W0245905

Applied Nonlinear Analysis

In honor of the 70th birthday of Professor Jindřich Nečas

Applied Nonlinear Analysis

Edited by

Adélia Sequeira

I.S.T. Technical University
Lisbon, Portugal

Hugo Beirão da Veiga

University of Pisa
Pisa, Italy

and

Juha Hans Videman

I.S.T. Technical University
Lisbon, Portugal

Springer Science+Business Media, LLC

Volume presented at the International Workshop on Nonlinear PDEs and Applications, December 13–17, 1999, i
Olomouc, Czech Republic, in honor of the 70th birthday of Professor Jindřich Nečas

ISBN 978-1-4757-8254-7 ISBN 978-0-306-47096-7 (eBook)
DOI 10.1007/978-0-306-47096-7

Originally published by Kluwer Academic/Plenum Publishers, New York in1999
Softcover reprint of the hardcover 1st edition 1999

10 9 8 7 6 5 4 3 2 1

A C.I.P. record for this book is available from the Library of Congress.

PREFACE

This book is meant as a present to honor Professor J. Nečas on the occasion of his 70th birthday.

It collects refereed contributions from sixty-one mathematicians from eleven countries. They cover many different areas of research related to the work of Professor Nečas, including Navier-Stokes equations, nonlinear elasticity, non-Newtonian fluids, regularity of solutions of parabolic and elliptic problems, operator theory and numerical methods.

The realization of this book could not have been made possible without the generous support of Centro de Matemática Aplicada (CMA/IST) and Fundação Calouste Gulbenkian.

Special thanks are due to Dr. Oldřich Ulrych for the careful preparation of the final version of this book.

Last but not least, we wish to express our gratitude to Dr. Šárka Matušů-Nečasová, for her invaluable assistance from the very beginning. This project could not have been successfully concluded without her enthusiasm and loving care for her father.

On behalf of the editors

ADÉLIA SEQUEIRA

JINDŘICH NEČAS

JINDŘICH NEČAS

Jindřich Nečas, honored by the Order of Merit of the Czech Republic by Václav Havel, President of the Czech Republic, on the October 28, 1998, Professor Emeritus of Mathematics at the Charles University in Prague, Presidential Research Professor at the Northern Illinois University and Doctor Honoris Causa at the Technical University of Dresden, has been enriching the Czech and world mathematics with his new ideas in the areas of partial differential equations, nonlinear functional analysis and applications of the both disciplines in continuum mechanics and hydrodynamics for more than forty years.

Born in Prague in December 14, 1929, Jindřich Nečas spent his youth in the nearby town of Mělník. He studied mathematics at the Faculty of Sciences of the Charles University in Prague between 1948–1952. After a short period at the Faculty of Civil Engineering of the Czech Technical University he joined the Mathematical Institute of the Czechoslovak Academy of Sciences where he headed the Department of Partial Differential Equations. Since 1977 he has been a member of the staff of the Faculty of Mathematics and Physics of the Charles University being in 1967–1971 the head of the Department of Mathematical Analysis, for many years the head of the Department of Mathematical Modelling and an active and distinguished member of the Scientific Council of the Faculty.

Let us go back to Nečas' first steps in mathematical research. He was the first PhD. student of I. Babuška, whom he still recalls with gratitude. As one of his first serious tasks he cooperated in the preparation of the pioneering monograph *Mathematical Methods of the Theory of Plane*

Elasticity by Babuška, Rektorys and Vyčichlo. It was mechanics which naturally directed him to applications of mathematics.

This period ended in 1957 with his defence of the dissertation *Solution of the Biharmonic Problem for Convex Polygons.* His interests gradually shifted to the functional analytic methods of solutions to partial differential equations. It was again I. Babuška who oriented him in this direction, introduced him to S. L. Sobolev and arranged his trip to Italy. His visits to Italy and France, where he got acquainted with the renowned schools of M. Picone, G. Fichera, E. Magenes and J. L. Lions deeply influenced the second period of Nečas' career.

Here we can find the fundamental contributions of Nečas to the linear theory: Rellich's identities and inequalities made it possible to prove the solvability of a wide class of boundary value problems for generalized data. They are important also for the application of the finite element method. This period culminated with the monograph *Les méthodes directes en théorie des équations elliptiques.* It became a standard reference book and found its way into the world of mathematical literature. We have only to regret that it has never been reedited (and translated into English). Its originality and richness of ideas was more than sufficient for J. Nečas to receive the Doctor of Science degree in 1966.

Without exaggeration, we can consider him the founder of the Czechoslovak school of modern methods of investigation of both boundary and initial value problems for partial differential equations. An excellent teacher, he influenced many students by his enthusiasm, never ceasing work in mathematics, organizing lectures and seminars and supervising many students to their diploma and Ph.D. thesis. Let us mention here two series of Summer Schools—one devoted to nonlinear partial differential equations and second interested in the recent results connected with Navier-Stokes equations. Both of them have had fundamental significance for the development of these areas.

While giving his monograph the final touch, J. Nečas already worked on another important research project. He studied and promoted the methods of solving nonlinear problems, and helped numerous young Czechoslovak mathematicians to start their careers in this domain. He also organized many international events and—last but not least—achieved many important results himself.

Nonlinear differential equations naturally lead to the study of nonlinear functional analysis and thus the monograph *Spectral Analysis of Nonlinear Operators* appeared in 1973. Among the many outstanding results let us mention the infinite dimensional version of Sard's theorem for analytic functionals which makes it possible to prove denumerability of the spectrum of a nonlinear operator. Theorems of the type of Fredholm's alternative represent another leading topic. The choice of the subject was extremely well-timed and many successors were appearing soon after the book had been published. This interest has not ceased till now and has resulted in deep and exact conditions of solvability of nonlinear boundary value problems. Svatopluk Fučík, who appeared as one of the co-authors of the monograph, together with Jan Kadlec, who worked primarily on problems characteristic for the previous period, and with younger Rudolf Švarc—were among the most talented and promising of Nečas' students. It is to be deeply regretted that the premature death of all three prevented them from gaining the kind of international fame as that of their teacher.

The period of nonlinearities, describing stationary phenomena, reached its top in the monograph *Introduction to the Theory of Nonlinear Elliptic Equations.* Before giving account of the next period, we must not omit one direction of his interest, namely, the problem of regularity of solutions to partial differential equations. If there is a leitmotif that can be heard through all of Nečas' work, then it is exactly this problem, closely connected to the solution of Hilbert's nineteenth problem.

In 1967 Nečas published his crucial work in this field, solving the problem of regularity of generalized solutions of elliptic equations of arbitrarily high order with nonlinear growth in a plane domain. His results allow a generalization for solutions to elliptic systems. In 1968 E. De Giorgi, E. Giusti and M. Miranda published counterexamples convincingly demonstrating that analogous theorems on regularity for systems fail to hold in space dimension greater then two. The series of papers by Nečas devoted to regularity in more dimensional domains can be divided into two groups. One of them can be characterized by the effort to find conditions guaranteeing regularity of weak solutions. Here an important result is an equivalent characterization of elliptic systems whose weak solutions are regular. This characterization is based on theorems of Liouville's type. The fact that Nečas' method can be applied to the study of regularity of solutions of both elliptic and parabolic systems demonstrates its general character. During this period Nečas collaborated also with many mathematicians (M. Giaquinta, B. Kawohl, J. Naumann). The other group of papers consists of those that aim at a deeper study of singularities of systems. J. Nečas is the author of numerous examples and counterexamples which help to map the situation.

In the next period, Nečas resumed his study of continuum mechanics. Again we can distinguish two fundamental groups of his interest. The former concerns the mechanics of elasto-plastic bodies. J. Nečas is the co-author of monographs *Mathematical Theory of Elastic and Elasto-plastic bodies: An Introduction* (with I. Hlaváček), *Solutions of Variational Inequalities in Mechanics* (with I. Hlaváček, J. Haslinger a J. Lovíšek). Let us also mention the theory of elastoplastic bodies admitting plastic flow and reinforcement, as well as the theory of contact problems with friction. It was J. Polášek who initiated Nečas' interest in transonic flow where he achieved remarkable results by using the method of entropic compactification and the method of viscosity. These results raised deep interest of the mathematical community. Nečas published the monograph *Écoulement de fluide, compacité par entropie.* In 1986 M. Padula presented her proof of the global existence of non-steady isothermal compressible fluids. This article led Nečas and Šilhavý to introduce a model of multipolar fluids satisfying the laws of thermodynamics. In this model the higher order stress tensor and its dependence on higher order velocity gradients are taking into account, the well-posedness of the model, the natural and logical construction of fundamental laws, and deep existence results were settled.

The most recent considerations are devoted to classical incompressible fluids, namely, to the Navier-Stokes fluids and to the power-law fluids. Essentially new existence, uniquenesss and regularity results are given for space periodic problem and for Dirichlet boundary value problem. Large time behaviour of solutions is analysed via the concept of short trajectories. A comprehensive survey of these results can be found in *Weak and Measure Valued Solutions to Evolutionary PDE's* (with J. Málek, M. Rokyta and M. Růžička).

The central theme in the mathematical theory of the Navier-Stokes fluids, i.e. the question of global existence of uniquely determined solution, has also become central in the research activities of J. Nečas in the past five years. Attention has been given to the proof that the possibility of constructing a singular solution in the self-similar form proposed by J. Leray in 1934, is excluded for the Cauchy problem. J. Nečas concentrates his energy to find the way of generalization of this result and to the resolution of the initial problem as well as to the study of influence of boundary conditions on the behaviour of the fluid described by Navier-Stokes equations.

A significant feature of Nečas' scientific work is his intensive and inspiring collaboration with many mathematicians ranging from the youngest to well-known and experienced colleagues from all over the world. Among them (without trying to get a complete list) we would like to mention: H. Bellout, F. Bloom, Ph. Ciarlet, A. Doktor, M. Feistauer, A. Friedman, M. Giaquinta, K. Gröger, Ch.P. Gupta, W. Hao, I. Hlaváček, R. Kodnár, V. Kondratiev, Y.C. Kwong, A. Lehtonen,

D.M. Lekveishvili, P.L. Lions, J. Lovíšek, D. Mayer, M. Müller, P. Neittaanmäki, I. Netuka, A. Novotný, O.A. Oleinik, M. Růžička, M. Rokyta, T. Roubíček, M. Šilhavý, M. Schönbeck, L. Trávníček.

We tried to collect some of the most important contributions of J. Nečas and to display the breadth of his interests and strivings, his encouragement of young people, his never ending enthusiasm, his deep and lively interest in mathematics. All these features of his personality have attracted students everywhere he has been working and have influenced many mathematicians.

OLDŘICH JOHN, JOSEF MÁLEK, JANA STARÁ

THE MOST SIGNIFICANT WORKS OF PROF. J. NEČAS

Monographs

[1] Svatopluk Fučík, Jindřich Nečas, and Vladimír Souček. *Einführung in die Variationsrechnung.* B. G. Teubner Verlagsgesellschaft, Leipzig, 1977. Mit englischen und russischen Zusammenfassungen, Teubner-Texte zur Mathematik.

[2] Svatopluk Fučík, Jindřich Nečas, and Vladimír Souček, Jiří Souček. *Spectral analysis of nonlinear operators.* Springer-Verlag, Berlin, 1973. Lecture Notes in Mathematics, Vol. 346.

[3] I. Hlaváček, J. Haslinger, J. Nečas, and J. Lovíšek. *Riešenie variačných nerovností v mechanike.* Alfa—Vydavatel'stvo Technickej a Ekonomickej Literatúry, Bratislava, 1982.

[4] I. Hlaváček, J. Haslinger, J. Nečas, and J. Lovíšek. *Solution of variational inequalities in mechanics*, volume 66 of *Applied Mathematical Sciences.* Springer-Verlag, New York, 1988. Translated from the Slovak by J. Jarník.

[5] J. Málek, J. Nečas, M. Rokyta, and M. Růžička. *Weak and measure-valued solutions to evolutionary PDEs*, volume 13 of *Applied Mathematics and Mathematical Computation.* Chapman & Hall, London, 1996.

[6] Jindřich Nečas. *Les méthodes directes en théorie des équations elliptiques.* Masson et Cie, Éditeurs, Paris, 1967.

[7] Jindřich Nečas. *Introduction to the theory of nonlinear elliptic equations*, volume 52 of *Teubner-Texte zur Mathematik [Teubner Texts in Mathematics]*. BSB B. G. Teubner Verlagsgesellschaft, Leipzig, 1983. With German, French and Russian summaries.

[8] Jindřich Nečas. *Introduction to the theory of nonlinear elliptic equations*. A Wiley-Interscience Publication. John Wiley & Sons Ltd., Chichester, 1986. Reprint of the 1983 edition.

[9] Jindřich Nečas. *Écoulements de fluide: compacité par entropie*, volume 10 of *RMA: Research Notes in Applied Mathematics*. Masson, Paris, 1989.

[10] Jindřich Nečas and Ivan Hlaváček. *Mathematical theory of elastic and elasto-plastic bodies: an introduction*, volume 3 of *Studies in Applied Mechanics*. Elsevier Scientific Publishing Co., Amsterdam, 1980.

Papers

[11] Hamid Bellout, Frederick Bloom, and Jindřich Nečas. Young measure-valued solutions for non-Newtonian incompressible fluids. *Comm. Partial Differential Equations*, 19(11-12):1763–1803, 1994.

[12] Hamid Bellout and Jindřich Nečas. Existence of global weak solutions for a class of quasilinear hyperbolic integro-differential equations describing viscoelastic materials. *Math. Ann.*, 299(2):275–291, 1994.

[13] Hamid Bellout, Frederick Bloom, and Jindřich Nečas. Existence of global weak solutions to the dynamical problem for a three-dimensional elastic body with singular memory. *SIAM J. Math. Anal.*, 24(1):36–45, 1993.

[14] Philippe G. Ciarlet and Jindřich Nečas. Problèmes unilatéraux en élasticité non linéaire tridimensionnelle. *C. R. Acad. Sci. Paris Sér. I Math.*, 298(8):189–192, 1984.

[15] Philippe G. Ciarlet and Jindřich Nečas. Injectivité presque partout, auto-contact, et non-interpénétrabilité en élasticité non-linéaire tridimensionnelle. *C. R. Acad. Sci. Paris Sér. I Math.*, 301(11):621–624, 1985.

[16] Philippe G. Ciarlet and Jindřich Nečas. Unilateral problems in nonlinear, three-dimensional elasticity. *Arch. Rational Mech. Anal.*, 87(4):319–338, 1985.

[17] Philippe G. Ciarlet and Jindřich Nečas. Injectivity and self-contact in nonlinear elasticity. *Arch. Rational Mech. Anal.*, 97(3):171–188, 1987.

[18] Miloslav Feistauer and Jindřich Nečas. Remarks on the solvability of transonic flow problems. *Manuscripta Math.*, 61(4):417–428, 1988.

[19] M. Giaquinta and J. Nečas. On the regularity of weak solutions to nonlinear elliptic systems via Liouville's type property. *Comment. Math. Univ. Carolin.*, 20(1):111–121, 1979.

[20] M. Giaquinta and J. Nečas. On the regularity of weak solutions to nonlinear elliptic systems of partial differential equations. *J. Reine Angew. Math.*, 316:140–159, 1980.

[21] M. Giaquinta, J. Nečas, O. John, and J. Stará. On the regularity up to the boundary for second order nonlinear elliptic systems. *Pacific J. Math.*, 99(1):1–17, 1982.

[22] K. Gröger and J. Nečas. On a class of nonlinear initial value problems in Hilbert spaces. *Math. Nachr.*, 93:21–31, 1979.

[23] K. Gröger, J. Nečas, and L. Trávníček. Dynamic deformation processes of elastic-plastic systems. *Z. Angew. Math. Mech.*, 59(10):567–572, 1979.

[24] Ivan Hlaváček and Jindřich Nečas. On inequalities of Korn's type. I. Boundary-value problems for elliptic system of partial differential equations. *Arch. Rational Mech. Anal.*, 36:305–311, 1970.

[25] Ivan Hlaváček and Jindřich Nečas. On inequalities of Korn's type. II. Applications to linear elasticity. *Arch. Rational Mech. Anal.*, 36:312–334, 1970.

[26] P.-L. Lions, J. Nečas, and I. Netuka. A Liouville theorem for nonlinear elliptic systems with isotropic nonlinearities. *Comment. Math. Univ. Carolin.*, 23(4):645–655, 1982.

[27] J. Nečas, O. John, and J. Stará. Counterexample to the regularity of weak solution of elliptic systems. *Comment. Math. Univ. Carolin.*, 21(1):145–154, 1980.

[28] J. Nečas and A. Novotný. Some qualitative properties of the viscous compressible heat conductive multipolar fluid. *Comm. Partial Differential Equations*, 16(2-3):197–220, 1991.

[29] J. Nečas, A. Novotný, and M. Šilhavý. Global solution to the compressible isothermal multipolar fluid. *J. Math. Anal. Appl.*, 162(1):223–241, 1991.

[30] J. Nečas, M. Růžička, and V. Šverák. On Leray's self-similar solutions of the Navier-Stokes equations. *Acta Math.*, 176(2):283–294, 1996.

[31] J. Nečas and M. Šilhavý. Multipolar viscous fluids. *Quart. Appl. Math.*, 49(2):247–265, 1991.

[32] Jindřich Nečas. Sur la coercivité des formes sesquilinéaires, elliptiques. *Rev. Roumaine Math. Pures Appl.*, 9:47–69, 1964.

[33] Jindřich Nečas. L'application de l'égalité de Rellich sur les systèmes elliptiques du deuxième ordre. *J. Math. Pures Appl. (9)*, 44:133–147, 1965.

[34] Jindřich Nečas. Sur l'appartenance dans la classe $C^{(k),\mu}$ des solutions variationnelles des équations elliptiques non-linéaires de l'ordre $2k$ en deux dimensions. *Comment. Math. Univ. Carolinae*, 8:209–217, 1967.

[35] Jindřich Nečas. Sur l'alternative de Fredholm pour les opérateurs non-linéaires avec applications aux problèmes aux limites. *Ann. Scuola Norm. Sup. Pisa (3)*, 23:331–345, 1969.

[36] Jindřich Nečas. Fredholm alternative for nonlinear operators and applications to partial differential equations and integral equations. *Časopis Pěst. Mat.*, 97:65–71, 94, 1972.

[37] Jindřich Nečas. Application of Rothe's method to abstract parabolic equations. *Czechoslovak Math. J.*, 24(99):496–500, 1974.

[38] Jindřich Nečas, Jiří Jarušek and Jaroslav Haslinger. On the solution of the variational inequality to the Signorini problem with small friction. *Boll. Un. Mat. Ital. B (5)*, 17(2):796–811, 1980.

[39] Jindřich Nečas, Ari Lehtonen, and Pekka Neittaanmäki. On the construction of Lusternik-Schnirelmann critical values with application to bifurcation problems. *Appl. Anal.*, 25(4):253–268, 1987.

[40] Jindřich Nečas, Michael Růžička, and Vladimir Šverák. Sur une remarque de J. Leray concernant la construction de solutions singulières des équations de Navier-Stokes. *C. R. Acad. Sci. Paris Sér. I Math.*, 323(3):245–249, 1996.

[41] Jindřich Nečas and Vladimír Šverák. On regularity of solutions of nonlinear parabolic systems. *Ann. Scuola Norm. Sup. Pisa Cl. Sci. (4)*, 18(1):1–11, 1991.

Contributions in Proceedings of Conferences

[42] J. Nečas. Entropy compactification of the transonic flow. In *Equadiff 6 (Brno, 1985)*, volume 1192 of *Lecture Notes in Math.*, pages 399–408. Springer, Berlin, 1986.

[43] J. Nečas. Theory of multipolar viscous fluids. In *The mathematics of finite elements and applications, VII (Uxbridge, 1990)*, pages 233–244. Academic Press, London, 1991.

[44] Jindřich Nečas. Theory of multipolar fluids. In *World Congress of Nonlinear Analysts '92, Vol. I–IV (Tampa, FL, 1992)*, pages 1073–1081. de Gruyter, Berlin, 1996.

[45] Jindřich Nečas and Luděk Trávníček. Variational inequalities of elastoplasticity with internal state variables. In *Theory of nonlinear operators (Proc. Fifth Internat. Summer School, Central Inst. Math. Mech. Acad. Sci. GDR, Berlin, 1977)*, volume 6 of *Abh. Akad. Wiss. DDR, Abt. Math. Naturwiss. Tech., 1978*, pages 195–204. Akademie-Verlag, Berlin, 1978.

CONTRIBUTING AUTHORS

Alliot Frédéric Laboratoire d'Analyse Numérique, Tour 55–65, 5ème étage, Université Pierre et Marie Curie, 4 Place Jussieu, 75252 Paris Cedex 05, France
email: `alliot@cermics.enpc.fr`

Amrouche Cherif Laboratoire de Mathématiques Appliqués, I.P.R.A., Avenue de l'Université, 64000 Pau, France
email: `Cherif.Amrouche@univ-pau.fr`

Chen G. Q. Center for Environmental Sciences, Peking University, Beijing, China

Daneček Josef Department of Mathematics, FAST VUT, Žižkova 17, 60200 Brno, Czech Republic
email: `mddan@fce.vutbr.cz`

Drábek Pavel University of West Bohemia, Americká 42, 306 14 Plzeň, Czech Republic
email: `pdrabek@kma.zcu.cz`

Eck Christof Institute of Applied Mathematics, University Erlangen-Nürnberg, Germany
email: `eck@am.uni-erlangen.de`

Egorov Yuri Vladimirovich Université Paul Sabatier, UFR MIG, MIP, 118 route de Narbonne, 31062 Toulouse, France
email: `egorov@mip.ups-tlse.fr`

Eisner Jan Mathematical Institute, Academy of Sciences of the Czech Republic, Žitná 25, 115 67 Praha 1, Czech Republic
email: `eisner@math.cas.cz`

Feistauer Miloslav Charles University Prague, Faculty of Mathematics and Physics, Malostranské nám. 25, 118 00 Praha 1, Czech Republic
email: feist@karlin.mff.cuni.cz

Fonseca Irene Department of Mathematical Sciences, Carnegie Mellon University, Pittsburgh, PA 15213, USA
email: fonseca@andrew.cmu.edu

Franců Jan Department of Mathematics, Technical University Brno, Technická 2, 616 69 Brno, Czech Republic
email: francu@fme.vutbr.cz

Galdi Giovanni Paolo Department of Mechanical Engineering and Department of Mathematics, University of Pittsburgh, USA
email: galdi@math.pitt.edu

Girault Vivette Laboratoire d'Analyse Numérique, Tour 55–65, 5ème étage, Université Pierre et Marie Curie, 4 Place Jussieu, 75252 Paris Cedex 05, France
email: girault@ann.jussieu.fr

Glowinski Roland Department of Mathematics, University of Houston,, Texas, USA
email: roland@math.uh.edu

Hecht Frédéric Laboratoire d'Analyse Numérique, Tour 55–65, 5ème étage, Université Pierre et Marie Curie, 4 Place Jussieu, 75252 Paris Cedex 05, France
email: hecht@ann.jussieu.fr

Hlaváček Ivan Mathematical Institute, Academy of Sciences of the Czech Republic, Žitná 25, 115 67 Praha 1, Czech Republic

Jarušek Jiří Mathematical Institute, Academy of Sciences of the Czech Republic, Žitná 25, 115 67 Praha 1, Czech Republic
email: jarusek@math.cas.cz

Kačur Josef Faculty of Mathematics and Physics, Comenius University, Mlynska dolina, 84215 Bratislava, Slovak Republic
email: kacur@fmph.uniba.sk

Kaplický Petr Charles University, Department of Mathematical Analysis, Sokolovská 83, 186 75 Praha 8, Czech Republic
email: kaplicky@karlin.mff.cuni.cz

Kawohl Bernd Mathematisches Institut, Universität zu Köln, D 50923 Köln, Germany
email: kawohl@thales.mi.uni-koeln.de

Klouček Petr Department of Computational and Applied Mathematics, Rice University, 6100 Main Street, Houston, TX 77005, USA
email: `kloucek@rice.edu`

Kondratiev Vladimir Alexandrovich Lomonossov University, Mehmat, Vorobievy Gory, Moscow, 119 899 Russia
email: `kondrat@vnmok.math.msu.su`

Křížek Michael Mathematical Institute, Academy of Sciences of the Czech Republic, Žitná 25, 115 67 Prague 1, Czech Republic
email: `krizek@math.cas.cz`

Kračmar Stanislav Czech Technical University, Faculty of Mechanical Engineering, Department of Technical Mathematics, Karlovo nám. 13, 12135 Prague, Czech Republic
email: `kracmar@fsik.cvut.cz`

Kučera Milan Mathematical Institute, Academy of Sciences of the Czech Republic, Žitná 25, 115 67 Praha 1, Czech Republic
email: `kucera@math.cas.cz`

Kufner Alois Mathematical Institute, Academy of Sciences of the Czech Republic, Žitná 25, 115 67 Praha 1, Czech Republic
email: `kufner@math.cas.cz`

Kutev Nikolay Mathematisches Institut, Universität zu Köln, D 50923 Köln, Germany
email: `kutev@mi.uni-koeln.de`

Leonardi Salvatore Dipartimento di Matematica, Viale A. Doria 6, 95125 Catania, Italy
email: `leonardi@dipmat.unict.it`

Lions Jacques-Louis Collège de France, 3 rue d'Ulm, 75005 Paris, France

Liu Liping Mathematical Institute, Academy of Sciences of the Czech Republic, Žitná 25, 115 67 Prague 1, Czech Republic
email: `liu@math.cas.cz`

Lovíšek Ján Faculty of Civil Engineering, Slovak Technical University, Bratislava, Slovak Republic
email: `lovisek@svf.stuba.sk`

Málek Josef Charles University, Mathematical Institute of Charles University, Sokolovská 83, 186 75 Praha 8, Czech Republic
email: `malek@karlin.mff.cuni.cz`

Malý Jan Department KMA, Charles University, 186 75 Praha 8, Czech Republic
email: maly@karlin.mff.cuni.cz

Matušů-Nečasová Šárka Czech Academy of Sciences, Mathematical Institute, Žitná 25, 115 67 Praha 1, Czech Republic
email: matus@math.cas.cz

Neittaanmäki Pekka Department of Mathematics, University of Jyväskylä, P. O. Box 35, FIN–40351 Jyväskylä, Finland
email: pn@mit.jyu.fi

Neustupa Jiří Czech Technical University, Faculty of Mechanical Engineering, Karlovo nám. 13, 121 35 Praha 2, Czech Republic
email: neustupa@marian.fsik.cvut.cz

Novotný Antonín Université de Toulon et du Var, Department of Mathematics, B.P.132, 83957 Toulon - La Garde, France
email: novotny@univ-tln.fr

Oliveira Paula de Departamento de Matemática da Universidade de Coimbra, 3000 Coimbra, Portugal
email: poliveir@mat.uc.pt

Padula Mariarosaria Dipartimento di Matematica, Università di Ferrara, via Machiavelli 35, 44100 Ferrara, Italy.
email: pad@dns.unife.it

Pan T.W. Department of Mathematics, University of Houston,, Texas, USA

Penel Patrick Université de Toulon et du Var, Department of Mathematics, B.P.132, 83957 Toulon - La Garde, France
email: penel@univ-tln.fr

Pironneau Olivier Laboratoire d'Analyse Numérique, Tour 55–65, 5ème étage, Université Pierre et Marie Curie, 4 Place Jussieu, 75252 Paris Cedex 05, France
email: pironneau@ann.jussieu.fr

Pokorný Milan Palacký University, Faculty of Science, Department of Mathematical Analysis and Applications of Mathematics, Tomkova 40, 779 00 Olomouc, Czech Republic
email: pokorny@risc.upol.cz

Rajagopal K. R. Department of Mechanical Engineering, Texas A&M University, College Station, Texas 77843-3123, USA
email: krajagopal@mengr.tamu.edu

Rautmann Reimund Fachbereich Mathematik und Informatik, Universitaet-GH Paderborn, Warburger Str. 100, 33098 Paderborn, Germany
email: `rautmann@plato.uni-paderborn.de`

Rodrigues José Francisco C.M.A.F. / Universidade de Lisboa, Av. Prof. Gama Pinto, 2, 1649-003 Lisboa, Portugal
email: `rodrigues@lmc.fc.ul.pt`

Rokyta Mirko Charles University, Department of Mathematical Analysis, Sokolovská 83, 186 75 Praha, Czech Republic
email: `rokyta@karlin.mff.cuni.cz`

Roubíček Tomáš Charles University, Mathematical Institute of Charles University, Sokolovská 83, 186 75 Praha 8, Czech Republic, and Institute of Information Theory and Automation, Academy of Sciences, Pod vodárenskou věží 4, CZ-182 08 Praha 8, Czech Republic
email: `roubicek@karlin.mff.cuni.cz`

Růžička Michael Institute of Applied Mathematics, University of Bonn, Beringstr. 4-6, D-53115 Bonn, Germany
email: `rose@iam.uni-bonn.de`

Santos José Departamento de Matemática da Universidade de Aveiro, 3810 Aveiro, Portugal
email: `jmss@mat.ua.pt`

Schonbek Maria Elena Department of Mathematics, University of California, Santa Cruz, CA 95060, USA
email: `schonbek@math.ucsc.edu`

Schwab Christoph Eidgenössische Technische Hochschule, Seminar für Angewandte Mathematik, CH – 8092 Zürich, Switzerland
email: `schwab@math.ethz.ch`

Sequeira Adélia Instituto Superior Técnico, Departamento de Matemática, Av. Rovisco Pais, 1, 1049-001 Lisboa, Portugal
email: `asequeir@math.ist.utl.pt`

Šilhavý Miloslav Mathematical Institute, Academy of Sciences of the Czech Republic, Žitná 25, 115 67 Praha 1, Czech Republic
email: `silhavy@matsrv.math.cas.cz`

Stará Jana Charles University, Department of Mathematical Analysis, Sokolovská 83, 186 75 Praha 8, Czech Republic
email: `stara@karlin.mff.cuni.cz`

Straškraba Ivan Mathematical Institute, Academy of Sciences of the Czech Republic, Žitná 25, 115 67 Prague 1, Czech Republic
email: strask@math.cas.cz

Tao Luoyi Department of Mechanical Engineering, Texas A&M University, College Station, Texas 77843-3123, USA
email: luoyitao@yahoo.com

Urbano José Miguel Departamento de Matemática, Universidade de Coimbra, 3000 Coimbra, Portugal
email: urbano@lmc.fc.ul.pt

Videman Juha Hans Instituto Superior Técnico, Departamento de Matemática, Av. Rovisco Pais, 1, 1049-001 Lisboa, Portugal
email: videman@math.ist.utl.pt

Viszus Eugen Department of Mathematical Analysis, MFF UK, Mlynska dolina, 84215 Bratislava, Slovak Republic
email: Eugen.Viszus@fmph.uniba.sk

Wolf Joerg Humboldt University, Institut fuer Mathematik, Mathematisch-Naturwissenschaftliche Fakultaet II, Unter den Linden 6, 10099 Berlin, Germany
email: jwolf@mathematik.hu-berlin.de

CONTENTS

ON THE REGULARITY AND DECAY OF THE WEAK SOLUTIONS TO THE STEADY-STATE NAVIER-STOKES EQUATIONS IN EXTERIOR DOMAINS

Frédéric Alliot, Cherif Amrouche

Abstract: In this article, we study the regularity properties of the weak solutions to the steady-state Navier-Stokes equations in exterior domains of $\mathbb{R}^3$. Our approach is based on a combination of the properties of Stokes problems in $\mathbb{R}^3$ and in bounded domains. We obtain in particular a decomposition result for the pressure and some sufficient conditions for the velocity to vanish at infinity.

Keywords: Exterior flows, Navier-Stokes, weak solutions, regularity, behaviour at infinity.

This paper is devoted to some mathematical questions related to the steady-state motion of an incompressible viscous fluid past a bounded body Ω'. In the three-dimensional space $\mathbb{R}^3$, let us denote by Ω the exterior of $\overline{\Omega'}$, which is filled by the fluid. Then, the velocity field u and the pressure π in the fluid satisfy the Navier-Stokes system:

$$(NS)\qquad \begin{aligned} -\nu\Delta u + u.\nabla u + \nabla \pi &= f \quad \text{in } \Omega,\\ \operatorname{div} u &= 0 \quad \text{in } \Omega,\\ u_{|\partial\Omega} &= \mathbf{0}, \end{aligned}$$

where f is a given external force-field and $\nu > 0$ stands for the kinematic viscosity of the fluid. The last equation of the system states that the fluid adheres at the surface of the body, which is the common no-slip condition. We shall moreover assume that the fluid is at rest at infinity and thus consider the additional condition:

$$\lim_{|x|\to+\infty} u(x) = \mathbf{0}. \tag{0.1}$$

Applied Nonlinear Analysis, edited by Sequcira *et al.*
Kluwer Academic / Plenum Publishers, New York, 1999.

Our purpose is to study some regularity properties of weak solutions to the problem (NS) (see Definition 1.1 below), keeping in mind that we wish the decay condition (0.1) to be fulfilled.

The paper is organized as follows: In Section 1, we recall a well-known result about existence of weak solutions for the problem (NS). The data and solutions will be chosen in weighted Sobolev spaces, in which distributions are well controlled at infinity. The second section is devoted to some regularity properties of the weak solutions u and the associated pressure π. We first obtain, with no additional assumption, the regularity of π, that leads to a "natural" decomposition of this term into a "viscous pressure" and a "convective pressure" (see Proposition 2.3 and Remark 2.4 below). Then, our main result establishes the L^p regularity of ∇u and π under some rather weak assumptions. Moreover, we deduce from this result some sufficient conditions on f such that each weak solution satisfies (0.1). The proof relies on the combination of the regularity properties of the Stokes problem in bounded domains and in $\mathbb{R}^3$. With similar arguments, we study the L^p regularity of higher-order derivatives of u and π and their decay at infinity. The last section is devoted to the regularity, in the Hardy space $\mathcal{H}^1$, of the second derivatives of the pressure in the whole space $\mathbb{R}^3$, and is based on sharp properties of the non-linear term.

We now conclude this introduction by giving some definitions and notation that we shall use throughout the paper.

Let us first settle the geometry of Ω. Let Ω' be a bounded open region of $\mathbb{R}^3$, not necessarily connected, with a Lipschitz-continuous boundary and let the fluid fill the complement of $\overline{\Omega'}$, denoted by Ω. We assume that Ω' has a finite number of connected components and that each connected component has a connected boundary, so that Ω is connected. In the sequel, such a set Ω will be referred to as an exterior domain.

We shall also denote by B_R the open ball of radius $R > 0$ centered at the origin. In particular, since Ω' is bounded, we can find some $R_0 > 0$ such that $\Omega' \subset B_{R_0}$ and we introduce, for any $R \geq R_0$, the sets

$$\Omega_R = \Omega \cap B_R \quad \text{and} \quad \Omega^R = \Omega - \overline{\Omega_R}.$$

Let $\mathcal{O}$ be an open region of $\mathbb{R}^3$. As usual, $\mathcal{D}(\mathcal{O})$ denotes the space of indefinitely differentiable functions with compact support in $\mathcal{O}$ and $\mathcal{D}'(\mathcal{O})$ denotes its dual space which is the space of distributions. For each $p \in]1, +\infty[$, the conjugate exponent p' is given by the relation $\frac{1}{p} + \frac{1}{p'} = 1$. We recall that $L^p(\mathcal{O})$ is the space of measurable functions such that $\int_{\mathcal{O}} |u|^p dx < \infty$. With its natural norm: $||u||_{L^p(\mathcal{O})} = (\int_{\mathcal{O}} |u|^p dx)^{1/p}$, it is a Banach space whose dual space is $L^{p'}(\mathcal{O})$. When $1 < p < 3$, we shall also use the Sobolev exponent of p that is $p^* = 3p/(3-p)$. Recall that the space $W^{1,p}(\mathcal{O})$ stands for the Sobolev space of functions $u \in L^p(\mathcal{O})$ with distributional derivatives in $L^p(\mathcal{O})$, endowed with its natural norm. Moreover, $\overset{\circ}{W}{}^{1,p}(\mathcal{O})$ is the closure of $\mathcal{D}(\mathcal{O})$ in $W^{1,p}(\mathcal{O})$ and $W^{-1,p'}(\mathcal{O})$ is the dual space of $\overset{\circ}{W}{}^{1,p}(\mathcal{O})$. When $p=2$, we shall also use the standard notation

$$H^1(\mathcal{O}) = W^{1,2}(\mathcal{O}), \quad H_0^1(\mathcal{O}) = \overset{\circ}{W}{}^{1,2}(\mathcal{O}), \quad H^{-1}(\mathcal{O}) = W^{-1,2}(\mathcal{O}).$$

Finally, we use bold type characters to denote vector distributions or spaces of vector distributions with 3 components. For instance, $f \in \mathbf{L}^p(\mathcal{O})$ means $(f_1, f_2, f_3) \in (L^p(\mathcal{O}))^3$.

1. EXISTENCE OF WEAK SOLUTIONS IN WEIGHTED SOBOLEV SPACES

The study of the steady-state Navier-Stokes problem in general domains was initiated by the fundamental works of J. Leray [13] who introduced the concept of weak solution:

Definition 1.1. A weak solution to the problem (NS) is a field $u \in \mathbf{H}^1_{loc}(\overline{\Omega})$ vanishing on $\partial\Omega$, with $\nabla u \in L^2(\Omega)$ and such that for all $\varphi \in \boldsymbol{\mathcal{V}}(\Omega) = \{v \in \boldsymbol{\mathcal{D}}(\Omega),\ \mathrm{div}\, v = 0\}$:

$$\nu \int_\Omega \nabla u \nabla \varphi dx + \int_\Omega u.\nabla u.\, \varphi dx = < f, \varphi > . \tag{1.1}$$

When Ω is an exterior domain, a weak solution u is only constrained at infinity by the condition $\nabla u \in L^2(\Omega)$. But such a condition is not sufficient to ensure that u satisfies (0.1), or even that u vanishes in a weaker sense at infinity. Hence, the general class of fields $u \in \mathbf{H}^1_{loc}(\overline{\Omega})$ vanishing on $\partial\Omega$, with $\nabla u \in L^2(\Omega)$ is too large for our purpose. It is more appropriate to control both ∇u and u itself at infinity, which can be achieved in a natural way in some weighted Sobolev spaces. Define the weight function $\rho(x) = (2 + |x|^2)^{1/2}$, then we can state the

Definition 1.2. Let Ω be either an exterior domain or $\Omega = \mathbb{R}^3$ and let p and α be real numbers with $1 < p < +\infty$. Then, we set $L^p_\alpha(\Omega) = \{u \in \mathcal{D}'(\Omega),\ \rho^\alpha u \in L^p(\Omega)\}$ and

$$W^{1,p}_\alpha(\Omega) = \{u \in \mathcal{D}'(\Omega),\ \rho^{\alpha-1} u \in L^p(\Omega),\ \rho^\alpha \nabla u \in \mathbf{L}^p(\Omega)\},\ \text{if } 3/p + \alpha \neq 1,$$

$$W^{1,p}_\alpha(\Omega) = \{u \in \mathcal{D}'(\Omega),\ \frac{\rho^{\alpha-1}}{\ln \rho} u \in L^p(\Omega),\ \rho^\alpha \nabla u \in \mathbf{L}^p(\Omega)\},\ \text{if } 3/p + \alpha = 1.$$

Each of these spaces is a reflexive Banach space when endowed with the norm:

$$\| u \|_{L^p_\alpha(\Omega)} = \| \rho^\alpha u \|_{L^p(\Omega)},$$

$$\| u \|_{W^{1,p}_\alpha(\Omega)} = (\| \rho^{\alpha-1} u \|^p_{L^p(\Omega)} + \| \rho^\alpha \nabla u \|^p_{\mathbf{L}^p(\Omega)})^{1/p} \text{ if } 3/p + \alpha \neq 1,$$

$$\| u \|_{W^{1,p}_\alpha(\Omega)} = (\| \frac{\rho^{\alpha-1}}{\ln \rho} u \|^p_{L^p(\Omega)} + \| \rho^\alpha \nabla u \|^p_{\mathbf{L}^p(\Omega)})^{1/p} \text{ if } 3/p + \alpha = 1.$$

In the definition above, the powers of the weight function ρ and the introduction of the logarithmic weight when $3/p + \alpha = 1$ are not anecdotal. Indeed, this definition allows to prove some weighted Poincaré inequalities which are the main interest of the spaces $W^{1,p}_\alpha$ (see Theorem 1.1 below).

Define now the space $\overset{\circ}{W}{}^{1,p}_\alpha(\Omega)$ as the closure of $\mathcal{D}(\Omega)$ for the norm $\| \cdot \|_{W^{1,p}_\alpha(\Omega)}$. Then, the dual space of $\overset{\circ}{W}{}^{1,p}_\alpha(\Omega)$, which we denote by $W^{-1,p'}_{-\alpha}(\Omega)$, is a space of distributions. When Ω is an exterior domain, and since each function of $W^{1,p}_\alpha(\Omega)$ locally belongs to the classical Sobolev space $W^{1,p}$, it is standard to check that

$$\overset{\circ}{W}{}^{1,p}_\alpha(\Omega) = \{v \in W^{1,p}_\alpha(\Omega), \gamma v = 0\}, \tag{1.2}$$

where γ stands for the trace operator on the Lipschitz-continuous boundary $\partial\Omega$. However, when $\Omega = \mathbb{R}^3$, we have $W^{1,p}_\alpha(\mathbb{R}^3) = \overset{\circ}{W}{}^{1,p}_\alpha(\mathbb{R}^3)$ (see [3], Th. 7.2).

We now recall a fundamental property of the spaces $W^{1,p}_\alpha$:

Theorem 1.1. (Amrouche-Girault-Giroire [3, 4]) *Let $\alpha \in \mathbb{R}$ and $1 < p < +\infty$.*
i) Let Ω be an exterior domain. There exists a constant $C = C(p, \alpha, \Omega) > 0$ such that

$$\forall u \in \overset{\circ}{W}{}^{1,p}_\alpha(\Omega), \quad \| u \|_{W^{1,p}_\alpha(\Omega)} \leq C \| \nabla u \|_{\mathbf{L}^p_\alpha(\Omega)}.$$

ii) There exists a constant $C = C(p, \alpha) > 0$ such that

$$\forall u \in W^{1,p}_\alpha(\mathbb{R}^3), \quad \| u \|_{W^{1,p}_\alpha(\mathbb{R}^3)} \leq C \| \nabla u \|_{\mathbf{L}^p_\alpha(\mathbb{R}^3)}, \quad \textit{if } 3/p + \alpha > 1,$$

$$\| u \|_{W^{1,p}_\alpha(\mathbb{R}^3)/\mathcal{P}_0} \leq C \| \nabla u \|_{\mathbf{L}^p_\alpha(\mathbb{R}^3)}, \quad \textit{otherwise},$$

where $\mathcal{P}_0$ stands for the subspace of constant functions in $W^{1,p}_\alpha(\mathbb{R}^3)$ when $3/p + \alpha \leq 1$.

Remark 1.2. Theorem 1.1 for instance states that the semi-norm $\|\nabla \ \|_{L^2(\Omega)}$ defines a norm on $\overset{\circ}{W}{}_0^{1,2}(\Omega)$ which is equivalent to the natural norm of this space.

We now turn to the question of existence of weak solutions to the exterior problem (NS). The key idea for proving existence, which has also been pointed out by J. Leray, is to find approximate solutions u_n that satisfy a uniform estimate:

$$\| \nabla u_n \|_{L^2(\Omega)} \leq M,$$

and then to pass to the limit. Following this idea, we state and prove the

Theorem 1.3. *Let $\Omega \subset \mathbb{R}^3$ be a Lipschitz exterior domain or $\Omega = \mathbb{R}^3$. Given a force $f \in \mathbf{W}_0^{-1,2}(\Omega)$, the problem (NS) has a weak solution $u \in \mathbf{W}_0^{1,2}(\Omega)$ such that:*

$$\nu \| \nabla u \|_{L^2(\Omega)} \leq \| f \|_{\mathbf{W}_0^{-1,2}(\Omega)}.$$

Besides, there exists a function $\pi \in L^2(\Omega_R)$ for all $R \geq R_0$, unique up to a constant, such that (u, π) solves problem (NS) in the sense of distributions.

Proof. Let $(R_n)_{n\geq 0}$ be an increasing sequence of real numbers with $R_0 > 0$ fixed in the introduction and such that $\lim_{n\to\infty} R_n = +\infty$. We approximate problem (NS) by the following sequence of problems on the bounded domains Ω_{R_n}:

Find $u_n \in \mathbf{H}_0^1(\Omega_{R_n})$ such that

$$\nu \int_{\Omega_{R_n}} \nabla u_n \nabla \varphi \, dx \; + \; \int_{\Omega_{R_n}} u_n . \nabla u_n . \varphi \, dx = < f, \varphi >, \qquad (1.3)$$
$$\forall \varphi \in \mathcal{D}(\Omega_{R_n}), \ \operatorname{div} \varphi = 0.$$

First remark that each function of $\overset{\circ}{\mathbf{W}}{}_0^{1,2}(\Omega)$ with support in $\overline{\Omega_{R_n}}$ also belongs to $\mathbf{H}_0^1(\Omega_{R_n})$. Then, since $f \in \mathbf{W}_0^{-1,2}(\Omega)$, its restriction to Ω_{R_n} satisfies

$$\| f \|_{\mathbf{H}^{-1}(\Omega_{R_n})} \leq \| f \|_{\mathbf{W}_0^{-1,2}(\Omega)}. \qquad (1.4)$$

Therefore, we know from [17](Th. 1.2, p. 164) that for each $n \geq 0$, problem (1.3) has a solution u_n such that

$$\nu \| \nabla u_n \|_{L^2(\Omega_{R_n})} \leq \| f \|_{\mathbf{H}^{-1}(\Omega_{R_n})}. \qquad (1.5)$$

We extend u_n by zero in Ω^{R_n}, and still denote u_n the extended function that belongs to $\overset{\circ}{\mathbf{W}}{}_0^{1,2}(\Omega)$. In view of (1.4) and (1.5), we thus have:

$$\nu \| \nabla u_n \|_{L^2(\Omega)} \leq \| f \|_{\mathbf{W}_0^{-1,2}(\Omega)}. \tag{1.6}$$

Hence, Theorem 1.1 (with $p = 2$, $\alpha = 0$) and (1.6) yield that u_n is bounded in $\overset{\circ}{\mathbf{W}}{}_0^{1,2}(\Omega)$, which is reflexive. Therefore, extracting subsequences if necessary, we have:

$$\begin{array}{l} u_n \rightharpoonup u \text{ in } \overset{\circ}{\mathbf{W}}{}_0^{1,2}(\Omega) \quad \text{and} \\ \nu \| \nabla u \|_{L^2(\Omega)} \leq \liminf \nu \| \nabla u_n \|_{L^2(\Omega)} \leq \| f \|_{\mathbf{W}_0^{-1,2}(\Omega)}. \end{array} \tag{1.7}$$

Let us now check that u is a weak solution. Let $\varphi \in \mathcal{V}(\Omega)$ and $N \geq 0$ be an integer such that $\operatorname{supp} \varphi \subset \Omega_{R_N}$. Then, we deduce from (1.3) that

$$\forall n \geq N, \qquad \nu \int_\Omega \nabla u_n \nabla \varphi dx + \int_\Omega u_n . \nabla u_n . \varphi dx = < f, \varphi > . \tag{1.8}$$

In view of (1.7), we can pass to limit in the first integral. Moreover, extracting a subsequence if necessary, we know that u_n converges strongly to u in $\mathbf{L}^2(\Omega_{R_N})$ since the imbedding $\mathbf{H}^1(\Omega_{R_N}) \subset \mathbf{L}^2(\Omega_{R_N})$ is compact. Hence, this convergence together with (1.7) ensures the convergence of the second integral of (1.8) and therefore $u \in \overset{\circ}{\mathbf{W}}{}_0^{1,2}(\Omega)$ satisfies (1.1).

Finally, existence of a pressure $\pi \in \mathcal{D}'(\Omega)$ such that (u, π) satisfies system (NS) in the sense of distributions follows from (1.1) and from a well-known consequence of a very general theorem of G. de Rham. Moreover, π is unique up to a constant because Ω is connected. Besides, the local regularity of π can be deduced from standard local properties of the distribution $f - u.\nabla u + \nu \Delta u$ and from a result of L. Tartar [16] (lemma 9, p. 30) and Girault-Raviart [10]. ◇

Remark 1.4. In this paper, we only focus on the regularity and decay of weak solutions in three-dimensional exterior domains. Let us nevertheless mention that many problems remain open for weak solutions that satisfy (0.1). For instance, it is not known whether such solutions are unique for "small" data, while such a property is established in bounded domains (See Temam [17], Ch. II and Girault-Raviart [10] for the case of bounded domains and Galdi [8], Ch. IX, for partial uniqueness properties in exterior domains).

The study of weak solutions in two-dimensional exterior domains is even more difficult. Although some existence results are known, the arguments developed in the proofs of our results below fail in two dimensions. As a matter of fact, the existence of weak solutions satisfying (0.1) for a large class of data is not established so far. We shall however give a positive answer to this problem for some particular data in a further work.

2. THE REGULARITY OF WEAK SOLUTIONS

Our approach relies on a localization argument which we develop in the paragraph below. This argument enables us to study on the one hand the regularity of a solution near infinity and on the other hand the regularity near the boundary.

2.1. Separating the regularity near infinity and near the boundary

Let Ω be an exterior domain. We introduce the following partition of unity: Let R_1 and R_2 be real numbers such that $R_2 > R_1 > R_0$ and choose some functions ψ_1 and ψ_2 such that:

$$\psi_1 \in C^\infty(\mathbb{R}^3), \quad \psi_1(x) = 0 \text{ if } |x| \leq R_1, \quad \psi_1(x) = 1 \text{ if } |x| \geq R_2, \tag{2.1}$$

$$\forall x \in \mathbb{R}^3, \quad \psi_1(x) + \psi_2(x) = 1. \tag{2.2}$$

Consider now a solution (u, π) to problem (NS) such that $u \in \overset{\circ}{\mathbf{W}}{}_0^{1,2}(\Omega)$ and π belongs to $L^2(\Omega_R)$ for all $R \geq R_0$ (think of a solution given by Theorem 1.3). Then, define (u^1, π^1) as follows:

$$(u^1, \pi^1) = (u\psi_1, \pi\psi_1) \text{ in } \Omega, \quad (u^1, \pi^1) = (\mathbf{0}, 0) \text{ in } \overline{\Omega'},$$

and set $(u^2, \pi^2) = (u\psi_2, \pi\psi_2)$ in Ω.

It is easy to check that $(u^1, \pi^1) \in \mathbf{W}_0^{1,2}(\mathbb{R}^3) \times L^2_{loc}(\mathbb{R}^3)$ (compute the weak derivatives of u^1 and use the fact that u^1 vanishes at the boundary $\partial\Omega$). We also note that (u^2, π^2) clearly belongs to $\mathbf{H}_0^1(\Omega_{R_2}) \times L^2(\Omega_{R_2})$. Moreover, further elementary calculations in the sense of distributions enable us to establish the equalities (respectively in $\mathcal{D}'(\mathbb{R}^3)$ if $i = 1$ and in $\mathcal{D}'(\Omega_{R_2})$ if $i = 2$):

$$-\nu\Delta u^i + \nabla\pi^i = f^i, \qquad \operatorname{div} u^i = g^i, \tag{2.3}$$

where

$$f^i = f\psi_i - 2\nu\nabla u\nabla\psi_i - \nu u\Delta\psi_i + \pi\nabla\psi_i - (u.\nabla u)\psi_i, \qquad g^i = -u.\nabla\psi_i. \tag{2.4}$$

Since ψ_1 is C^∞ on $\mathbb{R}^3$ with $\operatorname{supp}\psi_1 \subset \Omega$, we have naturally denoted by $f\psi_1$ the distribution on $\mathbb{R}^3$ given by:

$$\forall \varphi \in \mathcal{D}(\mathbb{R}^3), \qquad < f\psi_1, \varphi >_{\mathbb{R}^3} = < f, \varphi\psi_1 >_\Omega .$$

This notation also applies to each other term in the definition (2.4) with $i = 1$.

Finally, considering (2.3) and (2.4) with $i = 2$, the regularity of u and π near the boundary depends on the regularity of (f^2, g^2) and on the properties of the Stokes problem in the bounded domain Ω_{R_2}. Similarly, the regularity of u and π near infinity depends on the regularity of (f^1, g^1) and on the properties of the Stokes problem in $\mathbb{R}^3$.

Regularity properties for the Stokes problem in bounded domains have been first studied by L. Cattabriga [6] but we shall use more general results from [2] (see pp. 134–136).

Theorem 2.1. (Amrouche-Girault [2]) *Let $\mathcal{O} \subset \mathbb{R}^3$ be a bounded domain with $C^{1,1}$ boundary. Let $f \in \mathbf{W}^{-1,q}(\mathcal{O}), g \in L^q(\mathcal{O})$ with $1 < q < +\infty$ and assume that $\int_{\mathcal{O}} g(x)dx = 0$. Then, the problem: Find $(w, \tau) \in \mathbf{W}^{1,q}(\mathcal{O}) \times L^q(\mathcal{O})$ such that*

$$-\nu\Delta w + \nabla\tau = f, \ \operatorname{div} w = g \ \ in\ \mathcal{O}, \qquad w_{|\partial\mathcal{O}} = \mathbf{0},$$

has a unique solution such that $\int_{\mathcal{O}} \tau dx = 0$. If f and ∇g moreover belong to $\mathbf{L}^q(\mathcal{O})$, then $\nabla^2 w$ and $\nabla\tau$ also belong to $\mathbf{L}^q(\mathcal{O})$.

The Stokes problem in the whole space has been recently much studied in various functional spaces (see for instance Borchers-Miyakawa [5], Girault-Sequeira [9], Kozono-Sohr [11, 12] or Specovius Neugebauer [15]). The authors have also provided a rather complete study of this problem in weighted Sobolev spaces in [1]. For instance, as a particular case of the results established in the latter reference (section 3), we can state the:

Theorem 2.2. (Alliot-Amrouche [1]) *Let $l \leq 0$ be an integer and $1 < p < +\infty$ such that $3/p$ is not an integer smaller than or equal to $-l$. For each $(f, g) \in \mathbf{W}_l^{-1,p}(\mathbb{R}^3) \times L_l^p(\mathbb{R}^3)$, the Stokes problem:*

$$(S) \quad -\Delta v + \nabla\eta = f, \qquad \operatorname{div} v = g \quad in\ \mathbb{R}^3,$$

has a solution such that $(v, \eta) \in \mathbf{W}_l^{1,p}(\mathbb{R}^3) \times L_l^p(\mathbb{R}^3)$. If f and ∇g moreover belong to $L_{l+1}^p(\mathbb{R}^3)$ then $\nabla^2 v$ and $\nabla\eta$ also belong to $L_{l+1}^p(\mathbb{R}^3)$.

2.2. A decomposition result for the pressure

We have seen in Theorem 1.3 that we can associate with each weak solution u a pressure π that locally belongs to L^2. But, we do not have yet any information concerning the integrability at infinity of π. Our first result is dedicated to this question.

Proposition 2.3. *Let $\Omega \subset \mathbb{R}^3$ be an exterior domain or $\Omega = \mathbb{R}^3$ and let $f \in \mathbf{W}_0^{-1,2}(\Omega)$. The pressure π obtained in Theorem 1.3 has a representative such that*

$$\pi = \tau^1 + \tau^2 \quad \textit{with} \quad \tau^1 \in W_0^{1,3/2}(\Omega),\ \tau^2 \in L^2(\Omega).$$

Proof. Let $u \in \mathbf{W}_0^{1,2}(\Omega)$ be a weak solution to the problem (NS) given by Theorem 1.3 and let $\pi \in L^2(\Omega_R)$, $\forall R \geq R_0$ be the associated pressure. First recall the decomposition $\pi = \pi^1 + \pi^2$ introduced in paragraph 2.1. Since $\pi \in L^2(\Omega_{R_2})$, we obtain that $\pi^2 = \pi\psi_2$ belongs to $L^2(\Omega)$. Thus, the main part of the proof deals with the properties of π^1 and therefore of (f^1, g^1).

i) We first consider the term $(u.\nabla u)\psi_1$ of f^1. From Sobolev's imbedding theorem, we know that $\mathbf{W}_0^{1,2}(\Omega) \subset \mathbf{L}^6(\Omega)$. Then, we have $\nabla u \in L^2(\Omega)$ and $u \in \mathbf{L}^6(\Omega)$. Since ψ_1 is bounded and supported in Ω, Hölder's inequality yields:

$$(u.\nabla u)\psi_1 \in \mathbf{L}^{3/2}(\mathbb{R}^3). \tag{2.5}$$

But we have: $L^{3/2}(\mathbb{R}^3) \subset W_{-1}^{-1,3/2}(\mathbb{R}^3)$ which is the dual imbedding of $W_1^{1,3}(\mathbb{R}^3) \subset L^3(\mathbb{R}^3)$ (the latter is obvious from the definition of $W_1^{1,3}(\mathbb{R}^3)$). Hence, in view of Theorem 2.2 (with $p = 3/2, l = -1$), there exists $(v^1, \eta^1) \in \mathbf{W}_{-1}^{1,3/2}(\mathbb{R}^3) \times L_{-1}^{3/2}(\mathbb{R}^3)$ such that

$$-\nu\Delta v^1 + \nabla\eta^1 = -(u.\nabla u)\psi_1, \quad \operatorname{div} v^1 = 0, \quad \text{in } \mathbb{R}^3. \tag{2.6}$$

Considering (2.5), Theorem 2.2 yields besides that $\nabla\eta^1 \in \mathbf{L}^{3/2}(\mathbb{R}^3)$ and so we get that $\eta^1 \in W_0^{1,3/2}(\mathbb{R}^3)$.

ii) We consider now the other terms of f^1. Since ψ_1 is bounded and has bounded derivatives with compact support, it is easy to check that the terms $f\psi_1$, $\nabla u \nabla\psi_1$, $u\Delta\psi_1$ and $\pi\nabla\psi_1$ belong to $\mathbf{W}_0^{-1,2}(\mathbb{R}^3)$. Proving that $g^1 = -u.\nabla\psi_1 \in L^2(\mathbb{R}^3)$ is even simpler. Then, applying Theorem 2.2 (with $p = 2, l = 0$), we get the existence of $(v^2, \eta^2) \in \mathbf{W}_0^{1,2}(\mathbb{R}^3) \times L^2(\mathbb{R}^3)$ such that

$$\begin{aligned} -\nu\Delta v^2 + \nabla\eta^2 &= f\psi_1 - 2\nu\nabla u\nabla\psi_1 - \nu u\Delta\psi_1 + \pi\nabla\psi_1, \\ \operatorname{div} u^2 &= -u.\nabla\psi_1, \quad \text{in } \mathbb{R}^3. \end{aligned} \tag{2.7}$$

iii) Let us finally set $w = u^1 - v^1 - v^2$ and $\tau = \pi^1 - \eta^1 - \eta^2$. Subtracting (2.6) and (2.7) from (2.3) yields the relations:

$$-\nu\Delta w + \nabla\tau = \mathbf{0}, \quad \operatorname{div} w = 0, \quad \text{in } \mathbb{R}^3. \tag{2.8}$$

Then, computing the divergence of the first equation yields that τ is harmonic. Therefore, considering (2.8), Δw is also harmonic. Thus, w is a tempered biharmonic distribution on $\mathbb{R}^3$, and thus a polynomial. But this polynomial moreover belongs to $\mathbf{W}_0^{1,2}(\mathbb{R}^3) + \mathbf{W}_{-1}^{1,3/2}(\mathbb{R}^3)$ so that it has to be constant (a complete proof of this statement relies on some estimates of the L^p-mean on the sphere of radius R of the functions of $\mathbf{W}_\alpha^{1,p}(\mathbb{R}^3)$ when R tends to infinity ; see [1], Lemma 1.1). Since w is constant, we deduce from (2.8) that $\nabla\tau = \mathbf{0}$ and by the way the existence of a constant c such that $\pi^1 = \eta^1 + \eta^2 + c$. Hence, we have the equality $\pi = \pi^1 + \pi^2 = \eta^1 + (\eta^2 + \pi^2) + c$ in Ω and the proposition is proved setting $\tau^1 = \eta^1$ and $\tau^2 = \eta^2 + \pi\psi_2$. $\diamondsuit$

Remark 2.4. The decomposition of the pressure established in Proposition 2.3 allows to rewrite the first equation of the system (NS) as follows:

$$(-\nu\Delta u + \nabla\tau^2) + (u.\nabla u + \nabla\tau^1) = f.$$

Here, the first term belongs to $\mathbf{W}_0^{-1,2}$. The second term is more regular since it belongs to $L^{3/2}$. In a certain sense, the pressure τ^1 is associated with the viscosity term $\nu\Delta u$ while τ^1 is associated with the convection term $u.\nabla u$.

2.3. First L^p regularity results

From now on, we assume that the force f is more regular than needed in Theorem 1.3 and prove that weak solutions are also more regular. As in the previous paragraph, we consider separately the regularity near the boundary and near infinity. Let us begin with a few properties of the non-linear term.

Lemma 2.5. *Let $\Omega \subset \mathbb{R}^3$ be an exterior domain or $\Omega = \mathbb{R}^3$.*
i) Let $v \in \mathbf{W}_0^{1,2}(\Omega)$, then $(v.\nabla v) \in \mathbf{L}^{3/2}(\Omega) \cap \mathbf{W}_0^{-1,3}(\Omega)$.
ii) Let $v \in \mathbf{W}_0^{1,2}(\Omega) \cap \mathbf{W}_0^{1,3}(\Omega)$, then $(v.\nabla v) \in \mathbf{L}^{s_1}(\Omega) \cap \mathbf{W}_0^{-1,s_2}(\Omega)$, if $3/2 \le s_1 < 3$ and if $s_2 \ge 3$.

Proof. The proof relies on the Sobolev's imbedding theorem which implies that if $p < 3$ then $W_0^{1,p}(\Omega) \subset L^{p*}(\Omega)$, and therefore by duality that

$$\forall p < 3, \quad L^p(\Omega) \subset W_0^{-1,p^*}(\Omega). \tag{2.9}$$

i) If $v \in \mathbf{W}_0^{1,2}(\Omega)$, then v belongs to $\mathbf{L}^6(\Omega)$ and $\nabla v \in L^2(\Omega)$. Therefore, Hölder's inequality yields that $v.\nabla v \in \mathbf{L}^{3/2}(\Omega)$ which space is imbedded into $\mathbf{W}_0^{-1,3}(\Omega)$ in view of (2.9).

ii) Let $v \in \mathbf{W}_0^{1,2}(\Omega) \cap \mathbf{W}_0^{1,3}(\Omega)$. Since $\nabla v \in L^2(\Omega) \cap L^3(\Omega)$, we also have $\nabla v \in L^r(\Omega)$, $2 \le r \le 3$. Since v also belongs to $\mathbf{L}^6(\Omega)$, the Gagliardo-Nirenberg inequalities (see for instance, Nirenberg [14], p. 125, with $r = 3$, $q = 6$, $j = 0$ and $m = 1$) imply that $v \in \mathbf{L}^s(\Omega)$ provided that $6 \le s < +\infty$. Hence, Hölder's inequality yields that $(v.\nabla v) \in \mathbf{L}^{s_1}(\Omega)$ for all s_1 such that $3/2 \le s_1 < 3$. $\diamond$

We now prove the

Theorem 2.6. *Let $\Omega \subset \mathbb{R}^3$ be an exterior domain with $C^{1,1}$ boundary or $\Omega = \mathbb{R}^3$. Given $p \ge 3$ and $f \in \mathbf{W}_0^{-1,2}(\Omega) \cap \mathbf{W}_0^{-1,p}(\Omega)$, each weak solution $u \in \mathbf{W}_0^{1,2}(\Omega)$ to the problem (NS) also satisfies $u \in \mathbf{W}_0^{1,p}(\Omega)$. Moreover, the associated pressure π has a representative in $L^3(\Omega) \cap L^p(\Omega)$.*

Proof. We use once again the auxiliary problems introduced in paragraph 2.1. We first prove the case $p = 3$ and then consider the case $p > 3$.

i) The case $p = 3$: In view of Lemma 2.5, we know that $u.\nabla u \in \mathbf{W}_0^{-1,3}(\Omega)$ and therefore $(u.\nabla u)\psi_1 \in \mathbf{W}_0^{-1,3}(\mathbb{R}^3)$. Moreover, since $u \in \mathbf{H}^1_{loc}(\Omega)$, $\pi \in L^2(\Omega_{R_2})$ and since the derivatives of ψ_1 have compact support, we deduce from Sobolev injections theorem that

$$-2\nu \nabla u \nabla \psi_1 - \nu u \Delta \psi_1 + \pi \nabla \psi_1 \in \mathbf{W}_0^{-1,3}(\mathbb{R}^3), \qquad - u.\nabla \psi_1 \in L^3(\mathbb{R}^3).$$

Hence, the pair (f^1, g^1) (see (2.4)) belongs to $\mathbf{W}_0^{-1,3}(\mathbb{R}^3) \times L^3(\mathbb{R}^3)$. Then, there exists (Theorem 2.2 with $p = 3$, $l = 0$) some functions $(v, \eta) \in \mathbf{W}_0^{1,3}(\mathbb{R}^3) \times L^3(\mathbb{R}^3)$ such that:

$$-\nu \Delta v + \nabla \eta = f^1, \quad \operatorname{div} v = g^1 \quad \text{in } \mathbb{R}^3.$$

Subtracting these equalities from (2.3), we get:

$$-\nu \Delta(u^1 - v) + \nabla(\pi^1 - \eta) = \mathbf{0}, \quad \operatorname{div}(u^1 - v) = 0 \quad \text{in } \mathbb{R}^3. \tag{2.10}$$

Therefore, following the proof of Proposition 2.3 (*iii*), we prove that $u^1 - v$ is a polynomial. Since this polynomial belongs to $\mathbf{W}_0^{1,2}(\mathbb{R}^3) + \mathbf{W}_0^{1,3}(\mathbb{R}^3)$, it must be a constant polynomial $\mathbf{c}$. But constant polynomials belong to $\mathbf{W}_0^{1,3}(\mathbb{R}^3)$ (because of the logarithmic weight), so that

$$u^1 = v + \mathbf{c} \in \mathbf{W}_0^{1,3}(\mathbb{R}^3). \tag{2.11}$$

Besides, since $u^1 - v$ is constant, it follows from (2.10) that $\nabla(\pi^1 - \eta) = \mathbf{0}$ in $\mathbb{R}^3$. Therefore, there exists a constant function d such that

$$\pi^1 = \eta + d, \qquad \eta \in L^3(\mathbb{R}^3). \tag{2.12}$$

Let us now come to the regularity near the boundary. Recall that the auxiliary functions $(u^2, \pi^2) \in \mathbf{H}_0^1(\Omega_{R_2}) \times L^2(\Omega_{R_2})$ satisfy (2.3) with $i = 2$. Moreover, we can prove -as we proved that $(f^1, g^1) \in \mathbf{W}_0^{-1,3}(\mathbb{R}^3) \times L^3(\mathbb{R}^3)$, but applying local Sobolev's imbedding results- that $(f^2, g^2) \in \mathbf{W}^{-1,3}(\Omega_{R_2}) \times L^3(\Omega_{R_2})$. With such data, and since Ω_{R_2} has $C^{1,1}$ boundary, we can deduce from Theorem 2.1 that $(u^2, \pi^2) \in \mathbf{W}^{1,3}(\Omega_{R_2}) \times L^3(\Omega_{R_2})$, which immediately imply that

$$(u^2, \pi^2) \in \mathbf{W}_0^{1,3}(\Omega) \times L^3(\Omega). \tag{2.13}$$

Finally, since $u = u^1 + u^2$ et $\pi = \pi^1 + \pi^2$, our claim results from (2.11), (2.12) and (2.13). Note that we can also prove that the representative of π in $L^3(\Omega)$ is nothing but the representative obtained in Proposition 2.3.

ii) The case $p > 3$: Let $f \in \mathbf{W}_0^{-1,2}(\Omega) \cap \mathbf{W}_0^{-1,p}(\Omega)$. Owing to an interpolation argument, we can prove that $f \in \mathbf{W}_0^{-1,3}(\Omega)$ and since we have proved the theorem for $p = 3$, we know that $u \in \mathbf{W}_0^{1,2}(\Omega) \cap \mathbf{W}_0^{1,3}(\Omega)$ and we can choose $\pi \in L^3(\Omega)$. Then, Lemma 2.5 (*ii*) implies that $(u.\nabla u) \in \mathbf{W}_0^{-1,p}(\Omega)$ and therefore that $(u.\nabla u)\psi_1 \in \mathbf{W}_0^{-1,p}(\mathbb{R}^3)$. Besides, Sobolev's imbedding theorem yields that $u \in L^p(\Omega_{R_2})$ and so, as in the case $p = 3$, we prove that

$$(f^1, g^1) \in \mathbf{W}_0^{-1,p}(\mathbb{R}^3) \times L^p(\mathbb{R}^3) \quad \text{and} \quad (f^2, g^2) \in \mathbf{W}^{-1,p}(\Omega_{R_2}) \times L^p(\Omega_{R_2}).$$

Starting with this regularity, each argument used in the point (*i*) can be restated replacing the exponent 3 with p and so the proof is complete. ◇

Now, the existence of weak solutions to the problem (NS) that satisfy the decay condition (0.1) is a rather simple consequence of Theorem 2.6.

Corollary 2.7. *Assume that $f \in \mathbf{W}_0^{-1,2}(\Omega) \cap \mathbf{W}_0^{-1,p}(\Omega)$, $p > 3$. Then, each weak solution $u \in \mathbf{W}_0^{1,2}(\Omega)$ to the problem (NS) satisfies*

$$u \in \mathbf{L}^\infty(\Omega) \quad \textit{and} \quad \lim_{|x| \to \infty} u(x) = \mathbf{0}. \tag{2.14}$$

Proof. We know from Theorem 2.6 that $u \in \mathbf{W}_0^{1,2}(\Omega) \cap \mathbf{W}_0^{1,p}(\Omega)$ and therefore

$$u \in \mathbf{L}^6(\Omega) \quad \text{and} \quad \nabla u \in L^p(\Omega), \quad p > 3,$$

which property is known to imply (2.14). ◇

Remark 2.8. Let us mention a different version of Theorem 2.6 which focuses only on the properties at infinity of the solution. Owing to the partition of unit (2.1),(2.2), we have seen that the behaviour of the solution near the boundary and near infinity can be obtained separately. In fact, looking more carefully, we see that the properties of (u^1, π^1) only depend on the regularity of the restrictions of f and g to Ω^{R_1}. Therefore, if we only assume that $f \in \mathbf{W}_0^{-1,2}(\Omega^{R_1}) \cap \mathbf{W}_0^{-1,p}(\Omega^{R_1})$ with $p \geq 3$, we can still prove that each weak solution $u \in \mathbf{W}_0^{1,2}(\Omega)$ also satisfies $u \in \mathbf{W}_0^{1,p}(\Omega^{R_2})$ and that the associated pressure π has a representative such that $\pi \in L^p(\Omega^{R_2})$. The main interest of this version is that it requires no smoothness assumption on the boundary $\partial\Omega$ and therefore applies to a wider class of domains.

2.4. More regularity and decay

In this paragraph, we are interested in the L^p regularity of $\nabla^2 u$ and $\nabla\pi$. In particular we shall need the following imbedding results:

Lemma 2.9. *Let $\Omega \subset \mathbb{R}^3$ be an exterior domain or $\Omega = \mathbb{R}^3$. Assume that $\alpha, \beta \in \mathbb{R}$ and $1 < p < q < +\infty$ satisfy $3/q + \beta > 3/p + \alpha$. Then, the following relations hold*

$$L^q_\beta(\Omega) \subset L^p_\alpha(\Omega), \qquad W^{1,q}_\beta(\Omega) \subset W^{1,p}_\alpha(\Omega),$$

with continuous imbeddings.

Proof. *i)* Let $v \in L^q_\beta(\Omega)$. The assumption $3/q + \beta > 3/p + \alpha$ yields that

$$\alpha - \beta < 3(\frac{1}{q} - \frac{1}{p}). \tag{2.15}$$

Since $1 < p < q$, there exists a real number r such that $1 < r < +\infty$ and $1/r = 1/p - 1/q$. Then, the inequality (2.15) implies that $\rho^{\alpha-\beta} \in L^r(\Omega)$ and Hölder's inequality yields that

$$\| \rho^\alpha v \|_{L^p(\Omega)} \leq \| \rho^{\alpha-\beta} \|_{L^r(\Omega)} \| \rho^\beta v \|_{L^q(\Omega)},$$

which proves the first imbedding.

ii) The second imbedding is a straightforward consequence of the first one if $3/q + \beta \neq 1$ (there is no logarithmic weight in $W^{1,q}_\beta(\Omega)$). When $3/q + \beta = 1$, we remark that (2.15) also implies that $\rho^{\alpha-\beta} \ln \rho \in L^r(\Omega)$. Hence, Hölder's inequality yields the result because $\rho^{\alpha-1} v = (\rho^{\alpha-\beta} \ln \rho).(\rho^{\beta-1} v / \ln \rho)$. ◇

We now prove the following theorem:

Theorem 2.10. *Let $\Omega \subset \mathbb{R}^3$ be an exterior domain with $C^{1,1}$ boundary or $\Omega = \mathbb{R}^3$ and let $f \in \mathbf{W}_0^{-1,2}(\Omega)$.*

i) Assume that $f \in \mathbf{L}^p(\Omega) \cap \mathbf{W}_0^{-1,q}(\Omega)$, $q \geq p > 3$. Then, each weak solution $u \in \mathbf{W}_0^{1,2}(\Omega)$ to the problem (NS) satisfies $\nabla u \in L^q(\Omega)$, $\nabla^2 u \in L^p(\Omega)$ and the pressure π has a representative such that $\pi \in L^q(\Omega)$ and $\nabla\pi \in \mathbf{L}^p(\Omega)$.

ii) Assume that $f \in \mathbf{L}^p(\Omega)$ with $3/2 \leq p < 3$. Then, each weak solution $u \in \mathbf{W}_0^{1,2}(\Omega)$ to the problem (NS) satisfies $\nabla u \in L^{p*}(\Omega), \nabla^2 u \in L^p(\Omega)$ and the pressure π has a representative such that $\pi \in L^{p*}(\Omega)$ and $\nabla\pi \in \mathbf{L}^p(\Omega)$.

Proof. We first prove the first part of the theorem: Since $f \in \mathbf{W}_0^{-1,q}(\Omega)$ with $q > 3$, we know from Theorem 2.6 that

$$u \in \mathbf{W}_0^{1,2}(\Omega) \cap \mathbf{W}_0^{1,q}(\Omega), \quad \pi \in L^3(\Omega) \cap L^q(\Omega). \tag{2.16}$$

In particular, we have $\nabla u, \pi \in L^q(\Omega)$ and we now have to prove the regularity of $\nabla^2 u$ and $\nabla\pi$. But, (2.16) and obvious interpolation arguments imply that,

$$u \in \mathbf{W}_0^{1,r}(\Omega),\ 2 \leq r \leq q \quad \text{and} \quad \pi \in L^s(\Omega),\ 3 \leq s \leq q. \tag{2.17}$$

In particular, we have $\nabla u \in L^p(\Omega)$. Besides, Corollary 2.7 yields that $u \in \mathbf{L}^\infty(\Omega)$ so that we obtain

$$u.\nabla u \in \mathbf{L}^p(\Omega). \tag{2.18}$$

Since $q \geq p$, we can easily deduce from (2.16) and from (2.18) that:

$$(f^1, g^1) \in \mathbf{L}^p(\mathbb{R}^3) \times W_0^{1,p}(\mathbb{R}^3) \quad \text{and} \quad (f^2, g^2) \in \mathbf{L}^p(\Omega_{R_2}) \times W^{1,p}(\Omega_{R_2}). \tag{2.19}$$

Then, the regularity properties of the Stokes problem in bounded domains (Theorem 2.1) and the equalities (2.3) with $i = 2$ yield that

$$\nabla^2 u^2 \in L^p(\Omega_{R_2}), \quad \nabla\pi^2 \in \mathbf{L}^p(\Omega_{R_2}). \tag{2.20}$$

On the other hand, we can choose $r = s > p$ in (2.17) so that we have $3/r > 0 > 3/p - 1$. Then, Lemma 2.9 yields that $(u, \pi) \in \mathbf{W}_{-1}^{1,p}(\Omega) \times L_{-1}^p(\Omega)$, which implies that

$$(u^1, \pi^1) \in \mathbf{W}_{-1}^{1,p}(\mathbb{R}^3) \times L_{-1}^p(\mathbb{R}^3). \tag{2.21}$$

In view of (2.19) and (2.21), we can apply the regularity statement of Theorem 2.2 with $l = -1$. This yields that

$$\nabla^2 u^1 \in L^p(\mathbb{R}^3), \quad \nabla\pi^1 \in \mathbf{L}^p(\mathbb{R}^3), \tag{2.22}$$

which, together with (2.20), completes the proof, since $u = u^1 + u^2$ and $\pi = \pi^1 + \pi^2$.

ii) We now turn to the second point of the theorem. First remark that since $f \in \mathbf{L}^p(\Omega)$, with $p < 3$, then the imbedding (2.9) implies that $f \in \mathbf{W}_0^{-1,p*}(\Omega), p^* \geq 3$. In particular, all the arguments of the latter proof can be restated with p^* instead of q, except the proofs of (2.18) and (2.21) where some modifications occur. Indeed, in this case the relation (2.18) follows from Lemma 2.5 since we can set $r = 3$ in (2.17). The modified proof of (2.21) involves two cases. If $p > 3/2$, we can choose $3 < r = s < p*$ in (2.17) and then conclude with Lemma 2.9. In the remaining case $p = 3/2$, we use on the one hand the fact that $\mathbf{W}_0^{1,2}(\Omega) \subset \mathbf{W}_{-1}^{1,3/2}(\Omega)$ in view of Lemma 2.9. Therefore, we obtain that $u^1 \in \mathbf{W}_{-1}^{1,3/2}(\mathbb{R}^3)$. On the other hand, we recall that $\pi = \tau^1 + \tau^2$ with $\tau^1 \in W_0^{1,3/2}(\Omega)$ and $\tau^2 \in L^2(\Omega)$ (Proposition 2.3). Then, the imbedding $W_0^{1,3/2}(\Omega) \subset L_{-1}^{3/2}(\Omega)$ is obvious, and Lemma 2.9 proves that $L^2(\Omega) \subset L_{-1}^{3/2}(\Omega)$, so that $\pi^1 \in L_{-1}^{3/2}(\mathbb{R}^3)$. ◇

The following is an easy consequence of Theorem 2.10.

Corollary 2.11. *Let $\Omega \subset \mathbb{R}^3$ be an exterior domain with $C^{1,1}$ boundary or $\Omega = \mathbb{R}^3$ and let $f \in \mathbf{W}_0^{-1,2}(\Omega)$.*
i) Assume that $f \in \mathbf{L}^p(\Omega) \cap \mathbf{W}_0^{-1,q}(\Omega)$, $q \geq p > 3$. Then, each weak solution $u \in \mathbf{W}_0^{1,2}(\Omega)$ to the problem (NS) satisfies (0.1). *Moreover, $\nabla u, \pi \in L^\infty(\Omega)$ and*

$$\lim_{|x|\to+\infty} \nabla u(x) = 0, \quad \lim_{|x|\to+\infty} \pi(x) = 0 .$$

ii) Assume that $f \in L^p(\Omega)$, $3/2 < p < 3$. Then, each weak solution $u \in \mathbf{W}_0^{1,2}(\Omega)$ to the problem (NS) satisfies (0.1).

Proof. *i)* If $f \in \mathbf{L}^p(\Omega) \cap \mathbf{W}_0^{-1,q}(\Omega)$, $q \geq p > 3$, then Corollary 2.7 applies and so (0.1) holds. Besides, Theorem 2.10 yields that $\pi, \nabla u \in L^q(\Omega)$ and that $\nabla\pi, \nabla^2 u \in L^p(\Omega)$ with $p > 3$, which properties imply the result.

ii) If $f \in \mathbf{L}^p(\Omega)$, $3/2 < p < 3$, then (2.9) implies that $f \in \mathbf{W}_0^{-1,p^*}(\Omega)$ with $p^* > 3$ and thus Corollary 2.7 applies. ◇

Remark 2.12. *i)* The statement (*i*) in Theorem 2.10 still holds if $q > p = 3$. The adaptations of the proof to this case are straightforward. In contrast, the proof does not extend if $q = p = 3$. Indeed, we would have to apply Theorem 2.2 with $p = 3$ and $l = -1$, which case is excluded (we insist on the fact that the conclusions of Theorem 2.2 are false if $3/p$ is an integer smaller than or equal to $-l$).

ii) The method we used in this section also allows to prove more regularity properties. Assume for instance that the boundary $\partial\Omega$ is $C^{2,1}$. Then, if $f \in \mathbf{W}_0^{-1,2}(\Omega)$ and if one of the following conditions holds:

$$\begin{array}{ll} (a) \ f \in \mathbf{W}_0^{1,p}(\Omega), & 6/5 \leq p < 3/2, \\ (b) \ f \in \mathbf{W}^{1,p}(\Omega), & 3/2 < p < 3, \end{array}$$

then we can prove that each weak solution $u \in \mathbf{W}_0^{1,2}(\Omega)$ satisfies $\nabla^3 u \in L^p(\Omega)$. We obtain simultaneously that $\nabla^2 \pi \in L^p(\Omega)$. Besides, when assumption (b) holds, then we can establish that both $\nabla^2 u$ and $\nabla\pi$ are bounded and vanish at infinity.

2.5. Improved regularity for the pressure in $\mathbb{R}^3$

This last section is devoted to some sharp regularity properties of the pressure π when the domain Ω is the whole space $\mathbb{R}^3$. It is based on a result of R. Coifman, P.L.Lions, Y.Meyer and S. Semmes ([7], Th. II.1) that deals with the regularity of various non-linear quantities. This result is of particular interest to our problem since it establishes that

$$\text{if} \quad u \in \mathbf{W}_0^{1,2}(\mathbb{R}^3) \quad \text{then} \quad \operatorname{div}(u.\nabla u) \in \mathcal{H}^1(\mathbb{R}^3). \tag{2.23}$$

Here, the Hardy space $\mathcal{H}^1(\mathbb{R}^3)$ stands for the following subspace of $L^1(\mathbb{R}^3)$:

$$\mathcal{H}^1(\mathbb{R}^3) = \{u \in L^1(\mathbb{R}^3), \ R_j u \in L^1(\mathbb{R}^3), \ \forall j = 1,2,3\},$$

where the three-dimensional Riesz transforms R_j are given by :

$$R_j(f) = c\,\text{p.v.}\ (f * \frac{x_j}{|x|^4}), \ j = 1,\dots,3.$$

Therefore, we prove the following:

Theorem 2.13. *Let $f \in \mathbf{W}_0^{-1,2}(\mathbb{R}^3)$ and let $u \in \mathbf{W}_0^{1,2}(\mathbb{R}^3)$ be a weak solution to the problem (NS). If* $\operatorname{div} f \in \mathcal{H}^1(\mathbb{R}^3)$*, then the associated pressure π has a representative such that:*

$$\pi \in W_0^{1,3/2}(\mathbb{R}^3) \quad \text{and} \quad \nabla^2\pi \in \mathcal{H}^1(\mathbb{R}^3).$$

Proof.

i) Let us assume in view of Proposition 2.3 that $\pi \in W_0^{1,3/2}(\mathbb{R}^3) + L^2(\mathbb{R}^3)$. Since $\operatorname{div} u = 0$, we obtain by computing the divergence of the first equation of the problem (NS) that

$$\Delta\pi = \operatorname{div} f - \operatorname{div}(u.\nabla u) \ \text{in} \ \mathbb{R}^3. \tag{2.24}$$

In particular, if $\operatorname{div} f \in \mathcal{H}^1(\mathbb{R}^3)$ and in view of (2.23), we have

$$\operatorname{div} f - \operatorname{div}(u.\nabla u) \in \mathcal{H}^1(\mathbb{R}^3). \tag{2.25}$$

ii) We shall now obtain the regularity of π considering some regularity properties of the Laplacian in $\mathbb{R}^3$. First note that $\mathcal{H}^1(\mathbb{R}^3) \subset W_0^{-1,3/2}(\mathbb{R}^3) \perp \mathcal{P}_0$ where

$$W_0^{-1,3/2}(\mathbb{R}^3) \perp \mathcal{P}_0 = \{\varphi \in W_0^{-1,3/2}(\mathbb{R}^3), \quad < \varphi, 1 >_{W_0^{-1,3/2} \times W_0^{1,3}} = 0\}.$$

Now, we know from [3](Th. 5.1) that the Laplacian is an isomorphism from $W_0^{1,3/2}(\mathbb{R}^3)$ onto $W_0^{-1,3/2}(\mathbb{R}^3) \perp \mathcal{P}_0$. Therefore, there exists $\eta \in W_0^{1,3/2}(\mathbb{R}^3)$ such that

$$\Delta \eta = \operatorname{div} f - \operatorname{div}(u.\nabla u) \quad \text{in} \quad \mathbb{R}^3. \tag{2.26}$$

Moreover, since the Riesz transforms R_j, $j = 1, 2, 3$ are continuous from $\mathcal{H}^1(\mathbb{R}^3)$ into $\mathcal{H}^1(\mathbb{R}^3)$, the following identity

$$\frac{\partial^2 \eta}{\partial x_j \partial x_k} = -R_j R_k(\Delta \eta),$$

yields together with (2.23) that $\nabla^2 \eta \in \mathcal{H}^1(\mathbb{R}^3)$.

iii) Finally, we are going to prove that $\pi = \eta$, which completes the proof. Indeed, we obtain by subtracting (2.26) from (2.24) that $\pi - \eta \in L^2(\mathbb{R}^3) + W_0^{1,3/2}(\mathbb{R}^3)$ is an harmonic function. Then, $\pi - \eta$ is a polynomial that moreover belongs to $L^2(\mathbb{R}^3) + L^3(\mathbb{R}^3)$; and so it must be identically zero. $\diamondsuit$

Remark 2.14. We are not able to prove a similar result when Ω is an exterior domain. If we assume that, near infinity, $\operatorname{div} f$ is the restriction of a function belonging to $\mathcal{H}^1(\mathbb{R}^3)$, it seems difficult to establish that $\nabla^2 \pi$ enjoys the same regularity. For instance, we cannot use efficiently the cut-off procedure of Section 2. Indeed, it is easy to check that π^1 satisfies

$$\Delta \pi^1 = \operatorname{div} f^1 + \nu \Delta g^1,$$

but we cannot even prove that $\operatorname{div} f^1 + \nu \Delta g^1$ belongs to $L^1(\mathbb{R}^3)$.

References

[1] Alliot, F. and Amrouche, C. The Stokes Problem in $\mathbb{R}^n$: an approach in weighted Sobolev spaces. Math. Methods App. Sci. To appear.

[2] Amrouche, C. and Girault, V. (1994). Decomposition of vector spaces and application to the Stokes problem in arbitrary dimensions. *Czechoslovak Math. J.*, 44(119):109–140.

[3] Amrouche, C., Girault, V. and Giroire, J. (1994). Weighted Sobolev spaces for the Laplace equation in $\mathbf{R}^n$. *J. Math. Pures Appl.*, 20:579–606.

[4] Amrouche, C., Girault, V. and Giroire, J. (1997). Dirichlet and Neumann exterior problems for the n-dimensional Laplace operator. An approach in weighted Sobolev spaces. *J. Math. Pures Appl.*, 76(1):55–81.

[5] Borchers, W. and Miyakawa, T. (1992). On some coercive estimates for the Stokes problem in unbounded domains. In Springer-Verlag, editor, *Navier-Stokes equations: Theory and numerical methods*, volume 1530, pages 71–84. Heywood, J.G., Masuda, K., Rautmann, R., Solonnikov, V.A. Lecture Notes in Mathematics.

[6] Cattabriga, L. (1961). Su un problema al contorno relativo al sistema di equazioni di Stokes. *Rend. Sem. Mat. Univ. Padova*, 31:308–340.

[7] Coifman, R., Lions, P.L., Meyer, Y. and Semmes, S. (1993). Compensated compactness and Hardy spaces. *J. Math. Pures Appl.*, 72(3):247–286.

[8] Galdi, G.P. (1994). *An introduction to the mathematical theory of the Navier-Stokes equations*, volume II. Springer tracts in natural philosophy.

[9] Girault, V. and Sequeira, A. (1991). A well posed problem for the exterior Stokes equations in two and three dimensions. *Arch. Rational Mech. Anal.*, 114:313–333.

[10] Girault, V. and Raviart, P.A. (1986). *Finite Element Approximation of the Navier-Stokes Equations. Theory and Algorithms.* Springer-Verlag, Berlin.

[11] Kozono, H. and Sohr, H. (1991). New a priori estimates for the Stokes equations in exteriors domains. *Indiana Univ. Math. J.*, 40:1–25.

[12] Kozono, H. and Sohr, H. (1992). On a new class of generalized solutions for the Stokes equations in exterior domains. *Ann. Scuola Norm. Sup. Pisa, Ser. IV*, 19:155–181.

[13] Leray, J. (1934). Sur le mouvement d'un liquide visqueux emplissant l'espace. *Acta Math.*, 63:193–248.

[14] Nirenberg, L. (1959). On elliptic partial differential equations. *Ann. Scuola Norm. Sup. Pisa*, 13:116–162.

[15] Specovius Neugebauer, M. (1994). Weak Solutions of the Stokes Problem in Weighted Sobolev Spaces. *Acta Appl. Math.*, 37:195–203.

[16] Tartar, L. (1978). *Topics in non linear analysis.* Publications math matiques d'Orsay.

[17] Temam, R. (1977). *Navier-Stokes equations.* North-Holland, Amsterdam - New-York - Tokyo.

A NOTE ON TURBULENCE MODELING

G. Q. Chen, K. R. Rajagopal, Luoyi Tao

Abstract: The main thrust of this work is developing the basis for a mixed formulation of turbulence modeling, combining analytical theories and engineering modeling, which includes second order two-point correlations of velocity and pressure. Related issues such as the different outcomes that stem from differently chosen sets of ensemble averaging, the approximate nature and the advantages and disadvantages of such a formulation, and the choices of closure schemes are addressed.

Keywords: Closure, averaging Reynolds stress, dissipation.

1. INTRODUCTION

Analytical theories of turbulence, on the one side, deal with multi-point correlations of velocity restricting themselves to homogeneous turbulence (Orszag [11], Proudman and Reid [12], Tatsumi [17]). On the other side, engineering turbulence modeling deals with general turbulent flows restricting itself to single-point correlations of velocity and pressure (Launder [7], Rodi [13]). It seems worthwhile to develop a mixed formulation, combining these two methods, to obtain information on multi-point correlations of fluctuating velocity and pressure to general turbulent flows, a formulation that we discuss here.

The importance of multi-point correlations in turbulence, especially the two-point correlation of fluctuating velocity ($U_{i'j''}$), lies in that they provide some information on the basic structure of turbulent motions, such as the various scales of length (dissipative, integral), the direct interaction between the fluctuations of different positions and the distribution (transfer) of fluctuation energy on (between) different eddies (Batchelor [1], Hinze [6]). The mixed formulation contains the equations governing $U_{i'j''}$, and $P_{0'j''}$, the two-point correlation of fluctuating velocity and pressure. In such a formulation, some difficulties arise, compared with single-point correlation models, which include (i) how to model a two-point correlation of three fluctuating velocities ($U_{i'k''j''}$), (ii) how to prescribe physically sound initial and boundary conditions for $U_{i'j''}$ and boundary condition for $P_{0'j''}$, and (iii) how to solve the equation in the seven-dimensions space (six for space and one for time) in which the equations are formulated. Despite these difficulties, the formulation has great advantages: (i) $U_{i'j''}$ is obtained whose importance was mentioned above; (ii) only one quantity, $U_{i'k''j''}$, needs to be modeled; and (iii) the need to prescribe the initial and the boundary conditions for $U_{i'j''}$ and $P_{0'j''}$ implies that the formulation may have wider applicability than single-point correlation modeling since the latter cannot account for such information.

This formulation is based on ensemble averaging the solutions of the Navier-Stokes equations under proper conditions to be discussed. Various sets of solutions can be employed, depending on which pattern of turbulent flows is to be modeled. For example, a proper subset of the solutions can be chosen to get a large eddy simulation type model, or the whole set of solutions can be used to formulate a model without any fluctuations. In the case of the former, if the filtering scale is very small, the multi-point correlations may not be necessary and a lower level model of closure will yield reasonable results like the Smagorinsky eddy viscosity model for some flows (Smagorinsky [16]). However, in the latter example, the multi-point correlations are quite important since all scales of fluctuations are filtered out and the interaction among these fluctuations need to be taken into account by the correlations.

We will demonstrate that the present formulation is not simply an extension of analytical theories of homogeneous turbulence. As only $U_{i'j''}$ and $P_{0'j''}$ are included to simplify the modeling, some function may need to be introduced to ensure the divergence-free condition for $P_{0'j''}$ due to the incompressibility of the fluid. We will discuss the approximate nature of such a formulation, with the implication that an averaged model of general applicability may be out of reach and models appropriate to different classes of turbulent flows should be pursued. It will also be self-evident that the present formulation is not a simple extension of single-point correlation modeling since the former cannot reduce to the latter without the assumption to handle the reduction of the dimension of the space where the former is constructed. We will show that there are several schemes for modeling $U_{i'k''j''}$, and the choice of the scheme depends on whether we emphasize simplicity or comprehensiveness. Other issues are also to be discussed such as the consequences of the symmetries of the Navier-Stokes equations, realizability and molecular dissipation.

Our main concern at this moment is the basis for the formulation. A great deal of effort has to be expended yet to construct concrete models to solve problems.

2. FORMULATION

Suppose that the Navier-Stokes equations can describe the turbulent motion of incompressible Newtonian fluids in a flow domain $\mathcal{D}$, (here it is tacitly assumed whether a physical fluid is a Newtonian fluid or not is determined by its behavior in laminar states), that is, the velocity $\mathbf{u}$ and the pressure ρq are determined by, together with proper initial and boundary conditions,

$$u_{m,m} = 0, \tag{2.1}$$

$$\frac{\partial u_i}{\partial t} + (u_i u_m)_{,m} = -q_{,i} + \nu u_{i,mm}, \tag{2.2}$$

where ρ and ν are, respectively, the mass density and the kinetic viscosity of the fluid. Following the standard practice in turbulence modeling of averaging, we consider the ensemble of solutions $\mathcal{U}$ to (2.1) and (2.2) under "the same flow conditions" which are identified with some global (or large scale) quantities characterizing the flows (Monin and Yaglom [9]). Next, instead of carrying out the ensemble averaging on $\mathcal{U}$, we choose a suitable subset $\mathcal{S}$ of $\mathcal{U}$, which is to be discussed later, and introduce the ensemble averaging $\langle\ \rangle := \langle\ \rangle_{\mathcal{S}}$ on $\mathcal{S}$ to define

$$U_i := \langle u_i \rangle, \quad P := \langle q \rangle, \tag{2.3}$$

and the decomposition,

$$u_i = U_i + v_i, \quad q = P + p, \quad \langle v_i \rangle = 0, \quad \langle p \rangle = 0, \tag{2.4}$$

where $\mathbf{v}$ is the fluctuating velocity relative to U_i and ρp the fluctuating pressure relative to ρP. Consequently, equations (2.1) and (2.2) result in (Hinze [6])

$$U_{m,m} = 0, \tag{2.5}$$

$$\frac{\partial U_i}{\partial t} + U_m U_{i,m} = -P_{,i} + \nu U_{i,mm} - \langle v_i v_m \rangle_{,m}, \tag{2.6}$$

$$v_{m,m} = 0, \tag{2.7}$$

$$\frac{\partial v_i}{\partial t} + (v_i v_m + U_i v_m + v_i U_m - \langle v_i v_m \rangle)_{,m} = -p_{,i} + \nu v_{i,mm}, \tag{2.8}$$

and

$$p_{,nn} = -(v_m v_n + v_m U_n + v_n U_m - \langle v_m v_n \rangle)_{,mn}. \tag{2.9}$$

We now introduce the multi-point correlations such as

$$\begin{aligned} U_{i'j''} &:= U_{ij}(\mathbf{x}',\mathbf{x}'',t) := \langle v_{i'}v_{j''}\rangle, \quad v_{i'} := v_i(\mathbf{x}',t), \ \ldots \\ U_{i'j''k'''} &:= U_{ijk}(\mathbf{x}',\mathbf{x}'',\mathbf{x}''',t) := \langle v_{i'}v_{j''}v_{k'''}\rangle, \\ P_{0'i''} &:= P_{0i}(\mathbf{x}',\mathbf{x}'',t) := \langle p_{0'}v_{i''}\rangle, \quad p_{0'} := p(\mathbf{x}',t), \end{aligned} \tag{2.10}$$

which have the following symmetry properties (Proudman and Reid [12])

$$U_{i'j''} = U_{j''i'}, \quad U_{i'j''k'''} = U_{j''i'k'''} = U_{i'k'''j''}, \tag{2.11}$$

and

$$\frac{\partial}{\partial x'_i} U_{i'j''k'''} = 0, \quad \text{etc.} \tag{2.12}$$

Then, equations (2.7) and (2.8) yield (Hinze [6], Proudman and Reid [12])

$$\frac{\partial}{\partial x'_i} P_{0''i'} = 0, \tag{2.13}$$

$$\frac{\partial}{\partial x'_i} U_{i'j''} = 0, \tag{2.14}$$

$$\begin{aligned} &\left(\frac{\partial}{\partial t} + U_{m'}\frac{\partial}{\partial x'_m} + U_{m''}\frac{\partial}{\partial x''_m}\right)U_{i'j''} + U_{i',m'}\,U_{m'j''} \\ &+U_{j'',m''}\,U_{i'm''} + \frac{\partial}{\partial x'_k}U_{j''k'i'} + \frac{\partial}{\partial x''_k}U_{i'k''j''} \\ &= -\frac{\partial}{\partial x'_i}P_{0'j''} - \frac{\partial}{\partial x''_j}P_{0''i'} + \nu\left(\frac{\partial^2}{\partial \mathbf{x}'^2} + \frac{\partial^2}{\partial \mathbf{x}''^2}\right)U_{i'j''}, \end{aligned} \tag{2.15}$$

with $U_{i'} := U_i(\mathbf{x}',t)$ and $U_{j''} := U_j(\mathbf{x}'',t)$.

If we take U_i, $U_{i'j''}$, P and $P_{0'i''}$ as the primary field quantities and model $U_{i'k''j''}$ in terms of these primary fields appropriately, we will find out that there are 14 equations consisting of (2.5), (2.6) and (2.13) through (2.15), but there are only 13 primary quantities. Therefore, we may need to introduce a scalar function S with $S(\mathbf{x}',\mathbf{x}'',t) = S(\mathbf{x}'',\mathbf{x}',t)$ in to the formulation through equation (2.15), say, according to

$$\begin{aligned} &\left(\frac{\partial}{\partial t} + U_{m'}\frac{\partial}{\partial x'_m} + U_{m''}\frac{\partial}{\partial x''_m}\right)U_{i'j''} + U_{i',m'}\,U_{m'j''} \\ &+U_{j'',m''}\,U_{i'm''} + \frac{\partial}{\partial x'_k}U_{j''k'i'} + \frac{\partial}{\partial x''_k}U_{i'k''j''} \\ &= -\frac{\partial}{\partial x'_i}P_{0'j''} - \frac{\partial}{\partial x''_j}P_{0''i'} + \nu\left(\frac{\partial^2}{\partial \mathbf{x}'^2} + \frac{\partial^2}{\partial \mathbf{x}''^2}\right)U_{i'j''} + S\delta_{i'j''}. \end{aligned} \tag{2.16}$$

Here $\delta_{i'j''}$ is the Kronecker delta. We may associate this introduction of S with the constraint (2.13) in the sense that S is not needed if (2.13) is not enforced, since we have 13 equations for the 13 primary quantities. Therefore S could be considered physically as a force-like quantity resulting from the constraint (2.13) or a force-like quantity imposing (2.13) on the averaged field $P_{0'i''}$. Next, we justify the term S in (2.16) with the following argument. Assume that

$$\frac{\partial}{\partial x''_k} U_{i'k''j''} = -\hat{S}(\mathbf{x}', \mathbf{x}'', t)\delta_{i'j''} + \frac{\partial}{\partial x''_k} \hat{U}_{i'k''j''}$$

and

$$\frac{\partial}{\partial x'_k} U_{j''k'i'} = -\hat{S}(\mathbf{x}'', \mathbf{x}', t)\delta_{j''i'} + \frac{\partial}{\partial x'_k} \hat{U}_{j''k'i'}$$

where $\hat{S}(\mathbf{x}', \mathbf{x}'', t)$ is due to (2.13) and $\hat{U}_{i'k''j''}$ ($\hat{U}_{j''k'i'}$) is to be modeled in terms of those primary fields chosen previously. This treatment is analogous to the introduction of the hydrodynamic pressure to the Cauchy stress tensor of an incompressible material. One reason for modeling $\frac{\partial}{\partial x''_k} U_{i'k''j''}$ instead of $U_{i'k''j''}$ is to avoid the need for prescribing boundary conditions for $S(\mathbf{x}',\mathbf{x}'',t) := \hat{S}(\mathbf{x}',\mathbf{x}'',t) + \hat{S}(\mathbf{x}'',\mathbf{x}',t)$ and also for keeping the form simple. Now we have

$$\frac{\partial}{\partial x'_k} U_{j''k'i'} + \frac{\partial}{\partial x''_k} U_{i'k''j''} = \frac{\partial}{\partial x'_k} \hat{U}_{j''k'i'} + \frac{\partial}{\partial x''_k} \hat{U}_{i'k''j''} - S\delta_{i'j''}.$$

Substituting this relation into (2.15) and dropping the hat ˆ, we obtain (2.16). This replacement of $\hat{U}_{i'k''j''}$ ($\hat{U}_{j''k'i'}$) with $U_{i'k''j''}$ ($U_{j''k'i'}$) should not cause any confusion based on the fact that (i) both of the quantities have to be modeled; and (ii) in case that $\hat{U}_{i'k''j''}$ is constructed under (2.12), $S = 0$ results from (2.17) below and we can take $\hat{U}_{i'k''j''} = U_{i'k''j''}$.

It is easy to verify that (2.13), (2.14) and (2.16) yield

$$\frac{\partial^2 S}{\partial x'_k \partial x''_k} = \frac{\partial^2}{\partial x'_i \partial x''_j} \left(\frac{\partial}{\partial x'_k} U_{j''k'i'} + \frac{\partial}{\partial x''_k} U_{i'k''j''} \right) \tag{2.17}$$

and

$$\begin{aligned} \frac{\partial^2}{\partial \mathbf{x}'^2} P_{0'j''} &= \frac{\partial S}{\partial x'_j} - 2U_{n',m'} \frac{\partial}{\partial x'_n} U_{m'j''} \\ &- \frac{\partial}{\partial x'_n} \left(\frac{\partial}{\partial x'_k} U_{j''k'n'} + \frac{\partial}{\partial x''_k} U_{n'k''j''} \right). \end{aligned} \tag{2.18}$$

Equation (2.17) shows that S can be solved in terms of $\frac{\partial}{\partial x''_k} U_{i'k''j''}$ and $S = 0$ can occur when the model of $U_{i'k''j''}$ meets the constraint (2.12). This also implies that the introduction of S can be considered as part of the modeling of $U_{i'k''j''}$.

Thus, we have a determinate set of equations for U_i, $U_{i'j''}$, P, $P_{0'i''}$ (and S) consisting of (2.5), (2.6), (2.13), (2.14) and (2.16), provided that $U_{i'k''j''}$ is appropriately modeled in terms of these primary field quantities. The approximate nature of this truncation scheme is easily understood as follows. $U_{i'k''j''}$, together with S, though the latter is determined by the former, is supposed to account for the interaction between the lower order correlations (the primary fields chosen above) and the higher order correlations. Here, the modeling of $U_{i'k''j''}$ essentially serves to characterize this interaction through some specific structure in terms of the lower order correlations, and consequently it restricts the interaction to some special form, and some information related to the higher order correlations is left out. For example, a specific structure of $U_{i'k''j''}$ cannot accommodate all the possible initial and boundary conditions of itself and the higher order correlations because the structure is assumed to be fixed in terms of those lower order correlations. And thus the motion is completely determined from the model as long as proper initial and boundary conditions of the lower order correlations are prescribed, disregarding any initial and boundary conditions for the higher order correlations. A possible implication of this argument is that there may be no one general structure for $U_{i'k''j''}$ which can model optimally all turbulent flows.

Another limitation of the scheme, or any scheme based on averaging in fact, needs to be addressed, namely what is the class of turbulent flows for which an averaging scheme can be applied to produce physically meaningful results. We should restrict the model to flows where the fluctuation is relatively small, for instance,

$$U_{\alpha\alpha} \ll U_\alpha U_\alpha \text{ for large } |U_\alpha|; \quad U_{\alpha\alpha} \text{ small for small } |U_\alpha|. \tag{2.19}$$

(Here, we adopt the convention that the summation rule is suspended if Greek subscripts are used.) This restriction is physically essential, otherwise the large fluctuation will make the averaged velocity field practically useless.

On selecting $\mathcal{S}$, a set of solutions from $\mathcal{U}$, on which the averaging $\langle\ \rangle$ operates, we have two cases in mind. One is $\mathcal{S} = \mathcal{U}$, the assumption in standard engineering turbulence modeling, which supposedly smoothes out the fluctuation of all scales so that the resultant averaged equations are not of a chaotic nature. The other deals with a proper subset of $\mathcal{U}$. For the sake of demonstration, let us choose a length scale l, much smaller than the characteristic length of $\mathcal{D}$, and a subset $\mathcal{U}_l$ of $\mathcal{U}$, whose members display almost the same flow structures on the scales larger than l. Then, the ensemble averaging $\langle\ \rangle$ of relation (2.3) on $\mathcal{U}_l$ is to filter out the fluctuations on the length scales smaller than l, under the premise that $\mathcal{U}_l$ contains enough members so that the operation $\langle\ \rangle$ can effectively smooth the flow details on the scales smaller than l. Therefore we can, based on this argument, relate this case of averaging to large eddy simulation (LES) (Ferziger [3], [4]). This formulation has the advantage that both the commutativity of the averaging and differentiation and $(2.4)_{3,4}$ hold and it is clearly connected to the standard turbulence modeling of averaging (different subsets of $\mathcal{U}$). Its disadvantage may be that the fields do not seem to be defined as concretely as those of LES. In both cases of $\mathcal{S} = \mathcal{U}$ and $\mathcal{S} = \mathcal{U}_l$, each motion $\{u_i,\ q\}$ is decomposed into a main flow field $\{U_i,\ P\}$ and a fluctuating part $\{v_i,\ p\}$; the former varies relatively slowly in time and space and the latter changes rather randomly and rapidly in time and space. The averaging formulation seeks to reformulate the equations of motion within the frame of the characteristic time-space scale of the main flow field. The direct interaction between the fluctuations of different positions is described by the two-point correlations.

Now we present some justifications for introducing the above two-point correlation formulation. Firstly, this formulation is an extension of both the standard engineering turbulence modeling of single-point correlations and the analytic theory of homogeneous turbulence of multi-point correlations. Secondly, the model can be closed provided that the quantity $U_{i'k''j''}$ is formulated properly, while S is introduced. The interaction among fluctuations of different positions and the averaged velocity field can be characterized through $U_{i'j''}$. (The field $\{U_i,\ P\}$ is affected by $\{U_{i'j''},\ P_{0'j''}\}$ through U_{ij}, but a nontrivial U_i will influence $\{U_{i'j''},\ P_{0'j''}\}$ directly.) Finally, this two-point correlation model is much simpler than any formulation involved in higher order correlation quantities. For example, it follows from (2.7) and (2.8) that

$$
\begin{aligned}
&\left(\frac{\partial}{\partial t} + U_{m'}\frac{\partial}{\partial x'_m} + U_{m''}\frac{\partial}{\partial x''_m} + U_{m'''}\frac{\partial}{\partial x'''_m}\right)U_{i'j''k'''} \\
&\quad + U_{i',m'}U_{m'j''k'''} + U_{j'',m''}U_{i'm''k'''} + U_{k''',m'''}U_{i'j''m'''} \\
&= -\frac{\partial}{\partial x'_i}P_{0'j''k'''} - \frac{\partial}{\partial x'_m}\left(U_{i'm'j''k'''} - U_{i'm'}U_{j''k'''}\right) \\
&\quad -\frac{\partial}{\partial x''_j}P_{0''i'k'''} - \frac{\partial}{\partial x''_m}\left(U_{i'j''m''k'''} - U_{j''m''}U_{i'k'''}\right) \\
&\quad -\frac{\partial}{\partial x'''_k}P_{0'''i'j''} - \frac{\partial}{\partial x'''_m}\left(U_{i'j''k'''m'''} - U_{k'''m'''}U_{i'j''}\right) \\
&\quad +\nu\left(\frac{\partial^2}{\partial \mathbf{x}'^2} + \frac{\partial^2}{\partial \mathbf{x}''^2} + \frac{\partial^2}{\partial \mathbf{x}'''^2}\right)U_{i'j''k'''},
\end{aligned}
\tag{2.20}
$$

where

$$
\begin{aligned}
&U_{ij'k''l'''} := U_{ijkl}(\mathbf{x}, \mathbf{x}', \mathbf{x}'', \mathbf{x}''', t) := \langle v_i v_{j'} v_{k''} v_{l'''}\rangle, \\
&P_{0'i''j'''} := P_{0ij}(\mathbf{x}', \mathbf{x}'', \mathbf{x}''', t) := \langle p_{0'} v_{i''j'''}\rangle, \quad P_{0'i''j'''} = P_{0'j'''i''}.
\end{aligned}
\tag{2.21}
$$

We now have an extended model consisting of equations (2.5), (2.6), (2.12) to (2.14), (2.16) and (2.20) for U_i, $U_{i'j''}$, $U_{i'j''k'''}$, P, $P_{0'i''}$ and $P_{0'j''k'''}$ provided that $U_{i'm'j''k'''}$ is properly modeled. We notice that the divergence-free condition $\frac{\partial}{\partial x_j''} P_{0'j''k'''} = 0$ may not be enforced generally here unless the model of $U_{i'm'j''k'''}$ meets some symmetry constraints or a vectoral function is introduced in (2.20), similar to the previous introduction of S to handle (2.16). This treatment is not needed under some special form for $U_{i'm'j''k'''}$ like, in the case of $\mathcal{S} = \mathcal{U}$,

$$U_{ij'k''l'''} = U_{ij'}U_{k''l'''} + U_{ik''}U_{j'l'''} + U_{il'''}U_{j'k''}, \tag{2.22}$$

as suggested in the scheme of quasi-normal approximation (Proudman and Reid [12], Tatsumi [17]).[1] The advantage of this extended formulation is that $(2.11)_2$ and (2.12) will be satisfied, which has the following consequences: (i) $S = 0$ can be achieved (see equation (2.17)); (ii) In the case of $\mathcal{S} = \mathcal{U}$, in homogeneous turbulence we have from (2.16)

$$\frac{\partial}{\partial t}U_{kk} + (U_{k,m} + U_{m,k})\, U_{mk} - 2\nu \frac{\partial^2}{\partial \mathbf{r}^2} U_{kk}\bigg|_{\mathbf{r}=0} = 0, \tag{2.23}$$

with $\mathbf{r} := \mathbf{x}'' - \mathbf{x}'$, which reflects the fact that in homogeneous turbulence the mean turbulent kinetic energy is conserved by the non-linear terms of the Navier-Stokes equations (due to the absence of the correlation of three velocity components from the equation) and dissipated by molecular viscosity (Lesieur [8]); (iii) We have from (2.6) and (2.16),

$$\begin{aligned}
&\frac{\partial}{\partial t}\int_{\mathcal{V}} \frac{1}{2}(U_kU_k + U_{kk})\, dv + \int_{\partial\mathcal{V}} \frac{1}{2}(U_kU_k + U_{kk})\, U_m N_m da \\
&= \int_{\partial\mathcal{V}} \left[-PU_m - P_{0m} - U_kU_{km} - \frac{1}{2}U_{kmk} + \frac{1}{2}\nu\, (U_kU_k + U_{kk})_{,m}\right] N_m da \\
&-\nu \int_{\mathcal{V}} \left(U_{k,m}\, U_{k,m} + \frac{\partial^2}{\partial x'_m \partial x''_m} U_{k'k''}\bigg|_{\mathbf{x}''=\mathbf{x}'=\mathbf{x}} \right) dv,
\end{aligned} \tag{2.24}$$

with $\mathcal{V}$ being an arbitrary control volume in $\mathcal{D}$ and $\mathbf{N}$ the normal to $\mathcal{V}$. This equation shows that the direct effect of $U_{i'k''j''}$ on the kinetic energy of the fluid in $\mathcal{V}$ can be absorbed into a surface integral term. The disadvantages, however, include that (i) many more equations, (2.12) and (2.20), need to be solved which are involved in a high dimensional space; (ii) the initial and boundary conditions for $U_{i'j''k'''}$ and the boundary condition for $P_{0'j''k'''}$ need to be specified. These disadvantages will not only cause difficulties to the application of the model but also cause problems in calibrating the model in the first place. Certainly, similar disadvantages are also inherent in the two-point correlation formulation proposed in the present work, compared with any one-point correlation model. In the case of $\mathcal{S} = \mathcal{U}_l$, one-point correlation models may be appropriate if the scale l is very small, but the case of $\mathcal{S} = \mathcal{U}$ is involved in eliminating all scales of disturbance and the two-point correlations $U_{i'j''}$ and $P_{0i'}$ are needed to account for the interaction among different scales of fluctuations, and it is expected that this two-point correlation formulation is more general and appropriate than one-point correlation models.

There is no clear scheme for formulating $U_{i'k''j''}$ corresponding to $\mathcal{S} = \mathcal{U}$ or $\mathcal{S} = \mathcal{U}_l$ though the former is supposedly free from any chaotic behavior and the latter has a characteristic length l. In both cases $U_{i'k''j''}$ should be formulated so that (i) the solutions $U_{i'j''}$ and $P_{0'j''}$ of the model satisfy, from their definitions (2.10) and the Schwartz inequality (Schumann [14]),

$$U_{\alpha'\alpha'} \geq 0, \quad U_{\alpha'\alpha'}U_{\beta''\beta''} - (U_{\alpha'\beta''})^2 \geq 0, \tag{2.25}$$

which guarantees the positive (semi)-definiteness of U_{ij} to yield non-negative fluctuation energy and an estimate of the fluctuation around U_i, and

$$P_{0''\alpha'} = 0, \quad \text{under } U_{\alpha'\alpha'} = 0; \tag{2.26}$$

and (ii) the model is dissipative, that is, the source term in the equation on $(U_k U_k + U_{kk})/2$ has to be a sink due to the molecular dissipation, which reduces to

$$\left. \frac{\partial^2}{\partial x'_m \partial x''_m} U_{k'k''} \right|_{\mathbf{x}''=\mathbf{x}'} \geq 0$$

in case that $(2.11)_2$ and (2.12) are met. Next, the structure of $U_{i'k''j''}$ has to meet two invariance requirements. One is related to Galilean invariance: Though $U_{i'k''j''}$ is frame indifferent since $\mathbf{v}$ can be viewed as a velocity difference based on the ensemble averaging, whether such a restriction should be imposed on its modeling is another issue. Obviously, if we resort to the restriction, we would get a differential equation on $U_{i'j''}$, (2.16), which is not frame indifferent, while $U_{i'j''}$ is frame indifferent due to the same reason as that for which $U_{i'k''j''}$ is frame indifferent. To avoid such a dilemma, we will merely require that the model for $U_{i'k''j''}$ satisfy the principle of relativity of Galilei and Newton (Frisch [5], Sedov [15]), which is the very symmetry possessed by the original equations (2.1) and (2.2); Another invariance to consider is the scaling invariance possessed by equations (2.1) and (2.2): $\{\mathbf{x}, t, \nu, \mathbf{u}, q\} \to \{\lambda\mathbf{x},\ \lambda^{1-h}t,\ \lambda^{1+h}\nu,\ \lambda^h\mathbf{u},\ \lambda^{2h}q\}$, $\lambda > 0$, $h \in \mathbf{R}$; that is, if $\{\mathbf{u},\ q\}(\mathbf{x},\ t)$ is a solution to (2.1) and (2.2) with the viscosity ν, then $\{\lambda^h\mathbf{u},\ \lambda^{2h}q\}(\lambda\mathbf{x},\ \lambda^{1-h}t)$ is a solution with the viscosity $\lambda^{1+h}\nu$ (Carbone and Aubry [2], Frisch [5]). This same scaling invariance should also hold for the quantities from the ensemble averaging. Consequently we should not introduce any new dimensional constant, except ν and l, to model $U_{i'k''j''}$. Finally, in both cases chaotic motions up to some scale need to be smoothed out. One way to achieve this is through adopting a gradient structure for $U_{i'k''j''}$, which is motivated by the diffusive terms of molecular viscosity in the equations; an artificial viscosity is introduced both to simulate the intensified momentum transfer in turbulence and to dampen the fluctuation as intended by the averaging. This scheme has been widely used in turbulence modeling, such as the Smagorinsky eddy viscosity model in LES (Smagorinsky [16]), the Boussinesq assumption concerning the Reynolds stress in zero-equation models and two-equation models and the flux form in Reynolds stress equation models in engineering turbulence modeling (Rodi [13]). This treatment, however, is flawed in that the exact $U_{i'k''j''}$ corresponding to the choice $\mathcal{S} = \mathcal{U}$ just plays the role of redistributing the fluctuation kinetic energy U_{kk} among its components U_{11}, U_{22} and U_{33} and so conserves U_{kk} in homogeneous turbulence (see equation (2.23)). To remedy this flaw, the model for $U_{i'k''j''}$ has to meet both symmetry conditions of $(2.11)_2$ and (2.12), which will result in too complicated a structure for $U_{i'k''j''}$ that can be demonstrated by considering the modeling of $U_{i'j''k'''}$ as follows. Let $h_{i'j''k'''} := h_{ijk}(\mathbf{x}', \mathbf{x}'', \mathbf{x}''', t)$ be a non-zero basic form and $H_{i'j''k'''} := h_{i'j''k'''} + h_{i'k'''j''}$. Then

$$-U_{i'j''k'''} \sim H_{i'j''k'''} + H_{j''i'k'''} + H_{k'''j''i'}$$

meets $(2.11)_2$. Next, to satisfy (2.12), we propose

$$\begin{aligned} -U_{i'j''k'''} &= H_{i'j''k'''} + \frac{\partial}{\partial x'_i} R_{0'j''k'''} + H_{j''i'k'''} + \frac{\partial}{\partial x''_j} R_{0''i'k'''} \\ &+ H_{k'''j''i'} + \frac{\partial}{\partial x'''_k} R_{0'''i'j''} \end{aligned} \tag{2.27}$$

with

$$R_{0'j''k'''} = R_{0'k'''j''}, \quad R_{0'j''k'''} := R_{0jk}(\mathbf{x}', \mathbf{x}'', \mathbf{x}''', t), \tag{2.28}$$

in order that we have six equations (2.12) with six undetermined functions $R_{0'j''k'''}$. The following special solution of $R_{0'j''k'''}$ can be obtained under trivial boundary condition,

$$\begin{aligned} R_{0'j''k'''} = &\frac{1}{2}\int_{\mathcal{D}} G(\hat{\mathbf{x}}, \mathbf{x}') \times \\ &\times \Bigg[\frac{\partial^2}{\partial x''_j \partial x'''_k} \int_{\mathcal{D}^2} G(\bar{\mathbf{x}}, \mathbf{x}'') G(\tilde{\mathbf{x}}, \mathbf{x}''') \frac{\partial^3}{\partial \bar{x}_m \partial \tilde{x}_n \partial \hat{x}_l} (H_{\bar{m}\hat{l}\tilde{n}} + H_{\tilde{n}\hat{l}\bar{m}}) d\tilde{\mathbf{x}} d\bar{\mathbf{x}} \\ &+ \frac{\partial}{\partial x''_j} \int_{\mathcal{D}} G(\bar{\mathbf{x}}, \mathbf{x}'') \frac{\partial^2}{\partial \hat{x}_l \partial \bar{x}_m} (H_{\hat{l}k'''\bar{m}} + H_{\bar{m}k'''\hat{l}} + H_{k'''\hat{l}\bar{m}}) d\bar{\mathbf{x}} \end{aligned}$$

$$+ \quad \frac{\partial}{\partial x_k'''} \int_{\mathcal{D}} G(\tilde{\mathbf{x}}, \mathbf{x}''') \frac{\partial^2}{\partial \hat{x}_l \partial \tilde{x}_n} \left(H_{\hat{l}j''\tilde{n}} + H_{\tilde{n}j''\hat{l}} + H_{j''\hat{l}\tilde{n}} \right) d\tilde{\mathbf{x}}$$

$$+ \quad 2\frac{\partial}{\partial \hat{x}_l} \left(H_{\hat{l}j''k'''} + H_{j''\hat{l}k'''} + H_{k'''j''\hat{l}} \right) \Big] d\hat{\mathbf{x}}, \tag{2.29}$$

and G is the Green's function from, under some homogeneous boundary condition on $\mathcal{D}$,

$$\frac{\partial^2}{\partial \tilde{\mathbf{x}}^2} G(\tilde{\mathbf{x}}, \mathbf{x}') = -\delta(\tilde{\mathbf{x}} - \mathbf{x}'). \tag{2.30}$$

Then a relation for $U_{i'k''j''}$ can be derived from (2.27). One major problem with this relation is the explicit presence of the Green's function G which can be quite difficult to find if the flow domain is not simple enough. Also, it is not clear what homogeneous boundary conditions for G need to be adopted. The other problem is that multiple integrations and differentiations are involved. Thus we have to keep a balance between the simplicity and the comprehensiveness of the model for $U_{i'k''j''}$. To achieve the simplicity, we may need to allow the presence of dissipation caused by the artificial viscosity by relaxing the symmetry constraints, say, by removing $(2.11)_2$. This relaxation is formally allowable by observing that the term

$$W_{i'j''} = V_{i'j''} + V_{j''i'}, \quad V_{i'j''} := \frac{\partial}{\partial x_k''} U_{i'k''j''},$$

instead of $U_{i'k''j''}$, is present in (2.16), $W_{i'j''} = W_{j''i'}$ always holds regardless of whether the symmetry condition $(2.11)_2$ is satisfied by $U_{i'k''j''}$ or not, since in $V_{i'j''}$ the symmetry of $U_{i'k''j''}$ with respect to k'' and j'' is suppressed by the divergence operation, and the appropriate modeling of $V_{i'j''}$ should be the primary concern in order to get reasonable flow fields of $\{U_i, P, U_{i'j''}\}$.

Let us consider a possible model of $V_{i'j''}$ guided by the constraints mentioned above. The simple ones are

$$\frac{\partial}{\partial x_k''} U_{i'k''j''} = -\frac{\partial}{\partial x_k''} \left(\frac{\kappa''}{\sqrt{|\epsilon''|}} \Psi_{km}'' \frac{\partial}{\partial x_m''} U_{i'j''} \right) \tag{2.31}$$

for $\mathcal{S} = \mathcal{U}$ and

$$\frac{\partial}{\partial x_k''} U_{i'k''j''} = -\frac{\partial}{\partial x_k''} \left(l\sqrt{\kappa''} \Phi_{km}'' \frac{\partial}{\partial x_m''} U_{i'j''} \right) \tag{2.32}$$

for $\mathcal{S} = \mathcal{U}_l$ with

$$\kappa'' := U_{k''k''}, \quad \epsilon'' := \frac{\partial^2}{\partial x_k' \partial x_k''} U_{j'j''} \Big|_{\mathbf{x}'=\mathbf{x}''},$$

and both Ψ'' and Φ'' are dimensionless, positive (semi-)definite tensor functions as

$$\Psi'' := \Psi\left(U_{m''n''}, U_{i'j''}, \frac{\partial^2}{\partial x'_m \partial x''_n} U_{i'j''}\bigg|_{\mathbf{x}'=\mathbf{x}''}, \nu\right),$$
$$\Phi'' := \Phi(U_{m''n''}, U_{i'j''}, \nu, l)\,.$$

The specific forms of Ψ and Φ have to be fixed either directly or indirectly with the help of the experimental data, under the constraints of (2.25) and (2.26), the closed model being dissipative. If the data on $\{V_{i'j''}, U_i, U_{i'j''}\}$ or $\{V^d_{i'j''}, U^d_i, U^d_{i'j''}\}$ can be obtained directly, either by experimental measurement or by direct numerical simulation, the form on Ψ or Φ may be found by correlating $V^d_{i'j''}$ and $U^d_{i'j''}$ according to (2.31) or (2.32). Though it bypasses solving the equations as well as prescribing the initial and boundary conditions, this procedure does not necessarily guarantee a proper model for $\{U_i, U_{i'j''}\}$ due to the approximate nature of the model; for instance, if Ψ or Φ depends on $U_{i'j''}$, then $S \neq 0$ and the closed set of equations may not yield solutions for $\{U_i, U_{i'j''}\}$ compatible with that directly obtained for $\{U^d_i, U^d_{i'j''}\}$. To remedy this flaw, we may propose a form for $V_{i'j''}$ according to the criterion for producing reasonable solutions for $\{U_i, U_{i'j''}\}$ from the closed model, instead of focusing on the matching of $V_{i'j''}$ and $V^d_{i'j''}$. This scheme can be used even if $V^d_{i'j''}$ is not available. It remains to be resolved how to evaluate the appropriateness of the model.

3. SUMMARY

We have presented a mixed formulation for turbulence modeling, a combined version of analytical theories and engineering modeling. The equations for the two-point correlations $U_{i'j''}$ and $P_{0'j''}$ are discussed, and the related issues are addressed which include the necessity for introducing the function S, the approximate nature of any specific turbulence model of averaging, the different outcomes from differently chosen set of solutions on which the ensemble averaging is based, and the constraints on modeling $U_{i'k''j''}$ to make the equations determinate such as realizability and the symmetry properties of the Navier-Stokes equations. We have delineated the appealing side of such a formulation, and moreover, its disadvantages with regard to simplicity when compared with models of single-point correlations, like that involved in working in a higher dimensional space and the difficulty associated with prescribing initial and boundary conditions for $U_{i'j''}$ and the boundary conditions for $P_{0'j''}$. Future work has to resolve these problems and computational schemes have to be devised to help modeling.

Notes

1. This relation might be derived simply from (i) the properties of $U_{ij'k''l'''}$, such as, its symmetry with respect to its indices, formally linear in v_i, $(U_{ij'k''l'''})_{,m} = U_{(i,m)j'k''l'''}$ and so on; and (ii) the assumptions that $\mathbf{v}$ is of normal distribution at one point and that $U_{ij'k''l'''}$ depends on $U_{ij'}$, $U_{ij'k''}$ and the like. This relation unfortunately has the flaw of yielding negative energy spectra in isotropic turbulence at high Reynolds number $Re_\lambda = \sqrt{\langle \mathbf{v}\cdot\mathbf{v}\rangle}\lambda/\nu$ with λ being the dissipation length (Ogura [10], Orszag [11]). One reason for this failure might be explained by observing that

$$\langle v_j v_j v_{l'} v_{l'} \rangle = \langle v_j v_j \rangle \langle v_{l'} v_{l'} \rangle + 2 \langle v_j v_{l'} \rangle \langle v_j v_{l'} \rangle$$

from (2.22) where the left-hand side is physically expected to be small for large $|\mathbf{x} - \mathbf{x}'|$, but the right-hand side can be quite great under large fluctuations especially in homogeneous turbulence. Based on this observation, we may modify (2.22) to eliminate this flaw.

References

[1] Batchelor, G.K. (1982). *The Theory of Homogeneous Turbulence.* Cambridge University Press, Cambridge.

[2] Carbone, F. and Aubry, N. (1996). Hierachical order in wall-bounded shear turbulence. *Phys. Fluids*, A8:1061–1074.

[3] Ferziger, J.H. (1985). Large eddy simulation: its role in turbulence research. In *Theoretical Approaches to Turbulence.* (D. L. Dwoyer, M. Y. Hussaini, and R. G. Voigt, eds.) Springer-Verlag, New York.

[4] Ferziger, J.H. (1996). Large eddy simulation. In *Simulation and Modeling of Turbulent Flows.* (T. B. Gatski, M. Y. Hussaini, and J. L. Launder, eds.) pp. 109–154. Oxford University Press, Oxford.

[5] Frisch, U. (1995). *Turbulence: the legacy of A. N. Kolmogorov.* Cambridge University Press, New York.

[6] Hinze, J.O. (1995). *Turbulence.* McGraw-Hill, New York.

[7] Launder, B.E. (1996). An introduction to single-point closure methodology. In *Simulation and Modeling of Turbulent Flows.* (T.B. Gatski, M.Y. Hussaini, and J.L. Launder, eds.) pp. 243–310. Oxford University Press, Oxford.

[8] Lesieur, M. (1987). *Turbulence in Fluids.* Martinus Nijhoff Publishers, Boston.

[9] Monin, A.S. and Yaglom, A.M. (1971). *Statistical Fluid Mechanics: Mechanics of Turbulence,* 1. MIT Press, Cambridge, Mass.

[10] Ogura, Y. (1963). A consequence of the zero fourth cumulant approximation in the decay of isotropic turbulence. *J. Fluid Mech.*, 16:33–40.

[11] Orszag, S.A. (1970). Analytical theories of turbulence. *J. Fluid Mech.*, 41:363–386.

[12] Proudman, I. and Reid, W.H. (1954). On the decay of a normally distributed and homogeneous turbulent velocity field. *Philos. Trans. Roy. Soc. London Ser. A.*, A267:163–189.

[13] Rodi, W. (1980). *Turbulence Models and Their Application in Hydraulics – A State of the Art Review.* International Association for Hydraulic Research, Delft, the Netherlands.

[14] Schumann, U. (1977). Realizability of Reynolds stress turbulence models. *Phys. Fluids*, 20:721–725.

[15] Sedov, L.I. (1972). *A Course in Continuum Mechanics, Vol. II.* Wolters-Noordhoff Publishing Groningen, the Netherlands.

[16] Smagorinsky, J. (1963). General circulation experiments with the primitive equations: Part I, the basic experiment. *Monthly Weather Rev.*, 91:99–164.

[17] Tatsumi, T. (1980). Theory of homogeneous turbulence. *Adv. Appl. Mech.*, 20:39–133.

$L^{2,\lambda}$ - REGULARITY FOR NONLINEAR ELLIPTIC SYSTEMS OF SECOND ORDER

Josef Daněček, Eugen Viszus

Abstract: There is shown the $L^{2,\lambda}$ - regularity of the gradient of weak solutions of nonlinear elliptic systems.

Keywords: Nonlinear equations, regularity, Morrey spaces.

1. INTRODUCTION

In this paper we consider the problem of the regularity of the first derivatives of weak solutions to the nonlinear elliptic system

$$-D_\alpha a_i^\alpha(x,u,Du) = a_i(x,u,Du), \quad i = 1,\dots,N, \alpha = 1,\dots,n \tag{1.1}$$

where $a_i^\alpha(x,u,z)$, $a_i(x,u,z)$ are Caratheodorian mappings from $(x,u,z) \in \Omega \times \mathbf{R}^N \times \mathbf{R}^{nN}$ into $\mathbf{R}$. A function $u \in W^{1,2}_{loc}(\Omega,\mathbf{R}^N)$ is called a weak solution of (1.1) in Ω if

$$\int_\Omega a_i^\alpha(x,u,Du)D_\alpha\varphi^i\,dx = \int_\Omega a_i(x,u,Du)\varphi^i\,dx, \quad \forall\,\varphi \in C_0^\infty(\Omega,\mathbf{R}^N).$$

As it is known, in case of a general system (1.1) only partial regularity can be expected for $n > 2$ (see e.g.[2], [4], [7]). Under the assumptions below we will prove $L^{2,\lambda}$ - regularity $(0 < \lambda < n)$ of gradient of weak solutions for the system (1.1) whose coefficients $a_i^\alpha(x,u,Du)$ have the form

$$a_i^\alpha(x,u,Du) = A_{ij}^{\alpha\beta}(x)D_\beta u^j + g_i^\alpha(x,u,Du). \tag{1.2}$$

Here $A_{ij}^{\alpha\beta}$ is a matrix of functions, the following condition of strong ellipticity

$$A_{ij}^{\alpha\beta}(x)\xi_\alpha^i\xi_\beta^j \geq \nu\,|\xi|^2, \quad a.e.\ x \in \Omega,\ \forall \xi \in \mathbf{R}^{nN}; \nu > 0 \tag{1.3}$$

holds and $g_i^\alpha(x,u,z)$ are smooth functions with sublinear growth in z. In what follows, we formulate the conditions on the smoothness and the growth of the functions $A_{ij}^{\alpha\beta}(x)$, $g_i^\alpha(x,u,z)$ and $a_i(x,u,z)$ precisely.

Applied Nonlinear Analysis, edited by Sequeira *et al.*
Kluwer Academic / Plenum Publishers, New York, 1999.

Such result may open a way to prove BMO-regularity of gradient. In [3] the first author has proved $L^{2,\lambda}$ - regularity of gradient of weak solutions to (1.1) in the situation when the coefficients $A_{ij}^{\alpha\beta}$ are continuous. In this paper the coefficients $A_{ij}^{\alpha\beta}$ are discontinuous in general.

If we want to sketch our method of proof, we have to say that its crucial point is the assumption on $A_{ij}^{\alpha\beta}$: $A_{ij}^{\alpha\beta} \in L^\infty(\Omega) \cap \mathcal{L}_\Phi(\Omega)$ (for the definition see below). Taking into account higher integrability of gradient Du we obtain $L^{2,\lambda}$ - regularity of gradient.

2. NOTATIONS AND DEFINITIONS

We consider bounded open set $\Omega \subset \mathbf{R}^n$ with points $x = (x_1, \dots x_n)$, $n \geq 3$, $u\,\Omega \to \mathbf{R}^N$, $N \geq 1$, $u(x) = (u^1(x), \dots, u^N(x))$ is a vector-valued function, $Du = (D_1 u, \dots, D_n u)$, $D_\alpha = \partial/\partial x_\alpha$; we use the summation convention over repeated indices. The meaning of $\Omega_0 \subset\subset \Omega$ is that the closure of Ω_0 is contained in Ω, i.e. $\overline{\Omega}_0 \subset \Omega$. For the sake of simplicity we denote by $|\cdot|$ and $(.,.)$ the norm and scalar product in $\mathbf{R}^n$ as well as in $\mathbf{R}^N$ and $\mathbf{R}^{nN}$. If $x \in \mathbf{R}^n$ and r is a positive real number, we denote $B_r(x) = \{y \in \mathbf{R}^n \mid\, |y - x| < r\}$, i.e., the open ball in $\mathbf{R}^n$, $\Omega(x,r) = B_r(x) \cap \Omega$. Denote by $u_{x,r} = |\Omega(x,r)|_n^{-1} \int_{\Omega(x,r)} u(y)\,dy = \fint_{\Omega(x,r)} u(y)\,dy$ the mean value of the function $u \in L^1(\Omega, \mathbf{R}^N)$ over the set $\Omega(x,r)$, where $|\Omega(x,r)|_n$ is the n-dimensional Lebesgue measure of $\Omega(x,r)$. Beside the usually used space $C_0^\infty\left(\Omega, \mathbf{R}^N\right)$, Hölder spaces $C^{0,\alpha}\left(\Omega, \mathbf{R}^N\right)$, $C^{0,\alpha}\left(\overline{\Omega}, \mathbf{R}^N\right)$ and Sobolev spaces $W^{k,p}\left(\Omega, \mathbf{R}^N\right)$, $W^{k,p}_{loc}\left(\Omega, \mathbf{R}^N\right)$, $W^{k,p}_0\left(\Omega, \mathbf{R}^N\right)$ (see, e.g., [6]) we use the following Morrey spaces.

Definition 1. *Let $\lambda \in [0,n]$, $q \in [1,\infty)$. A function $u \in L^q(\Omega, \mathbf{R}^N)$ is said to belong to $L^{q,\lambda}(\Omega, \mathbf{R}^N)$ if*

$$||u||^q_{L^{q,\lambda}(\Omega,\mathbf{R}^N)} = \sup\left\{ r^{-\lambda} \int_{\Omega(x,r)} |u(y)|^q\,dy : x \in \Omega, r > 0 \right\} < \infty.$$

Remark. $u \in L^{q,\lambda}_{loc}(\Omega, \mathbf{R}^N)$ iff $u \in L^{q,\lambda}(\Omega_0, \mathbf{R}^N)$ for each $\Omega_0 \subset\subset \Omega$. For more details see [2], [4], [6] and [7].

The generalization of Campanato spaces $\mathcal{L}^{q,\lambda}(\Omega, \mathbf{R}^N)$ are the classes $\mathcal{L}_\Phi$ introduced by Spanne [8].

Definition 2. *A function $u \in L^2(\Omega, \mathbf{R}^N)$ is said to belong to $\mathcal{L}_\Phi(\Omega, \mathbf{R}^N)$ if*

$$[u]_{\Phi,\Omega} = \sup\left\{ \Phi(r)^{-1} \left(\fint_{\Omega(x,r)} |u(y) - u_{x,r}|^2\,dy \right)^{1/2} : x \in \Omega, r \in (0, diam\Omega] \right\} < \infty$$

and by $l_\Phi(\Omega, \mathbf{R}^N)$ *we denote subspace of all* $u \in \mathcal{L}_\Phi(\Omega, \mathbf{R}^N)$ *such that*

$$[u]_{\Phi,\Omega,r_0} =$$
$$\sup\left\{\Phi(r)^{-1}\left(\fint_{\Omega(x,r)}|u(y) - u_{x,r}|^2\,dy\right)^{1/2} : x \in \Omega, r \in (0, r_0]\right\} = o(1)$$

as $r_0 \searrow 0$, *where* $\Phi(r) = (1 + |\ln r|)^{-1}$.

Some basic properties of above mentioned spaces are formulated in the following proposition (for the proofs see [1], [2], [6] and [8]).

Proposition 1. *For a domain* $\Omega \subset \mathbf{R}^n$ *of the class* $C^{0,1}$ *we have the following*

(i) $L^{q,n}\left(\Omega, \mathbf{R}^N\right)$ *is isomorphic to the* $L^\infty(\Omega, \mathbf{R}^N)$.

(ii) $u \in W^{1,2}_{loc}(\Omega, \mathbf{R}^N)$ *and* $Du \in L^{2,\lambda}_{loc}(\Omega, \mathbf{R}^{nN})$, $n - 2 < \lambda < n$ *then* $u \in C^{0,\alpha}(\Omega, \mathbf{R}^N)$, $\alpha = (\lambda + 2 - n)/2$.

(iii) $\mathcal{L}_\Phi(\Omega, \mathbf{R}^N)$ *is a Banach space with norm* $\|u\|_{\mathcal{L}_\Phi(\Omega,\mathbf{R}^N)} = \|u\|_{L^2(\Omega,\mathbf{R}^N)} + [u]_{\mathcal{L}_\Phi(\Omega,\mathbf{R}^N)}$.

(iv) $C^0(\overline{\Omega}, \mathbf{R}^N)\backslash\mathcal{L}_\Phi(\Omega, \mathbf{R}^N)$ *and* $(L^\infty(\Omega, \mathbf{R}^N)\cap l_\Phi(\Omega, \mathbf{R}^N))\backslash C^0(\overline{\Omega}, \mathbf{R}^N)$ *are not empty.*

(v) For $p \in [1, \infty)$, $\Omega' \subset\subset \Omega$, $r_0 \in (0, dist(\Omega', \partial\Omega))$ *and* $u \in \mathcal{L}_\Phi(\Omega, \mathbf{R}^N)$ *set*

$$N_p(u; \Phi, \Omega', r_0) =$$
$$= \sup\left\{\Phi(r)^{-1}\left(\fint_{\Omega(x,r)}|u(y) - u_{x,r}|^p dy\right)^{1/p} x \in \Omega', r \in (0, r_0]\right\}.$$

Then we have for each $u \in \mathcal{L}_\Phi(\Omega, \mathbf{R}^N)$

$$N_1(u; \Phi, \Omega', r_0) \le N_p(u; \Phi, \Omega', r_0) \le c(p, n)[u]_{\Phi,\Omega,r_0}.$$

3. MAIN RESULTS

Suppose that for all $(x, u, z) \in \Omega \times \mathbf{R}^N \times \mathbf{R}^{nN}$ the following conditions hold:

$$|a_i(x, u, z)| \le f_i(x) + L\,|z|^{\gamma_0} \tag{3.1}$$
$$|g_i^\alpha(x, u, z)| \le f_i^\alpha(x) + L\,|z|^\gamma \tag{3.2}$$
$$g_i^\alpha(x, u, z)z_\alpha^i \ge \nu_1\,|z|^{1+\gamma} - f^2(x) \tag{3.3}$$

where L, ν_1 are positive constants, $1 \le \gamma_0 < (n+2)/n$, $0 \le \gamma < 1$, f, $f_i^\alpha \in L^{\sigma,\lambda}(\Omega)$, $\sigma > 2$, $0 < \lambda \le n$, $f_i \in L^{\sigma q_0, \lambda q_0}(\Omega)$, $q_0 = n/(n+2)$. We put $A = (A_{ij}^{\alpha\beta})$, $g = (g_i^\alpha)$, $a = (a_i)$, $\tilde{f} = (f_i)$, $\tilde{\tilde{f}} = (f_i^\alpha)$.

Theorem. *Let $u \in W_{loc}^{1,2}(\Omega, \mathbf{R}^N)$ be a weak solution to the system (1.1) and the conditions (1.2), (1.3), (3.1), (3.2) and (3.3) be satisfied. Suppose further that $A_{ij}^{\alpha\beta} \in L^\infty(\Omega) \cap \mathcal{L}_\Phi(\Omega)$, i, $j = 1,\dots,N$, α, $\beta = 1,\dots,n$. Then $Du \in L_{loc}^{2,\lambda}(\Omega, \mathbf{R}^{nN})$ for $\lambda < n$ and in the case $\lambda = n$ $Du \in L_{loc}^{2,\lambda'}(\Omega, \mathbf{R}^{nN})$ where $\lambda' < n$ is an arbitrary.*

Corollary. *Let the assumptions of Theorem be satisfied.*
If $n - 2 < \lambda < n$, then $u \in C^{0,(\lambda-n+2)/2}(\Omega, \mathbf{R}^N)$.

Proof. It follows from Proposition 1(ii).

4. SOME LEMMAS

In this section we present the results needed for the proof of Theorem. In $B(x,r) \subset \mathbf{R}^n$ we consider a linear elliptic system

$$-D_\alpha(A_{ij}^{\alpha\beta} D_\beta u^j) = 0 \tag{4.1}$$

with constant coefficients for which (1.3) holds.

Lemma 1. ([2] pp. 54–55) *Let $u \in W^{1,2}(B(x,r), \mathbf{R}^N)$ be a weak solution to the system (4.1). Then for each $t \in [0,1]$*

$$\int_{B_{tr}} |Du(y)|^2\,dy \le c\,t^n \int_{B_r} |Du(y)|^2\,dy$$

holds.

Lemma 2. ([4]) *Let $\Psi = \Psi(R)$, $R \in (0,d]$, $d > 0$ be a nonnegative function and let A, B, C, a, b be nonnegative constants. Suppose that for all $t \in (0,1]$ and all $R \in (0,d]$*

$$\Psi(tR) \le (At^a + B)\Psi(R) + CR^b$$

holds. Further, let $K \in (0,1)$ be such that $\varepsilon = AK^{a-b} + BK^{-b} < 1$. Then

$$\Psi(R) \le cR^b, \quad R \in (0,d]$$

where $c = \max\{C/K(1-\varepsilon),\ \sup_{R \in [Kd,d]} \Psi(R)/R^b\}$.

The following Lemma is the special case of Lemma 3.4 from the paper [3].

Lemma 3. ([3], pp.757–758) *Let $u \in W^{1,2}(\Omega, \mathbf{R}^N)$, $Du \in L^{2,\tau}(\Omega, \mathbf{R}^{nN})$, $0 \le \tau < n$ and (3.1) and (3.2) are satisfied with $f_i \in L^{2q_0,\lambda q_0}(\Omega)$, $f_i^\alpha \in L^{2,\lambda}(\Omega)$, $0 < \lambda \le n$.*

(i) Then $a_i \in L^{2q_0,\lambda_0}(\Omega)$ *and for each ball* $B_R(x) \subset \Omega$ *we have*

$$\int_{B_R(x)} |a_i(x,u,Du)|^{2q_0}\, dy \le c\, R^{\lambda_0} \tag{4.2}$$

where $c = c(n, L, \gamma_0, \operatorname{diam}\Omega, \|\widetilde{f}\|_{L^{2q_0,\lambda q_0}(\Omega,\mathbf{R}^N)}, \|Du\|_{L^2(\Omega,\mathbf{R}^{nN})})$ *and* $\lambda_0 = \min\{\lambda q_0, n - (n-\tau) q_0 \gamma_0\}$.

(ii) For each $\varepsilon \in (0,1)$ *and all* $B_R(x) \subset \Omega$

$$\int_{B_R(x)} |g_i^\alpha(x,u,Du)|^2\, dy \le c(L)\, \varepsilon \int_{B_R(x)} |Du|^2\, dy + c\, R^{\lambda_1} \tag{4.3}$$

Here $c = c(L, \varepsilon, \gamma, \operatorname{diam}\Omega, \|\widetilde{\widetilde{f}}\|_{L^{2,\lambda}(\Omega,\mathbf{R}^{nN})}, \|Du\|_{L^2(\Omega,\mathbf{R}^{nN})})$, $\lambda_1 = \lambda$ *for* $\lambda < n$ *and* $\lambda_1 < n$ *is an arbitrary for* $\lambda = n$.

Proof. For the proof (i) see [2], pp.106–107. According to (3.2) it follows that

$$\int_{B_R(x)} |g_i^\alpha(y,u,Du)|^2\, dy \le c\left(\|\widetilde{\widetilde{f}}\|^2_{L^{2,\lambda}(\Omega,\mathbf{R}^{nN})} R^\lambda + \int_{B_R(x)} |Du|^{2\gamma}\, dy \right).$$

By Young inequality we obtain

$$\int_{B_R(x)} |Du|^{2\gamma}\, dy \le \varepsilon \int_{B_R(x)} |Du|^2\, dy + c(n,\varepsilon,\gamma) R^n$$

for each $\varepsilon \in (0,1)$ and (4.3) easily follows.

In the following considerations we will use a result about higher integrability of gradient of weak solution of the system (1.1).

Proposition 4. ([4], p.138) *Suppose that (1.2), (1.3), (3.1)–(3.3) are fulfilled and let* $u \in W^{1,2}_{loc}(\Omega, \mathbf{R}^N)$ *be a weak solutions of (1.1). Then there exists an exponent* $r > 2$ *such that* $u \in W^{1,r}_{loc}(\Omega, \mathbf{R}^N)$. *Moreover there exists constant* $c = c(\nu, \nu_1, L, \|A\|_\infty)$ *and* $\widetilde{R} > 0$ *such that for all balls* $B_R(x) \subset \Omega$, $R < \widetilde{R}$ *the following inequality is satisfied*

$$\begin{aligned}\left(\fint_{B_{R/2}(x)} |Du|^r\, dy \right)^{1/r} &\le c\left\{ \left(\fint_{B_R(x)} |Du|^2)\, dy \right)^{1/2} \right.\\ &+ \left. \left(\fint_{B_R(x)} (|f|^r + |\widetilde{\widetilde{f}}|^r)\, dy \right)^{1/r} + R \left(\fint_{B_R(x)} |\widetilde{f}|^{rq_0} dy \right)^{1/rq_0} \right\}.\end{aligned}$$

5. PROOF OF THEOREM

Let $B_{R/2}(x_0) \subset B_R(x_0) \subset \Omega$ be an arbitrary ball and let $w \in W_0^{1,2}(B_{R/2}(x_0), \mathbf{R}^N)$ be a solution of the following system

$$\begin{aligned} &\int_{B_{R/2}(x_0)} (A_{ij}^{\alpha\beta})_{x_0,R} D_\beta w^j D_\alpha \varphi^i \, dx \\ &= \int_{B_{R/2}(x_0)} \left((A_{ij}^{\alpha\beta})_{x_0,R} - A_{ij}^{\alpha\beta}(x) \right) D_\beta u^j D_\alpha \varphi^i \, dx \\ &- \int_{B_{R/2}(x_0)} g_i^\alpha(x,u,Du) D_\alpha \varphi^i \, dx + \int_{B_{R/2}(x_0)} a_i(x,u,Du) \varphi^i \, dx \end{aligned} \tag{5.1}$$

for all $\varphi \in W_0^{1,2}(B_{R/2}(x_0), \mathbf{R}^N)$. It is known that under the assumption of theorem such solution exists and it is unique for all $R < R'$ (R' is sufficiently small).

We can put $\varphi = w$ in (5.1) and using ellipticity, Hölder and Sobolev inequalities we get

$$\begin{aligned} &\nu \int_{B_{R/2}(x_0)} |Dw|^2 \, dx \\ &\le c \Bigg(\int_{B_{R/2}(x_0)} |A_{x_0,R} - A(x)|^2 |Du|^2 \, dx + \int_{B_{R/2}(x_0)} |g(x,u,Du)|^2 \, dx \\ &+ \left(\int_{B_{R/2}(x_0)} |a(x,u,Du)|^{2q_0} \, dx \right)^{1/q_0} \Bigg) = c(I + II + III). \end{aligned}$$

Taking into account the properties of matrix $A = (A_{ij}^{\alpha\beta})$, Proposition 1(v), Proposition 4 with $r > 2$ and Hölder inequality ($r' = r/(r-2)$) we obtain

$$\begin{aligned} I &\le \left(\int_{B_{R/2}(x_0)} |A(x) - A_{x_0,R}|^{2r'} \, dx \right)^{1/r'} \left(\int_{B_{R/2}(x_0)} |Du|^r \, dx \right)^{2/r} \\ &\le \frac{c R^{n/r'}}{1 + |\ln R)|} \left((1 + |\ln R)|) \left(\fint_{B_{R/2}(x_0)} |A(x) - A_{x_0,R/2}|^{2r'} \, dx \right)^{1/r'} \right) \times \\ &\times \left(\int_{B_{R/2}(x_0)} |Du|^r \, dx \right)^{2/r} \\ &\le N_{2r'}(A; \Phi, B_{R/2}(x_0), R/2) \frac{R^{n/r'}}{1 + |\ln R|} \left(\int_{B_{R/2}(x_0)} |Du|^r \, dx \right)^{2/r} \\ &\le c(n, r, \|A\|_{\mathcal{L}_\Phi(\Omega, \mathbf{R}^{n^2 N^2})}) \frac{R^{n/r'}}{1 + |\ln R|} \left(\int_{B_{R/2}(x_0)} |Du|^r \, dx \right)^{2/r}. \end{aligned}$$

To the estimate the last integral in above inequality we use Proposition 4 and we get

$$\left(\int_{B_{R/2}(x_0)} |Du|^r\, dx\right)^{2/r}$$

$$\leq c\left\{\frac{1}{R^{n(1-2/r)}}\int_{B_R(x)} |Du|^2\, dy + \left(\int_{B_R(x)} (|f|^r + |\tilde{f}|^r)\, dy\right)^{2/r}\right.$$

$$\left. + R^{2(1-2/r)}\left(\int_{B_R(x)} |\tilde{f}|^{rq_0}\, dy\right)^{2/rq_0}\right\}$$

$$\leq c\left(\frac{1}{R^{n(1-2/r)}}\int_{B_R(x)} |Du|^2\, dy + R^{2\lambda/r} + R^{2(r-2+\lambda)/r)}\right)$$

where $c = c(r, \|f\|_{L^{r,\lambda}(\Omega)}, \|\tilde{\tilde{f}}\|_{L^{r,\lambda}(\Omega)}, \|\tilde{f}\|_{L^{rq_0,\lambda q_0}(\Omega)})$.

$$I \leq \frac{c}{1+|\ln R|}\int_{B_R(x_0)} |Du|^2 dx + c\left(R^{2\lambda/r} + R^{2(r-2+\lambda)/r)}\right) R^{n/r'}.$$

We can estimate II and III by means of Lemma 3 (with $\tau = 0$) and we have

$$\nu^2 \int_{B_{R/2}(x_0)} |Dw|^2\, dx \tag{5.2}$$

$$\leq c\left\{\left(\varepsilon + \frac{1}{1+|\ln R|}\right)\int_{B_R(x_0)} |Du|^2\, dx + R^{\mu}\right\}$$

where $\mu = \min\{(2\lambda + n(r-2))/r, (2\lambda + (n+2)(r-2))/r, \lambda, n+2-n\gamma_0\} = \min\{\lambda, n+2-n\gamma_0\}$ because $r > 2$.

The function $v = u - w \in W^{1,2}(B_{R/2}(x_0), \mathbf{R}^N)$ is the solution of the system

$$\int_{B_{R/2}(x_0)} (A_{ij}^{\alpha\beta})_{x_0,R/2} D_\beta v^j D_\alpha \varphi^i\, dx = 0 \tag{5.3}$$

for all $\varphi \in W_0^{1,2}(B_{R/2}(x_0), \mathbf{R}^N)$. From Lemma 1 we have for $t \in (0,1]$

$$\int_{B_{tR/2}(x_0)} |Dv(y)|^2\, dy \leq c\, t^n \int_{B_{R/2}(x_0)} |Dv(y)|^2\, dy.$$

By means of (5.2) and (5.3) we obtain for $t \in (0,1]$ and $\varepsilon \in (0,1)$

$$\int_{B_{tR/2}(x_0)} |Du|^2 dx \leq c\left\{\left(t^n + \varepsilon + \frac{1}{1+|\ln R|}\right)\int_{B_R(x_0)} |Du|^2\, dx + R^{\mu}\right\}.$$

For $t \in [1,2]$ the above inequality is trivial and we obtain for all $t \in [0,1]$

$$\int_{B_{tR}(x_0)} |Du|^2\,dx \le c_1\left(t^n + \varepsilon + \frac{1}{1+|\ln R|}\right)\int_{B_R(x_0)} |Du|^2\,dx + c_2\,R^\mu$$

where the constants c_1 and c_2 depends only above mentioned parameters.

Now from Lemma 2 we get the result of the following manner. If we put $\Psi(R) = \int_{B_R(x_0)} |Du|^2\,dx$, $A = c_1$, $B = c_1(\varepsilon + 1/(1+|\ln R|))$ and $C = c_2$ we can choose $0 < K < 1$ such that $AK^{n-\lambda} < 1/2$ (in the case $\lambda = n$ we have $AK^{n-\lambda_1} < 1/2$, where λ_1 is from Lemma 3(ii)). It is obvious that the constants $\varepsilon_0 > 0$, $R_0 > 0$ exist such that $BK^{-\lambda} < 1/2$ ($B = \varepsilon_0 + 1/(1+|\ln R_0|)$) and then for all $t \in (0,1)$, $R < R_0$ the assumptions of Lemma 2 are satisfied and therefore

$$\int_{B_R(x_0)} |Du|^2\,dx \le c\,R^\mu$$

If $\mu = \lambda$ Theorem is proved. If $\mu < \lambda$ the previous procedure can be repeated with $\tau = \mu$ in Lemma 3. It is clear that after a finite number of steps (since μ increases in each step as it follows from Lemma 3) we obtain $\mu = \lambda$.

References

[1] Acquistapace, P. (1992). On BMO regularity for linear elliptic systems, *Ann. Mat. Pura Appl.*, 161:231–269.

[2] Campanato, S. (1980). *Sistemi ellittici in forma divergenza. Regolarita all'interno.* Quaderni, Pisa.

[3] Daněček, J. (1986). Regularity for nonlinear elliptic systems, *Comment. Math. Univ. Carolin.*, 27:755–764.

[4] Giaquinta, M. (1983). *Multiple integrals in the calculus of variations and nonlinear elliptic systems*, Annals of Mathematics Studied N.105, Princenton university press, Princeton.

[5] Kadlec, J., Nečas, J. (1967). Sulla regularita delle soluzioni di equazioni ellitiche negli spazi $H^{k,\lambda}$, *Ann. Scuola Norm. Sup. Pisa*, 21:527–545.

[6] Kufner, A., John, O. and Fučík, S. (1977). *Function spaces*, Academia, Prague.

[7] Nečas J. (1983). *Introduction to the theory of nonlinear elliptic equations*, Teubner-Texte zur Mathematik, Leipzig.

[8] Spanne, S. (1965). Some function spaces defined using the mean oscillation over cubes, *Ann. Scuola Norm. Sup. Pisa*, 19:593–608.

ON THE FREDHOLM ALTERNATIVE FOR NONLINEAR HOMOGENEOUS OPERATORS

Pavel Drábek

Abstract: In this paper we discuss some issues concerning the generalization of the Fredholm alternative for nonlinear operators. We deal with both nonresonant and resonant cases which makes the situation fairly subtle. That is why we restrict ourselves on the case of one dimensional p–Laplacian.

Keywords: Nonlinear Fredholm alternative, p-Laplacian, eigenvalues, solvability, resonance.

1. INTRODUCTION

In this paper we want to discuss some aspects of the generalization of the Fredholm alternative for nonlinear operators. One of the first attempts to give a systematic treatment of this issue was done by Jindřich Nečas and his pupils and collaborators Svatopluk Fučík, Jiří Souček and Vladimír Souček in their book [FNSS] in early seventies. It follows from their very general results that if we drop the linear structure of the operator a lot of properties connected with the geometry of its range (and also the spaces considered) are lost or modified. The situation appears to be so complicated that in order to illustrate some of these phenomena we have to restrict our attention to a very special class of nonlinear operators.

Namely, we restrict ourselves to the second order o.d.e. operator $u \mapsto (|u'|^{p-2}u')'$, where $p > 1$ is a real number. This is one dimensional analogue of the p–Laplacian $u \mapsto \mathrm{div}(|\nabla u|^{p-2}\nabla u)$ which is frequently mentioned in many nonlinear mathematical models arising in various applications. From the theoretical point of view this (in general non additive for $p \neq 2$) operator plays a special role because it preserves *homogeneity* of order $p - 1$. This is very important because even if we loose linear structure of the operator for $p \neq 2$, we "stay not too far" from it for p "close to" 2.

The purpose of this paper is to illustrate that on one hand the operator $u \mapsto (|u'|^{p-2}u')'$ has very similar properties for $p = 2$ and $p \neq 2$ but on the other hand to show that the case $p = 2$ is singular in some sense.

The former case concerns the structure of the spectrum of nonlinear eigenvalue problem

$$\begin{aligned} &-(|u'|^{p-2}u')' - \lambda|u|^{p-2}u = 0 \text{ in } (0,1), \\ &u(0) = u(1) = 0. \end{aligned} \tag{EP}$$

It was proved that the set of all eigenvalues of (EP) and the properties of the corresponding eigenfunctions are very similar for any $p > 1$.

The latter case concerns the fact that the structure of the right hand sides f of the problem

$$\begin{aligned} &-(|u'|^{p-2}u')' - \lambda|u|^{p-2}u = f \text{ in } (0,1), \\ &u(0) = u(1) = 0, \end{aligned} \tag{RP}$$

where $\lambda > 0$ (λ being an eigenvalue or not), as well as the number of solutions to (RP), depend strongly on the fact whether $p = 2$ or not.

This paper is organized as follows. In section 2 we summarize the properties of all eigenvalues and eigenfunctions of the eigenvalue problem (EP). In section 3 we discuss existence and multiplicity of solutions to the problem (RP) for λ not an eigenvalue of (EP). We also point out some geometrical properties of the energy functional

$$J_f^\lambda : u \mapsto \frac{1}{p}\int_0^1 |u'|^p - \frac{\lambda}{p}\int_0^1 |u|^p - \int_0^1 fu$$

associated with (RP). The last section 4 is devoted exclusively to the case $\lambda = \lambda_1$ (the principal eigenvalue of (EP)). In this case the striking difference between $p = 2$ and $p \neq 2$ is shown. This difference concerns not only the structure of all right hand sides f for which (RP) is solvable, but it clarifies also the role of the conditions $p > 2$ and $p < 2$ in apriori estimates of the solutions as well as in the geometry of the energy functional $J_f^{\lambda_1}$.

This is a survey paper where the author was intended to summarize the research motivated by Professor J. Nečas more than twenty years ago. The author would like to express his gratitude to all his collaborators (Y. X. Huang, P. A. Binding, P. Takáč, M. del Pino, R. Manásevich) and to the following grants for the support during the work on this issue: the Grant Agency of the Czech Republic, grant # 201/97/03595, the Ministery of Education of the Czech Republic, grant # VS97156, NATO Collaborative Research Grant OUTR. CRG

961190, FONDAP de Matemáticas Aplicadas (Chile), the University of Calgary and the University of Rostock.

2. NONLINEAR HOMOGENEOUS EIGENVALUE PROBLEM

Let us consider eigenvalue problem

$$\begin{aligned} &(|u'|^{p-2}u')' + \lambda |u|^{p-2}u = 0 \text{ in } (0,1), \\ &u(0) = u(1) = 0 \end{aligned} \tag{2.1}$$

with a spectral parameter $\lambda \in \mathbb{R}$. Define

$$\pi_p = 2\int_0^1 \frac{ds}{(1-s^p)^{\frac{1}{p}}} = \frac{2\pi}{p\sin\frac{\pi}{p}}.$$

Then $\pi_2 = \pi$ and it was shown for instance in [DEM] that the set of all eigenvalues of (2.1) (i.e. the set of all values of λ for which (2.1) has a nonzero solution) is given by the sequence

$$\lambda_k = (p-1)(k\pi_p)^p \text{ for } k = 1, 2, \ldots$$

The set of eigenfunctions associated with $\lambda = \lambda_1$ corresponds precisely to that of constant multiples of the function $\sin_p(\pi_p t)$, where $\sin_p t$ is the solution of the initial value problem

$$\begin{aligned} &(|u'|^{p-2}u')' + (p-1)|u|^{p-2}u = 0 \text{ in } \mathbb{R}, \\ &u(0) = 0, u'(0) = 1, \end{aligned}$$

which for $t \in [0, \frac{\pi_p}{2}]$ can be described implicitly by the formula

$$t = \int_0^{\sin_p t} \frac{ds}{(1-s^p)^{\frac{1}{p}}}. \tag{2.2}$$

Furthermore, this function satisfies

$$\begin{aligned} &\sin_p(t) = \sin_p(\pi_p - t) \text{ for } t \in [\tfrac{\pi_p}{2}, \pi_p], \\ &\sin_p(t) = -\sin_p(-t) \text{ for } t \in [-\pi_p, 0] \end{aligned}$$

and can be uniquely extended as a $2\pi_p$ periodic function on the whole $\mathbb{R}$.

The set of eigenfunctions associated with $\lambda = \lambda_k, k = 2, 3, \ldots,$ are then constant multiples of the function $\sin_p(k\pi_p t)$.

For $t \in [0, \frac{\pi_p}{2})$ and $s \in [0, 1)$ setting

$$\cos_p t := \frac{d}{dt}\sin_p t, \quad \tan_p t := \frac{\sin_p t}{\cos_p t}, \quad \operatorname{arcsin}_p s := \sin_p^{-1} s,$$

we have the validity of formulas

$$\begin{aligned}&\sin_p^p t + \cos_p^p t = 1, \quad \cos_p' t = -\tan_p^{p-1} t \cos_p t,\\ &\tan_p' t = 1 + \tan_p^p t = \frac{1}{\cos_p^p t}, \quad \arcsin_p' s = \frac{1}{(1-s^p)^{\frac{1}{p}}}.\end{aligned} \tag{2.3}$$

These formulas fit with corresponding well known formulas for $p = 2$. Also the properties of eigenvalues ($\frac{\lambda_{k+1}^{1/p}}{\lambda_k^{1/p}} = \frac{k+1}{k}$, $\lambda_k \to \infty$ as $k \to \infty$, $\lambda_1 > 0$, etc.)
and eigenfunctions ($\sin_p(k\pi_p t)$ has $k-1$ equidistant nodes in (0,1), validity of formulas (2.3), etc.) for general $p > 1$ are similar to those already known before for $p = 2$.

Let us conclude this section by mentioning the paper of Nečas [N], where the eigenvalue problem of the type (2.1) was studied. Actually, the problem considered in [N] is more general (non autonomous) and so the results concerning the structure of the eigenvalues and associated eigenfunctions are not so accurate. The problem (2.1) was then studied by Drábek [D1], [D2] and Otani [O] where more accurate results were proved using autonomy of the equation in (2.1). The work of del Pino, Elgueta and Manásevich [DEM] provided then a very nice and transparent description of the eigenvalue problem (2.1) based on the generalization of "sin" function given implicitly by (2.2).

3. NONRESONANCE FOR THE P–LAPLACIAN

In this section we study the nonlinear boundary value problem

$$\begin{aligned}&-(|u'|^{p-2}u')' - \lambda|u|^{p-2}u = f \text{ in } (0,1),\\ &u(0) = u(1) = 0\end{aligned} \tag{3.1}$$

and associated energy functional $W_0^{1,p}(0,1) \to \mathbb{R}$,

$$J_f^\lambda(u) := \frac{1}{p}\int_0^1 |u'|^p - \frac{\lambda}{p}\int_0^1 |u|^p - \int_0^1 fu, \tag{3.2}$$

where λ is not and eigenvalue of (2.1) (i.e. $\lambda \neq \lambda_k, k = 1, 2\ldots$).

For simplicity we shall deal with $f \in C[0,1]$ and the solution of (3.1) will be such a function $u \in C^1[0,1]$ for which $|u'|^{p-2}u' \in C^1[0,1]$ and u satisfies the equation and the boundary conditions. It is not difficult to show that the critical points of J_f^λ are in one to one correspondence with the solutions of (3.1).

Due to the variational characterization of λ_1:

$$\lambda_1 = \min \frac{\int_0^1 |u'|^p}{\int_0^1 |u|^p}, \tag{3.3}$$

where minimum is taken over all nonzero elements of $W_0^{1,p}(0,1)$, and due to the monotonicity of the operators $A, B : W_0^{1,p}(0,1) \to (W_0^{1,p}(0,1))^*$ defined by

$$< Au, v >= \int_0^1 |u'|^{p-2}u'v', \quad < Bu, v >= \int_0^1 |u|^{p-2}uv$$

(here $< \cdot, \cdot >$ is the duality pairing between $(W_0^{1,p}(0,1))^*$ and $W_0^{1,p}(0,1)$) it is easy to prove that for $\lambda \le 0$ the energy functional J_f^λ has unique minimizer in $W_0^{1,p}(0,1)$ for arbitrary $f \in (W_0^{1,p}(0,1))^*$. In particular, it follows from here that given arbitrary $f \in C[0,1]$, the problem (3.1) has unique solution. So from this point of view, the situation is the same for $p = 2$ and $p \neq 2$.

The case $\lambda > 0$ is different. It is well known that for $p = 2$ and $\lambda \neq \lambda_k, k = 1, 2, \dots$, for any $f \in C[0,\pi]$ the problem (3.1) has unique solution, which follows e.g. from the Fredholm alternative. Let us consider now $p \neq 2$ and $0 < \lambda < \lambda_1$. Due to the variational characterization of λ_1 given by (3.3) the energy functional is still coercive but the monotone operators A and B "compete" because of the positive sign of λ. While in the linear case ($p = 2$) this fact does not affect uniqueness, for $p \neq 2$ the following interesting phenomenon is observed. There exist functions $f \in C[0,\pi]$ such that J_f^λ has at least two critical points. One of them corresponds to the global minimizer of J_f^λ on $W_0^{1,p}(0,1)$ (which does exist due to $\lambda < \lambda_1$) and the other is a critical point of "saddle" type. As an immediate consequence we obtain that for certain $f \in C[0,1]$ the problem (3.1) has at least two solutions.

The examples which illustrate these facts were given by Fleckinger, Hernández, Takáč and de Thélin [FHTT] for $1 < p < 2$ and del Pino, Elgueta and Manásevich [DEM] for $p > 2$. The result was generalized for general $\lambda > 0$ by Drábek and Takáč [DT].

As a summary of this section we point out that in the nonresonant case for the p–Laplacian we observe a lack of uniqueness for (3.1) with certain f when $p \neq 2$. This makes the nonlinear case qualitatively very different from the linear one.

4. RESONANCE FOR THE P–LAPLACIAN

The case $\lambda = \lambda_k$, $k = 1, 2, \dots$, in (3.1) is even more interesting and challenging. We shall restrict ourselves to $k = 1$, i.e. we study the nonlinear boundary value problem

$$\begin{aligned} &-(|u'|^{p-2}u')' - \lambda_1|u|^{p-2}u = f \text{ in } (0,1) \\ &u(0) = u(1) = 0, \end{aligned} \tag{4.1}$$

where $\lambda_1 = (p-1)\pi_p^p$ is the principal eigenvalue of (2.1).

For $p = 2$ this problem reduces to the simple linear problem

$$\begin{aligned} -u'' - \pi^2 u = f \text{ in } (0,1), \\ u(0) = u(1) = 0, \end{aligned} \tag{4.2}$$

whose solvability is fully described for instance by the linear Fredholm alternative. Namely, (4.2) is solvable if and only if

$$\int_0^1 f(t) \sin \pi t dt = 0. \tag{4.3}$$

In such case, the solution set is a continuum constituted by a one–dimensional linear manifold. Needless to say, such a nice characterization uses the underlying linear structure of the problem (4.2) in essential way.

It is natural to ask what is the role, if any, of the corresponding analogue of (4.3) for general $p > 1$, i.e.

$$\int_0^1 f(t) \sin_p \pi_p t dt = 0. \tag{4.4}$$

In fact, for instance it is shown in del Pino and Manásevich [DM] that no solution to (4.1) exists in the case $f \equiv$ constant $\neq 0$, case in which of course condition (4.4) is violated. However, in Binding, Drábek and Huang [BDH] an example is constructed which shows that (4.4) is not necessary for the existence of the solution to (4.1). This observation was refined by delPino, Drábek and Manásevich [DDM] in the following way.

Theorem 4.1. *Let $p \neq 2$. Then there exists an open cone $C \subset C[0,1]$ such that for all $f \in C$ problem (4.1) has at least two solutions. Moreover,*

$$\int_0^1 f(t) \sin_p \pi_p t dt \neq 0$$

for all $f \in C$.

So, if $p \neq 2$ the situation is completely different from the linear case.

On the other hand, as it follows from [DDM], if $f \in C^1[0,1]$ satisfies the orthogonality condition (4.4), linear in nature, then it is sufficient for solvability of (4.1) for any $p > 1$. In other words, the set of $f's$ for which (4.1) is solvable contains at least the linear space of all C^1 functions satisfying (4.4). More precisely we have

Theorem 4.2. *Let us assume that $f \in C^1[0,1]$, $f \not\equiv 0$, satisfies condition (4.4). Then the problem (4.1) has at least one solution. Moreover, if $p \neq 2$, then the set of all possible solutions is bounded in $C^1[0,1]$.*

We observe that also Theorem 4.2 reveals a striking difference between the cases $p \neq 2$ and $p = 2$, since in the latter case the solution set is an unbounded continuum.

A by product of the proof of Theorem 2 is the fact that the degree of the associated fixed–point operator with respect to a large ball in $C^1[0,T]$ becomes $+1$ if $p > 2$ while it equals -1 if $p < 2$.

Let us consider on $W_0^{1,p}(0,1)$ the energy functional

$$J_f^{\lambda_1}(u) = \frac{1}{p}\int_0^1 |u'|^p - \frac{\lambda_1}{p}\int_0^1 |u|^p - \int_0^1 fu$$

associated with (4.1). Assume that (4.4) holds. In case $p = 2$ this functional is bounded from below and it achieves its minimum in every solution of (4.1). The set of all solutions to (4.1) forms linear unbounded continuum in $W_0^{1,p}(0,1)$. In case $p \neq 2$ the situation is completely different. The following result is also proved in [DDM].

Theorem 4.3. *Assume that $f \in C^1[0,1]$, $f \not\equiv 0$ and f satisfies (4.4). Then*

(i) for $1 < p < 2$ the functional $J_f^{\lambda_1}$ is unbounded from below. The set of its critical points is nonempty and bounded;

(ii) for $p > 2$ the functional $J_f^{\lambda_1}$ is bounded from below and has a global minimizer. The set of its critical points is bounded, however $J_f^{\lambda_1}$ does not satisfy the Palais - Smale condition at the level 0.

Let us emphasize that changing p from $p < 2$ to $p > 2$ the following qualitative change of $J_f^{\lambda_1}$ occurs. The structure of $J_f^{\lambda_1}$ shifts from a saddle–point geometry to a global minima geometry for its level sets. The "singular value" $p = 2$ corresponds to a convex functional with a whole ray of minimizers. These facts open an interesting issue concerning the geometry of L^p–spaces and the structure of Poincaré–type inequalities.

References

[BDH] Binding, P.A., Drábek, P. and Huang, Y.X. On the Fredholm alternative for the p-Laplacian. Proc. Amer. Math. Soc., to appear.

[D1] Drábek, P. (1980). Ranges of a homogeneous operators and their perturbations. Časopis pěst. mat., 105:167–183.

[D2] Drábek, P. (1992). Solvability and Bifurcations of Nonlinear Equations. Research Notes in Mathematics 264, Longman, Harlow.

[DDM] delPino, M., Drábek, P. and Manásevich, R.F. The Fredholm alternative at the first eigenvalue for the one dimensional p-Laplacian, preprint.

[DEM] del Pino, M.A., Elgueta, M. and Manásevich, R.F. (1989). A homotopic deformation along p of a Leray-Schauder degree result and existence for $(|u'|^{p-2}u')' + f(t,u) = 0, u(0) = u(T) = 0, p > 1$. J. Differential Equations, 80:1–13.

[DM] del Pino, M.A. and Manásevich, R.F. (1991). Multiple solutions for the p-Laplacian under global nonresonance. Proc. Amer. Math. Soc., 112:131–138.

[DT] Drábek, P. and Takáč, P. A counterexample to the Fredholm alternative for the p-Laplacian, to appear.

[FHTT] Fleckinger, J., Hernández, J., Takáč, P. and deThélin, F. (1997). Uniqueness and positivity for solutions of equations with the p-Laplacian. Proc. of the Conference on Reaction-Diffusion Equations, Trieste, Italy, October 1995, Marcel Dekker, Inc., New York-Basel.

[FNSS] Fučík, S., Nečas, J., Souček, J. and Souček, V. (1973). Spectral Analysis of Nonlinear Operators. Lecture Notes in Mathematics Vol.346, Springer Verlag.

[N] Nečas, J. (1971). O diskretnosti spektra nelinejnoj zadaci Sturma - Liouvillja vtorogo porjadka. Dokl. Akad. Nauk SSSR, 201:1045–1048.

[O] Otani, M. (1984). A remark on certain nonlinear elliptic equations, Proc. Fac. Sci. Tokai Univ., Vol. XIX:23–28.

EXISTENCE OF SOLUTIONS TO A NONLINEAR COUPLED THERMO-VISCOELASTIC CONTACT PROBLEM WITH SMALL COULOMB FRICTION

Christof Eck, Jiří Jarušek

Abstract: The solvability of a coupled thermoviscoelastic contact problem with Coulomb friction is investigated. The heat generated by friction is described by a boundary term of quadratic order. The tensor of thermal conductivity is dependent on the temperature gradient and satisfies a certain growth condition.

Keywords: Dynamic contact problems, Signorini contact condition, nonlinear heat equation, viscoelasticity, Coulomb law of friction, existence of solutions.

1. INTRODUCTION

The investigation of contact problems with Coulomb friction started from an idea of Nečas to prove the existence of its solutions via the regularity of the solution of some approximate problem, [9]. This idea was also employed in the first existence results in dynamic contact problems derived for a viscoelastic body and a rigid undeformable support with a Signorini condition formulated in velocities, [5], [6]. As the friction represents an important heat source, the aspect of heat conduction and heat deformation must be included into the investigation. In [3] the linearized system of equations was treated. The linear character of the heat conduction equation forced us to limit there the growth of the heat generated by friction by a linear term. In the present paper the existence of a solution for a frictional thermoviscoelastic contact problem including the full quadratic growth of the heat generated by friction is proved for the first time. For

Applied Nonlinear Analysis, edited by Sequeira *et al.*
Kluwer Academic / Plenum Publishers, New York, 1999.

the viscoelastic material we employed the nonlinear constitutive law investigated in Nečas works [8], [10] and [11].

In the paper we use the standard notation W_p^k, $H^k \equiv W_2^k$, $\mathring{H}^k$ and $H^{-k} \equiv \mathring{H}^{k*}$ for isotropic Sobolev–Slobodeckii spaces with $k \geq 0$ and $p \geq 1$. If $k \in \mathbb{R}^2$, the first and second index corresponds to the time and the space variables, respectively. The spaces with range in $\mathbb{R}^N$ are denoted by bold letters.

2. DESCRIPTION OF THE PROBLEM

We consider a body occupying in some reference configuration the domain $\Omega \subset \mathbb{R}^N$ of dimension $N = 2$ or $N = 3$ with a Lipschitz boundary Γ composed of the three measurable, mutually disjoint parts Γ_U, Γ_F and Γ_C. Let $I_\mathcal{T} := [0, \mathcal{T}]$ be the considered time interval of the problem, let $Q_\mathcal{T} := I_\mathcal{T} \times \Omega$ denote the time–space domain and let $S_\mathcal{T} := I_\mathcal{T} \times \Gamma$ be its lateral boundary consisting of the parts $S_{X,\mathcal{T}} := I_\mathcal{T} \times \Gamma_X$ for $X = U, F, C$. For $\tau > 0$ we shall denote $I_\tau := [0, \tau]$ and analogously Q_τ, S_τ etc. The problem studied here consists of a dynamic contact problem with Coulomb friction coupled with a heat conduction equation. The contact problem is given by the set of relations

$$\ddot{u}_i - \sigma_{ij,j}(u) = f_i, \quad i = 1, \dots, N, \quad \text{in} \quad Q_\mathcal{T} \tag{2.1}$$

$$u = U \quad \text{on} \quad S_{U,\mathcal{T}}, \tag{2.2}$$

$$\sigma^{(n)}(u) = h \quad \text{on} \quad S_{F,\mathcal{T}}, \tag{2.3}$$

$$\left.\begin{array}{l} \dot{u}_n \leq 0, \quad \sigma_n \leq 0, \quad \sigma_n \dot{u}_n = 0, \\ \dot{u}_t = 0 \quad \Rightarrow \quad |\sigma_t| \leq \mathfrak{F}|\sigma_n|, \\ \dot{u}_t \neq 0 \quad \Rightarrow \quad \sigma_t = -\mathfrak{F}|\sigma_n| \dfrac{\dot{u}_t}{|\dot{u}_t|} \end{array}\right\} \quad \text{on} \quad S_{C,\mathcal{T}}, \tag{2.4}$$

$$u(0,x) = u_0(x), \quad \dot{u}(0,x) = u_1(x) \quad \text{for} \quad x \in \Omega. \tag{2.5}$$

Here and in the sequel, the summation convention is employed. By $v_{,i}$ we denote the derivative of a function v with respect to the space variable x_i. The respective time derivatives are denoted by dots. By u and Θ we denote the displacement field and the temperature, respectively. The strain–stress relation is given by a linear thermoviscoelastic law of the Kelvin–Voight type,

$$\sigma_{ij} \equiv \sigma_{ij}(u) = a^{(0)}_{ijk\ell} e_{k\ell}(u) + a^{(1)}_{ijk\ell} e_{k\ell}(\dot{u}) - b_{ij}\Theta, \quad i,j = 1, \dots, N,$$

with $e_{ij}(u) := \frac{1}{2}(u_{i,j} + u_{j,i})$. The tensors $\left\{a^{(0)}_{ijk\ell}\right\}$ and $\left\{a^{(1)}_{ijk\ell}\right\}$ are assumed to depend Lipschitz–continuously on the space variable and shall be symmetric, i.e. $a^{(\iota)}_{ijk\ell} = a^{(\iota)}_{jik\ell} = a^{(\iota)}_{k\ell ij}$, as well as bounded and elliptic, i.e.

$$a_0^{(\iota)}\xi_{ij}\xi_{ij} \le a_{ijk\ell}^{(\iota)}\xi_{ij}\xi_{k\ell} \le A_0^{(\iota)}\xi_{ij}\xi_{ij} \tag{2.6}$$

for all symmetric tensors $\{\xi_{ij}\} \in \mathbb{R}^{N,N}$ with constants $0 < a_0^{(\iota)} \le A_0^{(\iota)}$, $\iota = 0, 1$. The tensor $\{b_{ij}\}$ of thermal expansion shall be symmetric, Lipschitz with respect to the space variable and globally bounded. Moreover, n denotes the outer normal vector of the boundary, $\sigma_i^{(n)} = \sigma_{ij}n_j$ the components of the boundary traction; the subscripts $_n$ and $_t$ denote the normal and tangential components of the corresponding vectors. In particular, we have $\sigma_n = \sigma_{ij}n_in_j$ and $\sigma_t = \sigma^{(n)} - \sigma_n n$. Observe that the Signorini condition (the first row of (2.4)) is formulated in velocities.

The temperature field Θ satisfies the heat conduction problem

$$\begin{aligned}
\dot{\Theta} - (c_{ij}\Theta_{,j})_{,i} + b_{ij}\Theta\,\dot{u}_{i,j} &= 0 && \text{in } Q_T, && (2.7)\\
\Theta &= 0 && \text{on } S_{U,T}, && (2.8)\\
c_{ij}\Theta_{,j}n_i &= K\,(\Upsilon - \Theta) && \text{on } S_{F,T}, && (2.9)\\
c_{ij}\Theta_{,j}n_i = \mathfrak{F}|\sigma_n||\dot{u}_t| &+ K(\Upsilon - \Theta) && \text{on } S_{C,T}, && (2.10)\\
\Theta(0,x) &= 0 && \text{for } x \in \Omega. && (2.11)
\end{aligned}$$

The tensor of thermal conductivity c_{ij} is assumed to be symmetric and to depend locally Lipschitz–continuously on the temperature gradient such that it satisfies the growth condition

$$\check{c}_1\left(1 + |\nabla\Theta|^2\right)\xi_i\xi_i \le c_{ij}(\nabla\Theta)\xi_i\xi_j \le \check{c}_2\left(1 + |\nabla\Theta|^2\right)\xi_i\xi_i,\ \xi \in \mathbb{R}^N, \tag{2.12}$$

the strong monotonicity

$$\begin{aligned}
&\langle c_{ij}(\nabla\Theta)\Theta_{,j} - c_{ij}(\nabla\Xi)\Xi_{,j}, \Theta_{,i} - \Xi_{,i}\rangle_{Q_T}\\
&\quad \ge \check{c}_3\|\nabla(\Theta - \Xi)\|^4_{L_4(Q_T)} + \check{c}_4\|\nabla(\Theta - \Xi)\|^2_{L_2(Q_T)}
\end{aligned} \tag{2.13}$$

for each $\Theta, \Xi \in L_4\big(I_T; W_4^1(\Omega)\big)$, and the continuity relation

$$c_{ij}\big(\nabla\Theta^{(k)}\big)\Theta_{,j}^{(k)} \to c_{ij}(\nabla\Theta)\Theta_{,j} \text{ in } L_{\frac{4}{3}}(Q_T),\ i = 1, \dots, N, \tag{2.14}$$

for $\Theta^{(k)} \to \Theta$ strongly in $L_4(I_T; W_4^1(\Omega))$. An example for such a matrix–valued function is $c_{ij}(x;\Xi) = \delta_{ij}(d_0(x) + d_1(x)|\Xi|^2)$ with the Kronecker symbol δ_{ij} and measurable functions d_0 and d_1 such that $d_i \in [q_1, q_2]$, $i = 0, 1$, for some positive real constants q_1, q_2. In equation (2.7), the quadratic term describing the generation of heat by the viscosity (cf. [13]) has been neglected. Both the heat generated by friction and the heat exchange of the contact surface with the foundation is included into (2.10).

The variational formulation of the problem is given as follows. The sets of admissible functions for the viscoelastic equation of motion and the heat equation, respectively, are given by

$$\mathfrak{K} := \left\{v \in \boldsymbol{H}^{\frac{1}{2},1}(Q_T)\,;\, v = \dot{U} \text{ on } S_{U,T} \text{ and } v_n \le 0 \text{ on } S_{C,T}\right\}, \quad (2.15)$$

$$\mathfrak{U} := \{v \in L_2(I_T; \boldsymbol{H}^1(\Omega))\,;\, v = 0 \text{ on } S_{U,T}\}, \quad (2.16)$$

$$\mathfrak{V} := \{v \in L_4(I_T; W_4^1(\Omega))\,;\, v = 0 \text{ on } S_{U,T}\}. \quad (2.17)$$

A weak solution of the thermoviscoelastic contact problem is a pair of functions (u, Θ) with $\dot{u} \in \mathfrak{K}$, $\ddot{u} \in H^{\frac{1}{2}}(I_T; \boldsymbol{L}_2(\Omega))^*$, $\Theta \in \mathfrak{V}$, $\dot{\Theta} \in \mathfrak{V}^*$ which satisfy the initial conditions $u(0,\cdot) = u_0$, $\dot{u}(0,\cdot) = u_1$ and $\Theta(0,\cdot) = 0$ such that the variational inequality of the contact problem and the variational equation of the heat conduction problem are simultaneously satisfied. For a measurable set S let $\langle\cdot,\cdot\rangle_S$ denote the generalized $L_2(S)$–scalar product. Then the variational inequality of the thermoviscoelastic contact problem is given by

$$\begin{aligned}&\langle \ddot{u}, v - \dot{u}\rangle_{Q_T} + a^{(0)}(u, v - \dot{u}) + a^{(1)}(\dot{u}, v - \dot{u}) - \langle b_{ij}\Theta, e_{ij}(v - \dot{u})\rangle_{Q_T}\\ &+ \langle \mathfrak{F}|\sigma_n(u)|, |v_t| - |\dot{u}_t|\rangle_{S_{C,T}} \ge \mathcal{L}(v - \dot{u}) := \langle f_i, v_i\rangle_{Q_T} + \langle h_i, v_i\rangle_{S_{F,T}} \quad (2.18)\end{aligned}$$

for all $v \in \mathfrak{K}$ with the bilinear forms of elastic and viscoelastic energy

$$a^{(\iota)}(u,v) = \left\langle a^{(\iota)}_{ijk\ell} e_{k\ell}(u), e_{ij}(v)\right\rangle_{Q_T}, \quad \iota = 0, 1.$$

The variational equation of the heat conduction problem is defined by

$$\begin{aligned}\left\langle \dot{\Theta}, \varphi\right\rangle_{Q_T} &+ \langle c_{ij}(\nabla\Theta)\Theta_{,j}, \varphi_{,i}\rangle_{Q_T} + \langle b_{ij}\Theta\, \dot{u}_{i,j}, \varphi\rangle_{Q_T} + \\ &+ \langle K(\Theta - \Upsilon), \varphi\rangle_{S_{F,T}\cup S_{C,T}} = \langle \mathfrak{F}|\sigma_n||\dot{u}_t|, \varphi\rangle_{S_{C,T}} \quad (2.19)\end{aligned}$$

for all $\varphi \in \mathfrak{V}$.

The existence of solutions is proved in two steps: Using the penalty method, the contact problem (2.18) is replaced by an approximate variational inequality of the normal–compliance type. The existence of solutions to this problem is proved via a fixed point approach. In the second step we verify the convergence of solutions of the approximate problem to a solution of the original thermoviscoelastic contact problem.

3. APPROXIMATE CONTACT PROBLEM

Replacing the contact condition in (2.4) by the nonlinear boundary condition

$$\sigma_n(u) = -\tfrac{1}{\delta}[\dot{u}_n]_+ \text{ with } [\cdot]_+ := \max\{\cdot, 0\} \text{ and } \delta > 0,$$

we arrive at the following variational inequality:

Find a function u with $u(0,\cdot) = u_0$, $\dot{u}(0,\cdot) = u_1$ and $\dot{u} \in \dot{U} + \mathfrak{U}$ such that for all $v \in \dot{U} + \mathfrak{U}$ there holds

$$\langle \ddot{u}, v - \dot{u}\rangle_{Q_T} + a^{(0)}(u, v - \dot{u}) + a^{(1)}(\dot{u}, v - \dot{u}) - \left\langle b_{ij}\Theta^{(0)}, e_{ij}(v - \dot{u})\right\rangle_{Q_T}$$
$$+ \left\langle \tfrac{1}{\delta}[\dot{u}_n]_+, v_n - \dot{u}_n\right\rangle_{S_{C,T}} + \left\langle \mathfrak{F}\tfrac{1}{\delta}[\dot{u}_n]_+, |v_t| - |\dot{u}_t|\right\rangle_{S_{C,T}} \geq \mathcal{L}(v - \dot{u}). \quad (3.1)$$

Here, the temperature field Θ has been replaced by some given temperature $\Theta^{(0)}$. In the heat equation only the definition of the term describing the generation of heat by friction is changed; instead of $\mathfrak{F}|\sigma_n||\dot{u}_t|$ there appears $\mathfrak{F}\frac{1}{\delta}[\dot{u}_n]_+|\dot{u}_t|$. A solution of the coupled problem consisting both of the contact problem (3.1) and of the modified heat conduction problem (2.19) can be constructed as a fixed point: Let $\Theta^{(0)} \in \mathfrak{V}$ be a fixed temperature field. Then the solution of problem (3.1) with $\Theta^{(0)}$ defines a displacement field $u = u\big(\Theta^{(0)}\big)$ with $\dot{u} \in (\dot{U} + \mathfrak{U}) \cap \boldsymbol{H}^{\frac{1}{2},1}(Q_T)$. Solving the heat conduction problem with $u = u\big(\Theta^{(0)}\big)$, we obtain a function $\Theta = \Theta\big(u\left(\Theta^{(0)}\right)\big)$. If the problems (3.1) and (2.19) are uniquely solvable, then by this procedure an operator

$$\Phi : \Theta^{(0)} \mapsto \Theta \quad (3.2)$$

is defined and a solution of the approximate thermoviscoelastic contact problem is given by a fixed point of this operator and the corresponding solution u of the contact problem. In order to prove the existence of such a fixed point, we apply the fixed point theorem of Schauder, cf. [15]:

Theorem 3.1. *Let $\mathcal{X}$ be a Banach space, $C \subset \mathcal{X}$ be a bounded, convex, closed subset and $\Phi : C \to C$ be a completely continuous mapping from C into C. Then there exists at least one fixed point of Φ in C.*

Let us start with the investigation of the solvability of the approximate contact problem. In this section, all constants $\check{c}_\iota$, $\iota = 5, 6, \ldots$ may depend on the geometry of the domain Ω, on the given tensors $\{a^{(\iota)}_{ijk\ell}\}$, $\{b_{ij}\}$ and $\{c_{ij}\}$, and on the given data $\mathcal{L}$, U, Υ, K, u_0, u_1, $\mathfrak{F}$, but neither on the function $\Theta^{(0)}$ nor on the solution (u, Θ) of the problems to be solved. Some of the constants may also depend on the penalty parameter δ, this is then explicitly indicated by $\check{c} = \check{c}(\delta)$.

Proposition 1. *Let $\Theta^{(0)} \in H^{\frac{1}{4}}(I_T; L_2(\Omega))$ be a fixed temperature field. In addition to the assumptions concerning the regularity of the domain and the coefficients of the tensors $a^{(\iota)}_{ijk\ell}$ and b_{ij}, let $\mathcal{L} \in H^{\frac{1}{4}}\big(I_T; \boldsymbol{H}^1(\Omega)^*\big)$, $u_0, u_1 \in \boldsymbol{H}^{\frac{3}{2}}(\Omega)$ and $U \in \mathcal{H}_T$,*

$$\mathcal{H}_\tau := \left\{v \in \boldsymbol{H}^2(Q_\tau) \cap H^{\frac{5}{4}}(I_\tau; \boldsymbol{H}^1(\Omega))\,;\, v|_{S_{C,\tau}} = 0\right\}, \quad \tau > 0.$$

Let the compatibility conditions $U(0,\cdot) = u_0$ and $\dot{U}(0,\cdot) = u_1$ on Ω be satisfied. Let the coefficient of friction $\mathfrak{F}$ be bounded and nonnegative. Then the approximate contact problem (3.1) has a unique solution which satisfies the a-priori estimates

$$\|\dot{u}\|^2_{L_\infty(I_T;\boldsymbol{L}_2(\Omega))} + \|\dot{u}\|^2_{\boldsymbol{H}^{\frac{1}{2},1}(Q_T)} \leq \check{c}_5 \|\Theta^{(0)}\|^2_{L_2(Q_T)} + \check{c}_6, \tag{3.3}$$

$$\|\dot{u}\|^2_{H^{\frac{1}{4}}(I_T;\boldsymbol{H}^1(\Omega))} \leq \check{c}_7 \|\Theta^{(0)}\|^2_{H^{\frac{1}{4}}(I_T;L_2(\Omega))} + \check{c}_8(\delta). \tag{3.4}$$

The solution u depends continuously on the temperature: if $u^{(1)}$ and $u^{(2)}$ are two solutions with corresponding temperature fields $\Theta_1^{(0)}$ and $\Theta_2^{(0)}$, then

$$\|\dot{u}^{(1)} - \dot{u}^{(2)}\|_{\boldsymbol{H}^{\frac{1}{2},1}(Q_T)} \leq \check{c}_9 \|\Theta_1^{(0)} - \Theta_2^{(0)}\|_{L_2(Q_T)}. \tag{3.5}$$

Proof. All of the assertions have been proved in [3], see Proposition 1, except the boundedness of $\|\dot{u}\|_{H^{\frac{1}{4}}(I_T;\boldsymbol{H}^1(\Omega))}$. This is proved by the standard shift technique with respect to the time variable. For simplicity let us assume that variational inequality (3.1) is defined on the whole time axis $I_T = \mathbb{R}$. This can be achieved by a suitable localization technique, see. e.g. [6]. For a function $g(t,x)$, let $g_{-q}(t,x) := g(t+q,x)$ denote the shift with respect to the time variable and $\Delta_q g := g_{-q} - g$ the corresponding difference. We put the test function $v = \dot{u}_{-q}$ into the variational inequality, then we shift the inequality into the direction q and put the test function $v_{-q} = \dot{u}$ into the shifted inequality. Then we add both inequalities, multiply the result with $|q|^{-\frac{3}{2}}$ and integrate with respect to q. Using the technique described in [6], the inequality

$$\|\dot{u}\|^2_{H^{\frac{1}{4}}(\mathbb{R};\boldsymbol{H}^1(\Omega))} \leq \check{c}_{10} \|u\|^2_{H^{\frac{1}{4}}(\mathbb{R};\boldsymbol{H}^1(\Omega))} + \check{c}_{11} \|\Theta^{(0)}\|^2_{H^{\frac{1}{4}}(\mathbb{R};L_2(\Omega))} + \check{c}_{12}(\delta)$$

is obtained. This and the localization technique yield estimate (3.4). □

For the heat conduction problem (2.19) the following existence result is valid:

Proposition 2. *Let $\dot{u} \in (\dot{U} + \mathfrak{U}) \cap \boldsymbol{H}^{\frac{1}{2},1}(Q_T) \cap H^{\frac{1}{4}}(I_T;\boldsymbol{H}^1(\Omega))$ be a fixed displacement velocity and let the assumptions mentioned above concerning the regularity of the domain and the properties of the tensor of heat conduction $\{c_{ij}\}$ be valid. In addition, the Dirichlet part of the boundary for the heat equation shall have positive measure. Let the tensor of thermal expansion b_{ij} be bounded and symmetric. Let, moreover, the coefficient of heat exchange K and the coefficient of friction be bounded and non-negative. Finally, let $\Upsilon \in L_2(S_T)$. Then problem (2.19) (with $|\sigma_n(u)|$ replaced by $\frac{1}{\delta}[\dot{u}_n]_+$) has a unique solution* *which satisfies the a–priori estimate*

$$\|\Theta\|_{H^{\alpha,1}(Q_T)} + \|\Theta\|^3_{L_4(I_T;W^1_4(\Omega))} \leq \check{c}_{13}(\delta)\|\dot{u}\|^2_{\boldsymbol{H}^{\frac{1}{2},1}(Q_T)} + \check{c}_{14}(\delta) \quad (3.6)$$

for each $\alpha < \frac{1}{2}$. *The mapping* $\dot{u} \mapsto \Theta$ *is strongly continuous from* $(\dot{U} + \mathfrak{U}) \cap \boldsymbol{H}^{\frac{1}{2},1}(Q_T)$ *to* $L_4(I_T; W^1_4(\Omega)) \cap H^{\frac{1}{4}}(I_T; L_2(\Omega))$.

Proof. The existence of solutions is proved by the usual Galerkin method. Set $\mathcal{V} \equiv \{v \in W^1_4(\Omega);\ v = 0 \text{ on } \Gamma_U\}$, let $\{\mathcal{V}_m\}$ be an increasing sequence of m–dimensional subspaces such that $\bigcup_{m=1}^{+\infty} \mathcal{V}_m$ is dense in $\mathcal{V}$ (its existence is ensured by the separability of $\mathcal{V}$) and

$$\mathfrak{V}_m \equiv \Big\{ w : Q_T \to \mathbb{R};\ \exists c_i \in L_\infty(I_T),\ i = 1, \dots, m : \quad w(t,x) = \sum_{i=1}^m c_i(t) v_{im}(x) \Big\}$$

for an $L_2(\Omega)$–orthogonal basis $\{v_{im}\}_{i=1}^m$ of $\mathcal{V}_m$. Then, via the standard proof of the density of $\mathfrak{M} := \{\omega : I_T \to \mathcal{V};\ \text{card}\,(\omega(I_T)) < +\infty\}$ in $\mathfrak{V}$ and the easy approximability of functions from $\mathfrak{M}$ by elements from $\mathfrak{V}_0 \equiv \bigcup_{m=1}^{+\infty} \mathfrak{V}_m$, we can see that $\mathfrak{V}_0$ is dense in $\mathfrak{V}$. A Galerkin solution Θ_m of the heat conduction problem is a function from $\mathfrak{V}_m$ which satisfies for all test functions $\varphi \in \mathcal{V}_m$ and almost every $\tau \in I_T$ the Galerkin equations

$$\left\langle \dot{\Theta}_m, \varphi \right\rangle_\Omega + \langle c_{ij}(\nabla\Theta_m)\Theta_{m,j}, \varphi_{,i}\rangle_\Omega + \langle b_{ij}\Theta_m \dot{u}_{i,j}, \varphi\rangle_\Omega + \\ + \langle K(\Theta_m - \Upsilon), \varphi\rangle_{\Gamma_F \cup \Gamma_C} = \langle \mathfrak{F}\tfrac{1}{\delta}[\dot{u}_n]_+ |\dot{u}_t|, \varphi\rangle_{\Gamma_C} \quad (3.7)$$

and the initial condition $\Theta_m(0) = 0$. Equation (3.7) is equivalent to a system of ordinary differential equations. According to the well–known existence theorem about the Carathéodory solutions, this system has a solution. If we integrate the Galerkin equation (3.7) in time over $[0, \tau]$ with any $\tau \in \mathcal{T}$ and the test function $\varphi = \Theta_m(\tau)$, employ the monotonicity (2.13) of the tensor c_{ij} as well as the equivalence of the norms $\|\cdot\|_{W^1_4(\Omega)}$ and $v \mapsto \|\nabla v\|_{L_4(\Omega)}$ on $\mathcal{V}$, we derive the estimate

$$\|\Theta_m\|^2_{L_\infty(I_T;L_2(\Omega))} + \|\Theta_m\|^4_{L_4(I_T;W^1_4(\Omega))} \leq \check{c}_{15}\|\dot{u}\|_{L_2(I_T;\boldsymbol{H}^1(\Omega))} \|\Theta_m\|^2_{L_4(Q_T)} \\ + \check{c}_{16}\|\mathfrak{F}\|_{L_\infty(\Gamma_C)}\tfrac{1}{\delta}\,\|\dot{u}\|^2_{\boldsymbol{L}_{\frac{8}{3}}(S_{C,T})} \|\Theta_m\|_{L_4(S_{C,T})} + \check{c}_{17}\|\Upsilon\|^2_{L_2(S_T)}.$$

Due to the trace theorem and continuous embedding theorems there holds for dimension $N \leq 3$ $\|\dot{u}\|_{\boldsymbol{L}_{\frac{8}{3}}(S_{C,T})} \leq \check{c}_{18}\|\dot{u}\|_{\boldsymbol{H}^{\frac{1}{2},1}(Q_T)}$ and

$\|\Theta_m\|_{L_4(S_{C,T})} \leq \check{c}_{19} \|\Theta_m\|_{L_4(I_T;W^{1,4}(\Omega))}$. With the help of suitable Hölder inequalities we obtain

$$\|\Theta_m\|^2_{L_\infty(I_T;L_2(\Omega))} + \|\Theta_m\|^4_{L_4(I_T;W^1_4(\Omega))} \leq \check{c}_{20}(\delta)\|\dot{u}\|^{\frac{8}{3}}_{\boldsymbol{H}^{\frac{1}{2},1}(Q_T)} + \check{c}_{21}(\delta),$$

hence

$$\|\Theta_m\|^3_{L_4(I_T;W^1_4(\Omega))} \leq \check{c}_{22}(\delta)\|\dot{u}\|^2_{\boldsymbol{H}^{\frac{1}{2},1}(Q_T)} + \check{c}_{23}(\delta) \tag{3.8}$$

with $\check{c}_i(\delta)$, $i = 20, \ldots, 23$, independent of m, Θ_m and $\dot{u}$. Consequently, there exists a sequence $\{m_k\}$ converging to $+\infty$ and a corresponding sequence Θ_{m_k} of solutions to the Galerkin equations such that $\Theta_{m_k} \rightharpoonup \Theta$ in $L_4(I_T; W^1_4(\Omega))$ and $c_{ij}(\nabla\Theta_{m_k})\Theta_{m_k,j} \rightharpoonup C_i$ in $L_{\frac{4}{3}}(Q_T)$, $i = 1, \ldots, N$, with certain limits $C_i \in L_{\frac{4}{3}}(Q_T)$. Moreover, from (3.7) and (3.8) it follows that $\left\{\|\dot{\Theta}_m\|_{\mathfrak{V}^*_m}; m \in \mathbb{N}\right\}$ is bounded. Hence by a standard diagonal method we obtain a functional Λ and a subsequence indexed by m_k again such that for any $\varphi \in \mathfrak{V}_0$ there holds $\langle\dot{\Theta}_{m_k}, \varphi\rangle \to \Lambda(\varphi)$. Clearly, $|\Lambda(\varphi)| \leq \check{c}_{24}(\delta)\|\varphi\|_{\mathfrak{V}}$, $\varphi \in \mathfrak{V}_0$ and Λ is linear there. Hence $\Lambda \in \mathfrak{V}^*$ and its norm in $\mathfrak{V}^*$ is bounded by $\check{c}_{24}(\delta)$. It is standard to show that $\Lambda = \dot{\Theta}$. Passing to the limit $m_k \to +\infty$ we prove that Θ is a solution of the equation

$$\begin{aligned}\langle\dot{\Theta}, \varphi\rangle_{Q_T} + \langle C_i, \varphi_{,i}\rangle_{Q_T} + \langle b_{ij}\Theta\dot{u}_{i,j}, \varphi\rangle_{Q_T} + \langle K(\Theta - \Upsilon), \varphi\rangle_{S_{F,T}\cup S_{C,T}} \\ = \langle\mathfrak{F}\tfrac{1}{\delta}[\dot{u}_n]_+|\dot{u}_t|, \varphi\rangle_{S_{C,T}}, \quad \varphi \in \mathfrak{V}.\end{aligned} \tag{3.9}$$

It remains to prove $C_i = c_{ij}(\nabla\Theta)\Theta_{,j}$. Therefore we first investigate the regularity of Θ_m and Θ with respect to the time variable. For $0 \leq s_1, s_2 \leq T$, $s_1 \neq s_2$, we put the test function $\varphi = \Theta_m(s_2) - \Theta_m(s_1)$ into the Galerkin equations at time τ, we multiply the result by $|s_2 - s_1|^{-1-2\alpha}$ with a parameter $\alpha \in (0, \frac{1}{2})$ and integrate the result both with respect to τ from s_1 to s_2 and with respect to $s = (s_1, s_2)$ over I^2_T. Then we obtain the equation

$$\begin{aligned}&\int_{I^2_T}\int_{s_1}^{s_2} \frac{\langle\dot{\Theta}_m(\tau), \Theta_m(s_2) - \Theta_m(s_1)\rangle_\Omega}{|s_2 - s_1|^{1+2\alpha}}\, d\tau\, ds = \\ &-\int_{I^2_T}\int_{s_1}^{s_2} |s_2 - s_1|^{-1-2\alpha}\Big(\langle c_{ij}(\nabla\Theta_m(\tau))\Theta_{m,j}(\tau), \Theta_{m,i}(s_2) - \Theta_{m,i}(s_1)\rangle_\Omega \\ &+\langle b_{ij}\Theta_m(\tau)\dot{u}_{i,j}(\tau), \Theta_m(s_2) - \Theta_m(s_1)\rangle_\Omega + \langle K(\Theta_m - \Upsilon)(\tau), \Theta_m(s_2) \\ &-\Theta_m(s_1)\rangle_{\Gamma\setminus\Gamma_U} - \langle\mathfrak{F}\tfrac{1}{\delta}[\dot{u}_n(\tau)]_+|\dot{u}_t(\tau)|, \Theta_m(s_2) - \Theta_m(s_1)\rangle_{\Gamma_C}\Big)\, d\tau\, ds.\end{aligned}$$

After performing the integration with respect to τ one observes that the left hand side of this equation is equivalent to the square of the seminorm of Θ_m in the space $H^\alpha(I_T; L_2(\Omega))$. On the right hand side there are expressions of the type

$$\int_{I_T^2}\int_{s_1}^{s_2} \frac{|f(\tau)||g(s_2)-g(s_1)|}{|s_2-s_1|^{1+2\alpha}}\,d\tau\,ds,\ f\in L_p(I_T),\ g\in L_q(I_T),\ \tfrac{1}{p}+\tfrac{1}{q}=1.$$

Such expressions are bounded by $\check{c}_{25}\|f\|_{L_p(I_T)}\|g\|_{L_q(I_T)}$ with a constant $\check{c}_{25}$ independent of f and g if $\alpha < \frac{1}{2}$. Hence we obtain the estimate

$$\begin{aligned}\|\Theta_m\|^2_{H^\alpha(I_T;L_2(\Omega))} \le \check{c}_{26}\Big(&\|\Theta_m\|^4_{L_4(I_T;W_4^1(\Omega))}+\\ &+\|\Theta_m\|^2_{L_4(Q_T)}\|\dot{u}\|_{L_2(I_T;\boldsymbol{H}^1(\Omega))}+\|\Theta_m\|^2_{L_2(I_T;H^1(\Omega))}+\|\Upsilon\|^2_{L_2(S_T)}+\\ &+\big\|\mathfrak{F}\tfrac{1}{\delta}[\dot{u}_n]_+|\dot{u}_t|\big\|_{L_{\frac{4}{3}}(S_{C,T})}\|\Theta_m\|_{L_4(S_{C,T})}\Big)+\check{c}_{27}.\end{aligned} \tag{3.10}$$

This implies that Θ_m is bounded uniformly with respect to m in $H^\alpha(I_T; L_2(\Omega))$ for all $\alpha < \frac{1}{2}$. Interpolation with $\Theta_m \in L_2\big(I_T; H^1(\Omega)\big)$ gives $\Theta_m \in H^{\alpha(1-\varepsilon)}(I_T; H^\varepsilon(\Omega))$, $\varepsilon \in (0,1)$. This space is compactly embedded into $L_4(I_T; L_2(\Omega))$ for a suitable choice of ε, α. As a consequence, the sequence Θ_{m_k} mentioned above has an appropriate subsequence, denoted by Θ_{m_k} again, which converges also strongly in $L_4(I_T; L_2(\Omega))$ to the limit Θ. From the inequality

$$\|\Theta_{m_k}-\Theta\|^4_{L_4(Q_T)} \le \check{c}_{28}\|\Theta_{m_k}-\Theta\|^2_{L_4(I_T;L_\infty(\Omega))}\|\Theta_{m_k}-\Theta\|^2_{L_4(I_T;L_2(\Omega))}$$

there follows $\Theta_{m_k} \to \Theta$ strongly in $L_4(Q_T)$. With this strong convergence we are able to verify $C_i = c_{ij}(\nabla\Theta)\Theta_{,j}$. In fact, from the monotonicity condition (2.13) we derive $\liminf_{k\to+\infty}\langle c_{ij}(\nabla\Theta_{m_k})\Theta_{m_k,j}, \Theta_{m_k,i}\rangle_{Q_T} \ge \langle C_i, \Theta_{,i}\rangle_{Q_T}$. Employing both the Galerkin equation for Θ_{m_k} and the equation (3.9) for Θ we obtain $\limsup_{k\to+\infty}\langle c_{ij}(\nabla\Theta_{m_k})\Theta_{m_k,j}, \Theta_{m_k,i}\rangle_{Q_T} \le \langle C_i, \Theta_{,i}\rangle_{Q_T}$. Therefore we have $\lim_{k\to\infty}\langle c_{ij}(\nabla\Theta_{m_k})\Theta_{m_k,j}, \Theta_{m_k,i}\rangle_{Q_T} = \langle C_i, \Theta_{,i}\rangle_{Q_T}$ and from the limit in the monotonicity equation

$$\begin{aligned}c_0\|\Theta_{m_k}-\Theta\|^4_{L_4(I_T;W_4^1(\Omega))} &\le\\ &\le \langle c_{ij}(\nabla\Theta_{m_k})\Theta_{m_k,j}-c_{ij}(\nabla\Theta)\Theta_{,j}, \Theta_{m_k,i}-\Theta_{,i}\rangle_{Q_T}\end{aligned}$$

there follows that the convergence $\Theta_{m_k} \to \Theta$ is in fact strong in $L_4(I_T; W_4^1(\Omega))$. As a consequence

$$c_{ij}(\nabla\Theta_{m_k})\Theta_{m_k,j} \to c_{ij}(\nabla\Theta)\Theta_{,j}$$

strongly in $L_{\frac{4}{3}}(Q_T)$ and $C_i = c_{ij}(\nabla\Theta)\Theta_{,j}$. The *a priori* estimates (3.8) and (3.10) are also valid for the limit Θ what proves estimate (3.6).

In order to prove uniqueness, we assume that there are two solutions $\Theta^{(1)}$, $\Theta^{(2)}$ of the heat equation with the same velocity field $\dot{u}$. Let $\Xi := \Theta^{(1)} - \Theta^{(2)}$ denote the difference and let a test function ψ be defined by $\psi = \Xi$ for $\tau \leq \tau_0$ and $\psi = 0$ for $\tau \geq \tau_0$. Putting ψ into the equation with solution $\Theta^{(1)}$, putting $-\psi$ into the equation with solution $\Theta^{(2)}$ and adding the results we arrive at the inequality

$$\|\Xi(\tau_0)\|^2_{L_2(\Omega)} + \|\Xi\|^2_{L_2(0,\tau_0;\,H^1(\Omega))} \leq \check{c}_{29} \int_0^{\tau_0} \int_\Omega |\nabla \dot{u}||\Xi|^2 \, dx \, d\tau$$

$$\leq \check{c}_{30} \int_0^{\tau_0} \|\nabla \dot{u}\|_{L_2(\Omega;\mathbb{R}^{N^2})} \|\Xi\|^{\frac{1}{2}}_{L_2(\Omega)} \|\Xi\|^{\frac{3}{2}}_{L_6(\Omega)} \, d\tau$$

$$\leq \check{c}_{31}(\varepsilon) \int_0^{\tau_0} \|\nabla \dot{u}\|^4_{L_2(\Omega;\mathbb{R}^{N^2})} \|\Xi\|^2_{L_2(\Omega)} \, d\tau + \varepsilon \|\Xi\|^2_{L_2(0,\tau_0;L_6(\Omega))},$$

valid for any parameter $\varepsilon > 0$. For dimension $N \leq 3$ the space $H^{\frac{1}{4}}(I_T; H^1(\Omega))$ is embedded into $L_4(I_T; H^1(\Omega))$ and $H^1(\Omega)$ is embedded into $L_6(\Omega)$, hence from the previous equation there follows with the application of the Gronwall lemma the equation $\|\Xi(\tau_0)\|_{L_2(\Omega)} = 0$ for all $\tau_0 \geq 0$.

The continuity of the mapping $u \mapsto \Theta$ is seen as follows: Let $\dot{u}^{(k)}$ be a sequence of displacement velocities converging to $\dot{u}$ strongly in the space $\boldsymbol{H}^{\frac{1}{2},1}(Q_T)$ and let $\Theta^{(k)}$ be the sequence of corresponding solutions of the heat conduction problem. Since these solutions are uniformly bounded in $L_4(I_T; W_4^1(\Omega)) \cap H^\alpha(I_T; L_2(\Omega))$, $\alpha < \frac{1}{2}$, there exists a weakly convergent subsequence with limit Θ. As in the proof of the convergence of the Galerkin solutions above it is seen that Θ is a solution to the heat equation with corresponding displacement velocity field $\dot{u}$ and that the convergence $\Theta^{(k)} \to \Theta$ is in fact strong in $L_4(I_T; W_4^1(\Omega))$. The solution Θ is unique, hence every convergent subsequence of $\Theta^{(k)}$ has this limit. Due to the interpolation inequality

$$\left\|\Theta^{(k)} - \Theta\right\|_{H^{\frac{1}{4}}(I_T;L_2(\Omega))} \leq \check{c}_{32} \left\|\Theta^{(k)} - \Theta\right\|^{\frac{1}{4\alpha}}_{H^\alpha(I_T;L_2(\Omega))} \left\|\Theta^{(k)} - \Theta\right\|^{1-\frac{1}{4\alpha}}_{L_2(Q_T)}$$

valid for $\frac{1}{4} < \alpha < \frac{1}{2}$ the strong convergence in $H^{\frac{1}{4}}(I_T; L_2(\Omega))$ follows. □

With the help of Propositions 1 and 2 the existence of solutions to the approximate thermoviscoelastic contact problem can be formulated as follows:

Proposition 3. *Let the assumptions of Propositions 1 and 2 be valid. Then the thermoviscoelastic contact problem (3.1), (2.19) has a solution (u, Θ) which satisfies for all $\alpha < \frac{1}{2}$ the a-priori estimate*

$$\|\dot{u}\|_{\boldsymbol{H}^{\frac{1}{2},1}(Q_T)} + \|\Theta\|_{H^{\alpha,1}(Q_T)} + \|\Theta\|_{L_4(I_T;W_4^1(\Omega))} \leq \check{c}_{33}(\delta). \tag{3.11}$$

Proof. We must verify that the mapping Φ defined in (3.2) satisfies the assumptions of Theorem 3.1. Let $\mathcal{X} := H^{\frac{1}{4}}(I_T; L_2(\Omega))$ and let

$$\mathcal{C} \equiv \mathcal{C}(R) := \left\{ \Psi \in \mathcal{X};\ \|\Psi\|_{H^{\frac{1}{4}}(I_T;L_2(\Omega))} + \|\Psi\|^2_{L_2(Q_T)} \leq R \right\}$$

with $R > 0$ be the convex closed subset. If $\Theta^{(0)} \in \mathcal{C}(R)$ then, according to the estimates (3.3) and (3.6), the image $\Theta = \Phi(\Theta^{(0)})$ satisfies the inequality

$$\|\Theta\|^3_{L_4(I_T;W_4^1(\Omega))} \leq \check{c}_{34}(\delta) \left\|\Theta^{(0)}\right\|^2_{L_2(Q_T)} + \check{c}_{35}(\delta) \leq \check{c}_{36}(\delta) R + \check{c}_{37}(\delta).$$

Moreover, due to inequality (3.10) there holds

$$\|\Theta\|^2_{H^{\frac{1}{4}}(I_T;L_2(\Omega))} \leq \check{c}_{38} \|\Theta\|^4_{L_4(I_T;W_4^1(\Omega))} + \check{c}_{39}(\delta) \|\dot{u}\|^{\frac{8}{3}}_{\boldsymbol{H}^{\frac{1}{2},1}(Q_T)} + \check{c}_{40}$$

with the solution u of the approximate contact problem (3.1) to the temperature field $\Theta^{(0)}$. From the previous estimates and inequality (3.3) we derive

$$\|\Theta\|_{H^{\frac{1}{4}}(I_T;L_2(\Omega))} + \|\Theta\|^2_{L_2(Q_T)} \leq \check{c}_{41}(\delta) R^{\frac{2}{3}} + \check{c}_{42}(\delta) \tag{3.12}$$

with constants $\check{c}_{41}(\delta)$ and $\check{c}_{42}(\delta)$ independent of Θ, $\Theta^{(0)}$ and R. Hence there exists a value $R > 0$ such that Φ maps $\mathcal{C}(R)$ into itself. Due to the continuity results of Propositions 1 and 2 the mapping Φ is continuous in $\mathcal{X}$. Moreover, combining the *a priori* estimates of Propositions 1 and 2 we see that Φ maps bounded subsets of $H^{\frac{1}{4}}(I_T; L_2(\Omega))$ into bounded subsets of $H^{\alpha,1}(Q_T)$, $\frac{1}{4} < \alpha < \frac{1}{2}$, hence Φ is completely continuous. According to Theorem 3.1 Φ possesses a fixed point in $\mathcal{C}_R$. The *a priori* estimate (3.11) follows from the inequalities (3.12), (3.3) and (3.6). □

4. EXISTENCE OF THE SOLUTION TO THE THERMOVISCOELASTIC CONTACT PROBLEM

The existence of solutions to the original contact problem is proved by the investigation of the limit $\delta \to 0$ of the penalty parameter. Therefore it is necessary to have *a priori* estimates uniform with respect to the penalty parameter. For the contact problem such an estimate has been derived in [3]:

Proposition 4. *In addition to the assertions of Proposition 1 we assume* $\Gamma_C \in C_{1,1}$, $\Theta \in H^{\frac{1}{4},\frac{1}{2}}(Q_T)$ *and* $f \in L_2\left(I_T; \boldsymbol{H}^{\frac{1}{2}}(\Omega)\right)^* \cap H^{\frac{1}{4}}\left(I_T; \boldsymbol{H}^1(\Omega)^*\right)$. *The coefficient of friction shall depend on the space* *variable $x \in \Gamma_C$ such that $\operatorname{supp}\mathfrak{F} \subset \Gamma_{C,\eta} := \{x \in \Gamma_C;\, dist(x, \partial\Gamma_C) \geq \eta\}$ for some $\eta > 0$. Moreover, $\mathfrak{F}$ shall be uniformly bounded by the constant $C_{\mathfrak{F}}$ given in [6], Proposition 3.1 and formula (4.23) for anisotropic material and in [4] for isotropic material in two dimensions.*[1] *Then the solution to the approximate contact problem (3.1) satisfies the inequality*

$$\|\dot{u}\|_{\boldsymbol{H}^{\frac{1}{2},1}(Q_T)} + \|\dot{u}\|_{\boldsymbol{H}^{\frac{1}{2},1}(S_{C,\eta,T})} + \left\|\tfrac{1}{\delta}[\dot{u}_n]_+\right\|_{L_2(S_{C,\eta,T})} \leq \check{c}_{43}\|\Theta\|_{H^{\frac{1}{4},\frac{1}{2}}(Q_T)} + \check{c}_{44} \tag{4.1}$$

with $S_{C,\eta,T} := I_T \times \Gamma_{C,\eta}$ and constants $\check{c}_{43}$, $\check{c}_{44}$ independent of u, Θ and δ.

With the help of this Proposition it is possible to derive uniform estimates for the pair of solutions (u, Θ) of the coupled thermoviscoelastic contact problem.

Proposition 5. *Let the assumptions of Propositions 2 and 4 and $\Gamma \in C_{1,1}$ be valid. Then the temperature field Θ of the solution of the coupled approximate contact problem (3.1), (2.19) satisfies the inequality*

$$\|\Theta\|_{H^\alpha(I_T;L_2(\Omega))} + \|\Theta\|_{L_4(I_T;W_4^1(\Omega))} + \|\Theta\|_{L_p(S_T)} \leq \check{c}_{45} \tag{4.2}$$

for all $\alpha < \frac{1}{2}$ and all $p < 6$ with $\check{c}_{45}$ dependent on the given data but neither on the solution nor on the penalty parameter δ.

Proof. The usual *a priori* estimate of the heat equation (2.19) yields

$$\begin{aligned}\|\Theta\|^4_{L_4(I_T;W_4^1(\Omega))} \;\leq\;& \check{c}_{46}\left\|\tfrac{1}{\delta}[\dot{u}_n]_+\right\|_{L_2(S_{C,\eta,T})}\|\dot{u}\|_{L_{4-\varepsilon_1}(I_T;\boldsymbol{L}_2(\Gamma_C))} \\ &\cdot \|\Theta\|_{L_{4+\varepsilon_2}(I_T;L_\infty(\Gamma_C))} + \check{c}_{47}\|\dot{u}\|_{L_2(I_T;\boldsymbol{H}^1(\Omega))}\|\Theta\|^2_{L_4(Q_T)} + \check{c}_{48}\end{aligned}$$

with $\varepsilon_2 = \varepsilon_2(\varepsilon_1)$ such that $\varepsilon_2 \to 0$ for $\varepsilon_1 \to 0$. The constants $\check{c}_i$, $i = 46, \ldots$ employed here and in the sequel are independent both of the investigated solutions and of the penalty parameter δ. The estimate

---

[1] For anisotropic material this constant is given by

$$C_{\mathfrak{F}} := \sqrt{\frac{a_0^{(1)}}{2A_0^{(1)}}} \cdot \left\langle \begin{array}{ll} \sqrt{z/\left((1+\sqrt{8z})(z+1)\right)}, & z \geq 1, \\ \sqrt{\sqrt{z}/\left(2(1+\sqrt{8z})\right)}, & z \leq 1 \end{array}\right.$$

with $z = \frac{\pi\sqrt{a_0^{(1)}A_0^{(1)}}}{\sqrt{2}c^2_{d-1}(\frac{1}{2})}$, where $c_{d-1}(\frac{1}{2}) = 2\int_{\mathbb{R}} \frac{\sin^2 s}{s^2}\,ds \int_{\mathbb{R}^{d-2}}(1+|s|^2)^{-\frac{d}{2}}\,ds$ and $a_0^{(1)}, A_0^{(1)}$ are the constants of ellipticity and boundedness of the viscous part from formula (2.6).

(3.10) can be easily modified to

$$\|\Theta\|^2_{H^\alpha(I_T;L_2(\Omega))} \leq \check{c}_{49}\Big(\|\Theta\|^4_{L_4(I_T;W_4^1(\Omega))}+ +\|\Theta\|^2_{L_4(Q_T)}\|\dot{u}\|_{L_2(I_T;\boldsymbol{H}^1(\Omega))}+\|\Theta\|^2_{L_2(S_T)}+ +\big\|\tfrac{1}{\delta}[\dot{u}_n]_+\big\|_{L_2(S_{C,\eta,T})}\|\dot{u}\|_{L_{4-\varepsilon_1}(I_T;\boldsymbol{L}_2(\Gamma_C))}\|\Theta\|_{L_{4+\varepsilon_2}(I_T;L_\infty(\Gamma_C))}+1\Big).$$

The previous two estimates together with (3.3) lead after some calculation to the inequality

$$k\ \|\Theta\|^2_{H^\alpha(I_T;L_2(\Omega))}+\|\Theta\|^4_{L_4(I_T;W_4^1(\Omega))} \leq \check{c}_{50}\|\dot{u}\|_{L_{4-\varepsilon_1}(I_T;\boldsymbol{L}_2(\Gamma_C))} \cdot \big\|\tfrac{1}{\delta}[\dot{u}_n]_+\big\|_{L_2(S_{C,\eta,T})}\|\Theta\|_{L_{4+\varepsilon_2}(I_T;L_\infty(\Gamma))}+\check{c}_{51}\|\Theta\|^3_{L_4(I_T;W_4^1(\Omega))}+\check{c}_{52}, \quad (4.3)$$

valid for an appropriate $k>0$ independent of the penalty parameter. From the trace estimate $\|\dot{u}\|_{\boldsymbol{L}_2(\Gamma_C)} \leq \check{c}_{53}\|\dot{u}\|_{\boldsymbol{H}^{\frac{1}{2}+\varepsilon_3}(\Omega)}$ valid for all $\varepsilon_3>0$ and from the interpolation estimate $\|\dot{u}\|_{\boldsymbol{H}^{\frac{1}{2}+\varepsilon_3}(\Omega)} \leq \check{c}_{54}\|\dot{u}\|^{\frac{1}{2}-\varepsilon_3}_{\boldsymbol{L}_2(\Omega)} \cdot\|\dot{u}\|^{\frac{1}{2}+\varepsilon_3}_{\boldsymbol{H}^1(\Omega)}$ it is possible to derive

$$\begin{aligned}\|\dot{u}\|^{4-\varepsilon_1}_{L_{4-\varepsilon_1}(I_T;\boldsymbol{L}_2(\Gamma))} &\leq \check{c}_{55}\int_{I_T}\|\dot{u}\|^{2-\varepsilon_1}_{\boldsymbol{L}_2(\Omega)}\|\dot{u}\|^2_{\boldsymbol{H}^1(\Omega)}\,d\tau\\ &\leq \check{c}_{56}\|\dot{u}\|^{2-\varepsilon_1}_{L_\infty(I_T;\boldsymbol{L}_2(\Omega))}\|\dot{u}\|^2_{L_2(I_T;\boldsymbol{H}^1(\Omega))}. \end{aligned} \quad (4.4)$$

This together with the *a priori* estimate (3.3) of the contact problem yields

$$\|\dot{u}\|_{L_{4-\varepsilon_1}(I_T;\boldsymbol{L}_2(\Gamma_C))} \leq \check{c}_{57}\|\Theta\|_{L_2(Q_T)}+\check{c}_{58}.$$

The norm $\|\Theta\|_{L_{4+\varepsilon_2}(I_T;L_\infty(\Gamma_C))}$ can be estimated with the help of suitable embedding theorems by

$$\|\Theta\|^{4+\varepsilon_2}_{L_{4+\varepsilon_2}(I_T;L_\infty(\Gamma))} \leq \check{c}_{59}\int_{I_T}\|\Theta\|^{4+\varepsilon_2}_{W^{\alpha_0}_{4+\varepsilon_2}(\Gamma)}\,d\tau$$

for any $\alpha_0 \geq \frac{1}{2}$. Now we use the interpolation inequality

$$\|\varphi\|_{W^{\alpha_0}_{\beta_0}(M)} \leq \check{c}_{60}\|\varphi\|^{\lambda}_{W^{\alpha_1}_{\beta_1}(M)}\|\varphi\|^{1-\lambda}_{W^{\alpha_2}_{\beta_2}(M)} \quad (4.5)$$

valid for all $\alpha_0,\alpha_1,\alpha_2 \geq 0$, $\beta_0,\beta_1,\beta_2 \geq 1$ with $\alpha_0=\lambda\alpha_1+(1-\lambda)\alpha_2$ and $\frac{1}{\beta_0}=\frac{\lambda}{\beta_1}+\frac{1-\lambda}{\beta_2}$. This inequality is proved in [14], Theorem 1(d) in Section 2.4.2, relation (7) in Definition 1(d) on page 169 and Remark 2 on page 185 for arbitrary φ for $M=\mathbb{R}^N$, its generalization for M with C_∞-boundary is included into that book, too. The validity can be extended both to $M = \Omega$ and to its boundary $M = \Gamma$ by the usual localization technique and straightening of the boundary, provided $\Gamma \in C_{1,1}$ and $\alpha_0, \alpha_1, \alpha_2 \leq 1$. Employing this interpolation result for $\beta_0 = 4 + \varepsilon_2$, $\alpha_1 = \frac{3}{4}$, $\beta_1 = 4$, $\alpha_2 = 0$ and $\beta_2 = 6 - \vartheta$ with an arbitrarily small $\vartheta > 0$ and the usual trace theorem, we obtain

$$\begin{aligned}\|\Theta\|^{4+\varepsilon_2}_{L_{4+\varepsilon_2}(I_T;L_\infty(\Gamma))} &\leq \check{c}_{61} \int_{I_T} \|\Theta\|^{\lambda(4+\varepsilon_2)}_{W_4^{\frac{3}{4}}(\Gamma)} \|\Theta\|^{(1-\lambda)(4+\varepsilon_2)}_{L_{6-\vartheta}(\Gamma)}\, d\tau \\ &\leq \check{c}_{62} \left(\int_{I_T} \|\Theta\|^4_{W_4^1(\Omega)}\, d\tau\right)^{\frac{\lambda(4+\varepsilon_2)}{4}} \left(\int_{I_T} \|\Theta\|^{6-\vartheta}_{L_{6-\vartheta}(\Gamma)}\, d\tau\right)^{\frac{(1-\lambda)(4+\varepsilon_2)}{6-\vartheta}}.\end{aligned}$$

The parameter λ is the solution of the equation $\frac{1}{4+\varepsilon_2} = \frac{1}{4}\lambda + \frac{1}{6-\vartheta}(1-\lambda)$ and $\alpha_0 = \frac{3}{4}\lambda$. It holds $\lambda = 1 - \varepsilon_4(\varepsilon_2, \vartheta)$ with $\varepsilon_4 \to 0$ for $\vartheta, \varepsilon_2 \to 0$. Hence we obtain

$$\|\Theta\|_{L_{4+\varepsilon_2}(I_T;L_\infty(\Gamma))} \leq \check{c}_{63} \|\Theta\|^{1-\varepsilon_4}_{L_4(I_T;W_4^1(\Omega))} \|\Theta\|^{\varepsilon_4}_{L_{6-\vartheta}(S_T)}. \tag{4.6}$$

The estimate of $\|\Theta\|_{L_{6-\vartheta}(S_T)}$ is more complicated. The embedding theorem and the trace theorem for Sobolev spaces yield for arbitrary $\varphi \in L_p(S_T)$

$$\|\varphi\|_{L_p(S_T)} \leq \check{c}_{64} \|\varphi\|_{W^{\gamma}_{\beta_0}(I_T;W^{\alpha_0-1/\beta_0}_{\beta_0}(\Gamma))} \leq \check{c}_{65} \|\varphi\|_{W^{\gamma}_{\beta_0}(I_T;W^{\alpha_0}_{\beta_0}(\Omega))},$$

if $\gamma > \frac{1}{\beta_0} - \frac{1}{p}$ and $\frac{1}{N-1}\left(\alpha_0 - \frac{1}{\beta_0}\right) > \frac{1}{\beta_0} - \frac{1}{p}$. This gives

$$\|\Theta\|^{\beta_0}_{L_p(S_T)} \leq \check{c}_{66} \left(\int_{I_T^2} \frac{\|\Theta(s_1) - \Theta(s_2)\|^{\beta_0}_{W^{\alpha_0}_{\beta_0}(\Omega)}}{|s_1 - s_2|^{1+\beta_0\gamma}}\, ds + \int_{I_T} \|\Theta(s_1)\|^{\beta_0}_{W^{\alpha_0}_{\beta_0}(\Omega)}\, ds_1\right)$$

with $s = (s_1, s_2)$. Now, we use the interpolation inequality (4.5) with $\alpha_1 = 1$, $\beta_1 = 4$, $\alpha_2 = 0$, $\beta_2 = 2$ and $\lambda = \alpha_0$ and the Hölder inequality with $q = \frac{4}{\alpha_0\beta_0}$ and $q' = \frac{2}{(1-\alpha_0)\beta_0}$. Here $\frac{1}{q} + \frac{1}{q'} = 1$ shall be valid; this is equivalent to $\beta_0 = \frac{4}{2-\alpha_0}$. Then we obtain

$$\begin{aligned}\|\Theta\|^{\beta_0}_{L_p(S_T)} \leq \check{c}_{67} &\left(\int_{I_T^2} \frac{\|\Theta(s_1) - \Theta(s_2)\|^4_{W_4^1(\Omega)}}{|s_1 - s_2|^{1-\eta}}\, ds + \|\Theta\|^4_{L_4(I_T;W_4^1(\Omega))}\right)^{\frac{1}{q}} \cdot \\ &\cdot \left(\int_{I_T^2} \frac{\|\Theta(s_1) - \Theta(s_2)\|^2_{W_2^0(\Omega)}}{|s_1 - s_2|^{2-\eta}}\, ds + \|\Theta\|^2_{L_2(Q_T)}\right)^{\frac{1}{q'}}\end{aligned}$$

with $\eta = \frac{1}{q'} - \beta_0\gamma = \beta_0\left(\frac{1-\alpha_0}{2} - \gamma\right)$ which must be positive. Choosing the values $\alpha_0 = \frac{2}{3}$, $\beta_0 = 3$ and some $\gamma \in \left(0, \frac{1}{6}\right)$ it follows

$$\|\Theta\|_{L_p(S_T)} \leq \check{c}_{68}\|\Theta\|^{\frac{2}{3}}_{L_4(I_T;W^1_4(\Omega))}\|\Theta\|^{\frac{1}{3}}_{H^{\frac{1}{2}-\tilde{\eta}}(I_T;L_2(\Omega))} \tag{4.7}$$

for all $p < 6$ with $\tilde{\eta} = \tilde{\eta}(p) > 0$ and the estimate is done. We insert this estimate into (4.6) and then both (4.6) and (4.4) into (4.3) and then, after use of the interpolation

$$\begin{aligned} \left\|\tfrac{1}{\delta}[\dot{u}_n]_+\right\|_{L_2(S_{C,\eta,T})} &\leq \check{c}_{69}\|\Theta\|_{H^{\frac{1}{4},\frac{1}{2}}(Q_T)} + \check{c}_{70} \\ &\leq \check{c}_{71}\left(\|\Theta\|^{\lambda}_{H^{\alpha}(I_T;L_2(\Omega))}\|\Theta\|^{1-\lambda}_{L_2(Q_T)} + \|\Theta\|_{L_2(I_T;H^1(\Omega))} + 1\right) \end{aligned}$$

which is valid for $\lambda = 1/(4\alpha)$, we arrive at the estimate

$$\begin{aligned} &k\ \|\Theta\|^2_{H^{\alpha}(I_T;L_2(\Omega))} + \|\Theta\|^4_{L_4(I_T;W^1_4(\Omega))} \\ &\leq \check{c}_{72}\|\Theta\|^{\frac{5}{2}-\varepsilon_5}_{L_4(I_T;W^1_4(\Omega))}\|\Theta\|^{\frac{1}{2}+\varepsilon_5}_{H^{\alpha}(I_T;L_2(\Omega))} + \check{c}_{73}\|\Theta\|^3_{L_4(I_T;W^1_4(\Omega))} + \check{c}_{74}. \end{aligned} \tag{4.8}$$

Here, $\varepsilon_5 > 0$ can be arbitrarily small, therefore an easy Hölder inequality yields

$$\|\Theta\|^4_{L_4(I_T;W^1_4(\Omega))} \leq \check{c}_{75}\|\Theta\|^{\frac{10}{3}+\varepsilon}_{L_4(I_T;W^1_4(\Omega))} + \check{c}_{76},\ \varepsilon > 0 \text{ arbitrarily small.}$$

Hence, $\|\Theta\|^2_{H^{\alpha}(I_T;L_2(\Omega))} + \|\Theta\|^4_{L_4(I_T;W^1_4(\Omega))} \leq \check{c}_{77}$ and then using (4.7) we obtain (4.2). Therefore the proposition is proved. □

From the inequalities (4.1) and (4.2) we can derive that there exist sequences $\{\delta_k\}$, $\{u^{(k)}\}$ and $\{\Theta^{(k)}\}$ such that $\dot{u}^{(k)} \rightharpoonup \dot{u}$ in $\boldsymbol{H}^{\frac{1}{2},1}(Q_T)$ and strongly both in $\boldsymbol{L}_2(S_{C,T})$ and in $\boldsymbol{L}_{\tilde{p}}(S_{C,\eta,T})$ for any $\tilde{p} < 4$, $\sigma_n(u^{(k)}) = -\frac{1}{\delta_k}\left[\dot{u}^{(k)}_n\right]_+ \rightharpoonup \sigma_n(u)$ in $L_2(S_{C,\eta,T})$ and $\Theta^{(k)} \rightharpoonup \Theta$ in $L_4(I_T; W^1_4(\Omega))$ and strongly both in $L_4(Q_T)$ and in $L_p(S_{C,\eta,T})$ for $p < 6$. Regarding these convergence properties we can prove the convergence in the energy term $\langle c_{ij}(\nabla\Theta)\Theta_{,j}, \varphi_{,i}\rangle_{Q_T}$ like in the proof of Proposition 2 and, in particular, the strong convergence of $\Theta^{(k)}$ in $L_4\big(I_T; W^1_4(\Omega)\big)$. Passing to the limit $k \to +\infty$ in the variational inequality (3.1) with test function $v \in \mathfrak{K}$ and in the penalized version of variational equation (2.19) for test functions $\varphi \in C_\infty(Q_T) \cap \mathfrak{V}$ therefore shows that the limit functions (u, Θ) solve the original non–smoothed problem (2.18), (2.19) for all test functions $v \in \mathfrak{K}$ and $\varphi \in C_\infty(Q_T) \cap \mathfrak{V}$. The estimates (4.1) and (4.2) remain valid for the limit functions u and Θ. Using appropriate embedding theorems it is possible to verify that the linear functionals $\varphi \mapsto \langle \mathfrak{F}|\sigma_n(u)||\dot{u}_t|, \varphi\rangle_{S_{C,\eta,T}}$ and $\varphi \mapsto \langle b_{ij}\Theta\,\dot{u}_{i,j}, \varphi\rangle_{Q_T}$ are bounded in the dual space of $L_4(I_{\mathcal{T}}; W_4^1(\Omega))$. Since $C_\infty(Q_{\mathcal{T}}) \cap \mathfrak{V}$ is dense in $\mathfrak{V}$ we see that variational equation (2.19) is satisfied for all test functions $\varphi \in \mathfrak{V}$.

Let us collect the assumptions for the final result derived above:

Assumptions: Ω is a bounded domain with a boundary Γ of the class $C_{1,1}$ consisting of the measurable parts Γ_U, Γ_F and Γ_C with $(\text{mes}\,\Gamma_U \cdot \text{mes}\,\Gamma_C) > 0$. The time interval is given by $I_{\mathcal{T}} := [0, \mathcal{T}]$ with $0 < \mathcal{T} < +\infty$. The given data satisfy the properties $u_0 \in \boldsymbol{H}^{\frac{3}{2}}(\Omega)$, $u_1 \in \boldsymbol{H}^{\frac{3}{2}}(\Omega)$, $U \in \mathcal{H}_{\mathcal{T}}$, $U(0,\cdot) = u_0$, $\dot{U}(0,\cdot) = u_1$ on Ω, $f \in L_2\big(I_{\mathcal{T}}; \boldsymbol{H}^{\frac{1}{2}}(\Omega)^*\big) \cap H^{\frac{1}{4}}(I_{\mathcal{T}}; \boldsymbol{H}^1(\Omega)^*)$, $h \in H^{\frac{1}{4}}\big(I_{\mathcal{T}}; \boldsymbol{H}^{\frac{1}{2}}(\Gamma_F)^*\big)$, and $\Upsilon \in L_2(S_{\mathcal{T}})$. The coefficients $a_{ijkl}^{(\iota)}$, $\iota = 0, 1$, c_{ij} and b_{ij} are symmetric, depend Lipschitz–continuously on the space variable, are globally bounded and satisfy the conditions (2.6), (2.12), (2.13) and (2.14). The coefficient of friction is nonnegative, bounded by the constant $C_{\mathfrak{F}}$ and vanishes outside the set $\Gamma_{C,\eta}$ for some $\eta > 0$ as in Proposition 4. The coefficient K in the heat exchange is non–negative and belongs to $L_\infty(\Gamma)$.

Theorem 4.1. *Let the assumptions collected above be valid. Then the thermoviscoelastic contact problem with Coulomb friction (2.18), (2.19) has at least one solution.*

Remark. The assumptions of vanishing initial conditions $\Theta(0, x) = 0$ and Dirichlet–conditions $\Theta = 0$ on $S_{U,\mathcal{T}}$ is not essential here. The result is also true for non–vanishing data $\Theta(0, x) = \Theta_0(x)$ and $\Theta = \Theta_1$ on $S_{U,\mathcal{T}}$ with $\Theta_0 \in L_2(\Omega)$ and $\Theta_1 \in H^{\frac{1}{2},1}(Q_{\mathcal{T}}) \cap L_4(I_{\mathcal{T}}; W_4^1(\Omega))$. The changes in the proofs are not substantial. For simplicity of the presentation we have omitted this case.

Acknowledgments

The work presented here was partially supported by the Czech Academy of Sciences under grant A 1075707 and by the Sonderforschungsbereich 404 of the German Research Foundation (DFG).

References

[1] Duvaut, G. and Lions, J.L. (1972). *Les inéquations en mécanique et en physique.* Dunod, Paris.

[2] Eck, C. (1996). Existenz und Regularität der Lösungen für Kontaktprobleme mit Reibung. PhD–Thesis, University of Stuttgart.

[3] Eck, C. and Jarušek, J. (1998). The solvability of a coupled thermoviscoelastic contact problem with small Coulomb friction and linearized growth of frictional heat. SFB 404, Preprint 98/04, University of Stuttgart. To appear in *Math. Meth. Appl. Sci.*

[4] Eck, C. and Jarušek, J. (1998). Existence of solutions for the dynamic frictional contact problem of isotropic elastic bodies. Preprint No. 242, Institute for Applied Mathematics, University Erlangen–Nürnberg.

[5] Jarušek, J. and Eck, C. (1996). Dynamic contact problems with friction in linear viscoelasticity. *C. R. Acad. Sci. Paris, Sér. I*, 322:507–512.

[6] Jarušek, J. and Eck, C. (1999). Dynamic contact problems with small Coulomb friction for viscoelastic bodies. Existence of solutions. *Math. Models & Meth. Appl. Sci.*, 9(1):11–34.

[7] Lions, J.L. and Magenes, E. (1968). *Problèmes aux limites non-homogènes et applications.* Dunod, Paris.

[8] Nečas, J. (1989). Dynamics of thermoelastic systems with strong viscosity. In: *Proc. confer. partial diff. eq., 25.-29.3.1988, Holzhau, GDR.* Teubner Texte Math., Leipzig.

[9] Nečas, J., Jarušek, J. and J. Haslinger. (1980). On the solution of the variational inequality to the Signorini problem with small friction. *Boll. Un. Mat. Ital.*, 5(17–B):796–811.

[10] Nečas, J., Novotný, A. and Šverák, V. (1990). Uniqueness of solutions to the system of thermoelastic bodies with strong viscosity. *Math. Nachr.*, 150:319–324.

[11] Nečas, J. and Růžička, M. (1991). A dynamic problem of thermoelasticity. *Z. Anal. Anwendungen*, 10(3):358–368.

[12] Nečas, J. (1967). *Les méthodes directes en équations elliptiques.* Academia, Praha.

[13] Sprekels, J. (1991). Global solutions in onedimensional magneto–thermo-viscoelasticity. *European J. Appl. Math.*, 2:83–96.

[14] Triebel, H. (1978). *Interpolation Theory, Function Spaces, Differential Operators.* North Holland, Amsterdam - New York - Oxford.

[15] Zeidler, E. (1990). *Nonlinear Functional Analysis and its Applications. II B: Monotone Operators*, Springer–Verlag, New York.

ON SOME GLOBAL EXISTENCE THEOREMS FOR A SEMILINEAR PARABOLIC PROBLEM

Yuri Vladimirovich Egorov, Vladimir Alexandrovich Kondratiev

Abstract: Some conditions are obtained sufficient for solutions to a non-linear parabolic equation of second order with non-linear boundary conditions to be bounded or to tend to infinity at a finite time.

Keywords: Parabolic equations, semilinear boundary value problem, blow-up.

Let Ω be a bounded domain in $I\!R^n$ with a smooth boundary Γ. Set $Q_T =]0,T[\times\Omega, S_T =]0,T[\times\Gamma, Q = \Omega\times]0,\infty[, S = \Gamma\times]0,\infty[$. Consider the boundary value problem

$$\frac{\partial\varphi(u)}{\partial t} = Lu - a(x)f(u) \quad \text{in} \quad Q_T, \tag{1}$$

$$\frac{\partial u}{\partial N} - b(x)g(u) = 0 \quad \text{on} \quad S_T, \tag{2}$$

$$u(0,x) = u_0(x) \quad \text{in} \quad \Omega, \tag{3}$$

where

$$Lu = \sum_{i,j=1}^{n} a_{ij}(t,x)u_{x_i x_j} + \sum_{i=1}^{n} a_i(t,x)u_{x_i} + c(t,x)u + d(t,x),$$

$$\sum_{i,j=1}^{n} a_{ij}(t,x)\xi_i\xi_j \geq 0 \ \text{ for } \xi \in I\!R^n; \ \frac{\partial u}{\partial N} = \sum_{i,j=1}^{n} \cos(\nu, x_i)a_{ij}(t,x)\frac{\partial u}{\partial x_j},$$

ν is the outer normal unit vector, $\varphi(u), f(u), g(u)$ are increasing smooth non-linear functions, positive for positive u, tending to $+\infty$ as $u \to +\infty$, $f(0) = 0, g(0) = 0$. We suppose that the coefficients are smooth enough and the solution to the problem (1)-(3) can be extended for all $t > 0$ or until the moment T when $\sup_\Omega u(t,x)$ becomes equal to ∞. In particular, we assume that $\partial u_0/\partial N - b(x)g(u_0) = 0$ on Γ. We are interested in the

Applied Nonlinear Analysis, edited by Sequeira *et al.*
Kluwer Academic / Plenum Publishers, New York, 1999.

study of the asymptotic behavior of the classical solution u as $t \to \infty$ or as $t \to T$ in the case of explosion.

A function $u = u(t,x)$ is said to be a classical solution in Q_T if u is bounded and uniformly continuous in $\overline{Q}_T$, the functions $u_t, \nabla_x u, \nabla_x \nabla_x u$ are continuous in Q_T. A global solution is a function, defined in $\overline{Q}$, whose restriction to Q_T is a classical solution for any $T > 0$. The existence of the solution to the considered problems can be usually proved locally, on a finite interval of T. We use in this paper the term "blowup" as a pseudonym for "global nonexistence", i.e. "a solution blowups" means that the maximal interval of its existence is bounded.

The following theorems generalize the results of H.A. Levine, L.E. Payne [3], W. Walter [4], J. Esher [5], J.L. Gómez, V. Márquez, N. Wolanski [8], M. Chipot, M. Fila, P. Quittner [9], P. Quittner [10], authors [11] and others.

Theorem 1. *Suppose that* $b(x) \geq b_0 > 0, 0 \leq a(x) \leq a_0$. *Let* $\varphi(z)$, $f(z)$, $g(z)$, $g'(z)$, $g'(z)/\varphi'(z)$ *be continuous, positive functions for* $z > 0$. *Let the functions* $\varphi(z), g(z), g'(z)/\varphi'(z)$ *be increasing for* $z > 0$. *Assume that there exists a unit vector* $\gamma \in I\!R^n$ *and two positive constants* δ, A *such that*

$$\begin{matrix} \sum_{i,j=1}^n a_{ij}(t,x)\gamma_i\gamma_j \geq \delta, & \sum_{i=1}^n a_i(t,x)\gamma_i > -A, \\ c(t,x) > -A, & d(t,x) > -A \end{matrix} \qquad \text{in } Q.$$

Let $\varepsilon > 0$ *be so small that* $\varepsilon \sum_{i=1}^{n} \frac{\partial x_i}{\partial N}\gamma_i < b_0$ *on* Γ. *Let for* $z \geq c_0$

$$\frac{\delta\varepsilon^2}{2} g(z)g'(z) > A(\varepsilon g(z) + z + 1) + a_0 f(z), \quad \int_1^\infty \frac{d\varphi(z)}{g(z)g'(z)} < \infty. \tag{4}$$

Then there exist a T_0 *and a positive function* $v(t,x)$, *continuous in* $\overline{\Omega}$ *for* $0 \leq t < T_0$ *and tending to infinity as* $t \to T_0, x \to x_0 \in \overline{\Omega}$ *such that for any classical solution* u *of the problem* (1) – (3) *in* $Q_{T_0 - t_0}$ *satisfying* $u(0,x) > v(t_0,x)$ *we have* $u(t,x) > v(t+t_0,x)$ *in* Ω *and*

$$\lim_{t \to T_0 - t_0 -} \max_{x \in \Omega} u(t,x) = \infty.$$

Proof. It suffices to prove that there exists a positive function $v(t,x)$, satisfying the conditions

$$\frac{\partial\varphi(v)}{\partial t} - Lv + a(x)f(v) < 0 \quad \text{in} \quad \Omega \quad \text{for} \quad t \geq 0,$$

$$\frac{\partial v}{\partial N} - b(x)g(v) < 0 \quad \text{on} \quad \Gamma \quad \text{for} \quad t \geq 0,$$

continuous in $[0,T] \times \overline{\Omega}$ and tending to infinity as $t \to T_0$, $x \to x_0 \in \overline{\Omega}$. The function v constructed in below depends only on a_0, b_0, φ, f, g and the diameter of Ω.

We are looking for the function v in the form

$$v = \alpha(\tau), \quad \text{where } \tau = h(t) + c + \varepsilon \sum_{i=1}^{n} \gamma_i x_i,$$

where c is such that $c + \varepsilon \sum_{i=1}^{n} \gamma_i x_i > 0$ in Ω, α is a solution to the equation

$$\alpha'(t) = g(\alpha(t)), \quad \alpha(0) = c_0,$$

where c_0 is a positive constant.

Assume at first that $\int_1^{\infty} dz/g(z) = \infty$. Then $\alpha(t)$ is defined for all $t > 0$ and $\alpha(t) \to +\infty$ as $t \to +\infty$.

Let h satisfy the relation

$$\varphi'(\alpha(h(t)))h'(t) = \frac{\delta\varepsilon^2}{2} g'(\alpha(h(t))), \quad h(0) = 0.$$

Then $h(t)$ is defined for all $t > 0$, $h'(t) > 0$ for $t > 0$. We have

$$\frac{\partial\varphi(v)}{\partial t} - Lv + a(x)f(v) = \varphi'(\alpha(\tau))\alpha'(\tau)h'(t) - \varepsilon^2\alpha''(\tau) \sum_{i,j=1}^{n} a_{ij}(t,x)\gamma_i\gamma_j -$$

$$-\varepsilon g(\alpha(\tau)) \sum_{i=1}^{n} a_i(t,x)\gamma_i - c(t,x)\alpha(\tau) - d(t,x) + a(x)f(\alpha(\tau)) <$$
$$< \varphi'(\alpha(\tau))\alpha'(\tau)h'(t) - \delta\varepsilon^2 g(\alpha(\tau))g'(\alpha(\tau)) + \varepsilon A g(\alpha(\tau)) +$$
$$+ A\alpha(\tau) + A + a_0 f(\alpha(\tau)) \le$$
$$\le \varphi'(\alpha(\tau))\alpha'(\tau)h'(t) - \delta\varepsilon^2 g(\alpha(\tau))g'(\alpha(\tau))/2 =$$
$$= \frac{\delta\varepsilon^2}{2}\varphi'(\alpha(\tau))g(\alpha(\tau)) \left[\frac{g'(\alpha(h(t)))}{\varphi'(\alpha(h(t)))} - \frac{g'(\alpha(\tau))}{\varphi'(\alpha(\tau))}\right] \le 0$$

in Q_{T_0}, the value T_0 will be indicated in below.

On the other hand,

$$\frac{\partial v}{\partial N} - b(x)g(v) = \alpha'(\tau)\varepsilon \sum_{i=1}^{n} \gamma_i \frac{\partial x_i}{\partial N} - b(x)g(\alpha(\tau))$$

$$\le g(\alpha(\tau))(\varepsilon \sum_{i=1}^{n} \gamma_i \frac{\partial x_i}{\partial N} - b_0) < 0 \quad \text{on} \quad S_{T_0}.$$

Remark that the above stated inequalities are valid for the functions $\alpha(h(t + t_0) + c + \sum_{i=1}^{n} \gamma_i x_i)$ if $0 < t_0 < T_0$. By the maximum principle, the condition $u(0,x) > v(t_0,x)$ implies the relation

$$u(t,x) > v(t + t_0, x) \quad \text{in} \quad \Omega, \quad \text{for} \quad t \le T_0 - t_0.$$

On the other hand, the function h is such that

$$\delta\varepsilon^2 t = \int_0^h \frac{2\varphi'(\alpha(y))g(\alpha(y))dy}{g(\alpha(y))g'(\alpha(y))} = \int_{c_0}^{\alpha(h)} \frac{2d\varphi(s)}{g'(s)g(s)}.$$

Here we have used the substitution $s = \alpha(y)$. Now we obtain the result, since the last integral converges as $h \to \infty$. We can put

$$T_0 = \frac{2}{\delta\varepsilon^2}\int_0^\infty \frac{d\varphi(s)}{g'(s)g(s)} - c_1,$$

where $c_1 = \sup_\Omega(c + \varepsilon \sum \gamma_i x_i)$.

If the integral $\int_1^\infty dz/g(z)$ is convergent, then the function $\alpha(t)$ is defined for $0 \le t < T'$ and $\alpha(T') = \infty$. The function $\alpha(h(t))$ is defined for $0 \le t < T_0' = T_0 + c_1$ and $\alpha(h(T_0')) = \infty$. Therefore, the function $h(t)$ is defined for $0 \le t \le T_0', h(T_0') = T'$, the function $v(t,x)$ is defined for $0 \le t \le T_0' - c_1 = T_0$ and $\lim_{t\to T_0} \sup_\Omega v(t,x) = \infty$.

Then the rest of the above proof goes through. ∎

Corollary 1. *Let $L = \Delta$, $a(x) = a = const > 0$, $b(x) = b = const > 0$, $\varphi(u) = u, f(u) = u^p, g(u) = u^q$ and $1 < p < 2q - 1$. Then there exists a constant $b_0 > 0$ such that all solutions of the problem*

$$\frac{\partial u}{\partial t} = \Delta u - au^p \quad in \quad Q_T, \tag{5}$$

$$\frac{\partial u}{\partial \nu} - bu^q = 0 \quad on \quad S_T, \tag{6}$$

$$u(0,x) = u_0(x) \quad in \quad \Omega, \tag{7}$$

blow-up if $u_0(x) \ge b_0$.

Theorem 2. *Let $b_1 \ge b(x) \ge 0, a(x) \ge a_0 \ge 0$. Let $\varphi(z), \varphi'(z)$, $f(z)$, $g(z)$, $g'(z)$ be continuous, positive functions for $z > 0$. Suppose that there exists $A > 0$ such that*

$$|a_{ij}| < A,\ |a_i| < A,\ c < A,\ d < A.$$

Suppose also that there exists a function $\psi(x) \in C^2(\overline{\Omega})$ such that $0 \ge \psi(x) \ge -c_1$ in Ω and $\partial\psi/\partial N > b_1$ on Γ. Suppose that there exists $c_0 > 0$ such that

$$\int_{c_0}^\infty dz/g(z) > c_1.$$

Let

$$\sum_{i,j=1}^n a_{ij}\frac{\partial\psi}{\partial x_i}\frac{\partial\psi}{\partial x_j} \le A, \quad \sum_{i,j=1}^n a_{ij}\frac{\partial^2\psi}{\partial x_i \partial x_j} + \sum_{j=1}^n a_j\frac{\partial\psi}{\partial x_j} \le A \qquad in\ \Omega.$$

Put $\Phi(v) = [g(v)g'(v) + g(v) + v - a_0 f(v)/A]_+$, *where* $z_+ = max(z, 0)$.

If the function

$$\varphi'(v)g(v)/(\Phi(v) + 1)$$

is decreasing for $v > 0$, *and*

$$\int_1^\infty \frac{d\varphi(\tau)}{\Phi(v) + 1} = \infty,$$

then the solution u *to the problem* (1) – (3) *can be defined for all* $t \geq 0$. *Moreover, there exists a function* $w(t, x)$, *continuous in* $[0, \infty[\times\overline{\Omega}$, *such that for any classical solution* u *of the problem* (1) – (3) *satisfying* $u(0, x) < w(t_0, x)$ *with* $t_0 \geq 0$ *the relation*

$$u(t, x) \leq w(t + t_0, x)$$

holds in Ω. *The function* w *depends only on* f, g, φ *and* Ω.

If there exists $v_0 > 0$ *such that*

$$g(v)g'(v) + g(v) + v + 1 \leq a_0 f(v)/A \quad \text{for} \quad v \geq v_0, \tag{8}$$

then all solutions of the problem (1) – (3) *are bounded.*

Proof. Let the function $\alpha(\tau)$ be the solution to the problem

$$\alpha'(\tau) = g(\alpha(\tau)), \quad \alpha(0) = c_0,$$

where c_0 is a positive constant.

Let at first $\int_{c_0}^\infty dz/g(z) = \infty$. Then the function $\alpha(s)$ is defined for $0 \leq s < \infty$ and grows monotonically at infinity. Moreover, $\alpha'(s)$ is also monotone. Put now

$$w(t, x) = \alpha(\tau), \; \tau = h(t) + \psi(x),$$

where $h(0) = c_1$ and $h'(t) > 0$. Then

$$\frac{\partial w}{\partial N} - b(x)g(w) \geq g(w)(\frac{\partial \psi}{\partial N} - b_1) > 0 \quad \text{on} \quad S.$$

Moreover,

$$\frac{\partial \varphi(w)}{\partial t} - Lw + a(x)f(w) = \varphi'(\alpha(\tau))\alpha'(\tau)h'(t) - \alpha'(\tau)\sum_{i,j=1}^n a_{ij}\frac{\partial^2\psi}{\partial x_i \partial x_j} -$$
$$-\alpha''(\tau)\sum_{i,j=1}^n a_{ij}\frac{\partial\psi}{\partial x_i}\frac{\partial\psi}{\partial x_j} - g(\alpha(\tau))\sum_{i=1}^n a_i(t, x)\frac{\partial\psi}{\partial x_i} - c(t, x)\alpha(\tau) -$$
$$-d(t, x) + a(x)f(\alpha(\tau)) >$$
$$> \varphi'(\alpha(\tau))\alpha'(\tau)h'(t) - Ag(\alpha(\tau))g'(\alpha(\tau)) - Ag(\alpha(\tau)) -$$
$$-A\alpha(\tau) - A + a_0 f(\alpha(\tau)) \geq \varphi'(\alpha(\tau))\alpha'(\tau)h'(t) - A\Phi(\alpha(\tau)) - A$$

in Q.

Let the function h be the solution to the Cauchy problem

$$\varphi'(\alpha(h(t)))g(\alpha(h(t)))h'(t) = A\Phi(\alpha(h(t)) + A, \ h(0) = c_1.$$

Then

$$\frac{\partial\varphi(w)}{\partial t} - Lw + a(x)f(w)$$

$$> A\varphi'(\alpha(\tau))g(\alpha(\tau))\left[\frac{\Phi(\alpha(h(t)) + 1}{\varphi'(\alpha(h(t)))g(\alpha(h(t)))} - \frac{\Phi(\alpha(\tau)) + 1}{\varphi'(\alpha(\tau))g(\alpha(\tau))}\right] \geq 0$$

in Ω for $t > 0$. Now if t_0 is determined for a given solution u in such a way that

$$u(0,x) < \alpha(\psi(x) + h(t_0)) \quad \text{in} \quad \overline{\Omega},$$

then by the maximum principle the function $\alpha(\psi(x) + h(t + t_0))$ is an upper bound for u. Therefore, the function $w(t,x) = \alpha(\psi(x) + h(t))$ has the properties stated in Theorem 2.

If $\int_{c_0}^{\infty} dz/g(z) = T < \infty$ then the function $\alpha(s)$ is defined for $0 \leq s < T$ and grows monotonically to infinity as $t \to T_0$. The function h is defined as above and $c_1 \leq h(t) < T_0$ for $t > 0$. Then $0 \leq w(t,x) < \alpha(h(t))$ for $t > 0$ and the first statement of Theorem 2 is proved.

Let the condition (8) be fulfilled. Let $\sup u(0,x) = b_0$. Then

$$\frac{\partial\varphi(w)}{\partial t} - Lw + a(x)f(w) > \varphi'(\alpha(\tau))\alpha'(\tau)h'(t)$$

for $t \geq t_1$ where t_1 is such that $\alpha(t_1) \geq v_0$. Set

$$b = \inf_{\tau\geq 0} \varphi'(\alpha(\tau))g(\alpha(\tau)), \ m_1 = \sup_{0\leq\tau\leq v_0} A[\Phi(\alpha(\tau)) + 1].$$

Now put

$$h(t) = \frac{m_1}{b}t + b_0 + 1$$

for $0 < t < t_1$ and

$$h(t) = \frac{m_1}{b}t_1 + b_0 + 1$$

for $t \geq t_1$.

Then

$$\frac{\partial\varphi(w)}{\partial t} - Lw + a(x)f(w) > 0$$

for all $t \geq 0$ and the rest of the above proof goes through. ∎

Corollary 2. *Let $L = \Delta$, $a(x) = a = const > 0$, $b(x) = b = const > 0$, $\varphi(u) = u, f(u) = u^p, g(u) = u^q$ and $p > 2q - 1 > 1$ or $p = 2q - 1$ and $a > b$. Then all solutions of the problem (5) – (7) are defined for all $t \geq 0$ and are bounded.*

Theorem 3. *Let u be a positive solution to the problem*

$$\frac{\partial u}{\partial t} = Lu - a(t,x)f(u) \quad in \quad Q_T, \ Lu = \sum_{i,j=1}^{n} \frac{\partial}{\partial x_i}(a_{ij}(t,x)\frac{\partial u}{\partial x_j}),$$

$$\frac{\partial u}{\partial N} - b(t,x)g(u) = 0 \quad on \quad S_T,$$

$$u(0,x) = u_0(x) > 0 \quad in \quad \Omega.$$

If $f'(u) > 0, g(u) \geq kf(u)$ *for* $u \geq 0$,

$$\int_0^\infty \frac{ds}{f(s)} < \infty \ \ and \ \int_0^\infty [\int_\Gamma b(t,x)dS_x - \int_\Omega a(t,x)dx]dt = \infty.$$

Then there exists $T_0 > 0$ *such that*

$$\lim_{t\to T_0} \sup_{x\in\Omega} u(t,x) = \infty.$$

Proof. Put

$$v = \int_\infty^u \frac{ds}{f(s)} \equiv F(u).$$

Then $F(u) < 0, \lim_{u\to\infty} F(u) = 0$ and $F'(u) > 0$. We have

$$\frac{\partial v}{\partial t} = Lv + \sum_{i,j=1}^n a_{ij} \frac{f'(u)}{f^2(u)} \frac{\partial u}{\partial x_i} \frac{\partial u}{\partial x_j} - a(t,x) \quad \text{in} \quad Q_T,$$

$$\frac{\partial v}{\partial N} = b(t,x)\frac{g(u)}{f(u)} \geq kb(t,x) \quad \text{on} \quad S_T,$$

$$v(0,x) = F(u_0(x)) < 0 \quad \text{in} \quad \Omega.$$

Put

$$W(t) = \int_\Omega v(t,x)dx.$$

Then

$$W'(t) \geq k\int_\Gamma b(t,x)dS_x - \int_\Omega a(t,x)dx, \ W(0) < 0.$$

Therefore,

$$W(t) \geq W(0) + \int_0^t [k\int_\Gamma b(s,x)dS_x - \int_\Omega a(s,x)dx]ds.$$

The function in the right-hand side tends to $+\infty$, when $t \to \infty$. Therefore, there exists $T_0 > 0$ such that $W(T_0) = 0$. Then there exists $x' \in \Omega$ such that $v(T_0,x') = 0$ and therefore $u(T_0,x') = \infty$. ∎

Corollary 3. *Let* $L = \Delta$, $a(x) = a = const > 0$, $b(x) = b = const > 0$, $\varphi(u) = u, f(u) = u^p, g(u) = u^q$ *and* $p > q > 1$ *or* $p = q$ *and* $b|\Gamma| > a|\Omega|$. *Then all positive solutions of the problem* (5)–(7) *blow-up at some finite moment of* t.

Consider now the boundary value problem (5)-(7). As we saw before, all solutions u satisfying the condition

$$u_0(x) \geq H \tag{9}$$

with a sufficiently large constant $H > 0$ are blowing-up at a finite moment T_0 if $p < 2q-1$. At the same time, if $p = q$, then by Theorem 3 all solutions are blowing-up if just $u_0(x) > 0, b|\Gamma| > a|\Omega|$. Now we can precise this result:

Theorem 4. *If $2q-1 > p \geq q > 1$ and the constant b/a is large enough, then all solutions of the problem (6) – (8) are blowing-up if $u_0(x) > 0$.*

This theorem generalizes the results of W. Walter [4], M.Chipot, M.Fila, P.Quittner [9], P.Quittner [10], and some others in the spirit of H.Fujita, see [1], [2], [7].

In particular, in [9] it is shown that for small values of the constant b/a the stationary problem

$$\Delta u = au^p \quad \text{in} \quad \Omega, \ \frac{\partial u}{\partial \nu} - bu^q = 0 \quad \text{on } \Gamma$$

admits non-trivial positive solutions, so that the solution of the problem (5)-(7) with positive initial data u_0 can tend to one of them as $t \to \infty$.

Our proof is based on the following lemma of S.Kamin, L.A.Peletier, J.L.Vazquez (see [6], Lemma 3.1, p.608).

Lemma 2. *Let u be a positive solution of the equation (5) bounded in Q. Then there exists a positive constant C such that*

$$u(t,x) \leq C(d(x)^{2/(1-p)} + t^{1/(1-p)}),$$

where $d(x)$ is the distance of the point x from Γ.

Therefore, for any $\sigma \in]0,(p-1)/2[$ $\sigma \in (0,(p-1)/2)$ and $t \geq 1$ we have

$$\int_\Omega u(t,x)^\sigma dx \leq C_1,$$

where the constant C_1 depends on p, n and Ω.

Proof of Theorem 4. Consider the integral

$$0 = \int_\Omega (\frac{\partial u}{\partial t} - \Delta u + au^p)u^{-q} dx$$

$$= \frac{1}{1-q}\frac{\partial}{\partial t}\int_\Omega u^{1-q}dx - q\int_\Omega \frac{|\nabla u|^2}{u^{q+1}}dx + b|\Gamma| - a\int_\Omega u^{p-q}dx.$$

Since $0 \leq p-q < (p-1)/2$, we have by Lemma 2 that

$$\int_\Omega u^{p-q} dx \leq C_2$$

and therefore,

$$\frac{1}{q-1}\int_\Omega u(0,x)^{1-q}dx = \frac{1}{q-1}\int_\Omega u(T,x)^{1-q}dx + q\int_0^T\int_\Omega \frac{|\nabla u|^2}{u^{q+1}}dxdt$$

$$+b|\Gamma|T - a\int_0^T\int_\Omega u^{p-q}dxdt \geq (b|\Gamma| - aC_2)T$$

and we have a contradiction as $T \to \infty$ if $b|\Gamma| > aC_2$. ∎

Other our result is concerning the solutions of the equation

$$\frac{\partial u}{\partial t} = \Delta u - a(x)f(u) \quad \text{in} \quad Q_T, \tag{10}$$

satisfying the boundary condition

$$\frac{\partial u}{\partial \nu} - b(x)g(u) = 0 \quad \text{on} \quad S_T. \tag{11}$$

We suppose that $a(x) \geq 0, b(x) \geq b_0 > 0$, the functions f and g are positive increasing.

As we showed before, the solutions of the problem (10), (11), (7) are blowing-up at a finite time, if

$$g(s)g'(s) - cf(s) > 0, \int_0^\infty \frac{ds}{g(s)g'(s) - cf(s)} < \infty, \tag{12}$$

We show that this condition is sufficient for explosion even if the boundary condition (11) is fulfilled on a part of Γ only.

Theorem 5. *Let Γ_0 be a part of Γ having positive Hausdorff $(n-1)$-measure which is a smooth connected $(n-1)$-dimensional manifold and u be a positive classical solution of* (10) *satisfying the boundary condition*

$$\frac{\partial u}{\partial \nu} - b(x)g(u) = 0 \quad on \quad \Sigma_T = [0,T] \times \Gamma_0.$$

Let the function $g(u)/u$ be increasing and tending to ∞ as $u \to \infty$,

$$g(0) = 0, \int_M^\infty \frac{ds}{g(s)} < \infty, \ M > 0.$$

Let also

$$f(s) < \beta s \ \textit{for}\ s > M, \ \textit{and} \ \lim_{s\to\infty} f(s)/g(s) = 0.$$

If $u(0,x) \geq H$ and H is large enough, then there exists a positive T_0 such that

$$\lim_{t \to T_0-} \sup_\Omega u(t,x) = \infty.$$

Moreover, for any $T_0 > 0$ there exists a constant H such that the explosion happens before T_0 if $u_0(x) \geq H$.

Let $\Gamma_1 \subset\subset \Gamma_0$, $meas_{n-1}\Gamma_1 > 0$. Let Γ_2 be a smooth $(n-1)$-dimensional manifold containing in Ω such that $\Gamma_0 \cup \Gamma_2$ serves as a smooth boundary of a domain $\Omega_1 \subset \Omega$. Our proof uses the following lemmas.

Lemma 3. *There exists a positive constant k and a positive in Ω_1 function $\Phi(x)$ satisfying the equation*

$$\Delta\Phi = \alpha a(x)\Phi(x) \quad \text{for } x \in \Omega_1$$

such that $\Phi(x) = 0$ on Γ_2, $\partial\Phi/\partial\nu - k^2\Phi = 0$ on Γ_1 and $\partial\Phi/\partial\nu = 0$ on $\Gamma_0 \setminus \Gamma_1$. This function is larger than a positive constant on Γ_1.

Proof. Put

$$\lambda(k) = \inf_{w \in H^1(\Omega_1), w=0 \text{ on } \Gamma_2} \frac{\int_{\Omega_1} |\nabla w|^2 dx + \alpha \int_{\Omega_1} a(x) w^2 dx - k^2 \int_{\Gamma_1} w^2 dS}{\int_{\Omega_1} w^2 dx}.$$

Evidently, $\lambda(0) > 0$ and $\lambda(k) < 0$ if k is large, because the functional takes negative values if a function w is fixed and k is big. By continuity, there exists a k such that $\lambda(k) = 0$. Obviously there exists a positive function $w = \Phi$ on which the functional takes its minimal value. This function satisfies the equation $\Delta\Phi = \alpha a(x)\Phi(x)$ and by the maximum principle cannot vanish on Γ_0. ∎

We can assume that $0 < \Phi(x) \leq 1$ in Ω_1, $\Phi(x) \geq c_0 > 0$ on Γ_1.

Lemma 4. *Let u be a function satisfying the conditions of Theorem 5 in Ω for $0 < t < T$. Let m be a constant such that $0 < m < M, g(s)/s > k^2$ for $s > m$, where k is the constant found in Lemma 3. If $u > m$ on Γ_1 for $0 < t < T$, then $u(t, x) > M\Phi(x)$ in Ω_1 for $0 < t < T$.*

Proof. We have

$$\frac{\partial(u - M\Phi)}{\partial t} - \Delta(u - M\Phi) = -a(x)f(u) + \alpha(x)M\Phi(x) \text{ in } Q_1 = [0,T] \times \Omega_1,$$

$$u - M\Phi > 0 \quad \text{on} \quad [0,T] \times \Gamma_2,$$

$$\frac{\partial(u - M\Phi)}{\partial\nu} > 0 \quad \text{on} \quad [0,T] \times \Gamma_0 \subset \Gamma_1,$$

$$\frac{\partial(u - M\Phi)}{\partial\nu} = b(x)[g(u) - k\Phi] > b(x)k(u - M\Phi) > 0 \quad \text{on} \quad [0,T] \times \Gamma_1.$$

Therefore, $u > M\Phi$ in $[0,T] \times \Omega_1$. ∎

Lemma 5. *Let u be a function satisfying the conditions of Theorem 5 in Ω for $t > 0$ and m be the constant defined in Lemma 3. Let $c_0M > m$. Then $u > m$ on Γ_1 for $0 < t < T$ and therefore, $u(t,x) > M\Phi(x)$ in Ω_1 for $0 < t < T$.*

Proof. We have $u > M > m$ on Γ_1 at $t = 0$. Let $u(t, x) > m$ on Γ_1 for $0 < t < t_0$ but $u(t_0, x_0) = m$ and $x_0 \in \Gamma_1$. By Lemma 4 we have $u(t_0, x_0) \geq M\Phi(x_0) \geq c_0M > m$, a contradiction. ∎

Proof of Theorem 5. Let Ω' be a subdomain of Ω_1 and the part Γ' of its boundary common with Γ contains Γ_1 and is contained in Γ_0. Let ψ be a positive function minimizing the functional

$$\frac{\int_{\Omega'} |\nabla u|^2 dx}{\int_{\Omega'} u^2 dx}$$

in the class of functions from $H^1(\Omega')$, vanishing on $\Gamma'' = \partial\Omega' \setminus \Gamma'$. Then $\partial\psi/\partial\nu = 0$ on Γ', $\psi > 0$ in Ω', and $\Delta\psi + \lambda\psi = 0, \lambda > 0$.

Let G be such a function that $G'(u) = -1/g(u), G(u) \geq 0$ and $G \to 0$ as $u \to \infty$.

Consider the integral

$$0 = \int_0^T \int_{\Omega'} (\frac{\partial}{\partial t} - \Delta u + a(x)f(u))\frac{\psi(x)}{g(u)} dxdt$$

$$= \int_{\Omega'} \psi(x)G(u)(0,x)dx - \int_{\Omega'} \psi(x)G(u)(T,x)dx -$$

$$- \int_0^T \int_{\Omega'} \frac{\psi(x)|\nabla u|^2 g'(u)}{g(u)^2} dxdt + \int_0^T \int_{\Omega'} a(x)\psi(x)\frac{f(u)}{g(u)} dxdt +$$

$$+ \int_0^T \int_{\Omega'} \frac{\nabla\psi(x)\nabla u dx}{g(u)} dt - \int_0^T \int_{\Gamma'} b(x)\psi(x)dSdt$$

$$= \int_{\Omega'} \psi(x)G(u)(0,x)dx - \int_{\Omega'} \psi(x)G(u)(T,x)dx -$$

$$- \int_0^T \int_{\Omega'} \frac{\psi(x)|\nabla u|^2 g'(u)}{g(u)^2} dxdt + \int_0^T \int_{\Omega'} \Delta\psi(x)G(u)dxdt -$$

$$- \int_0^T \int_{\Gamma''} \frac{\partial\psi}{\partial\nu} G(u)dSdt - \int_0^T \int_{\Gamma'} b(x)\psi(x)dSdt$$

$$+ \int_0^T a(x)\psi(x) \int_{\Omega'} \frac{f(u)}{g(u)} dxdt \leq A$$

$$- \int_0^T \int_{\Gamma'} b(x)\psi(x)dSdt - \int_0^T \int_{\Gamma''} \frac{\partial\psi}{\partial\nu} G(u)dSdt + \int_0^T \int_{\Omega'} a(x)\psi(x)\frac{f(u)}{g(u)} dxdt,$$

where A is a constant independent of T. We have used here the fact that $\Delta\psi(x) < 0$.

By Lemma 5, $u(t,x) \geq c_1 M$ on Γ'', where $c_1 = \inf_{\Omega'} \Phi(x)$. We have

$$\int_0^T \int_{\Gamma'} b(x)\psi(x)dSdt \geq C_1 T,$$

$$\int_0^T \int_{\Gamma''} \frac{\partial\psi}{\partial\nu} G(u)dSdt \leq C_2 T G(c_0 M),$$

$$|\int_0^T \int_{\Omega'} a(x)\psi(x)\frac{f(u)}{g(u)} dxdt| \leq C_3 T \gamma(M),$$

where the constants C_1, C_2, C_3 are independent of T and M, $\gamma(M) = \max_{u>M} f(u)/g(u)$. It implies that

$$C_2TG(c_0M) + C_3T\gamma(M) \geq C_1T - A.$$

This leads to contradiction if M and T are large enough. Therefore, the solution $u(t,x)$ cannot exist for big t. ∎

Acknowledgments

This work was supported by the project INTAS-96-1060.

References

[1] Fujita, H. (1996). *On the blowing up of the solutions for* $u_t = \Delta u + u^{1+\alpha}$. J. Fac. Sci. Univ. Tokyo, Sect. I, 13:109–124.

[2] Fujita, H. (1970) *On some nonexistence and nonuniqueness theorems for nonlinear parabolic equations.* Proc. of Symposia in Pure Mathematics, XVIII,4:105–113.

[3] Levine, H.A. and Payne, L.E. (1974). *Some nonexistence theorems for initial-boundary value problems with nonlinear boundary constraints.* Proc. of AMS, 46,7:277–284.

[4] Walter, W. (1975). *On existence and nonexistence in the large of solutions of parabolic differential equations with a nonlinear boundary condition.* SIAM J. Math. Anal., 6:5–90.

[5] Esher, J. (1989). *Global existence and non-existence of semilinear parabolic equations with nonlinear boundary conditions.* Math. Ann., 284:285–305.

[6] Kamin, S., Peletier, L.A. and Vazquez, J.L. (1989). *Classification of singular solutions of a nonlinear heat equation.* Duke Math. J., 58:243–263.

[7] Levine, H.A. (1990). *The role of critical exponents in blowup theorems.* SIAM Review, 32:262–288.

[8] Gómez,J.L., Márquez,V. and Wolanski, N. (1991). *Blow-up results and localization of blow-up points for the heat equation with a nonlinear boundary condition.* J. Differential Equations, 92:384–401.

[9] Chipot,M., Fila, M. and Quittner, P. (1991). *Stationary solutions, blow-up and convergence to stationary solutions for semilinear parabolic equations with nonlinear boundary conditions.* Acta Math. Univ. Comenian., LX,1:35–103.

[10] Quittner, P. (1991). *On global existence and stationary solutions for two classes of semilinear parabolic problems.* Comment. Math. Univ. Carolin., 34:105–124.

[11] Egorov, Yu. V. and Kondratiev, V.A. (1996). *On a nonlinear boundary problem for a heat equation.* C. R. Acad. Sci. Paris, Série I, 322:55–58.

[12] Egorov, Yu. V. and Kondratiev, V.A. (1998). *Two theorems on blow-up solutions for semilinear parabolic equations of second order* C. R. Acad. Sci. Paris, Série I, 327:47–52.

BIFURCATION OF SOLUTIONS TO REACTION-DIFFUSION SYSTEMS WITH JUMPING NONLINEARITIES

Jan Eisner, Milan Kučera

Abstract: Bifurcation of stationary solutions to reaction-diffusion systems of activator-inhibitor type with jumping nonlinearities are located. The result can be understood as a certain destabilizing effect of jumping terms.

Keywords: Jumping nonlinearities, bifurcation, reaction-diffusion systems, spatial patterns.

1. INTRODUCTION

Let us consider a boundary value problem

$$d_1\Delta u + b_{11+}(x)u^+ - b_{11-}(x)u^- + b_{12}v + n_1(u,v) = 0, \tag{1.1}$$
$$d_2\Delta v + b_{21}u + b_{22+}(x)v^+ - b_{22-}(x)v^- + n_2(u,v) = 0 \quad \text{in } \Omega,$$
$$u = v = 0 \quad \text{on } \partial\Omega, \tag{1.2}$$

where Ω is a bounded domain in $\mathbb{R}^n$ with a lipschitzian boundary $\partial\Omega$, b_{11+}, b_{11-}, b_{22+}, $b_{22-} \in L^\infty(\Omega)$, b_{12}, b_{21} are given reals, u^+, v^+ and u^-, v^- denote respectively the positive and negative parts of u, v (i.e., $u = u^+ - u^-$, $v = v^+ - v^-$), $n_j(\xi,\eta) = o(|\xi| + |\eta|)$, d_1, d_2 are positive parameters. Our aim is to locate bifurcation points under certain assumptions.

The problem (1.1) can be written as

$$d_1\Delta u + b_{11}u + b_{12}v - b_{1+}(x)u^+ + b_{1-}(x)u^- + n_1(u,v) = 0, \tag{1.3}$$
$$d_2\Delta v + b_{21}u + b_{22}v - b_{2+}(x)v^+ + b_{2-}(x)v^- + n_2(u,v) = 0 \quad \text{in } \Omega$$

with $b_{jj} = \max\{\operatorname{ess\,sup}_{x\in\Omega} b_{jj+}(x), \operatorname{ess\,sup}_{x\in\Omega} b_{jj-}(x)\}$, $b_{j+}(x) = b_{jj} - b_{jj+}(x)$, $b_{j-}(x) = b_{jj} - b_{jj-}(x)$, $j = 1, 2$. We will suppose that

$$b_{11} > 0,\ b_{12} < 0,\ b_{21} > 0,\ b_{22} < 0,\ b_{11} + b_{22} < 0,\ \det b_{ij} > 0 \tag{1.4}$$

Applied Nonlinear Analysis, edited by Sequeira *et al.*
Kluwer Academic / Plenum Publishers, New York, 1999.

and consider a fixed $p \in \mathbb{N}$ such that the following condition holds: there is an eigenfunction e_p corresponding to the p-th eigenvalue κ_p of the Laplacian with (1.2) such that

$$\begin{aligned} &b_{1+} > 0 \text{ a.e. in } \Omega^+_{e_p},\ b_{1+} = 0 \text{ a.e. in } \Omega^-_{e_p},\\ &b_{1-} > 0 \text{ a.e. in } \Omega^-_{e_p},\ b_{1-} = 0 \text{ a.e. in } \Omega^+_{e_p},\\ &b_{2+} > 0 \text{ a.e. in } \Omega^-_{e_p},\ b_{2+} = 0 \text{ a.e. in } \Omega^+_{e_p},\\ &b_{2-} > 0 \text{ a.e. in } \Omega^+_{e_p},\ b_{2-} = 0 \text{ a.e. in } \Omega^-_{e_p},\\ &\text{meas}(\Omega \setminus (\Omega^+_{e_p} \cup \Omega^-_{e_p})) = 0 \end{aligned} \tag{1.5}$$

where $\Omega^+_{e_p} = \{x \in \Omega;\ e_p(x) > 0\}$, $\Omega^-_{e_p} = \{x \in \Omega;\ e_p(x) < 0\}$. (The last condition in (1.5) is automatically fulfilled in reasonable situations.)

As a consequence of our general result, we will obtain the following assertion:

Let $d_2 = \tilde{d}_2$ be fixed and let d_1 be the bifurcation parameter. Then there exists a bifurcation point $\tilde{d}_1$ of (1.1) *greater than the largest bifurcation point of the stationary problem corresponding to* (1.7), (1.8) *below. Precisely, for any δ small enough there exist d_1^δ, u_δ, $v_\delta \in W_0^{1,2}(\Omega)$ satisfying* (1.1), (1.2), $\|u_\delta\|^2 + \|v_\delta\|^2 = \delta$, $d_1^\delta \to \tilde{d}_1$ *for* $\delta \to 0_+$.

All solutions are understood in a weak sense. In fact, we will consider (1.1) with d_1, d_2 changing along a given general curve σ described by a parameter $s \in \mathbb{R}$ and s will be a bifurcation parameter.

Notice that in the case $p = 1$, the eigenvalue κ_1 is simple, the corresponding eigenfunctions do not change their sign and if b_{jj+}, b_{jj-} are constant then the validity of the condition (1.5) with b_{11}, b_{22}, b_{1+}, b_{1-}, b_{2+}, b_{2-} defined as above is ensured if

$$\text{either } b_{11+} > b_{11-},\ b_{22-} > b_{22+} \quad \text{or } b_{11-} > b_{11+},\ b_{22+} > b_{22-}. \tag{1.6}$$

(The first and the second possibility corresponds respectively to the choice of the negative and positive eigenfunction e_1 in (1.5)).

The assumption (1.4) ensures that the diffusion driven instability occurs for the problem

$$u_t = d_1 \Delta u + b_{11} u + b_{12} v + n_1(u, v), \tag{1.7}$$
$$v_t = d_2 \Delta v + b_{21} u + b_{22} v + n_2(u, v) \quad \text{in } [0, +\infty) \times \Omega,$$
$$u = v = 0 \quad \text{on } [0, +\infty) \times \partial\Omega \tag{1.8}$$

as well as for (1.7) with the Neumann boundary conditions. More precisely, the trivial solution is stable as a solution of the ordinary differential equations $u_t = b_{11}u + b_{12}v + n_1(u,v)$, $v_t = b_{21}u + b_{22}v + n_2(u,v)$ but as a solution of (1.7), (1.8), the trivial solution is stable only for some parameters d_1, d_2 (the domain of stability) and unstable for the other d_1, d_2 (the domain of instability) – see Proposition 2.4, Fig. 1. Spatially nonhomogeneous stationary solutions (spatial patterns) of (1.7), (1.8) bifurcate from the trivial solutions at the boundary of the domain of stability, bifurcation is excluded in the domain of stability. The same holds for (1.7) with the Neumann boundary conditions. See, e.g., [15].

The bifurcation point of (1.1), (1.2) we obtain can lie in the domain of stability of the problem (1.7), (1.8) where bifurcation for (1.7), (1.8) is excluded. This can be understood as a destabilizing effect of the terms with positive and negative parts in (1.3) (Remark 2.7).

Our system can correspond to a reaction which changes by a jump, or it can describe an additional source which is switched on (off) if the concentration crosses the value of the basic steady state (which need not be trivial in general – see a note below).

If b_{jj+}, b_{jj-} are constant and (1.4), (1.6) hold then the bifurcation point of (1.1), (1.2) can be located in an elementary way – see Remark 3.12. In the general case of the assumption (1.5), the method is the same as in the papers [11] and [3] where a destabilizing effect (with respect to spatial patterning) of unilateral boundary conditions described respectively by variational inequalities and inclusions was proved. The main idea is taken from [7]. Notice that P. Quittner found a simpler method for the proof of existence of bifurcations for unilateral problems (see [18]). Unfortunately, it is not clear how to use it in our situation or in the situation of [11], [3] where unilateral conditions are prescribed for both u and v.

In fact, our bifurcation result remains valid if the assumption (1.4) is replaced by

$$b_{11} > 0,\ b_{12} < 0,\ b_{21} > 0,\ \det\ b_{ij} > 0. \tag{1.9}$$

In the proof, only this weaker condition will be used. In this case the picture of hyperbolas in the first quadrant described below (Fig. 1) and the domains of stability and instability can be more complicated. Diffusion driven instability does not occur in general.

We consider here bifurcation from the branch of trivial solutions. In fact, from the point of view of applications, rather the situation when a fixed constant positive solution $[\overline{u}, \overline{v}]$ loses the stability for a parameter crossing a critical value is of interest and bifurcation of spatially nonconstant solutions is considered. Of course, in this case it is possible to transfer the basic constant solution to the trivial one and consider bifurcation from the trivial solutions again. Hence, our results can be reformulated for bifurcation from a constant positive solution.

2. MAIN RESULT

Notation 2.1. Define the inner product on the space $W_0^{1,2}(\Omega)$ (the space of functions from the usual Sobolev space with zero traces) by

$$\langle u, \varphi \rangle = \int_\Omega \nabla u \cdot \nabla \varphi \, dx.$$

Then the corresponding norm $\| \cdot \|$ is equivalent to the usual Sobolev norm. Set $\mathtt{V} = W_0^{1,2}(\Omega) \times W_0^{1,2}(\Omega)$. The symbols $\langle \cdot, \cdot \rangle$ and $\| \cdot \|$ will be used also for the inner product and the norm in $\mathtt{V}$, respectively, i.e.,

$$\langle U, W \rangle = \langle u, w \rangle + \langle v, z \rangle, \ \|U\|^2 = \|u\|^2 + \|v\|^2$$

for $U = [u, v], W = [w, z] \in \mathtt{V}$. We will always suppose that there is $c > 0$ such that $|n_j(\xi, \eta)| \leq c(1 + |\xi|^{q-1} + |\eta|^{q-1})$ for all $\xi, \eta \in \mathtt{R}$ with some $q \geq 1$ for $\mathtt{n} \leq 2$ and $1 \leq q \leq \frac{2\mathtt{n}}{\mathtt{n}-2}$ for $\mathtt{n} > 2$. Then it follows from the compactness of the embedding $W^{1,2}(\Omega) \subset L^q(\Omega)$ and from the Nemytskij theorem (see, e.g., [5]) that the operators $A : W_0^{1,2}(\Omega) \to W_0^{1,2}(\Omega)$, $N_j : \mathtt{V} \to W_0^{1,2}(\Omega)$, $j = 1, 2$, defined by

$$\langle Au, \varphi \rangle = \int_\Omega u\varphi \, dx, \quad \langle N_j(u, v), \varphi \rangle = \int_\Omega n_j(u, v)\varphi \, dx$$

for all $u, v, \varphi \in W_0^{1,2}(\Omega)$ have the following properties:

A is linear, symmetric, positive and completely continuous, (2.1)

$$N_j \text{ are completely continuous, } \lim_{\|U\| \to 0} \frac{\|N_j(U)\|}{\|U\|} = 0$$

(see Appendix in [12] for the detailed proof of the last condition). Set

$$AU = [Au, Av], \ N(U) = [N_1(u), N_2(v)] \text{ for all } U = [u, v] \in \mathtt{V},$$

$$D(d) = \begin{pmatrix} d_1 & 0 \\ 0 & d_2 \end{pmatrix}, \ B = \begin{pmatrix} b_{11} & b_{12} \\ b_{21} & b_{22} \end{pmatrix}, \ B^* = \begin{pmatrix} b_{11} & b_{21} \\ b_{12} & b_{22} \end{pmatrix}.$$

We will denote by $\to$ and $\rightharpoonup$ the strong convergence and the weak convergence, respectively.

A weak solution of the stationary problem corresponding to (1.7), (1.8) is a solution of the system of operator equations

$$\begin{aligned} d_1 u - b_{11} Au - b_{12} Av - N_1(u, v) &= 0, \\ d_2 v - b_{21} Au - b_{22} Av - N_2(u, v) &= 0 \end{aligned}$$

which can be written in the vector form as

$$D(d)U - BAU - N(U) = 0. \tag{2.2}$$

Notation 2.2. Let us define operators $\beta_1, \beta_2 : W_0^{1,2}(\Omega) \to W_0^{1,2}(\Omega)$ by

$$\langle \beta_1 u, \varphi \rangle = \int_\Omega \left[b_{1+} u^+ - b_{1-} u^- \right] \varphi \, dx$$

for all $u, v, \varphi \in W_0^{1,2}(\Omega)$.

$$\langle \beta_2 v, \varphi \rangle = \int_\Omega \left[b_{2+} v^+ - b_{2-} v^- \right] \varphi \, dx$$

Set $\beta U = [\beta_1 u, \beta_2 v]$.

A weak solution of the problem (1.1), (1.2), (i.e., of (1.3), (1.2)) can be introduced as a solution of the equation

$$D(d)U - BAU - N(U) + \beta U = 0. \tag{2.3}$$

We will need also the "linearized" equation

$$D(d)U - BAU + \beta U = 0. \tag{2.4}$$

Of course (2.4) is nonlinear again. Our problem cannot be approximated by a linear equation and therefore standard methods of the bifurcation theory cannot be used.

Notation 2.3. $\mathbb{R}_+$ — the set of all positive reals, $\mathbb{R}_+^2 = \mathbb{R}_+ \times \mathbb{R}_+$;

κ_j, e_j, $j = 1, 2, \ldots$ — the eigenvalues and orthogonal eigenfunctions of $-\Delta u = \lambda u$ with (1.2);

$C_j = \{d = [d_1, d_2] \in \mathbb{R}_+^2; d_2 = \frac{b_{12} b_{21}/\kappa_j^2}{d_1 - b_{11}/\kappa_j} + \frac{b_{22}}{\kappa_j}\}$, $j = 1, 2, 3, \ldots$ (Fig. 1);

D_S — domain of stability of the problem (1.7), (1.8) – the set of all $d \in \mathbb{R}_+^2$ lying to the right from all C_j, $j = 1, 2, 3, \ldots$ (see Fig. 1);

D_U — domain of instability of the problem (1.7), (1.8) – the set of all $d \in \mathbb{R}_+^2$ lying to the left from C_j for at least one j (see Fig. 1);

bifurcation point of (2.7) *or* (2.8) — a parameter $s_1 \in \mathbb{R}$ such that in any neighbourhood of $[s_1, 0]$ in $\mathbb{R} \times \mathbb{V}$ there is $[s, U] = [s, u, v]$, $\|U\| \neq 0$, satisfying (2.7) or (2.8), respectively.

Recall that if Re $\lambda \leq -\varepsilon < 0$ for all eigenvalues of the problem

$$\begin{aligned} d_1 \Delta u + b_{11} u + b_{12} v &= \lambda u, \\ d_2 \Delta v + b_{21} u + b_{22} v &= \lambda v \quad \text{in } \Omega \end{aligned} \tag{2.5}$$

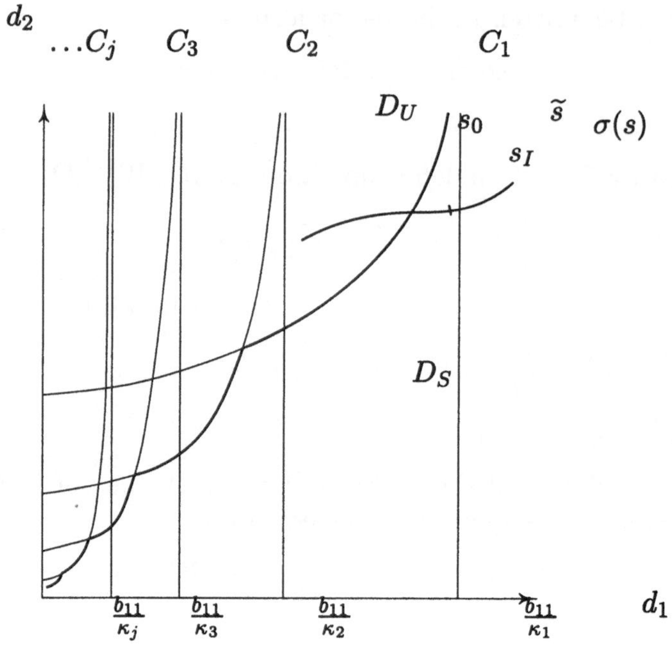

Fig. 1

with the boundary conditions (1.2) then the trivial solution of (1.7), (1.8) is stable (e.g. with respect to the L^2-norm) and if there is an eigenvalue of (2.5), (1.2) satisfying Re $\lambda > 0$ then the trivial solution of (1.7), (1.8) is unstable (see, e.g., [6], [20]). Hence, our definition of the domain of stability and instability in Notation 2.3 is justified by the following statement.

Proposition 2.4. *Let the assumption* (1.4) *be fulfilled. Then* $\bigcup_{j=1}^{\infty} C_j$ *is the set of all* $d \in \mathbb{R}^2_+$ *such that* $\lambda = 0$ *is an eigenvalue of* (2.5), (1.2). *Further, for any* $d \in D_S$ *there is* $\varepsilon > 0$ *such that all eigenvalues of* (2.5), (1.2) *satisfy* Re $\lambda < -\varepsilon < 0$ *and for any* $d \in D_U$ *there exists at least one positive (real) eigenvalue of* (2.5), (1.2).

Proof. See e.g. [14], [16] for Neumann boundary conditions, $\mathbf{n} = 1$, or [2] for the general case.

Consider a differentiable curve $\sigma : \mathbb{R} \to \mathbb{R}^2_+$, $\sigma = [\sigma_1, \sigma_2]$, such that

$$\begin{aligned} &\sigma \text{ intersects transversally } C_p \text{ at a point } \sigma(s_0) = d^0 \in C_p, \\ &\sigma \text{ intersects the line } d_1 = \tfrac{b_{11}}{\kappa_1} \text{ at a point } \sigma(\widetilde{s}),\ \widetilde{s} > s_0, \\ &\sigma(s) \text{ lies to the right from } C_p \text{ and } \sigma_1(s) < \tfrac{b_{11}}{\kappa_1} \text{ for all } s \in (s_0, \widetilde{s}), \\ &\sigma_1(s) > \tfrac{b_{11}}{\kappa_1} \text{ for } s \in (\widetilde{s}, \widetilde{s} + \zeta_0) \text{ with some } \zeta_0 > 0 \text{ small.} \end{aligned} \tag{2.6}$$

(Note that the line $d_1 = \frac{b_{11}}{\kappa_1}$ is the asymptote to C_1 – see Fig. 1.) We will study the problems (2.2), (2.3) only on the curve σ, i.e., the problems

$$D(\sigma(s))U - BAU - N(U) = 0, \tag{2.7}$$

$$D(\sigma(s))U - BAU - N(U) + \beta U = 0 \tag{2.8}$$

with the single bifurcation parameter $s \in \mathbb{R}$.

Consequence 2.5. *If $\sigma(s) \in D_S$ for $s \in (s_0, \tilde{s})$ then there is no bifurcation point of (2.7) in the interval $(s_0, \tilde{s}+\zeta_0)$. This follows directly from Proposition 2.4 and the fact that $\mathrm{Ker}(D(\sigma(s))I - BA) \neq \{0\}$ for any bifurcation point s of (2.7).*

Note that s_0 is a bifurcation point of (2.7) under reasonable assumptions (cf. [14], [16]).

Theorem 2.6. *Let us suppose that (1.4) holds and there exists an eigenfunction e_p corresponding to the eigenvalue κ_p of the Laplacian such that (1.5) is fulfilled. Consider a differentiable curve σ satisfying (2.6). Then there is a bifurcation point $s_I \in (s_0, \tilde{s}]$ of (2.8). For any $\delta \in (0, \delta_0)$ ($\delta_0 > 0$ small) there exist s_δ, u_δ, v_δ satisfying (2.8) (i.e., (1.1), (1.2) with $d_1 = \sigma_1(s_\delta)$, $d_2 = \sigma_2(s_\delta)$ in the weak sense), $\|u_\delta\|^2 + \|v_\delta\|^2 = \delta$, $s_\delta \to s_I$ for $\delta \to 0_+$.*

Note that the assumption (1.5) can be generalized – see the end of Remark 3.7.

Remark 2.7. Consider the case when d^0 from (2.6) is in the part of C_p lying to the right from all C_j, $C_j \neq C_p$ (i.e., on the boundary of the domain of stability of (1.7), (1.8)). Then Theorem 2.6 asserts that bifurcation of nontrivial solutions to (2.8) (i.e., nontrivial weak solutions to (1.1), (1.2)) occurs in the domain where bifurcation for the problem (2.7) (i.e., that of stationary solutions of (1.7), (1.8)) is excluded by Consequence 2.5. This means that spatial patterns for (1.1), (1.2) arise in a certain sense sooner than for (1.7), (1.8) (see [8] – Introduction, or [12] – Interpretation). This can be understood as a destabilizing effect of the "jumping terms" in (1.3) – cf. e.g. [8], [9], [11], [3] where a destabilizing effect of unilateral boundary conditions was proved. Notice that also a stabilizing effect of unilateral conditions can be shown in some cases (see e.g. [10], [12]).

3. PROOF OF THE MAIN RESULT

In the sequel we will always suppose that the eigenvalues κ_j and the eigenfunctions e_j of the Laplacian with (1.2) are numbered so that e_p

is the eigenfunction from the assumption (1.5), $\kappa_p = \ldots = \kappa_{p+k-1}$, k is the multiplicity of κ_p.

Let us consider the eigenvalue problem

$$D^{-1}(\sigma(s))BAU - U = \mu U. \tag{3.1}$$

Proposition 3.1. *Let σ be a curve satisfying (2.6). If $d^0 \in C_p$, $C_p = \ldots = C_{p+k-1}$, k is the multiplicity of κ_p, $d^0 \notin C_q$ for all $C_q \neq C_p$ then there is an eigenvalue $\mu_p(s) = \ldots = \mu_{p+k-1}(s)$ of (3.1) (depending continuously on s) losing positiveness as s crosses s_0 in the following sense: there exists $\eta_0 > 0$ such that $\mu_p(s)$ is positive in one of the one-sided neighbourhoods $(s_0 - \eta_0, s_0)$, $(s_0, s_0 + \eta_0)$ and is either negative or complex in the other one. For any $s \in (s_0 - \eta_0, s_0 + \eta_0)$, this eigenvalue has multiplicity k and the corresponding eigenvectors have the form*

$$U_j(s) = \left[\frac{\sigma_2(s)\kappa_j - b_{22} + \mu_j(s)\sigma_2(s)\kappa_j}{b_{21}} e_j, e_j\right], \quad j = p, \ldots, p+k-1.$$

We have

$$\mathrm{Ker}(D^{-1}(d^0)BA - I) = \mathrm{Lin}\{U_0, U_{p+1}(s_0), \ldots, U_{p+k-1}(s_0)\} \tag{3.2}$$

where we denote $U_0 = [u_0, v_0] = U_p(s_0)$ to emphasize the importance of this vector for our considerations.

Further, let $d^0 \in C_p \cap C_q$ for some $C_q \neq C_p$, $C_q = \ldots = C_{q+l-1}$ where κ_q has multiplicity l. If σ only touches (but does not intersect) C_q at d^0 then the same assertion as above remains valid. If σ really crosses also C_q at d^0 then there exist precisely two different eigenvalues $\mu_p(s) = \ldots = \mu_{p+k-1}(s)$, $\mu_q(s) = \ldots = \mu_{q+l-1}(s)$ losing positiveness in the sense explained above. (In fact, in this case these eigenvalues are always real and change their sign as s crosses s_0.) For $s = s_0$, $\mu_p(s_0) = \mu_q(s_0) = 0$ has multiplicity $k + l$. The eigenvectors corresponding to $\mu_p(s)$, $\mu_q(s)$ are

$$U_i(s) = \left[\frac{\sigma_2(s)\kappa_i - b_{22} + \mu_i(s)\sigma_2(s)\kappa_i}{b_{21}} e_i, e_i\right], \quad i = p, \ldots, p+k-1.$$

$$U_j(s) = \left[\frac{\sigma_2(s)\kappa_j - b_{22} + \mu_j(s)\sigma_2(s)\kappa_j}{b_{21}} e_j, e_j\right], \quad j = q, \ldots, q+l-1.$$

Setting $U_0 = U_p(s_0)$ again, we have

$$\mathrm{Ker}\left(D^{-1}(d^0)BA - I\right) = \mathrm{Lin}\{U_0, U_j(s_0)\}_{j=p+1\ldots,p+k-1,q,\ldots,q+l-1}. \tag{3.3}$$

Proof. See [3], Observations 4.1, 4.2.

Remark 3.2. Set

$$U_0^* = [u_0^*, v_0^*] = \left[\frac{b_{21}}{b_{12}} u_0, v_0\right] = \left[\frac{\sigma_2(s_0)\kappa_p - b_{22}}{b_{12}} e_p, e_p\right]$$

where $U_0 = [u_0, v_0]$ is introduced in Proposition 3.1. Clearly

$$D(d^0)U_0^* - B^* A U_0^* = 0. \tag{3.4}$$

A jump of the Leray-Schauder degree of a certain mapping will be essential for our considerations (see Proof of Theorem 2.6 for details). If $\dim \operatorname{Ker}(D^{-1}(d^0)BA - I) = 1$ then there is only one eigenvalue losing positiveness in the sense of Proposition 3.1 and the jump of the degree follows – see Proof of Theorem 2.6. In the general case, the number of the eigenvalues losing positiveness can be greater than one (see Proposition 3.1) and we must consider a suitable perturbation of (3.1) such that the assertion of Lemma 3.5 below holds. This is the reason for introducing the operator L below.

Notation 3.3. Set $I(d^0) = \{i \neq p;\ d^0 \in C_i\}$. Note that $I(d^0) = \emptyset$ if $\dim \operatorname{Ker}(D^{-1}(d^0)BA - I) = 1$. Let η_0 be from Proposition 3.1. Let us choose $\eta \in (0, \min\{\eta_0, \tilde{s} - s_0\})$. Hence, $\sigma_1(s) < \frac{b_{11}}{\kappa_1}$ for $s \in (s_0, s_0 + \eta)$ by (2.6). Let χ be a continuous function such that $\chi(s_0) = 1$, $\chi(s) \in (0,1)$ for $s \in (s_0 - \eta, s_0) \cup (s_0, s_0 + \eta)$ and $\chi(s) = 0$ for $s \notin (s_0 - \eta, s_0 + \eta)$. Set

$$\begin{aligned} L(s)U &= \chi(s) \cdot \sum_{i \in I(d^0)} \frac{\langle U_i(s), U\rangle}{\|U_i(s)\|^2} \cdot U_i(s) \quad \text{for } s \in (s_0 - \eta, s_0 + \eta) \\ &= 0 \quad \text{for } s \notin (s_0 - \eta, s_0 + \eta) \end{aligned}$$

(cf. [3]). For any $s \in \mathbb{R}$ fixed, $L(s) : \mathbb{V} \to \mathbb{V}$ is a linear completely continuous operator.

Remark 3.4. We have $L(s) \equiv 0$ for $s \in \mathbb{R}$ if $I(d^0) = \emptyset$, i.e.,if $\dim \operatorname{Ker}(D^{-1}(d^0)BA - I) = 1$. Further, it follows from the orthogonality of e_i and the form of $U_i(s)$, U_0^* (see Proposition 3.1 and Remark 3.2) that $\langle U_i(s), U_j(s)\rangle = 0$ for all $j \neq i$, $s \in \mathbb{R}$ and

$$L(s)U_0 = 0,\ L(s)U_i(s) = \chi(s)U_i(s) \text{ for any } i \in I(d^0),\ s \in \mathbb{R} \tag{3.5}$$

$$\langle D(\sigma(s))L(s)U, U_0\rangle = \langle D(\sigma(s))L(s)U, U_0^*\rangle = 0 \text{ for any } s \in \mathbb{R},\ U \in \mathbb{V}.$$

Lemma 3.5. *We can choose $\varrho_0 > 0$ and $\eta \in (0, \tilde{s} - s_0)$ (see Notation 3.3) such that for any $\varrho \in (0, \varrho_0)$ we have*

$$\operatorname{Ker}(D^{-1}(\sigma(s_0))BA - \varrho L(s_0) - I) = \operatorname{Lin}\{U_0\}. \tag{3.6}$$

Further, $\nu(s_0 - \varepsilon) - \nu(s_0 + \varepsilon) = \pm 1$ for all $\varepsilon \in (0, \eta)$ where $\nu(s)$ is the sum of algebraic multiplicities of all positive eigenvalues of the operator $D^{-1}(\sigma(s))BA - \varrho L(s) - I$.

Proof. follows from Proposition 3.1 and the properties of L – see [3], Lemma 4.1.

Notation 3.6. Set $K_j = \{\varphi \in W_0^{1,2}(\Omega);\ \beta_j\varphi = 0\}$, $j = 1,2$, $K = K_1 \times K_2$. Clearly

$$K_1 = \{\varphi \in W_0^{1,2}(\Omega); \varphi \le 0 \text{ a.e. in } \Omega_{ep}^+, \varphi \ge 0 \text{ a.e. in } \Omega_{ep}^-\},$$
$$K_2 = \{\varphi \in W_0^{1,2}(\Omega); \varphi \ge 0 \text{ a.e. in } \Omega_{ep}^+, \varphi \le 0 \text{ a.e. in } \Omega_{ep}^-\}$$

by (1.5). The set K is a closed convex cone in $\mathbb{V}$ with its vertex at the origin.

Remark 3.7. Let $U_0 = [u_0, v_0]$ and $U_0^* = [u_0^*, v_0^*]$ be from Proposition 3.1 and Remark 3.2. It follows from (1.4) that $U_0^* \in K$ and $\pm U_0 \notin K$ (see Notation 3.6). Moreover, let us show that $W \notin K$ for any $W = [w, z] = \sum_{i=1}^r a_i U_i(s_0)$, $a_i \in \mathbb{R}$, $r \in \mathbb{N}$, $\|W\| \neq 0$. If $W \in K$ is of the form considered then $w = \sum_{i=1}^r a_i\alpha_i e_i \le 0$ a.e. in Ω_{ep}^+, $w \ge 0$ a.e. in Ω_{ep}^- , and $z = \sum_{i=1}^r a_i e_i \ge 0$ a.e. in Ω_{ep}^+, $z \le 0$ a.e. in Ω_{ep}^-, with $\alpha_i = \frac{d_2^0\kappa_i - b_{22}}{b_{21}} > 0$ (see Proposition 3.1). Multiplying w by z, integrating over Ω and using the L^2-orthogonality of e_i and the last condition in (1.5) we obtain

$$0 \ge \int_{\Omega_{ep}^+} w(x)z(x)\mathrm{d}x + \int_{\Omega_{ep}^-} w(x)z(x)\mathrm{d}x = \int_\Omega w(x)z(x)\mathrm{d}x =$$
$$= \int_\Omega \left(\sum_{i=1}^r a_i\alpha_i e_i(x)\right)\left(\sum_{i=1}^r a_i e_i(x)\right)\mathrm{d}x = \sum_{i=1}^r a_i^2\alpha_i \int_\Omega e_i^2(x)\mathrm{d}x.$$

This is possible only for $a_i = 0$, $i = 1, 2, \ldots, r$, which is a contradiction with $\|W\| \neq 0$.

Note that Theorem 2.6 remains valid if we replace the positiveness of $b_1^+, b_1^-, b_2^+, b_2^-$ in (1.5) by the positiveness only on some subsets $\Omega_1^+ \subset \Omega_{ep}^+$, $\Omega_1^- \subset \Omega_{ep}^-$, $\Omega_2^+ \subset \Omega_{ep}^-$, $\Omega_2^- \subset \Omega_{ep}^+$ such that $\mathrm{meas}(\Omega_1^+ \cup \Omega_1^-) > 0$, $\mathrm{meas}(\Omega_2^+ \cup \Omega_2^-) > 0$, $\mathrm{Ker}(D(d^0)I - BA) \cap K = \{0\}$. In fact, only this is used in our proof.

Observation 3.8. *The operator β has the following properties:*

$$\beta \text{ is completely continuous,} \tag{3.7}$$
$$\beta U = 0 \text{ if and only if } U \in K;\ \langle \beta U, U\rangle > 0 \text{ for all } U \notin K, \tag{3.8}$$
$$\beta(tU) = t\beta U \text{ for all } t > 0, U \in \mathbb{V}, \tag{3.9}$$
$$\langle \beta U - \beta V, U - V\rangle \ge 0 \text{ for all } U, V \in \mathbb{V}, \tag{3.10}$$
$$\langle \beta V, U\rangle \le 0 \text{ for all } V \in \mathbb{V}, U \in K. \tag{3.11}$$

Observation 3.9. *We will show that $\langle \beta W, U_0^* \rangle < 0$ for all $W \notin K$. In particular, if $\langle \beta W, U_0^* \rangle = 0$ for some $W \in \mathbb{V}$ then $W \in K$, i.e., $\beta W = 0$. A special choice $W = \pm U_0$ gives $\langle \beta U_0, U_0^* \rangle < 0$ and $\langle \beta(-U_0), U_0^* \rangle < 0$ by Remark 3.7. Indeed, $W = [w, z] \notin K$ means that there is a set $E \subset \Omega$ such that meas$E > 0$ and one of the following four cases occurs:*

$$E \subset \Omega_{e_p}^+ \text{ and } w > 0 \text{ a.e. in } E, \quad E \subset \Omega_{e_p}^- \text{ and } w < 0 \text{ a.e. in } E,$$
$$E \subset \Omega_{e_p}^+ \text{ and } z < 0 \text{ a.e. in } E, \quad E \subset \Omega_{e_p}^- \text{ and } z > 0 \text{ a.e. in } E.$$

Hence, we obtain by using (1.4), (1.5), Notation 2.2 and Remark 3.2 that

$$\langle \beta W, U_0^* \rangle = \langle \beta_1 w, u_0^* \rangle + \langle \beta_2 z, v_0^* \rangle = \frac{b_{21}}{b_{12}} \langle \beta_1 w, u_0 \rangle + \langle \beta_2 z, v_0 \rangle =$$
$$\int_{\Omega_{e_p}^+} \frac{b_{21}}{b_{12}} b_{1+} w^+ u_0 - b_{2-} z^- v_0 \, dx + \int_{\Omega_{e_p}^-} -\frac{b_{21}}{b_{12}} b_{1-} w^- u_0 + b_{2+} z^+ v_0 \, dx < 0.$$

Lemma 3.10. *(Cf. [4].) Let (1.4) and (1.5) hold. If $d = [d_1, d_2] \in \mathbb{R}_+^2$, $d_1 > \frac{b_{11}}{\kappa_1}$ (i.e., if $d \in Z_0$ in the notation of [4]), $\xi \in [0, +\infty)$ then the only solution of*

$$D(d)U - BAU + \xi \beta U = 0 \tag{3.12}$$

is trivial. (Let us recall that the line $d_1 = \frac{b_{11}}{\kappa_1}$ is the asymptote to C_1 – see Fig. 1.)

Proof. Let $d = [d_1, d_2]$ be fixed, $d_1 > \frac{b_{11}}{\kappa_1}$. We can write (3.12) in the form

$$\begin{aligned} d_1 u - b_{11} Au - b_{12} Av + \xi \beta_1 u &= 0, \\ d_2 v - b_{21} Au - b_{22} Av + \xi \beta_2 v &= 0. \end{aligned} \tag{3.13}$$

We have

$$\frac{d_1}{b_{11}} > \frac{1}{\kappa_1} = \sup_{\|\varphi\| \neq 0} \frac{\langle A\varphi, \varphi \rangle}{\|\varphi\|^2} \tag{3.14}$$

and it follows by using (3.10) that

$$\langle d_1 \varphi - b_{11} A\varphi + \xi \beta_1 \varphi - d_1 \psi + b_{11} A\psi - \xi \beta_1 \psi, \varphi - \psi \rangle \geq c \cdot \|\varphi - \psi\|^2$$

for all $\varphi, \psi \in W_0^{1,2}(\Omega)$ with $c = d_1 - \frac{b_{11}}{\kappa_1} > 0$. It follows that $(d_1 I - b_{11} A + \xi \beta_1)^{-1}$ is well-defined (see, e.g., [13]). From the first equation of (3.13), we can express

$$u = b_{12}(d_1 I - b_{11} A + \xi \beta_1)^{-1} Av \tag{3.15}$$

and put it into the second equation to obtain

$$d_2 v - b_{21} b_{12} A(d_1 I - b_{11} A + \xi\beta_1)^{-1} Av - b_{22} Av + \xi\beta_2 v = 0. \qquad (3.16)$$

It follows from (2.1), (3.14), (3.8) that $\langle A(d_1 I - b_{11} A + \xi\beta_1)^{-1} Av, v\rangle \geq 0$. Hence, multiplying (3.16) by v we obtain by using (1.4), (2.1), (3.8) that

$$\langle d_2 v - b_{21} b_{12} A(d_1 I - b_{11} A + \xi\beta_1)^{-1} Av - b_{22} Av + \xi\beta_2 v, v\rangle > 0$$

provided $\|v\| \neq 0$. This implies $v = 0$ and (3.15) gives also $u = 0$.

Similarly as in [11] and [3], for the proof of Theorem 2.6 we will choose a small $\varrho > 0$ and consider a penalty equation

$$D(\sigma(s))U - BAU - \frac{\tau}{1+\tau} N(U) + \varrho \frac{D(\sigma(s))}{1+\tau} L(s)U + \frac{\tau}{1+\tau} \beta U = 0 \quad (3.17)$$

together with the norm condition

$$\|U\|^2 = \frac{\delta\tau}{1+\tau}. \qquad (3.18)$$

Here $\delta > 0$ is a given small number, τ is an additional parameter. The equation (3.17) can be understood as a homotopy joining the linearization of (2.7) perturbed by the operator $\varrho D(\sigma(s))L(s)$ (obtained for $\tau = 0$) with our equation with the jumping nonlinearity (2.8) (obtained for $\tau \to +\infty$). The common idea of the papers mentioned is to prove the existence of a branch of $[s, U, \tau]$ satisfying (3.17), (3.18) starting at $[s_0, 0, 0]$ which is unbounded in τ. The limiting process $\tau \to +\infty$ along this branch gives a solution U_δ of (2.8) with some s_δ, $\|U_\delta\|^2 = \delta$. Any accumulation point of s_δ for $\delta \to 0_+$ is a bifurcation point of (2.8).

Lemma 3.11. *(Cf. [4].) Let ζ_0 be from the assumption* (2.6). *For any $\zeta \in (0, \zeta_0)$ there exists $\delta_0 > 0$ such that there is no nontrivial solution U of* (3.17) *with $s = \tilde{s} + \zeta$, $\tau \in [0, +\infty)$ and $\|U\|^2 < \delta_0$.*

Proof. Suppose by contradiction that there exist $\zeta_1 \in (0, \zeta_0)$ and $U_n \to 0$, $\frac{U_n}{\|U_n\|} \rightharpoonup W$, $\frac{\tau_n}{1+\tau_n} \to \xi \in [0,1]$ such that U_n, τ_n satisfy (3.17) with $s = \tilde{s} + \zeta_1$. Note that $L(\tilde{s} + \zeta_1) \equiv 0$ (see (2.6) and Notation 3.3). Multiplying (3.17) by $\|U_n\|^{-1}$ and passing to the limit we obtain by using (2.1), (3.7), (3.9) that $\frac{U_n}{\|U_n\|} \to W$, W is a nontrivial solution of (3.12) with $d = [d_1, d_2] = \sigma(\tilde{s} + \zeta_1)$, $d_1 > \frac{b_{11}}{\kappa_1}$ (see (2.6)). This is a contradiction with Lemma 3.10.

Proof of Theorem 2.6. Let us choose ϱ and η small enough so that the assertion of Lemma 3.5 holds. Let $\zeta \in (0, \zeta_0)$ be fixed where ζ_0 is from the assumption (2.6). Let $\delta_0 > 0$ be the corresponding number from Lemma 3.11. Consider $\delta \in (0, \delta_0)$ fixed. The equations (3.17), (3.18) are equivalent to

$$x - T(s)x + G(s,x) = 0 \tag{3.19}$$

in the space $\mathbf{X} = \mathbf{V} \times \mathbf{R}$ (with points $x = [U, \tau]$ and the norm $|||x||| = \|U\| + |\tau|$) where

$$T(s)x = \Big[D^{-1}(\sigma(s))BAU - \varrho L(s)U, 0\Big] \text{ for all } s \in \mathbf{R}, x = [U,\tau] \in \mathbf{X}$$

$$G(s,x) = \left[\frac{\tau}{1+\tau} D^{-1}(\sigma(s))\Big(-N(U) + \beta U\Big) - \varrho \frac{\tau}{1+\tau} L(s)U, -\frac{1+\tau}{\delta}\|U\|^2\right]$$
$$\text{for all } s \in \mathbf{R}, x = [U,\tau] \in \mathbf{X}.$$

The operators T, G have the following properties:

$$T : \mathbf{R} \times \mathbf{X} \to \mathbf{X},\ G : \mathbf{R} \times \mathbf{X} \to \mathbf{X} \text{ are completely continuous,} \tag{3.20}$$

$$T(s) : \mathbf{X} \to \mathbf{X} \text{ is linear for any fixed } s \in \mathbf{R}, \tag{3.21}$$

$$\lim_{|||x||| \to 0} \frac{|||G(s,x)|||}{|||x|||} = 0 \text{ uniformly on bounded } s\text{-intervals.} \tag{3.22}$$

If T is a compact linear operator in a Banach space, Ker $(I - T) = \{0\}$ then $\operatorname{ind}(I - T) = (-1)^{\nu(T)}$ where ind denotes the Leray-Schauder index, $\nu(T)$ is the sum of the algebraic multiplicities of all positive eigenvalues of $T - I$ (see, e.g., [17]). It is easy to see that λ is an eigenvalue of multiplicity k of $T(s)$ (for some $s \in \mathbf{R}$) and $x = [U, \tau]$ is the corresponding eigenvector if and only if λ is an eigenvalue of multiplicity k of $D^{-1}(\sigma(s))BA - \varrho L(s)$ and U is the corresponding eigenvector, $\tau = 0$. Thus, it follows from Lemma 3.5 that

$$\operatorname{Ker}(I - T(s_0)) = \operatorname{Lin}\{x_0\} \text{ with } x_0 = [U_0, 0] \tag{3.23}$$

and there is $\varepsilon_0 > 0$ such that

$$\operatorname{ind}(I - T(s_0 - \varepsilon)) \neq \operatorname{ind}(I - T(s_0 + \varepsilon)) \text{ for all } \varepsilon \in (0, \varepsilon_0). \tag{3.24}$$

(Recall that the vector U_0 is from Proposition 3.1.) Set

$$\mathcal{C} = \overline{\{[s,x] \in \mathbf{R} \times \mathbf{X};\, |||x||| \neq 0,\ (3.19) \text{ holds}\}}$$
$$= \overline{\{[s,U,\tau];\, \tau \neq 0,\ (3.17),\ (3.18) \text{ hold}\}}$$

and let $\mathcal{C}_0$ be the component of $\mathcal{C}$ containing $[s_0, 0, 0]$. Analogously as in [1] we can define subcontinua $\mathcal{C}_0^+$ and $\mathcal{C}_0^-$ of $\mathcal{C}_0$ starting at $[s_0, 0, 0]$ in the direction x_0 and $-x_0$, respectively, with $x_0 = [U_0, 0]$. See [1], [7] for details. In particular, $\mathcal{C}_0^+$ and $\mathcal{C}_0^-$ contain $[s_0, 0, 0]$ and

$$\text{there are } [s_n, U_n, \tau_n] \in \mathcal{C}_0^+, [s_n, U_n, \tau_n] \to [s_0, 0, 0], \frac{U_n}{\|U_n\|} \to \frac{U_0}{\|U_0\|}, \tag{3.25}$$

there are $[s_n, U_n, \tau_n] \in \mathcal{C}_0^-, [s_n, U_n, \tau_n] \to [s_0, 0, 0], \frac{U_n}{\|U_n\|} \to -\frac{U_0}{\|U_0\|}$. (3.26)

Under the assumptions (3.20)–(3.24), considerations from the proof of the global Dancer's bifurcation theorem ([1], Theorem 2) can be used and an analogue of this theorem for the equation (3.19) can be proved (cf. [7], Theorem 4.1). That means

$$\text{either } \mathcal{C}_0^+ \cap \mathcal{C}_0^- \neq \{[s_0, 0, 0]\} \text{ or both } \mathcal{C}_0^+ \text{ and } \mathcal{C}_0^- \text{ are unbounded} \tag{3.27}$$

(cf. [1] or [7] for details in the classical case or in our situation, respectively).

We will write $\mathcal{C}_\delta, \mathcal{C}_{\delta,0}, \mathcal{C}_{\delta,0}^+, \mathcal{C}_{\delta,0}^-$ instead of $\mathcal{C}, \mathcal{C}_0, \mathcal{C}_0^+, \mathcal{C}_0^-$ in order to emphasize the role of δ from the norm condition (3.18). We shall prove successively that $\delta_0 > 0$ could be chosen such that the following statements hold if $\delta \in (0, \delta_0)$:

$$\text{if } [s_n, U_n, \tau_n] \in \mathcal{C}_{\delta,0}^+, [s_n, U_n, \tau_n] \to [s_0, 0, 0], \tfrac{U_n}{\|U_n\|} \to \tfrac{U_0}{\|U_0\|} \text{ then } \lim \tfrac{s_n - s_0}{\tau_n} < 0, \tag{3.28}$$

$$\text{if } [s_n, U_n, \tau_n] \in \mathcal{C}_{\delta,0}^-, [s_n, U_n, \tau_n] \to [s_0, 0, 0], \tfrac{U_n}{\|U_n\|} \to -\tfrac{U_0}{\|U_0\|} \text{ then } \lim \tfrac{s_n - s_0}{\tau_n} > 0 \tag{3.29}$$

(the branch $\mathcal{C}_{\delta,0}^+$ starts downwards from s_0, $\mathcal{C}_{\delta,0}^-$ starts upwards from s_0),

$$[s_0, U, \tau] \in \mathcal{C}_{\delta,0} \Longrightarrow \tau = 0 \tag{3.30}$$

($\mathcal{C}_{\delta,0}^+$, $\mathcal{C}_{\delta,0}^-$ cannot intersect the level s_0 with the exception of the starting point),

$$[s, U, \tau] \in \mathcal{C}_{\delta,0} \Longrightarrow s \leq \widetilde{s} + \zeta \tag{3.31}$$

(where $\widetilde{s}$ is from (2.6)). Suppose for a moment that (3.28) – (3.31) hold. It follows from (3.28), (3.29), (3.30) and the definition of $\mathcal{C}_{\delta,0}^+$ and $\mathcal{C}_{\delta,0}^-$ (see [1], [7] for details) that $\mathcal{C}_{\delta,0}^+$ and $\mathcal{C}_{\delta,0}^-$ remain below and above s_0, respectively, with the exception of $[s_0, 0, 0]$ and therefore $\mathcal{C}_{\delta,0}^+ \cap \mathcal{C}_{\delta,0}^- = \{[s_0, 0, 0]\}$. Hence (3.27) implies that $\mathcal{C}_{\delta,0}^+, \mathcal{C}_{\delta,0}^-$ are unbounded. However, (3.31) together with (3.18) imply the boundedness of $\mathcal{C}_{\delta,0}^-$ in s and $\|U\|$ and therefore $\mathcal{C}_{\delta,0}^-$ is unbounded in τ. It is easy to see that $\tau \geq 0$ for all $[s, U, \tau] \in \mathcal{C}_{\delta,0}$. In particular, there exists a sequence $[s_n, U_n, \tau_n] \in \mathcal{C}_{\delta,0}^-$ satisfying (3.17), (3.18) with $s_n \in (s_0, \widetilde{s} + \zeta]$, $\tau_n \to +\infty$. We can suppose $s_n \to s_\delta, U_n \rightharpoonup U_\delta$ with some $s_\delta \in [s_0, \widetilde{s} + \zeta]$, U_δ and it follows from (3.17) by using the compactness of the operators A, L and β that $U_n \to U_\delta$, $\|U_\delta\|^2 = \delta$, U_δ is a solution of (2.8) with $s = s_\delta$. We would like to know that

$$s_\delta \geq s_0 + \varepsilon \text{ for all } \delta > 0 \text{ small enough with some } \varepsilon > 0. \tag{3.32}$$

Suppose by contradiction that there are $\delta_n \to 0$ and $s_{\delta_n} \to s_0$, $\|U_{\delta_n}\| \to 0$, $\frac{U_{\delta_n}}{\|U_{\delta_n}\|} \rightharpoonup W$,

$$D(\sigma(s_{\delta_n}))U_{\delta_n} - BAU_{\delta_n} - N(U_{\delta_n}) + \beta U_{\delta_n} = 0. \tag{3.33}$$

Dividing (3.33) by $\|U_{\delta_n}\|$ and using (2.1), (3.7), (3.9) we obtain $\frac{U_{\delta_n}}{\|U_{\delta_n}\|} \to W$ and W is a solution of (2.4) with $d = d_0$. Multiply (2.4) (with $U = W$ and $d = d^0$) by U_0^*, (3.4) by W and subtract. We obtain $\langle \beta W, U_0^* \rangle = 0$. Observation 3.9 implies $\beta W = 0$ and consequently, $W \in K$ by (3.8). It follows from (2.4) with $U = W$ and $d = d^0$ that $W \in \operatorname{Ker}(D(d^0)I - BA)$. Simultaneously, we obtain from Remark 3.7 and (3.2) or (3.3) in Proposition 3.1 that $W \notin K$. This is a contradiction and (3.32) is proved. Moreover, ζ could be chosen arbitrarily small and therefore, any accumulation point s_I of s_δ for $\delta \to 0_+$ (which is simultaneously a bifurcation point of (2.8)) satisfies $s_I \in [s_0 + \varepsilon, \tilde{s}]$.

To complete the proof it is sufficient to show that (3.28) – (3.31) hold.

Proof of (3.28): Multiply the equation

$$\begin{aligned} D(\sigma(s_n))U_n - BAU_n - \frac{\tau_n}{1+\tau_n}N(U_n) + \\ + \varrho\frac{D(\sigma(s_n))}{1+\tau_n}L(s_n)U_n + \frac{\tau_n}{1+\tau_n}\beta U_n = 0 \end{aligned} \tag{3.34}$$

by $\frac{U_0^*}{\|U_n\|}$, (3.4) by $\frac{U_n}{\|U_n\|}$ and subtract. A simple calculus using (2.1), (3.9), (3.5) yields

$$R_n(s_n - s_0) + \frac{\tau_n}{1+\tau_n}\left\langle -\frac{N(U_n)}{\|U_n\|} + \beta\frac{U_n}{\|U_n\|}, U_0^* \right\rangle = 0,$$

$$R_n = \sigma_1'(\bar{s}_n)\left\langle \frac{u_n}{\|U_n\|}, u_0^* \right\rangle + \sigma_2'(\tilde{s}_n)\left\langle \frac{v_n}{\|U_n\|}, v_0^* \right\rangle$$

with some $\bar{s}_n, \tilde{s}_n$ between s_0, s_n where $U_n = [u_n, v_n], U_0 = [u_0, v_0]$. Further, it follows from the formula for u_0, u_0^*, v_0, v_0^* (see Proposition 3.1 and Remark 3.2), the equation defining C_p and the assumption (2.6) (the transversality and orientation of σ) that

$$\begin{aligned} R_n \to R &= \sigma_1'(s_0)\left\langle \frac{u_0}{\|U_0\|}, u_0^* \right\rangle + \sigma_2'(s_0)\left\langle \frac{v_0}{\|U_0\|}, v_0^* \right\rangle \\ &= \left[\frac{(\sigma_2(s_0)\kappa_p - b_{22})^2}{b_{12}b_{21}}\sigma_1'(s_0) + \sigma_2'(s_0)\right] \cdot \frac{\|v_0^*\|^2}{\|U_0\|} < 0. \end{aligned}$$

Hence, it follows by using (2.1) and Observation 3.9 that

$$\lim_{n\to\infty} \frac{s_n - s_0}{\tau_n} = -\frac{1}{R\|U_0\|}\langle \beta U_0, U_0^* \rangle < 0.$$

Proof of (3.29) is the same but we have $\frac{U_n}{\|U_n\|} \to -\frac{U_0}{\|U_0\|}$ and $R > 0$.

Proof of (3.30) for δ small enough: suppose by contradiction that there are $\delta_n \to 0$ and $[s_0, U_n, \tau_n] \in \mathcal{C}_{\delta_n,0}, \tau_n > 0$. Then $\|U_n\| \to 0$, (3.34) holds with $s_n = s_0$. We can suppose without loss of generality that $\frac{U_n}{\|U_n\|} \rightharpoonup W$, $\tau_n \to \tau \in [0, +\infty]$ and $\frac{\tau_n}{1+\tau_n} \to \xi \in [0, 1]$. We obtain from (3.34) (with $s_n = s_0$) divided by $\|U_n\|$ (by using (2.1), (3.7), (3.9)) that $\frac{U_n}{\|U_n\|} \to W$,

$$\begin{aligned} &D(\sigma(s_0))W - BAW + \varrho\frac{D(\sigma(s_0))}{1+\tau}L(s_0)W + \xi\beta W = 0 \quad \text{if } \tau < +\infty, \\ &D(\sigma(s_0))W - BAW + \xi\beta W = 0, \quad \xi = 1 \quad \text{if } \tau = +\infty. \end{aligned} \tag{3.35}$$

Multiply (3.35) by U_0^*, (3.4) by W and subtract. We get $\xi\langle\beta W, U_0^*\rangle = 0$ by using (3.5). If $\xi \neq 0$ then $W \in K$ by Observation 3.9. Hence, (3.35) gives $W \in \operatorname{Ker}(D^{-1}(\sigma(s_0))BA - \frac{\varrho}{1+\tau}L(s_0) - I)$ if $\tau < \infty$ or $W \in \operatorname{Ker}(D^{-1}(\sigma(s_0))BA - I)$ if $\tau = \infty$. One of the possibilities (3.6), (3.2), (3.3) holds (see Proposition 3.1, Lemma 3.5) and it follows from Remark 3.7 that $W \notin K$, which is a contradiction. Thus $\xi = 0$, i.e., $\tau_n \to 0$. Multiply (3.34) (with $s_n = s_0$) by $\frac{U_0^*}{\|U_n\|}$, (3.4) by $\frac{U_n}{\|U_n\|}$ and subtract. We obtain

$$\left\langle \frac{-\tau_n}{1+\tau_n}\frac{N(U_n)}{\|U_n\|} + \varrho\frac{D(d^0)}{1+\tau_n}L(s_0)\frac{U_n}{\|U_n\|} + \frac{\tau_n}{1+\tau_n}\beta\frac{U_n}{\|U_n\|}, U_0^* \right\rangle = 0.$$

Using (2.1), (3.5), dividing by τ_n and letting $n \to \infty$ we get $\langle\beta W, U_0^*\rangle = 0$ and this leads to a contradiction as above.

Proof of (3.31) follows directly from the conectedness of $\mathcal{C}_{\delta,0}$, the fact that $[s_0, 0, 0] \in \mathcal{C}_{\delta,0}$, Lemma 3.11 and the choice of δ_0 at the beginning of the proof.

Remark 3.12. Consider the particular case when b_{jj+}, b_{jj-} are constants, (1.4) and the first condition from (1.6) are fulfilled. The problem (1.1) can be written as (1.3) with $b_{11} = b_{11+}$, $b_{22} = b_{22-}$, $b_{1-} = b_{11} - b_{11-}$, $b_{2+} = b_{22} - b_{22+}$, $b_{1+} = b_{2-} = 0$. We have $b_{22}v - b_{2+}v^+ = b_{22+}v - b_{2+}v^-$ and therefore, all positive solutions of

$$\begin{aligned} &d_1\Delta u + b_{11}u + b_{12}v + n_1(u, v) = 0, \\ &d_2\Delta v + b_{21}u + b_{22+}v + n_2(u, v) = 0 \quad \text{in } \Omega \end{aligned} \tag{3.36}$$

are simultaneously solutions of (1.3) (i.e., also of (1.1)). Define $\widetilde{C}_1$ in the same way as C_1 but with the coefficient b_{22} replaced by b_{22+}. The first part of Proposition 2.4 and the formulas (3.2), (3.3) in Proposition 3.1 remain valid if we replace the assumption (1.4) by (1.9) (see the proof in [2]). It follows that the linearization of (3.36), (1.2) (i.e., (3.36),

(1.2) with $n_j \equiv 0$) has a nontrivial solution $\widetilde{U}_0 = \left[\frac{d_2\kappa_1 - b_{22+}}{b_{21}} e_1, e_1\right]$ where e_1 is the principal eigenfunction of the Laplacian which does not change its sign. If we take e_1 positive then clearly $\widetilde{U}_0$ is simultaneously a solution of (1.3), (1.2) (i.e., (1.1), (1.2)) without the terms n_j. If σ is a curve intersecting $\widetilde{C}_1$ at $d = \sigma(s_1)$ then it follows from the classical bifurcation results ([19] or [1]) that there is a global bifurcation branch of solutions to (3.36), (1.2) starting at s_1 in the direction $\widetilde{U}_0$. It is not hard to show by using a regularity argument that the positiveness of both components of bifurcating solutions lying on this branch is preserved in a neighbourhood of s_1. Hence, these solutions are simultaneously solutions of (1.3), (1.2). Similarly, in the case of the second condition in (1.6), we obtain a nontrivial solution $\widetilde{U}_0 = \left[\frac{d_2\kappa_1 - b_{22-}}{b_{21}} e_1, e_1\right]$ of the linearization of (3.36) with b_{22+} replaced by b_{22-} for $d \in \widetilde{C}_1$ defined in the same way as C_1 but with the coefficient b_{22} replaced now by b_{22-}. If we choose e_1 negative then $\widetilde{U}_0$ is simultaneously a solution of (1.3), (1.2) (i.e., (1.1), (1.2)) without the terms n_j. The solutions of (3.36), (1.2) with b_{22+} replaced by b_{22-} lying in the bifurcation branch starting in the direction $\widetilde{U}_0$ are negative near s_1 and are simultaneously solutions of our original problem (1.1), (1.2). Hence, in the particular case under consideration, some bifurcation points of our problem with jumping nonlinearities can be obtained elementarily as a bifurcation of a modified classical problem. Note that in both cases $\widetilde{C}_1$ lies below C_1 and if σ is the curve from the assumptions of Theorem 2.6 then its assertion is an easy consequence of our considerations.

Acknowledgments

The authors are supported by the grant No. 201/98/1453 of the Grant Agency of the Czech Republic and by the grant No. VS 97156 of the Ministry of Education of the Czech Republic.

References

[1] Dancer E.N. (1974). On the structure of solutions of non-linear eigenvalue problems. *Indiana Univ. Math. J.*, 23:1069–1076 .

[2] Drábek P. and Kučera M. (1988). Reaction-diffusion systems: Destabilizing effect of unilateral conditions. *Nonlinear Anal.*, 12:1173–1192.

[3] Eisner J. and Kučera M. (1997). Spatial patterns for reaction-diffusion systems with conditions described by inclusions. *Appl. Math.*, 42:421–449.

[4] Eisner J. (1996). Critical and bifurcation points of reaction-diffusion systems with conditions given by inclusions. Preprint Math. Inst. Acad. Sci. Czech Rep., 118. Submitted to *Nonlinear Anal.*

[5] Fučík S. and Kufner A. (1980). *Nonlinear Differential Equations.* Elsevier, Amsterdam.

[6] Kielhöfer H. (1974). Stability and semilinear evolution equations in Hilbert space. *Arch. Rational Mech. Anal.*, 57:150–165.

[7] Kučera M. (1982). Bifurcation points of variational inequalities. *Czechoslovak Math. J.*, 32:208–226.

[8] Kučera M. and Bosák M. (1993). Bifurcation for quasi-variational inequalities of reaction-diffusion type. *SAACM*, 3:111–127.

[9] Kučera M. (1995). Bifurcation of solutions to reaction-diffusion systems with unilateral conditions. In *Navier-Stokes Equations and related Topics* (Edited by A. Sequeira), pp. 307–322. Plenum Press, New York.

[10] Kučera M. (1996). Reaction-diffusion systems: Bifurcation and stabilizing effect of conditions given by inclusions. *Nonlinear Anal.*, 27:249–260.

[11] Kučera M. (1997). Bifurcation of solutions to reaction-diffusion systems with conditions described by inequalities and inclusions. *Nonlinear Anal.*, 30:3683–3694.

[12] Kučera M. (1997). Reaction-diffusion systems: Stabilizing effect of conditions described by quasivariational inequalities. *Czechoslovak Math. J.*, 47:469–486.

[13] Lions J. L. (1969). *Quelques méthodes de resolution de problèmes aux limites non linéaires.* Paris.

[14] Mimura M., Nishiura Y. and Yamaguti M. (1979). Some diffusive prey and predator systems and their bifurcation problems. *Ann. New York Acad. Sci.*, 316:490–521.

[15] Murray J. D. (1993). *Mathematical Biology.* Biomathematics Texts, Vol. 19. Springer-Verlag Berlin, Heidelberg.

[16] Nishiura Y. (1982). Global structure of bifurcating solutions of some reaction-diffusion systems. *SIAM J. Math. Anal.*, 13:555–593.

[17] Nirenberg L. (1974). *Topics in Nonlinear Functional Analysis.* Academic Press, New York.

[18] Quittner P. (1987). Bifurcation points and eigenvalues of inequalities of reaction-diffusion type. *J. Reine Angew. Math.*, 380:1–13.

[19] Rabinowitz P.H. (1971). Some global results for nonlinear eigenvalue problems. *J. Funct. Anal.*, 7:487–513.

[20] Sattinger D.H. (1973). *Topics in Stability and Bifurcation Theory*, Lecture Notes in Mathematics 309. Springer-Verlag, Berlin-Heidelberg-New York.

COUPLED PROBLEMS FOR VISCOUS INCOMPRESSIBLE FLOW IN EXTERIOR DOMAINS

Miloslav Feistauer, Christoph Schwab

Abstract: The formulation of the *fluid flow in an unbounded exterior domain* Ω is not always convenient for computations and, therefore, the problem is often truncated to a bounded domain $\Omega^- \subset \Omega$ with an artificial exterior boundary Γ. Then the problem of the choice of suitable "transparent" boundary conditions on Γ appears. Another possibility is to simulate the presence of the fluid in the domain Ω^+ exterior to Γ with the use of a suitable (preferably linear) approximation of the equations describing the flow. The interior and exterior problems are coupled with the aid of *transmission conditions* on the interface Γ.

Here we briefly describe the formulation and analysis of the coupling of the interior Navier–Stokes problem and the exterior Stokes problem and Oseen problem. At the end we give the reformulation of the coupled problems with the aid of integral equations on the artificial interface.

Keywords: Stokes problem, Oseen problem, Navier-Stokes equations, coupled procedures, boundary integral equations.

1. COUPLING OF INTERIOR NAVIER-STOKES PROBLEM WITH EXTERIOR STOKES PROBLEM

Let $\Omega \subset I\!R^3$ be an unbounded domain which is the complement of the closure of a bounded open set (representing, e. g., a body emerged into a moving fluid). We set $\Gamma_0 = \partial\Omega$ and introduce an artificial interface Γ dividing Ω into two subdomains, a bounded interior domain Ω^- with boundary $\partial\Omega^- = \Gamma_0 \cup \Gamma$ and an unbounded domain Ω^+ with $\partial\Omega^+ = \Gamma$.

Classical formulation of the coupled problem: Find the velocity $\boldsymbol{u}^\pm = (u_1^\pm, u_2^\pm, u_3^\pm) : \overline{\Omega^\pm} \to I\!R^3$ and the pressure $p^\pm : \overline{\Omega^\pm} \to I\!R$ such

Applied Nonlinear Analysis, edited by Sequeira *et al.*
Kluwer Academic / Plenum Publishers, New York, 1999.

that

$$\left.\begin{array}{ll}
\text{a)} & u_i^\pm \in C^2(\overline{\Omega^\pm}), \quad i=1,\,2,\,3, \quad p^\pm \in C^1(\overline{\Omega^\pm}),\\
\text{b)} & -2\nu \sum_{j=1}^{3} \dfrac{\partial D_{ij}(\boldsymbol{u}^-)}{\partial x_j} + \sum_{j=1}^{3} u_j^- \dfrac{\partial u_i^-}{\partial x_j} + \dfrac{\partial p^-}{\partial x_i} = f_i,\\
& \qquad\qquad\qquad\qquad\qquad i = 1,\,2,\,3, \text{ in } \Omega^-,\\
\text{c)} & \operatorname{div} \boldsymbol{u}^- = 0 \quad \text{in } \Omega^-,\\
\text{d)} & \boldsymbol{u}^-|_{\Gamma_0} = 0,\\
\text{e)} & -\nu\,\Delta\,\boldsymbol{u}^+ + \nabla p^+ = 0 \quad \text{in } \Omega^+,\\
\text{f)} & \operatorname{div} \boldsymbol{u}^+ = 0 \quad \text{in } \Omega^+,\\
\text{g)} & \lim_{|x|\to\infty} \boldsymbol{u}^+(x) = \boldsymbol{u}_\infty,\\
\text{h)} & \boldsymbol{u}^- = \boldsymbol{u}^+ \quad \text{on } \Gamma,\\
\text{i)} & \left(p^- + \dfrac{1}{2}\,|\boldsymbol{u}^-|^2\right)\boldsymbol{n} + 2\nu\,\mathbb{D}(\boldsymbol{u}^-)\,\boldsymbol{n} = \sigma_n(\boldsymbol{u}^+,\,p^+) \text{ on } \Gamma\,,\\
\text{j)} & \sigma_n(\boldsymbol{u}^+,\,p^+) := \sigma[\boldsymbol{u}^+,\,p^+]\boldsymbol{n}, \quad \sigma[\boldsymbol{u},\,p] = -p\mathbf{1} + 2\nu\mathbb{D}(\boldsymbol{u}).
\end{array}\right\} \quad (1.1)$$

$\sigma[\boldsymbol{u},\,p]$ denotes the hydrostatic stress tensor for the Stokes problem.

We prescribe the following data: $\nu > 0$ – constant viscosity, $\boldsymbol{f} = (f_1, f_2, f_3)$ – volume force with support $\operatorname{supp}\boldsymbol{f} \subset \Omega^-$, $\boldsymbol{u}_\infty \in \mathbb{R}^3$ – the free-stream velocity at ∞. By $\boldsymbol{n}$ we denote the unit outer normal to $\partial\Omega^-$ on Γ (pointing from Ω^- into Ω^+). $\mathbb{D}(\boldsymbol{u})$ is the velocity deformation tensor with components $\mathbb{D}_{ij}(\boldsymbol{u}) = (\partial u_i/\partial x_j + \partial u_j/\partial x_i)/2$.

In the domain Ω^- and Ω^+ the Navier–Stokes system and the Stokes system are considered, respectively. The coupling conditions on Γ representing the continuity of the velocity and the normal stress, augmented in Ω^- by the kinetic energy, were chosen in accordance with [16], [1].

1.1. Weak formulation

In order to reformulate the above problem in a weak sense, we introduce the following function spaces ([4], [6], [10], [12], [13]): $H^1(\Omega^-)$ – the Sobolev space equipped with the standard norm $\|\cdot\|_{1,\Omega^-}$, $H^{1/2}(\Gamma)$ – Sobolev–Slobodetskii space of traces $\gamma_0 u$ on Γ of functions $u \in H^1(\Omega^-)$ equipped with norm $\|\cdot\|_{1/2,\Gamma}$, $H^{-1/2}(\Gamma)$ – dual of $H^{1/2}(\Gamma)$, $W^1(\Omega^+)$ – weighted Sobolev space $= \{u;\, (1+|x|^2)^{-1/2}\, u \in L^2(\Omega^+)$, $\partial u/\partial x_i \in L^2(\Omega^+),\ i = 1,\,2,\,3\}$, equipped with the norm

$$\|u\|_{1,\Omega^+} = \left\{ \int_{\Omega^+} (1+|x|^2)^{-1/2}\, |u(x)|^2\, dx + |u|^2_{1,\Omega^+} \right\}^{1/2},$$

where the seminorm

$$|u|_{1,\Omega^+} = \Big(\int_{\Omega^+} |\nabla u|^2 \, \mathrm{d}x \Big)^{1/2}$$

is a norm equivalent to the norm $\| \cdot \|_{1,\Omega^+}$. We set $W_0^1(\Omega^+) = \{v \in W^1(\Omega^+);\ \gamma_0 \, v = 0 \text{ on } \Gamma\}$. The space $H^{1/2}(\Gamma)$ can be interpreted as the space of traces $\gamma_0 \, u$ of all $u \in W^1(\Omega^+)$ on Γ. By $\langle \cdot, \cdot \rangle$ we denote the duality between $H^{-1/2}(\Gamma)$ and $H^{1/2}(\Gamma)$ induced by the $L^2(\Gamma)$ – scalar product.

If X is a Banach space with a norm $\| \cdot \|$, then we define the space $\boldsymbol{X} = X \times X \times X$ equipped with the norm

$$\|\boldsymbol{u}\| = \Big(\sum_{i=1}^{3} \|u_i\|^2 \Big)^{1/2}, \qquad \boldsymbol{u} = (u_1, u_2, u_3) \in \boldsymbol{X}.$$

Now we set

$$\begin{aligned}
\boldsymbol{V}(\Omega^-) &= \{\boldsymbol{v} \in \boldsymbol{H}^1(\Omega^-);\ \ \boldsymbol{v}|_{\Gamma_0} = 0,\ \operatorname{div} \boldsymbol{v} = 0 \text{ in } \Omega^-\}, \\
\boldsymbol{W}_0^1(\Omega^+) &= \{\boldsymbol{v} \in \boldsymbol{W}^1(\Omega^+);\ \ \gamma_0 \boldsymbol{v} = 0 \text{ on } \Gamma\}, \\
\boldsymbol{W}(\Omega^+) &= \{\boldsymbol{v} \in \boldsymbol{W}^1(\Omega^+);\ \ \operatorname{div} \boldsymbol{v} = 0 \text{ in } \Omega^+\}, \\
\boldsymbol{V}_0(\Omega^+) &= \{\boldsymbol{v} \in \boldsymbol{W}(\Omega^+);\ \ \gamma_0 \boldsymbol{v} = 0 \text{ on } \Gamma\}, \\
\boldsymbol{H}_0^{1/2}(\Gamma) &= \{\boldsymbol{v} \in \boldsymbol{H}^{1/2}(\Gamma);\ \ \int_\Gamma \boldsymbol{v} \cdot \boldsymbol{n} \, \mathrm{d}s = 0\}\,.
\end{aligned}$$

We have $\gamma_0 \, \boldsymbol{v} \in \boldsymbol{H}_0^{1/2}(\Gamma)$ for $\boldsymbol{v} \in \boldsymbol{V}(\Omega^-)$.

It is possible to show that for any $\boldsymbol{u}_0 \in \boldsymbol{H}_0^{1/2}(\Gamma)$ there exists its extension $\mathbb{R}\, \boldsymbol{u}_0 \in \boldsymbol{W}(\Omega^+)$ such that $\gamma_0(\mathbb{R}\, \boldsymbol{u}_0) = \boldsymbol{u}_0$.

For the weak formulation we introduce the following forms:

$$\begin{aligned}
a_0(\boldsymbol{u}, \boldsymbol{v}) &= 2\nu \int_{\Omega^-} \sum_{i,j=1}^{3} D_{ij}(\boldsymbol{u}) \, D_{ij}(\boldsymbol{v}) \, \mathrm{d}x, \\
a_1(\boldsymbol{u}, \boldsymbol{w}, \boldsymbol{v}) &= \int_{\Omega^-} \sum_{i,j=1}^{3} u_j \frac{\partial w_i}{\partial x_j} v_i \, \mathrm{d}x, \\
a_2(\boldsymbol{u}, \boldsymbol{w}, \boldsymbol{v}) &= -\frac{1}{2} \int_\Gamma (\boldsymbol{u} \cdot \boldsymbol{w}) \, (\boldsymbol{v} \cdot \boldsymbol{n}) \, \mathrm{d}s, \qquad (1.2) \\
a(\boldsymbol{u}, \boldsymbol{v}) &= a_0(\boldsymbol{u}, \boldsymbol{v}) + a_1(\boldsymbol{u}, \boldsymbol{u}, \boldsymbol{v}) + a_2(\boldsymbol{u}, \boldsymbol{u}, \boldsymbol{v}), \\
&\qquad \boldsymbol{u},\ \boldsymbol{v},\ \boldsymbol{w} \in \boldsymbol{H}^1(\Omega^-), \\
a^+(\boldsymbol{z}, \boldsymbol{v}) &= \int_{\Omega^+} \sum_{i,j=1}^{3} \frac{\partial z_i}{\partial x_j} \, \frac{\partial v_i}{\partial x_j} \, \mathrm{d}x, \qquad \boldsymbol{z},\ \boldsymbol{v} \in \boldsymbol{W}^1(\Omega^+)\,.
\end{aligned}$$

Let us assume that $\boldsymbol{f} \in \boldsymbol{V}^*(\Omega^-)$ (= dual of $\boldsymbol{V}(\Omega^-)$) and denote by $\langle\cdot,\cdot\rangle_{\Omega^-}$ the duality between $\boldsymbol{V}^*(\Omega^-)$ and $\boldsymbol{V}(\Omega^-)$.

Starting from the classical formulation (1.1), using suitable (smooth) test functions (with compact supports) and Green's theorem, we arrive at the following weak formulations:

Weak formulation in Ω^-. Assume that $\sigma_n(\boldsymbol{u}^+,p^+) \in \boldsymbol{H}^{-1/2}(\Omega)$ is given. Find $\boldsymbol{u}^- \in \boldsymbol{V}(\Omega^-)$ such that

$$a(\boldsymbol{u}^-,\boldsymbol{v}) - \langle\sigma_n(\boldsymbol{u}^+,p^+),\, \gamma_0\,\boldsymbol{v}\rangle = \langle\boldsymbol{f},\,\boldsymbol{v}\rangle_{\Omega^-} \quad \forall \boldsymbol{v} \in \boldsymbol{V}(\Omega^-)\,. \tag{1.3}$$

Weak formulation in Ω^+. Assume that $\boldsymbol{u}_0 \in \boldsymbol{H}_0^{1/2}(\Gamma)$ is given. Find $\boldsymbol{u}^+$ satisfying the following conditions:

$$\begin{aligned} &\text{a)} \quad (\boldsymbol{u}^+ - \boldsymbol{u}_\infty) - I\!R(\boldsymbol{u}_0 - \boldsymbol{u}_\infty) \in \boldsymbol{V}_0(\Omega^+),\\ &\text{b)} \quad a^+(\boldsymbol{u}^+ - \boldsymbol{u}_\infty,\, \boldsymbol{v}) = 0 \quad \forall \boldsymbol{v} \in \boldsymbol{V}_0(\Omega^+). \end{aligned} \tag{1.4}$$

Using the Lax–Milgram lemma and results from [4] and [10], it is possible to establish

Theorem 1.1. *There exists a unique solution $\boldsymbol{u}^+$ of problem (1.4). This solution is independent of the choice of the extension $I\!R(\boldsymbol{u}_0 - \boldsymbol{u}_\infty) \in \boldsymbol{W}(\Omega^+)$ of $\boldsymbol{u}_0 - \boldsymbol{u}_\infty$ from Γ onto Ω^+. The velocity $\boldsymbol{u}^+$ can be associated with a uniquely determined pressure $p^+ \in L^2(\Omega^+)$ such that*

$$a^+(\boldsymbol{u}^+ - \boldsymbol{u}_\infty,\, \boldsymbol{v}) - \int_{\Omega^+} p^+ \operatorname{div} \boldsymbol{v}\, \mathrm{d}x = 0 \quad \forall \boldsymbol{v} \in \boldsymbol{W}_0^1(\Omega^+)\,. \tag{1.5}$$

■

Assuming that it is possible to define a generalization $\sigma_n(\boldsymbol{u}^+,p^+) \in \boldsymbol{H}^{-1/2}(\Gamma)$ of the normal stress for $\boldsymbol{u}^+$, p^+ from Theorem 1.1, we arrive at the **weak formulation of the coupled problem**:
Find $\boldsymbol{u}^-$, $\boldsymbol{u}^+$ satisfying (1.3) and (1.4) with

$$\boldsymbol{u}_0 = \gamma_0\,\boldsymbol{u}^- \quad \text{on } \Gamma\,. \tag{1.6}$$

1.2. Abstract problem

Let us assume for now that $\boldsymbol{u}^-$ is known. Then we solve problem (1.4) with the Dirichlet boundary condition (1.6). If the solution $\boldsymbol{u}^+$ and the pressure p^+ associated with $\boldsymbol{u}^+$ by Theorem 1.1 allow to express $\sigma_n(\boldsymbol{u}^+,p^+) \in \boldsymbol{H}^{-1/2}(\Gamma)$, we see that $\sigma_n(\boldsymbol{u}^+,p^+)$ is a function of $\boldsymbol{u}_0 = \gamma_0\,\boldsymbol{u}^-$:

$$\sigma_n(\boldsymbol{u}^+,p^+) = -\Lambda(\boldsymbol{u}_0). \tag{1.7}$$

The operator $\Lambda : \boldsymbol{H}_0^{1/2}(\Gamma) \to \boldsymbol{H}^{-1/2}(\Gamma)$ converting Dirichlet data into "Neumann" data via the solution of the exterior Stokes problem

(1.4), is called the **Steklov–Poincaré operator**. It allows us to reformulate the coupled problem as the **abstract problem**: Given $\Lambda : \boldsymbol{H}_0^{1/2}(\Gamma) \to \boldsymbol{H}^{-1/2}(\Gamma)$ and $\boldsymbol{f} \in \boldsymbol{V}^*(\Omega^-)$, find $\boldsymbol{u}^-$ such that

$$\begin{aligned} &\text{a)} \quad \boldsymbol{u}^- \in \boldsymbol{V}(\Omega^-), \\ &\text{b)} \quad a(\boldsymbol{u}^-, \boldsymbol{v}) + \langle \Lambda(\gamma_0, \boldsymbol{u}^-), \gamma_0 \boldsymbol{v}\rangle + \langle \boldsymbol{f}, \boldsymbol{v}\rangle_{\Omega^-} \quad \forall \boldsymbol{v} \in \boldsymbol{V}(\Omega^-). \end{aligned} \tag{1.8}$$

The investigation of problem (1.8) yields the following result:

Theorem 1.2. *Let the operator* Λ *be* **weakly sequentially continuous** *and* **weakly noncoercive**, *i. e.,*

$$\begin{aligned} &\boldsymbol{z}^n,\ \boldsymbol{z} \in \boldsymbol{H}_0^{1/2}(\Gamma),\ \boldsymbol{z}^n \to \boldsymbol{z} \text{ weakly in } \boldsymbol{H}^{1/2}(\Gamma) \text{ as } n \to \infty \implies \\ &\implies \langle \Lambda(\boldsymbol{z}^n), \boldsymbol{w}\rangle \longrightarrow \langle \Lambda(\boldsymbol{z}), \boldsymbol{w}\rangle \quad \forall \boldsymbol{w} \in \boldsymbol{H}^{1/2}(\Gamma) \text{ as } n \to \infty, \end{aligned} \tag{1.9}$$

and there exist constants $c_3 \in \mathbb{R}$, $c_4 \geq 0$ *such that*

$$\langle \Lambda(\boldsymbol{z}), \boldsymbol{z}\rangle \geq c_3 - c_4 \|\boldsymbol{z}\|_{1/2,\Gamma} \quad \forall \boldsymbol{z} \in \boldsymbol{H}_0^{1/2}(\Gamma), \tag{1.10}$$

respectively. Then problem (1.8) has at least one solution.

Proof. of this theorem is carried out by the Galerkin method, similarly as, e. g., in [6, Par. 8.4.20] or [12, Theorem 1.2, page 280] with the aid of the compact imbeddings $H^1(\Omega^-) \hookrightarrow\hookrightarrow L^2(\Omega^-)$, $H^1(\Omega^-) \hookrightarrow\hookrightarrow L^3(\Gamma)$, Korn's inequality and the relation $a_1(\boldsymbol{u}, \boldsymbol{v}, \boldsymbol{v}) + a_2(\boldsymbol{v}, \boldsymbol{v}, \boldsymbol{u}) = 0$ valid for all $\boldsymbol{u}$, $\boldsymbol{v} \in \boldsymbol{V}(\Omega^-)$. ∎

1.3. Properties of the Steklov–Poincaré operator Λ

It remains to establish the existence of the operator Λ and its properties (1.9) and (1.10):

Theorem 1.3. *Let* $\boldsymbol{u}^+$ *be the solution of the exterior problem (1.4) and* p^+ *be the associated pressure by relation (1.5). Then, for all* $\boldsymbol{w} \in \boldsymbol{H}^{1/2}(\Gamma)$ *and* $\boldsymbol{v} \in \boldsymbol{W}^1(\Omega^+)$ *such that* $\gamma_0 \boldsymbol{v} = \boldsymbol{w}$, *the formula*

$$\langle \sigma_n(\boldsymbol{u}^+, p^+), \boldsymbol{w}\rangle = -2\nu \int_{\Omega^+} \sum_{i,j=1}^{3} D_{ij}(\boldsymbol{u}^+)\, D_{ij}(\boldsymbol{v})\, \mathrm{d}x + \int_{\Omega^+} p^+ \operatorname{div} \boldsymbol{v}\, \mathrm{d}x \tag{1.11}$$

determines a unique element $\sigma_n(\boldsymbol{u}^+, p^+) \in \boldsymbol{H}^{-1/2}(\Gamma)$. *If* $\boldsymbol{u}^+$ *and* p^+ *are sufficiently regular, then this element can be identified with the function* $\sigma_n(\boldsymbol{u}^+, p^+)$ *defined in (1.1, j). Further, the Steklov-Poincaré operator* Λ *defined by (1.7) has properties (1.9) and (1.10).* ∎

The results of Theorem 1.1 – 1.3 imply the existence of a weak solution of the coupled problem (1.1). All details can be found in [8].

2. COUPLING OF INTERIOR NAVIER-STOKES PROBLEM WITH EXTERIOR OSEEN PROBLEM

In this section we are concerned with the modelling of viscous incompressible flow in an unbounded exterior domain with the aid of the coupling of the nonlinear Navier–Stokes equations considered in a bounded domain with the linear Oseen system in an exterior domain.

Similarly as in the case of the coupling of the Navier–Stokes problem with the Stokes problem, an important question is the choice of transmission conditions on the artificial interface Γ. The transmission condition used in Section 1 is not suitable in the case of the exterior Oseen problem and, therefore, we propose its modification resembling a "natural" boundary condition from [3]. We arrive then at the following **classical formulation of the coupled problem:** Find $\boldsymbol{u}^{\pm} = (u_1^{\pm}, \dots, u_N^{\pm}) = \overline{\Omega}^{\pm} \to I\!R^N$, $p^{\pm} : \overline{\Omega}^{\pm} \to I\!R$ such that

$$\left.\begin{array}{ll}
\text{a)} & u_i^{\pm} \in C^2(\overline{\Omega^{\pm}}),\ i = 1, \dots, N,\ p^{\pm} \in C^1(\overline{\Omega^{\pm}}), \\
\text{b)} & -\nu \Delta \boldsymbol{u}^- + (\boldsymbol{u}^- \cdot \nabla)\, \boldsymbol{u}^- + \nabla p^- = \boldsymbol{f} \ \text{ in } \Omega^-, \\
\text{c)} & \operatorname{div} \boldsymbol{u}^- = 0 \ \text{ in } \Omega^-, \\
\text{d)} & \boldsymbol{u}^-|_{\Gamma_0} = 0, \\
\text{e)} & -\nu \Delta \boldsymbol{u}^+ + (\boldsymbol{u}_\infty \cdot \nabla)\, \boldsymbol{u}^+ + \nabla p^+ = 0 \ \text{ in } \Omega^+, \\
\text{f)} & \operatorname{div} \boldsymbol{u}^+ = 0 \ \text{ in } \Omega^+, \\
\text{g)} & \lim_{|x| \to \infty} \boldsymbol{u}^+(x) = \boldsymbol{u}_\infty, \\
\text{i)} & \boldsymbol{u}^- = \boldsymbol{u}^+ \ \text{ on } \Gamma, \\
\text{j)} & -p^- \boldsymbol{n} + \nu \dfrac{\partial \boldsymbol{u}^-}{\partial \boldsymbol{n}} - \dfrac{1}{2} (\boldsymbol{u}^- \cdot \boldsymbol{n})\, \boldsymbol{u}^- = \sigma_n(\boldsymbol{u}^+, p^+) \text{ on } \Gamma\,.
\end{array}\right\} \quad (2.1)$$

Here, and throughout we understand $\sigma_n(\boldsymbol{u}, p)$ in the context of the Oseen problem as

$$\sigma_n(\boldsymbol{u}^+, p^+) := \sigma[\boldsymbol{u}^+, p^+]\boldsymbol{n} \text{ where } \sigma[\boldsymbol{u}, p] := -p\mathbf{1} + 2\nu I\!D(\boldsymbol{u}) - \tfrac{1}{2}\, \boldsymbol{u}\boldsymbol{u}_\infty^\top$$

denotes the hydrostatic stress tensor for the Oseen problem.

Other than that, we use the same notation as in Section 1.

Remark 2.1. For simplicity we consider the terms $\partial \boldsymbol{u}^{\pm}/\partial \boldsymbol{n}$ in (2.1,j), corresponding naturally to equations (2.1,b) and e). If we use the relations

$$\Delta u_i = \sum_{i=1}^{N} \frac{\partial D_{ij}(\boldsymbol{u})}{\partial x_j}, \qquad D_{ij}(\boldsymbol{u}) = \frac{1}{2}\Big(\frac{\partial u_i}{\partial x_j} + \frac{\partial u_j}{\partial x_i}\Big),$$

valid for $\boldsymbol{u} \in C^2(\Omega^\pm)$ with $\operatorname{div} \boldsymbol{u} = 0$, then $\partial \boldsymbol{u}^\pm/\partial \boldsymbol{n}$ can be replaced by $\sum_{j=1}^{N} D_{ij}(\boldsymbol{u}^\pm)\, n_j$ as in Section 1.

2.1. Weak formulation

In what follows **we will assume** that $\partial\Omega^- = \Gamma_0 \cup \Gamma$ is Lipschitz-continuous. If $\tilde{\Omega} \subset \Omega$ is a domain, then by $L^p(\tilde{\Omega})$ and $W^{k,p}(\tilde{\Omega})$ we denote the Lebesgue and Sobolev spaces, respectively, defined over $\tilde{\Omega}$ (cf., [13]). For a bounded domain $\tilde{\Omega}$ we set $W_0^{1,2}(\tilde{\Omega}) = \{v \in W^{1,2}(\Omega);\ v|_{\partial\tilde{\Omega}} =$ the trace of v on $\partial\tilde{\Omega} = 0\}$. In $W_0^{1,2}(\tilde{\Omega})$ we can use two equivalent norms

$$\|v\|_{W_0^{1,2}(\tilde{\Omega})} = \Big(\int_{\tilde{\Omega}} (|v|^2 + |\nabla v|^2)\, \mathrm{d}x\Big)^{1/2}$$

and

$$|v|_{W_0^{1,2}(\tilde{\Omega})} = \Big(\int_{\Omega} |\nabla v|^2\, \mathrm{d}x\Big)^{1/2}.$$

It is well-known that

$$W_0^{1,2}(\tilde{\Omega}) = \text{closure of } C_0^\infty(\tilde{\Omega}) \text{ in } W^{1,2}(\tilde{\Omega}),$$

where $C_0^\infty(\tilde{\Omega})$ is the space of all infinitely continuously differentiable functions with compact supports in $\tilde{\Omega}$: $\operatorname{supp} v \subset \tilde{\Omega}$ for $v \in C_0^\infty(\tilde{\Omega})$.

For the unbounded domain Ω we define the weighted Sobolev space

$$W^1(\Omega) = \Big\{u;\, (1+|x|^2)^{-1/2}\,\sigma_N\, u \in L^2(\Omega),\ \frac{\partial u}{\partial x_i} \in L^2(\Omega)\Big\},$$

where $\sigma_N(x) = 1$ for $N = 3$ and $\sigma_N(x) = |\ln(1+|x|)|^{-1}$ for $N = 2$, equipped with the norm

$$\|u\|_{W^1(\Omega)} = \Big\{\int_{\Omega} [(1+|x|^2)^{-1}\sigma_N^2\, |u|^2 + |\nabla u|^2]\, \mathrm{d}x\Big\}^{1/2},$$

which is equivalent to the seminorm

$$|u|_{W^1(\Omega)} = \Big\{\int_{\Omega} |\nabla u|^2\, \mathrm{d}x\Big\}^{1/2}.$$

(See, e. g., [4, Theorem 1, page 118] or [10, Vol. I, page 60].)

Further, we put

$$W_0^1(\Omega) = \text{closure of } C_0^\infty(\Omega) \text{ in } W^1(\Omega).$$

Then

$$W_0^1(\Omega) = \left\{ v \in W^1(\Omega); v|_{\Gamma_0} = 0 \right\}.$$

We write $v \in W^{k,p}_{\text{loc}}(\Omega)$, if $v|_{\tilde{\Omega}} = W^{k,p}(\tilde{\Omega})$ for every bounded domain $\tilde{\Omega} \subset \Omega$.

Let us define subspaces of $\boldsymbol{W}^1(\Omega)$:

$$\begin{aligned} \boldsymbol{\mathcal{V}}(\Omega) &= \{\boldsymbol{v} \in \boldsymbol{C}_0^\infty(\Omega); \operatorname{div} \boldsymbol{v} = 0 \text{ in } \Omega\}, \\ \boldsymbol{V}(\Omega) &= \text{closure of } \boldsymbol{\mathcal{V}}(\Omega) \text{ in } \boldsymbol{W}^1(\Omega). \end{aligned}$$

For functions $\boldsymbol{v}$ from subspaces of Sobolev spaces, the restrictions $\boldsymbol{v}|_\Gamma$, $\boldsymbol{v}|_{\Gamma_0}$ etc. will be understood in the sense of traces.

For $\boldsymbol{v} \in \boldsymbol{V}(\Omega)$, the limit at ∞ is zero and $\boldsymbol{v}|_{\Gamma_0} = 0$. In order to realize condition (2.1, g) in the weak formulation, we introduce a function $\boldsymbol{\phi}_\infty$ defined in the following way. Let $\mathcal{B}$ be a sufficiently large ball with centre at the origin such that $\overline{\Omega}^- \subset \mathcal{B}$. Then $\Omega^* := (\mathcal{B} \cap \Omega) - \overline{\Omega}^- \subset \Omega^+$ and $\partial\Omega^* = \Gamma \cup \Gamma^*$, where Γ and Γ^* is the interior and exterior component of $\partial\Omega^*$, respectively. Since $\int_\Gamma \boldsymbol{u}_\infty \cdot \boldsymbol{n}\, \mathrm{d}S = 0$, in virtue of [12, Lemma 2.2, page 24], there exists a function $\boldsymbol{\phi}^*$ such that

$$\boldsymbol{\phi}^* \in \boldsymbol{W}^{1,2}(\Omega^*), \quad \boldsymbol{\phi}^*|_\Gamma = 0, \quad \boldsymbol{\phi}^*|_{\Gamma^*} = \boldsymbol{u}_\infty, \quad \operatorname{div} \boldsymbol{\phi}^* = 0 \text{ in } \Omega^*.$$

Now we define $\boldsymbol{\phi}_\infty : \overline{\Omega} \to I\!R^N$:

$$\boldsymbol{\phi}_\infty = \begin{cases} 0 & \text{in } \overline{\Omega}^-, \\ \boldsymbol{\phi}^* & \text{in } \Omega^*, \\ \boldsymbol{u}_\infty & \text{in } \Omega^+ - \Omega^*. \end{cases}$$

Obviously, $\boldsymbol{\phi}_\infty \in \boldsymbol{W}^{1,2}_{\text{loc}}(\tilde{\Omega})$ and $\operatorname{div} \boldsymbol{\phi}_\infty = 0$ a. e. (= almost everywhere) in Ω.

Let us assume that $\boldsymbol{u}^\pm$, $p^\pm$ form a classical solution of the coupled problem (2.1). Let $\boldsymbol{v} \in \boldsymbol{\mathcal{V}}(\Omega)$. Multiplying equation (2.1, b) by $\boldsymbol{v}|_{\Omega^-}$ and (2.1, e) by $\boldsymbol{v}|_{\Omega^+}$, integrating over Ω^- and Ω^+, respectively, summing these integrals, applying Green's theorem and using the fact that $\operatorname{div} \boldsymbol{v} = 0$ in Ω and $\boldsymbol{v}|_{\Gamma_0} = 0$, and putting

$$\boldsymbol{u} = \begin{cases} \boldsymbol{u}^- & \text{in } \overline{\Omega}^-, \\ \boldsymbol{u}^+ & \text{in } \overline{\Omega}^+. \end{cases}$$

we obtain the identity

$$\nu \int_{\Omega^-} \sum_{i,j=1}^N \frac{\partial u_i}{\partial x_j} \frac{\partial v_i}{\partial x_j}\, \mathrm{d}x + \nu \int_{\Omega^+} \sum_{i,j=1}^N \frac{\partial u_i}{\partial x_j} \frac{\partial v_i}{\partial x_j}\, \mathrm{d}x + \int_{\Omega^-} \sum_{i,j=1}^N u_j \frac{\partial u_i}{\partial x_j} v_i\, \mathrm{d}x$$

$$+\int_{\Omega^+} \sum_{i,j=1}^{N} u_{\infty j} \frac{\partial u_i}{\partial x_j} v_i \, \mathrm{d}x - \frac{1}{2} \int_{\Gamma} [(\boldsymbol{u} - \boldsymbol{u}_\infty) \cdot \boldsymbol{n}] \, [\boldsymbol{u} \cdot \boldsymbol{v}] \, \mathrm{d}s = \int_{\Omega^-} \boldsymbol{f} \cdot \boldsymbol{v} \, \mathrm{d}x \, .$$

Let us introduce the forms

$$\left.\begin{aligned}
a_0(\boldsymbol{u}, \boldsymbol{v}) &= \nu \int_{\Omega^-} \sum_{i,j=1}^{N} \frac{\partial u_i}{\partial x_j} \frac{\partial v_i}{\partial x_j} \, \mathrm{d}x, \\
a_1(\boldsymbol{u}, \boldsymbol{w}, \boldsymbol{v}) &= \int_{\Omega^-} \sum_{i,j=1}^{N} u_j \frac{\partial w_i}{\partial x_j} v_i \, \mathrm{d}x, \\
a_2(\boldsymbol{u}, \boldsymbol{w}, \boldsymbol{v}) &= -\frac{1}{2} \int_{\Gamma} [(\boldsymbol{u} - \boldsymbol{u}_\infty) \cdot \boldsymbol{n}] \, [\boldsymbol{w} \cdot \boldsymbol{v}] \, \mathrm{d}s, \\
b_0(\boldsymbol{u}, \boldsymbol{v}) &= \nu \int_{\Omega^+} \sum_{i,j=1}^{N} \frac{\partial u_i}{\partial x_j} \frac{\partial v_i}{\partial x_j} \, \mathrm{d}x, \\
b_1(\boldsymbol{u}, \boldsymbol{v}) &= \int_{\Omega^+} \sum_{i,j=1}^{N} u_{\infty j} \frac{\partial u_i}{\partial x_j} v_i \, \mathrm{d}x, \\
L(\boldsymbol{v}) &= \int_{\Omega^-} \boldsymbol{f} \cdot \boldsymbol{v} \, \mathrm{d}x, \\
a(\boldsymbol{u}, \boldsymbol{u}, \boldsymbol{v}) &= a_0(\boldsymbol{u}, \boldsymbol{v}) + a_1(\boldsymbol{u}, \boldsymbol{u}, \boldsymbol{v}) + a_2(\boldsymbol{u}, \boldsymbol{u}, \boldsymbol{v}) \\
b(\boldsymbol{u}, \boldsymbol{v}) &= b_0(\boldsymbol{u}, \boldsymbol{v}) + b_1(\boldsymbol{u}, \boldsymbol{v}), \\
\text{for } \boldsymbol{u}, \boldsymbol{v} : \Omega &\to I\!\!R^N, \ \boldsymbol{u}, \boldsymbol{w} \in \boldsymbol{W}^{1,2}_{\mathrm{loc}}(\Omega), \ \boldsymbol{v} \in \boldsymbol{C}_0^\infty(\Omega).
\end{aligned}\right\} \tag{2.2}$$

On the basis of the above considerations we come to the following concept:

Definition 2.2. We call a vector valued function $\boldsymbol{u} : \Omega \to I\!\!R^N$ a **weak solution** of the coupled problem (2.1), if the following conditions are satisfied:

$$\begin{aligned}
&\text{a)} \quad \boldsymbol{u} - \boldsymbol{\phi}_\infty \in \boldsymbol{V}(\Omega), \\
&\text{b)} \quad a(\boldsymbol{u}, \boldsymbol{u}, \boldsymbol{v}) + b(\boldsymbol{u}, \boldsymbol{v}) = L(\boldsymbol{v}) \quad \forall \boldsymbol{v} \in \boldsymbol{\mathcal{V}}(\Omega).
\end{aligned} \tag{2.3}$$

Remark 2.3. From above it follows that the classical solution yields the weak solution. In (2.2, a), conditions (2.1, c, d, f, g) are hidden and $\boldsymbol{u} \in \boldsymbol{W}^{1,2}_{\mathrm{loc}}(\Omega)$. Since $\boldsymbol{v} \in \boldsymbol{\mathcal{V}}(\Omega)$ has compact support, all integrals over Ω in (2.2) have sense. Moreover, also the form a_2 is well defined as follows from the trace theorem for functions from $\boldsymbol{W}^{1,2}(\widetilde{\Omega})$, where $\widetilde{\Omega} \subset \Omega$ is a bounded domain with $\Gamma \subset \partial\widetilde{\Omega}$. However, it is not possible to use $\boldsymbol{v} \in \boldsymbol{V}(\Omega)$ as test functions in (2.3, b), because the form $b_1(\boldsymbol{u}, \boldsymbol{v})$ is not defined for $\boldsymbol{u} \in \boldsymbol{W}^{1,2}_{\mathrm{loc}}(\Omega)$ and $\boldsymbol{v} \in \boldsymbol{V}(\Omega)$ in general (cf. [10]). This is the reason that we cannot carry out the existence treatment as in Section 1. We apply now a completely different approach for proving the existence of a solution of problem (2.3). In fact, this new technique can also be applied to the coupling of the interior Navier–Stokes problem with the exterior Stokes problem. (Details will appear in [9].)

Remark 2.4. On the basis of results from [10], Chap. VII, the weak solution $\boldsymbol{u}$ of problem (2.3) can be associated with the pressure $p \in L^2_{\mathrm{loc}}(\Omega)$ such that

$$a(\boldsymbol{u},\boldsymbol{v}) - (p, \operatorname{div} \boldsymbol{v}) = L(\boldsymbol{v}) \quad \forall \boldsymbol{v} \in \boldsymbol{C}_0^\infty(\Omega). \tag{2.4}$$

2.2. Existence of a weak solution

First we prove some important properties of the forms a_0, a_1, a_2 defined in (2.2). These forms have sense, of course, also for functions from the space $\boldsymbol{W}^{1,2}(\Omega^-)$, as follows from the continuous imbedding $\boldsymbol{W}^{1,2}(\Omega^-) \hookrightarrow \boldsymbol{L}^4(\Omega^-)$ and the continuity of the trace operator from the space $\boldsymbol{W}^{1,2}(\Omega^-)$ into $\boldsymbol{L}^3(\Gamma)$. (We simply write $\boldsymbol{W}^{1,2}(\Omega^-) \hookrightarrow \boldsymbol{L}^3(\Gamma)$.)

Let us set

$$\begin{aligned}
\boldsymbol{V}(\Omega^-) &= \left\{ \boldsymbol{v} \in \boldsymbol{W}^{1,2}(\Omega^-); v|_{\Gamma_0} = 0, \operatorname{div} \boldsymbol{v} = 0 \ \text{ a.e. in } \Omega^- \right\}, \\
\mathcal{V}_0(\Omega^-) &= \left\{ \boldsymbol{v} \in \boldsymbol{C}^\infty(\overline{\Omega}^-); \operatorname{supp} \boldsymbol{v} \subset \Omega^- \cup \Gamma, \operatorname{div} \boldsymbol{v} = 0 \ \text{ in } \Omega^- \right\}, \\
\tilde{a}(\boldsymbol{u},\boldsymbol{v}) &= -\frac{1}{2} \int_\Gamma (\boldsymbol{u} \cdot \boldsymbol{n})\, |\boldsymbol{v}|^2 \, ds, \quad \boldsymbol{u}, \boldsymbol{v} \in \boldsymbol{W}^{1,2}(\Omega^-).
\end{aligned}$$

Lemma 2.5. *a_0 is a continuous bilinear form on $\boldsymbol{W}^{1,2}(\Omega^-)$. Further, a_1 and a_2 are continuous trilinear forms on $\boldsymbol{W}^{1,2}(\Omega^-)$.*

For $\boldsymbol{u}$, $\boldsymbol{v}$, $\boldsymbol{w} \in \boldsymbol{V}(\Omega^-)$ we have

$$a_1(\boldsymbol{u},\boldsymbol{v},\boldsymbol{w}) = -a_1(\boldsymbol{u},\boldsymbol{w},\boldsymbol{v}) - \tilde{a}(\boldsymbol{u},\boldsymbol{v}+\boldsymbol{w}) + \tilde{a}(\boldsymbol{u},\boldsymbol{v}) + \tilde{a}(\boldsymbol{u},\boldsymbol{w}). \tag{2.5}$$

Let us define the form

$$d(\boldsymbol{u},\boldsymbol{v},\boldsymbol{w}) = a_1(\boldsymbol{u},\boldsymbol{v},\boldsymbol{w}) + a_2(\boldsymbol{u},\boldsymbol{v},\boldsymbol{w}), \quad \boldsymbol{u}, \boldsymbol{v}, \boldsymbol{w} \in \boldsymbol{W}^{1,2}(\Omega^-). \tag{2.6}$$

Then it holds: If $\boldsymbol{z}$, $\boldsymbol{v}$, $\boldsymbol{z}_n \in \mathcal{V}_0(\Omega^-)$, $n = 1, 2, \ldots$, and if

$$\begin{aligned}
&\text{a)} \quad |\boldsymbol{z}_n|_{\mathbf{W}^{1,2}(\Omega^-)} \le C, \quad n = 1, 2, \ldots, \\
&\text{b)} \quad \boldsymbol{z}_n \longrightarrow \boldsymbol{z} \ \textit{ strongly in } \boldsymbol{L}^2(\Omega^-) \\
&\text{c)} \quad \boldsymbol{z}_n|_\Gamma \longrightarrow \boldsymbol{z}|_\Gamma \ \textit{ strongly in } \boldsymbol{L}^3(\Gamma) \ \textit{ as } n \to \infty,
\end{aligned} \tag{2.7}$$

then

$$d(\boldsymbol{z}_n, \boldsymbol{z}_n, \boldsymbol{v}) \longrightarrow d(\boldsymbol{z}, \boldsymbol{z}, \boldsymbol{v}) \quad \textit{as } n \to \infty. \tag{2.8}$$

The solvability of the coupled problem in the unbounded domain Ω is established with the aid of coupled problems considered on a monotone sequence of bounded subdomains. For any positive integer n we denote by $\mathcal{B}_n$ the ball with radius n and centre at the origin. We will consider $n \geq n_0$ with fixed n_0 such that $\mathcal{B} \subset \mathcal{B}_{n_0}$ $(\subset \mathcal{B}_n)$, where $\mathcal{B}$ is the ball used in the definition of the function $\boldsymbol{\phi}_\infty$. Hence, $\partial\mathcal{B}_n \subset \Omega^+$ and $\boldsymbol{\phi}_\infty|_{\partial\mathcal{B}_n} = \boldsymbol{u}_\infty$ for $n \geq n_0$. We set $\Omega_n = \Omega \cap \mathcal{B}_n$ and $\Omega_n^+ = \Omega^+ \cap \mathcal{B}_n$. Then for $n \geq n_0$, we have $\Omega^- \subset \Omega_n$, $\Omega_n = \Omega^- \cup \Gamma \cup \Omega_n^+$, $\partial\Omega_n = \Gamma_0 \cup \Gamma_n$ and $\partial\Omega_n^+ = \Gamma \cup \Gamma_n$. Moreover, $\Omega_n \subset \Omega_{n+1}$ and $\bigcup_{n=n_0}^{\infty} \Omega_n = \Omega$. Γ_n is the exterior component of $\partial\Omega_n$ and $\partial\Omega_n^+$.

For $n \geq n_0$ we define the forms

$$\begin{aligned} b_0^n(\boldsymbol{u}, \boldsymbol{v}) &= \nu \int_{\Omega_n^+} \sum_{i,j=1}^{N} \frac{\partial u_i}{\partial x_j} \frac{\partial v_i}{\partial x_j} \, \mathrm{d}x, \\ b_1^n(\boldsymbol{u}, \boldsymbol{v}) &= \int_{\Omega_n^+} \sum_{i,j=1}^{N} \phi_{\infty j} \frac{\partial u_i}{\partial x_j} v_i \, \mathrm{d}x, \\ a^n(\boldsymbol{u}, \boldsymbol{v}) &= a_0(\boldsymbol{u}, \boldsymbol{v}) + a_1(\boldsymbol{u}, \boldsymbol{u}, \boldsymbol{v}) + a_2(\boldsymbol{u}, \boldsymbol{u}, \boldsymbol{v}) + b_0^n(\boldsymbol{u}, \boldsymbol{v}) + b_1^n(\boldsymbol{u}, \boldsymbol{v}), \\ & \qquad \boldsymbol{u}, \ \boldsymbol{v} \in \boldsymbol{W}^{1,2}(\Omega_n). \end{aligned} \tag{2.9}$$

For every $n \geq n_0$ we introduce the spaces

$$\begin{aligned} \boldsymbol{\mathcal{V}}(\Omega_n) &= \{\boldsymbol{v} \in \boldsymbol{C}_0^\infty(\Omega_n);\ \operatorname{div} \boldsymbol{v} = 0 \text{ in } \Omega_n\}, \\ \boldsymbol{V}(\Omega_n) &= \text{closure of } \boldsymbol{\mathcal{V}}(\Omega_n) \text{ in } \boldsymbol{W}^{1,2}(\Omega_n) \\ &= \left\{\boldsymbol{v} \in \boldsymbol{W}_0^{1,2}(\Omega_n);\ \operatorname{div} \boldsymbol{v} = 0 \text{ in } \Omega_n\right\}, \end{aligned}$$

and consider the following **auxiliary problem** in Ω_n: Find $\boldsymbol{u}_n : \Omega_n \to \mathbb{R}^N$ such that

$$\begin{aligned} &\text{a)} \quad \boldsymbol{u}_n - \boldsymbol{\phi}_\infty|_{\Omega_n} \in \boldsymbol{V}(\Omega_n), \\ &\text{b)} \quad a^n(\boldsymbol{u}_n, \boldsymbol{v}) = L(\boldsymbol{v}) \quad \forall \boldsymbol{v} \in \boldsymbol{V}(\Omega_n) \end{aligned} \tag{2.10}$$

(the form $L(\boldsymbol{v})$ has sense for $\boldsymbol{v} \in \boldsymbol{V}(\Omega_n)$ extended by zero on Ω). Conditions (2.10) represent the weak formulation of a coupled "Navier–Stokes – Oseen" problem in the bounded domain $\Omega_n = \Omega^- \cup \Gamma \cup \Omega_n^+$.

The solution of problem (2.3) can be written in the form

$$\boldsymbol{u} = \boldsymbol{\phi}_\infty + \boldsymbol{z}, \quad \boldsymbol{z} \in \boldsymbol{V}(\Omega). \tag{2.11}$$

Hence, (2.3) is equivalent to finding $\boldsymbol{z} : \Omega \to \mathbb{R}^N$ such that

$$\begin{aligned} &\text{a)} \quad \boldsymbol{z} \in \boldsymbol{V}(\Omega), \\ &\text{b)} \quad a(\boldsymbol{\phi}_\infty + \boldsymbol{z}, \boldsymbol{v}) = L(\boldsymbol{v}) \quad \forall \boldsymbol{v} \in \boldsymbol{\mathcal{V}}(\Omega). \end{aligned} \tag{2.12}$$

Similarly we can reformulate problem (2.10): Find $\boldsymbol{z}_n : \Omega_n \to \mathbb{R}^N$ such that

$$\begin{aligned} &\text{a)} \quad \boldsymbol{z}_n \in \boldsymbol{V}(\Omega_n), \\ &\text{b)} \quad a^n(\boldsymbol{\phi}_\infty + \boldsymbol{z}_n, \boldsymbol{v}) = L(\boldsymbol{v}) \quad \forall \boldsymbol{v} \in \boldsymbol{V}(\Omega_n). \end{aligned} \tag{2.13}$$

Then $\boldsymbol{u}_n = \boldsymbol{\phi}_\infty + \boldsymbol{z}_n$. From the definition of $\boldsymbol{\phi}_\infty$ it follows that $\boldsymbol{u}_n = \boldsymbol{z}_n$ in $\overline{\Omega}^-$.

The solvability of the above auxiliary problems is proved with the aid of the following results:

Lemma 2.6. *For each $\boldsymbol{z} \in \boldsymbol{V}(\Omega_n)$ we have*

$$a_1(\boldsymbol{z}, \boldsymbol{z}, \boldsymbol{z}) + a_2(\boldsymbol{z}, \boldsymbol{z}, \boldsymbol{z}) + b_1^n(\boldsymbol{\phi}_\infty + \boldsymbol{z}, \boldsymbol{z}) = \int_{\Omega_n^+} \sum_{i,j=1}^N u_{\infty j} \frac{\partial \phi_{\infty i}}{\partial x_j} z_i \, dx \, .$$

Theorem 2.7. *For each $n \geq n_0$ problem (2.13) has at least one solution $\boldsymbol{z}_n$. There exists a constant $K > 0$ independent of n such that*

$$|\boldsymbol{z}_n|_{\mathbf{W}^{1,2}(\Omega_n)} \leq K, \quad n \geq n_0. \tag{2.14}$$

Proof. is carried out with the aid of the Galerkin method in a standard way as, e. g., in [12], Theorem 1.2, page 280, [17], Chap. II, or [6], Par. 8.4.20. ■

The main result of this section reads:

Theorem 2.8. *There exists at least one solution $\boldsymbol{u}$ of problem (2.3) This $\boldsymbol{u}$ is a weak solution of the coupled problem (2.1).*

Proof. As was stated above, problem (2.3) is equivalent to problem (2.12). In order to prove the solvability of problem (2.12), we extend the solution $\boldsymbol{z}_n$ of problem (2.13) ($n \geq n_0$) by zero from the domain Ω_n onto Ω. For simplicity, we will denote this extension again by $\boldsymbol{z}_n$. Hence, we have a sequence $\{\boldsymbol{z}_n\}_{n=n_0}^\infty$ such that

$$\begin{aligned} &\boldsymbol{z}_n \in \boldsymbol{V}(\Omega), \quad n \geq n_0, \\ &|\boldsymbol{z}_n|_{\mathbf{W}^1(\Omega)} = |\boldsymbol{z}_n|_{\mathbf{W}^{1,2}(\Omega_n)} \leq K, \quad n \geq n_0. \end{aligned} \tag{2.15}$$

Since the space $\boldsymbol{V}(\Omega)$ is reflexive and the sequence $\{\boldsymbol{z}_n\}_{n=n_0}^\infty$ is bounded in $\boldsymbol{V}(\Omega)$, there exists $\boldsymbol{z} \in \boldsymbol{V}(\Omega)$ and a subsequence of $\{\boldsymbol{z}_n\}_{n=n_0}^\infty$ (let us denote it again by $\{\boldsymbol{z}_n\}$) such that

$$\boldsymbol{z}_n \longrightarrow \boldsymbol{z} \quad \text{weakly in } \boldsymbol{V}(\Omega) \text{ as } n \to \infty. \tag{2.16}$$

Our goal is to show that z is a solution of problem (2.12).

Let $v \in \mathcal{V}(\Omega)$. Then there exists $n^* \geq n_0$ such that $\operatorname{supp} v \subset \Omega_{n^*}$ and, in virtue of (2.13), (2.2) and (2.9) we have $v|_{\Omega_n} \in V(\Omega_n)$ for $n \geq n^*$ and

$$a(\phi_\infty + z_n, v) = a^{n^*}(\phi_\infty + z_n, v) = a^n(\phi_\infty + z_n, v) = L(v), \quad n \geq n^*. \tag{2.17}$$

Taking into account that $|z_n|_{W^{1,2}(\Omega_{n^*})} \leq |z_n|_{W^1(\Omega)}$, from (2.15) we see that the sequence $\{z_n|_{\Omega_{n^*}}\}$ is bounded in $W^{1,2}(\Omega_{n^*})$. Thus, we can suppose that

$$z_n|_{\Omega_{n^*}} \longrightarrow z|_{\Omega_{n^*}} \quad \text{weakly in } W^{1,2}(\Omega_{n^*}) \quad \text{as } n \to \infty. \tag{2.18}$$

This and the compact imbeddings $W^{1,2}(\Omega_{n^*}) \hookrightarrow\hookrightarrow L^2(\Omega_{n^*})$ and $W^{1,2}(\Omega_{n^*}) \hookrightarrow\hookrightarrow L^3(\Gamma)$ imply that

$$\begin{aligned} z_n|_{\Omega_{n^*}} &\longrightarrow z|_{\Omega_{n^*}} \text{ strongly in } L^2(\Omega_{n^*}), \\ z_n|_\Gamma &\longrightarrow z|_\Gamma \text{ strongly in } L^3(\Gamma), \quad \text{as } n \to \infty. \end{aligned} \tag{2.19}$$

Now we are ready to carry out the limit process in (2.17) for $n \to \infty$. Linearity and continuity of the forms $a_0(\phi_\infty + \cdot, v) = a_0(\cdot, v)$, $b_0(\phi_\infty + \cdot, v)$ and $b_1^{n^*}(\phi_\infty + \cdot, v)$ (let us remind that $\phi_\infty = 0$ in Ω^-) imply that

$$\begin{aligned} a_0(\phi_\infty + z_n, v) = a_0(z_n, v) &\longrightarrow a_0(z, v) = a_0(\phi_\infty + z, v), \\ b_0^{n^*}(\phi_\infty + z_n, v) &\longrightarrow b_0^{n^*}(\phi_\infty + z, v), \\ b_1^{n^*}(\phi_\infty + z_n, v) &\longrightarrow b_1^{n^*}(\phi_\infty + z, v) \text{ as } n \to \infty. \end{aligned} \tag{2.20}$$

From (2.15) and (2.19) we see that the sequence $\{z_n\}_{n=n_0}^{\infty}$ satisfies conditions (2.7, a–c). This and Lemma 2.5 imply that

$$a_1(z_n, z_n, v) + a_2(z_n, z_n, v) \longrightarrow a_1(z, z, v) + a_2(z, z, v) \quad \text{as } n \to \infty. \tag{2.21}$$

Now, from (2.17), (2.20) and (2.21) we conclude that the function $z \in V(\Omega)$ satisfies the identity

$$a(\phi_\infty + z, v) = L(v) \quad \text{for all } v \in \mathcal{V}(\Omega),$$

which means that z is a solution of problem (2.12) and $u = \phi_\infty + z$ is a solution of problem (2.3), which we wanted to prove. ∎

3. FORMULATION OF THE COUPLED PROBLEMS WITH THE AID OF BOUNDARY INTEGRAL EQUATIONS

The fact that the Stokes equations as well as the Oseen equations possess fundamental solutions allows us to reformulate the exterior

Stokes and Oseen problem as integral equation on the coupling interface Γ. This may be used to reduce the coupled problems on the unbounded domain analyzed above to equivalent problems in the bounded domain Ω^- which are equipped with nonlocal boundary conditions on Γ. In this section, we derive explicit representations of the nonlocal boundary operators in terms of the Calderón Projector of the linear exterior problem which describes the far-field. The nonlocal boundary operators for the Navier-Stokes equations coupled with the exterior Stokes and Oseen problems will turn out to be strongly elliptic boundary integral operators which can be discretized by Galerkin boundary element methods. This approach was used for the solution of a number of elliptic problems in exterior domains in e.g., [2, 7, 11, 16].

As it is well-known, there are generally many possible approaches to reformulate exterior boundary value problems in terms of boundary integral equations. Correspondingly there are many ways to represent the Poincaré-Steklov operators. For the exterior Stokes problem of Section 1, we present a formulation in terms of single layer potentials based on indirect boundary reduction by potentials. The resulting representation of the Poincaré Steklov operator requires the inversion of a coercive, self-adjoint boundary integral operator of order -1 on Γ.

For the Oseen problem, there is an analogous formulation; however, the coercivity of the first kind boundary operator to be inverted is open — only a weaker Gårding-Inequality can be established then. Therefore, we present a different formulation based on a pure double layer ansatz for the exterior velocity field $\boldsymbol{u}^+$ in the Oseen problem [5]. Contrary to the Stokes problem, this is admissible in the Oseen case due to the different decay behaviour of the Oseen fundamental solution as $|x| \to \infty$. Here, the boundary reduction is direct, via the Faxén-formulas on Γ.

3.1. Exterior Stokes Problem

For the integral equation of the exterior Stokes problem, we shall require hydrodynamic potentials that are defined in terms of fundamental solutions of the Stokes operator (1.1, e). We shall in particular require the **velocity fundamental tensor $\boldsymbol{E}(z)$** given by

$$E_{ij}(z) = (\delta_{ij}\,\Delta - \partial_i \partial_j)\,\Phi(z), \quad z \in I\!R^3 \backslash \{0\} \tag{3.1}$$

where $1 \le i, j \le 3$ and $\Phi(z) = \Phi_{St}(z) := |z|/(8\pi\nu)$.

Further, we shall also use the **pressure fundamental vector $\boldsymbol{e}(z)$** given by

$$e_i(z) = -\frac{1}{4\pi}\,\partial_i\Big(\frac{1}{|z|}\Big) = \frac{1}{4\pi}\,\frac{z_i}{|z|^3} \quad \text{where } 1 \le i \le 3\,. \tag{3.2}$$

To obtain an expression of Λ in (1.8) in terms of boundary integral operators, we require a certain factor space of $\boldsymbol{H}^{-1/2}(\Gamma)$: we set

$$\boldsymbol{T} := \boldsymbol{H}^{-1/2}(\Gamma)/\mathcal{R} \tag{3.3}$$

where $\mathcal{R}$ denotes the equivalence relation

$$\boldsymbol{t} \sim \boldsymbol{t}' \Longleftrightarrow \boldsymbol{t} = \boldsymbol{t}' + \lambda \boldsymbol{n} \tag{3.4}$$

for some $\lambda \in I\!R$ (recall that $\boldsymbol{n}$ denotes the exterior unit normal to Ω^-, pointing into Ω^+). Then there holds

Theorem 3.1. *Assume that the coupling boundary Γ is smooth. The solution of the exterior Stokes Problem (1.1, e)–(1.1, h) in Ω^+ can be represented in the form of the Odqvist hydrodynamic potentials*

$$u_i^+(x) = u_{\infty,i} + \sum_{k=1}^{3} \int_{y\in\Gamma} E_{ki}(x-y) t_k(y)\, ds_y, \quad x \in \Omega^+, \quad i = 1,2,3, \tag{3.5}$$

$$p^+(x) = \sum_{k=1}^{3} \int_{y\in\Gamma} e_k(x-y)\, t_k(y) ds_y, \quad x \in \Omega^+ \tag{3.6}$$

for some boundary densities $\boldsymbol{t} \in \boldsymbol{H}^{-1/2}(\Gamma)$ which are the unique solutions of the first kind boundary integral equations:

$$u_{\infty,i} + \sum_{k=1}^{3} \int_{y\in\Gamma} t_k(y)\, E_{ki}(x-y) ds_y = u_i^+(x), \; i = 1,2,3, \; x \in \Gamma\,, \tag{3.7}$$

or, more precisely, in variational form: find $\boldsymbol{t} \in \boldsymbol{T}$ such that

$$b(\boldsymbol{t}, \boldsymbol{t}') = \langle \boldsymbol{t}', \boldsymbol{u}^+ - \boldsymbol{u}_\infty \rangle \quad \forall \boldsymbol{t}' \in \boldsymbol{T} \tag{3.8}$$

where the bilinear form $b(\boldsymbol{t}, \boldsymbol{t}')$, given by

$$b(\boldsymbol{t}, \boldsymbol{t}') = \sum_{i,j=1}^{3} \int_\Gamma \int_\Gamma t_i(x)\, E_{ij}(x-y)\, t_j'(y) ds_y\, ds_x\,, \tag{3.9}$$

is symmetric and coercive on $\boldsymbol{T}$: there exists $\beta > 0$ such that

$$b(\boldsymbol{t}, \boldsymbol{t}) \geq \beta\, \|\boldsymbol{t}\|^2_{H^{-\frac{1}{2}}(\Gamma)} \quad \forall \boldsymbol{t} \in \boldsymbol{T}. \tag{3.10}$$

For the proof, we refer to [4], Chap. VI, Theorem 1. We remark that the symmetry and coercivity of the bilinear form $b(\cdot,\cdot)$ in (3.9) gives, upon discretization with a Galerkin boundary element method on Γ, a symmetric and positive-definite stiffness matrix corresponding to the hydrodynamic single layer operator $\boldsymbol{S}$ on the left hand side of (3.7). The operator $\boldsymbol{S}$ is continuous from $\boldsymbol{H}^{-1/2}(\Gamma) \to \boldsymbol{H}^{1/2}(\Gamma)$. Using (3.7) and (1.6), we get the nonlocal boundary condition

$$\boldsymbol{t} \in \boldsymbol{T}: \ \langle \boldsymbol{t}', \boldsymbol{S}\,\boldsymbol{t}\rangle = \langle \boldsymbol{t}', \gamma_0 \boldsymbol{u}^- - \boldsymbol{u}_\infty\rangle \quad \forall \boldsymbol{t}' \in \boldsymbol{T} \tag{3.11}$$

where $\langle \cdot, \cdot \rangle$ denotes the $\boldsymbol{H}^{-1/2}(\Gamma) \times \boldsymbol{H}^{1/2}(\Gamma)$ duality pairing. By (3.10), $\boldsymbol{S}$ is invertible on $\boldsymbol{T}$ and we get that

$$\boldsymbol{t} = \boldsymbol{S}^{-1}(\boldsymbol{u}^- - \boldsymbol{u}_\infty)\,. \tag{3.12}$$

Having obtained $\boldsymbol{t}$ by (3.8), the exterior Stokes flow $(\boldsymbol{u}^+, p^+)$ is given by (3.5). In particular, we get with normal stress operator $\sigma_n(\boldsymbol{u}^+, p^+)$, applied to (3.5), (3.6), for a point $x_0 \in \Gamma$ with the jump relations of the Odqvist potentials that

$$\begin{aligned} \sigma_n(\boldsymbol{u}^+, p^+)(x_0) &= \lim_{\varepsilon \to 0+} \sigma_n(\boldsymbol{u}^+, p^+)(x_0 + \varepsilon \boldsymbol{n}) \\ &= \frac{1}{2}\,\boldsymbol{t}(x_0) + \text{p.v.} \int_{y \in \Gamma} \boldsymbol{n}(x_0)\, \sigma_{x_0}[\boldsymbol{E}, \boldsymbol{e}](x_0 - y)\, \boldsymbol{t}(y)\, ds_y \\ &= \Big(\Big(\frac{1}{2}\,\boldsymbol{I} + \boldsymbol{K}'\Big)\,\boldsymbol{t}\Big)(x_0) \end{aligned} \tag{3.13}$$

where the integral over Γ has to be understood in the Cauchy principal value sense and the subscript x_0 indicates that the differentiations are with respect to x_0. The expression $\boldsymbol{n}\,\sigma[\boldsymbol{E}, \boldsymbol{e}]\,\boldsymbol{t}$ is interpreted as the vector with components $\sum_{j,k=1}^{3} n_j\, \sigma_{ijk}\, t_k$, $i = 1, 2, 3$, where σ_{ijk}, $i, j = 1, 2, 3$, are components of the tensor $\sigma(\boldsymbol{E}_k, e_k)$, using the notation $\boldsymbol{E}_k$ and e_k for the k–th row of $\boldsymbol{E}$ and the k–th component of $\boldsymbol{e}$, respectively.

We therefore obtain with the weak formulation (1.3) in Ω^- the following, nonlocal boundary problem in $\Omega^- \cup \Gamma$ which is equivalent to the weak formulation of the coupled problem (1.8):
Find $\boldsymbol{u}^- \in \boldsymbol{V}(\Omega^-)$, $\boldsymbol{t} \in \boldsymbol{T}$ such that

$$\begin{aligned} a(\boldsymbol{u}^-, \boldsymbol{v}) - \Big\langle \Big(\frac{1}{2}\,\boldsymbol{I} + \boldsymbol{K}'\Big)\,\boldsymbol{t},\, \gamma_0\,\boldsymbol{v}\Big\rangle &= \langle \boldsymbol{f}, \boldsymbol{v}\rangle_{\Omega^-} \quad \forall \boldsymbol{v} \in \boldsymbol{V}(\Omega^-) \\ -\langle \boldsymbol{t}', \gamma_0\,\boldsymbol{u}^- - \boldsymbol{u}_\infty\rangle + \langle \boldsymbol{t}', \boldsymbol{S}\boldsymbol{t}\rangle &= 0 \qquad\qquad\quad \forall \boldsymbol{t}' \in \boldsymbol{T}\,. \end{aligned} \tag{3.14}$$

With (3.12) and (3.13) we obtain the representation of the Steklov–Poincaré operator in terms of boundary integral operators

$$\Lambda(\boldsymbol{u}^-) = -\sigma_n(\boldsymbol{u}^+, p^+) = -\Big(\frac{1}{2}\,\boldsymbol{I} + \boldsymbol{K}'\Big)\,\boldsymbol{S}^{-1}(\boldsymbol{u}^- - \boldsymbol{u}_\infty)\,. \tag{3.15}$$

Naturally, in a numerical implementation of the nonlocal boundary condition (1.1, i) in (1.3), the discrete inverse of $\boldsymbol{S}$ should not be explicitly calculated, but rather realized numerically by a fast algorithm.

3.2. Exterior Oseen Problem

We consider now the exterior Oseen Problem (2.1, e) - (2.1, j). We will use once again the Odqvist hydrodynamic potentials to reduce the coupled problem (2.1) to a nonlocal boundary value problem in $\Omega^- \cup \Gamma$. We shall now, however, not use a single layer ansatz (the so-called "indirect" method of boundary reduction), but rather the "direct" method based on the Faxén representation formula on Γ, leading to the "one integral equation" approach of [11].

To do so, we require once more for the exterior Oseen problem the velocity fundamental tensor $\boldsymbol{E}(z)$ and the pressure fundamental vector $\boldsymbol{e}(z)$. To define them, we assume without loss of generality that

$$\boldsymbol{u}_\infty = (u_\infty, 0, 0)^\top \,. \tag{3.16}$$

Then $\boldsymbol{E}$ and $\boldsymbol{e}$ are once more defined by (3.1), (3.2), however now with $\Phi(z)$ given by [15]

$$\Phi_{Os}(z) := \frac{1}{4\pi u_\infty} \int\limits_0^{u_\infty s(z)/2\nu} (1 - e^{-\alpha})\alpha^{-1} d\alpha, \quad ; s(z) := |z| - z_1 \,. \tag{3.17}$$

We recall further that for the Oseen problem the hydrostatic stress is given by

$$\sigma[\boldsymbol{u}, p] := -p\,\mathbf{1} + 2\nu\, \mathbb{D}(\boldsymbol{u}) - \frac{1}{2}\,\boldsymbol{u}\boldsymbol{u}_\infty^\top \,. \tag{3.18}$$

We shall also require the **adjoint stress operator**

$$\sigma^*[\boldsymbol{v}, q] := q\,\mathbf{1} + 2\nu\, \mathbb{D}(\boldsymbol{v}) + \frac{1}{2}\,\boldsymbol{v}\boldsymbol{u}_\infty^\top \,. \tag{3.19}$$

Then there holds the **Faxén representation formula**:
Any $(\boldsymbol{u}^+ - \boldsymbol{u}_\infty, p^+) \in \boldsymbol{H}^2_{\text{loc}}(\Omega^+) \times H^1_{\text{loc}}(\Omega^+)$ solving (2.1, e) - (2.1, j) can be represented in the form: for any $x \in \Omega^+$

$$\begin{aligned} \boldsymbol{u}^+(x) - \boldsymbol{u}_\infty = \int\limits_{y\in\Gamma} \{(\boldsymbol{u}^+(y) - \boldsymbol{u}_\infty)\, \sigma^*_y[\boldsymbol{E}, \boldsymbol{e}](x - y)\,\boldsymbol{n} - \\ - \boldsymbol{E}(x - y)\sigma_y[\boldsymbol{u}^+ - \boldsymbol{u}_\infty, p]\,\boldsymbol{n})\} ds_y, \end{aligned} \tag{3.20}$$

$$p^+(x) = \int\limits_{y\in\Gamma} \{(\boldsymbol{u}^+ - \boldsymbol{u}_\infty)\, \sigma^*_y[\boldsymbol{e}, e^*](x - y)\,\boldsymbol{n} - \boldsymbol{e}(x - y)\,\sigma_y[\boldsymbol{u}^+ - \boldsymbol{u}_\infty, p]\,\boldsymbol{n})\} ds_y, \tag{3.21}$$

where $p(x)$ is determined only mod $\mathbb{R}$, σ_y, σ^*_y are as in (3.18), (3.19), with the subscript y indicating that the differentiations are with respect to y and where

$$e^*(z) = \frac{u_\infty}{4\pi}\,\partial_1\Big(\frac{1}{|z|}\Big) = -u_\infty\, e_1(z) \tag{3.22}$$

is the pressure corresponding to the velocity field $\boldsymbol{e}(z)$. The expression $(\boldsymbol{u}^+ - \boldsymbol{u}_\infty)\,\sigma^*[\boldsymbol{E},\boldsymbol{e}]\,\boldsymbol{n}$ is interpreted in an analogous way as $\boldsymbol{n}\,\sigma[\boldsymbol{E},\boldsymbol{e}]\,\boldsymbol{t}$ in Par. 3.1.

We observe that the leading singularities of $\boldsymbol{E}(z)$ and of $\boldsymbol{e}(z)$ at $|z| = 0$ in the Oseen and the Stokes case are identical. More precisely, for small $|z|$

$$\boldsymbol{E}(z) = (\delta_{ij}\Delta - \partial_i\partial_j)\,\Phi_{Os}(z) = \frac{1}{8\pi\nu}\,(\delta_{ij}\Delta - \partial_i\partial_j)\,|z| + O(1)\,. \quad (3.23)$$

The hydrodynamic potentials admit therefore the same jump relations in the Stokes and the Oseen case. We reduce the exterior Oseen problem to Γ by passing with x in (3.20) to $x_0 \in \Gamma$: For any $x_0 = \lim\limits_{\varepsilon\to 0+} x_0 + \varepsilon\,\boldsymbol{n}(x_0) \in \Gamma$, the jump relations give

$$\begin{aligned} &\lim_{\epsilon\to 0+} \boldsymbol{u}^+(x_0+\varepsilon\boldsymbol{n}) - \boldsymbol{u}_\infty = \boldsymbol{u}^+(x_0) - \boldsymbol{u}_\infty = \\ &-\int\limits_{y\in\Gamma} \boldsymbol{E}(x_0-y)\sigma_n(\boldsymbol{u}^+ - \boldsymbol{u}_\infty, p^+)(y)ds_y - \frac{1}{2}\,(\boldsymbol{u}^+(x_0) - \boldsymbol{u}_\infty) \\ &-\text{p.v.} \int\limits_{y\in\Gamma} \{(\boldsymbol{u}^+(y) - \boldsymbol{u}_\infty)\cdot\sigma_y^*[\boldsymbol{E},\boldsymbol{e}](x_0-y)\,\boldsymbol{n}(y)\}ds_y \end{aligned}$$

or, symbolically,

$$\gamma_0\boldsymbol{u}^+(x_0) - \boldsymbol{u}_\infty = -(\boldsymbol{S}\,\sigma_n(\boldsymbol{u}^+ - \boldsymbol{u}_\infty, p^+))(x_0) + \Big(\frac{1}{2}\,\boldsymbol{I} + \boldsymbol{K}\Big)(\boldsymbol{u}^+ - \boldsymbol{u}_\infty)(x_0)\,, \quad (3.24)$$

where $\boldsymbol{K}$ denotes the hydrodynamic double layer operator, or, equivalently

$$\Big(\frac{1}{2}\,\boldsymbol{I} - \boldsymbol{K}\Big)(\boldsymbol{u}^+ - \boldsymbol{u}_\infty) = -\boldsymbol{S}\,\sigma_n\,. \quad (3.25)$$

We emphasize that now $\boldsymbol{S}$ is neither symmetric nor coercive, generally. Using the continuity of the velocities (1.6), and casting (3.25) in weak form, we find the integral equation for the hydrodynamic normal stress σ_n corresponding to the exterior Oseen problem due to the velocity $\boldsymbol{u}^-$ on Γ:

$$\boldsymbol{\sigma}_n \in \boldsymbol{H}^{-1/2}(\Gamma) : \langle\boldsymbol{\tau}, \boldsymbol{S}\boldsymbol{\sigma}_n\rangle + \Big\langle\boldsymbol{\tau}, \Big(\frac{1}{2}\,\boldsymbol{I} - \boldsymbol{K}\Big)(\boldsymbol{u}^- - \boldsymbol{u}_\infty)\Big\rangle = 0 \quad (3.26)$$

for all $\boldsymbol{\tau} \in \boldsymbol{H}^{-1/2}(\Gamma)$.

The hydrodynamic single layer potential $\boldsymbol{S} : \boldsymbol{H}^{-1/2}(\Gamma) \to \boldsymbol{H}^{1/2}(\Gamma)$ is continuous and satisfies, in virtue of (3.23) and (3.10), the Gårding inequality: there is $c > 0$ such that

$$\forall\boldsymbol{\tau} \in \boldsymbol{H}^{-1/2}(\Gamma) : \langle\boldsymbol{\tau}, \boldsymbol{S}\boldsymbol{\tau}\rangle \geq c\,\|\boldsymbol{\tau}\|^2_{H^{-1/2}(\Gamma)} - k(\boldsymbol{\tau},\boldsymbol{\tau})$$

where $k(\cdot,\cdot)$ is a compact form on $\boldsymbol{H}^{-1/2}(\Gamma)$. Equation (3.26) gives now, together with the (formal) weak form (1.3) of the Navier-Stokes system in Ω^- the desired nonlocal boundary value problem in $\Omega^- \cup \Gamma$: Find $\boldsymbol{u}^- \in \boldsymbol{V}(\Omega^-)$, $\boldsymbol{\sigma}_n \in \boldsymbol{H}^{-1/2}(\Gamma)$ such that

$$\begin{aligned} a(\boldsymbol{u}^-, \boldsymbol{v}) - \langle \boldsymbol{\sigma}_n, \gamma_0 \boldsymbol{v}\rangle &= \langle \boldsymbol{f}, \boldsymbol{v}\rangle_{\Omega^-} \quad && \forall \boldsymbol{v} \in \boldsymbol{V}(\Omega^-)\,, \\ \Big\langle \boldsymbol{\tau}, \Big(\frac{1}{2}\boldsymbol{I} - \boldsymbol{K}\Big)(\boldsymbol{u}^- - \boldsymbol{u}_\infty)\Big\rangle + \langle \boldsymbol{\tau}, \boldsymbol{S}\boldsymbol{\sigma}_n\rangle &= 0 && \forall \boldsymbol{\tau} \in \boldsymbol{H}^{-1/2}(\Gamma)\,. \end{aligned} \tag{3.27}$$

Here the nonlinear form $a(\cdot,\cdot)$ is as in (1.2).

Whereas the nonlocal problem (3.14) and the corresponding one (3.27) for the exterior Stokes equation are mathematically on solid ground due to Theorem 1.3 and 3.1, in the Oseen case research is in progress on the following questions:

a) Existence of solutions to the nonlocal problems (3.14), (3.27) in the exterior Oseen case,

b) Coercivity of $\boldsymbol{S}$ in the Oseen case,

c) Convergence of Galerkin-discretizations of (3.14), (3.27) in the Stokes and Oseen case (note that the nonlinearity is not of the type treated in [11]).

Acknowledgments

The research of M. Feistauer has been supported under the Grant No. 201/97/0217 of the Czech Grant Agency. The authors gratefully acknowledge this support.

References

[1] Bégue, C., Conca, E., Murat, F. and Pironneau O. (1988). The Stokes and Navier–Stokes equations with boundary conditions involving pressure. Publications du Laboratoire d'Analyse Numérique, Université P. et. M. Curie, Paris VI, R 88027, Novembre 1988.

[2] Berger, H., Warnecke, G. and Wendland W. (1997). Analysis of a FEM/BEM coupling method for transonic flow computations. *Math. Comp.*, 66:1407–1440.

[3] Bruneau, Ch. and Fabrie P. (1994). Effective downstream boundary conditions for incompressible Navier–Stokes equations. *Int. J. Numer. Meth. Fluids*, 19:693–705.

[4] Dautray, R. and Lions, J.L. (1993). *Mathematical Analysis and Numerical Methods for Science and Technology*. Vol. 4, Springer–Verlag, Berlin – Heidelberg – New York.

[5] Farwig, R. (1997). The stationary Navier–Stokes equations in a 3D–exterior domain. TH Darmstadt, Fachbereich Mathematik, Preprint - Nr. 1911.

[6] Feistauer, M. (1993). *Mathematical Methods in Fluid Dynamics.* Pitman Monographs and Surveys in Pure and Applied Mathematics 67, Longman Scientific & Technical, Harlow.

[7] Feistauer, M., Hsiao, G.C., Kleinman, R.E. and Tezaur, R. (1995). Analysis and numerical realization of coupled BEM and FEM for nonlinear exterior problems. In: Methoden und Verfahren der Mathematischen Physik, Band 40, "Inverse Scattering and Potential Problems in Mathematical Physics" (R. E. Kleinman, R. Kress, E. Martensen - eds.), P. Lang, Frankfurt am Main, 47–73.

[8] Feistauer, M. and Schwab C. (1998). On coupled problems for viscous flow in exterior domains. Math. Models Methods Appl. Sci., 8:657–684.

[9] Feistauer, M. and Schwab C. (1998). Coupling of an interior Navier–Stokes problem with an exterior Oseen problem. Research Report No. 98-01, February 1998, SAM ETH Zürich (submitted).

[10] Galdi, G.P. (1994). *An Introduction to the Mathematical Theory of the Navier–Stokes Equations.* Springer Tracts in Natural Philosophy, Volume 38, Springer–Verlag, New York.

[11] Gatica, G.N. and Hsiao, G.C. (1995). Boundary field equation methods for a class of nonlinear problems, Pitman Res. Notes in Math. No.331, Pitman Publ. Harlow, UK.

[12] Girault, V. and Raviart, P.A. (1986). *Finite Element Methods for Navier–Stokes Equations.* Springer Series in Computational Mathematics 5, Springer–Verlag, Berlin.

[13] Kufner, A., John, O. and Fučík, S. (1977). *Function Spaces.* Academia, Praha, 1977.

[14] Lions, J.L. (1969). *Quelques méthodes de résolution des problèmes aux limites non linéaires.* Dunod, Paris.

[15] Oseen, C.W. (1927). *Neuere Methoden und Ergebnisse in der Hydrodynamik.* Akad. Verlagsgesellschaft, Leipzig.

[16] Sequeira, A. (1983). The coupling of boundary integral and finite element methods for the bidimensional exterior steady problems. *Math. Methods Appl. Sci.,* 5:356–375.

[17] Temam R. (1977). *Navier–Stokes Equations.* North–Holland, Amsterdam - New York - Oxford.

REMARKS ON THE DETERMINANT IN NONLINEAR ELASTICITY AND FRACTURE MECHANICS

Irene Fonseca, Jan Malý

Abstract: The role of the determinant in ensuring local invertibility of Sobolev functions in $W^{1,N}(\Omega;\mathbb{R}^N)$ is studied. Weak continuity of minors of gradients of functions in $W^{1,p}(\Omega;\mathbb{R}^N)$ for $p < N$ is fully characterized. Properties of the determinant are addressed within the framework of functions of bounded variation, and a change of variables formula is obtained. These results are relevant in the study of equilibria, cavitation, and fracture of nonlinear elastic materials.

Keywords: Topological degree, weak lower semicontinuity, functions of bounded variation.

1. INTRODUCTION

Remarkable advances in industry and technology have motivated the study of instabilities in certain advanced materials. The need to understand and predict macroscopic behavior from microscopic and mesoscopic data, as well and the analysis of questions related to phase transformations, defectiveness, the onset of microstructures in smart materials, and other issues related to optimal design and homogenization of composite materials and very thin films, have opened new areas of mathematics virtually unexplored until recently. As it turns out, mathematical models for equilibria and dynamical evolution of phase boundaries for these materials fall outside the scope of classical theories, mostly due to the facts that the underlying energies are nonconvex, or that the admissible fields may exhibit discontinuities. Indeed, spaces of discontinuous mappings have proven to be useful in the modeling of deformations of continua which may undergo fracture or develop defects (see e.g. [13], [14], [15], [27]). A natural space for the underlying deformations is that of functions of bounded variation, BV, or its

Applied Nonlinear Analysis, edited by Sequeira *et al.*
Kluwer Academic / Plenum Publishers, New York, 1999.

subclass introduced by De Giorgi and Ambrosio [24] of functions of special bounded variation, SBV.

It is well known that in nonlinear elasticity interpenetration of matter is prevented by assuming that energy densities blow up as the jacobian of the deformation gradient approaches zero. Precisely, if $\Omega \subset \mathbb{R}^N$ represents the reference configuration of an elastic body, and if the bulk energy of an admissible deformation $u : \Omega \to \mathbb{R}^N$ is given by

$$E(u) := \int_\Omega W(\nabla u)\, dx,$$

where $W : \mathbb{R}^N \times \mathbb{R}^{N\times N} \to [0, +\infty)$ is the energy density, then it is commonly assumed that $\det \nabla u > 0$ a.e. in Ω, and $W(\xi) \to +\infty$ as $\det \xi \to 0^+$. Under this degeneracy hypothesis, a question that has challenged mathematicians for several years concerns the search for physically reasonable hypotheses on W and on the class of admissible fields u guaranteeing weak lower semicontinuity of the energy E.

Ciarlet and Nečas [18] introduced a "local invertibility condition"

$$\int_\Omega \det \nabla u\, dx \leq \mathcal{L}^N(u(\Omega))$$

where $\mathcal{L}^N$ stands for the Lebesgue measure in $\mathbb{R}^N$, which, together with the condition $\det \nabla u > 0$ a.e. in Ω, ensures local invertibility of u in appropriate Sobolev spaces.

In this paper, we start by recalling a local inverse function theorem obtained by Fonseca and Gangbo [28] for mappings u in $W^{1,N}(\Omega; \mathbb{R}^N)$ such that $\det \nabla u(x) > 0$ for a.e. $x \in \Omega$.

If we require less regularity of the admissible fields, as it happens in the study of cavitation in rubber-like materials, then weak continuity properties of minors become very challenging. To this end, we fully characterize weak convergence of det in Sobolev spaces $W^{1,p}(\Omega; \mathbb{R}^N)$ for $p < N$. This analysis was carried out in Fonseca and Malý [32].

Finally, and drawing on the work of Choksi and Fonseca [17], we show how properties of $\det \nabla u$ may ensure a change of variables formula for functions of bounded variation, where ∇u denotes the part of the distributional derivative Du which is absolutely continuous with respect to $\mathcal{L}^N$.

In what follows, if ξ is a $d \times N$ matrix then $\mathbb{M}(\xi)$ stands for the list of all minors of ξ, and $\mathbb{M}_k(\xi)$ denotes all minors of order k, with $k \leq \min\{d, N\}$. If $d = N$ then $\operatorname{adj} \xi$ is the adjugate of ξ, i.e. the matrix of minors of order $N-1$ so that

$$\det \xi\, \mathbb{I} = (\operatorname{adj} \xi)^T \xi.$$

If μ is a Radon measure in Ω and if $B \subset \Omega$ is a Borel set, then $\mu\lfloor B$ stands for the restriction of μ to B, i.e. $\mu\lfloor B(X) := \mu(X \cap B)$ for all Borel set $X \subset \Omega$.

2. LOCAL INVERTIBILITY IN SOBOLEV SPACES

The study of equilibria for defective crystals has motivated the introduction of variational problems where the domain of integration is varying as well as its deformations. In particular, Fonseca and Parry [35] proposed a model where the underlying energy is given by

$$E(u,v) := \int_{v(\Omega)} W(\nabla(u \circ v^{-1}))\, dx,$$

where u denotes a deformation of Ω and v represents the slip or plastic deformation, with $\det \nabla v = 1$. Invertibility of v will guarantee that the energy may be reformulated as

$$E(u,v) := \int_{\Omega} W(\nabla u (\nabla v)^{-1})\, dx$$

where now the domain of integration is held fixed.

The theorem below provides sufficient conditions under which we may ensure local invertibility (see [28]).

Theorem 2.1. *Let Ω be a bounded, open subset of $\mathbb{R}^N$, and let $u \in W^{1,N}(\Omega;\mathbb{R}^N)$ be such that $\det \nabla u(x) > 0$ for a.e. $x \in \Omega$. Then for $\mathcal{L}^N$ almost every $x_0 \in \Omega$ the function u is locally almost invertible in a neighborhood of x_0, i.e., there exist $r = r(x_0) > 0$, an open set $D = D(x_0) \subset\subset \Omega$, and a function $w : B(y_0;r) \to D$, with $y_0 := u(x_0)$, such that*

$$\begin{gathered} w \in W^{1,1}(B(y_0,r);\mathbb{R}^N), \quad w \circ u(x) = x \text{ a.e. } x \in D, \\ u \circ w(y) = y \text{ a.e. } y \in B(y_0,r), \end{gathered} \tag{2.1}$$

and

$$\nabla w(y) = (\nabla u)^{-1}(w(y)) \text{ a.e. } y \in B(y_0,r).$$

If, in addition, $\left|\frac{\operatorname{adj} \nabla u}{\det \nabla u}\right|^s \det \nabla u \in L^1(\Omega)$ for some $1 \le s < +\infty$, then $w \in W^{1,s}(B(y_0,r);D)$.

The latter part of this result was independently obtained by Šverák [59].

If u is more regular, precisely if $u \in W^{1,p}(\Omega;\mathbb{R}^N)$ with $p \ge N(N-1)$, then $u : D \to u(D)$ and $w : u(D) \to D$ are homeomorphisms, and there exists a set $\mathcal{N} \subset \Omega$ with $\mathcal{L}^N(\mathcal{N}) = 0$ such that $u : \Omega \setminus \mathcal{N} \to \mathbb{R}^N$

is an open mapping (see also [44], where a stronger version of this result was established). Also, Heinonen and Koskela [43] showed that if $u \in W^{1,p}(\Omega; \mathbb{R}^N)$ with $p > N(N-1)$, $\det \nabla u > 0$ a.e. in Ω, and $N \geq 3$, then u is open and discrete.

We note that there are Lipschitz homeomorphisms which do not satisfy $\det \nabla u \neq 0$ a.e. in Ω. Indeed, Martio and Ziemer [50] proved that for every bounded, open set $\Omega \subset \mathbb{R}^N$, there exist a measurable set $A \subset \Omega$ with $\mathcal{L}^N(A) > 0$, and a homeomorphism $u \in W^{1,\infty}(\Omega; \mathbb{R}^N)$ such that $\det \nabla u = 0$ for every $x \in A$.

The proof of Theorem 2.1 uses properties of degree theory, and departs considerably from earlier work on this subject (see e.g. [8], [18], [59], [60]). We remark that we cannot expect to prove this result by approximating u by a sequence of smooth functions with positive determinant, since Ball [9] has provided an example of a mapping $v \in W^{1,\infty}(\Omega; \mathbb{R}^2)$ with $\det \nabla v = 1$ a.e. in Ω, and for which there is no sequence $\{v_n\} \in C^1(\bar{\Omega}; \mathbb{R}^2)$ such that $v_n \to u$ uniformly and $\det \nabla u_n > 0$ in Ω.

Theorem 2.1 illustrates how in $W^{1,N}(\Omega; \mathbb{R}^N)$ the determinant still behaves, essentially, as the Jacobian of a smooth deformation. There are situations, however, where we are led to the study of properties of minors for deformations in $W^{1,p}(\Omega; \mathbb{R}^N)$ for $p < N$. This question is addressed in the next section.

3. WEAK CONTINUITY OF THE JACOBIAN INTEGRAL

In [32] we search for minimal conditions ensuring weak convergence of minors in Sobolev spaces.

Certain energies for nonlinear elastic materials may be represented as

$$E(u) = \int_\Omega W(\nabla u)\, dx$$

where W is a *polyconvex integrand*, i.e. W is a convex function of $\mathbb{M}(\nabla u)$. It follows that E is $W^{1,p}$-sequentially weakly lower semicontinuous, i.e.

$$u_n \rightharpoonup u \text{ in } W^{1,p}(\Omega; \mathbb{R}^d) \Rightarrow E(u) \leq \liminf_{n\to\infty} E(u_n), \tag{3.1}$$

if

$$u_n \rightharpoonup u \text{ in } W^{1,p}(\Omega; \mathbb{R}^d) \Rightarrow \mathbb{M}(\nabla u_n) \overset{*}{\rightharpoonup} \mathbb{M}(\nabla u) \text{ in the sense of measures.}$$

We recall that if $g_n, g \in L^1(\Omega)$, then we say that $g_n \overset{*}{\rightharpoonup} g$ *in the sense of measures* if

$$\int_\Omega g_n \varphi\, dx \to \int_\Omega g \varphi\, dx$$

for all $\varphi \in C_c^\infty(\Omega)$.

Weak continuity of minors has been studied in depth by Murat and Tartar (see [56], [61]) within the framework of compensated compactness, and lower semicontinuity for polyconvex functions was undertaken by several authors (e.g. see [2], [7], [8], [10], [16], [20], [21], [23], [30], [36], [37], [38], [45], [51], [52], [55], [61]).

Here, and for simplicity, we are going to restrict the study to the higher order minor, det, and when $d = N$, although most of the results may be extended to lower order minors. As it is usual, the admissible fields are assumed to be in $W^{1,N}(\Omega;\mathbb{R}^N)$ because in this space we may integrate by parts and write the determinant as a divergence operator. Precisely, if $u \in W^{1,N}(\Omega;\mathbb{R}^N)$ then

$$\det \nabla u = \mathrm{Det}\nabla u, \tag{3.2}$$

where

$$\mathrm{Det}\nabla u := \frac{1}{N}\mathrm{div}((\mathrm{adj}\nabla u)^T u).$$

The relation (3.2) may fail if u is not sufficiently regular. As an example, consider

$$u(x) := \sqrt[N]{a^N + |x|^N}\,\frac{x}{|x|}, \quad \Omega := B(0,1).$$

Then $u \in W^{1,p}(B(0,1);\mathbb{R}^N)$ for all $p < N$, $\det \nabla u = 1$ a.e. in $B(0,1)$, but

$$\mathrm{Det}\nabla u = \mathcal{L}^N\lfloor B(0,1) + \omega_N\, a^N \delta_0, \qquad \omega_N := \mathcal{L}^N(B(0,1))$$

Moreover, it was proven by Müller [54] (see also [53]) that if S is a closed set with Hausdorff dimension $\alpha \in (0,N)$, then there exists $u \in W^{1,p}(B(0,1);\mathbb{R}^N) \cap C^0(\overline{\Omega})$ for all $p < N$, such that

$$\mathrm{Det}\nabla u = \det \nabla u\, \mathcal{L}^N\lfloor B(0,1) + \mu_s, \tag{3.3}$$

where μ_s is a positive Radon measure, singular with respect to the Lebesgue measure $\mathcal{L}^N$, and such that $\mathrm{supp}\,\mu_s = S$.

Further results by Müller, Tang and Yan [55] established that if $u \in W^{1,q}(\Omega;\mathbb{R}^N)$ with $q \geq N^2/(N+1)$ then (3.3) holds, and $\mu_s = 0$ if

$$u \in W^{1,N-1}(\Omega;\mathbb{R}^N), \quad \mathrm{adj}\,\nabla u \in L^{N/(N-1)}.$$

The exploitation of spaces such as BMO and Hardy spaces allows one to refine these results along the lines of the work of Müller [52], [53], [54], and Coifman, Lions, Meyers and Semmes [19], and, in particular, it can be shown that if $u \in W^{1,N-1}(\Omega;\mathbb{R}^N)$ is such that $\det \nabla u \geq 0$ and $\mathrm{adj}\,\nabla u \in L^{N/(N-1)}$ then

$$\det \nabla u \ln(2 + \det \nabla u) \in L^1_{\mathrm{loc}}.$$

In earlier works, Ball [8], Dal Maso and Celada [16], Dacorogna and Murat [22], Dal Maso and Sbordone [23], Giaquinta, Modica and Souček [38], [39], [40], (see also [41]), and Reshetnyak [57] established that

$$u_n \rightharpoonup u \text{ in } W^{1,N}(\Omega;\mathbb{R}^N) \Rightarrow \det \nabla u_n \overset{*}{\rightharpoonup} \det \nabla u,$$

where the convergence $\overset{*}{\rightharpoonup}$ is in the sense of measures. Moreover, if $u_n, u \in W^{1,N}(\Omega;\mathbb{R}^N)$ and $u_n \rightharpoonup u$ in $W^{1,p}(\Omega;\mathbb{R}^N)$ with $p > \frac{N^2}{N+1}$, then we still have

$$\det \nabla u_n \overset{*}{\rightharpoonup} \det \nabla u \quad \text{in the sense of distributions.} \tag{3.4}$$

Also, if $p > N-1$ and if $\sup_n \|u_n\|_\infty < +\infty$, with $\{\det \nabla u_n\}$ equi-integrable, then (3.4) is still valid (see [21]).

A complete characterization of weak convergence of the determinant has been obtained by Fonseca and Malý in [32], where the results below may be found.

Theorem 3.1. *If the sequence $\{u_n\} \subset W^{1,N}(\Omega;\mathbb{R}^N)$ converges to u in $L^1(\Omega;\mathbb{R}^N)$, if $\{\mathbb{M}(\nabla u_n)\}$ is bounded in L^1, and if $\{\det \nabla u_n\}$ is equi-integrable, then*

$$\det \nabla u_n \rightharpoonup \det \nabla u \text{ in } L^1.$$

Remarks 3.2. (i) This result was proven earlier by Giaquinta, Modica and Souček, see [41], Theorem II.3.2.1, under the additional assumption that all minors of all orders are equi-integrable. Their proof relies of tools from Geometric Measure Theory. The proof presented in [32] is entirely analytical.

(ii) Note that from the hypotheses of Theorem 3.1, one can only guarantee apriori that $u \in BV$, so ∇u must be understood as the Radon Nikodym derivative of the distributional derivative Du with respect to the Lebesgue measure $\mathcal{L}^N$.

(iii) In Theorem 3.1 equi-integrability is necessary. Indeed, it is possible to construct a sequence $u_n \in W^{1,N}(\Omega;\mathbb{R}^N)$ such that $u_n \rightharpoonup u$ weakly in $W^{1,N-1}$, u is affine, $\{\det \nabla u_n\}$ is bounded in L^1, and still

$$\det \nabla u_n \overset{*}{\rightharpoonup} \det \nabla u + \mu_s,$$

where μ_s is a nonzero Radon measure singular with respect to $\mathcal{L}^N$. Let us consider a nonincreasing smooth function $\psi : [0,\infty) \to \mathbb{R}$ such that

$$\psi(r) = \begin{cases} \frac{\pi}{2} & \text{for } 0 \le r \le 1, \\ -\frac{\pi}{2} & \text{for } r \ge 2, \end{cases}$$

and set

$$(u_n)_1(x) := \frac{x_1 x_3}{\rho} \cos\psi(k\rho),$$
$$(u_n)_2(x) := \frac{x_2 x_3}{\rho} \cos\psi(k\rho),$$
$$(u_n)_3(x) := x_3 \sin\psi(k\rho),$$

where $\rho := \sqrt{x_1^2 + x_2^2}$. Then $u_n \rightharpoonup u$, where $u(x) = (0, 0, x_3)$, and the weak* limit of $\{\det \nabla u_n\}$ is equal to $\det \nabla u \mathcal{L}^N \lfloor B(0,1) + \mu_s$, where the density of the singular measure μ_s is supported on the x_3-axis.

Theorem 3.3. *If the sequence $\{u_n\} \subset W^{1,N}(\Omega;\mathbb{R}^N)$ converges to u in $L^1(\Omega;\mathbb{R}^N)$, if $\{u_n\}$ is bounded in $W^{1,N-1}(\Omega;\mathbb{R}^N)$, and if* $\det \nabla u_n \overset{*}{\rightharpoonup} \mu$ *for some Radon measure μ, then*

$$\frac{d\mu}{d\mathcal{L}^N} = \det \nabla u \quad a.e.\ x \in \Omega.$$

The two theorems above are sharp, in that they are complemented by the following result.

Theorem 3.4. *Let $u \in W^{1,N}(\Omega;\mathbb{R}^N)$.*

(i) Let μ be a Radon measure on Ω, and let $1 \le p < N-1$. Then there exists a sequence $\{u_n\} \subset C^1(\overline{\Omega};\mathbb{R}^N)$ such that

$$u_n \to u \text{ in } W^{1,p}(\Omega;\mathbb{R}^N) \quad \text{and } \det \nabla u_n \overset{*}{\rightharpoonup} \mu.$$

(ii) Let $f \in L^q(\Omega;\mathbb{R})$, $1 \le q < +\infty$. Then there exists a sequence $\{u_n\} \subset C^1(\overline{\Omega};\mathbb{R}^N)$ such that

$$u_n \rightharpoonup u \text{ in } W^{1,p}(\Omega;\mathbb{R}^N) \quad \text{and } \det \nabla u_n \to f \text{ in } L^q.$$

Finally, we notice that the argument of the proof of Theorem 3.1 may be used to provide an alternative proof for the following lower semicontinuity result of Dal Maso and Sbordone [23] and Celada and Dal Maso [16] (see [32], and also the paper by Fusco and Hutchinson [36] for analytical proofs which do not need Geometric Measure Theory tools such as currents).

Theorem 3.5. *Let $u_n \in W^{1,N}(\Omega;\mathbb{R}^N)$, $u_n \to u$ in $L^1(\Omega;\mathbb{R}^N)$, and let $\{\mathbb{M}(\nabla u_n)\}$ be bounded in L^1. If g is a nonnegative, convex function, then*

$$\int_\Omega g(\det \nabla u)\,dx \le \liminf_{n\to\infty} \int_\Omega g(\det \nabla u_n)\,dx.$$

We end this section with a brief overview of some recently obtained relaxation results for quasiconvex and polyconvex energies (see [1], [12],

[30], [33], [31], [46], [49]). If $f : \Omega \times \mathbb{R}^{d\times N} \to \mathbb{R}$ is a Carathéodory function, then the effective (or relaxed) energy is defined as

$$\mathcal{F}_{p,q}(u;\Omega) := \inf_{\{u_n\}} \left\{ \liminf_{n\to\infty} \int_\Omega f(x,\nabla u_n)\,dx : \; u_n \in W^{1,q}_{\text{loc}},\, u_n \rightharpoonup u \text{ in } W^{1,p} \right\}.$$

First we consider the case where

$$f(x,\xi) := g(\det \xi),$$

$g : \mathbb{R} \to [0,+\infty)$ is a convex function. We have shown that (see [32])

$$\mathcal{F}_{p,N}(u;\Omega) \geq \int_\Omega g(\det \nabla u)\,dx \quad \text{if } p \geq N-1,$$

and if $p > N-1$ then (see [12], [31])

$$\mathcal{F}_{p,N}(u;\Omega) = \int_\Omega g(\det \nabla u)\,dx + \mu_s(\Omega)$$

for some Radon measure μ, singular with respect to the Lebesgue measure $\mathcal{L}^N$. For a general f, and under the growth condition

$$0 \leq f(x,\xi) \leq C(1+|\xi|^q)$$

and

$$p > \frac{N-1}{N} q,$$

we have

$$\mathcal{F}_{p,q}(u;\Omega) = h_u \mathcal{L}^N \lfloor \Omega + \mu_s,$$

where (see [1])

$$h_u \leq Qf(x,\nabla u)$$

and μ_s is a singular measure. In the case where $f = f(\xi)$, we have

$$h_u = Qf(\nabla u), \tag{3.5}$$

where Qf stands for the *quasiconvexification* of f, precisely,

$$Qf(\xi) := \inf \left\{ \int_{(0,1)^N} f(\xi + \nabla\varphi(x))\,dx : \varphi \in C_c^\infty(\Omega;\mathbb{R}^N). \right\}$$

This may no longer be valid when f depends also on x and $p < q$. Indeed, Gangbo [37] has constructed an example where

$$f(x,\xi) = \chi_K(x)\,|\det \xi|,$$

and $h_u = f$ if and only if $\mathcal{L}^N(\partial K) = 0$. Hence, in general, (3.5) fails and

$$f^{**}(x,\nabla u) \leq h_u$$

is the only known lower estimate.

4. A CHANGE OF VARIABLES FORMULA IN BV

In the previous two sections we dealt with properties of $\det \nabla u$ for $u \in W^{1,p}(\Omega;\mathbb{R}^N)$ and for certain ranges of p. Here we go outside the Sobolev spaces framework to handle situations relevant to the study of phase transitions of incoherent phase deformations, to the analysis of fracture, or to tackle problems where the growth of bulk energy densities is at most linear. In the latter case minimizing sequences may convergence to macroscopic states which are only in BV (see [6], [34]).

Given $u \in BV(\Omega;\mathbb{R}^N)$, its distributional derivative may be written as

$$D\varphi = \nabla u\, \mathcal{L}^N \lfloor \Omega + (u^+ - u^-) \otimes \nu H^{N-1} \lfloor S(u) + C(u),$$

where H^{N-1} denotes the $(N-1)$-dimensional Hausdorff measure, $S(u)$ is the jump set of u, which differs from the complement of the set of Lebesgue points, $\Gamma(u)$, by a set of H^{N-1} measure zero, $C(u)$ is the Cantor part, u^+ and u^- are the traces of u on $S(u)$ at x, $\nu(x)$ is the unit normal vector to $S(u)$ pointing towards the side of u^+ (see [25], [62]). Here, and in the sequel, we assume that u is appropriately represented, namely that

$$u(x) = \lim_{r\to 0+} \mathcal{L}^N(B(x,r))^{-1} \int_{B(x,r)} u\, dy$$

whenever the limit on the right exists.

We remark that ∇u does not have the structure of a gradient, i.e. it is not necessarily curl-free. Indeed, according to a result by Alberti [3] ∇u may be any L^1 function. In spite of this degeneracy, the usual change of variables formula still holds (see [29]).

It is well known that the change of variables formula

$$\int_A v(x)|Ju(x)| = \int_{\mathbb{R}^N} \Big(\sum_{x\in A\cap G\cap u^{-1}(y)} v(x)\Big)\, dy \tag{4.1}$$

holds for a set G of full measure in Ω, provided u is an almost everywhere approximately differentiable function (hence, if $u \in BV$), u has the N-property on G, and v is a measurable function on Ω such that $v|Ju| \in L^1(\Omega)$. Here Ju is the Jacobian computed from the approximate derivative, which, in turn, coincides a.e. with the absolutely continuous part ∇u of Du. We recall that u is said to have the N-*property* on G if $\mathcal{L}^N(u(E)) = 0$ for each set $E \subset G$ with $\mathcal{L}^N(E) = 0$.

In the case where $v = 1$, equation (4.1) is often called the area formula.

Federer (see [26], 3.2.1) showed that u has the N-property on G if

$$G \subset \left\{ x : \operatorname{ap} \limsup_{y\to x} \frac{|u(y)-u(x)|}{|y-x|} < \infty \right\}.$$

In [17] another class of admissible domains was proposed. We denote by $M(Du)(x)$, the Hardy-Littlewood maximal function of the total variation measure $|Du|$ computed using the balls contained in Ω. As it is usual, it μ is a finite, positive Radon measure on Ω, we define the *maximal function of* μ as (see [58])

$$M(\mu)(x) := \sup\left\{\frac{\mu(B(x,r))}{\mathcal{L}^N(B(x,r))} : 0 < r < \operatorname{dist}(x, \partial\Omega)\right\},$$

and it can be shown that (see [4])

$$\mathcal{L}^N(\{x \in \Omega : M(\mu)(x) = +\infty\}) = 0.$$

Theorem 4.1. *Let $u \in BV(\Omega;\mathbb{R}^N)$ and let*

$$G \subset \{x : M(|Du|)(x) < \infty\}$$

be a measurable set with $\mathcal{L}^N(G) = \mathcal{L}^N(\Omega)$. Then the change of variables (4.1) holds on G for any measurable function v such that $v|Ju| \in L^1(\Omega)$.

Proof. Set

$$(u)_{x,r} := \mathcal{L}^N(B(x,r))^{-1}\int_{B(x,r)} u\,dy.$$

By Poincaré's inequality,

$$\begin{aligned}\left|(u)_{x,r/2} - (u)_{x,r}\right| &\le C\mathcal{L}^N(B(x,r))^{-1}\int_{B(x,r)} |u(y) - (u)_{x,r}|\,dy \\ &\le Cr\mathcal{L}^N(B(x,r))^{-1}\int_{B(x,r)} |Du| \\ &\le CrM(|Du|)(x).\end{aligned} \tag{4.2}$$

Iterating this inequality for r, $r/2$, $r/4, \dots$, we obtain

$$|u(x) - (u)_{x,r}| \le CrM(|Du|)(x)$$

for all $x \in G$. Using Poincaré's inequality once again on a ball $B(x,R)$, where $R = 2\operatorname{dist}(x,y)$ and $B(x,R)$ is contained in Ω, we obtain the Bojarski-Hajłasz inequality (see [11])

$$|u(y) - u(x)| \le C|y - x|(M(|Du|)(x) + M(|Du|)(y)).$$

Hence the graph of $u\lfloor G$ can be covered by a countable union of graphs of Lipschitz functions and thus from the N-property of Lipschitz functions we conclude the N-property of u on G. □

A consequence of the above theorem results in conservation of volume.

Corollary 4.2. *Let $u \in BV(\Omega;\mathbb{R}^N)$ be such that $|\det \nabla u(x)| = 1$ for a.e. $x \in \Omega$, and let*

$$G \subset \{x : M(|Du|)(x) < \infty\}$$

be a measurable set such that

$$\mathcal{L}^N(u(G)) = \mathcal{L}^N(G) = \mathcal{L}^N(\Omega).$$

Then for any measurable set $A \subset G$ we have

$$\mathcal{L}^N(u(A)) = \mathcal{L}^N(A).$$

Next we present another another version of change of variable formula for BV functions. Namely, we show that if the absolutely continuous part of ∇u is in L^N, we may essentially enlarge the set G for which (4.1) is valid. This is a generalization of change of variable formula for $W^{1,N}$ functions by Malý and Martio [48] and Malý [46], following some ideas from the above mentioned works (see also [43], [44], [47], [50]).

If a function $u \in BV(\Omega;\mathbb{R}^N)$ then we write

$$\begin{aligned} G_1(u) &:= \{x \in \Omega : M(|(Du)_s|)(x) < \infty\}, \\ G_2(u) &:= \Big\{x \in \Omega : \text{ there exists } \alpha \in (0,1) \text{ such that} \\ &\qquad \limsup_{r\to 0+} r^{-N-\alpha} \int_{B(x,r)} |u(y) - u(x)|\, dy < \infty\Big\}. \end{aligned}$$

We recall that for $\mathcal{L}^N$ a.e. $x \in \Omega$

$$\lim_{r\to 0+} r^{-N-1} \int_{B(x,r)} |u(y) - u(x) - \nabla u(x)\cdot(y-x)|\, dy = 0,$$

therefore for all $\alpha \in (0,1)$

$$\limsup_{r\to 0+} r^{-N-\alpha} \int_{B(x,r)} |u(y) - u(x)|\, dy = 0;$$

hence

$$\mathcal{L}^N(G_1(u) \cap G_2(u)) = \mathcal{L}^N(\Omega).$$

Theorem 4.3. *Let $u \in BV(\Omega;\mathbb{R}^N)$ be such that $\nabla u \in L^N(\Omega)$. Let*

$$G \subset G_1(u) \cap G_2(u)$$

be a measurable set with $\mathcal{L}^N(G) = \mathcal{L}^N(\Omega)$. Let v be a measurable function on Ω such that $v\,|Ju| \in L^1(\Omega)$. Then the change of variables formula (4.1) *holds on G.*

Proof. We will verify the N-property of u on G. Let $E \subset G$ be a set of measure zero, and choose an open set $U \subset \Omega$ containing E. For $x \in E$ write

$$r_k := 2^{-k} r_0, \quad B_k := B(x, r_k), \text{ with } B(x, r_0) \subset U.$$

Fix a point $x \in E$, and find an $\alpha \in (0,1)$ according to the definition of G_2. With each $k = 1, 2, \ldots$ we associate a value $\rho_k = \rho_k(x) \geq 0$ such that

$$\begin{aligned} |B_k \cap \{y : |u(y) - u(x)| \leq \rho_k\}| &\geq \mathcal{L}^N(B_k)/2, \\ |B_k \cap \{y : |u(y) - u(x)| \geq \rho_k\}| &\geq \mathcal{L}^N(B_k)/2. \end{aligned} \tag{4.3}$$

Notice that, by Chebychev's inequality and the definition of G_2,

$$\rho_k \leq C(x) r_k^\alpha \tag{4.4}$$

for some $C(x) > 0$. We write

$$\begin{aligned} V_k &:= \{y \in B_k : |u(y) - u(x)| \leq \rho_k\}, \\ b_k &:= \int_{V_k} (1 + |\nabla u|^N)\, dy, \end{aligned}$$

and set

$$\begin{aligned} I_1 &:= \{k : b_k \geq r_k^{N\alpha}\}, \\ I_2 &:= \{k : 2 r_k^{-N\alpha} b_k \geq \sup_{j \geq k} r_j^{-N\alpha} b_j\}, \\ I = I(x) &:= I_1 \cup I_2. \end{aligned}$$

If I_1 is finite, then the sequence $\{r_k^{-N\alpha} b_k\}$ is bounded and thus I_2 is infinite. It follows that $I_1 \cup I_2$ is infinite. We claim that

$$k \in I \implies \rho_k^N \leq K(x) \int_{V_k} (1 + |\nabla u|^N)\, dx, \tag{4.5}$$

where $K = K(x)$ is a positive constant which may depend on x and u but not on k. Let η_k be an auxiliary smooth function with values

$$\eta_k := \begin{cases} (\rho_k - \rho_{k+1})^+ & \text{on } (-\infty, \rho_{k+1}), \\ 0 & \text{on } (\rho_k, +\infty), \end{cases} \tag{4.6}$$

and satisfying

$$0 \leq \eta_k \leq (\rho_k - \rho_{k+1})^+, \quad 0 \leq -\eta_k' \leq 2.$$

Set

$$u_k(y) := \eta_k(|u(y) - u(x)|).$$

By the chain rule (see Ambrosio and Dal Maso [5]), $u \in BV(B_k; \mathbb{R}^N)$ and

$$|Du_k| \leq 2|(Du)_s| + 2|\nabla u| \chi_{V_k}. \tag{4.7}$$

In order to estimate $(\rho_k - \rho_{k+1})^+$, without loss of generality we may assume that $\rho_k \geq \rho_{k+1}$. We have

$$\begin{aligned} |\{y : u_k(y) = \rho_k - \rho_{k+1}\}| &\geq 2^{-N-1} \mathcal{L}^N(B_k), \\ |\{y : u_k(y) = 0\}| &\geq 2^{-1} \mathcal{L}^N(B_k), \end{aligned}$$

and thus by a Poincaré type inequality and (4.7) we obtain

$$\begin{aligned}\rho_k - \rho_{k+1} &\le C r_k^{-N+1} \int_{B_k} |Du_k| \le \\ &\le C r_k M(|(Du)_s|)(x) + C r_k^{-N+1} \int_{V_k} |\nabla u|\, dy \le K b_k^{1/N},\end{aligned} \tag{4.8}$$

where we have used the fact that $M(|(Du)_s|)(x) < +\infty$ and that $b_k \ge \mathcal{L}^N(B_k)/2$. Now, if $k \in I_1$ then by (4.4) we have

$$\rho_k^N \le K r_k^{N\alpha} \le K b_k,$$

and this asserts (4.5). If $k \in I_2$, then by (4.8)

$$\rho_k \le \sum_{j \ge k} (\rho_j - \rho_{j+1})^+ \le K \sum_{j \ge k} b_j^{1/N} \le K \sum_{j \ge k} (r_j/r_k)^\alpha b_k^{1/N} \le K b_k^{1/N},$$

which yields (4.5) as well. We conclude the proof using Vitali's covering argument. We write the set E as

$$E = \bigcup_{m=1}^{\infty} E_m, \quad E_m := \{x \in E : K(x) \le m\}.$$

It is enough to show that $\mathcal{L}^N(u(E_m)) = 0$ for a fixed $m \in \{1, 2, \dots\}$. We cover $u(E_m)$ with balls $B(u(x), \rho_k + r_k)$ where $x \in E$ and $k \in I(x)$. Notice that we add r_k to ρ_k only to avoid degeneracy when $\rho_k = 0$. We have

$$r_k^N \le C\mathcal{L}^N(B_k) \le 2C\mathcal{L}^N(V_k). \tag{4.9}$$

Since $I(x)$ is infinite, by virtue of (4.4) we obtain a fine covering of E; hence; hence we may extract a disjoint subcover $\{B(n)\}$ of $u(E)$ up to a set of measure zero. If $B(n) = B(u(x), \rho_k + r_k)$ is one of selected balls, then the corresponding radius $\rho_k + r_k$ is denoted by $\rho(n)$, and the corresponding set V_k by $V(n)$. Notice that the sets $V(n)$ are disjoint. By (4.5) and (4.9) we obtain

$$\begin{aligned}\mathcal{L}^N(u(E_m)) &\le C \sum_n \rho(n)^N \le C\, m \sum_n \int_{V(n)} (1 + |\nabla u|^N)\, dy \\ &\le C\, m \int_U (1 + |\nabla u|^N)\, dy.\end{aligned}$$

A suitable choice of the set U allows us to deduce that the right hand side on the above set of inequalities may be rendered as small as we want. Hence the N-property is verified, which, in turn, implies that the (4.1) holds as well. □

Acknowledgments

The research of I. Fonseca was partially supported by the National Science Foundation through the Center for Nonlinear Analysis, and under Grants No. DMS-9500531 and DMS-9731957.

The research of J. Malý was supported by grants GAČR 201/96/0311 and GAUK 170/1999 and by Research Project J13/98113200007 of MŠMT (Czech Republic).

References

[1] Acerbi, E.G., Bouchitté, G. and Fonseca, I. Relaxation of convex functionals and the Lavrentiev phenomenon. In preparation.

[2] Acerbi, E. and Dal Maso., G. (1994). New lower semicontinuity results for polyconvex integrals case. *Calc. Var. Partial Differential Equations*, 2:329–372.

[3] Alberti, G. (1991). A Lusin type theorem for gradients. *J. Funct. Anal.*, 100:110–118.

[4] Ambrosio, L. (1994). On the lower semicontinuity of quasiconvex integrals in $SBV(\Omega;\mathbb{R}^k)$. *Nonlinear Anal*, 23:691–702.

[5] Ambrosio, L. and Dal Maso, G. (1990). A general chain rule for distributional derivatives. *Proc. Amer. Math. Soc.*, 108:691–702.

[6] Ambrosio, L. and Dal Maso, G. (1992). On the representation in $BV(\Omega;\mathbb{R}^m)$ of quasiconvex integrals. *J. Funct. Anal.*, 109:76–97.

[7] Acerbi, E. and Fusco, N. (1984). New lower semicontinuity results for polyconvex integrals case. *Arch. Rational Mech. Anal.*, 86:125–145.

[8] Ball, J.M. (1977). Convexity conditions and existence theorems in nonlinear elasticity. *Arch. Rational Mech. Anal.*, 63:337–403.

[9] Ball, J.M. (1988). Global invertibility of Sobolev functions and the interpenetration of the matter. *Proc. Roy. Soc. Edinburgh*, 88A:315–328.

[10] Ball, J.M. and Murat, F. (1984). $W^{1,p}$ quasiconvexity and variational problems for multiple integrals. *J. Funct. Anal.*, 58:225–253.

[11] Bojarski, B. and Hajłasz, P. (1993). Pointwise inequalities for Sobolev functions and some applications. *Studia Math.*, 106:77–92.

[12] Bouchitté, G., Fonseca, I. and Malý, J. (1998). The effective bulk energy of the relaxed energy of multiple integrals below the growth exponent. *Proc. Roy. Soc. Edinburgh*, 128A:463–479.

[13] Braides, A. (1998). *Approximation of Free-Discontinuity Problems.* Lecture Notes in Mathematics, Springer Verlag, Berlin.

[14] Braides, A. and Coscia, A. (1994). The interaction between bulk energy and surface energy in multiple integrals. *Proc. Royal Soc. Edinburgh*, 124A:737–756.

[15] Braides, A. and Chiadò Piat, V. (1996). Integral representation results for functionals defined on $SBV(\Omega;\mathbb{R}^m)$. *J. Math. Pures Appl.*, 75:595–626.

[16] Celada, P. and Dal Maso, G. (1994). Further remarks on the lower semicontinuity of polyconvex integrals. *Ann. Inst. H. Poincaré, Anal. Non Linéaire*, 11:661–691.

[17] Choksi, R. and Fonseca, I. (1997). A change of variables formula for mappings in *BV*. *Proc. Amer. Math. Soc.*, 125:2065–2072.

[18] Ciarlet, P.G. and Nečas, J. (1987). Injectivity and self contact in non linear elasticity. *Arch. Rational Mech. Anal.*, 97:171–188.

[19] Coifman, R. Lions, P.-L., Meyer, Y. and Semmes, S. (1989). Compacité par compensation et espaces de Hardy. *CRAS Paris*, 309:945–949.

[20] Dacorogna, B. (1989). *Direct Methods in Calculus of Variations*. Appl. Math. Sciences 78, Springer–Verlag, Berlin.

[21] Dacorogna, B. and Marcellini, P. (1990). Semicontinuité pour des intégrandes polyconvexes sans continuité des determinants. *C. R. Acad. Sci. Paris Sér. I Math.*, 311, 6:393–396.

[22] Dacorogna, B. and Murat, F. (1992). On the optimality of certain Sobolev exponents for the weak continuity of determinants. *J. Funct. Anal.*, 105:42–62.

[23] Dal Maso, G. and Sbordone, C. (1995). Weak lower semicontinuity of polyconvex integrals: a borderline case. *Math. Z.*, 218:603–609.

[24] De Giorgi, E. and Ambrosio, L. (1988). Un nuovo funzionale del calcolo delle variazioni. *Atti Accad. Naz. Lincei Rend. Cl. Sci. Fis. Mat. Nat.*, 82:199–210.

[25] Evans, L.C. and Gariepy, R.F. (1992). *Measure Theory and Fine Properties of Functions*. CRC Press.

[26] Federer, H. (1996). *Geometric Measure Theory*. Springer, (Second edition 1996).

[27] Fonseca, I. and Francfort, G. (1995). Relaxation in BV versus quasiconvexification in $W^{1,p}$: a model for the interaction between damage and fracture. *Calc. Var. Partial Differential Equations*, 3:407–446.

[28] Fonseca, I. and Gangbo, W. (1995). Local invertibility of Sobolev functions. *SIAM J. Math. Anal.*, 26,2:280–304.

[29] Fonseca, I. and Gangbo, W. (1995). *Degree Theory in Analysis and Applications*. Clarendon Press, Oxford.

[30] Fonseca, I. and Leoni, G. On lower semicontinuity and relaxation. To appear.

[31] Fonseca, I. and Malý, J. (1997). Relaxation of Multiple Integrals below the growth exponent. *Ann. Inst. H. Poincaré. Anal. Non Linéaire*, 14,3:309–338.

[32] Fonseca, I. and Malý, J. Weak continuity of jacobian integrals. In preparation.

[33] Fonseca, I. and Marcellini, P. (1997). Relaxation of multiple integrals in subcritical Sobolev spaces. *J. Geom. Anal.*, 7,1:57–81.

[34] Fonseca, I. and Müller, S. (1993). Relaxation of quasiconvex functionals in $BV(\Omega, \mathbb{R}^p)$ for integrands $f(x, u, \nabla u)$. *Arch. Rational Mech. Anal.*, 123:1–49.

[35] Fonseca, I. and Parry, G. (1992). Equilibrium configurations of defective crystals. *Arch. Rational Mech. Anal.*, 120:245–283.

[36] Fusco, N. and Hutchinson, J.E. (1995). A direct proof for lower semicontinuity of polyconvex functionals. *Manuscripta Math.*, 85:35–50.

[37] Gangbo, W. (1994). On the weak lower semicontinuity of energies with polyconvex integrands. *J. Math. Pures Appl.*, 73, 5:455–469.

[38] Giaquinta, M., Modica, G. and Souček, J. (1990). Cartesian currents, weak dipheomorphisms and existence theorems in nonlinear elasticity. *Arch. Rational Mech. Anal.* **106** *(1989), 97–159. Erratum and addendum. Arch. Rational Mech. Anal.*, 109:385–592.

[39] Giaquinta, M., Modica, G. and Souček, J. (1989). Cartesian Currents and Variational Problems for Mappings into Spheres. *Ann. Scuola Norm. Sup. Pisa Cl. Sci.*, 16:393–485.

[40] Giaquinta, M., Modica, G. and Souček, J. (1990). Liquid crystals: relaxed energies, dipoles, singular lines and singular points. *Ann. Scuola Norm. Sup. Pisa Cl. Sci.*, 17,3:415–437.

[41] Giaquinta, M., Modica, G. and Souček, J. (1995). *Cartesian Currents in the Calculus of Variations I, II.* Ergebnisse der Mathematik und Ihrer Grenzgebiete Vol. 38, Springer, Berlin–Heidelberg.

[42] Hajłasz, P. Note on weak approximation of minors. Ann. Inst. H. Poincaré, to appear.

[43] Heinonen, J. and Koskela, P. (1993). Sobolev mappings with integral dilatations. *Arch. Rational Mech. Anal.*, 125:81–97.

[44] Iwaniec, T. and Šverák, V. (1993). On mappings with integrable dilatation. *Proc. Amer. Math. Soc.*, 118,1:181–188.

[45] Malý, J. (1993). Weak lower semicontinuity of polyconvex integrals. *Proc. Roy. Soc. Edinburgh*, 123A:681–691.

[46] Malý, J. (1994). Weak lower semicontinuity of quasiconvex integrals. *Manuscripta Math.*, 85:419–428.

[47] Malý, J. (1994). The area formula for $W^{1,n}$-mappings. *Comment. Math. Univ. Carolin.*, 35,2:291–298.

[48] Malý, J. and Martio, O. (1995). Lusin's condition (N) and mappings of the class $W^{1,n}$. *J. Reine Angew. Math.*, 458:19–36.

[49] Marcellini, P. (1985). Approximation of quasiconvex functions and lower semicontinuity of multiple quasiconvex integrals. *Manuscripta Math.*, 51:1–28.

[50] Martio, O. and Ziemer, W.P. (1992). Lusin's condition (N) and mappings with non-negative jacobians. *Michigan Math. J.*, 39:495–508.

[51] Morrey, C.B. (1996). *Multiple Integrals in the Calculus of Variations.* Springer.

[52] Müller, S. (1988). Weak continuity of determinants and nonlinear elasticity. *C. R. Acad. Sci. Paris Ser. I Math.*, 307:501–506.

[53] Müller, S. (1990). Det=det. A Remark on the distributional determinant. *C. R. Acad. Sci. Paris Ser. I Math.*, 311:13–17 .

[54] Müller, S. (1993). On the singular support of the distributional determinant. *Ann. Inst. H. Poincare Anal. Non-Lineaire*, 10:657–696.

[55] Müller, S., Tang, Q. and Yan, S.B. (1994). On a new class of elastic deformations not allowing for cavitation. *Ann.-Inst.-H.-Poincare-Anal.-Non-Lineaire*, 11:217–243.

[56] Murat, F. (1981). Compacité par compensation: condition necessaire et suffisante de continuité faible sous une hypothése de rang constant. *Ann. Scuola Norm. Sup. Pisa*, 4,8:68–102.

[57] Reshetnyak, Y. (1968). Weak convergence and completely additive vector functions on a set. *Sibir. Math.*, 9:1039–1045.

[58] Stein, E. (1970). *Singular Integrals and Differentiability Properties of Functions.* Princeton Univ. Press.

[59] Šverák, V. (1988). Regularity properties of deformations with finite energy. *Arch. Rational Mech. Anal.*, 100:105–127.

[60] Tang, Q. (1988). Almost-everywhere injectivity in nonlinear elasticity. *Proc. Roy. Soc. Edinburgh*, 109:79–95.

[61] Tartar, L. (1979). Compensated compactness and applications to partial differential equations. *Nonlinear Analysis and Mechanics: Heriot-Watt Symposium, vol. IV (ed. R. Knops), Pitman Res. Notes in Math.*, 39:136–212.

[62] Ziemer, W.P. (1970). *Weakly Differentiable Functions.* Springer-Verlag, Berlin 1989.

ON MODELLING OF CZOCHRALSKI FLOW, THE CASE OF NON PLANE FREE SURFACE

Jan Franců

Abstract: The flow of the melt during the industrial production of single crystal from melt by Czochralski method is called Czochralski flow. The mathematical description of the flow consists of a coupled system of six P.D.E. in cylindrical coordinates containing Navier-Stokes equations (with the stream function vorticity and swirl), heat convection-conduction equation, convection-diffusion equation for oxygen impurity and an equation describing magnetic field effect.

The paper deals with analysis of the system in the form used for numerical simulation. Weak formulation and existence of the weak solution to stationary and evolution problem is studied. The results from paper [J. Franců: Modelling of Czochralski flow, Abstract and Applied Analysis, 3 (1998) No.1–2, pp. 1–39] are extended to the case of non-plane free surface of the melt.

Keywords: Navier-Stokes equations, Czochralski method, single crystal growth, operator equation, existence theorem, weighted Sobolev spaces, Rothe method.

1. INTRODUCTION

Czochralski method is one of the most important methods for industrial production of silicon single crystals. It consists in pulling up the single crystal from silicon melt in a device called Czochralski device. Since impurities in the melt (mostly oxygen atoms from the silica (SiO_2) walls of the pot) build in the single crystal, the producers are interested in character of the melt flow. The flow is not visible, it is very hard to measure during the procedure, therefore producers are interested in mathematical modelling of the flow on computers.

Applied Nonlinear Analysis, edited by Sequeira *et al.*
Kluwer Academic / Plenum Publishers, New York, 1999.

We shall call the flow of the melt in the Czochralski device during the single crystal growth *Czochralski flow.* Mathematical model of the flow used for numerical simulation is represented by a system of six coupled partial differential equations (2.1)–(2.6) with boundary and initial conditions.

A brief derivation of the system, weak formulation and proof of existence to stationary (4.3) and evolution problem (4.4) is introduced in [3]. In this paper we extend the result in the following way. We remove the assumption of plane free surface of the melt, in this paper we assume that the free surface of the melt is known and axially symmetric. (Let us remark that free surface does not mean free boundary, the problem is not "free boundary problem" since the shape of the free surface of the melt is considered to be known.) This is a non-trivial generalization. We derive conditions on free surface Γ_s for swirl Ω and vorticity S. In later we neglect a term corresponding to curvature. On the other hand the curvature was not taken into account in (2.8). In weak formulation due to special type of boundary conditions (2.22) on the free surface Γ_s we must change bilinear form for Ω to (3.7) to enable application of existence result for weakly continuous operators [2] and to follow Rothe method in evolution problem [4].

This research was initiated by professor Nečas. In a small group leaded by professor Litzman at Masaryk University in Brno we were developing numerical simulations of the Czochralski flow for Tesla Rožnov company. In 1990 in a conference I referred on the model and its numerical computation. In discussion professor Nečas proclaimed that he thought that it was possible to prove existence of the weak solution to this problem.

He was true but it took several years to overcome many troubles connected with the problem. The existence results were first proved only for small material constants $\alpha_T, \alpha_C, \beta_T$, β_C, see [1] in 1992. A discussion with Dr. Knobloch inspired me to find a way of removing these restrictions. In 1996 professor Tobiska inspired me to try to generalize the result to the case of non-plane free surface of the melt.

Most paper dealing with modelling of Czochralski flow were devoted to numerical experiments and schemes for numerical computations. On the other hand there is an extensive bibliography dealing with the Navier-Stokes system and its analysis, e. g. [4]. But the Navier-Stokes system is usually uncoupled, formulated in terms of velocity vector (not in terms of the flow function) in Cartesian coordinates (not in the cylindrical coordinates) and mostly with homogeneous Dirichlet boundary conditions. Only in [6] there is a mathematical analysis including existence proof and numerical experiments to the model of Czochralski flow formulated in Cartesian coordinates.

In the paper we give a precise weak formulation of the problem and prove existence of the weak solution. We investigate the system in the form which is used for numerical simulation. Thus we use cylindrical coordinates, Navier-Stokes equations with the flow function and derived variables Svanberg vorticity S, swirl Ω etc.

The problem is rather complicated. Special difficulties arise from the so-called "wet axis", in the cylindrical coordinates the coefficients have singularities, which involves use of weighted Sobolev spaces, see [5]. The Navier-Stokes equations are formulated in terms of stream function, vorticity and swirl. They are coupled with heat convection-conduction equation and oxygen concentration convection-diffusion equation. The last equation in the system describes the effect of the axial magnetic field. The system is evolutionary but not in all unknowns, it is elliptic in χ.

In this short paper we follow notation of [3]. The parts formulated in details in [3] are only mentioned. More space is devoted only to parts which differ from the comprehensive paper [3].

2. MATHEMATICAL MODEL

We shall deal with modelling of melt flow during single crystal growth by the Czochralski method in a device called crystal puller or Czochralski device.

2.1. Czochralski device

The heart of the device consists of a melting pot (crucible) set on a turning base. Polycrystalline silicon is put into the pot (crucible) and heated by electric heaters around the pot. When the silicon is melted, a single crystal nucleus tightened in a turning hanger touches the surface of the melt. The single crystal starts "growing" as the silicon melt contacts the silicon solid. Both the pot and the hanger rotate around the common vertical axis to obtain the axially symmetric single crystal. It grows in a protective inert atmosphere and often in an axial magnetic field produced by an electromagnetic coil.

Our modelling is confined to the region V of the melt in the melting pot. We assume axial symmetry of the problem. Derivation of the system of partial differential equations with corresponding boundary conditions is in [3].

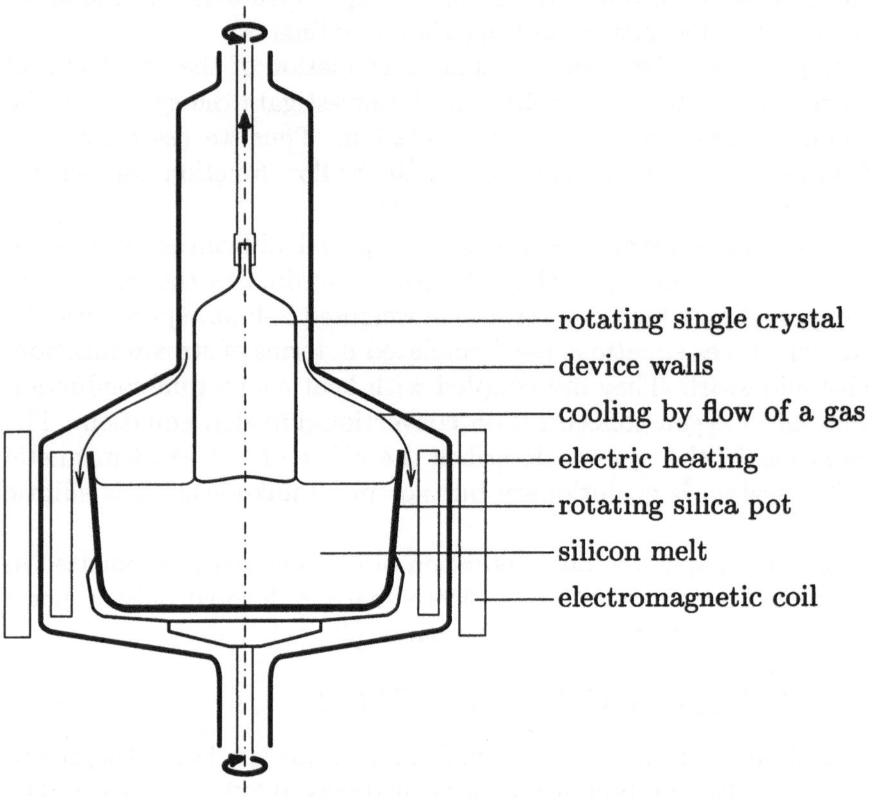

Fig. 1. Czochralski device

2.2. Geometry of the problem

We assume that the region occupied by the melt is constant and known. It is denoted by V in Cartesian coordinates. In the cylindrical coordinates (r, φ, z) the region V corresponds (up to a zero measure set) to $G \times (0, 2\pi)$. Thus the domain G represents a radial cross-section of V in the r, z–half plane $(r > 0)$.

Due to axial symmetry of the problem all variables are independent of φ. The problem is considered in the domain G. Boundary Γ of the domain G is divided into four parts, see Fig. 2:

Γ_p — contact with the bottom and wall of the melting pot,
Γ_s — free surface of the melt,
Γ_c — contact with the crystal and
Γ_a — axis of the symmetry.

In [3] we assumed that the free surface of the melt has a plane shape, in this paper we admit non-plane but known and axially symmetric free surface of the melt.

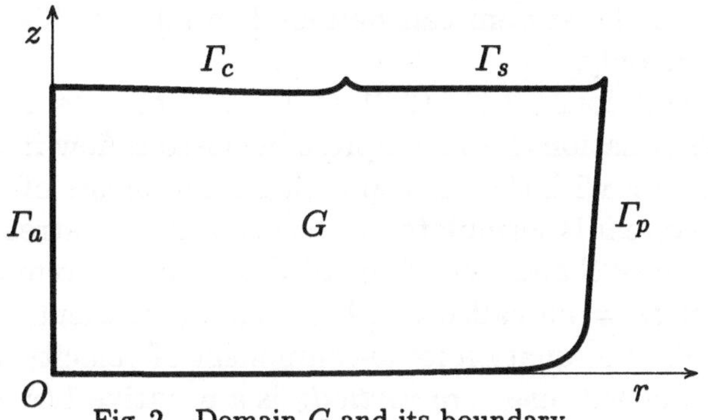

Fig. 2. Domain G and its boundary

2.3. System of differential equations

The mathematical model of Czochralski flow usually used in the literature dealing with its numeric computations consists of the following coupled system of differential equations:

$$\frac{\partial S}{\partial t} + \frac{1}{r}\frac{\partial}{\partial r}(ruS) + \frac{\partial}{\partial z}(wS) + \frac{\partial}{\partial z}\left(\frac{\Omega^2}{r^4}\right) = \tag{2.1}$$

$$= \nu\left[\frac{1}{r}\frac{\partial}{\partial r}\left[\frac{1}{r}\frac{\partial}{\partial r}\left(r^2 S\right)\right] + \frac{\partial^2 S}{\partial z^2}\right] + \alpha_T\frac{1}{r}\frac{\partial T}{\partial r} + \alpha_C\frac{1}{r}\frac{\partial C}{\partial r} + \alpha_m\frac{1}{r^2}\frac{\partial^2 \psi}{\partial z^2},$$

$$\frac{\partial \Omega}{\partial t} + \frac{1}{r}\frac{\partial}{\partial r}(ru\Omega) + \frac{\partial}{\partial z}(w\Omega) = \tag{2.2}$$

$$= \nu\left[\frac{1}{r}\frac{\partial}{\partial r}\left[r^3\frac{\partial}{\partial r}\left(\frac{\Omega}{r^2}\right)\right] + \frac{\partial^2 \Omega}{\partial z^2}\right] - \alpha_m\frac{\partial \chi}{\partial z},$$

$$\frac{\partial T}{\partial r} + \frac{1}{r}\frac{\partial}{\partial r}(ruT) + \frac{\partial}{\partial z}(wT) = \nu_T\left[\frac{1}{r}\frac{\partial}{\partial r}\left(r\frac{\partial T}{\partial r}\right) + \frac{\partial^2 T}{\partial z^2}\right], \tag{2.3}$$

$$\frac{\partial C}{\partial r} + \frac{1}{r}\frac{\partial}{\partial r}(ruC) + \frac{\partial}{\partial z}(wC) = \nu_C\left[\frac{1}{r}\frac{\partial}{\partial r}\left(r\frac{\partial C}{\partial r}\right) + \frac{\partial^2 C}{\partial z^2}\right], \tag{2.4}$$

$$\frac{\partial}{\partial r}\left(\frac{1}{r}\frac{\partial \psi}{\partial r}\right) + \frac{1}{r}\frac{\partial^2 \psi}{\partial z^2} = -rS, \tag{2.5}$$

$$\frac{\partial}{\partial r}\left(\frac{1}{r}\frac{\partial \chi}{\partial r}\right) + \frac{1}{r}\frac{\partial^2 \chi}{\partial z^2} = \frac{1}{r}\frac{\partial \Omega}{\partial z} \tag{2.6}$$

for unknown functions $S, \Omega, T, C, \psi, \chi$, where $u = u(\psi), w = w(\psi)$, see (2.7).

2.4. Comments to the equations

Derivation of the system can be found in [3]. In this paper we give brief comments only.

The first equations (2.1), (2.2) with (2.5) represent the system of Navier-Stokes equations for incompressible viscous flow in the cylindrical coordinates r, φ, z with the corresponding components of velocity vector u, v, w. The system is formulated in terms of Stokes stream function ψ, Svanberg vorticity S and swirl Ω instead of velocity components u, v, w. The unknown $\Omega = rv$ called swirl is angular moment. The equation (2.2) represents the equation for φ-component of velocity v rewritten for Ω. Variable S called *Svanberg vorticity* is a negative $1/r$ multiple of the φ-component of vorticity ω

$$S = -\frac{1}{r}\omega_\varphi \equiv \frac{1}{r}\left[\frac{\partial w}{\partial r} - \frac{\partial u}{\partial z}\right].$$

The equation (2.1) represents equations for r, z components of velocity u, w rewritten for S. Continuity equation enabled to introduce the Stokes stream function ψ replacing the velocity components u, w by

$$u = \frac{1}{r}\frac{\partial \psi}{\partial z}, \qquad w = -\frac{1}{r}\frac{\partial \psi}{\partial r}. \tag{2.7}$$

Combining the last two equalities we obtain (2.5) — the relation between S, ψ.

The equation (2.3) is heat conduction and convection equation for the unknown temperature T. The equation (2.4) models diffusion and convection of oxygen in the melt; the unknown C is oxygen concentration. The last equation (2.6) describes the effect of magnetic field. The unknown χ is the stream function for induced electric current in the melt.

The equations are coupled: convection term with the stream function ψ appears in (2.1), (2.2), (2.3) and (2.4), stream function χ is in (2.2). Variables T, C in (2.1) describe natural convection caused by buoyance due to density gradient of the melt in gravitational field.

In the equations two types of operators appear: a generalized Laplace operator and a convection operator. Denoting the former by A_k and the latter by B

$$A_k(f) = -\frac{k}{r}\frac{\partial f}{\partial r} - \frac{\partial^2 f}{\partial r^2} - \frac{\partial^2 f}{\partial z^2}, \quad B(\psi, f) = \frac{1}{r}\left(\frac{\partial \psi}{\partial z}\frac{\partial f}{\partial r} - \frac{\partial \psi}{\partial r}\frac{\partial f}{\partial z}\right) \tag{2.8}$$

and inserting from (2.7) we can rewrite the system as follows:

$$\frac{\partial S}{\partial t} + \nu A_3(S) + B(\psi, S) + \frac{1}{r^4}\frac{\partial}{\partial z}\left(\Omega^2\right) = \tag{2.9}$$

$$= \alpha_T \frac{1}{r}\frac{\partial T}{\partial r} + \alpha_C \frac{1}{r}\frac{\partial C}{\partial r} + \alpha_m \frac{1}{r^2}\frac{\partial^2 \psi}{\partial z^2},$$

$$\frac{\partial \Omega}{\partial t} + \nu A_{-1}(\Omega) + B(\psi, \Omega) = -\alpha_m \frac{\partial \chi}{\partial z}, \tag{2.10}$$

$$\frac{\partial T}{\partial t} + \nu_T A_1(T) + B(\psi, T) = 0, \tag{2.11}$$

$$\frac{\partial C}{\partial t} + \nu_C A_1(C) + B(\psi, C) = 0, \tag{2.12}$$

$$A_{-1}(\psi) = r^2 S, \tag{2.13}$$

$$A_{-1}(\chi) = -\frac{\partial \Omega}{\partial z}. \tag{2.14}$$

2.5. Boundary conditions

The system of differential equations is completed with boundary conditions:

— at the melting pot wall Γ_p:

$$\Omega = r^2 o_p, \quad T = T_p, \quad C = C_p, \quad \psi = 0, \quad \nabla\psi = 0, \quad \chi = 0, \tag{2.15}$$

— at the crystal interface Γ_c:

$$\Omega = r^2 o_c, \quad T = T_c, \quad \frac{\partial C}{\partial n} = 0, \quad \psi = 0, \quad \nabla\psi = 0, \quad \chi = 0, \tag{2.16}$$

— at the free surface Γ_s:

$$S = \beta_T \frac{1}{r}\frac{\partial T}{\partial r} + \beta_C \frac{1}{r}\frac{\partial C}{\partial r}, \quad \frac{\partial \Omega}{\partial n} - \frac{2}{r}\Omega n_r = 0, \tag{2.17}$$

$$\frac{\partial T}{\partial n} = g_T - \gamma_T T, \quad \frac{\partial C}{\partial n} = g_C - \gamma_C C, \quad \psi = 0, \quad \chi = 0,$$

— and at the symmetry axis Γ_a:

$$\Omega = 0, \quad \frac{\partial T}{\partial r} = 0, \quad \frac{\partial C}{\partial r} = 0, \quad \psi = 0, \quad \nabla\psi = 0, \quad \chi = 0. \tag{2.18}$$

2.6. Comments to the boundary conditions

The forced convection is caused by rotation of the melting pot and by rotation or counter-rotation of the crystal. Denoting the angular velocity of the pot by o_p and of the crystal by o_c we obtained nonslip conditions for u, v, w on $\Gamma_p \cup \Gamma_c$ which yield the conditions for Ω on Γ_p and Γ_c. On Γ_s and Γ_a the normal component of the velocity vector equals to zero. Thus the stream function ψ has zero tangent derivatives on the whole boundary and we put $\psi = 0$ on Γ. Moreover $\nabla \psi = 0$ on $\Gamma_p \cup \Gamma_c \cup \Gamma_a$. Similarly, due to insulating boundary the stream function for induced electric current satisfies $\chi = 0$ on Γ.

We assume that the temperature T is known at the pot walls and crystal interface. At the free surface we consider a linearized law for heat flow. At the axis Γ_a the symmetry conditions are assumed. Concerning oxygen concentration C we assume that it is known at the pot walls, is symmetric at the axis and no segregation occurs at the crystal interface. On the free surface we consider a linearized law for oxygen flow due to evaporating. This condition is often replaced by $C = 0$.

2.7. The surface tension on the free surface

In contrast to [3] we assume the non-plane free surface of the melt. We shall deal with the conditions in details.

On the free surface of the melt the surface tension variations occur due to temperature and concentration gradients. This surface tension variations produce shear stress which generates a surface flow — the so-called Marangoni effect.

We assume linear dependence of the surface tension A on T and C

$$A = A_o[1 - \mathrm{const}_T(T - T_o) - \mathrm{const}_C(C - C_o)] . \tag{2.19}$$

The shear stress is given by the surface gradient of A and it represents the only tangential surface force acting on the free surface. Denoting the stress tensor by τ we have

$$\mathbf{t} \cdot \nabla A = \mathbf{t} \cdot \tau \mathbf{n} \tag{2.20}$$

for any tangential $\mathbf{t}$ and the normal vector $\mathbf{n} = (n_r, 0, n_z)$ to the surface. Between the stress tensor τ and the stretching tensor $\varepsilon(\mathbf{v}) = (\nabla \mathbf{v} + (\nabla \mathbf{v})^T)/2$ we assume linear dependence (Newton law) $\tau = 2\nu\rho\varepsilon(\mathbf{v})$. Combining these relations we obtain

$$\mathbf{t} \cdot \nabla A = \nu\rho\,\mathbf{t} \cdot [\nabla \mathbf{v} + (\nabla \mathbf{v})^T]\,\mathbf{n} \tag{2.21}$$

In our case of axially symmetric free surface the curve Γ_s can be described by functions σ_r, σ_z satisfying $(\sigma_r')^2 + (\sigma_z')^2 = 1$ as follows

$$\Gamma_s = \{(r,z) \mid r = \sigma_r(s), z = \sigma_z(s) \quad s \in (s_a, s_b)\}\,.$$

In cylindric coordinates we have

$$\mathbf{n} \equiv (n_r, n_\varphi, n_z) = (\sigma_z', 0, -\sigma_r')\,.$$

We choose a tangent vector, transform the relation (2.21) into cylindric coordinates and rewrite it for our unknowns.

First in (2.21) we take the tangent vector $\mathbf{t}_1$, in cylindrical coordinates $\mathbf{t}_1 = (0,1,0)$. Since A, u, w are independent of φ on the plane surface after some computation we obtain

$$0 = -\nu\rho \left(\frac{\partial v}{\partial r} n_r + \frac{\partial v}{\partial z} n_z - \frac{1}{r} v\, n_r \right)$$

which rewritten for Ω yields the condition

$$\frac{\partial \Omega}{\partial n} - \frac{2}{r} \Omega\, n_r = 0\,, \qquad \text{on } \Gamma_s\,. \tag{2.22}$$

Then in (2.21) we take the tangent vector $\mathbf{t}_2$ in the r, z plane, in cylindrical coordinates $\mathbf{t}_2 = (\sigma_r', 0, \sigma_z') = (-n_z, 0, n_r)$. After some computation we obtain

$$-\frac{\partial A}{\partial s} = \nu\rho \left[2\left(\frac{\partial u}{\partial r} - \frac{\partial w}{\partial z} \right) n_r\, n_z + \left(\frac{\partial u}{\partial z} + \frac{\partial w}{\partial r} \right) \left(n_z^2 - n_r^2 \right) \right]\,.$$

Since normal component of velocity is zero, its tangent derivative is zero, too. Neglecting the second derivatives of σ_r, σ_z the condition $\partial/\partial s(u\, n_r + w\, n_z) = 0$ yields

$$2\left(\frac{\partial w}{\partial z} - \frac{\partial u}{\partial r} \right) n_r\, n_z + 2\frac{\partial u}{\partial z} n_r^2 - 2\frac{\partial w}{\partial r} n_z^2 = 0\,.$$

Combining these relations we obtain

$$\frac{\partial u}{\partial z} - \frac{\partial w}{\partial r} \neq -\frac{\partial A}{\partial s}\,.$$

The left hand side equals to $-Sr$. Inserting for A we obtain the condition for S in (2.17) with material constants β_T and β_C.

2.8. Summary of the data

The material constants used in the system are:
$\nu > 0$ — silicon melt viscosity,
$\nu_T > 0$ — thermal diffusivity of the silicon melt,
$\nu_C > 0$ — oxygen diffusion coefficient in the silicon melt,
α_T, α_C — coefficient of buoyance caused by thermal and oxygen volume expansion in the gravitation field.
The other "data" in the system are:
β_T, β_C — coefficients of condition describing the surface flow in the free surface,
$\gamma_T \geq 0$, g_T, $\gamma_C \geq 0$, g_C — data in conditions describing the linearized heat and oxygen flow on the free surface. They depend also on the surrounding walls and on the flow of cooling gas.

3. INTEGRAL IDENTITIES

The integral identities are the base for weak formulation of the problems. Since the identities except for (3.2) are same as in [3] their derivation will be only outlined.

Lemma. *Let the functions* $S, \Omega, T, C, \psi, \chi$ *satisfy the system of equations (2.9)–(2.14) with the boundary conditions (2.15)–(2.18). Then they satisfy the following integral identities*

$$a_{-1}\left(\frac{\partial \psi}{\partial t}, \tilde{\psi}\right) + \nu\, a(\psi, \tilde{\psi}) - b\left(\psi, \tilde{\psi}; \frac{1}{r^2} A_{-1}(\psi)\right) +$$

$$+ \int_G \frac{1}{r^3} \frac{\partial}{\partial z}\left(\Omega^2\right) \tilde{\psi}\, \mathrm{d}G + \alpha_m \left(\frac{1}{r}\frac{\partial \psi}{\partial z}, \frac{1}{r}\frac{\partial \tilde{\psi}}{\partial z}\right) = \qquad (3.1)$$

$$= \left(\alpha_T \frac{\partial T}{\partial r} + \alpha_C \frac{\partial C}{\partial r}, \frac{\tilde{\psi}}{r}\right) - \nu \left\langle \beta_T \frac{\partial T}{\partial r} + \beta_C \frac{\partial C}{\partial r}, \frac{\partial \tilde{\psi}}{\partial z}\right\rangle_{\Gamma_s},$$

$$\left(\frac{\partial}{\partial t}\frac{\Omega}{r}, \frac{\tilde{\Omega}}{r}\right) + \nu\, a^*_{-1}(\Omega, \tilde{\Omega}) + b\left(\psi, \Omega; \frac{\tilde{\Omega}}{r^2}\right) = \alpha_m \left(\frac{\chi}{r}, \frac{1}{r}\frac{\partial \tilde{\Omega}}{\partial z}\right), \qquad (3.2)$$

$$\left(\frac{\partial T}{\partial t}, \tilde{T}\right) + \nu_T\, a_1(T, \tilde{T}) + b(\psi, T; \tilde{T}) = \nu_T \int_{\Gamma_s} r(g_T - \gamma_T T)\tilde{T}\, \mathrm{d}\Gamma, \qquad (3.3)$$

$$\left(\frac{\partial C}{\partial t}, \tilde{C}\right) + \nu_C\, a_1(C, \tilde{C}) + b(\psi, C; \tilde{C}) = \nu_C \int_{\Gamma_s} r(g_C - \gamma_C C)\tilde{C}\, \mathrm{d}\Gamma, \qquad (3.4)$$

$$a_{-1}(\chi, \tilde{\chi}) = -\left(\frac{1}{r}\frac{\partial \Omega}{\partial z}, \frac{\tilde{\chi}}{r}\right) \qquad (3.5)$$

with test functions $\tilde{\psi}, \tilde{\Omega}, \tilde{T}, \tilde{C}, \tilde{\chi}$ satisfying

$$\begin{aligned} \tilde{\psi} &= 0 \quad \text{on } \Gamma\,, \\ \tilde{\Omega} &= 0 \quad \text{on } \Gamma - \Gamma_s\,, \\ \tilde{T} &= 0 \quad \text{on } \Gamma_p \cup \Gamma_c\,, \\ \tilde{C} &= 0 \quad \text{on } \Gamma_p\,, \\ \tilde{\chi} &= 0 \quad \text{on } \Gamma\,. \end{aligned} \tag{3.6}$$

In the identities (u, v) means the scalar product with weight r

$$(u, v) = \int_G r\, u\, v\, \mathrm{d}G\,,$$

the bilinear forms corresponding to operators A_k are defined by

$$a_k(u, v) = \int_G r^k \left(\frac{\partial u}{\partial r}\frac{\partial v}{\partial r} + \frac{\partial u}{\partial z}\frac{\partial v}{\partial z} \right) \mathrm{d}G\,, \qquad k = 1, -1\,,$$

$$a^*_{-1}(\Omega, \tilde{\Omega}) = \int_G \left[r^3 \frac{\partial}{\partial r}\left(\frac{\Omega}{r^2}\right) \frac{\partial}{\partial r}\left(\frac{\tilde{\Omega}}{r^2}\right) + \frac{1}{r}\frac{\partial \Omega}{\partial z}\frac{\partial \tilde{\Omega}}{\partial z} \right] \mathrm{d}G\,, \tag{3.7}$$

$$a(\psi, \tilde{\psi}) =$$

$$= \int_G r \left[\frac{\partial}{\partial r}\left(\frac{1}{r}\frac{\partial \psi}{\partial r}\right) \frac{\partial}{\partial r}\left(\frac{1}{r}\frac{\partial \tilde{\psi}}{\partial r}\right) + 2\frac{1}{r}\frac{\partial^2 \psi}{\partial r\,\partial z}\frac{1}{r}\frac{\partial^2 \tilde{\psi}}{\partial r\,\partial z} + \frac{1}{r}\frac{\partial^2 \psi}{\partial z^2}\frac{1}{r}\frac{\partial^2 \tilde{\psi}}{\partial z^2} \right] \mathrm{d}G$$

$$\langle u\,,\, v \rangle_{\Gamma_s} = \int_{\Gamma_s} u\, v\; \mathrm{d}\Gamma$$

and the trilinear form corresponding to operator $B(u, v)$ is given by

$$b\,(u, v; w) \equiv \int_G r\, B(u, v) w\, \mathrm{d}G = \int_G \left(\frac{\partial u}{\partial z}\frac{\partial v}{\partial r} - \frac{\partial u}{\partial r}\frac{\partial v}{\partial z} \right) w\, \mathrm{d}G$$

On the other hand if the functions ψ, Ω, T, C, χ satisfy the derived integral identities for all smooth test functions $\tilde{\psi}, \tilde{\Omega}, \tilde{T}, \tilde{C}, \tilde{\chi}$ satisfying (3.6) and the functions ψ, Ω, T, C, χ are sufficiently smooth, then they also satisfy the system of differential equations (2.9)–(2.14) with boundary conditions (2.15)–(2.18) where S is given by (2.13).

3.1. Comment to derivation of the identities

All identities are derived in usual way. Each equation of (2.10), (2.11), (2.12), (2.14) is multiplied with a test function satisfying the conditions (3.6) and the weight r in case of (3.3), (3.4). In the other equations the weight is $1/r$. The second order operators A_k $(k = 1, -1)$ are converted using integration by parts in the plane to the corresponding bilinear form a_k:

$$\int_G r^k A_k(u)v \, dG = a_k(u,v) - \int_\Gamma r^k \frac{\partial u}{\partial n} v \, d\Gamma .$$

In the curve integral we use boundary condition for the unknowns and the test functions.

The equations (2.9) and (2.13) are "processed" together since on boundary $\Gamma - \Gamma_s$ the second order equation (2.13) for ψ has two boundary conditions $\psi = 0$, $\partial\psi/\partial n = 0$ while equation (2.9) for S has no boundary condition. Thus we express S by ψ using equation (2.13)

$$S \equiv S(\psi) = r^{-2} A_{-1}(\psi)$$

and insert it into equation (2.9) to obtain a fourth order equation for ψ which has two boundary conditions along the whole Γ.

The equation (2.9) is multiplied by $r\tilde{\psi}$ and integrated over domain G. The first integral can be rewritten to the form of bilinear form $a_{-1}(u,v)$. The second integral with operator $A_3\left(r^{-2}A_{-1}(\psi)\right)$ is transformed to the bilinear form $\mathcal{a}$ by double "integration by parts" and conversion of two "mixed" terms with integrand of type $a\tilde{b}$ to the form $c\tilde{c}$. In the integrals over Γ_s the boundary conditions are used.

3.2. Identity for swirl Ω

As in the other cases we multiplied the equation (2.10) by $\tilde{\Omega}/r$ and integrated over G. The problem was with the integral containing $A_1(\Omega)$. In case of plane free surface the Newton condition on Γ_s reduces to Neuman condition and one can use the bilinear form $a_1(u,v)$. In our case of non plane free surface the condition (2.17) is not convenient since T has opposite sign and it would cause troubles in proof of coercivity.

Thus we shall use other integration by parts

$$\int_G \frac{1}{r} A_{-1}(\Omega)\tilde{\Omega} \, dG = - \int_G \frac{1}{r}\left[\frac{1}{r}\frac{\partial}{\partial r}\left[r^3 \frac{\partial}{\partial r}\left(\frac{\Omega}{r^2}\right)\right] + \frac{\partial^2 \Omega}{\partial z^2}\right]\tilde{\Omega} \, dG =$$

$$= a^*_{-1}(\Omega, \tilde{\Omega}) - \int_\Gamma \left[r^3 \frac{\partial}{\partial r}\left(\frac{\Omega}{r^2}\right) n_r \frac{\tilde{\Omega}}{r^2} + \frac{1}{r}\frac{\partial \Omega}{\partial z} n_z \tilde{\Omega}\right] d\Gamma ,$$

where we obtained other bilinear form $a^*_{-1}(\Omega, \tilde{\Omega})$ defined by (3.7). Since

$$r^2 \frac{\partial}{\partial r}\left(\frac{\Omega}{r^2}\right) n_r = \frac{\partial \Omega}{\partial r} n_r - \frac{2}{r}\Omega \, n_r$$

we can use the boundary condition to eliminate the boundary integral. The other terms are converted in usual way.

4. SOLVABILITY OF THE PROBLEMS

The following weak formulations and solvability of the problems are similar to those in [3]. We choose convenient function spaces, the problem is reformulated into an operator equation with a vector of unknowns. We shift the problem to homogeneous boundary conditions. In the stationary problem existence proof consists in verifying assumption of an abstract existence theorem. In the evolution problem we follow Rothe method.

4.1. Function spaces

For the unknowns we introduce weighted Sobolev spaces with weight r or $1/r$ such that the bilinear forms are continuous. Thus the forms $a_1(u,v)$, $a_{-1}(u,v)$, $a^*_{-1}(u,v)$ and $\mathfrak{a}(u,v)$ define weighted Sobolev spaces $W^1_r(G)$, $W^1_{1/r}(G)$, $W^{1*}_{1/r}(G)$ and $W^{2*}_{1/r}(G)$ respectively. Namely

$$W^{1*}_{1/r}(G) = \left\{ u \mid a^*_{-1}(u,u) < \infty, \quad \int_G \frac{1}{r} u^2 \, \mathrm{d}G < \infty \right\}.$$

The weighted Sobolev spaces are studied in [5].

Then according to various boundary conditions for the test functions we introduce function spaces denoted by subscript of the unknown

$$\begin{aligned}
V_\psi &= \{\psi \in W^{2*}_{1/r}(G) \mid \psi = 0 \quad \text{on } \Gamma, \quad \nabla\psi = 0 \text{ on } \Gamma - \Gamma_s\}, \\
V_\Omega &= \{\Omega \in W^{1*}_{1/r}(G) \mid \Omega = 0 \quad \text{on } \Gamma - \Gamma_s\}, \\
V_T &= \{T \in W^1_r(G) \mid T = 0 \quad \text{on } \Gamma_p \cup \Gamma_c\}, \\
V_C &= \{C \in W^1_r(G) \mid C = 0 \quad \text{on } \Gamma_p\}, \\
V_\chi &= \{\chi \in W^1_{1/r}(G) \mid \chi = 0 \quad \text{on } \Gamma\}.
\end{aligned}$$

The bilinear forms define equivalent norms on these subspaces.

Since Ω, T, C have nonhomogeneous boundary conditions we introduce auxiliary functions Ω_b, T_b, C_b satisfying these boundary conditions. We assume

$$\Omega_b \in W^{1*}_{1/r}(G), \quad T_b \in W^1_r(G), \quad C_b \in W^1_r(G). \tag{4.1}$$

$$\nu, \nu_T, \nu_C > 0, \quad \gamma_T, \gamma_C, \alpha_m \geq 0, \quad g_T, g_C \in L^2(\Gamma_s). \tag{4.2}$$

4.2. Vector formulation

We gather all the unknowns into a vector U of unknown functions and all the test functions into a vector V of test functions

$$\mathrm{U} = (\psi, \Omega, T, C, \chi), \qquad \mathrm{V} = (\tilde{\psi}, \tilde{\Omega}, \tilde{T}, \tilde{C}, \tilde{\chi}).$$

We introduce a basic space $\mathbb{W}$ for the vector functions U and V

$$\mathbb{W} = W^{2*}_{1/r}(G) \times W^{1*}_{1/r}(G) \times W^1_r(G) \times W^1_r(G) \times W^1_{1/r}(G),$$

its subspace $\mathbb{V}$ of functions with prescribed zero traces for the test vector V by

$$\mathbb{V} = V_\psi \times V_\Omega \times V_T \times V_C \times V_\chi$$

and a vector of functions determining the nonhomogeneous boundary conditions $\mathrm{U}_b = (0, \Omega_b, T_b, C_b, 0)$.

4.3. Stationary problem

We take the integral identities (3.1), (3.2), (3.3), (3.4), (3.5) without the first evolution terms with time derivative. We multiply them with positive constants $k_\psi, k_\Omega, k_T, k_C, k_\chi$, respectively and sum them up. We choose $k_\psi = k_\Omega = 1$, the constants k_T, k_C, k_χ will be chosen later such that they ensure coerciveness of the operator.

Summing the identities we obtain an identity containing 17 terms, we associate them into four groups, the first three represent defining formulae for operators $\mathscr{A}, \mathscr{B}, \mathscr{C} : \mathbb{W} \to \mathbb{V}^*$:

— the first *principal* linear operator $\mathscr{A}$ contains all scalar product type terms,

— the second *convective* nonlinear operator $\mathscr{B}$ consists of all trilinear forms,

— the third *coupling* operator $\mathscr{C}$ contains remaining bilinear terms,

— the remaining terms form a functional $\mathscr{F}_0$ on $\mathbb{V}$, [3].

Thus the problem is converted into an operator equation

$$\mathscr{A}(\mathrm{U}) + \mathscr{B}(\mathrm{U}) + \mathscr{C}(\mathrm{U}) = \mathscr{F}_0, \qquad \mathrm{U} - \mathrm{U}_b \in \mathbb{V}.$$

Finally we get rid of nonhomogeneous boundary conditions. In the equation we replace the unknown U by $\mathrm{U} + \mathrm{U}_b$ with $\mathrm{U} \in \mathbb{V}$ and rewrite the equation $\mathscr{A}(\mathrm{U} + \mathrm{U}_b) + \mathscr{B}(\mathrm{U} + \mathrm{U}_b) + \mathscr{C}(\mathrm{U}) = \mathscr{F}_0$ to the form

$$\mathscr{A}(\mathrm{U}) + \mathscr{B}(\mathrm{U}) + \mathscr{C}(\mathrm{U}) + \mathscr{D}(\mathrm{U}) = \mathscr{F}_s, \tag{4.3}$$

where the operator $\mathscr{D}$ contains all the new terms linear in the unknown U and $\mathscr{F}_s$ the other terms. Now we formulate the stationary problem:

Stationary problem. *The function* $\mathrm{U}+\mathrm{U}_b$ *is called the* weak solution *to the stationary problem iff* $\mathrm{U} \in \mathbb{V}$ *and the operator equation (4.3) holds on* $\mathbb{V}^*$.

Theorem 4.1. *Let the assumption (4.1) and (4.2) be satisfied. Then the* Stationary problem *is well defined and admits a weak solution.*

The proof of existence of the solution consists in verifying assumptions of the following existence theorem, see [2]:

The operator equation $T(u) = b$ *on a separable reflexive Banach space* V *with coercive and weakly continuous* ($u_n \rightharpoonup u$ *implies* $T(u_n) \rightharpoonup T(u)$) *operator* $T : V \to V^*$ *and* $b \in V^*$ *admits a solution.*

Weak continuity of the operator can be proved using linearity in the highest derivatives and compact imbedding which makes lower order terms converging strongly. Coerciveness is ensured by the first term with $\mathscr{A}$. The second term $\langle \mathscr{B}(\mathrm{U}), \mathrm{U} \rangle = 0$ and the other terms can be made arbitrary small by a special choice of auxiliary functions in U_b with a cut off function when the convective operator $\mathscr{D}$ is estimated and by a special choice of constants k_T, k_C, k_χ when the coupling operator $\mathscr{C}$ is estimated. For detailed proof see [3].

4.4. Evolution problem

Using the same vector and operator formulation starting from integral identities including their evolution terms which forms another operator $\mathscr{E}$ we obtain the operator equation

$$\tfrac{\mathrm{d}}{\mathrm{d}t}\mathscr{E}(\mathrm{U}(t)) + (\mathscr{A}(t) + \mathscr{B} + \mathscr{C}(t) + \mathscr{D}(t))\,(\mathrm{U}(t)) = \mathscr{F}_e(t)\,. \tag{4.4}$$

We complete it with initial conditions.

The problem admits a solution. The proof follows the Rothe method, see [4]. The evolution problem is semidiscretized to a sequence of stationary problems whose solvability is ensured in previous paragraph. Using the corresponding sequence of solutions we construct the Rothe piecewise constant function and the continuous piecewise linear function. A priori estimates yield existence of a weakly converging subsequence. Justification of the limit procedure ensures that the limit solves the evolution problem and completes the proof.

Acknowledgments

This research was supported by grant No. 201/97/0153 of Grant Agency of Czech Republic.

References

[1] Franců, J. (1992). On modelling of Czochralski flow. (Czech), Thesis, Masaryk University, Brno.

[2] Franců, J. (1994). Weakly continuous operators. Applications to differential equations *Appl. Math.*, 39:45–56.

[3] Franců, J. (1998). Modelling of Czochralski flow. *Abstr. Appl. Anal.*, 3,1–2:1–39.

[4] Feistauer, M. (1993). Mathematical Methods in fluid dynamics. Longman Scientific&Technical.

[5] Kufner, A. (1980). Weighted Sobolev spaces. Teubner-Texte zur Mathematik – Band 31, Teubner Leipzig.

[6] Knobloch, P. (1996). Solvability and finite element discretization of a mathematical model related to Czochralski crystal growth, *Otto-von-Guericke-Universität Magdeburg, Preprint MBI-96-5.*

SYMMETRIC STATIONARY SOLUTIONS TO THE PLANE EXTERIOR NAVIER-STOKES PROBLEM FOR ARBITRARY LARGE REYNOLDS NUMBER

Giovanni Paolo Galdi

Abstract: We show that the two-dimensional stationary exterior Navier-Stokes problem is solvable for arbitrary large Reynolds number, in the class of symmetric solutions, provided the corresponding homogeneous problem has only the zero solution.

Keywords: Navier-Stokes equations, existence, stationary solution, two-dimensional exterior problem.

1. INTRODUCTION

In his celebrated paper of 1933, J.Leray studied the solvability of an exterior boundary-value problem related to the Navier-Stokes equations. In a suitable dimensionless form, the problem can be written as follows

$$\left.\begin{aligned} \Delta v &= v \cdot \nabla v + \nabla p \\ \nabla \cdot v &= 0 \end{aligned}\right\} \text{ in } \Omega \qquad (1.1)$$
$$v = 0 \ \text{ at } \Sigma \equiv \partial\Omega$$

along with the condition at infinity

$$\lim_{|x|\to\infty} v(x) = v_\infty. \qquad (1.2)$$

As is known, these equations describe the steady motion of a viscous liquid $\mathcal{F}$ around a body $\mathcal{B}$, translating with a prescribed constant velocity $-v_\infty$, when the motion is viewed from a frame attached to $\mathcal{B}$. In our dimensionless form, the magnitude of v_∞ can be identified with the "Reynolds number" of the problem. The vector field v and the scalar field p are velocity and pressure, respectively, associated with the flow of

Applied Nonlinear Analysis, edited by Sequeira *et al.*
Kluwer Academic / Plenum Publishers, New York, 1999.

$\mathcal{F}$. Ω is the complement of $\mathcal{B}$ and represents the relevant region occupied by $\mathcal{F}$, and it can be a domain of $\mathbb{R}^2$ or of $\mathbb{R}^3$, according to whether the motion of $\mathcal{F}$ may be considered plane or fully threedimensional.

Undoubtedly, the most significative contribution of Leray to problem (1.1), (1.2) was to show that, for *any* prescribed nonzero v_∞ there is at least one solution to (1.1). Moreover, in the case $\Omega \subset \mathbb{R}^3$, he also showed the validity of (1.2), in a suitable generalized sense. [(1)] Therefore, in the threedimensional case, Leray proved the solvability of (1.1), (1.2) for *arbitrary* Reynolds number. However, for $\Omega \subset \mathbb{R}^2$, namely, in the case of plane flow, he was not able to get the same type of result, since he could not establish (1.2), in any sense.

The question of whether Leray's solution satisfies (1.2), and, more generally, the question of the solvability of (1.1), (1.2) for $\Omega \subset \mathbb{R}^2$ has become the object of deep researches by many authors. Among others, D.Gilbarg and H.Weinberger [9], [10] have shown that the solution constructed by Leray is always bounded and that it converges at large distances, in the mean square over the angle, to a certain vector v_0. More detailed information about convergence can be given if the solution is symmetric. Specifically, a pair $v \equiv (u, w), p$ is said *symmetric* if u and p are even in x_2 and w is odd in x_2. If $\mathcal{B}$ is symmetric around the x_1-axis, and $v_\infty = \lambda e$, with e unit vector along x_1, Leray's construction leads to a symmetric solution. In such a case, C.Amick [1] has shown that v tends to v_0 uniformly pointwise. However, it is *not* known whether or not $v_\infty = v_0$ (v_0 may be even zero!) and, consequently, the question of whether Leray's solution satisfies (1.2) remains open.

Solvability of problem (1.1), (1.2), with methods completely different than Leray's, was considered by R.Finn and D.R.Smith [4], [5], [13], and, more recently, by me [6], [7]. In these papers it is shown that the problem has one solution, at least for Reynolds number of *restricted* size (small translational velocity). Moreover, the solution is *physically reasonable* in the sense of Finn [13], and it is locally unique.

In view of all the above considerations, the fundamental question that remains still open is whether or not (1.1), (1.2) is solvable for arbitrary *large* Reynolds number.

The objective of this note is to give a contribution along this direction. We shall limit ourselves to give the main ideas of proofs, referring the reader to a forthcoming full detailed paper, that will appear elsewhere. Denote by $(\mathrm{NS})_0$ the problem (1.1), (1.2) with $v_\infty = 0$, where (1.2) is understood in the sense of pointwise, uniform convergence. Clearly,

---

(1) The validity of (1.2), pointwise and uniformly, was first proved by R.Finn [3].

the zero solution $v = 0$, p=const is a solution to $(\mathrm{NS})_0$. Furthermore, assume $\mathcal{B}$ symmetric around the x_1-axis (say) and denote by $\mathcal{C}$ the class of symmetric pairs v, p with v having a finite Dirichlet integral. Our result states that *if* the zero solution is the *only* solution to $(\mathrm{NS})_0$ in the class $\mathcal{C}$, then problem (1.1), (1.2) is solvable in $\mathcal{C}$, for *arbitrary large* Reynolds numbers and the corresponding solutions are physically reasonable. In particular, denoting by M the set of λ for which (1.1), (1.2) has at least one symmetric solution associated to a given $v_\infty = \lambda e$, we show that M contains an *unbounded* set of the positive real axis. [(2)]

Stated in a different way, for our result to be true it is sufficient that every symmetric solution to the homogeneous problem $(\mathrm{NS})_0$ with v having a finite Dirichlet integral is identically zero. We wish to emphasize that the problem here is not related to local regularity of solutions to $(\mathrm{NS})_0$ (they are real-analytic in Ω) but to their behavior at large distances.

The paper is organized as follows. In section 2 we collect some known results concerning Leray's solutions. Section 3 is devoted to sketch the proof of a result which is crucial for the proof of our main theorem. The result shows, in particular, that the velocity field of *any* symmetric solution constructed by Leray's method is bounded from below by a constant (depending only on Σ) times a suitable power of λ (see Theorem 3.1). To our knowledge, this is the first *explicit* relation between these two quantities. Finally, in Section 4, we give a proof of our main theorem.

Throughout the paper we shall use standard notations for function spaces. So, for instance, $L^q(\mathcal{A})$, $W^{m,q}(\mathcal{A})$, etc., will denote the usual Lebesgue and Sobolev spaces on the domain $\mathcal{A}$, with norms $\|\cdot\|_{q,\mathcal{A}}$ and $\|\cdot\|_{m,q,\mathcal{A}}$, respectively. Whenever confusion will not arise, we shall omit the subscript $\mathcal{A}$. Moreover, we represent a function, say u, at a point in Ω by either $u(x_1, x_2)$ or $u(r, \theta)$ in polar coordinates. This latter is used as a notation for $u(r\cos\theta, r\sin\theta)$.

2. SYMMETRIC LERAY SOLUTIONS AND RELATED PROPERTIES

In this section we shall consider Leray solutions and recall some related properties that we shall use later in the paper. Even though several of the results we shall state continue to hold for more general situations, we shall restrict our attention to the class of *symmetric solutions*. To this end, we denote by $\mathcal{B}$ a smooth compact, connected set of $\mathbb{R}^2$, symmetric

(2) Without loss of generality, we may take $\lambda > 0$.

around the x_1-axis. We also denote by Σ its boundary, and by Ω its complement. We take the origin of coordinates in the interior of $\mathcal{B}$ and assume, without loss, $\mathcal{B} \subset \{|x| \leq 1\}$. Finally, we set

$$\Omega_R = \{x \in \Omega : |x| < R, \quad R \geq 1\}.$$

Let $\lambda > 0$ and let e be the unit vector $(1, 0)$. Then, by a symmetric solution to the exterior Navier-Stokes problem in Ω corresponding to λ, we mean a pair constituted by a vector field $v = (u, w)$ and a scalar field p, where u and p even in x_2 and w odd in x_2, satisfying the following equations

$$\left.\begin{aligned} \Delta v = v \cdot \nabla v + \nabla p \\ \nabla \cdot v = 0 \end{aligned}\right\} \text{ in } \Omega \qquad v = 0 \text{ at } \Sigma \tag{2.1}$$

along with the condition at infinity

$$\lim_{|x| \to \infty} v(x) = \lambda e. \tag{2.2}$$

A solution to (2.1)-(2.2) was sought by Leray [12] by means of the following procedure of "invading domains". Let $\{R_k\}_{k \in \mathbb{N}}$ be an unbounded, increasing sequence of positive numbers, with $R_k > 2$. For each k, consider the sequence of problems:

$$\left.\begin{aligned} \Delta v_k = v_k \cdot \nabla v_k + \nabla p_k \\ \nabla \cdot v_k = 0 \end{aligned}\right\} \text{ in } \Omega_{R_k} \qquad v_k = 0 \text{ at } \Sigma, \qquad v_k = \lambda e \text{ at } |x| = R_k. \tag{2.3}$$

Combining the work of Leray with classical regularity theory (see, *e.g.*, [7]) we obtain the following result.

Lemma 2.1. *There exist a subsequence of* $\{v_k, p_k\}$ *–that we still denote by* $\{v_k, p_k\}$*– and two fields* $v = (u, w)$ *and* p *such that*

(i) $\int_{\Omega_{R_k}} |\nabla v_k|^2 \leq M$, *for some* M *depending only on* Σ *and* λ;

(ii) $v \in C^\infty(\Omega) \cap C^2(\overline{\Omega'})$, $p \in C^\infty(\Omega) \cap C^1(\overline{\Omega'})$, *for all bounded subdomains* Ω';

(iii) $\|v_k - v\|_{C^2(\overline{\Omega'})} + \|p_k - p\|_{C^1(\overline{\Omega'})} \longrightarrow 0$ *as* $k \to \infty$;

(iv) v, p *is symmetric and satisfies* (2.1),

(v) $\nabla v \in L^2(\Omega)$ *and the following* energy inequality *holds*

$$\int_\Omega |\nabla v|^2 \le -\lambda\, e \cdot \int_\Sigma T(v,p)\cdot n, \tag{2.4}$$

where $T(v,p) = \nabla v + (\nabla v)^\top - pI$.

Throughout this paper, solutions v, p to (2.1)-(2.2) described in Lemma 2.1 will be referred to as *symmetric Leray solutions.*

As is well-known, Leray's result recalled in Lemma 2.1 does not establish the validity of (2.2) for v. In fact, it does not even ensure the boundedness of v or of the approximating velocity field v_k. These issues were later considered by several authors. We will collect the results we need in the following lemma, where part (i) is due to Gilbarg and Weinberger [9], [10], part (ii) is due to Amick [1], and part (iii) is proved by Amick [2] in conjunction with a result of Smith [13]; for part (iii), see also [7], Section X.6, and [8].

Lemma 2.2. (i) *There exists a positive constant* C_0 *depending only on* λ *and* Σ *such that*

$$|v_k(x)| \le C_0, \quad \text{for all } x \in \Omega_{3R_k/4}.$$

Thus, as a consequence of Lemma 2.1(iii), v *is bounded.*

(ii) *There is* $v_\infty \in \mathbb{R}^2$ *such that*

$$\lim_{|x|\to\infty} v(x) = v_\infty, \quad \textit{uniformly.}$$

(iii) *If* $v_\infty \neq 0$, *we have, for all sufficiently large* $|x|$

$$|v(x) - v_\infty| \le c|x|^{-1/2}, \quad |\nabla v(x)| \le c|x|^{-1}\log^2|x|, \quad |p(x)| \le c|x|^{-1}\log|x|,$$

with c *independent of* x.

This lemma does not ensure that $v_\infty = \lambda e$. However, one can prove the following result.

Lemma 2.3. *Let* v_∞ *be as in* Lemma 2.2(ii). *Then,* $v_\infty = \mu e$, *where* $\mu = \alpha\lambda$, *for some* $\alpha \in [0,1]$.

3. A KEY RESULT

Throughout this section, we denote by $\{v_k = (u_k, w_k), p_k\}$ a symmetric solution to (2.3) and by $\omega_k \equiv \partial u_k/\partial x_2 - \partial w_k/\partial x_1$ the corresponding vorticity. We also set

$$D_k \equiv \int_{\Omega_{R_k}} |\nabla v_k|^2.$$

and

$$\delta \equiv \left(\int_{\Omega_3} |\nabla v|^2\right)^{1/4},$$

where v, p is a symmetric Leray solution. Furthermore, we put

$$\|f\|_m \equiv \|f\|_{C^m(\overline{\Omega}_2)} = \max_{\Omega_2} \sum_{0 \le \alpha_1+\alpha_2 \le m} \left| \frac{\partial^{\alpha_1+\alpha_2} f(x)}{\partial x_1^{\alpha_1} \partial x_2^{\alpha_2}} \right|.$$

Finally, we indicate by c a constant depending at most on Σ, and whose numerical value is not essential to our aims. In particular, c may have several different values in a single computation. For example, we may have, in the same line, $2c \le c$.

The main objective of this section is to sketch a proof of the following key result.

Theorem 3.1. *Let v, p be a symmetric Leray solution corresponding to a given λ. Then, there exists a homogeneous polynomial $P = P(\delta)$ with coefficients depending only on Σ, such that*

$$\lambda^2 \le P(\delta).$$

Remark 3.1. The proof of this theorem is a consequence of several intermediate steps. Before doing this, however, we wish to point out a particular, immediate consequence of our result, namely, that a symmetric Leray solution corresponding to $\lambda > 0$ can never be trivial, *i.e*, $v \equiv 0$, p=const. This was proved for the first time by Amick [1], Theorem 29. It is not known if the same result is true for non-symmetric solutions.

Lemma 3.1. *The following inequality holds, for all $\rho_k \in (R_k/2, 3R_k/4)$*

$$\lambda^2 \le c\left(D_k + \int_0^{2\pi} |v(\rho_k, \theta)|^2 d\theta\right).$$

We recall that the vorticity ω_k satisfies the following equation

$$\Delta\omega_k - v \cdot \nabla\omega_k = 0. \tag{3.1}$$

By using suitable "cut-off " and "energy" arguments, from (3.1) and (2.3) we can show the following lemmas.

Lemma 3.2. *The following inequality holds*

$$\int_{\Omega_{3R_k/4}} |\nabla\omega_k|^2 \le c\left(\|v_k\|_2^2 + \|v_k\|_2^3 + R_k^{-1} M(C_0+1)\right) \equiv A_k,$$

where M and C_0 are the constants introduced in Lemma 2.1(i) *and* Lemma 2.2(i), *respectively.*

Lemma 3.3. *There exists $\widetilde{R}_k \in (R_k/2, 3R_k/4)$ such that*

(i) $\int_0^{2\pi} |v_k(\widetilde{R}_k, \theta)|^2 d\theta \leq c(|v(\widetilde{x}_k)|^2 + A_k + D_k)$, *for all $\widetilde{x}_k$ with $|\widetilde{x}_k| = \widetilde{R}_k$;*

(ii) $\int_0^{2\pi} \left| \frac{\partial p_k(\widetilde{R}_k, \theta)}{\partial \theta} \right|^2 d\theta \leq c(1 + \max_\theta |v(\widetilde{R}, \theta)|^2)(A_k + D_k)$;

(iii) $\max_\theta |v(\widetilde{R}_k, \theta)| \leq \lambda + c(A_k + D_k)^{1/2}$,

where the quantity A_k is defined in Lemma 3.2.

Lemma 3.4. *Let $\widetilde{R}_k$ be the number defined in* Lemma 3.3. *The following inequality holds*

$$\max_\theta |p_k(\widetilde{R}_k, \theta)| \leq c\left((1 + \lambda)(A_k + D_k)^{1/2} + \|p_k\|_0 + A_k + D_k\right)$$

Using a result of Amick [1], §4.2, we can show the following one.

Lemma 3.5. *The following inequality holds*

$$\lambda^2 \leq c\left((1 + \lambda)(A_k + D_k)^{1/2} + A_k + D_k + \|p_k\|_0 + \|v_k\|_0^2\right). \qquad (3.2)$$

The following two results are based on energy estimates for (2.3), and on classical local estimates for the Stokes problem, respectively.

Lemma 3.6. *Let v, p be a symmetric Leray solution corresponding to λ. Then, the following inequality holds*

$$\lambda^2 \leq c\left(S^3 + S^2 + S^{3/2} + S\right),$$

where

$$S = \|p\|_0 + \|v\|_2.$$

Lemma 3.7. *Let v, p be as in* Lemma 3.6. *Then,*

$$\|p\|_0 + \|v\|_2 \leq c\left((\chi^2 + \chi)^2 + \chi^2 + \chi\right),$$

where

$$\chi = \|\nabla v\|_{\Omega_3}^2 + \|\nabla v\|_{\Omega_3}.$$

We are now in a position to give a *proof of* Theorem 3.1. In fact, it is an immediate consequence of Lemma 3.6 and Lemma 3.7.

4. EXISTENCE OF SYMMETRIC SOLUTIONS FOR ARBITRARY LARGE REYNOLDS NUMBER

We begin to introduce a suitable regularity class. Specifically we denote by $\mathcal{C}$ the class of pairs constituted by a vector field $u = (u_1, u_2)$ and scalar field ϕ such that:

(i) *Symmetry*:

$$u_1(x_1, x_2) = u_1(x_1, -x_2), \quad u_2(x_1, x_2) = -u_2(x_1, -x_2)$$
$$\phi(x_1, x_2) = \phi(x_1, -x_2),$$

(ii) *Finite Dirichlet Integral*: $\displaystyle\int_\Omega |\nabla u|^2 < \infty.$

The objective of this section is to prove the following result.

Theorem 4.1. *Let $\mathcal{B}$ be symmetric around the x_1-axis. Assume that the following problem*

$$\left.\begin{aligned} &\Delta u = u \cdot \nabla u + \nabla \phi \\ &\nabla \cdot u = 0 \end{aligned}\right\} \text{ in } \Omega$$
$$u = 0 \text{ at } \Sigma, \qquad \lim_{|x|\to\infty} u(x) = 0, \text{ uniformly} \tag{4.1}$$

has only the zero solution in the class $\mathcal{C}$. Then, there is a set M with the following properties:

(i) *$M \subset [0, \infty)$;*

(ii) *$M \supset [0, c)$ for some $c = c(\Sigma) > 0$;*

(iii) *M is unbounded;*

(iv) *For any $\mu \in M$, the problem*

$$\left.\begin{aligned} &\Delta v = v \cdot \nabla v + \nabla p \\ &\nabla \cdot v = 0 \end{aligned}\right\} \text{ in } \Omega$$
$$v = 0 \text{ at } \Sigma, \qquad \lim_{|x|\to\infty} v(x) = \mu e \tag{4.2}$$

has at least one solution in the class $\mathcal{C}$.

Before we give the proof of Theorem 4.1, we wish to make the following remarks.

Remark 4.1. In the case when $\mathcal{B} = \emptyset$ (an unrealistic assumption in our present situation) the uniqueness of the zero solution to (4.1) is a simple consequence of the maximum principle applied to the vorticity equation; see [10], Theorem 2.

Remark 4.2. All possible solutions to (4.1) in the class $\mathcal{C}$ are smooth and satisfy the following asymptotic conditions ([7], Theorem X.3.2, Theorem X.3.3)

$$\lim_{|x|\to\infty} D^m u(x) = \lim_{|x|\to\infty} D^m \phi(x) = 0,$$

where D^m represents a derivative of arbitrary order $m \geq 0$. Furthermore, denoting by ω the vorticity field associated to u, we have ([10], Theorem 6)

$$|\omega(x)| \leq c(|x| + 1)^{-3/4}.$$

Let us now come back to the proof of Theorem 4.1. To this end, we need the following auxiliary result, that can be obtained by means of a suitable variation of the classical Hopf extension method..

Lemma 4.1. *Let $\mu_0 > 0$ and let v, p be a solution to* (4.2) *corresponding to $\mu \in (0, \mu_0]$. Then, there exists a positive constant κ, depending only on Σ and μ_0, such that*

$$\int_\Omega |\nabla v|^2 \leq \kappa$$

We are now in a position to give the *proof of* Theorem 4.1. Let us denote by M the set of those $\mu > 0$ for which problem (4.2) has a corresponding solution v, p. From the work of Finn and Smith [5] and Galdi [6], we know that $M \supset [0, c]$, for some positive $c = c(\Sigma)$. We shall now show that $M \supset M_0$ where M_0 enjoys the properties:

(i) $M_0 \subset (0, \infty)$;

(ii) M_0 is unbounded.

Actually, let M_0 be defined as follows:

$$\mu \in M_0 \quad \text{if and only if} \quad \lim_{|x|\to\infty} v(x) = \mu\, e, \quad \text{uniformly}$$

where v, p is a symmetric Leray solution corresponding to a given $\lambda > 0$. Clearly, $M_0 \subset M$. Also, by Lemma 2.2, $M_0 \neq \emptyset$. Furthermore, $M_0 \subset (0, \infty)$. In fact, by Lemma 2.3, $M_0 \subset [0, \infty)$. However, $0 \notin M_0$, because, otherwise, v, p satisfy the homogeneous equation (4.1), and this, by assumption, would imply $v \equiv 0$. So, by Theorem 3.1, we would conclude $\lambda = 0$, which leads to a contradiction. This proves property (i) of M_0. Let us show (ii). Assuming M_0 bounded means $0 < \mu \leq \mu_0 < \infty$, for some $\mu_0 > 0$. Thus, from Lemma 4.1, we have

$$\int_\Omega |\nabla v|^2 \leq c(\mu_0), \tag{4.3}$$

for all symmetric Leray solutions corresponding to *arbitrary* $\lambda > 0$. Using (4.3) into Theorem 3.1, we obtain

$$\lambda \leq c_1(\mu_0),$$

for *arbitrary* $\lambda > 0$, which gives a contradiction. The theorem is, therefore, completely proved.

References

[1] Amick, C.J. (1988). *On Leray's Problem of Steady Navier-Stokes Flow Past a Body in the Plane.* Acta Math., 161:71–130.

[2] Amick, C.J. (1991). On the Asymptotic Form of Navier-Stokes Flow Past a Body in the Plane. *J. Diff. Equations,* 91:149–167.

[3] Finn, R. (1959). On Steady-State Solutions of the Navier-Stokes Partial Differential Equations. *Arch. Rational Mech. Anal.,* 3:381–396.

[4] Finn, R. and Smith, D.R. (1967). On the Linearized Hydrodynamical Equations in Two Dimensions. *Arch. Rational Mech. Anal.,* 25:1–25.

[5] Finn, R. and Smith, D.R. (1967). On the Stationary Solution of the Navier-Stokes Equations in Two Dimensions. *Arch. Rational Mech. Anal.,* 25:26–39.

[6] Galdi, G.P. (1993). Existence and Uniqueness at Low Reynolds Number of Stationary Plane Flow of a Viscous Fluid in Exterior Domains. *Recent Developments in Theoretical Fluid Mechanics.* Galdi, G.P. and Nečas, J. (Eds.), Pitman Research Notes in Mathematics, Longman Scientific and Technical, 291:1–33.

[7] Galdi, G.P. (1998). *An Introduction to the Mathematical Theory of the Navier-Stokes Equations: Nonlinear Steady Problems.* Springer Tracts in Natural Philosophy, Vol. 39, Springer-Verlag, Revised Edition.

[8] Galdi, G.P. and Sohr, H. (1995). On the Asymptotic Structure of Plane Steady Flow of a Viscous Fluid in Exterior Domains. *Arch. Rational Mech. Anal.*, 131:101–119.

[9] Gilbarg, D. and Weinberger, H.F. (1974). Asymptotic Properties of Leray's Solution of the Stationary Two-Dimensional Navier-Stokes Equations. *Russian Math. Surveys,* 29:109–123.

[10] Gilbarg, D. and Weinberger, H.F. (1978). Asymptotic Properties of Steady Plane Solutions of the Navier-Stokes Equations with Bounded Dirichlet Integral. *Ann. Scuola Norm. Sup. Pisa,* 5(4):381–404.

[11] Hamel, G. (1916). Spiralförmige Bewegungen zäher Flüssigkeiten. *Jber. Dtsch. Math. Ver.*, 25:34–60; English Trans.: NACA Tech. Memo. 1342, 1953.

[12] Leray, J. (1933). Etude de Diverses Équations Intégrales non Linéaires et de Quelques Problèmes que Pose l'Hydrodynamique. *J. Math. Pures Appl.,* 12:1–82.

[13] Smith, D.R. (1965). Estimates at Infinity for Stationary Solutions of the Navier-Stokes Equations in Two Dimensions. *Arch. Rational Mech. Anal.*, 20:341–372.

A FICTITIOUS-DOMAIN METHOD WITH DISTRIBUTED MULTIPLIER FOR THE STOKES PROBLEM

Vivette Girault, Roland Glowinski, T.W. Pan

Abstract: This article is devoted to the numerical analysis of a fictitious domain method for the Stokes problem, where the boundary condition is enforced weakly by means of a multiplier defined in a portion of the domain. In practice, this is applied for example to the sedimentation of many particles in a fluid. It is found that the multiplier is divergence-free. We present here sufficient conditions on the relative mesh sizes for convergence of the discrete method. Also, we show how the constraint on the divergence of the discrete multiplier can be relaxed when such a sedimentation problem is discretized.

Keywords: Fictitious domain, distributed multiplier, particle sedimentation.

The fictitious-domain method presented here is motivated by the numerical simulation of an incompressible flow around moving rigid bodies, when the rigid-body motions are caused by hydrodynamical forces and for example, gravity. One example is the problem of sedimentation of particles. Our method consists in filling the moving bodies by the surrounding fluid and imposing weakly the rigid-body motions, in this region, by means of a distributed Lagrange multiplier. This leads to a modified flow problem in the entire region. The advantage of this approach is that a single uniform mesh is used for the entire region and the particles are meshed independently, once and for all. On the other hand, as we shall see here, the corresponding scheme has a low order of convergence. The numerical analysis of this problem is difficult, and to simplify, we shall mostly consider the case of a single particle with a known rigid-body motion, immersed in a fluid whose equation of motion is a steady Stokes system of equations in two dimensions.

In the next section, we shall state the fictitious-domain formulation of an exterior Stokes problem with a non-homogeneous Dirichlet boundary

Applied Nonlinear Analysis, edited by Sequeira *et al.*
Kluwer Academic / Plenum Publishers, New York, 1999.

condition, interpret the new problem and show that it is well-posed. We shall see that the Lagrange multiplier is determined by a divergence-free condition. Section 2 is devoted to the numerical discretization of this problem, characterized by the fact that the fluid mesh and the Lagrange multiplier mesh are unrelated, up to a mesh-length ratio. Particular emphasis is placed on the proof of a discrete uniform inf-sup condition. Section 3 studies briefly the particular case where the Dirichlet boundary condition is given by a rigid-body motion. In this case, the discrete divergence-free constraint on the multiplier can be relaxed.

The fictitious-domain method discussed in this article has been generalized to the solution of the full time-dependent Navier-Stokes equations modelling incompressible viscous flow in regions with moving boundaries. The corresponding computational methods are described in [9], [10], [11]. In [10], the motion of the moving boundary is known in advance, while [9] and [11] discuss the simulation of particulate flow, with up to 500 particles. Simulations involving 10^3 (resp. 10^2) particles in two dimensions (resp. three dimensions) have become routines for Newtonian viscous fluids. In articles to appear or in preparation, one describes further applications to particulate flow with more than 10^3 particles in three dimensions (for Newtonian fluids) and of the order of 10 particles in two dimensions for visco-elastic liquids of the Oldroyd-B type. Via parallel computing, one expects being able to simulate in a near future particulate flow with more than 10^3 particles in three dimensions for visco-elastic liquids.

1. A FICTITIOUS-DOMAIN FORMULATION OF THE STOKES PROBLEM

We consider the case of a single particle occupying a bounded plane domain B, immersed in a fluid contained in a rectangular domain Ω, so that the original domain of interest is $\Omega \setminus \bar{B}$. We assume that the boundary ∂B of B is Lipschitz-continuous, that Ω is large enough so that ∂B is far enough from the boundary Γ of Ω, and Γ has sides parallel to the axes. We do not suppose from the onset that particle B has a rigid-body motion, but we assume that the given velocity on ∂B is the trace of a known function defined in B, with zero divergence. If this function were not known, and a lifting had been explicitly constructed, the method discussed here would lose much interest. We introduce the spaces, on a domain $\mathcal{O}$

$$W(\mathcal{O}) = \{\mathbf{v} \in H^1(\mathcal{O})^2 \,;\, \operatorname{div} \mathbf{v} = 0 \text{ in } \mathcal{O}\},$$

$$L_0^2(\mathcal{O}) = \{q \in L^2(\mathcal{O}) \,;\, \int_{\mathcal{O}} q \, d\mathbf{x} = 0\},$$

(we refer the reader to [15] or [1] for the properties of Sobolev spaces). Then, for $\mathbf{f}$ given in $L^2(\Omega \setminus \bar{B})^2$ and $\mathbf{g}$ given in $W(B)$, we want to derive an equivalent variational formulation for the following Stokes problem: Find $\mathbf{u}$ in $H^1(\Omega \setminus \bar{B})^2$ and p in $L_0^2(\Omega \setminus \bar{B})$, solution of

$$-\nu \Delta \mathbf{u} + \nabla p = \mathbf{f} \quad \text{in } \Omega \setminus \bar{B}, \tag{1.1}$$

$$\operatorname{div} \mathbf{u} = 0 \quad \text{in } \Omega \setminus \bar{B}, \tag{1.2}$$

$$\mathbf{u} = \mathbf{0} \quad \text{on } \Gamma, \ \mathbf{u} = \mathbf{g} \quad \text{on } \partial B, \tag{1.3}$$

where $\nu > 0$ is the given viscosity constant. Since by assumption, $\mathbf{g}$ belongs to $W(B)$, it satisfies the compatibility condition

$$\int_{\partial B} \mathbf{g} \cdot \mathbf{n} \, d\sigma = 0,$$

where $\mathbf{n}$ denotes the unit normal to ∂B, directed inside B. This Stokes problem has a unique solution. Note that the boundary condition (1.3) on Γ is chosen according to convenience and can be replaced by another one.

As explained in the introduction, we propose to impose weakly (1.3) on ∂B by means of a Lagrange multiplier λ defined in B, and to set our problem in the whole domain Ω. The reader will see below that it is reasonable to take λ with H^1 regularity. Since the pressure p is the multiplier associated with the divergence constraint, we consider the following bilinear form

$$b(\mathbf{v}, (\lambda, p)) = -\int_\Omega p \operatorname{div} \mathbf{v} \, d\mathbf{x} - \int_B \nabla \lambda : \nabla \mathbf{v} \, d\mathbf{x} - \int_B \lambda \cdot \mathbf{v} \, d\mathbf{x}. \tag{1.4}$$

The next lemma shows that we must impose on the volume multiplier a zero divergence constraint, because otherwise it is not determined.

Lemma 1.1. *If* $\operatorname{div} \lambda = 0$, *the equation: Find* (λ, p) *in* $H^1(B)^2 \times L_0^2(\Omega)$ *such that*

$$\forall \mathbf{v} \in H_0^1(\Omega)^2, \ b(\mathbf{v}, (\lambda, p)) = 0, \tag{1.5}$$

has the only solution $(\lambda, p) = (\mathbf{0}, 0)$. *If* $\operatorname{div} \lambda \neq 0$, *then (1.5) has an infinity of nontrivial solutions.*

Thus, we choose $W(B)$ as space for the volume multiplier. Then, we extend $\mathbf{f}$ in B by a function $\tilde{\mathbf{f}} \in L^2(\Omega)^2$ (for instance, $\tilde{\mathbf{f}} = \mathbf{0}$ in B), and we propose the following fictitious-domain variational formulation of (1.1)–(1.3): Find $(\mathbf{u}, p, \lambda)$ in $H_0^1(\Omega)^2 \times L_0^2(\Omega) \times W(B)$, solution of

$$\forall \mathbf{v} \in H_0^1(\Omega)^2 \,,\ \nu \int_\Omega \nabla \mathbf{u} : \nabla \mathbf{v}\, d\mathbf{x} - \int_\Omega p \operatorname{div} \mathbf{v}\, d\mathbf{x} = \int_\Omega \tilde{\mathbf{f}} \cdot \mathbf{v}\, d\mathbf{x} + \int_B \nabla \lambda : \nabla \mathbf{v}\, d\mathbf{x} + \int_B \lambda \cdot \mathbf{v}\, d\mathbf{x}, \tag{1.6}$$

$$\forall q \in L_0^2(\Omega) \,,\ \int_\Omega q \operatorname{div} \mathbf{u}\, d\mathbf{x} = 0, \tag{1.7}$$

$$\forall \mu \in W(B) \,,\ \int_B \nabla(\mathbf{u} - \mathbf{g}) : \nabla \mu\, d\mathbf{x} + \int_B (\mathbf{u} - \mathbf{g}) \cdot \mu\, d\mathbf{x} = 0. \tag{1.8}$$

To interpret this problem, let $(\mathbf{u}, p, \lambda)$ be a solution of (1.6)–(1.8). We easily derive that:

$$\mathbf{u} = \mathbf{g} \quad \text{in } B, \tag{1.9}$$

$$\operatorname{div} \mathbf{u} = 0 \quad \text{in } \Omega,$$

$$-\nu \Delta \mathbf{u} + \nabla p = \mathbf{f} \quad \text{in } \Omega \setminus \bar{B} \,,\ -\nu \Delta \mathbf{u} + \nabla p = \tilde{\mathbf{f}} + \lambda - \Delta \lambda \quad \text{in } B,$$

$$[\frac{\partial}{\partial \mathbf{n}}(\nu \mathbf{u} - \lambda) - p\mathbf{n}]_{\partial B} = \mathbf{0},$$

where

$$[v]_{\partial B} = (v|_{\Omega \setminus \bar{B}} - v|_B)|_{\partial B}$$

denotes the jump of v across ∂B and λ is extended by zero in $\Omega \setminus \bar{B}$ in order to define the jump of $\frac{\partial \lambda}{\partial \mathbf{n}}$. Hence, the restriction of $(\mathbf{u}, p)$ is a solution of the original Stokes problem (1.1)–(1.3), $\mathbf{u}$ satisfies (1.9), and the pair $(\nu \mathbf{g} - \lambda, p)$ is the solution of a Stokes problem in B with a Neumann boundary condition:

$$-\Delta(\nu \mathbf{g} - \lambda) + \nu \mathbf{g} - \lambda + \nabla p = \tilde{\mathbf{f}} + \nu \mathbf{g} \quad \text{in } B, \tag{1.10}$$

$$\operatorname{div}(\nu \mathbf{g} - \lambda) = 0 \quad \text{in } B, \tag{1.11}$$

$$\frac{\partial}{\partial \mathbf{n}}(\nu \mathbf{g} - \lambda) - p\mathbf{n} = (\nu \frac{\partial \mathbf{u}}{\partial \mathbf{n}} - p\mathbf{n})|_{\Omega \setminus \bar{B}} \quad \text{on } \partial B. \tag{1.12}$$

This problem has a unique solution because $(\mathbf{u}, p)$ is known in $\Omega \setminus \bar{B}$. Note also that it simplifies when $\Delta \mathbf{g}$ belongs to $L^2(B)^2$, and this always holds in the applications we have in mind.

Conversely, if $(\mathbf{u}, p)$ is a solution of (1.1)–(1.3), then extending $\mathbf{u}$ by $\mathbf{g}$ in B, $\mathbf{f}$ by $\tilde{\mathbf{f}}$ in B and defining the pair (λ, p) in B by (1.10)–(1.12), the triple $(\mathbf{u}, p, \lambda)$ satisfies (1.6)–(1.8), except that the mean-value of p is generally not zero in Ω.

Problem (1.6)–(1.8) is a mixed variational problem. Since the bilinear form $\nu \int_\Omega \nabla \mathbf{u} : \nabla \mathbf{v}\, d\mathbf{x}$ is elliptic on $[H_0^1(\Omega)^2]^2$, to show that this problem is well-posed, we must establish the following inf-sup condition (cf. [12], [6]).

Theorem 1.2. *There exists a constant $\beta > 0$ such that, for all (λ, p) in $W(B) \times L_0^2(\Omega)$,*

$$\sup_{\mathbf{v}\in H_0^1(\Omega)^2} \frac{1}{|\mathbf{v}|_{H^1(\Omega)}} b(\mathbf{v}, (\lambda, p)) \geq \beta(\|p\|^2_{L^2(\Omega)} + \|\lambda\|^2_{H^1(B)})^{1/2}. \tag{1.13}$$

Proof. For an arbitrary $p \in L_0^2(\Omega)$, we first construct $\mathbf{z}$ in $H^1(B)^2$ satisfying

$$\operatorname{div} \mathbf{z} = p \quad \text{in } B,$$

$$\forall \mathbf{w} \in W(B)\,, \int_B \nabla \mathbf{z} : \nabla \mathbf{w}\, d\mathbf{x} + \int_B \mathbf{z} \cdot \mathbf{w}\, d\mathbf{x} = 0\,, \tag{1.14}$$

that depends continously on p:

$$\|\mathbf{z}\|_{H^1(B)} \leq C_1 \|p\|_{L^2(B)}\,.$$

Next, as

$$\int_B p\, d\mathbf{x} = -\int_{\partial B} \mathbf{z} \cdot \mathbf{n}\, d\sigma\,,$$

we extend $\mathbf{z}$ in $\Omega \setminus \bar{B}$ by constructing $\tilde{\mathbf{z}}$ in $H^1(\Omega \setminus \bar{B})^2$ such that

$$\operatorname{div} \tilde{\mathbf{z}} = p \quad \text{in } \Omega \setminus \bar{B}\,,$$

$$\tilde{\mathbf{z}} = \mathbf{z} \quad \text{on } \partial B\,,\ \tilde{\mathbf{z}} = \mathbf{0} \quad \text{on } \Gamma\,,$$

$$\|\tilde{\mathbf{z}}\|_{H^1(\Omega\setminus\bar{B})} \leq C_2(\|p\|_{L^2(\Omega\setminus\bar{B})} + \|\mathbf{z}\|_{H^{1/2}(\partial B)})\,.$$

By construction, the extended function $\mathbf{z}$ belongs to $H_0^1(\Omega)^2$, $\operatorname{div} \mathbf{z} = p$ in Ω, $\mathbf{z}$ satisfies(1.14) and depends continuously on p:

$$\|\mathbf{z}\|_{H^1(\Omega)} \leq C_3 \|p\|_{L^2(\Omega)}\,. \tag{1.15}$$

Finally, we extend λ to Ω so that the extended function λ belongs to $H_0^1(\Omega)^2$, has zero divergence and

$$|\lambda|_{H^1(\Omega)} \leq C_4 \|\lambda\|_{H^1(B)}\,. \tag{1.16}$$

The choice $\mathbf{v} = -(\mathbf{z} + \lambda)$ verifies

$$b(\mathbf{v}, (\lambda, p)) = \|p\|^2_{L^2(\Omega)} + \|\lambda\|^2_{H^1(B)}\,,$$

$$|\mathbf{v}|_{H^1(\Omega)} \leq (C_3^2 + C_4^2)^{1/2}(\|p\|^2_{L^2(\Omega)} + \|\lambda\|^2_{H^1(B)})^{1/2}\,.$$

■

Remark 1. As a consequence, problem (1.6)–(1.8) has a unique solution $(\mathbf{u}, p, \lambda)$ in $H_0^1(\Omega)^2 \times L_0^2(\Omega) \times W(B)$ that depends continuously on the data $\tilde{\mathbf{f}}$ and $\mathbf{g}$. However, even if $\mathbf{g}$ is smooth (which is the case in practice), it is unlikely that globally $(\mathbf{u}, p)$ have a stronger regularity than $H^{3/2-\varepsilon}(\Omega)^2 \times H^{1/2-\varepsilon}(\Omega)$, because of the jump of $\nu \frac{\partial \mathbf{u}}{\partial \mathbf{n}} - p\,\mathbf{n}$ across ∂B (cf. [13]). In contrast, it is possible that λ belongs to $H^2(B)^2$.

2. DISCRETIZATION

To simplify the discussion, we assume from now on that the boundary ∂B is a polygon. The case of a curved boundary is more technical but brings no essential difficulty. Since the mesh of the fictitious domain Ω and the particle B are unrelated, we choose two independent discretization parameters, $h > 0$ and $\eta > 0$, that will tend to zero. Let $\mathcal{T}_h$ be a uniform triangulation of $\overline{\Omega}$, composed of squares divided into two triangles along the same diagonal, let $\mathcal{S}_\eta$ be a regular triangulation of $\overline{B}$ (cf. [7]) and define the two finite-element spaces

$$X_h = \{\mathbf{v}_h \in \mathcal{C}^0(\overline{\Omega})^2\,;\ \forall T \in \mathcal{T}_h,\ \mathbf{v}_h|_T \in \mathbb{P}_1^2,\ \mathbf{v}_h|_\Gamma = \mathbf{0}\} \subset H_0^1(\Omega)^2\,, \tag{2.1}$$

$$M_h = \{q_h \in L_0^2(\Omega)\,;\ \forall T \in \mathcal{T}_h,\ q_h|_T \in \mathbb{P}_0\} \subset L_0^2(\Omega)\,, \tag{2.2}$$

where the step-size of M_h is sufficiently large with respect to the step-size of X_h, so that this pair satisfies a uniform discrete inf-sup condition (cf. [12]): there exists a constant $\delta^* > 0$, independent of h, such that

$$\inf_{q_h \in M_h} \sup_{\mathbf{v}_h \in X_h} \frac{\int_\Omega q_h \operatorname{div} \mathbf{v}_h \, d\mathbf{x}}{\|q_h\|_{L^2(\Omega)} |\mathbf{v}_h|_{H^1(\Omega)}} \geq \delta^*\,. \tag{2.3}$$

Here we shall impose this condition by asking that the support of the basis functions of M_h be spread over an adequate number of triangles of $\mathcal{T}_h$, so that the mesh-size of M_h, say h_p, is for instance the double of that of X_h. To be precise, we should use a different index h for X_h and M_h, but for the sake of simplicity, we only make this distinction when it is necessary.

Let Λ_η be a standard discretization of $H^1(B)^2$:

$$\Lambda_\eta = \{\mu_\eta \in \mathcal{C}^0(\overline{B})^2\,;\ \forall S \in \mathcal{S}_\eta,\ \mu_\eta|_S \in \mathbb{P}_1^2\} \subset H^1(B)^2\,. \tag{2.4}$$

In order to approximate $W(B)$, we observe that in (1.10) and (1.11), the restriction of p to B is the Lagrange multiplier associated with the divergence constraint on λ. This suggests to retain the same structure at the discrete level. But, we cannot ask that $\int_B q_h \operatorname{div} \mu_\eta \, d\mathbf{x} = 0$ for q_h in M_h without considering the intersection of the support of q_h with B. If this support is too small, we run the risk of imposing too many conditions on λ_η. Hence we discretize $W(B)$ by

$$\begin{aligned} W_\eta = \{\mu_\eta \in \Lambda_\eta\,;\ &\int_B q_h \operatorname{div} \mu_\eta \, d\mathbf{x} = 0 \text{ for all basis functions} \\ &q_h \in M_h \text{with } |T \cap \bar{B}| \geq \gamma |T|, \text{ where } T \text{ is the support of } q_h\}\,, \end{aligned} \tag{2.5}$$

where $\gamma \in (0, 1/2)$ is an adequate parameter. Then we approximate (1.6)–(1.8) by: Find $(\mathbf{u}_h, p_h, \lambda_\eta) \in X_h \times M_h \times W_\eta$ solution of

$$\forall \mathbf{v}_h \in X_h \ , \ \nu \int_\Omega \nabla \mathbf{u}_h : \nabla \mathbf{v}_h \, d\mathbf{x} - \int_\Omega p_h \operatorname{div} \mathbf{v}_h \, d\mathbf{x} = \int_\Omega \tilde{\mathbf{f}} \cdot \mathbf{v}_h \, d\mathbf{x} \\ + \int_B \nabla \lambda_\eta : \nabla \mathbf{v}_h \, d\mathbf{x} + \int_B \lambda_\eta \cdot \mathbf{v}_h \, d\mathbf{x} , \tag{2.6}$$

$$\forall q_h \in M_h \ , \ \int_\Omega q_h \operatorname{div} \mathbf{u}_h \, d\mathbf{x} = 0 , \tag{2.7}$$

$$\forall \mu_\eta \in W_\eta \ , \ \int_B \nabla(\mathbf{u}_h - \mathbf{g}) : \nabla \mu_\eta \, d\mathbf{x} + \int_B (\mathbf{u}_h - \mathbf{g}) \cdot \mu_\eta \, d\mathbf{x} = 0 . \tag{2.8}$$

Again, this is a mixed problem and since the bilinear form $\nu \int_\Omega \nabla \mathbf{u}_h : \nabla \mathbf{v}_h \, d\mathbf{x}$ is elliptic on $X_h \times X_h$, we must check that the bilinear form b satisfies a uniform discrete inf-sup condition. It is convenient to split it as follows: for each $(\lambda_\eta, p_h) \in W_\eta \times M_h$, there exists $\mathbf{z}_h \in X_h$ such that

$$b(\mathbf{z}_h, (\lambda_\eta, p_h)) \geq \beta_1^* (\|p_h\|^2_{L^2(\Omega)} + \|\lambda_\eta\|^2_{H^1(B)}) , \tag{2.9}$$

$$|\mathbf{z}_h|_{H^1(\Omega)} \leq \beta_2^* (\|p_h\|_{L^2(\Omega)} + \|\lambda_\eta\|_{H^1(B)}) , \tag{2.10}$$

where $\beta_1^* > 0$ and $\beta_2^* > 0$ are two constants independent of h, η and (λ_η, p_h).

Theorem 2.1. *Assume (2.3) and suppose S_η is uniformly regular. There exists a constant $\kappa > 1$, independent of h and η, such that, if*

$$\frac{\eta}{h} \geq \kappa , \tag{2.11}$$

and if

$$\left(\frac{h_p}{\eta}\right)^{1/2-\varepsilon} > \sqrt{\frac{2\gamma}{1-\gamma}} , \tag{2.12}$$

then there exist two constants $\beta_1^* > 0$ and $\beta_2^* > 0$, independent of h, η and γ such that (2.9) and (2.10) are satisfied for all $(\lambda_\eta, p_h) \in W_\eta \times M_h$.

Proof. Let $(\lambda_\eta, p_h) \in W_\eta \times M_h$ be arbitrary. First, we extend λ_η to $\Omega \setminus \bar{B}$, so that the extended function $\mathbf{v}$ satisfies for all supports T of the basis functions q_h in M_h:

$$\int_T \operatorname{div} \mathbf{v} \, d\mathbf{x} = 0 .$$

Let $\{T_i\}_{i=1}^I$ be the set of supports of the basis functions $q_h \in M_h$ for which

$$0 < |T_i \cap \bar{B}| < \gamma |T_i|, \tag{2.13}$$

set

$$D_s = \cup_{i=1}^I (T_i \cap \bar{B}),$$

and define a piecewise constant function d_v in $\Omega \setminus \bar{B}$ by

$$d_v = -\frac{1}{|T_i \setminus (T_i \cap \bar{B})|} \int_{T_i \cap \bar{B}} \operatorname{div} \lambda_\eta \, d\mathbf{x}, \text{ in } T_i \setminus (T_i \cap \bar{B}), \, 1 \le i \le I,$$
$$d_v = 0 \text{ elsewhere}.$$

In view of (2.5), we have

$$\int_{\Omega \setminus \bar{B}} d_v \, d\mathbf{x} = \sum_{i=1}^I \int_{T_i \setminus (T_i \cap \bar{B})} d_v \, d\mathbf{x} = -\int_{D_s} \operatorname{div} \lambda_\eta \, d\mathbf{x} = \int_{\partial B} \lambda_\eta \cdot \mathbf{n} \, d\sigma.$$

Furthermore, d_v belongs to $H^{1/2-\varepsilon}(\Omega \setminus \bar{B})$ and λ_η belongs in particular to $H^{1-\varepsilon}(\partial B)^2$. Therefore, we choose $\mathbf{v} = \mathcal{K}(d_v, \lambda_\eta)$, where $\mathcal{K}$ is the operator defined by Theorem 7.1 of [2]; more precisely, we have

$$\operatorname{div} \mathbf{v} = d_v \text{ in } \Omega \setminus \bar{B}, \quad \mathbf{v}|_{\partial B} = \lambda_\eta|_{\partial B}, \quad \mathbf{v}|_\Gamma = \mathbf{0}.$$

Thus, $\mathbf{v}$ belongs to $H^{3/2-\varepsilon}(\Omega \setminus \bar{B})^2$ and, with the same constant C_1, we have

$$\|\mathbf{v}\|_{H^1(\Omega \setminus \bar{B})} \le C_1 \{ \|\lambda_\eta\|_{H^{1/2}(\partial B)} + \|d_v\|_{L^2(\Omega \setminus \bar{B})} \},$$
$$\|\mathbf{v}\|_{H^{3/2-\varepsilon}(\Omega \setminus \bar{B})} \le C_1 \{ \|\lambda_\eta\|_{H^{1-\varepsilon}(\partial B)} + \|d_v\|_{H^{1/2-\varepsilon}(\Omega \setminus \bar{B})} \}.$$

On one hand, (2.13) implies that

$$\|d_v\|_{L^2(\Omega \setminus \bar{B})} \le \sqrt{\frac{2\gamma}{1-\gamma}} |\lambda_\eta|_{H^1(D_s)} \quad \text{where} \quad \sqrt{\frac{2\gamma}{1-\gamma}} < \sqrt{2}.$$

Hence

$$\|\mathbf{v}\|_{H^1(\Omega \setminus \bar{B})} \le C_2 \big(1 + (\frac{2\gamma}{1-\gamma})^{1/2}\big) \|\lambda_\eta\|_{H^1(B)}. \tag{2.14}$$

On the other hand, since d_v belongs to $H^{1/2-\varepsilon}(\Omega \setminus \bar{B})$ and vanishes on all elements except on the sets $T_i \setminus (T_i \cap \bar{B})$, whose measure is bounded below by $(1-\gamma)|T_i| > \frac{1}{2}|T_i|$, and since $\mathcal{S}_\eta$ is uniformly regular, we can apply the following inverse inequality (cf. [4]):

$$\|d_v\|_{H^{1/2-\varepsilon}(\Omega \setminus \bar{B})} \le \frac{C_3}{h_p^{1/2-\varepsilon}} \|d_v\|_{L^2(\Omega \setminus \bar{B})}, \tag{2.15}$$

where the constant C_3 is independent of h_p and γ. Therefore,

$$\|\mathbf{v}\|_{H^{3/2-\varepsilon}(\Omega\setminus\bar{B})} \leq C_4\big(\|\lambda_\eta\|_{H^{3/2-\varepsilon}(B)} + (\frac{2\gamma}{1-\gamma})^{1/2}\frac{1}{h_p^{1/2-\varepsilon}}|\lambda_\eta|_{H^1(D_s)}\big) . \tag{2.16}$$

Then, we set $\mathbf{v} = \lambda_\eta$ in B. By construction, $\mathbf{v}$ belongs to $H_0^1(\Omega)^2$ and to $H^{3/2-\varepsilon}(\Omega)^2$; it satisfies

$$\forall q_h \in M_h \ , \ \int_\Omega q_h \operatorname{div} \mathbf{v}\, d\mathbf{x} = 0 , \tag{2.17}$$

and it satisfies in Ω inequalities similar to (2.14) and (2.16).

Next, we construct an approximation of $\mathbf{v}$ in X_h that preserves (2.17). For this, we choose an approximation operator $r_h \in \mathcal{L}(H_0^1(\Omega)^2; X_h)$ such that (cf. [8], [3], [16]):

$$\forall \mathbf{v} \in H_0^1(\Omega)^2 \ , \ |r_h(\mathbf{v})|_{H^1(\Omega)} \leq C_5 |\mathbf{v}|_{H^1(\Omega)} ,$$
$$\forall \mathbf{v} \in (H^2(\Omega) \cap H_0^1(\Omega))^2 \ , \ |r_h(\mathbf{v}) - \mathbf{v}|_{H^1(\Omega)} \leq C_6 h\, |\mathbf{v}|_{H^2(\Omega)} .$$

Then, by virtue of the inf-sup condition (2.3), there exists an operator $\Pi_h \in \mathcal{L}(H_0^1(\Omega)^2; X_h)$ satisfying for all $\mathbf{v} \in H_0^1(\Omega)^2$, (cf. [12]):

$$\forall q_h \in M_h \ , \ \int_\Omega q_h \operatorname{div}(\Pi_h(\mathbf{v}) - \mathbf{v})\, d\mathbf{x} = 0 , \tag{2.18}$$

$$|\Pi_h(\mathbf{v}) - \mathbf{v}|_{H^1(\Omega)} \leq (1 + \frac{\sqrt{2}}{\delta^*})|r_h(\mathbf{v}) - \mathbf{v}|_{H^1(\Omega)} . \tag{2.19}$$

Finally, owing again to (2.3), there exists $\mathbf{w}_h \in X_h$ such that

$$\|p_h\|^2_{L^2(\Omega)} = \int_\Omega p_h \operatorname{div} \mathbf{w}_h\, d\mathbf{x} \ , \ |\mathbf{w}_h|_{H^1(\Omega)} \leq \frac{1}{\delta^*}\|p_h\|_{L^2(\Omega)} .$$

Then we consider the linear combination, for an adequate parameter $\xi > 0$:

$$\mathbf{z}_h = -\xi \mathbf{w}_h - \Pi_h(\mathbf{v}) \in X_h .$$

By construction, and using (2.17) and (2.18), we obtain

$$-\int_\Omega p_h \operatorname{div} \mathbf{z}_h\, d\mathbf{x} = \xi\, \|p_h\|^2_{L^2(\Omega)} . \tag{2.20}$$

Now, on one hand, for any $\alpha > 0$, we can write

$$|\int_B \nabla \lambda_\eta : \nabla \mathbf{w}_h\, d\mathbf{x} + \int_B \lambda_\eta \cdot \mathbf{w}_h\, d\mathbf{x}| \leq \frac{1}{2}(\alpha\|\lambda_\eta\|^2_{H^1(B)} + \frac{1}{\alpha}\|\mathbf{w}_h\|^2_{H^1(B)})$$
$$\leq \frac{1}{2}(\alpha\|\lambda_\eta\|^2_{H^1(B)} + \frac{1+\mathcal{P}^2}{\alpha(\delta^*)^2}\|p_h\|^2_{L^2(\Omega)}) , \tag{2.21}$$

where $\mathcal{P}$ is the constant of Poincaré's inequality. On the other hand, we derive from the definition of $\mathbf{v}$:

$$\int_B \nabla \lambda_\eta : \nabla \Pi_h(\mathbf{v})\, d\mathbf{x} + \int_B \lambda_\eta \cdot \Pi_h(\mathbf{v})\, d\mathbf{x} = \|\lambda_\eta\|^2_{H^1(B)}$$
$$+ \int_B \nabla \lambda_\eta : \nabla(\Pi_h(\mathbf{v}) - \mathbf{v})\, d\mathbf{x} + \int_B \lambda_\eta \cdot (\Pi_h(\mathbf{v}) - \mathbf{v})\, d\mathbf{x} \tag{2.22}$$
$$\geq \|\lambda_\eta\|^2_{H^1(B)} - \|\lambda_\eta\|_{H^1(B)} \|\Pi_h(\mathbf{v}) - \mathbf{v}\|_{H^1(B)}\,.$$

By using (2.19), the above approximation properties of r_h and interpolation between spaces (cf. [14]), we obtain

$$\|\Pi_h(\mathbf{v}) - \mathbf{v}\|_{H^1(B)} \leq C_7 (1 + \frac{\sqrt{2}}{\delta^*}) h^{1/2-\varepsilon} \|\mathbf{v}\|_{H^{3/2-\varepsilon}(\Omega)}\,.$$

Owing to (2.16) and applying again the inverse inequality (2.15), we obtain

$$\|\Pi_h(\mathbf{v}) - \mathbf{v}\|_{H^1(B)} \leq C_8 \Big((\frac{h}{\eta})^{1/2-\varepsilon} \|\lambda_\eta\|_{H^1(B)} +$$
$$+ (\frac{2\gamma}{1-\gamma})^{1/2} (\frac{h}{h_p})^{1/2-\varepsilon} |\lambda_\eta|_{H^1(D_s)} \Big)\,. \tag{2.23}$$

Collecting (2.20)–(2.23), we derive the lower bound:

$$b(\mathbf{z}_h, (\lambda_\eta, p_h)) \geq \xi (1 - \frac{1 + \mathcal{P}^2}{2\,\alpha (\delta^*)^2}) \|p_h\|^2_{L^2(\Omega)}$$
$$+ (1 - \frac{\alpha\,\xi}{2} - C_8 ((\frac{h}{\eta})^{1/2-\varepsilon} + \sqrt{\frac{2\gamma}{1-\gamma}} (\frac{h}{h_p})^{1/2-\varepsilon}) \|\lambda_\eta\|^2_{H^1(B)}\,.$$

Let us choose for example

$$\alpha = \frac{1 + \mathcal{P}^2}{(\delta^*)^2}\,, \quad \xi = \frac{1}{5\,\alpha}\,.$$

This choice and (2.12) imply

$$b(\mathbf{z}_h, (\lambda_\eta, p_h)) \geq \frac{\xi}{2} \|p_h\|^2_{L^2(\Omega)} + (\frac{9}{10} - 2C_8 (\frac{h}{\eta})^{1/2-\varepsilon}) \|\lambda_\eta\|^2_{H^1(B)}\,.$$

Thus, by choosing κ such that

$$\kappa^{1/2-\varepsilon} = \frac{5}{2} C_8\,, \tag{2.24}$$

we derive (2.9) with $\beta_1^* = \min(\frac{1}{10}, \frac{\xi}{2})$. The proof of the estimate (2.10) is straightforward. ■

Remark 2. Let V_h denote the set of all $\mathbf{w}_h \in X_h$ satisfying

$$\forall q_h \in M_h\,, \int_\Omega q_h \operatorname{div} \mathbf{w}_h\, d\mathbf{x} = 0\,,$$

$$\forall \mu_\eta \in W_\eta\,, \quad \int_B \nabla\, \mu_\eta : \nabla\, \mathbf{w}_h \, d\mathbf{x} + \int_B \mu_\eta \cdot \mathbf{w}_h \, d\mathbf{x} = 0\,.$$

Then, the inf-sup condition (2.9), (2.10) implies in particular:

$$\inf_{\mathbf{v}_h \in V_h} |\mathbf{u} - \mathbf{v}_h|_{H^1(\Omega)} \le C \inf_{\mathbf{v}_h \in X_h} |\mathbf{u} - \mathbf{v}_h|_{H^1(\Omega)}\,, \tag{2.25}$$

with a constant C independent of $\mathbf{u}$, h, η and γ.

It remains to lift the constraint on the functions of W_η. For this, it suffices to find a constant β_3^*, independent of h_p and η, such that for all q_h in M_h satisfying

$$q_h = 0 \text{ on all } T \text{ for which } |T \cap \bar{B}| < \gamma |T|\,, \tag{2.26}$$

we have

$$\sup_{\lambda_\eta \in \Lambda_\eta} \frac{1}{\|\lambda_\eta\|_{H^1(B)}} \int_B q_h \operatorname{div} \lambda_\eta \, d\mathbf{x} \ge \beta_3^* \|q_h\|_{L^2(B)}\,. \tag{2.27}$$

Lemma 2.2. *There exists a constant $C > 0$, independent of h_p, η and γ, such that if*

$$\left(\frac{h_p}{\eta}\right)^{1/2-\varepsilon} \ge \frac{C}{\sqrt{\gamma}}\,, \tag{2.28}$$

then the inf-sup condition (2.27) holds, with a constant β_3^* that is proportional to $\sqrt{\gamma}$, for all $q_h \in M_h$ satisfying (2.26).

Proof. Let us choose once and for all a domain $\mathcal{O} \subset \Omega$ with smooth boundary $\partial\mathcal{O}$, that contains the support of all $q_h \in M_h$ satisfying (2.26). (Here we assume that Ω is large enough with respect to B). We take $q_h \in M_h$ satisfying (2.26) and in $\mathcal{O}$, we solve the problem:

$$\Delta\, \varphi = q_h \text{ in } \mathcal{O}\,, \quad \frac{\partial \varphi}{\partial \mathbf{n}} = 0 \text{ on } \partial\mathcal{O}\,.$$

As $\mathcal{O}$ contains the support of q_h, then $\int_{\mathcal{O}} q_h \, d\mathbf{x} = 0$ and this problem has a unique solution $\varphi \in H^1(\mathcal{O}) \cap L_0^2(\mathcal{O})$. Since $\partial\mathcal{O}$ is smooth, φ belongs to $H^2(\mathcal{O})$,

$$\|\varphi\|_{H^2(\mathcal{O})} \le C_1 \|q_h\|_{L^2(\mathcal{O})}\,.$$

The smoothness of $\partial\mathcal{O}$ also implies that

$$\|\varphi\|_{H^{5/2-\varepsilon}(\mathcal{O})} \le C_2 \|q_h\|_{H^{1/2-\varepsilon}(\mathcal{O})} \le \frac{C_3}{h_p^{1/2-\varepsilon}} \|q_h\|_{L^2(\mathcal{O})}\,, \tag{2.29}$$

where we have applied again the inverse inequality (2.15). In addition, in view of (2.26),

$$\|q_h\|_{L^2(\mathcal{O})} \le \frac{1}{\sqrt{\gamma}} \|q_h\|_{L^2(B)}\,. \tag{2.30}$$

Next, we set $\mathbf{v} = \nabla\varphi$ and we choose $\lambda_\eta = r_\eta(\mathbf{v}) \in \Lambda_\eta$, where $r_\eta \in \mathcal{L}(H^1(B)^2; \Lambda_\eta)$ is the analogue of the approximation operator r_h. Thus

$$\|\mathrm{div}(r_\eta(\mathbf{v})-\mathbf{v})\|_{L^2(B)} \le C_4\eta^{1/2-\varepsilon}\|\mathbf{v}\|_{H^{3/2-\varepsilon}(B)} \le \frac{C_5}{\sqrt{\gamma}}(\frac{\eta}{h_p})^{1/2-\varepsilon}\|q_h\|_{L^2(B)}\,.$$

Therefore,

$$\int_B q_h \,\mathrm{div}\,\lambda_\eta \,d\mathbf{x} = \int_B q_h \,\mathrm{div}\,\mathbf{v}\,d\mathbf{x} + \int_B q_h \,\mathrm{div}(r_\eta(\mathbf{v}) - \mathbf{v})\,d\mathbf{x}$$
$$\ge (1 - \frac{C_5}{\sqrt{\gamma}}(\frac{\eta}{h_p})^{1/2-\varepsilon})\|q_h\|^2_{L^2(B)}\,.$$

If we choose for example

$$(\frac{h_p}{\eta})^{1/2-\varepsilon} > \frac{5}{4}\frac{C_5}{\sqrt{\gamma}}\,, \tag{2.31}$$

then $\int_B q_h \,\mathrm{div}\,\lambda_\eta\,d\mathbf{x} \ge \frac{1}{5}\|q_h\|^2_{L^2(B)}$ and $\|r_\eta(\mathbf{v})\|_{H^1(B)} \le \frac{C_6}{\sqrt{\gamma}}\|q_h\|_{L^2(B)}$, so that (2.27) holds with $\beta_3^* = \frac{\sqrt{\gamma}}{5C_6}$. ■

Remark 3. Summing up, with the choices (2.31) and (2.12), we obtain the condition

$$(\frac{h_p}{\eta})^{1/2-\varepsilon} > \max(\sqrt{\frac{2\gamma}{1-\gamma}}, \frac{5}{4}\frac{C_5}{\sqrt{\gamma}})\,.$$

Note that the largest bound is likely to be the second one. However, (2.24) and (2.28) are both theoretical bounds and are usually ignored in practice: η is often taken slightly larger than h and h_p is often the double of h. Nevertheless, there are cases where this simple choice leads to an unstable solution in some elements near the boundary ∂B.

Remark 4. The inf-sup condition (2.27) implies in particular

$$\inf_{\mu_\eta \in W_\eta} \|\lambda - \mu_\eta\|_{H^1(B)} \le (1 + \frac{\sqrt{2}}{\beta_3^*}) \inf_{\mu_\eta\in\Lambda_\eta} \|\lambda - \mu_\eta\|_{H^1(B)}\,, \tag{2.32}$$

and observe that the constant of (2.32) is large if $\sqrt{\gamma}$ is small.

As the inf-sup condition (2.9), (2.10) is satisfied, problem (2.6)–(2.8) has a unique solution and standard results on mixed methods (cf. [12], [6], [5]), together with (2.25) and (2.32) yield the following error estimate.

Theorem 2.3. *Assume that S_η is uniformly regular and that (2.11), (2.12) and (2.28) hold. Then we have the following error bounds*

$$|\mathbf{u}-\mathbf{u}_h|_{H^1(\Omega)} \leq 2C \inf_{\mathbf{v}_h\in X_h} |\mathbf{u}-\mathbf{v}_h|_{H^1(\Omega)} + \frac{\sqrt{2}}{\nu} \inf_{q_h\in M_h} \|p-q_h\|_{L^2(\Omega)} + \frac{1}{\nu}(1+\frac{\sqrt{2}}{\beta_3^*})(1+\mathcal{P}^2)^{1/2} \inf_{\mu_\eta\in\Lambda_\eta} \|\lambda-\mu_\eta\|_{H^1(B)}, \tag{2.33}$$

$$\|p-p_h\|_{L^2(\Omega)} + \|\lambda-\lambda_\eta\|_{H^1(B)} \leq \nu\frac{\sqrt{2}}{\beta^*}C \inf_{\mathbf{v}_h\in X_h} |\mathbf{u}-\mathbf{v}_h|_{H^1(\Omega)} + (1+\frac{2}{\beta^*}) \inf_{q_h\in M_h} \|p-q_h\|_{L^2(\Omega)} + (1+\frac{\sqrt{2}(1+\mathcal{P}^2)^{1/2}}{\beta^*})(1+\frac{\sqrt{2}}{\beta_3^*}) \inf_{\mu_\eta\in\Lambda_\eta} \|\lambda-\mu_\eta\|_{H^1(B)}, \tag{2.34}$$

where C is the constant of (2.25) and $\beta^* = \frac{\beta_1^*}{\sqrt{2}\beta_2^*}$.

This theorem states that the accuracy of (2.6)–(2.8) depends on the global regularity of $\mathbf{u}$, p and λ. But we have mentioned in Remark 1 that this regularity is not high. Thus, we have the following corollary.

Corollary 2.4. *Under the assumptions of Theorem 2.3 and if $\mathbf{u}\in H^{3/2-\varepsilon}(\Omega)^2$, $p\in H^{1/2-\varepsilon}(\Omega)$, $\lambda\in H^2(B)^2$, then there exists a constant C, such that*

$$|\mathbf{u}-\mathbf{u}_h|_{H^1(\Omega)} + \|p-p_h\|_{L^2(\Omega)} + \|\lambda-\lambda_\eta\|_{H^1(B)} \leq C(h^{1/2-\varepsilon}\|\mathbf{u}\|_{H^{3/2-\varepsilon}(\Omega)} + h_p^{1/2-\varepsilon}\|p\|_{H^{1/2-\varepsilon}(\Omega)} + \eta\|\lambda\|_{H^2(B)}).$$

3. THE CASE OF A RIGID BODY

Here, we retain the setting and assumptions of the preceding section. When $\mathbf{g}$ is a rigid-body motion:

$$\mathbf{g} = (a + b\,x_2, c + d\,x_1)\ ,\ a,b,c,d\in\mathbb{R}, \tag{3.1}$$

the constraint (2.5) on W_η can be relaxed. For this, we first construct an adequate lifting of $\mathbf{g}$.

Lemma 3.1. *Assume that Ω is large enough and $\mathcal{T}_h$ is such that there exists a rectangle R containing B and strictly contained in Ω such that* *the support T of any basis function $q_h \in M_h$ satisfies either $T \subset \bar{R}$ or $T \subset \Omega \setminus \bar{R}$. Then, for any $\mathbf{g}$ of the form (3.1), there exists $\mathbf{w}_h \in X_h$ satisfying*

$$\forall q_h \in M_h \ , \ \int_\Omega q_h \operatorname{div} \mathbf{w}_h \, d\mathbf{x} = 0 \, , \tag{3.2}$$

$$\forall \mu_\eta \in \Lambda_\eta \ , \ \int_B \{\nabla(\mathbf{w}_h - \mathbf{g}) : \nabla \mu_\eta + (\mathbf{w}_h - \mathbf{g}) \cdot \mu_\eta\} d\mathbf{x} = 0 \, . \tag{3.3}$$

Proof. Since $\mathbf{g} \in \mathbb{P}_1^2$ and $\operatorname{div} \mathbf{g} = 0$, it suffices to correct $\mathbf{g}$ so that it vanishes on Γ. First note that $\mathbf{g} = \mathbf{curl}\, \psi$, where $\psi = a\,x_2 + \frac{1}{2} b\, x_2^2 - c\, x_1 - \frac{1}{2} d\, x_1^2$. Next, we choose once and for all a smooth truncating function θ such that $\theta = 1$ in R, θ has compact support in Ω, and we set

$$\mathbf{G} = \mathbf{curl}(\theta \psi) \, .$$

Let I_h be the standard Lagrange interpolation operator in X_h. Then $I_h(\mathbf{G}) \in X_h$, and the assumption on R implies that $I_h(\mathbf{G}) = \mathbf{g}$ in R, because I_h preserves $\mathbb{P}_1^2$ in each triangle. Consequently,

$$\int_{\Omega \setminus \bar{R}} \operatorname{div} I_h(\mathbf{G}) \, d\mathbf{x} = \int_{\partial R} I_h(\mathbf{G}) \cdot \mathbf{n} \, d\sigma = \int_{\partial R} \mathbf{g} \cdot \mathbf{n} \, d\sigma = 0 \, ,$$

where $\mathbf{n}$ is the unit normal to ∂R, pointing inside R. The above assumption on R and the imbedding of the mesh of X_h into that of M_h imply that the inf-sup condition (2.3) also holds for the restrictions of X_h and M_h to $\Omega \setminus \bar{R}$. Thus there exists a function $\mathbf{z}_h \in X_h$ that vanishes in R, such that

$$\forall q_h \in M_h \ , \ \int_{\Omega \setminus \bar{R}} q_h \operatorname{div} \mathbf{z}_h \, d\mathbf{x} = \int_{\Omega \setminus \bar{R}} q_h \operatorname{div} I_h(\mathbf{G}) \, d\mathbf{x} \, .$$

The desired lifting is $\mathbf{w}_h = I_h(\mathbf{G}) - \mathbf{z}_h \in X_h$. ■

As a consequence, we consider the following version without constraint of (2.6)–(2.8): Find $(\mathbf{u}_h', p_h', \lambda_\eta') \in X_h \times M_h \times \Lambda_\eta$ such that

$$\forall \mathbf{v}_h \in X_h \ , \ \nu \int_\Omega \nabla \mathbf{u}_h' : \nabla \mathbf{v}_h \, d\mathbf{x} - \int_\Omega p_h' \operatorname{div} \mathbf{v}_h \, d\mathbf{x} = \int_\Omega \tilde{\mathbf{f}} \cdot \mathbf{v}_h \, d\mathbf{x} + \int_B \nabla \lambda_\eta' : \nabla \mathbf{v}_h \, d\mathbf{x} + \int_B \lambda_\eta' \cdot \mathbf{v}_h \, d\mathbf{x} \, , \tag{3.4}$$

$$\forall q_h \in M_h \ , \ \int_\Omega q_h \operatorname{div} \mathbf{u}_h' \, d\mathbf{x} = 0 \, , \tag{3.5}$$

$$\forall \mu_\eta \in \Lambda_\eta \ , \ \int_B \nabla(\mathbf{u}'_h - \mathbf{g}) : \nabla \mu_\eta \, d\mathbf{x} + \int_B (\mathbf{u}'_h - \mathbf{g}) \cdot \mu_\eta \, d\mathbf{x} = 0 \, . \quad (3.6)$$

Theorem 3.2. *Let $\mathcal{T}_h$ be as in Lemma 3.1. Then problem (3.4)–(3.6) has at least one solution. The multipliers p'_h and λ'_η are not necessarily unique, but $\mathbf{u}'_h = \mathbf{u}_h$, the unique solution of (2.6)–(2.8).*

Proof. From the existence of a lifting $\mathbf{w}_h$ satisfying (3.2) and (3.3), it is easy to prove that (3.4)–(3.6) has at least one solution $(\mathbf{u}'_h, p'_h, \lambda'_\eta)$. In addition, the inf-sup condition (2.9), (2.10) allows us to construct $p_h \in M_h$ and $\lambda_\eta \in W_\eta$ such that

$$\forall \mathbf{v}_h \in X_h \ , \ b(\mathbf{v}_h, (\lambda_\eta, p_h)) = b(\mathbf{v}_h, (\lambda'_\eta, p'_h)) \, .$$

In view of (3.4), we find that $(\mathbf{u}'_h, p_h, \lambda_\eta)$ is a solution of (2.6)–(2.8) and the uniqueness of this solution shows that $\mathbf{u}'_h = \mathbf{u}_h$. In particular, this means that $\mathbf{u}_h$ satisfies

$$\forall \mu_\eta \in \Lambda_\eta \ , \ \int_B \nabla(\mathbf{u}_h - \mathbf{g}) : \nabla \mu_\eta \, d\mathbf{x} + \int_B (\mathbf{u}_h - \mathbf{g}) \cdot \mu_\eta \, d\mathbf{x} = 0 \, , \quad (3.7)$$

and hence (2.8) holds for all $\mu_\eta \in \Lambda_\eta$. ∎

It is well-known that in this situation, even if the multipliers are not unique, we can use a gradient algorithm to solve efficiently the systems (3.4)–(3.6).

References

[1] Adams, R.A. (1975). *Sobolev Spaces.* Academic Press, New York.

[2] Arnold, D.N., Scott, L.R. and Vogelius, M. (1988). Regular inversion of the divergence operator with Dirichlet boundary conditions on a polygon. *Ann. Scuola Norm. Sup. Pisa–Serie IV,* XV:169–192.

[3] Bernardi, C.. and Girault, V. (1998). A local regularization operator for triangular and quadrilateral finite elements. *SIAM J. Numer. Anal.*, 35:1893–1916.

[4] Bramble, J., Pasciak, J. and Xu, J. (1991). The analysis of multigrid algorithms with non nested spaces and noninherited quadratic forms. *Math. Comp.*, 56:1–34.

[5] Brenner, S. and Scott, L.R. (1994). *The Mathematical Theory of Finite Element Methods.* Springer-Verlag, New York.

[6] Brezzi, F. and Fortin, M. (1991). *Mixed and Hybrid Finite Element Methods.* Springer-Verlag, New York.

[7] Ciarlet, P.G. (1978). *The Finite Element Method for Elliptic Problems.* North-Holland, Amsterdam.

[8] Clément, P. (1975). Approximation by finite element functions using local regularization. *RAIRO, Anal. Numér.*, R-2:77–84.

[9] Glowinski, R., Hesla, T., Joseph, D.D., Pan, T.W. and Périaux, J. (1997). Distributed Lagrange multiplier methods for particulate flow, in: *Computational Science for the 21st Century*, pp.270–279, M.O. Bristeau, G. Etgen, W. Fitzgibbon, J.L. Lions, J. Périaux, M.F. Wheeler (Eds.).

[10] Glowinski, R., Pan, T.W. and Périaux, J. (1998). Distributed Lagrange multiplier methods for incompressible viscous flow around moving rigid bodies. *Comp. Meth. Appl. Mech. Engrg.*, 151:181–194.

[11] Glowinski, R., Pan, T.W., Hesla, T., Joseph, D.D. and Périaux, J. (1998). A fictitious-domain method with distributed Lagrange multipliers for the numerical simulation of particulate flow, in: *Domain Decomposition Methods*, 10:121–137, J. Mandel, C. Favhat, X.C. Cai (Eds.), AMS, Providence, R.I.

[12] Girault, V. and Raviart, P.A. (1986). *Finite Element Methods for Navier-Stokes Equations. Theory and Algorithms.* Springer-Verlag, New York.

[13] Grisvard, P. (1985). *Elliptic Problems in Nonsmooth Domains*, Pitman, Boston.

[14] Lions, J.L. and Magenes, E. (1968). *Problèmes aux Limites non Homogènes et Applications, I.* Dunod, Paris.

[15] Nečas, J. (1967). *Les Méthodes directes en théorie des équations elliptiques*, Masson, Paris.

[16] Scott, L.R. and Zhang, S. (1990). Finite element interpolation of non-smooth functions satisfying boundary conditions, *Math. Comp.*, 54:483–493.

RELIABLE SOLUTION OF A UNILATERAL CONTACT PROBLEM WITH FRICTION, CONSIDERING UNCERTAIN INPUT DATA

Ivan Hlaváček

Abstract: A Signorini contact problem with an approximate model of friction is analyzed, when Lamé's coefficients, body forces and friction coefficients are uncertain, being prescribed in a given set of admissible functions. Three kinds of criteria, characterizing the stress intensity, are chosen to define three maximization problems. Approximate problems are proposed on the basis of a mixed finite element method. Some theoretical convergence analysis is presented.

Keywords: Uncertain data, unilateral contact, Coulomb friction.

Introduction

Mathematical models involve data (coefficients, right-hand side, boundary values), which cannot be sometimes determined uniquely, but only in some intervals, which result from experimental measurements and inverse (identification) problems.

Assume that the main aim of the computations is to find maximal value of certain functional, which depends on the solution of the mathematical model. Then we can formulate a corresponding maximization problem and employ methods of Optimal Design. Such a general approach has been proposed in [6] and applied to nonlinear elliptic problems in heat conduction [7], elasto-plasticity [8] - [10] and to parabolic problems [11].

The aim of the present paper is to apply the general approach to a unilateral contact problem for an elastic body, with an approximate Coulomb friction. Problems of this kind with uncertain input data occur e.g. in modelling of plate tectonics, based on the global geodynamics. If a litospheric plate is obducting with time onto the oceanic litospheric plate, the model can be represented by a Signorini problem with uncertain Lamé's coefficients, body forces and a coefficient of the friction — see, e.g., [14]. Another example is a model of interaction between a tunnel wall and the rock [13], where the Lamé's coefficients of the rock and the friction coefficients are uncertain.

In the first Section we introduce a unilateral Signorini problem with approximate friction and define a set of admissible uncertain data. Then Maximization Problems are formulated for three kinds of the criterion: (i) the intensity of shear stresses, (ii) the normal component of the surface traction or (iii) a norm of the surface traction. The existence of a maximizing data is proved in Section 2 on the basis of the continuous dependence of the solution of the contact problem on the data.

We introduce approximate maximization problems in Section 3, using a mixed finite element method to solve the contact problem [1, 3, 4] and prove the solvability of the approximate problems. In Section 4 we show the existence of sequences of approximate solutions, which tend to a solution of the original Maximization Problem, when the mesh-sizes of the discretizations tend to zero and that the approximate maxima tend to the maximum of the original problem.

1. SETTING OF MAXIMIZATION PROBLEMS

Let an elastic piecewise homogeneous isotropic body occupy a bounded domain $\Omega \subset \mathbf{R}^2$ with Lipschitz boundary $\partial\Omega$. We assume that

$$\partial\Omega = \Gamma_K \cup \Gamma_u \cup \Gamma_p$$

is a disjoint decomposition, meas $\Gamma_u > 0$, meas $\Gamma_K > 0$. We introduce the bilinear form

$$a(A; u, v) = \int_\Omega (\lambda \text{ div } u \text{ div } v + 2\mu e_{ij}(u) e_{ij}(v)) dx,$$

where

$$\begin{aligned} \text{div } u &= \partial u_i / \partial x_i, \\ e_{ij}(u) &= \frac{1}{2}(\partial u_i/\partial x_j + \partial u_j/\partial x_i), \qquad i = 1, 2 \end{aligned}$$

and the virtual work of external forces

$$L(A; v) = \int_\Omega F_i v_i dx + \int_{\Gamma_p} P_i v_i ds.$$

The Coulomb friction will be approximated by the functional

$$j(A; v) = \int_{\Gamma_K} g|v_t| ds,$$

($v_t = v_i t_i$ is the tangential component of the displacement).

Here the repeated index implies a summation within $\{1, 2\}$. The input data are

$$A \equiv (\lambda, \mu, F_1, F_2, g) \in U_{ad} = U_{ad}^{\lambda} \times U_{ad}^{\mu} \times U_{ad}^{F_1} \times U_{ad}^{F_2} \times U_{ad}^{g},$$

where the sets of admissible data will be defined as follows.

Let

$$\begin{aligned} \overline{\Omega} &= \bigcup_{j=1}^{J} \overline{\Omega}_j, \quad \Omega_j \cap \Omega_k = \emptyset \quad \text{for} \quad j \neq k, \\ \overline{\Gamma}_K &= \bigcup_{q=1}^{Q} \overline{\Gamma}_q, \quad \Gamma_p \cap \Gamma_q = \emptyset \quad \text{for} \quad p \neq q \end{aligned} \tag{1.1}$$

be a given partition of the domain $\overline{\Omega}$ and of $\overline{\Gamma}_K$, respectively.

We define

$$\begin{aligned} U_{ad}^{\lambda} &= \{\lambda \in L^{\infty}(\Omega) : \lambda_{\min}^{j} \leq \lambda|_{\Omega_j} = \text{ const.} \leq \lambda_{\max}^{j}, \ 1 \leq j \leq J\} \\ U_{ad}^{\mu} &= \{\mu \in L^{\infty}(\Omega) : \mu_{\min}^{j} \leq \mu|_{\Omega_j} = \text{ const.} \leq \mu_{\max}^{j}, \ 1 \leq j \leq J\} \\ U_{ad}^{F_i} &= \{f \in L^{\infty}(\Omega) : F_{i,\min}^{j} \leq f|_{\Omega_j} = \text{ const.} \leq F_{i,\max}^{j}, \ 1 \leq j \leq J\}, \\ &\qquad i = 1, 2 \\ U_{ad}^{g} &= \{g \in L^{\infty}(\Gamma_K) : g|_{\overline{\Gamma}_q} \in C^{(0),1}(\overline{\Gamma}_q); \ 0 \leq g(s) \leq g_{\max}^{(q)}, \\ &\qquad |dg/ds| \leq C_g \text{ a.e. in } \Gamma_q, \quad 1 \leq q \leq Q\}, \end{aligned}$$

where

$$\begin{aligned} 0 \leq \lambda_{\min}^{j} < \lambda_{\max}^{j}, \quad 0 < \mu_0 \leq \mu_{\min}^{j} < \mu_{\max}^{j}, \\ F_{i,\min}^{j} \leq F_{i,\max}^{j}, \ g_{\max}^{(q)} > 0 \quad \text{and} \quad C_g > 0 \end{aligned}$$

are given constants.

Furthermore, let $P_i \in L^2(\Gamma_p)$ be a given surface load and let a function $\overline{u} \in [H^1(\Omega)]^2$ be given, such that $\overline{u}_\nu = \overline{u}_i \nu_i = 0$ on Γ_K, where ν denotes the unit outward normal to the boundary.

We introduce the set

$$K = \{v \in [H^1(\Omega)]^2 : v = \overline{u} \quad \text{on} \quad \Gamma_u, \ v_\nu \leq 0 \quad \text{on} \quad \Gamma_K\} \tag{1.2}$$

and the following (state) problem for any given $A \in U_{ad}$:

$$\begin{aligned} &\text{find} \quad u(A) \in K \quad \text{such that} \\ &a(A; u(A), v - u(A)) + j(A; v) - j(A; u(A)) \geq L(A; v - u(A)) \\ &\text{holds for all} \quad v \in K. \end{aligned} \tag{1.3}$$

Proposition 1.1. *There exists a unique solution $u(A)$ of the problem* (1.3) *for any $A \in U_{ad}$.*

Proof is based on the equivalence of the variational inequality (1.3) with the minimization of the following functional

$$L(u) \equiv \frac{1}{2}a(A;u,u) + j(A;u) - L(A;u)$$

over the set K. ■

Let us choose a finite number of (small) subdomains $G_j \subset \Omega$ (adjacent to Γ_K, for example), $j = 1, \ldots, N$ and define

$$\psi_j(A;u) = (\text{meas } G_j)^{-1} \int_{G_j} I_2^2(\tau(A;u))dx, \tag{1.4}$$

where $I_2(\tau)$ denotes the "intensity of shear stress", i.e., an invariant of the stress tensor deviator τ^D. We have the formulae

$$\begin{aligned} I_2^2(\tau) &= \tau_{ij}^D \tau_{ij}^D = \\ &= \frac{2}{3}[\tau_{11}^2 + \tau_{22}^2 + \tau_{33}^2 - (\tau_{11}\tau_{22} + \tau_{11}\tau_{33} + \tau_{22}\tau_{33}) + 3\tau_{12}^2] \\ \tau_{ij}(A;u) &= \lambda\delta_{ij} \text{ div } u + 2\mu e_{ij}(u), \quad i,j = 1,2, \\ \tau_{33}(A;u) &= \lambda \text{ div } u, \end{aligned}$$

corresponding to the *plane strain* elasticity.

Let us consider the functional

$$\Phi(A;u) = \max_{1 \le j \le N} \psi_j(A;u) \tag{1.5}$$

and the following *Maximization Problem:* find

$$A^0 = \arg \max_{A \in U_{ad}} \Phi(A; u(A)). \tag{1.6}$$

If Γ_K is polygonal and the friction can be neglected (as in [14]), we set $g \equiv 0$ and define (instead of (1.4)) for instance

$$\psi_j(A;u) = (\text{meas } G_j)^{-1} \int_{G_j} (-\tau_{ij}(A;u)\nu_i\nu_j)dx. \tag{1.7}$$

If a norm of the surface traction vector

$$T_i = \tau_{ij}\nu_j \tag{1.8}$$

on a part Γ_0 of the boundary is the most important aim of computations, we can extend the formula (1.8), which holds for enough smooth stress tensors only, to cover the general case, when

$$\tau_{ij} \in L^2(\Omega) \quad \text{and} \quad \partial\tau_{ij}/\partial x_j \in L^2(\Omega), \qquad i,j = 1,2$$

is the only assumed regularity of the stress field. To this end, we find a solution $z(A;u)$ of an auxiliary elliptic boundary value problem — see, e.g., [1], [5]. Then we define

$$\Phi(A;u) = a(A;u,z(A;u)) - \int_\Omega F_i z_i(A,u)dx, \tag{1.9}$$

since this functional equals to the square of a norm of the surface traction vector.

2. EXISTENCE OF A SOLUTION TO MAXIMIZATION PROBLEMS

To prove the solvability of the problem (1.6), we have to verify a continuity of the mapping $A \mapsto u(A)$ on the set U_{ad}. We introduce the space

$$U = [\mathbf{R}^J]^4 \times \sqcap_{q=1}^{Q} C(\overline{\Gamma}_q)$$

and prove the crucial

Proposition 2.1. *Let $A_n \in U_{ad}$, $A_n \to A$ in U as $n \to \infty$. Then*

$$u(A_n) \to u(A) \quad in \quad [H^1(\Omega)]^2.$$

Proof is based on the following observation: if $A_n \in U_{ad}$, $A_n \to A$ in U and $u_n \rightharpoonup u$ (weakly) in $[H^1(\Omega)]^2$, then

$$a(A_n;u_n,v) \to a(A;u,v) \quad \text{for all} \quad v \in [H^1(\Omega)]^2,$$
$$L(A_n;u_n) \to L(A;u), \quad j(A_n;u_n) \to j(A;u).$$

First, we show that the sequence $u(A_n)$ is bounded, so that a weak cluster point ω exists. Second, we verify that $\omega = u(A)$ and the uniqueness (Proposition 1.1) implies that the whole sequence $\{u(A_n)\}$ tends to $u(A)$ weakly. Third, we prove the strong convergence. For detailed proof — see the paper [12].

Lemma 2.2. *Let the criterion-functional Φ be defined either by* (1.5) (*with* (1.4) *or* (1.7)), *or by* (1.9). *Let $A_n \in U_{ad}$, $A_n \to A$ in U and $u_n \to u$ in $[H^1(\Omega)]^2$, as $n \to \infty$.*

Then

$$\Phi(A_n;u_n) \to \Phi(A;u).$$

Theorem 2.3. *There exists at least one solution of the Maximization Problem* (1.6).

Proof. By Lemma 2.2 and Proposition 2.1, the functional $A \mapsto \Phi(A;u(A))$ is continuous on the set U_{ad}. Since the set U_{ad} is compact in U, the existence of a maximizer follows.

3. APPROXIMATE MAXIMIZATION PROBLEMS

Assume that Ω is a *polygonal* domain and that $\overline{\Gamma}_K \cap \overline{\Gamma}_u$ consists of a finite number of points. Let $M > 1$ be an integer and let every Γ_q be partitioned into M equal segments $\Delta_m^{(q)}$, $m = 1, \ldots, M$; $q = 1, \ldots, Q$. Denote the above partition of $\overline{\Gamma}_K$ by T_M. We define

$$U_{ad}^{g^M} = \{g \in U_{ad}^g : g|_{\Delta_m}^{(q)} \in P_1(\Delta_m^{(q)}), \quad 1 \le q \le Q, \quad 1 \le m \le M\},$$

(i.e., piecewise linear functions) and denote

$$U_{ad}^M = U_{ad}^\lambda \times U_{ad}^\mu \times U_{ad}^{F_1} \times U_{ad}^{F_2} \times U_{ad}^{g^M}, \quad (U_{ad}^M \subset U_{ad}).$$

The state problem (1.3) can be solved by various methods — see, e.g., [1–4]. Here we choose a mixed finite element method, which removes the unpleasant presence of the nondifferentiable term $j(v)$.

Let T_h be a triangulation of the domain $\overline{\Omega}$, consistent with the partitions (1.1), with the decomposition of the boundary and with the boundaries ∂G_j from (1.4), (1.7). Let h denote the length of the maximal side of all triangles in T_h. We introduce a finite-dimensional subspace V_h of piecewise linear vector functions

$$V_h = \{v_h \in [C(\overline{\Omega})]^2 : v_h|_T \in [P_1(T)]^2 \quad \forall\, T \in T_h, \ v_h = 0 \quad \text{on} \quad \Gamma_u\}$$

and a subset

$$K_{0h} = \{v_h \in V_h : v_{hj}\nu_j(a_i) \le 0 \quad \text{for all nodes} \quad a_i \in \overline{\Gamma}_K \setminus \overline{\Gamma}_u\}.$$

The nodes a_i need not coincide with those of the partition T_M.

Let any segment Γ_q be divided into equal subsegments e_r. We define

$$H = \max_{e_r \in \overline{\Gamma}_K} (\text{meas } e_r),$$

$$\Lambda_H = \{\eta_H \in L^2(\Gamma_K) : \eta_H|_{e_r} \in P_0(e_r), \quad |\eta_H| \le 1, \quad \forall\, e_r \in \overline{\Gamma}_K\},$$

(i.e., a set of bounded piecewise constant functions) and denote the above partition of Γ_K by T_H.

By a *mixed finite element approximation* of the problem (1.3) we call the problem of finding a *saddle-point* of the following Lagrangian

$$\mathscr{L}(A; v, \eta) = \frac{1}{2} a(A; v, v) - L(A; v) + \int_{\Gamma_K} g\eta v_t ds$$

on the set $(\overline{u} + K_{0h}) \times \Lambda_H$, i.e., a couple $(w_h(A), \chi_H(A))$, satisfying the inequalities

$$\mathscr{L}(A; w_h(A), \eta_H) \le \mathscr{L}(A; w_h(A), \chi_H(A)) \le \mathscr{L}(A; v_h, \chi_H(A)) \qquad (3.1)$$

for all $v_h \in \overline{u} + K_{0h}$ and $\eta_H \in \Lambda_H$.

The problem (3.1) has a solution for all $A \in U_{ad}$. The first component $w_h(A)$ is uniquely determined. For the proof — see [3 - §2.5.41, Theorem 5.5] or [4 - Theorem 9.2].

Proposition 3.1. *If $A_m \in U_{ad}$, $A_m \to A$ in U, then*

$$w_h(A_m) \to w_h(A), \quad \textit{as} \quad m \to \infty.$$

Proof — see [12].

We introduce the following *Approximate Maximization Problem:* given a triangulation T_h, a partition T_H and the set U_{ad}^M, find

$$A_M^0(h,H) = \arg \max_{A_m \in U_{ad}^M} \Phi(A_M; w_h(A_M)). \tag{3.2}$$

In case of the criterion (1.9) we define a finite element approximation $z_h(A; w_h)$ of $z(A; u)$, the functional

$$\Phi_h(A; w) = a(A; w, z_h(A; w)) - \int_\Omega F_i z_{hi}(A; w) dx \tag{3.3}$$

and replace the functional Φ in (3.2) by Φ_h.

Lemma 3.2. *If $A_m \in U_{ad}$, $A_m \to A$ in U and $w^m \to w$ in $[H^1(\Omega)]^2$, as $m \to \infty$, then*

$$\Phi_h(A_m; w^m) \to \Phi_h(A; w).$$

Theorem 3.3. *The Approximate Maximization Problem* (3.2) *has at least one solution for any T_h, T_H and U_{ad}^M.*

Proof follows from the compactness of the set U_{ad}^M in U and the continuity of the mapping $A \mapsto \Phi(A; w_h(A))$, which is a consequence of Proposition 3.1 and Lemma 2.2 or Lemma 3.2.

4. SOME CONVERGENCE ANALYSIS

We will study the behavior of $A_M^0(h,H)$, $w_h(A_M^0(h,H))$ and $\Phi(A_M^0(h,H); w_h(A_M^0(h,H)))$, when the mesh-sizes h, H tend to zero and M tends to infinity. To this end, we need the following result (see [3 - §2.5.41, Theorems 5.7, 5.4]).

Proposition 4.1. *Let $\{T_h\}$, $h \to 0+$, be a regular family of triangulations, let $H \to 0+$ and $A \in U_{ad}$ be fixed. Then*

$$w_h(A) \to u(A) \quad \textit{in} \quad [H^1(\Omega)]^2,$$

where $u(A)$ is the solution of the state problem (1.3).

Lemma 4.2. *For any $A \in U_{ad}$ there exists a sequence $\{A_M\}$, $M \to \infty$, such that $A_M \in U_{ad}^M$ and $A_M \to A$ in U.*

Proof is based on the Lagrange linear interpolate of any $g \in U^g_{ad}$ on the partition T_M.

Theorem 4.3. *Let $\{T_h\}$, $h \to 0+$, be a regular family of triangulations. Let $\{A^0_M(h,H)\}$, $h \to 0+$, $H \to 0+$, $M \geq \xi(h,H)$ be a sequence of solutions of the Approximate Maximization Problem* (3.2), *where $\xi : (\mathbf{R}^+)^2 \to \mathbf{R}^+$ is some function, such that*

$$\lim \xi(h,H) \to +\infty, \quad \textit{as} \quad h \to 0+ \quad \textit{and} \quad H \to 0+ .$$

Then there exists a subsequence $\{A^0_{M_n}(h_n, H_n)\}$, such that

$$A^0_{M_n}(h_n, H_n) \to A^0 \quad \textit{in} \quad U, \tag{4.1}$$

$$w_{h_n}(A^0_{M_n}(h_n, H_n)) \to u(A^0) \quad \textit{in} \quad [H^1(\Omega)]^2, \tag{4.2}$$

$$\Phi(A^0_{M_n}(h_n, H_n); w_{h_n}(A^0_{M_n}(h_n, H_n))) \to \Phi(A^0; u(A^0)), \tag{4.3}$$

where A^0 is a solution of the Maximization Problem (1.6) *for the functional* (1.5) *with* (1.4) *or* (1.7).

Proof. Let $A \in U_{ad}$ be arbitrary. By Lemma 4.2, there is a sequence $\{A_M\}$, $A_M \in U^M_{ad}$, $A_M \to A$ in U, as $M \to \infty$. By definition, we have

$$\Phi(A^0_M(h,H); w_h(A^0_M(h,H))) \geq \Phi(A_m; w_h(A_M)) \tag{4.4}$$

for all triples $(h, H; M)$ under consideration. Since $U^M_{ad} \subset U_{ad}$ and U_{ad} is compact, there exists $A^0 \in U_{ad}$ and a subsequence $\{A^0_{M_n}(h_n, H_n)\}$ such that (4.1) holds. Using Proposition 3.1 and 4.1, we obtain that (4.2) holds provided ξ is "sufficiently fast growing" function. In the same way, we deduce that

$$w_{h_n}(A_{M_n}) \to u(A) \quad \text{in} \quad [H^1(\Omega)]^2. \tag{4.5}$$

Let us consider (4.4) for triples $(h_n, H_n; M_n)$ and pass to the limit with $n \to \infty$. Using (4.1), (4.2), (4.5) and Lemma 2.2, we arrive at (4.3) and

$$\Phi(A^0; u(A^0)) \geq \Phi(A; u(A)),$$

so that A^0 is a solution of the problem (1.6). ∎

Remark 4.4. The most important result is the convergence (4.3). In fact, whereas the "most dangerous" data A^0 are not required in practice, the maximal stress intensity is the main aim of computations.

Remark 4.5. An analogous convergence result can be derived for the case of the criterion (1.9) and (3.3) (see [12 - Theorem 4.2]).

Acknowledgments

The support of the Grant Agency of the Czech Republic under Grant 201/97/0217 if gratefully acknowledged.

References

[1] Haslinger, J. and Hlaváček, I. (1982). Approximation of the Signorini problem with friction by a mixed finite element method. J. Math. Anal. Appl., 86:99–122.

[2] Haslinger, J. and Tvrdý, M. (1983). Approximation and numerical realization of contact problems with friction. Apl. Mat., 28:55–71.

[3] Hlaváček, I., Haslinger, J., Nečas, J. and Lovíšek, J. (1988). Solution of variational inequalities in Mechanics. Springer-Verlag, Appl. Math. Sciences, vol. 66, New York.

[4] Haslinger, J., Hlaváček, I. and Nečas, J. (1996). Numerical methods for unilateral problems in Solid Mechanics. Handbook of Numer. Anal., vol. IV, ed. by P.G. Ciarlet and J.L. Lions, pp. 313–485, Elsevier, Amsterdam.

[5] Hlaváček, I. (1987). Shape optimization in two-dimensional elasticity by the dual finite element method. RAIRO Model Math. Anal. Numer., 21:63–92.

[6] Hlaváček, I. (1997). Reliable solutions of elliptic boundary value problems with respect to uncertain data. Proceedings of the WCNA-96, Nonlin. Analysis, Theory, Methods and Appl., 30:3879–3890.

[7] Hlaváček, I. (1997). Reliable solution of a quasilinear nonpotential elliptic problem of a nonmonotone type with respect to the uncertainty in coefficients. J. Math. Anal. Appl. 212:452–466.

[8] Hlaváček, I. (1996). Reliable solutions of problems in the deformation theory of plasticity with respect to uncertain material function. Appl. Math., 41:447–466.

[9] Hlaváček, I. Reliable solution of an elasto-plastic torsion problem. To appear.

[10] Hlaváček, I. (1998). Reliable solution of an elasto-plastic Reissner-Mindlin beam for the Hencky's model with uncertain yield function. Appl. Math., 43:223–237.

[11] Hlaváček, I. (1999). Reliable solution of linear parabolic problems with uncertain coefficients. Z. Angew. Math. Mech., 79:291–301.

[12] Hlaváček, I. Reliable solution of a Signorini contact problem with friction, considering uncertain data. To appear.

[13] Janovský, V. and Procházka, P. (1980). Contact problem of two elastic bodies. Apl. Mat., 25:87–148.

[14] Nedoma, J., Haslinger, J. and Hlaváček, I. (1989). The problem of an obducting litospheric plate in the Aleutian arc system. A finite element analysis in the frictionless case. Math. Comput. Modelling, 12:61–75.

DOMAIN DECOMPOSITION ALGORITHM FOR COMPUTER AIDED DESIGN

Frédéric Hecht, Jacques-Louis Lions, Olivier Pironneau

Abstract: We present a decomposition algorithm similar to Schwarz for the numerical solution of partial differential equations in complex domains. The method is well suited to domains described by Constructive Solid Geometry (CSG), i.e. by set operations on simple shapes, a data structure often used in image synthesis and Virtual Reality. This work extends the algorithms presented in [1], [2], [3] which were based on "virtual controls", whereas here compactness is used for convergence proofs.

Keywords: Partial differential equations, domain decomposition, virtual reality, finite element method, chimera.

1. INTRODUCTION

In many areas, such as architecture, style departments, image synthesis, one has to solve Partial Differential Equations (PDE) in domains Ω of $\mathbb{R}^2$ or of $\mathbb{R}^3$ which are described by set operations on simple shapes, but the number of elementary shapes is large.

For such situations, which are referred to as Constructive Solid Geometry (CSG), and which are often used in image synthesis and Virtual Reality (VR) (cf [8] for instance), it is difficult to construct a global triangulation of the domain while it is simple to triangulate each individual domains.

Thus DDM (Domain Decomposition Methods) is certainly a very natural approach for such problems.

In a series of notes, two of the authors (cf. [1],[2],[3]) have introduced a systematic method to address DDM, based on the idea of *Virtual Control.* In the third note [3], our motivation is explained at length, namely

Applied Nonlinear Analysis, edited by Sequeira *et al.*
Kluwer Academic / Plenum Publishers, New York, 1999.

1. To compute with the data structures of VR without having to translate CSG data.
2. But also to extend and improve on the Chimera method [4].

In this paper we do not use virtual control but rather an alternative, based on a fixed-point algorithm. It is less flexible and less general than the "virtual control" approach, but as in the classical Schwarz algorithm – to which the method presented here is an alternative – it has the advantage of not requiring the computation of boundary integrals. It is even better than Schwarz' in that it does not require the computation of interpolations on boundaries.

In each elementary shape Ω_i of Ω we compute, iteratively, a function u_i of the solution u of the PDE (in fact the boundary value problem we want to solve). In order to proceed with the iterations (cf. (3.3), Section 3 below) each u_i may have to be interpolated on any Ω_j, $j(\neq i)$, of course at a reasonable cost. We present here such an interpolator (explained in [4] with more details) which is efficient even on multiply connected domains.

We consider four geometrical cases, the general case immediately following. The convergence proof (presented in Section 3) is general. The error estimates are presented in a partly formal fashion in Section 3, the method being rigorous for other geometrical cases.

2. THE MODEL PROBLEM

Let Ω be a bounded open set of $\mathcal{R}^d$. We wish to solve the following: find

$$u \in V \equiv H_0^1(\Omega) \quad : \qquad a(u, \hat{u}) = (f, \hat{u}) \quad \forall \hat{u} \in V, \tag{2.1}$$

where a is a bilinear coercive form on V and $(f, .)$ is a continuous linear form on V, for instance, with $a_{ij} \in L^\infty(\Omega)$:

$$a(u,v) = \sum_{i,j=1}^{d} \int_\Omega a_{ij} \frac{\partial u}{\partial x_i} \frac{\partial v}{\partial x_j}, \qquad (f,v) = \int_\Omega fv. \tag{2.2}$$

The domain Ω is obtained by sets operations on a family of bounded open sets Ω_k. The sets operations are:

- Union: $\Omega_i \cup \Omega_j$
- Difference: $\Omega_i \backslash \overline{\Omega_j}$ provided that $\Omega_j \subset \Omega_i$
- Extrusion: $\Omega_i \backslash (\overline{\Omega_j \cap \Omega_i})$.
- Intersection: $\Omega_i \cap \Omega_j$.

We analyze the four cases independently, then the general case will be straightforward.

There are also other cases which ought to receive a separate treatment but which we will not investigate here, such as the case of two tangent objects, for example a book on a table.

3. THE DOMAIN IS THE UNION OF TWO OVERLAPPING SETS

Assume that $\Omega = \Omega_1 \cup \Omega_2$ and that $\Omega_1 \cap \Omega_2 \neq \emptyset$; denote by $S_1 = \partial\Omega_1 \cap \Omega_2$ and $S_2 = \partial\Omega_2 \cap \Omega_1$ and set

$$V_i = \{v \in L^2(\Omega) : \quad v|_{\Omega_i} \in H_0^1(\Omega_i), \quad v|_{\Omega-\Omega_i} = 0\} \tag{3.1}$$

Note that the Schwarz domain decomposition algorithm can be used here:

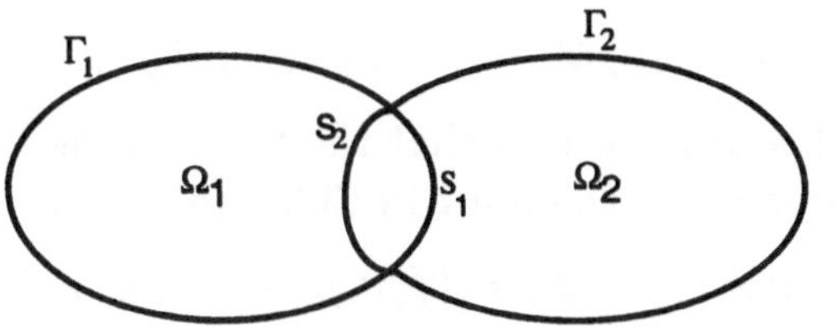

Figure 1.

3.1 Algorithm 1. (Schwarz)

Choose $u_i^0 \in V_i$, set $n = 0$.

Begin loop

Find u_i^{n+1} such that $u_i^{n+1} - u_j^n \in V_i, \quad i,j = 1,2, j \neq i$ by solving

$$a(u_i^{n+1}, \hat{u}_i) = (f, \hat{u}_i) \quad \forall \hat{u}_i \in V_i \tag{3.2}$$

End loop

The convergence has been analyzed by P.L. Lions [6]. In search for precision, we present the following alternative

3.2 Algorithm 2. (fixed-point)

Let $b(,)$ be an equivalent scalar product on $L^2(\Omega)$, for instance $b(u,v) = (\beta u, v)$ for some positive scalar β, and choose two arbitrary functions u_i^0 in V_i.

Once $u_i^0 \in V_i$ are chosen, set $n = 0$.

Begin loop

Find $u_i^{n+1} \in V_i$ by solving

$$\begin{aligned} b(u_1^{n+1} - u_1^n, \hat{u}_1) + a(u_1^{n+1} + u_2^n, \hat{u}_1) &= (f, \hat{u}_1) \quad \forall \hat{u}_1 \in V_1 \\ b(u_2^{n+1} - u_2^n, \hat{u}_2) + a(u_1^n + u_2^{n+1}, \hat{u}_2) &= (f, \hat{u}_2) \quad \forall \hat{u}_2 \in V_2 \end{aligned} \tag{3.3}$$

End loop

Remark 1. When $\beta = 0$ Algorithm 2 is identical to Algorithm 1 with u_i^{n+1} replaced by $u_i^{n+1} - u_j^n, \ i,j = 1,2, \ j \neq i$.

Let A be the second order operator associated with a, i.e.

$$a(u,v) = (Au, v) \; \forall u, v \in V.$$

Theorem 1. *When $a(\cdot,\cdot)$ is symmetric, Algorithm 2 converges in the sense that $u_i^n \to u_i^*$ weakly in $H_0^1(\Omega)$ with $u_1^* + u_2^* = u$ solution of (2.1) and the decomposition is uniquely defined in $\Omega_1 \cap \Omega_2$ by*

$$(\beta + A)u_1 = \frac{1}{2}(\beta + A)(u + u_1^0 - u_2^0), \; u_1|_{S_1} = 0, \quad u_1|_{S_2} = u, \tag{3.4}$$

$$(\beta + A)u_2 = \frac{1}{2}(\beta + A)(u + u_2^0 - u_1^0), \; u_2|_{S_2} = 0, \; u_2|_{S_1} = u., \; u_i|_{\partial\Omega_i} = 0$$

Proof. Assume for the moment that the algorithm converges: $u_i^n \to u_i$ weakly in V_i. Note that the decomposition

$$V = V_1 + V_2, \quad (i.e. \; \forall u \in V, \; \exists u_1 \in V_1, \; u_2 \in V_2 \; : \; u = u_1 + u_2) \tag{3.5}$$

is not unique and so it is natural that the limits u_i depend on the initial guesses u_i^0. By passing to the limit in (3.3) we obtain

$$a(u_1 + u_2, \hat{u}_i) = (f, \hat{u}_i), \qquad \forall \hat{u}_i \in V_i, \tag{3.6}$$

and since any $\hat{u} \in V$ can be decomposed into $\hat{u}_1 + \hat{u}_2$, $\hat{u}_i \in V_i$, equation (2.1) follows by summing the two equations of (3.6). It remains to see (assuming that weak convergence holds) towards which of the decomposition (3.5) convergence takes place.
Let φ be given in $\mathcal{D}(\Omega_1 \cap \Omega_2)$ (the set of C^∞ functions with compact support in $\Omega_1 \cap \Omega_2$). We can take $\hat{u}_i = \varphi$ in (3.3). Subtracting, we obtain

$$b(u_1^{n+1} - u_1^n - (u_2^{n+1} - u_2^n), \varphi) + a(u_1^{n+1} - u_1^n - (u_2^{n+1} - u_2^n), \varphi) = 0. \tag{3.7}$$

Summing (3.7) with respect to n, we obtain

$$b(u_1^{n+1} - u_2^{n+1}, \varphi) + a(u_1^{n+1} - u_2^{n+1}, \varphi) = b(u_1^0 - u_2^0, \varphi) + a(u_1^0 - u_2^0, \varphi). \tag{3.8}$$

Hence letting $n \to \infty$,

$$b(u_1 - u_2, \varphi) + a(u_1 - u_2, \varphi) = b(u_1^0 - u_2^0, \varphi) + a(u_1^0 - u_2^0, \varphi), \; \forall \varphi \in \mathcal{D}(\Omega_1 \cap \Omega_2). \tag{3.9}$$

This is equivalent to $(\beta + A)(u_1 - u_2) = (\beta + A)(u_1^0 - u_2^0)$ in $\Omega_1 \cap \Omega_2$. Since we already know that (3.5) holds true, it is equivalent to

$$(\beta + A)u_1 = \frac{1}{2}(\beta + A)(u + u_1^0 - u_2^0) \quad \text{in} \quad \Omega_1 \cap \Omega_2, \quad u_1|_{S_1} = 0, \; u_1|_{S_2} = u, \tag{3.10}$$

which defines uniquely u_1 (hence u_2) in $\Omega_1 \cap \Omega_2$. Therefore, if Algorithm 2 converges, it converges to the unique decomposition (3.5), (3.9).

Let us prove weak convergence in the symmetric case $a_{ij} = a_{ji}$, $\forall i,j$. To simplify we take $b(u,v) = \int_\Omega u.v$. Moreover we will show that

$$\sum_n (|u_1^{n+1} - u_1^n|^2 + |u_2^{n+1} - u_2^n|^2) < \infty, \tag{3.11}$$

where $|\cdot|$ is the $L^2(\Omega)$ norm (recall that the functions with subscript i are extended by 0 outside Ω_i).
For the proof let us consider an arbitrary decomposition

$$u = w_1 + w_2, \quad w_i \in V_i. \tag{3.12}$$

We introduce

$$w_i^n = u_i^n - w_i. \tag{3.13}$$

One has

$$\begin{aligned} \beta(w_1^{n+1} - w_1^n, \hat{u}_1) + a(w_1^{n+1} + w_2^n, \hat{u}_1) = 0 \\ \beta(w_2^{n+1} - w_2^n, \hat{u}_2) + a(w_2^{n+1} + w_1^n, \hat{u}_2) = 0 \end{aligned} \tag{3.14}$$

Taking $\hat{u}_i = w_i^{n+1}$ and writing $a(\hat{u})$ for $a(\hat{u},\hat{u})$ we obtain

$$\begin{aligned} &\frac{\beta}{2}[\,|w_1^{n+1} - w_1^n|^2 + |w_1^{n+1}|^2 - |w_1^n|^2 + |w_2^{n+1} - w_2^n|^2 + |w_2^{n+1}|^2 - |w_2^n|^2] \\ &+a(w_1^{n+1}) + a(w_2^n, w_1^{n+1}) + a(w_1^n, w_2^{n+1}) + a(w_2^{n+1}) = 0. \end{aligned} \tag{3.15}$$

Owing to the symmetry of a, one has $a(w,\hat{w}) \le (a(w) + a(\hat{w}))/2$ and it follows from (3.15) that

$$|w_1^{n+1}|^2 + |w_2^{n+1}|^2 + |w_1^{n+1} - w_1^n|^2 + |w_2^{n+1} - w_2^n|^2 \tag{3.16}$$

$$+\frac{1}{\beta}[a(w_1^{n+1}) + a(w_2^{n+1})] \le |w_1^n|^2 + |w_2^n|^2 + \frac{1}{\beta}[a(w_1^n) + a(w_2^n)]. \tag{3.17}$$

Summing up in n it follows that

$$\begin{aligned} &|w_1^{n+1}|^2 + |w_2^{n+1}|^2 + \sum_{k=0}^{n}(|w_1^{k+1} - w_1^k|^2 + |w_2^{k+1} - w_2^k|^2) \\ &+\frac{1}{\beta}[a(w_1^{n+1}) + a(w_2^{n+1})] \le |w_1^0|^2 + |w_2^0|^2 + \frac{1}{\beta}[a(w_1^0) + a(w_2^0)]. \end{aligned} \tag{3.18}$$

Together with the coercivity of a, and the uniqueness of the limit, this proves that the whole sequence u_i^n converges weakly in V_i. Furthermore (3.11) follows from (3.18).

4. DISCRETIZATION

For clarity we assume that the Ω_i are polygonal and that $a_{ij} = \delta_{ij}$. Let V_{1h} and V_{2h} be two Lagrange conforming continuous finite element approximation spaces of order p of $V_1 = H_0^1(\Omega_1)$ and $V_2 = H_0^1(\Omega_2)$. Then the discrete version of Algorithm 2 is:

Find $u_{1h}^{n+1} \in V_{1h}$ such that

$$\int_{\Omega_1} \beta(u_{1h}^{n+1} - u_{1h}^n) v_{1h} + \int_{\Omega_1} \nabla u_{1h}^{n+1} \nabla v_{1h} + \int_{\Omega_1} \nabla u_{2h}^n \nabla v_{1h} = \int_{\Omega_1} f v_{1h}, \ \forall v_{1h} \in V_{1h} \tag{4.1}$$

Find $u_{2h}^{n+1} \in V_{2h}$ such that

$$\int_{\Omega_2} \beta(u_{2h}^{n+1} - u_{2h}^n) v_{2h} + \int_{\Omega_2} \nabla u_{2h}^{n+1} \nabla v_{2h} + \int_{\Omega_2} \nabla u_{1h}^n \nabla v_{2h} = \int_{\Omega_2} f v_{2h}, \ \forall v_{2h} \in V_{2h}. \tag{4.2}$$

5. ERROR ESTIMATE

We state the result with $p \geq 1$ but when it comes to quadrature errors we will control the error for linear elements only ($p = 1$).

Theorem 2. *Assume that the solution of (2.2) is in $H^{p+1}(\Omega)$ for some $p \geq 1$. Assume that every element u of $H^{p+1}(\Omega)$ can be decomposed in*

$$u = u_1 + u_2 \tag{5.1}$$

where u_i restricted to Ω_i is in $H^{p+1}(\Omega_i)$ and $u_i = 0$ outside Ω_i.
If $u_h = \lim(u_{1h}^n + u_{2h}^n)$ is computed with Lagrange conforming finite elements of order p, then

$$\|u - u_h\|_{1,\Omega} \leq C h^p (\|u_1\|_{p+1,\Omega_1} + \|u_2\|_{p+1,\Omega_2}). \tag{5.2}$$

Proof. The proof of convergence is the same as for the continuous case, so there exists $u_{ih} \in V_{ih}$ such that

$$\int_\Omega \nabla(u_{1h} + u_{2h}) \nabla(v_{1h} + v_{2h}) = \int_\Omega f(v_{1h} + v_{2h}), \ \forall v_{ih} \in V_{ih}, i = 1,2 \tag{5.3}$$

or equivalently

$$\int_\Omega \nabla(u_{1h} - u_1 + u_{2h} - u_2) \nabla(v_{1h} + v_{2h}) = 0, \ \forall v_{ih} \in V_{ih}, i = 1,2 \tag{5.4}$$

This means that u_{ih} also solves

$$\min_{u_{ih} \in V_{ih}} \int_\Omega |\nabla(u_1 + u_2 - u_{1h} - u_{2h})|^2. \tag{5.5}$$

Hence if Π_{ih} denotes the finite element interpolator from V_i to V_{ih}

$$(\int_\Omega |\nabla(u_1 + u_2 - u_{1h} - u_{2h})|^2)^{\frac{1}{2}} \leq (\int_\Omega |\nabla(u_1 + u_2 - \Pi_{1h} u_1 - \Pi_{2h} u_2)|^2)^{\frac{1}{2}} \tag{5.6}$$

$$\leq C(||u_1||_{\Omega_1, p+1} + ||u_2||_{\Omega_2, p+1}) h^p \tag{5.7}$$

Remark 2. Let us comment on the decomposition $u_1 + u_2$. Assume that $\theta \in H^{p+1}(\Omega_1 \cap \Omega_2)$ is one on $(\partial\Omega_1) \cap \Omega_2$ and zero on $(\partial\Omega_2) \cap \Omega_1)$, and has $\partial^k\theta/\partial n^k = 0, k = 1,..,p$ on $\partial(\Omega_1 \cap \Omega_2)$. Then it can be extended in $\Omega_1 \cup \Omega_2$ by one and zero and

$$\begin{aligned} u_1 &= u\theta \text{ in } \Omega_1 \\ u_2 &= u(1-\theta) \text{ in } \Omega_2 \end{aligned}$$

has the desired property.
If Ω is decomposed into slices then such functions θ exist.

5.1. Quadratures

As such the scheme is too costly to implement because it requires the intersection of triangulations. Integrals of piecewise constant functions g are computed exactly by

$$\int_{\Omega_i} g = \sum_{k=1}^{n_i} |T_k^i| g(\xi_k^i) \tag{5.8}$$

where n_i is the number of triangles of the triangulation of Ω_i and ξ_k^i is the chosen quadrature point in triangle T_k^i (its center for instance).
To compute integrals involving products of functions on two triangulations like $\int \nabla u_{1h} \nabla v_{2h}$ we propose the following formula

$$\int_{\Omega_1 \cap \Omega_2} g \approx \frac{1}{2} \sum_{\{k:\xi_k^1 \in \Omega_1 \cap \Omega_2\}} |T_k^1| g(\xi_k^1) + \frac{1}{2} \sum_{\{k:\xi_k^2 \in \Omega_1 \cap \Omega_2\}} |T_k^2| g(\xi_k^2) \tag{5.9}$$

This can be summarized by saying that when $u \in V_{ih}$ and $v \in V_{jh}$, $i \neq j$, then $a(\cdot,\cdot)$ is replaced by $a_h(\cdot,\cdot)$ with

$$a_h(u,v) = \sum_{k=1}^{n_1} \left(\frac{|T_k^1| \nabla u \cdot \nabla v}{I_{\Omega_1} + I_{\Omega_2}} \right) |_{x=\xi_k^1} + \sum_{k=1}^{n_2} \left(\frac{|T_k^2| \nabla u \cdot \nabla v}{I_{\Omega_1} + I_{\Omega_2}} \right) |_{x=\xi_k^2} \tag{5.10}$$

and $a_h(u,v) = a(u,v)$ when $u, v \in V_{ih}$.

Applied to $g = \nabla u_{1h} \nabla u_{2h}$ this formula requires the computation of ∇u_{ih} on the mesh of Ω_i and then its computation at ξ_k^i which in turn requires to identify the position of this point in the mesh of Ω_j, $j \neq i$.

With such definitions we propose to solve the discrete problems:

- Find $u_{ih}^{n+1} \in V_{ih}$ such that $\forall v_{ih} \in V_{ih}$

$$b(u_{1h}^{n+1} - u_{1h}^n, \hat{u}_{1h}) + a_h(u_{1h}^{n+1} + u_{2h}^n, \hat{u}_{1h}) = (f, \hat{u}_{1h}) \quad \forall \hat{u}_{1h} \in V_{1h} \tag{5.11}$$

$$b(u_{2h}^{n+1} - u_{2h}^n, \hat{u}_{2h}) + a_h(u_{1h}^n + u_{2h}^{n+1}, \hat{u}_{2h}) = (f, \hat{u}_{2h}) \quad \forall \hat{u}_{2h} \in V_{2h} \tag{5.12}$$

Clearly these equations define u_{ih}^{n+1} uniquely. At convergence the problem solved is

- Find $u_{ih} \in V_{ih}$ such that $\forall \hat{u}_{ih} \in V_{ih}$

$$a_h(u_{1h} + u_{2h}, \hat{u}_{1h} + \hat{u}_{2h}) = (f, \hat{u}_{1h} + \hat{u}_{2h}). \tag{5.13}$$

The bilinear form is symmetric but this discrete problem may not have a solution because the form may not be coercive. For this there is clearly a compatibility condition between the triangulation which we haven't found yet but which will be investigate in the future; if it is too restrictive the same idea can be used with more quadrature points. So we state only a partial result below, but it circumscribes the difficulty.

5.1.1. Quadrature error. We prove here a partial result which has the merit of showing were lies the difficulty, and hope to solve it later.

Proposition. *Assume that the triangulations of Ω_1 and Ω_2 are compatible in the sense that they give a coercive bilinear form. Then the error between the approximate problem (5.13) and the continuous problem is*

$$\|u - u_h\| < Ch(\|u_1\|_{2,\Omega_1} + \|u_2\|_{2,\Omega_2}$$

Proof. Recall Strang's Lemma (see Ciarlet, [9,p186])

$$\|u - u_h\| < \|u - \Pi_h u\| + \sup_{w_h \in V_h} \frac{a(\Pi_h u, w_h) - a_h(\Pi_h u, w_h)}{\|w_h\|}$$

where by $\Pi_h u$ we mean the interpolation of u_1 on V_h^1 plus that of u_2 on V_h^2 for some decomposition of u into $u^1 + u^2$.
Here, with linear elements on triangles and $a(,)$ defined from the Laplace equation, quadrature errors are only on mixed integrals.
For a decomposition of u into $u^1 + u^2$, another way of writing $a_h(\Pi_h u, w_h^1 + w_h^2)$ is

$$\begin{aligned}
a_h(\Pi_h^1 u^1 + \Pi_h^2 u^2, w_h^1 + w_h^2) &= a(\Pi_h^1 u^1, w_h^1) + a(\Pi_h^2 u^2, w_h^2) \\
&+ a(\Pi_h^2 u^1, w_h^2) \\
&- a_h(\Pi_h^2 u^1 - \Pi_h^1 u^1, w_h^2) \\
&+ a(\Pi_h^1 u^2, w_h^1) \\
&- a_h(\Pi_h^1 u^2 - \Pi_h^2 u^2, w_h^1)
\end{aligned}$$

$$\begin{aligned}
&= a(\Pi_h^1 u^1 + \Pi_h^2 u^2, w_h^1 + w_h^2) \\
&- (a_h - a)(\Pi_h^2 u^1 - \Pi_h^1 u^1, w_h^2) - (a_h - a)(\Pi_h^1 u^2 - \Pi_h^2 u^2, w_h^1)
\end{aligned}$$

because $a(\Pi_h^2 u^1, w_h^2)$ is equal to $a_h(\Pi_h^2 u^1, w_h^2)$. Hence

$(a-a_h)(\Pi_h u, w_h) = (a_h-a)(\Pi_h^2 u^1 - \Pi_h^1 u^1, w_h^2) + (a_h-a)(\Pi_h^1 u^2 - \Pi_h^2 u^2, w_h^1)$

Now $\Pi_h^2 u^1 - \Pi_h^1 u^1$ can be bounded like an interpolation error by rewriting it as

$$\Pi_h^2 u^1 - \Pi_h^1 u^1 = \Pi_h^2 u^1 - u^1 + u^1 - \Pi_h^1 u^1.$$

So

$$\begin{aligned} (a-a_h)(\Pi_h u, w_h) &= (a-a_h)(\Pi_h^1 u^1 - u^1, w_h^2) - \\ -(a-a_h)(\Pi_h^2 u^1 - u^1, w_h^2) &+ (a-a_h)(\Pi_h^1 u^2 - u^2, w_h^1) - \\ &- (a-a_h)(\Pi_h^2 u^2 - u^2, w_h^1) \end{aligned}$$

Each four pieces are bounded independently; the parts that involve $a(,)$ are easy to bound, the parts that involve $a_h(,)$ are treated as follows. By definition

$$\begin{aligned} a_h((\Pi_h^2 u^1 - u^1), w_h^2) &= \sum_m |T_m^1| \nabla(\Pi_h^2 u^1 - u^1)(\xi_m^1) \nabla w_h^2(\xi_m^1) \\ &\leq C\|\Pi_h^2 u^1 - u^1\| \|w_h^2\| \leq Ch\|w_h^2\| \end{aligned}$$

This proves that

$$|(a-a_h)(\Pi_h u, w_h)| \leq Ch(\|w_h^1\| + \|w_h^2\|)$$

Since $\|w_h^1\| + \|w_h^2\|$ is an equivalent H^1 norm in V_h, the quadrature error is bounded by h.

6. OTHER CASES

6.1. The domain is the difference of two sets

Now assume that $\Omega = \Omega' \setminus C$ where $C \subset \Omega'$.

We take a larger set Ω_2' containing C and inside Ω'. For Ω_1 we choose a set Ω_1' containing C but inside Ω_2:

$$C \subset \Omega_1' \subset \Omega_2' \subset \Omega' \tag{6.1}$$

Then we take

$$\Omega_1 = \Omega' \setminus \Omega_1', \qquad \Omega_2 = \Omega_2' \setminus C \tag{6.2}$$

Obviously we have $\Omega = \Omega_1 \cup \Omega_2, \Omega_1 \cap \Omega_2 \neq 0$ so we can apply Algorithm 1.

Remark 4. This idea is borrowed from the Chimera method except that the latter is framed in the context of Schwarz algorithm.

Remark 5. In the discrete case, the domains Ω_i are found automatically by finding first all the triangles of Ω_{1h} which are touching C then taking one or two layers of triangles around it; this determines the boundary S_1. Then surrounding C with a boundary S_2 of the same type as ∂C which contains S_1 in its interior and is contained in Ω_1. This may not be possible if the triangles of Ω' are too large.

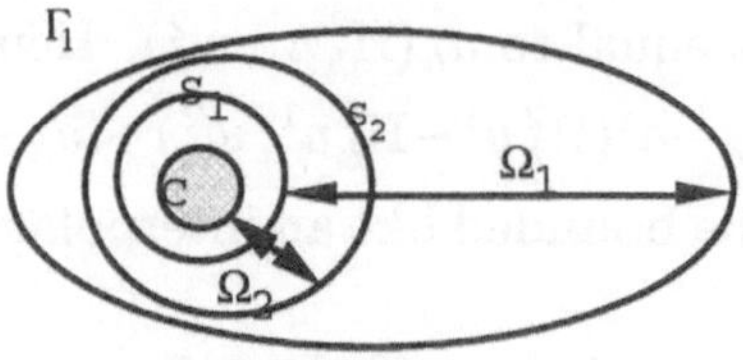

Figure 2.

6.2. The domain is obtained by extrusion

Consider the case where a portion of volume, C, is extruded from the primary volume Ω':

$$\Omega = \Omega_1 \backslash C \cap \Omega' \tag{6.3}$$

As before we construct an auxiliary domain Ω_2 which is around C and an auxiliary domain Ω_1 which is exterior to C but intersect Ω_2. Let S_1 be the part of $\partial\Omega'$ in Ω_2 ans S_2 the part of $\partial\Omega_2$ in Ω_1 The boundary conditions on u_i^{n+1} in Algorithm 1 will be

$$u_1^{n+1} = 0 \quad \text{on } S_2 \cup \partial\Omega_1, \qquad u_2^{n+1} = 0 \quad \text{on } S_1 \cup \partial\Omega_1 \tag{6.4}$$

Note that a condition on u_2 is on a boundary strictly inside Ω_2. If this causes a difficulty then the fictitious domain method may be used to impose this condition.

6.3. The domain is the intersection of two sets

Extending the idea used for the extrusion we simply compute the u_i^{n+1} in Ω_i with homogeneous Dirichlet conditions on *both* boundaries $\partial\Omega_i$, i.e. on $\partial\Omega_1 \cup \partial\Omega_2$. Again the fictitious domain method will avoid the need for intersecting both domains.

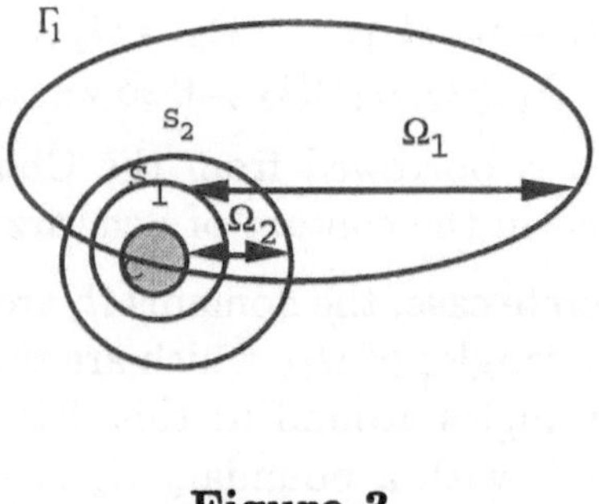

Figure 3.

7. A FAST FINITE ELEMENT INTERPOLATOR

In practice one may discretize the variational equations by the Finite Element method. Then there will be one mesh for Ω_1 and another one for Ω_2. The computation of integrals of products of functions defined on different meshes is difficult. Quadrature formulae and interpolations from one mesh to another at quadrature points are needed. We present below the interpolation operator which we have used and which is new, to the best of our knowledge.

Let $\mathcal{T}_h^0 = \cup_k T_k^0, \mathcal{T}_h^1 = \cup_k T_k^1$ be two triangulations of a domain Ω. Let

$$V(\mathcal{T}_h^i) = \{C^0(\Omega_h^i) \ : \ f|_{T_k^i} \in P^1\}, \quad i = 0, 1 \tag{7.1}$$

be the spaces of continuous piecewise affine functions on each triangulation.

Let $f \in V(\mathcal{T}_h^0)$. The problem is to find $g \in V(\mathcal{T}_h^1)$ such that

$$g(q) = f(q) \quad \forall q \text{ vertex of } \mathcal{T}_h^1 \tag{7.2}$$

Although this is a seemingly simple problem, finding an efficient algorithm is difficult in practice. We propose an algorithm which is of complexity $N^1 \log N^0$, where N^i is the number of vertices of $\mathcal{T}_h^i$, and which is very fast for most practical 2D applications.

Algorithm 4. The method has 5 steps. First a quadtree is built containing all the vertices of mesh $\mathcal{T}_h^0$ such that in each terminal cell there are at least one, and at most 4, vertices of $\mathcal{T}_h^0$.

For each q^1, vertex of $\mathcal{T}_h^1$ do:

Step 1 Find the terminal cell of the quadtree containing q^1.

Step 2 Find the the nearest vertex q_j^0 to q^1 in that cell.

Step 3 Choose one triangle $T_k^0 \in \mathcal{T}_h^0$ which has q_j^0 for vertex.

Step 4 Compute the barycentric coordinates $\{\lambda_j\}_{j=1,2,3}$ of q^1 in T_k^0.

– if all barycentric coordinates are positive, go to Step 5

– else if one barycentric coordinate λ_i is negative replace T_k^0 by the adjacent triangle opposite q_i^0 and go to Step 4.

– else two barycentric coordinates are negative so take one of the two randomly and replace T_k^0 by the adjacent triangle as above.

Step 5 compute $g(q^1)$ on T_k^0 by linear interpolation of f:

$$g(q^1) = \sum_{j=1,2,3} \lambda_j f(q_j^0) \tag{7.3}$$

End
Two problems need to be solved:

- *What if q^1 is not in Ω_h^0 ?* Then Step 5 will stop with a boundary triangle. So we add a step which tests the distance of q^1 to the two adjacent boundary edges and select the nearest, and so on till the distance grows.

- *What if Ω_h^0 is not convex and the marching process of Step 4 locks on a boundary?*

By construction Delaunay-Voronoi mesh generators always triangulate the convex hull of the vertices of the domain. So we make sure that this information is not lost when $\mathcal{T}_h^0, \mathcal{T}_h^1$ are constructed and we keep the triangles which are outside the domain in a special list. Hence in step 5 we can use that list to step over holes if needed.

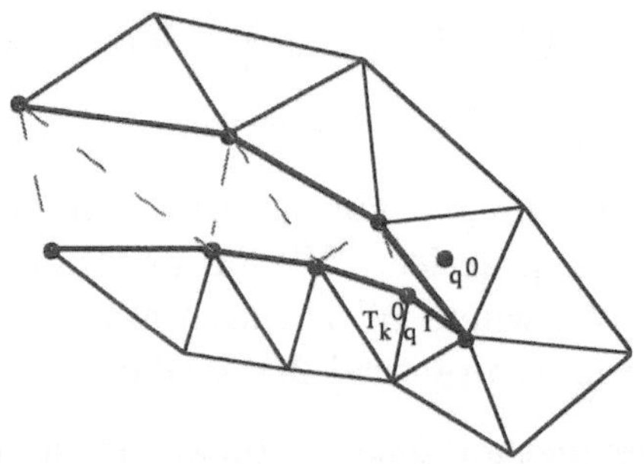

Figure 4. *To interpolate a function at q^0 the knowledge of the triangle which contains q^0 is needed. The algorithm may start at $q^1 \in T_k^0$ and stall on the boundary (thick line) because the line q^0q^1 is not inside Ω. But if the holes are triangulated too (doted line) then the problem does not arise.*

Remark. Step 3 requires an array of pointers such that each vertex points to one triangle of the triangulation.

8. NUMERICAL EXAMPLES

The test case is geared to reproduce the situation of scientific computing with CAD data. The temperature equation is solved for an object, a stylized spanner, described by set operations on 4 elementary shapes, A,B,C,D. A is a rectangle, B is a circle, C is a trapezoidal quadrangle and D is a circle. The spanner is the union of B and C with A extruded and D removed.
We do not have yet the software to treat extrusions so A is intersected with B first so as to reduce the case to a set difference rather than an

extrusion. Then A and D are surrounded by artificial domains, some elements of B and C are removed so that the final domain becomes the union of 4 sets feasible for Schwarz algorithm.

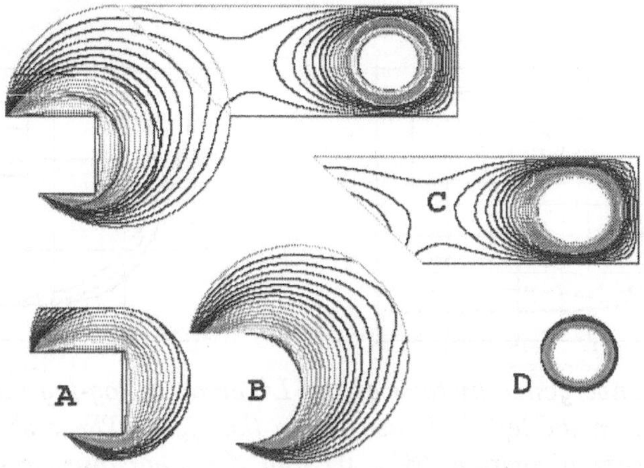

Figure 5. *The top picture shows the temperature level lines in the spanner as computed by standard FEM. The other pictures show the same lines but computed by domain decomposition after 26 iterations. Note that the spanner is reconstructed by sliding horizontally A into B, and vertically C into B and D.*

The geometry is prepared with the software "freefem+" (see [5] for more details).
The PDE is a simple Laplacian with non-homogeneous Dirichlet data: the temperature is 100° in the mouth of the spanner and in the hole in the handle and zero in the remaining boundaries.
Freefem contains also a PDE solver capable of handling several meshes within one program and which uses triangular conforming finite elements of degree 1 and Gauss factorizations to solve the linear systems. Mixed integrals are computed with quadrature points on the mid-edges of the triangles, which is similar to formula (3.9) when the function is piecewise constant. Naturally the interpolation operator for the computation of integrals is the one presented above.
We have tested the method for different values of β and for two cases:
- I. In the first case the boundaries of the subdomains are edges of the triangulations of the other domains.
- II. In the second case the triangulations are completely independent from one another.
Figure 5 shows the convergence behavior for these two cases (I on the left, II on the right) for $\beta = 0, 1, 5$ and 10. (recall that $\beta = 0$ is Schwarz' algorithm).

Figure 6 shows the solution with the standard finite element method (top) compared with the solution on each domain.

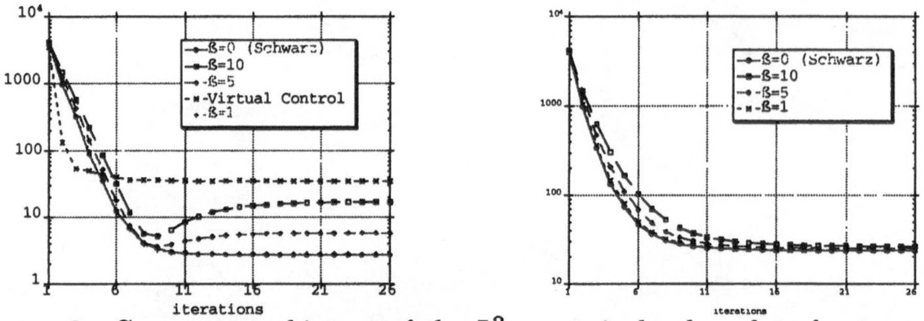

Figure 6. *Convergence history of the L^2 error in log-log plots for two cases. Case I. on the left and case II. on the right. These show that the classical Schwarz algorithm ($\beta = 0$) and the algorithm presented here have similar performances except if β is too large and if the meshes match. Comparison with the method of "virtual controls" described in [1] is also made and shows that the latter is faster but gives a less perfect matching in this case.*

Acknowledgments

We wish to thank Yves Achdou and Vivette Girault for their very helpful suggestions.

References

[1] Lions, J.L. and Pirroneau, O. (1998). Algorithmes parallèles pour la solution de problèmes aux limites. C.R.A.S., 327:947–352, Paris.

[2] Lions, J.L. and Pironneau, O. (1998). Sur le contrôle des sytèmes distribués. C.R.A.S., 327:993–998, Paris.

[3] Lions J.L., Pironneau O. (1999). Domain decomposition methods for CAD. C.R.A.S., 328:73–80, Paris.

[4] Steger, J.L. (1991). The Chimera method of flow simulation. Workshop on applied CFD, Univ of Tennessee Space Institute, August 1991.

[5] Hecht, F., Pironneau O. (1999). Multiple meshes and the implementation of freefem+. INRIA report March, 1999. Also on the web at ftp://ftp.ann.jussieu.fr/pub/soft/pironneau.

[6] Lions, P.L. (1988,1989,1990). On the Schwarz alternating method. I,II,III. Int Symposium on Domain decomposition Methods for Partial Differential Equations. SIAM, Philadelphia.

[7] Lions, J.L. and Magenes, E. (1968). Problèmes aux limites non-homogènes et applications. Vol 1, Dunod 1968.

[8] Burden, G. and Coiffe, Ph. (1994). Virtual Reality Technology. New York, Wiley.

[9] Ciarlet, P.G. (1997). The Finite Element Method. Prentice Hall.

SOLUTION OF CONVECTION-DIFFUSION PROBLEMS WITH THE MEMORY TERMS

Jozef Kačur

Abstract: A new numerical scheme is proposed for solving a contaminant transport problem with adsorption. Both, equilibrium and nonequilibrium sorption modes with Freundlich and langmiur type isotherms are included in the considered mathematical model. The approximation scheme is based on a relaxation scheme and on the method of characteristics. The convergence of approximation scheme is proved and some numerical experiments are presented.

Keywords: Method of characteristics, convection-diffusion, contaminant transport, memory effects.

1. INTRODUCTION

In this paper an approximation solution of the following convection diffusion problem is discussed

$$\begin{aligned} \partial_t b(u) + \operatorname{div}(\bar{F}(t,x,u) - k(t,x,u)\nabla u) &= f(t,x,u,s), \\ s(t,x) = \textstyle\int_0^t K(t,z)\psi(u(z,x))dz & \end{aligned} \quad \text{in } (0,T)\times\Omega, \tag{1.1}$$

where $\Omega \subset \mathbb{R}^N$ is a bounded domain with a Lipschitz continuous boundary $\partial\Omega$, $T < \infty$.

We consider the mixed boundary conditions

$$\begin{aligned} u = 0 \quad &\text{on } I\times\Gamma_1, \\ -k(t,x,u)\nabla u\cdot\nu = g(t,x,u) \quad &\text{on } I\times\Gamma_2 \end{aligned} \tag{1.2}$$

where $I \equiv (0,T)$, $\Gamma_1, \Gamma_2 \subset \partial\Omega$, $\Gamma_1 \cap \Gamma_2 = \emptyset$ and $\text{mes}_{N-1}\Gamma_1 + \text{mes}_{N-1}\Gamma_2 = \text{mes}_{N-1}\partial\Omega$. Together with (1.1), (1.2) we consider the initial condition

$$b(u(0,x)) = b(u_0(x)) \quad \text{in } \Omega. \tag{1.3}$$

We assume that $b(s)$ is strictly increasing in s, $\bar{F}(t,x,s)$ is Lipschitz continuous in x, s and f, g, ψ are sublinear in u. Problem (1.1)-(1.3) has

Applied Nonlinear Analysis, edited by Sequeira *et al.*
Kluwer Academic / Plenum Publishers, New York, 1999.

been studied in [9] for a special case when $f \equiv f(t,x,u)$, i.e., without memory term.

As an example we present a model of contaminant transport in porous media intensively studied in the last years, see [1, 3, 4, 5, 10, 11, 12]

$$\begin{aligned} \partial_t(\theta C + \rho S) + \operatorname{div}(\bar{v}C - D\nabla C) &= 0 \\ \rho\partial_t S &= d(\psi(C) - S) \end{aligned} \tag{E}$$

where C is the concentration of the contaminant, $\bar{v}$ is (Darcy) velocity field of water, D is diffusion tensor, ρ is bulkdensity, ψ is sorption isotherm of the porous media with porosity θ. Here S is the mass of contaminant adsorbed by unit mass of porous medium. Coefficient d describes the rate of adsorption. If $d \to \infty$ then equilibrium sorption process occurs and consequently $S = \psi(C)$. Then $b(s) = \theta s + \rho\psi(s)$ generates the parabolic term in (1.1) with $f \equiv 0$. Moreover, when $\psi(s) = cs^p$, $0 < p$ (so called Freundlich isotherm) then $b'(0) = \infty$ for $p < 1$ which occurs in most practical situations. In that case (1.1) is of porous media type with convective term and the support of contaminant develops with the finite speed. In the nonequilibrium case ($d < \infty$) we can eliminate S from ODE and we obtain

$$b(z) \equiv z, \quad f(t,x,u,s) \equiv s - d\psi(u)$$

and

$$s = da\int_0^t e^{-a(t-z)}\,\psi(u(z,x))dz \qquad \text{with } a = \frac{d}{\rho} \quad (S(0) = 0)\,.$$

In the case of Freundlich isotherm with $p < 1$ the function ψ is not Lipschitz continuous.

Numerical analysis of the model with the equilibrium sorption process (i.e. $s \equiv 0$) is included in our previous paper [9]. The contribution of the present paper is the numerical analysis of the mathematical model (1) which includes both equilibrium and nonequilibrium sorption process in (E). The degeneracy $b' = 0, \infty$ in some points is included and thus convective term can be strongly dominant. Numerical solution of (1.1)-(1.3) thus represents a delicate problem. We extend our concept of approximation introduced in [9] for the case $s \equiv 0$. We prove the convergence of the approximate solution. The existence and uniqueness of the variational solution is discussed in [12] where $\bar{F}(t,x,u) \equiv \bar{v}(x)u$.

Our concept of approximation is based on the relaxation schemes developed by W. Jäger and J. Kačur in [6, 7] and on the method of characteristics initiated by O. Pironneau [14] and J. Douglas and T. Russel [2].

2. APPROXIMATION SCHEME

The transport part of (1.1) is of the form

$$\partial_t b(u) + \mathrm{div}\bar{F}(t, x, u) = 0$$

which formally we can rewrite into the form

$$\partial_t u + \frac{\bar{F}'_u(t, x, u)}{b'(u)} \cdot \nabla u = -\frac{\mathrm{div}_x \bar{F}(t, x, u)}{b'(u)}$$

and hence the corresponding velocity field $\bar{v} = \frac{\bar{F}'_u(t,x,u)}{b'(u)}$ is depending on the unknown u and cannot be expected to be smooth. Thus we regularize b by b_n with the properties listed below. The transport part with the velocity field $\bar{v}$ can be realized by means of characteristics $X(s; t, x)$ governed by ODE

$$\frac{dX}{ds} = \bar{v}(X, s); \quad X(t; t, x) = x.$$

Their Euler backwards type approximation between time levels $t = t_i$, $t = t_{i+1}$ is given by $\varphi^{i+1}(x) = x - (t_{i+1} - t_i)\bar{v}(x, t_{i+1})$. Then concentration profile $u_i(x)$ at $t = t_i$ after the transport along approximated characteristics prolonging $t_{i+1} - t_i$ became $u_i \circ \varphi^{i+1}$. The transport can be realized if characteristics X or their approximations φ do not intersect each other which requires the boundedness of $|\nabla \bar{v}|_\infty \leq c$ and small time step $\tau = t_{i+1} - t_i$. In our case we cannot guarantee it. As it was proven in [9] a smoothing (or averaging) of velocity field can guarantee that the corresponding characteristic will not intersect provided the time step $\tau = t_{i+1} - t_i$ is small. Applying the method of characteristics the points $\varphi^{i+1}(x)$ can cross the boundary $\partial\Omega$ and in such case we understand by $u_i \circ \varphi^i := \tilde{u}_i \circ \varphi^i$ where $\tilde{u}_i$ is an extension of $u_i \in W_2^1(\Omega)$ to $\tilde{u}_i \in W_2^1(\Omega^*)$ with $\Omega^* \supset \bar{\Omega}$ so that $\|\tilde{u}_i\|_{W_2^1(\Omega^*)} \leq \|u_i\|_{W_2^1(\Omega)}$.

We realize smoothing of $\bar{v}$ by convolution $\omega_h * \bar{v}$ where ω_h is standard mollifier with $\omega_h(x) = \frac{1}{h^N}\omega_1(\frac{x}{h})$ where $\omega_1(x) = \kappa \exp(\frac{|x|^2}{|x|^2 - 1})$, $\int_{\mathbb{R}^N} \omega_1 dx = 1$.

We consider nonstandard time discretization of (1.1) with time step $\tau \equiv \tau_n = T/n$ $(n \in \mathbb{N})$ and u_i is an approximation of $u(x, t_i)$ at time level $t_i = \tau.i$, $i = 1, \ldots, n$. We have to determine u_i from linear elliptic equation coupled with a relaxation parameter $0 < \lambda_i \in L_\infty(\Omega)$

$$\lambda_i(u_i - u_{i-1} \circ \varphi^i_{\mu_i}) - \mathrm{div}(k_i \nabla u_i) = \tau H_i + \tau f(t_i, x, u_{i-1} s_i) \qquad (2.1)$$

$$u_i = 0 \quad \text{on } \Gamma_1, \quad -k_i \nabla u_i \cdot \nu = g_i \equiv g(t_i, u_{i-1}) \quad \text{on } \Gamma_2,$$

where

$$s_i = \sum_{j=1}^{i-1} \alpha_{ij}\psi(u_j)\tau, \quad \alpha_{ij} = \tfrac{1}{\tau}\int_{t_{j-1}}^{t_j} K(t_i,s)ds,$$
$$H_i := \mathrm{div}_x \bar{F}(x,t_i,u_{i-1}),$$
$$\varphi_\nu^i(x) := x - \tau\omega_h * \left(\tfrac{F_u'(t_i,x,u_{i-1})}{\nu}\right)$$
$$\text{with } 0 < \nu \in L_\infty(\Omega), \; h = \tau^\omega, \; \omega \in (0,1)$$

and the following "convergence conditions" (2.2), (2.3) have to be satisfied

$$\|\mu_i - G_i(\mu_i)\|_0 < \tau^\alpha, \quad \alpha \in (0,1) \tag{2.2}$$

where

$$G_i(\nu) = \int_0^1 b_n'(u_{i-1} + s(u_{i-1}\circ\varphi_\nu^i - u_{i-1}))ds \equiv \frac{b_n(u_{i-1}) - b_n(u_{i-1}\circ\varphi_\nu^i)}{u_{i-1} - u_{i-1}\circ\varphi_\nu^i}$$

($\|\cdot\|_0$ is L_2 norm and b_n is a regularization of b) and

$$\left\|\lambda_i - \frac{b_n(u_i) - b_n(u_{i-1}\circ\varphi_{\mu_i}^i)}{u_i - u_{i-1}\circ\varphi_{\mu_i}^i}\right\|_0 < \tau^\beta, \quad \beta \in (0,1)\,. \tag{2.3}$$

The scheme is implicit and to guarantee (2.2), (2.3) we propose the iterations

$$\lambda_{i,k-1}(u_{i,k} - u_{i-1}\circ\varphi_{\mu_i}^i) - \tau\mathrm{div}(k_i\nabla u_{i,k}) = \tau H_i + \tau f_i \tag{2.4$_k$}$$

$$u_{i,k} = 0 \text{ on } \Gamma_1, \quad -k_i\nabla u_i\cdot\nu = g_i \text{ on } \Gamma_2, \; f_i := f(t_i,x,u_{i-1},s_i)$$
$$\lambda_{i,k} := \tfrac{b_n(u_{i,k}) - b_n(u_{i-1}\circ\varphi_{\mu_i}^i)}{u_{i,k} - u_{i-1}\circ\varphi_{\mu_i}^i}, \quad \lambda_{i,0} := b_n'(u_{i-1})\,.$$

These iterations are not coupled with (2.2). If

$$\|\lambda_{i,k_0} - \lambda_{i,k_0-1}\|_0 < \tau^\beta$$

then we put $u_i := u_{i,k_0}$, $\lambda_i := \lambda_{i,k_0-1}$.

To obtain $\varphi_{\mu_{i+1}}^{i+1}$ we propose fixed point type iterations

$$\mu_{i,l} = G_{i+1}(\mu_{i,l-1}), \quad l = 1, \ldots \tag{2.5$_l$}$$

and when $\|\mu_{i,l_0} - \mu_{i,l_0-1}\|_0 < \tau^\alpha$ then we put $\mu_{i+1} := \mu_{i,l_0-1}$ and obtain (2.2) (with $i+1$ in the place of i). Then we continue (2.1) on the next time level $t = t_{i+1}$.

3. ASSUMPTIONS AND CONVERGENCE OF (2.5_l)

By c we denote generic positive constants. We shall assume

(H_1) b is increasing, absolutely continuous function satisfying $b(0) = 0$. We assume that there exist $b_n \in C^1(\mathbb{R})$, $b_n(0) = 0$ $(\tau = \tau_n = T/n)$ with $b_n'(s)$ locally Lipschitz continuous such that:

(i) $b_n(s) \to b(s)$ locally uniformly;

(ii) $cn^{-d} \le b_n'(s) \le cn^{\gamma} \quad \forall s \in \mathbb{R};\ d, \gamma \in (0,1)$;

(iii) $\sup_{|z|<K} |b_n(z)| \le c(K) < \infty \quad 0 < K < \infty$;

(iv) $\min\{b'(s), \varepsilon\} \le cb_n'(s)$ for some $\varepsilon > 0$;

(v) $|b_n''(s)| \le cn^{\rho},\ \rho \in (0,1)$;

(H_2) $k(t,x,s) : I \times \Omega \times \mathbb{R} \to \mathbb{R}^{N\times N}$ is continuous and

$$c_1|\xi|^2 \le (k(t,x,s)\xi, \xi) \le c_2|\xi|^2 \ ;$$

(H_3) $\bar{F}(t,x,s),\ \bar{F}_s' \equiv \partial_s \bar{F}(t,x,s) : I \times \Omega \times \mathbb{R}^1 \to \mathbb{R}^N$ are continuous and

$$|\partial_s \bar{F}| \le c,\ |\partial_s^2 \bar{F}| \le c,\ |\partial_x \bar{F}(t,x,s)| + |\partial_x \bar{F}_s'(t,x,s)| \le c(L(t,x) + |s|)$$
$$\text{for a.e. } (t,x) \in Q_T \equiv I \times \Omega,\ s \in \mathbb{R}\ .$$

We also assume that $\bar{F}(t,x,s)$ can be extended to $\Omega^* \supset \bar{\Omega}$ so that the estimates hold true for $x \in \Omega^*$ and $L \in L_2(\Omega^* \times I)$;

(H_4) $f(t,x,s,\eta),\ g(t,x,s)$ are continuous in their variables and

$$|f(t,x,s,\eta)| \le c(1 + |s| + |\eta|),\ |g(t,x,s)| \le c(1+|s|)$$

$\psi(s) : \mathbb{R} \to \mathbb{R}$ is continuous and $|\psi(s)| \le c(1 + |s|)$; $K(t,s) \in L_\infty(I \times I)$.

(H_5) $u_0 \in W_2^1(\Omega) \cap L_\infty(\Omega)$.

We denote the standard functional spaces by $L_2 = L_2(\Omega)$, $L_\infty(\Omega)$, $V = \{v \in W_2^1(\Omega); v = 0 \text{ on } \Gamma_1\}$, $L_2(I,V)$ - see [13]. By $\|\cdot\|_0$, $\|\cdot\|_\infty$, $\|\cdot\|$, $\|\cdot\|_{\Gamma_2}$ we denote the norms in L_2, $L_\infty(\Omega)$, $W_2^1(\Omega)$, $L_2(\Gamma_2)$, respectively. We denote by $(u,v) = \int_\Omega uvdx$, $(u,v)_{\Gamma_2} = \int_{\Gamma_2} uvdx$ and V^* the dual space to V. In the sequel we drop the variable x in the terms $k, f, g, \bar{F}, \lambda, \mu$.

We use the concept of variational solution. Let $< u, v >$ represents the duality between V^* and V.

Definition 3.1. *$u \in L_2(I,V)$ is a variational solution of (1.1)-(1.3) iff* *(i)* $b(u) \in L_\infty(I, L_1(\Omega)),\ \partial_t b(u) \in L_2(I, V^*)$;

(ii) $\int_I < \partial_t b(u), v > + \int_I (div\bar{F}(t,u), v > + \int_I (k(t,u)\nabla u, \nabla v) +$
$+ \int_I (g(t,u), v)_{\Gamma_2} = \int_I (f(t,u,s), v) \qquad \forall v \in L_2(I,V)$,
$s(x,t) = \int_0^t K(t,s)\psi(u(x,s))ds \qquad a.e.(t,s) \in I \times \Omega;$

(iii) $\int_I < \partial_t b(u), v >= \int_I (b(u) - b(u_0), \partial_t v) \quad \forall v \in L_2(I,V) \cap L_\infty(I \times \Omega)$, $\partial_t v \in L_\infty(Q_T),\ v(T) = 0$

Also $u_i \in V$ in (2.1) is a variational solution of (2.1)

$$(\lambda_i(u_i - u_{i-1} \circ \varphi^i_{\mu_i}), v) + \tau(k_i \nabla u_i, \nabla v_i) + \tau(g_i, v)_{\Gamma_2} = \\ = \tau(H_i, v) + \tau(f(t_i, u_{i-1}, s_i) \qquad \forall v \in V \tag{3.1}$$

If $u_{i-1},\ u_{i-1} \circ \varphi^i_{\mu_i} \in L_2$ then the existence of $u_i \in V$ in (3.1) is guaranteed by Lax-Milgram lemma. To obtain the a priori estimates for u_i and to prove that $u_{i-1} \circ \varphi^i \in L_2$ the crucial role place the estimate (see [9] Lemma 11)

$$\frac{1}{2}|x-y| \le |\varphi^i_{\mu_i}(x) - \varphi^i_{\mu_i}(y)| \le 2|x-y| \tag{3.2}$$

uniformly for $i = 1, \dots, n$, provided $\mu_i = G_i(\nu)$ for any $0 < \nu \in L_\infty(\Omega)$ and $\omega + d < 1,\ \tau \le \tau_0$. Then $u_{i-1} \circ \varphi^i_{\mu_i} \in L_2$ provided $u_{i-1} \in L_2$. Hence $\{u_i\}_{i=1}^n \in V$ satisfying (3.1) is guaranteed.

In the following we shall assume that

$$\|u_i\|_\infty \le c \quad \text{uniformly for } n,\ i = 1, \dots, n \tag{3.3}$$

without any structural restrictions on $b, \bar{F}, g, f$. If $f \equiv f(t,x,u)$ is not dependent on s (i.e. the memory term is not considered in (1.1)) then (3.3) has been proved in [9], Lemma 15 under the following structural restrictions:

(i) $f, g, \mathrm{div}_x \bar{F}(t,x,s) \equiv 0$; or

(ii) $g \equiv 0,\ \lambda_i \ge q > 0,\ L \in L_\infty(I \times \Omega)$ in H3); or

(iii) $f(t,x,s)s \ge 0,\ g(t,x,c)s \ge 0, \quad H(t,x,s)s \ge 0$

(in the case *(iii)* $f_i \equiv f(t_i,x,u_i),\ g_i = g(t_i,x,u_i),\ H_i = H(t_i,x,u_i)$ in (3.1)).

In our situation the result (3.3) hold true in the case *(ii)*. Indeed as in [9] we obtain estimate

$$\|u_i\|_\infty \le \|u_{i-1}\|_\infty + \tau \left(\left\| \frac{H_i}{\lambda_i} \right\|_\infty + \left\| \frac{f_i}{\lambda_i} \right\|_\infty \right)$$

where we estimate H_i, f_i in terms of $\|u_j\|_\infty$ using (H_3), (H_4). Then we obtain

$$\|u_i\|_\infty \leq \|u_{i-1}\|_\infty + c\tau\left(\|u_{i-1}\|_\infty + \sum_{j=1}^{i-1}\|u_j\|_\infty \tau + 1\right)$$
$$\|u_i\|_\infty \leq (1+c\tau)\|u_{i-1}\|_\infty + c\tau\sum_{j=1}^{i-1}\|u_j\|_\infty \tau + c\tau \leq$$
$$\leq (1+c\tau)^i\left(c_1 + c_2\sum_{j=1}^{i-1}\|u_j\|_\infty \tau\right)$$

since $(1+c\tau)^i \leq e^{cT}$, Gronwall argument implies the estimate (3.3).

The convergence of iterations (2.5_l) has been proved in [9] under the assumptions:

$$\|\nabla u_i\|_0 \leq c,\ \forall n,\ i=1,\dots,n \text{ and } \rho + d + \frac{N}{2}\omega < 1,\ \tau \leq \tau_0$$

where $N = \dim\Omega$.

Remark 3.2. *The convergence of iterations in (2.4_k) has been analyzed in [8].*

Remark 3.3. *The regularization $b_n(s)$ of $b(s)$ in (H_1) is not so much restrictive with respect to asymptotic behaviour of b'_n, b''_n. For example, let us consider $b(s) = |s|^p \operatorname{sgn} s$, $p \in (0,1)$. We can take for $b_n(s)$*

$$b_n(s) = \begin{cases} \left(s + n^{-\delta}\right)^p - n^{-p\delta} & \text{for } s_1 \geq s \geq 0, \\ -\left|s - n^{-\delta}\right|^p + n^{-p\delta} & \text{for } -s_1 < s \leq 0, \end{cases} \qquad s_1 = s_1(n) = n^\delta - n^{-\delta}$$

and $b_n(s) = b_n(s_1) + b'_n(s_1)(s - s_1)$ for $s \geq s_1$ and similarly for $s \leq -s_1$. Then we have $c_1 n^{-(1-p)\delta} \leq b'_n(s) \leq c_2 n^{(1-p)\delta}$ (i.e. $d = \gamma = (1-p)\delta$ for any $\delta > 0$) and $b''_n(s) \leq cn^\rho$ with $\rho = (2-p)\delta$. We can verify easily (i)-(v) in (H_1).

4. CONVERGENCE OF THE METHOD

To obtain a priori estimates for $\{u_i\}_{i=1}^n$ we follow [9] and only sketch the additional terms on R.H.S. in (3.1) concerning memory. We obtain

Lemma 4.1. *Under the assumptions (H_1)-(H_5), (3.3) and $\omega + d < 1$, $\gamma + 2d + \rho < \omega$, $\alpha > d$ the following a priori estimates hold true*

$$\max_{1\leq i\leq n}\int_\Omega B_n(u_j)dx \leq c,\ \sum_{i=1}^n \|u_i\|^2 \tau \leq c$$

uniformly for u, where $B_n(s) := b_n(s)s - \int_0^s b(z)dz$.

Sketch of the proof. We put $v = u_i$ into (3.1) and sum it up for $i = 1, \dots, j$. Similarly as in [9] we obtain

$$\max_{1\le j\le i} \int_\Omega B_n(u_j)dx + \sum_{i=1}^j \|u_i\|^2 \tau \le$$
$$\le c\left(\varepsilon + \tfrac{c}{\varepsilon}\tau^{2\omega-2\gamma-4d-2\rho}\right) \sum_{i=1}^j \|u_i\|^2 \tau + c_\varepsilon + c\sum_{i=1}^j \sum_{k=1}^i \|u_k\|_0^2 \tau ,$$

where the last term arises in estimation of memory terms. Since $\|u_i\|_0 \le c\|u_i\|_\infty \le c$ we estimate the last term by a constant. Then we take ε sufficiently small and consequently $\tau \le \tau_0(\varepsilon)$ we obtain the required a priori estimate.

By means of $\{u_i\}_{i=1}^n$ we construct Rothe's functions

$$\bar{u}^n(t) := u_i \qquad \text{for } t \in (t_{i-1}, t_i >,\ i = 1, \dots, n$$
$$u^n(t) := u_{i-1} + \tfrac{t-t_{i-1}}{\tau}(u_i - u_{i-1})$$

with $\bar{u}^n(0) = u_0$.

Lemma 4.2. *$\{\bar{u}^n\}$ is compact in $L_s(I\times\Omega)$ $\forall s > 1$, i.e. there exists $u \in L_\infty(I\times\Omega)$ and $\{\bar{n}\} \subset \{n\}$ such that $\bar{u}^{\bar{n}} \to u$ in $L_s(I\times\Omega)$.*

The proof is based on the a priori estimate

$$\sum_{i=1}^{n-k} (b_n(u_{i+k}) - b_n(u_i), u_{i+k} - u_i)\tau \le ck\tau$$

which can be obtained in the same way as in [9] estimating the memory terms using (3.3). This a priori estimates can be rewritten in the form

$$\int_0^{T-z} (b_n(\bar{u}^n(t+z) - b_n(\bar{u}^n(t)),\ \bar{u}^n(t+z) - \bar{u}^n(t))dt \le cz$$

uniformly for $z \in (0, z_0)$. From this and the estimate $\int_0^T \|\bar{u}^n(t)\|^2 dt \le c$ (see Lemma 4.1) we deduce that $\bar{u}^n(t,x) \to u(t,x)$ a.e. in $I\times\Omega$ because b is strictly increasing and because of the regularization properties of b_n — see [7]. Then (3.3) implies L_s convergence.

As a consequence we obtain $u \in L_2(I,V)$ and $\bar{u}^{\bar{n}} \rightharpoonup u$ in $L_2(I,V)$.

Now we can prove our main result.

Theorem 4.3. *Let the assumptions (H_1)-(H_5), (3.3) and $\alpha > d$, $\gamma + 2d + \rho < \omega$ be satisfied. Then $\bar{u}^{\bar{n}} \to u$ in $L_s(I\times\Omega)$ $\forall s > 1$ and $\bar{u}^{\bar{n}} \rightharpoonup u$ in $L_2(I,V)$ where $\bar{u}^n$ is from (2.1)-(2.3) and u is a variational solution of (1.1). If the variational solution u is unique then the original sequence $\{\bar{u}^n\}$ is convergent.*

Sketch of the proof. We rewrite (3.1) in the form

$$\begin{aligned}(b_n(u_i) - b_n(u_{i-1}), v) &= -(b_n(u_{i-1}) - b_n(u_{i-1} \circ \varphi^i_{\mu_i}), v) + \\ &+ \tau^\beta(\kappa_i(u_i - u_{i-1} \circ \varphi^i_{\mu_i}), v) - \tau(k_i \nabla u_i, \nabla v) - \tau(g_i, v)_{\Gamma_2} + \\ &+ \tau(H_i, v) + \tau(f_i, v) \qquad \forall v \in V\end{aligned} \tag{4.1}$$

where $\|\kappa_i\|_0 \le 1$. Then we consider $v \in L_2(I, V) \cap C^1(I, C^\infty(\bar{\Omega}))$ with $v(x,t) = 0$ for t in a neighbourhood of T and put it into (4.1). Then we integrate it over I and denote the corresponding terms by $J_{1,n} - J_{7,n}$. Similarly as in [9] we obtain

$$J_{1,n} \to -(b(u_0), v(0)) - \textstyle\int_I (b(u), \partial_t v)dt \qquad \text{for } n \to \infty$$

$$J_{2,n} \to \textstyle\int_I (\bar{F}'_u(t,u) \cdot \nabla u, v)dt \qquad \text{for } n \to \infty$$

where we have rearranged

$$b_n(u_{i-1}) - b_n(u_{i-1} \circ \varphi^i_{\mu_i}) = \tfrac{b_n(u_{i-1}) - b_n(u_{i-1} \circ \varphi^i_{\mu_i})}{u_{i-1} - u_{i-1} \circ \varphi^i_{\mu_i}} \left(u_{i-1} - u_{i-1} \circ \varphi^i_{\mu_i}\right)$$

$$u_{i-1} - u_{i-1} \circ \varphi^i_{\mu_i} = \textstyle\int_0^1 \nabla u_{i-1}(x + s(\bar{\varphi}^n - x))ds \left[\mathcal{F}_i + (\omega_h * \mathcal{F}_i - \mathcal{F}_i)\right]$$

with $\mathcal{F}_i := \frac{\bar{F}'_s(t_i, u_{i-1})}{\mu_i}$. Here we use also $I^n \to 1$, $\bar{M}^n \to 0$ in $L_2(I \times \Omega)$ for $n \to \infty$ where

$$I^n(t) = \frac{b_n(u_{i-1}) - b_n(u_{i-1} \circ \varphi^i_{\mu_i})}{u_{i-1} - u_{i-1} \circ \varphi^i_{\mu_i}} \cdot \frac{1}{\mu_i}, \qquad \bar{M}^n(t) := \omega_h * \mathcal{F}_i - \mathcal{F}_i$$

for $t \in (t_{i-1}, t_i)$, $i = 1, \ldots, n$. Similarly we obtain

$$J_{3,n} \to 0 \qquad \text{for } n \to \infty\,.$$

Easily we deduce

$$J_{4,n} \to \int_I (k(t,u)\nabla u, \nabla v)dt \qquad \text{for } n \to \infty$$

and

$$\left.\begin{aligned} J_{5,n} &\to \textstyle\int_I (g(t,u), v)_{\Gamma_2} dt\,, \\ J_{6,n} &\to \textstyle\int_I (\mathrm{div}_x \bar{F}(t,x,u), v)dt \end{aligned}\right. \qquad \text{for } n \to \infty\,.$$

For the last term we have to use the following facts

$$\bar{u}^n \to u \quad \text{a.e. in } I \times \Omega,\ \|\bar{u}^n\|_\infty \le c\,,$$

$$\bar{s}^n(x,t) = \textstyle\int_0^{t-\tau} K(t,s)\psi(\bar{u}^n_\tau(s))ds \to \int_0^t K(t,s)\psi(u(s))ds$$

$$\text{a.e. in } (t,x) \in I \times \Omega \text{ and } \|\bar{s}^n\|_\infty \le c \quad \forall n\,.$$

Hence as a special case we obtain

$$\bar{s}^n \to s \equiv \int_0^t K(t,z)\psi(u(z))dz \qquad \text{in } L_2(I\times\Omega)$$

and consequently

$$J_{7,n} \to \int_I (f(t,u,s),v)dt\ .$$

Then we take the limit $n\to\infty$ in

$$J_{1,n} = -J_{2,n} + J_{3,n} - J_{4,n} - J_{5,n} + J_{6,n} + J_{7,n}$$

and obtain

$$\begin{aligned}-\textstyle\int_I(b(u),\partial_t v)dt - (b(u_0),v(0)) = -\int_I(\mathrm{div}\bar{F}(t,x,u),v)dt-\\ -\textstyle\int_I(k(t,u)\nabla u,\nabla v)dt + \int_I(g(t,u),v)_{\Gamma_2}dt + \int_I(f(t,u,s),v)dt\end{aligned}$$

where we have added together terms J_2 and J_6. Hence we deduce that there exists $\partial_t b(u)\in L_2(I,V^*)$ similarly as in [9]. Then we conclude that u is a variational solution of (1.1)-(1.3).

The uniqueness of variational solution has been studied in [12].

The more strong convergence results we obtain under the regularity assumptions on b.

Theorem 4.4. *Let the assumptions of Theorem 4.3 be satisfied. Suppose that (2.2), (2.3) are satisfied with the norm* $\|.\|_\infty$ *in the place of* $\|.\|_0$ *and let* $0<\varepsilon\le b'(s)\le M<\infty$ *a.e. in* $\mathbb{R}$ *and* $d=\gamma=0$ *in* (H_1). *Then* $\bar{u}^{\bar{n}}\to u$ *in* $L_2(I,V)$.

The proof of Theorem 4.4 is the same as that one in [9] (Theorem 48) and the presence of the memory term represents no substantial difficulties.

5. NUMERICAL IMPLEMENTATION

The numerical implementation of (2.1)-(2.3) is rather costly also without the presence of memory term - see [9]. The additional difficulties arise including the memory terms since at the time level $t=t_i$ we need for evaluation of s_i the values of u_j for all $j=0,\dots,i-1$. In the numerical realization of (2.1) we project it into finite dimensional space $V_\lambda\subset V$ using FEM. We assume that $V_\lambda\to V$ for $\lambda\to 0$ in canonical sense (λ being the discretization parameter). Then instead of $u_i\in V$ we obtain $u_i^\lambda\in V_\lambda$ as a solution of projected equation (2.1). Then by means of $\{u_i^\lambda\}_{i=1}^n$ we construct Rothe's function u_α ($\alpha=(\tau,\lambda)$) as our approximate solution. For the convergence $u_\alpha\to u$ for $\alpha\to 0$ we obtain the same results as in Theorems 4.3, 4.4. In the projected problem (2.1) we assume that u_0^λ is a projection of $u_0\in W_2^1$ into V_λ and that

$u_0^\lambda \to u_0$ in $L_2(\Omega)$. The evaluation of $u_{i-1} \circ \varphi^i$ is the most costly from the numerical point of view. As an alternative to the standard back tracing we can use the following procedure. On each time level t_i we construct a new basis $\{\psi_j^i\}_{j=1}^{m_{\lambda,i}}$ which we obtain by shifting along characteristics of the basis $\{\psi_j^{i-1}\}_{j=1}^{m_{\lambda,i-1}}$ corresponding to $t = t_{i-1}$. The new basis elements are locally completed by new elements or locally reduced with respect to the density of the grid points which is changing by means of characteristics. This process is simply realizable when, e.g., piecewise linear elements are used. Then, in the place of back tracing in $u_{i-1} \circ \varphi_{\mu_i}^i$ we obtain immediately the values in nodal points for the new basis on time level t_i. Additional treatment needs the extension of u_{i-1} outside Ω (using boundary conditions).

We shall discuss now the treatment of memory terms. For realistic contaminant transport problem (E) with sorption isotherms $\psi(C)$ (e.g. $\psi(C) = k_1 C^p$ for Freunlich type, $\psi(C) = \frac{k_2 C}{k_2 + k_3 C}$ for Langmuir type etc.) we can express

$$S(t) = S(0)\,\mathrm{e}^{-at} + a\int_0^t \mathrm{e}^{-a(t-s)}\,\psi(C(s))ds \qquad \left(a = \frac{d}{\rho}\right)$$

and when we insert $\partial_t S$ into transport equation we obtain a memory term with $f(t,x,u,s) \equiv s - \psi(u)$ and

$$s = ad\int_0^t \mathrm{e}^{-a(t-s)}\,\psi(C(s))ds, \quad (K(t,s) \equiv \mathrm{e}^{-a(t-s)}) \quad \text{provided } S(0) = 0\ .$$

In that case we can verify that $\alpha_{i+1,j} = \mathrm{e}^{-a\tau}\,\alpha_{i,j}$ since $\alpha_{i,j} = \frac{1}{\tau}\int_{t_{j-1}}^{t_j} \mathrm{e}^{-a(t_i-s)}\,ds$ and

$$s_{i+1} = \mathrm{e}^{-a\tau}\,s_i + \alpha_{i+1,i}\psi(u_i) \qquad \text{for } i = 1,\dots,n\ .$$

As a consequence we do not need to store the values of u_j ($j = 1,\dots,i-1$) for evaluation of s_i.

Example 1. We apply the proposed method in numerical solution for the problem (E) with equilibrium sorption isotherm i.e. $d \to \infty$ which implies $S = \psi(C) \equiv \kappa C^p$. Then (E) with specific data (in 1D) reduces to

$$\partial_t\left(\frac{1}{2}u + 1.5u^p\right) + \partial_x(3u - 0.05\partial_x u) = 0\ .$$

We shall consider $p = 1.2,\ 0.8,\ 0.6,\ 0.4,\ \Omega = (0,100),\ T = 30$ and the Dirichlet boundary condition

(i) $u(0,t) = 1$

(ii) $u(0,t) = 0$

with the corresponding initial conditions

(i) $u(x,0) = u_0^1(x)$

(ii) $u(x,0) = u_0^2(x)$.

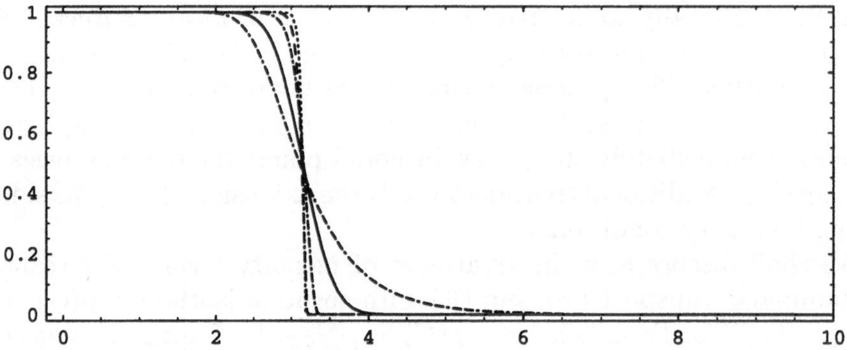

Figure 1

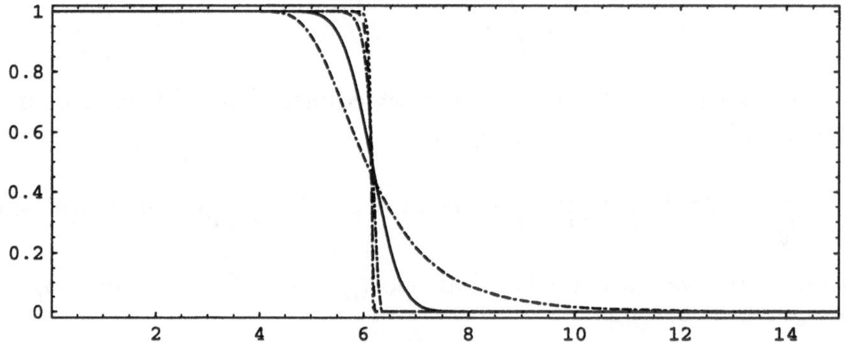

Figure 2

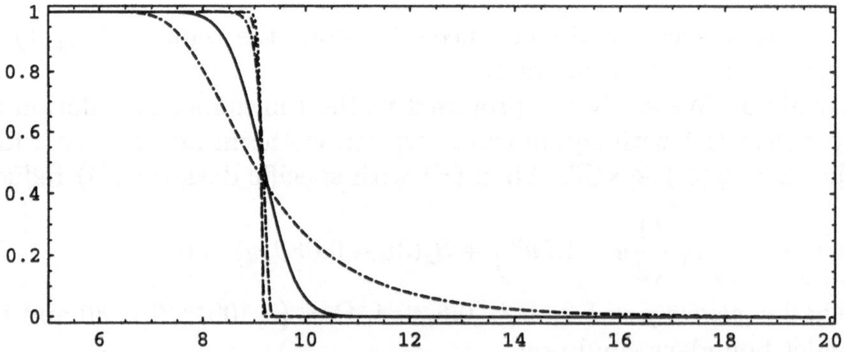

Figure 3

Here $u_0^i(x)$ $(i = 1,2)$ are piecewise linear functions of the following form: $u_0^1(x) = 1$ for $x \in (0, 0.1)$, $u_0^1(0.2) = 0, u_0^1(x) = 0$ for $x > 0.2$;

$u_0^2(x) = 0$ for $x \in (0, 100) \setminus (0.1, 0.5)$, $u_0^2(0.2) = u_0^2(0.4) = 1$. The solutions are drawn in three time moments for various p ($p = 1.2$ - dash-dash-dotted line, $p = 1.0$ - full line, $p = 0.8$ - dash-dotted line, $p = 0.6$ - dashed line, $p = 0.4$ - dotted line) in Figure 1 for $t = 2$, in Figure 2 for $t = 4$ and in Figure 3 for $t = 6$ in the case *(i)*. The case *(ii)* is drawn in Figure 4 for $t = 1$, in Figure 5 for $t = 2$ and in Figure 6 for $t = 6$.

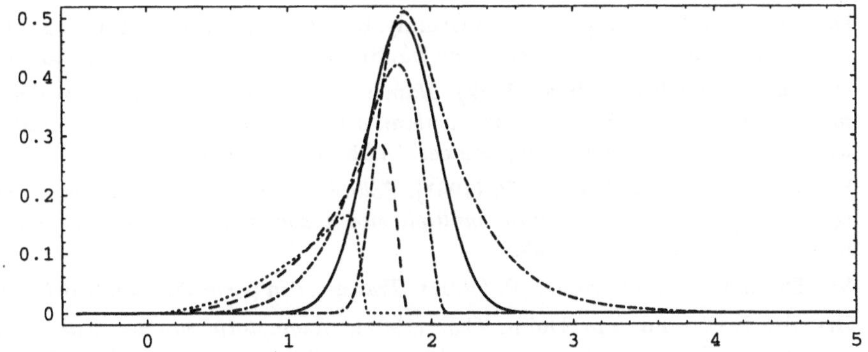

Figure 4

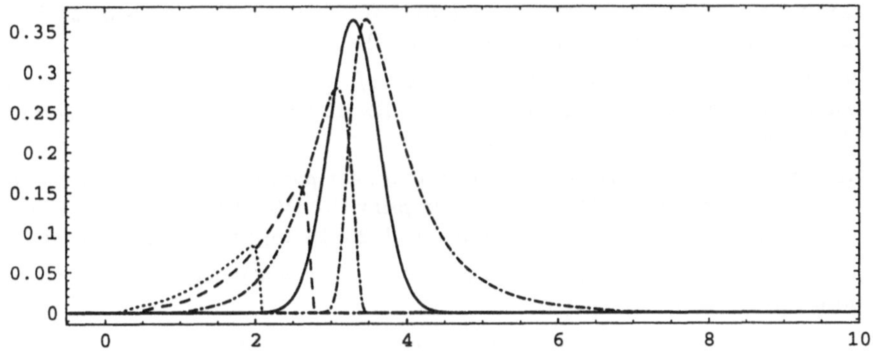

Figure 5

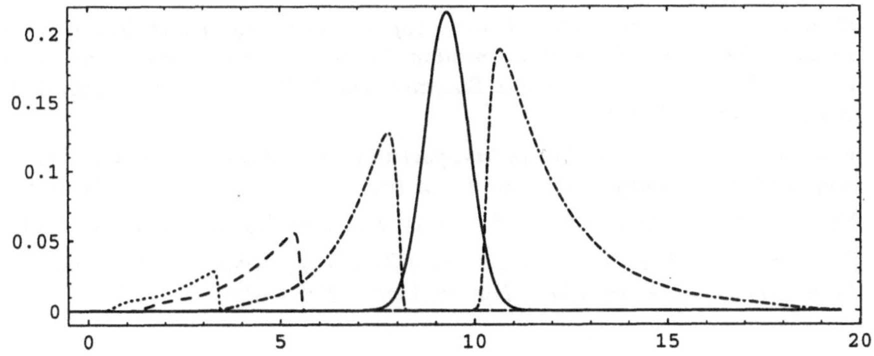

Figure 6

Acknowledgments

I want to express my thanks to D. Kostecky for his help with numerical experiments.

This research was partially supported by scientific grant VEGA and by grant 201/97/0153 of Grant Agency of Czech Republic.

References

[1] Dawson, C.N., Van Duijn, C.J. and Grundy, R.E. (1996). *Large time asymptotics in contaminant transport in porous media.* SIAM J. Appl. Math. 56(4):965–993.

[2] Douglas, J. and Russel, T.F. (1982). *Numerical methods for convection dominated diffusion problems based on combining the method of the characteristics with finite elements or finite differences.* SIAM J. Numer. Anal., 19:871–885.

[3] Van Duijn, C.J. and Knabner, P. (1991). *Solute transport in porous media with equilibrium and non-equilibrium multiple-site adsorption: Traveling waves.* J. Reine Angew. Math., 415:1–49.

[4] Van Duijn, C.J. and Knabner, P. (1992). *Transport in porous media,* 8:167–226.

[5] Grundy, R.E., and Van Duijn, C.J. (1994). *Asymptotic profiles with finite mass in one-dimensional contaminant transport through porous media: The fast reaction case.* Quart. J. Mech. Appl. Math., 47:69–106.

[6] Jäger, W. and Kačur, J. (1991). *Solution of porous medium systems by linear approximation scheme.* Numer. Math., 60:407–427.

[7] Jäger, W. and Kačur, J. (1995). *Solution of doubly nonlinear and degenerate parabolic problems by relaxation schemes.* RAIRO Model. Math. Anal. Numer., 29(5):605–627.

[8] Kačur, J. (1996). *Solution to strongly nonlinear parabolic problems by a linear approximation scheme.* Mathematics Preprint No. IV-M1-96, Comenius University Faculty of Mathematics and Physics, 1–26, to appear in IMA J. Num.

[9] Kačur, J. *Solution of degenerate convection- diffusion problems by the method of characteristics.* To appear.

[10] Knabner, P. (1991). Mathematische Modelle fur Transport und Sorption geloster Stoffe in porosen Medien. *Meth. Verf. Math. Phys.,*36, Verlag Peter D. Lang, Frankfurt am Main.

[11] Knabner, P. (1992). *Finite-Element-Approximation of Solute Transport in Porous Media with General Adsorption Processes.* "Flow and Transport in Porous Media" Ed. Xiao Shutie, Summer school, Beijing, 8–26 August 1988, World Scientific, 223–292.

[12] Knabner, P. and Otto, F. *Solute transport in porous media with equilibrium and non-equilibrium multiple-site adsorption: Uniqueness of the solution.* To appear.

[13] Kufner, A., John, O. and Fučík, S. (1977). *Function spaces.* Noordhoff, Leiden.

[14] Pironneau, O. (1982). *On the transport-diffusion algorithm and its applications to the Navier-Stokes equations.* Numer. Math., 38:309–332.

ON GLOBAL EXISTENCE OF SMOOTH TWO-DIMENSIONAL STEADY FLOWS FOR A CLASS OF NON-NEWTONIAN FLUIDS UNDER VARIOUS BOUNDARY CONDITIONS

Petr Kaplický, Josef Málek, Jana Stará

Abstract: We study steady two-dimensional flows of shear dependent fluids in a bounded domain subjected to three kinds of boundary conditions: (i) general nonhomogeneous Dirichlet, (ii) nonhomogeneous Dirichlet with zero normal component at the boundary (fixed wall) and (iii) free-stick (slippery boundary). The existence of a $C^{1,\alpha}$-solution is proved: while condition (i) requires smallness of a given function at boundary, conditions (ii) provide smooth solutions for all choice of data. Some results regarding a special construction of an extension operator are interesting on their own.

Keywords: Non Newtonian fluids, shear dependent viscosity, regularity, Hölder continuity of gradients, non homogeneous Dirichlet boundary condition, free stick boundary condition.

1. INTRODUCTION

Non-newtonian fluid mechanics involves all problems described by the system of equations where the constitutive relation between the stress tensor T and the symmetric part of the velocity gradient $D(v)$ is not linear. One of the simplest examples of a nonlinear relation between T and $D(v)$ is given by the form

$$T = 2\nu(|D(v)|^2)D(v), \tag{1.1}$$

where $\nu : \mathbb{R}_0^+ \to \mathbb{R}_0^+$ is a nonlinear, typically monotone function of the modulus of $D(v)$.

Model (1.1) has the ability to shear thicken (when the generalized viscosity function ν is increasing) or to shear thin (when ν is decreasing),

Applied Nonlinear Analysis, edited by Sequeira *et al.*
Kluwer Academic / Plenum Publishers, New York, 1999.

however it cannot capture further phenomena of nonlinear fluids as normal stress differences, stress relaxation, nonlinear creep or yield stress. On the other hand, the ability to shear thin is exhibited by a significant amount of materials, as polymers, dyes, chemical solutions, glaciers, geological materials, blood, etc. (See [17] for explanations of the notions, and [10], [11] and [17] for further references.)

When restricted to isothermal process of incompressible materials in a bounded set $\Omega \subset \mathbb{R}^d$, then governing equations read[1]

$$\begin{aligned} \operatorname{div} v &= 0 \\ \rho \frac{\partial v}{\partial t} + \rho v_k \frac{\partial v}{\partial x_k} - \operatorname{div} T &= \nabla \pi + \rho f. \end{aligned} \tag{1.2}$$

Here, $v = (v_1, \cdots, v_d)$ is the unknown velocity field, π is the unknown pressure, ρ is a given positive constant expressing the density of the fluid and $f = (f_1, \cdots, f_d)$ stands for the vector of external body forces.

Although (1.2) together with (1.1) represents the simplest deviation from the Navier-Stokes system (that can be obtained by setting $\nu(|D(v)|^2) = \nu_0$, ν_0 being a positive constant), the mathematical analysis of (1.2) with (1.1) differs from the analysis of the Navier-Stokes equations tremendously. We are going to illustrate it discussing the question of full regularity in two space dimensions, when equations (1.2) are supplemented by the Dirichlet boundary condition, and by the initial condition v_0, i.e.,

$$\begin{cases} v(0,x) = v_0(x) & \text{for all } x \in \Omega, \\ v(t,x) = 0 & \text{for all } t \in (0,t^*) \text{ and } x \in \partial\Omega. \end{cases}$$

Here, $\operatorname{div} v_0 = 0$ on Ω and $v_0 = 0$ at $\partial\Omega$.

Set $f = 0$ for simplicity in the following exposition.

It is nowadays well-known that the Navier-Stokes equations in two dimensions possess global uniquely defined weak solution provided that v_0 is L^2-integrable. It is also standard to observe that if data (v_0 and Ω) are smooth then the solution is also smooth.

Completely different situation concerns the system (1.2) with (1.1) even in the case when ν is bounded (but nonconstant) function of $|D(v)|^2$: while the global-in-time existence of weak solution is available, and also L^2-integrability of the second derivatives is known, the question of higher regularity (up to the boundary) is completely open. The essential step is how to pass from L^2-integrability of the second spatial derivatives and first time derivatives to the boundedness of the first gradient. To our knowledge, except for the paper of Seregin [18], who proved recently interior $C^{1,\alpha}$-regularity of solution to (1.1), (1.2) with ν

[1]The summation convention is used throughout the whole paper.

bounded, there are no other results. Closely related is a paper by Nečas, Šverák [16], where the nonlinear parabolic systems are studied and local $C^{1,\alpha}$-regularity is proved.

In order to understand better both the passage from "$W^{2,2}(\Omega)$ regularity to $C^{1,\alpha}(\Omega)$ regularity" and the analysis near the boundary, we have started to treat the steady case first. The presented paper is a continuation of our previous studies, where we treated the space periodic case (see [8]) and the homogeneous Dirichlet problem (see [9]).

In this paper, we deal with steady problem in two dimensions and we consider three kinds of boundary conditions:

1. nonhomogeneous Dirichlet boundary conditions with two types of restrictions on boundary data

   (a) $$v = \varphi \quad \text{at } \partial\Omega \text{ with } \varphi \text{ sufficiently small;} \tag{1.3}$$

   (b) $$v = \varphi, \quad \varphi \cdot \nu = 0 \quad \text{at } \partial\Omega \tag{1.4}$$

   (where ν denotes the outer normal vector to $\partial\Omega$) without any restrictions on smallness of φ in tangential direction.

2. free-stick boundary conditions

   $$v \cdot \nu = 0 \quad \text{and} \quad T_{ij}(|Dv|)\nu_i \tau_j = 0 \quad \text{at } \partial\Omega \tag{1.5}$$

   (where τ denotes a tangent vector to $\partial\Omega$). See [3] for more details on this condition.

2. DEFINITION OF THE PROBLEM AND MAIN RESULTS

We investigate the following problem: to construct $v = (v_1, v_2) : \Omega \to \mathbb{R}^2$ and $\pi : \Omega \to \mathbb{R}$ solving

$$\begin{aligned} v_k \frac{\partial v}{\partial x_k} - \operatorname{div}(T(D(v))) + \nabla\pi = f \quad &\text{in } \Omega\,, \\ \operatorname{div} v = 0 \quad \text{in } \Omega \end{aligned} \tag{2.1}$$

together with one of the boundary conditions (1.3)-(1.5) and satisfying for a certain $\epsilon > 0$

$$v \in W^{2,2+\epsilon}(\Omega)\,, \quad \pi \in W^{1,2+\epsilon}(\Omega).$$

Note that higher regularity of v and π (corresponding to the smoothness of data) is a consequence of linear theory, see [8] for more details if needed.

Let $\mathbf{S}$ be the set of symmetric matrices of the type 2×2. Recall that $D(v) \in \mathbf{S}$ denotes the symmetrized gradient of v with components $\frac{1}{2}(\frac{\partial v_i}{\partial x_j} + \frac{\partial v_j}{\partial x_i}), i, j = 1, 2$.

Throughout the whole paper we suppose that:

(A_1) $\Omega \subset \mathbb{R}^2$ is a bounded open set with the boundary $\partial\Omega$;

(A_2) function $T : \mathbf{S} \to \mathbf{S}$ satisfies for a given $p \in (1,\infty)$:[2]

- there exists $F : \mathbb{R}_0^+ = \langle 0,\infty) \to \mathbb{R}_0^+$ such that $F \in C^2\left(\mathbb{R}_0^+\right)$ and for all $i,j = 1,2$

$$\begin{aligned} T_{ij}(\eta) = \partial_{ij}F(|\eta|^2) = 2F'(|\eta|^2)\eta_{ij}, \quad \forall \eta \in \mathbf{S}, \\ F(0) = \partial_{ij}F(0) = 0; \end{aligned} \tag{2.2}$$

- there exist $C_1, C_2 > 0$ such that

$$\partial_{ij}\partial_{kl}F(|\eta|^2)\xi_{ij}\xi_{kl} \geq C_1(1+|\eta|^2)^{\frac{p-2}{2}}|\xi|^2, \quad \forall \eta,\xi \in \mathbf{S}; \tag{2.3}$$

and (for all $i,j,k,l = 1,2$)

$$\left|\partial_{ij}\partial_{kl}F(|\eta|^2)\right| \leq C_2\left(1+|\eta|^2\right)^{\frac{p-2}{2}}, \quad \forall \eta \in \mathbf{S}. \tag{2.4}$$

(A_3) $f = (f_1, f_2) : \Omega \to \mathbb{R}^2$ is given: $f \in L^{p'}(\Omega)$, $p' = p/(p-1)$, for $p \in (1,2)$ and $f \in L^{2+\varepsilon_0}(\Omega)$ with a certain $\varepsilon_0 > 0$ for $p \geq 2$.

Now, we are going to give an overview of our main results stated in Theorems 2.1, 2.2, 2.3 and 2.4; proofs of Theorems 2.1 and 2.2 treating the nonhomogeneous Dirichlet boundary condition will be given in Section 2, proofs of Theorems 2.3 and 2.4 dealing with free-stick boundary conditions are contained in Section 3.

Theorem 2.1. *Let $p > \frac{3}{2}$, $\Omega \in C^3$ and (A_1)–(A_3) hold. Suppose that*

$$\varphi \text{ is a trace of } \Phi \in W^{3,q}(\Omega),\ q > 2, \quad (\text{shortly } \varphi = Tr(\Phi)) \tag{2.5}$$

$$\varphi \cdot \nu = 0 \quad \text{at } \partial\Omega\,. \tag{2.6}$$

Then there exist $\epsilon > 0$ and at least one solution $v \in W^{2,2+\epsilon}(\Omega)$, $\pi \in W^{1,2+\epsilon}(\Omega)$, $\int_\Omega \pi\, dx = 0$, satisfying (2.1) and (1.4).

Theorem 2.2. *Let $p > \frac{3}{2}$, $\Omega \in C^2$ and (A_1)–(A_3) hold. Then there exists a constant δ (δ may depend on f) so that for any boundary condition φ satisfying (2.5) and*

$$\|\Phi\|_{3,q} \leq \delta\,, \quad (q > 2)$$

there exist $\epsilon > 0$ and $v \in W^{2,2+\epsilon}(\Omega)$, $\pi \in W^{1,2+\epsilon}(\Omega)$, $\int_\Omega \pi\, dx = 0$, satisfying (2.1) and (1.3). Moreover, if $p > \frac{6}{5}$, then for sufficiently

[2] Notation: if $Z : \mathbf{S} \to \mathbb{R}$ (or $\mathbf{S}$), then $\partial_{ij}Z(\eta) \equiv \frac{\partial Z}{\partial \eta_{ij}}(\eta)$, $i,j = 1,2$, denotes partial derivative of Z with respect to the independent variable η_{ij} at the point η.

small Φ *there exist a positive* ϵ *and* $v \in W^{1,p}(\Omega) \cap W^{2,2+\epsilon}_{loc}(\Omega)$, $\pi \in L^{p'}(\Omega) \cap W^{1,2+\epsilon}_{loc}(\Omega)$ *satisfying (2.1) and (1.3).*

Theorem 2.3. *Let* $p > \frac{3}{2}$, Ω *be a non circular* C^2 *domain and* (A_1)–(A_3) *hold. Then there exist a positive* ϵ *and* $v \in W^{2,2+\epsilon}(\Omega)$, $\pi \in W^{1,2+\epsilon}(\Omega)$, $\int_\Omega \pi\, dx = 0$, *satisfying (2.1) and (1.5). Moreover, if* $p > \frac{6}{5}$, *then there exist a positive* ϵ *and* $v \in W^{1,p}(\Omega) \cap W^{2,2+\epsilon}_{loc}(\Omega)$, $\pi \in L^{p'}(\Omega) \cap W^{1,2+\epsilon}_{loc}(\Omega)$ *satisfying (2.1) and (1.5).*

Theorem 2.4. *Let* $p > \frac{3}{2}$, Ω *be a circle and* (A_1)–(A_3) *hold. Then there exist a positive* ϵ *and* $v \in W^{2,2+\epsilon}(\Omega)$, $\pi \in W^{1,2+\epsilon}(\Omega)$, $\int_\Omega \pi\, dx = 0$, *satisfying (2.1), (1.5) and*

$$\int_\Omega \left(\frac{\partial v_2}{\partial x_1} - \frac{\partial v_1}{\partial x_2}\right) dx = 0.$$

Moreover, if $p > \frac{6}{5}$, *then there exist a positive* ϵ *and* $v \in W^{1,p}(\Omega) \cap W^{2,2+\epsilon}_{loc}(\Omega)$, $\pi \in L^{p'}(\Omega) \cap W^{1,2+\epsilon}_{loc}(\Omega)$ *satisfying (2.1) and (1.5).*

The idea of the proof of Theorems 2.1 - 2.4 has similar structure[3]:

- First we show that for $p = 2$ there exists at least one weak solution of the problem and that all such solutions are smooth.
- Then we approximate function T with a growth $p - 1 (\neq 1)$ by functions T^λ with linear growth defining for $\lambda \in (0,1)$

$$T^\lambda(\xi) := (1 + \lambda |\xi|^2)^{\frac{2-p}{2}} T(\xi) \quad \text{for all } \xi \in \mathbf{S}. \tag{2.7}$$

  Thus T^λ satisfies assumptions (1.2)–(1.4) above with $p = 2$. It means that we can use previous step and obtain existence of aproximating regular solutions.
- The last step consists in finding estimates for approximating solutions in $W^{1,p}(\Omega)$ and in $W^{2,2+\epsilon}(\Omega)$ uniform with respect to λ. It is easy to see that T^λ satisfies (for $p \in (1,2)$)

$$\left|\partial_{ij} T^\lambda_{kl}(\xi)\right| \le C_1, \text{ for all } \xi \in \mathbf{S} \tag{2.8}$$

$$\partial_{ij} T^\lambda_{kl}(\eta) \xi_{ij} \xi_{kl} \ge C_2 (1 + |\eta|^2)^{\frac{p-2}{2}} |\xi|^2, \text{ for all } \xi, \eta \in \mathbf{S} \tag{2.9}$$

  with constants C_1 and C_2 that do not depend on λ. It allows us to obtain the above mentioned uniform estimates and then

[3]It is worth remarking that we work only with divergence-free test functions to gain estimates for the velocity field. The properties of the pressure are obtained only at the final stage by a variant of the De Rham theorem.

choose a sequence denoted v_n converging weakly to a function v in $W^{2,2+\epsilon}(\Omega)$. The final passage from the approximating problems to the original problem for $\lambda \to 0$ is then easy.

This structure of the proof is successfully applied and explained in [9] in detail, where homogeneous Dirichlet boundary conditions are considered. The scheme carry on the method developed by Nečas and Stará to prove regularity of weak solutions to nonlinear elliptic equations and systems of an arbitrary growth $p \in (1, \infty)$ in two dimensions, see [13], [14] and [20].

Here we extend the results from [9] to boundary conditions (1.3)-(1.5). Using the paper [9] as a reference, we concentrate on and present only those parts of the proofs that are different or new. Thus the most difficult step 3 is not discussed here at all, as it follows [9] line by line.

A final remark concerns interior regularity. As stated in above Theorems 2.2, 2.3 and 2.4, $C^{1,\alpha}$-regularity of v holds for larger range of p, namely $p > \frac{6}{5}$. As the introduction of λ-approximating problems differs from case to case we prefer not to formulate one general theorem on interior regularity for all kinds of boundary condition. In fact, the method of the proof of Theorem 2.1 does not allow to show interior regularity for $p \leq \frac{3}{2}$. On the other hand, when the existence of approximations is achieved, the regularity procedure coincides with homogeneous Dirichlet problem performed in [9] in detail. This is why we completely skip this part here.

3. PROOFS OF THEOREMS 2.1 AND 2.2

Before starting with the proofs we give two auxiliary lemmas on the properties of an extension of boundary conditions.

Lemma 3.1. *Let $\Omega \in C^2$ bounded and φ be a trace of $\Phi \in W^{1,2}(\Omega)$ on $\partial\Omega$ and $\eta > 0$. Then there exists a prolongation $\Phi^\eta \in W^{1,2}(\Omega)$ such that φ is a trace of Φ^η on $\partial\Omega$ and*

$$\left| \int_\Omega v_i \frac{\partial \Phi_j^\eta}{\partial x_i} v_j \, dx \right| < \eta \, \|v\|_{1,2}^2, \quad \forall v \in W_0^{1,2}(\Omega). \tag{3.1}$$

Proof. A construction following Hopf can be found in [21] or [5]. ∎

The following lemma generalizes this result for $p \neq 2$ and gives estimates of Φ^η in $W^{1,p}(\Omega)$.

Lemma 3.2. *Let $p \in (1, 2\rangle, \eta > 0, \Omega \in C^3, \varphi = Tr(\Phi), \Phi \in W^{3,q}(\Omega)$ for some $q > 2$, and φ satisfy (1.4). Then there exists $\Phi^\eta \in W^{1,\infty}(\Omega)$ such* *that*

$$\operatorname{div}\Phi^\eta = 0 \ \ in\ \Omega \quad and \quad Tr(\Phi^\eta) = \varphi \ \ at\ \partial\Omega, \tag{3.2}$$

$$\left| \int_\Omega v_i \frac{\partial \Phi_j^\eta}{\partial x_i} v_j \, dx \right| < \eta^{\frac{3p-4}{p}} \|v\|_{1,p}^2 \quad \forall v \in W_0^{1,p}(\Omega), \tag{3.3}$$

$$\begin{aligned} \|\Phi^\eta\|_q &< C\eta^{\frac{1}{q}} \\ \|\Phi^\eta\|_{1,q} &< C\eta^{\frac{1}{q}-1} \end{aligned} \quad \forall q \in \langle 1, \infty). \tag{3.4}$$

Proof. An analogous lemma for $p = 2$ (following Miranville and Wang) is sketched in [22]. We use the same construction here, what differs are the estimates.

Since Ω is bounded and regular enough, there exists $\delta_0 > 0$ such that the interior normals to $\partial\Omega$ do not intersect in the neighborhood of $\partial\Omega$ of width $2\delta_0$. We denote this neighborhood of $\partial\Omega$ located inside Ω by $\mathbf{O}_{2\delta_0}$. Moreover, due to regularity of $\partial\Omega$, for every $x = (x_1, x_2) \in \mathbf{O}_{\delta_0}$, there exists a unique point $b(x_1, x_2) \in \partial\Omega$ such that

$$d(x) := \operatorname{dist}((x_1, x_2), \partial\Omega) = \operatorname{dist}((x_1, x_2), b(x_1, x_2)).$$

Let $\tau_{b(x_1,x_2)} = (\tau_1, \tau_2)$ denote the clockwise tangent vector to $\partial\Omega$ at the point $b(x_1, x_2)$. Then $\nu = \nu_{b(x_1,x_2)} = (-\tau_2, \tau_1)$ represents the outer normal to $\partial\Omega$.

We consider a smooth function $\rho : \langle 0, \infty) \to \langle -1, 1\rangle$ such that

$$\rho(0) = 1, \quad \rho(s) = 0 \text{ for all } s \geq 1, \text{ and } \int_0^1 \rho(s)\, \mathrm{d}s = 0,$$

and we set ($\eta < \delta_0$)

$$\Psi^\eta(x) = \begin{cases} \varphi(b(x)) \cdot \tau_{b(x_1,x_2)} \displaystyle\int_0^{\operatorname{dist}(x)} \rho(s/\eta)\, \mathrm{d}s & \text{if } (x_1, x_2) \in \mathbf{O}_{\delta_0} \\ 0, & \text{elsewhere.} \end{cases}$$

We finally set

$$\Phi(x_1, x_2) = \Phi^\eta(x_1, x_2) = \operatorname{rot}\Psi^\eta(x_1, x_2) = \begin{pmatrix} -\frac{\partial \Psi^\eta}{\partial x_2}(x_1, x_2) \\ \frac{\partial \Psi^\eta}{\partial x_1}(x_1, x_2) \end{pmatrix}. \tag{3.5}$$

Thus in $\mathbf{O}_{\delta_0}$ we can compute

$$\Phi(x) \cdot \nu_{b(x)} = \\ \varphi(b(x)) \cdot \tau_{b(x)} \rho\left(\tfrac{d(x)}{\eta}\right) \tfrac{\partial}{\partial \tau}(d(x)) + \tfrac{\partial}{\partial \tau}(\varphi(b(x)) \cdot \tau_{b(x)}) \textstyle\int_0^{d(x)} \rho(\frac{s}{\eta})\, \mathrm{d}s$$

and

$$\Phi(x) \cdot \tau_{b(x)} = \\ -\varphi(b(x)) \cdot \tau_{b(x)} \rho\left(\tfrac{d(x)}{\eta}\right) \tfrac{\partial}{\partial \nu}(d(x)) - \tfrac{\partial}{\partial \nu}(\varphi(b(x)) \cdot \tau_{b(x)}) \textstyle\int_0^{d(x)} \rho(\frac{s}{\eta})\, \mathrm{d}s\,.$$

Since $\frac{\partial}{\partial \nu}(d(x)) = -1$, $\frac{\partial}{\partial \tau}(d(x)) = 0$ and $\frac{\partial}{\partial \nu}(\varphi(b(x)) \cdot \tau_{b(x)}) = 0$, these relations reduce to

$$\left.\begin{aligned} \Phi(x) \cdot \nu_{b(x)} &= \tfrac{\partial}{\partial \tau}(\varphi(b(x)) \cdot \tau_{b(x)}) \textstyle\int_0^{d(x)} \rho(\frac{s}{\eta})\, \mathrm{d}s \\ \Phi(x) \cdot \tau_{b(x)} &= \varphi(b(x)) \cdot \tau_{b(x)} \rho\left(\tfrac{d(x)}{\eta}\right), \end{aligned}\right\} \text{ in } \mathbf{O}_{\delta_0}, \tag{3.6}$$

from which we can easily observe that at the boundary, where $d(x) = 0$ and $b(x) = x$, we have

$$\Phi(x) \cdot \nu_x = 0 \quad \text{and} \quad \Phi(x) \cdot \tau_x = \varphi(x) \cdot \tau_x = \varphi(x) \quad \text{because of (1.4).}$$

Thus (3.2) holds.

Next, as $\Phi^\eta = (\Phi^\eta \cdot \nu)\nu + (\Phi^\eta \cdot \tau)\tau$, and ν and τ depend smoothly on $(x_1, x_2) \in \mathbf{O}_{\delta_0}$, one can obtain (3.4) looking at the values of $\Phi \cdot \nu$, $\Phi \cdot \tau$ and their tangential and normal derivatives.

Indeed, as $\Phi \cdot \nu$, $\Phi \cdot \tau$ are bounded uniformly with respect to η and they have the support in an η-neighbourhood of the boundary, we get $(3.4)_1$.

Computing tangential and normal derivatives of $\Phi \cdot \nu$ and $\Phi \cdot \tau$ we see that all of them are bounded uniformly with respect to η except for

$$\frac{\partial}{\partial \nu}(\Phi(x) \cdot \tau_{b(x)}) = \varphi(b(x)) \cdot \tau_{b(x)} \frac{1}{\eta} \rho'(\frac{d(x)}{\eta}) \frac{\partial}{\partial \nu}(d(x)) + \text{ bounded terms},$$

which implies

$$|\nabla \Phi^\eta| \leq \frac{C}{\eta} \qquad \text{for all } x \in \mathbf{O}_{\delta_0}, \tag{3.7}$$

which is $(3.4)_2$.

Finally, with help of (3.7) we also have

$$\begin{aligned} \left| \int_\Omega v_i \frac{\partial \Phi_i^\eta}{\partial x_j} v_j \, \mathrm{d}x \right| &\leq \left(\int_\Omega \frac{|v(x)|^2}{d(x)^{\frac{4}{p'}}} \, \mathrm{d}x \right) \left\| d^{\frac{4}{p'}} \nabla \Phi^\eta \right\|_\infty \\ &\leq C \eta^{\frac{4}{p'}-1} \left(\int_\Omega \frac{|v(x)|^2}{d(x)^{\frac{4}{p'}}} \, \mathrm{d}x \right) \equiv C \eta^{\frac{4}{p'}-1} I_1, \end{aligned} \tag{3.8}$$

where constant C does not depend on η.

Using Hölder inequality with $\delta = 1 - 4/(p'p)$ we estimate I_1 as

$$I_1 \leq \int_\Omega \left| \frac{v}{d(x)} \right|^{\frac{4}{p'}} |v|^{2-\frac{4}{p'}} \, \mathrm{d}x \leq \left(\int_\Omega \left| \frac{v}{d(x)} \right|^p \, \mathrm{d}x \right)^{\frac{4}{p'p}} \left(\int_\Omega |v|^{\frac{2-4/p'}{\delta}} \, \mathrm{d}x \right)^\delta$$

Since $(2 - 4/p')/\delta = 2p/(2-p)$ we can use imbedding theorems to the second integral and the Hardy inequality to the first integral at the right hand side to obtain

$$I_1 \leq \|\nabla v\|_p^{\frac{4}{p'}} \|\nabla v\|_p^{2-\frac{4}{p'}} \leq \|\nabla v\|_p^2 \tag{3.9}$$

Inserting (3.9) into (3.8) leads to (3.3) (realizing that $\frac{4}{p'} - 1 = \frac{3p-4}{p}$). The proof is complete. ∎

Now, we come to the proofs of Theorems 2.1, 2.2. They follow the scheme sketched above and detailed in [9] with two important modifications clarified in the following two lemmas. The first one concerns the coerciveness needed for the existence of solutions in case $p = 2$, for which the Hopf construction of prolongation of the boundary value φ (see Lemma 3.1) is used. It then allows us to show that all solutions of the quadratic problem (2.7) are smooth as stated in the following lemma.

Lemma 3.3. *Let $p = 2, q > 2$, $f \in L^q(\Omega)$, $\Omega \in C^3$, $\varphi = Tr(\phi)$, $\phi \in W^{2,q}$. Then every solution v, π of (2.1) and (1.4) satisfies $v \in W^{2,2+\epsilon}(\Omega)$, $\pi \in W^{1,2+\epsilon}(\Omega)$ for some $\epsilon > 0$.*

Proof. Proof of this Lemma with $\phi \equiv 0$ can be found in [9], Section 3. Here we proceed analogously. The only difference is that for the use of Leray - Lions theorem as well as for the first apriori estimate we have to apply Lemma 3.1 in order to obtain a bound of $\|v\|_{1,2}$ depending only on the data. The rest of the proof follows closely [9]. ∎

The second modification is needed in order to show that approximating problems parametrized by λ are uniformly bounded in $W^{1,p}(\Omega)$. For this purpose we use two ingredients (as explained in Lemma 2.4 below): (i) the generalized estimates with $p \neq 2$ for the prolongation of φ constructed in Lemma 3.2 (recall that it requires the restriction to assumption (1.4)); (ii) the procedure of Blavier and Mikelic' (see [2]) to obtain uniform bounds in $W^{1,p}(\Omega)$.

Once the uniform estimates in $W^{1,p}(\Omega)$ are achieved, one can derive the uniform estimates in $W^{2,2+\epsilon}(\Omega)$ as in [9].

Lemma 3.4. *Let the assumptions of Theorem A hold. Then for every $\lambda \in (0,1)$ there exists at least one solution v^λ, π^λ of the approximated system*

$$\begin{aligned} v_k \frac{\partial v}{\partial x_k} - \operatorname{div}(T^\lambda(D(v))) + \nabla \pi = f \quad & in\ \Omega\,, \\ \operatorname{div} v = 0 \quad & in\ \Omega \end{aligned} \tag{3.10}$$

with T^λ given by (2.7) and a constant C independent of λ such that

$$\left\|v^\lambda\right\|_{1,p} < C, \left\|\pi^\lambda\right\|_{p'} < C.$$

Proof. We will construct solutions via Galerkin approximations.

Let $\{\psi^i\}_{i=1}^{\infty}$ be basis of smooth functions of $V_p \equiv \{\psi \in W_0^{1,p}(\Omega),\ \operatorname{div}\psi = 0\}$. Let $\Phi^\eta \in V_p$ be a function with trace φ which is a suitable prolongation of φ (the existence of Φ^η is ensured by Lemma 3.2) with small η, specified precisely later. We look for $c^N \in \mathbb{R}^N$ such that $v^N := \Phi^\eta + \sum_{i=1}^N c_i^N \psi^i$ solve the following system of algebraic equations

$$\int_\Omega \{T^\lambda(D(v^N))D(\psi^i) + v_k^N \frac{\partial v_j^N}{\partial x_k}\psi_j^i\}\,dx = \int_\Omega f\psi^i\,dx, \quad \forall i \in \{1,\dots,N\}. \tag{3.11}$$

The existence of at least one solution $c^N \in \mathbb{R}^N$ of (3.11) can be shown by Brouwer Fixed Point Theorem.

Let us define a continuous mapping on $\mathbb{R}^N$:

$$P_i^\lambda(c^N) := \textstyle\int_\Omega \{T^\lambda(D(v^N))D(\psi^i) + v_k^N \frac{\partial v_j^N}{\partial x_k}\psi_j^i - f\psi^i\}\,dx, i = 1,2,\dots,N,$$
$$\left\|c^N\right\|_{\mathbb{R}^N} := \left\|D\left(\textstyle\sum_{i=1}^N c_i^N \psi^i\right)\right\|_p .$$

If we show that there exists $K > 0$ such that

$$(P(c^N), c^N) > 0 \quad \text{for all } \|c^N\|_{\mathbb{R}^N} = K\,, \tag{3.12}$$

then we obtain the existence of a solution c^N such that $\|c^N\|_{\mathbb{R}^N} < K$ by Brouwer Fixed Point Theorem. We shall prove that K can be found independent of λ and N. By standard monotone operator theory we can then pass to the limit and obtain the existence of a solution v^λ to (3.10) with $\left\|v^\lambda\right\|_{1,p}$ uniformly bounded, which then implies the statement of Lemma 3.4.

Let us now prove (3.12). First we write some estimates. (Note that constants C's do not depend on λ in what follows.) ($v := v^N, v' := v^N - \Phi^\eta$)

$$\begin{aligned}
&\textstyle\int_\Omega T^\lambda(D(v))D(v')\,dx \\
&\quad \geq C\textstyle\int_\Omega (1+|D(v)|^2)^{\frac{p-2}{2}}\,|D(v)|^2\,dx - C\,\|D(\Phi^\eta)\|_2^2 \\
&\quad \geq C\textstyle\int_\Omega |D(v)|^p\,dx - C\,|\Omega| - C\,\|D(\Phi^\eta)\|_2^2 \\
&\quad \geq C2^{1-p}\,\|Dv'\|_p^p - C\,\|D\Phi^\eta\|_p^p - C\,|\Omega| - C\,\|D(\Phi^\eta)\|_2^2\,, \\
&\left|\textstyle\int_\Omega v_k \frac{\partial v_j}{\partial x_k} v_j'\,dx\right| \\
&\quad \leq \left|\textstyle\int_\Omega v_k' \frac{\partial \Phi_j^\eta}{\partial x_k} v_j'\,dx\right| + \left|\textstyle\int_\Omega \Phi_k^\eta \Phi_j^\eta \frac{\partial v_j'}{\partial x_k}\,dx\right| \\
&\quad \leq C\eta^{\frac{3p-4}{p}}\,\|\nabla v'\|_p^2 + \delta\,\|\nabla v'\|_p^p + C(\delta)\,\|\Phi^\eta\|_{2p'}^{2p'} \\
&\textstyle\int_\Omega v'\,dx \leq \delta\,\|\nabla v'\|_p^p + C(\delta)\,\|f\|_{p'}^{p'}.
\end{aligned}$$

Here, we used the fact that $|a+b|^p \geq 2^{1-p}(|a|^p - |b|^p)$, the Hölder, Korn, Young inequalities and properties of function Φ^η from Lemma 3.2.

All together we obtain ($c := c^N$)

$$(P(c), c) \geq a\,\|Dv'\|_p^p - b\,\|Dv'\|_p^2 - e\,, \tag{3.13}$$

where $a = C_1 2^{-p}$ (choice of δ), $b = C_2\eta^{\frac{3p-4}{p}}$ and $e = C_2(|\Omega| + \|\Phi^\eta\|_{2p'}^{2p'} + \|D\Phi^\eta\|_2^2 + \|\nabla\Phi^\eta\|_p^p + \|f\|_{p'}^{p'})$. Denoting $g(x) = ax^p - bx^2 - e$, we want to show that $g(K) > 0$ for a positive K.

The function g attains its maximum at the point $\tilde{x} = (\frac{2b}{ap})^{\frac{1}{p-2}}$ and

$$g(\tilde{x}) = (\frac{2b}{ap})^{\frac{p}{p-2}} a(1 - \frac{p}{2}) - e = \left(\frac{2C_2}{ap}\right)^{\frac{p}{p-2}} a(1 - \frac{p}{2})\eta^{\frac{3p-4}{p-2}} - e.$$

Denoting $\left(\frac{2C_2}{ap}\right)^{\frac{p}{p-2}} a(1 - \frac{p}{2})$ by A and realizing that $e \le B\eta^{-1}$ for $\eta < 1$ due to (3.4) we see that

$$g(\tilde{x}) \ge A\eta^{\frac{3p-4}{p-2}} - B\eta{-}1.$$

As $\frac{3p-4}{p-2} < -1$ for $p > 3/2$ we see that it is possible to choose η so small that $g(\tilde{x}) > 0$ holds.

The lemma is proved. ∎

Corollary 3.5. *Let v^λ, π^λ be the solution from Lemma 3.4. Then $v^\lambda \in W^{2,2+\epsilon}(\Omega)$ and $\pi^\lambda \in W^{1,2+\epsilon}(\Omega)$ for some $\epsilon > 0$. Moreover, there exists a constant C that does not depend on λ so that*

$$\left\|v^\lambda\right\|_{1,p} < C, \left\|\pi^\lambda\right\|_{p'} < C.$$

The uniform apriori estimates of v^λ in $W^{2,2+\epsilon}(\Omega)$ needed for the proof of the Theorem 2.1 can the be proved as in [9].

Regarding the proof of Theorem 2.2 it is enough to notice that to achieve the positive maximum of function g in Lemma 3.4 we can use sufficiently small norm of Φ in estimates of convective term instead of Lemma 3.2. The rest of the proof is analogous.

4. PROOFS OF THEOREMS 2.3 AND 2.4

In this part we mean by V_p the space

$$V_p \equiv \{u \in W^{1,p}(\Omega), \quad u \cdot \nu = 0 \quad \text{at } \partial\Omega, \quad \operatorname{div} u = 0 \text{ on } \Omega\}.$$

A function $v \in V_p$ is a weak solution of (2.1), (1.5) iff for any test function $\varphi \in V_p$, the equality

$$\int_\Omega \left(T(D(v))D(\varphi) + v_k \frac{\partial v_i}{\partial x_k}\varphi_i\right) \mathrm{d}x = \int_\Omega f \cdot \varphi \, \mathrm{d}x \tag{4.1}$$

holds. We introduce the operator $A : V_p \to V_p'$ given by the main part of (4.1), i.e.,

$$\langle Av, \varphi\rangle \equiv \int_\Omega \{T(D(v))D(\varphi) + v_k \frac{\partial v}{\partial x_k}\varphi\} \, \mathrm{d}x.$$

Let us start with smoothness of solutions for quadratically approximated systems (3.10) with T^λ given by (2.7). Then A is continuous and

$$\langle Av^\lambda, v^\lambda\rangle = \int_\Omega T(D(v^\lambda))D(v^\lambda) \geq \|Dv^\lambda\|_2^2 .$$

As $\partial\Omega$ is not a circle, we can use the Korn inequality (see [7]) and get coerciveness of A (i.e. $\nabla v^\lambda \in L^2(\Omega)$). As the first part of A is strongly monotone, we can obtain by Leray-Lions argument the existence of a weak solution $v^\lambda \in V_2$. Moreover, for any $\lambda \in (0,1)$ there is a C_λ such that

$$\left\|v^\lambda\right\|_{1,2} + \left\|\pi^\lambda\right\|_2 \leq C_\lambda \|f\|_2 .$$

Now, we are going to prove that for any λ the corresponding solution $\{v^\lambda, \pi^\lambda\}$ belongs to $C^{1,\alpha}(\overline{\Omega}) \times C^{0,\alpha}(\overline{\Omega})$ (in fact, we will discuss in detail the step "from $W^{1,2}$ to $W^{2,2}$") and in the second step we find estimates of $\{v^\lambda, \pi^\lambda\}$ in $W^{2+\epsilon,2}(\Omega) \times W^{1+\epsilon,2}(\Omega)$ that do not depend on approximation parameter λ.

Let λ be fixed. We want to show that any solution $v^\lambda \in W^{1,p}(\Omega)$ belongs to $W^{2,2+\epsilon}(\Omega)$. As interior regularity is same as in the case of the homogeneous Dirichlet problem (see [9], Section 3) we prove here the estimates near the boundary only.

We shall use the local description of the boundary $\partial\Omega$ related to the domain $\Omega \in C^2$. This means that $\partial\Omega$ is locally described by C^2-maps $a_1, a_2, \ldots, a_k$, $k \in \mathbb{N}$. In the corresponding ℓ-th coordinate system ($\ell = 1, 2, \ldots, k$), we suppose that for a fixed positive α and $x_1 \in (-\alpha, \alpha)$

(*i*) $(x_1, x_2) \in \partial\Omega$ if and only if $x_2 = a_\ell(x_1)$,
(*ii*) $U_\ell^+ = \{(x_1, x_2); x_1 \in (-\alpha, \alpha),\ x_2 \in (a_\ell(x_1), a_\ell(x_1) + \alpha)\} \subset \Omega$,
(*iii*) $U_\ell^- = \{(x_1, x_2); x_1 \in (-\alpha, \alpha),$
$x_2 \in (a_\ell(x_1) - \alpha, a_\ell(x_1))\} \subset \mathbb{R}^2 \setminus \overline{\Omega}$.

We also assume that

$$\frac{\partial a_\ell(0)}{\partial x_1} = 0 .$$

Further, we denote

$$U_\ell = \{(x_1, x_2); x_1 \in (-\alpha, \alpha), x_2 \in (a_\ell(x_1) - \alpha, a_\ell(x_1) + \alpha)\}$$

and choose an open smooth set U_0 so that

$$\bigcup_{\ell=0}^{k} U_\ell \supset \Omega .$$

In this description the tangent derivative on U^ℓ is defined by

$$\frac{\partial}{\partial\tau} = \frac{\partial}{\partial x_1} + (a^\ell)' \frac{\partial}{\partial x_2}$$

and the outer normal vector (up to a multiplicative constant) is given by

$$\nu = (-(a^\ell)', 1).$$

In what follows we shall omit indices ℓ (counting coverings of $\partial\Omega$) and λ (approximation parameter).

We shall test (4.1) by

$$\Phi_{(\varphi)} := \xi\Big(-\frac{\partial \varphi}{\partial \tau} + (0, a''\varphi_1)\Big) + \psi^\varphi \tag{4.2}$$

where $\varphi \in V_2$, ξ is a cut-off function in $C_0^\infty(U)$ and ψ^φ is chosen so that

$$\operatorname{div} \Phi_{(\varphi)} = 0.$$

Function ψ^φ is defined as a solution of

$$\begin{aligned} \operatorname{div} \psi^\varphi &= \operatorname{div}\big(\xi(\tfrac{\partial \varphi}{\partial \tau} - (0, a''\varphi_1))\big) \ \text{ in } U \\ \psi^\varphi &= 0 \text{ at } \partial U. \end{aligned} \tag{4.3}$$

By Amrouche, Girault (see [1]) we know that for any $s \in (1, \infty)$ there exists a ψ^φ

$$\|\psi^\varphi\|_{1,s} \le C(s)\, \|\varphi\|_{1,s} \qquad \text{and} \qquad \|\psi^\varphi\|_{0,s} \le C(s)\, \|\varphi\|_{0,s} \,.$$

As $\varphi \in V_2$ we have $\varphi \cdot \nu = 0$ at $\partial\Omega$ and also $\frac{\partial}{\partial \tau}(\varphi \cdot \nu) = 0$ at $\partial\Omega$. Thus

$$\frac{\partial \varphi}{\partial \tau} \cdot \nu - a''\varphi_1 = 0 \quad \text{ at } \partial\Omega\,.$$

Finally we obtain

$$\Phi_{(\varphi)} \cdot \nu = -\xi \frac{\partial \varphi}{\partial \tau} \cdot \nu + \xi a''\varphi_1\nu_2 + \psi^\varphi \cdot \nu = 0 \quad \text{ at } \partial\Omega.$$

Let us denote

$$u := \xi\Big(-\frac{\partial v}{\partial \tau} + (0, a''v_1)\Big) + \psi^v,$$

where ψ^v is defined by (4.3). After integration by parts and rearrangement, we see that u solves the equation:

$$\begin{aligned} &\int_\Omega \partial_{ij} T_{kl}(D(v)) D_{kl}(u) D_{ij}(\varphi)\, dx = \\ &\int_\Omega P_{ij} \tfrac{\partial \varphi_i}{\partial x_j} + P_i \varphi_i \, dx + \int_\Omega Q_{ij} \tfrac{\partial \psi_i^\varphi}{\partial x_j} + \\ &\qquad Q_i \psi_i^\varphi - v_k \tfrac{\partial v_i}{\partial x_k} (\Phi_{(u)})_i \, dx, \quad \forall \varphi \in V_2 \end{aligned} \tag{4.4}$$

with $\Phi_{(u)}$ and ψ^φ given by (4.2) and (4.3). The terms P_{ij}, P_i, Q_{ij}, Q_i that are not written explicitly here can be estimated by $C(|f| + |\nabla v|)$. Inserting $\varphi = u$ in (4.4) we obtain (using ellipticity condition (2.4))

$$\int_\Omega |Du|^2\, dx \le C\, \|Du\|_2 \,(\|f\|_2 + 1) + \Big|\int_\Omega v_k \frac{\partial v}{\partial x_k} \Phi_{(u)}\, dx\Big| \tag{4.5}$$

The last term in (4.5) can be estimated by $C(\varepsilon) + \varepsilon \|\nabla^2 v\|_2$ for an arbitrarily small $\varepsilon > 0$.

The Korn inequality ([7]) implies that we can estimate symmetric gradient by gradient in (4.5) and for all $\varepsilon > 0$ there exists $K(\varepsilon)$ such that

$$\forall \ell = 1, \ldots, k : \quad \left\| \frac{\partial v}{\partial \tau} \xi_\ell \right\|_{1,2;U_\ell} \leq K(\varepsilon) + \varepsilon \left\| \nabla^2 v \right\|_2 . \tag{4.6}$$

Using condition (2.4) we realize easily that to estimate normal derivatives of ∇v, it is enough to estimate $\frac{\partial^2 v}{\partial x_2^2}$. As $\operatorname{div} v = 0$ we obtain $\frac{\partial^2 v_2}{\partial x_2^2} = -\frac{\partial^2 v_1}{\partial x_1 \partial x_2}$ and it is controlled by (4.6) for a sufficiently fine covering $\{U_\ell\}_{\ell=1}^k$. The estimate of $\frac{\partial^2 v_1}{\partial x_2^2}$ can be extracted directly from (4.1). Taking curl of (4.1)[4] (in distributional sense) we get (the empty brackets () abbreviates $(|D(v)|^2)$)

$$\begin{aligned} \frac{\partial}{\partial x_1}\frac{\partial}{\partial x_1}\partial_{21}F() + \frac{\partial}{\partial x_1}\frac{\partial}{\partial x_2}\partial_{22}F() - \frac{\partial}{\partial x_2}\frac{\partial}{\partial x_1}\partial_{11}F() - \frac{\partial}{\partial x_2}\frac{\partial}{\partial x_2}\partial_{12}F() \\ = \operatorname{curl}(v_k \frac{\partial v}{\partial x_k}) - \operatorname{curl} f \, . \end{aligned} \tag{4.7}$$

Set $G \equiv \frac{\partial}{\partial x_2}\partial_{12}F(|D(v)|^2)$. Then clearly ($\ell = 1, 2, \ldots, k$)

$$\|\xi_\ell G\|_{-1,2} \leq C\|\partial_{12}F()\|_2 \leq C\|Dv\|_2 \leq C \, .$$

Next, by (2.4),

$$\left\| \frac{\partial}{\partial x_1} (\xi_\ell G) \right\|_{-1,2} \leq \left\| \frac{\partial}{\partial x_1} \partial_{12}F() \right\|_2 \leq C + \sum_{i=1}^{2} \left\| \frac{\partial^2 v}{\partial x_1 \partial x_i} \right\|_2 .$$

Finally, from (4.7), we have

$$\begin{aligned} &\left\| \frac{\partial}{\partial x_2} (\xi_\ell G) \right\|_{-1,2} \\ &\quad \leq C \left(\left\| \frac{\partial}{\partial x_1} (\partial_{21}F + \partial_{22}F - \partial_{11}F) \right\|_2 + \left\| v_k \frac{\partial v}{\partial x_k} \right\|_2 + \left\| f \right\|_2 + 1 \right) \\ &\quad \leq C + \textstyle\sum_{i=1}^2 \left\| \frac{\partial^2 v}{\partial x_1 \partial x_i} \right\|_2 . \end{aligned}$$

Now, we can apply the Nečas' theorem on negative norms (cf. [15] or [10] for a formulation of theorem) to obtain

$$\|\xi_\ell G\|_2 \leq C(\|\xi_\ell G\|_{-1,2} + \|\nabla \xi_\ell G\|_{-1,2} \leq C + \sum_{i=1}^{2} \left\| \frac{\partial^2 v}{\partial x_1 \partial x_i} \right\|_2 . \tag{4.8}$$

From the definition of G, we observe

$$\partial_{12}\partial_{12}F() \, D_{12}(\frac{\partial v}{\partial x_2}) = \frac{G}{2} - \frac{1}{2}\partial_{11}\partial_{12}F() \, \frac{\partial D_{11}(v)}{\partial x_2} - \frac{1}{2}\partial_{22}\partial_{12}F() \, \frac{\partial D_{22}(v)}{\partial x_2} . \tag{4.9}$$

[4]Recall that $\operatorname{curl} z = \frac{\partial z_2}{\partial x_1} - \frac{\partial z_1}{\partial x_2}$ for $z : \Omega \to \mathbb{R}^2$.

Using the fact that $\frac{\partial^2 v_2}{\partial x_2^2} = -\frac{\partial^2 v_1}{\partial x_1 \partial x_2}$ for the last term, and the facts that $D_{12}(\frac{\partial v}{\partial x_2}) = \frac{\partial^2 v_1}{\partial x_2^2} + \frac{\partial^2 v_2}{\partial x_1 \partial x_2}$ and $\partial_{12}\partial_{12}F(|D(v)|^2) \geq C_1$ by (2.3), we can conclude from (4.9) and (4.8) that

$$\left\|\xi_\ell \frac{\partial^2 v_1}{\partial x_2^2}\right\|_2 \leq C + \sum_{i=1}^{2} \left\|\frac{\partial^2 v}{\partial x_1 \partial x_i}\right\|_2 . \tag{4.10}$$

Then (4.10) together with (4.1)-(4.6) lead to

$$\sum_{j=1}^{2} \left\|\xi_\ell \frac{\partial^2 v_j}{\partial x_2^2}\right\|_2 \leq C + \left\|\frac{\partial \nabla v}{\partial \tau}\right\|_2 + \tilde{c} \sup_{x_1 \in (-\alpha,\alpha)} \left|\frac{\partial a(x_1)}{\partial x_1}\right| \sum_{j=1}^{2} \left\|\xi_\ell \frac{\partial^2 v_j}{\partial x_2^2}\right\|_2 . \tag{4.11}$$

If α is chosen so that

$$\tilde{c} \max_\ell \sup_{x_1 \in (-\alpha,\alpha)} \left|\frac{\partial a(x_1)}{\partial x_1}\right| \leq \frac{1}{2} .$$

the last term in (4.11) can be moved to the left-hand side and

$$\sum_{j=1}^{2} \left\|\xi_\ell \frac{\partial^2 v_j}{\partial x_2^2}\right\|_2 \leq K(\varepsilon) + \varepsilon \left\|\nabla^2 v\right\|_2 \qquad \text{for } \ell = 1, ..., k.$$

Adding these estimates and choosing ε sufficiently small we obtain

$$\left\|\frac{\partial^2 v_j}{\partial x_2^2}\right\|_2 \leq K,$$

which means that $v \in W^{2,2}(\Omega)$. Remind that we omitted index λ so that by this way we proved $v^\lambda \in W^{2,2}(\Omega)$

It implies that the right-hand side in (4.4) belongs certainly to $L^q(\Omega)$ for $q > 2$. We can thus complete the first part of the proof combining the scheme from [9] with L^p-theory for the Stokes system with the boundary conditions of the type (1.5) proved in [19]. (They can be also deduced from the results of Grubb [6] studying the evolutionary Stokes system.) The first part of the proof (dealing with λ fixed) of Theorem C is complete.

To obtain estimates uniform with respect to λ we repeat this procedure again using the estimates (2.8), (2.9) (uniform with respect to λ) instead of (2.3) and (2.4). Testing equation (2.1) by v and using the p-version of the Korn inequality (see [12]), we obtain

$$\exists C \quad \forall \lambda \in (0,1) \quad \|v\|_{1,p} \leq C.$$

The rest of the proof is analogous to [9]. The only substantial changes are in using the L^p-theory for Stokes problem for boundary conditions

(1.5) (performed in [19]) instead for homogeneous Dirichlet problem. Theorem 2.3 is proved.

In order to prove Theorem 2.4, it is possible to proceed as in the proof above provided that the Korn inequality holds. As Ω is a circle we modify the definition of the space V_p and set

$$V_p \equiv \{u \in W^{1,p}(\Omega); \\ u \cdot \nu = 0 \text{ at } \partial\Omega, \ \operatorname{div} u = 0 \text{ in } \Omega, \int_\Omega \left(\frac{\partial v_2}{\partial x_1} - \frac{\partial v_1}{\partial x_2}\right) \, dx = 0\}.$$

Then, by [7] and [12], we see that the Korn inequality $\|\nabla u\|_p \le c\|D(u)\|_p$ holds for all $u \in V_p$. The rest of the proof follows lines of the proof above. ■

Acknowledgments

The research was supported by the grants No. 201/96/0228 (J. Málek) and No. 201/96/0311 (J. Stará) of the Grant Agency of the Czech Republic and by the grant 189/96 (J. Stará) of the Grant Agency of the Charles University.

References

[1] Amrouche, Ch. and Girault, V. (1994). *Decomposition of vector spaces and application to the Stokes problem in arbitrary dimension.* Czechoslovak Math. J., 44:109–141.

[2] Blavier, E. and Mikelić, A. (1995). *On the stationary Quasi-Newtonian Flow Obeying a Power-law.* Math. Meth. Appl. Sci., 18:927–948.

[3] Ebin, D.G. (1983). *Viscous fluids in a domain with frictionless boundary.* Global Analysis-Analysis on Manifolds (eds. H. Kurke, J. Meele, H. Triebel, R. Thiele) 57:93–110.

[4] Galdi, G.P. (1994). *An introduction to the mathematical theory of the Navier-Stokes equations, Volume I, Linearized Steady Problems.* Springer-Verlag, New York, Berlin, Heidelberg.

[5] Galdi, G.P. (1994). *An introduction to the mathematical theory of the Navier-Stokes equations, Volume II, Nonlinear Steady Problems.* Springer-Verlag, New York, Berlin, Heidelberg.

[6] Grubb, G. (1995). *Nonhomogeneous time-dependent Navier-Stokes problems in L_p Sobolev spaces.* Differential and Integral Equations 8:1013–1046.

[7] Hlaváček, I. and Nečas, J. (1970). *On inequalities of Korn's type.* Arch. Rational Mech. Anal., 36:305–333, Part I&II.

[8] Kaplický, P., Málek, J. and Stará, J. (1997). *Full regularity of weak solutions to a class of nonlinear fluids in two dimensions - stationary, periodic problem.* Comment. Math. Univ. Carolinae, 38(4):681–695.

[9] Kaplický, P., Málek, J. and Stará, J. (1999). *$C^{1,\alpha}$-solutions to a class of nonlinear fluids in two dimensions – stationary Dirichlet problem.* Zapiski naukhnych seminarov POMI 259. To appear.

[10] Málek, J., Nečas, J., Rokyta, M. and Růžička, M. (1996). *Weak and measure-valued solutions to the evolutionary PDE's.* Chapman & Hall, London.

[11] Málek, J., Rajagopal, K.R. and Růžička, M. *Existence and regularity of solutions and the stability of the rest state for fluids with shear dependent viscosity.* Math. Models Methods in Appl. Sci., 6(5):789–812.

[12] Mosolov, P.P. and Mjasnikov, V.P. (1971). *Dokazatelstvo neravenstva Korna.* Doklady Akademii Nauk SSSR, 1:36–39.

[13] Nečas, J. (1968). *Sur la régularitè des solutions faibles des équations elliptiques non linéaires.* Comment. Math. Univ. Carolinae, 9(3):365–413.

[14] Nečas, J. (1967). *Sur la régularité des solutions variationnelles des équations elliptiques non-linéaires d'ordre 2k en deux dimensions.* Annali della Scuola Normale Superiore di Pisa XXI Fasc. III:427–457.

[15] Nečas, J. (1969). *Sur les normes équivalentes dans $W^{k,p}(\Omega)$ et sur la coercivité des formes formellement positives.* Les Presses de l'Université de Montréal, Montréal, 102–128.

[16] Nečas, J. and Šverák, V. (1991). *On regularity of solutions of nonlinear parabolic systems.* Ann. Scuola Norm. Sup. Pisa Cl. Sci., 18:1–11.

[17] Rajagopal, K.R. (1993). *Mechanics of Non-Newtonian Fluids* Recent development in theoretical fluid mechanics. (Eds. G.P. Galdi and J. Nečas) 129–162, Pitman Research Notes in Mathematics Series 291, Longman Scientific & Technical, Essex, England.

[18] Seregin, G.A. (1996). *On regularity for solution to initial-boundary value problems describing the motion of non-newtonian fluids in dimension two.* Preprint SFB256, No.482. To appear in St. Petersburg Mathematical J..

[19] Solonnikov, V.A. and Sčadilov, V.E. (1973). *On one boundary-value problems to stationary Navier-Stokes equations.* Trudy MIAN 125:196–210.

[20] Stará, J. (1971). *Regularity Results for Non-linear Elliptic Systems in two Dimensions.* Annali della Scuola Normale Superiore di Pisa XXV Fasc. I, 163–190.

[21] Temam, R. (1984). *Navier-Stokes Equations, Theory and Numerical Analysis.* North Holland, Amsterdam, Third revised edition.

[22] Temam, R. (1997). *Infinite-Dimensional Dynamical Systems in Mechanics and Physics.* Springer-Verlag, New York, 122–123. Second edition.

VISCOSITY SOLUTIONS FOR DEGENERATE AND NONMONOTONE ELLIPTIC EQUATIONS

Bernd Kawohl, Nikolay Kutev

Abstract: Motivated by the theory of viscosity solutions we suggest and discuss criteria to choose a particular solution from possibly many solutions in situations where there is nonuniqueness or discontinuity. Particular examples include the Cahn-Hilliard equation.

Keywords: Viscosity solution, Perron method, Cahn Hilliard equation, gradient blow-up, discontinuity.

1. INTRODUCTION AND MOTIVATION

The aim of this paper is to extend the notion of continuous viscosity solutions of M.Crandall and P.L.Lions [8], [21] to a wide class of degenerate nonlinear elliptic equations which are not proper, i.e. equations without the fundamental monotonicity condition with respect to the solution $u(x)$. Originally viscosity solutions were introduced for first order equations by the method of vanishing viscosity. Later on for second order fully nonlinear elliptic and parabolic equations the existence of viscosity solutions was proved by means of the Perron method. That is why the unique viscosity solution obtained by Perron's procedure is automatically a continuous function. However there are many examples of equations which have either discontinuous solutions or more than one continuous viscosity solution. This is our motivation to introduce a notion of discontinuous solutions which are stable under small perturbations and are still unique in the class of discontinuous solutions, and which we will call later on limit solutions. This new definition is a natural extension of the classical one of M.Crandall and P.L.Lions. In fact, if the problem has a unique continuous viscosity solution, then this solution is also the unique limit solution of the problem according to the new definition.

The most important example which motivates such considerations is the Cahn-Hilliard equation. It is well known from the abundant literature that this problem has discontinuous solutions. The reason for this phenomenon is the fact that the equation is not proper, combined with the high degeneracy of the equation. One can see from the analysis of the solutions of the regularized problem

$$-\varepsilon^2 u''(x) + u^3(x) - u(x) = 0 \qquad \text{in } (0, b) \tag{1.1}$$

that in some sense the discontinuity of the solution as $\varepsilon \to 0$ is a consequence of "an interior gradient blow up" of the perturbed solutions $u^\varepsilon(x)$

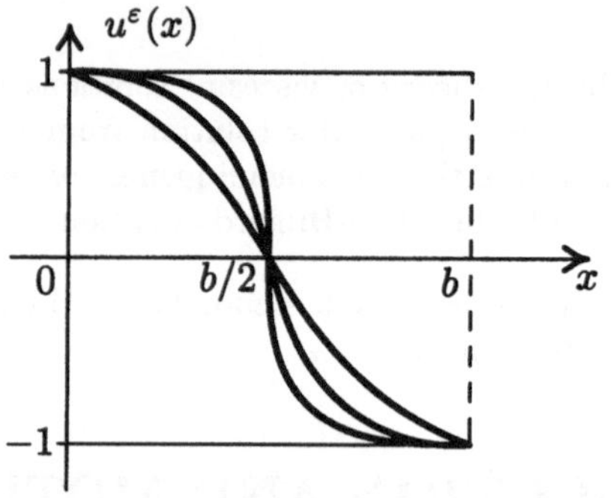

Figure 1. Solutions of (1.1) for different ε

This is more clear in the problem considered by M.Bertsch and R.dal Passo [4], arising in the theory of phase transitions where the corresponding free-energy functional has a linear growth rate with respect to the gradient, i.e.

$$u_t = [\varphi(u)\psi(u_x)]_x \quad \text{in } \mathbf{R} \times \mathbf{R}_+, \quad u(x,0) = u_0(x) \quad \text{in } \mathbf{R} \tag{1.2}$$

where φ, ψ are smooth, $\varphi > 0$, $\psi' > 0$, ψ is odd and ψ has the following asymptotic behaviour at infinity:

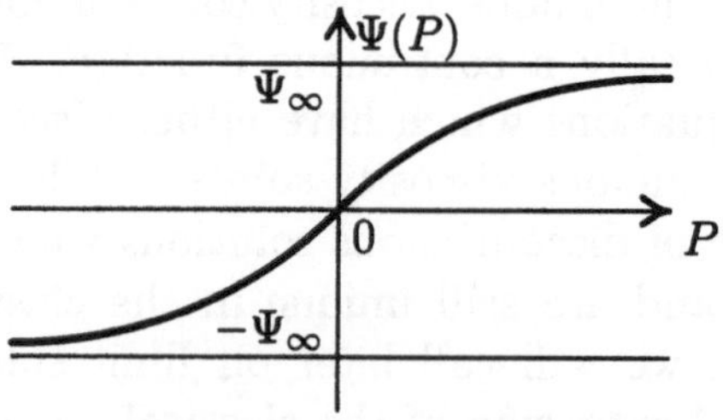

Figure 2. Shape of ψ

Since the equation, in general, is nonmonotone and highly degenerate, the gradient of the solution blows up after a finite time and then the solution becomes discontinuous, where the discontinuity is only as in Figure 3a) but not as in Figure 3b).

The paper of Angenent and Fila [2] treats strictly parabolic nonmonotone equations, a typical example of which is the problem

$$\begin{aligned} u_t &= u_{xx} + u|u_x|^{m-1}u_x && \text{in } (-1,1)\times \mathbf{R}_+ \\ u(x,0) &= \varphi(x) && \text{in } (-1,1), \\ u(\pm 1,t) &= A_\pm && \text{for } t>0 \end{aligned} \tag{1.3}$$

for $m > 2$, for constants $A_- < 0 < A_+$ and for sufficiently large $A_+ - A_-$.

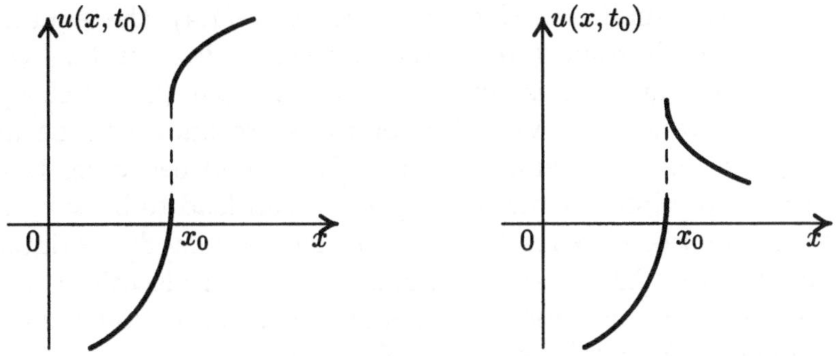

Figure 3a. Possible Discontinuity **3b.** Impossible Discontinuity

The authors prove that after a finite time the gradient of the solution will blow up at some interior nodal point of u, provided the oscillation of the data, $A_+ - A_-$, is large enough. They did not investigate the behaviour of the solution after the blow up time but one can prove that the solution will be discontinuous as in Figure 3a). The reason for the interior gradient blow up and the discontinuity of the solution is the supergrowth of the gradient term ($m > 2$) and the large oscillation of the data. Boundary gradient blow up can be excluded due to the concavity (or convexity) of the solution u on the boundary $(1,t)$ (or $(-1,t)$), see [2]. In this case the nodal line of the solution, where the gradient blows up, is called interior boundary of the equation, see [11], and it should not be confused with the topological boundary of the domain. If we interpret the lower order term $|u_x|^{m-1}u_x \cdot u$ as a zero order term with coefficient $|u_x|^{m-1}u_x$ of the wrong sign, we see that the maximum principle does not apply when $|u_x| \to \infty$. In some sense this equation can be considered as an implicit degenerate parabolic equation, because if we divide the equation by $(1+u_x^2)^\beta$, $\beta > 0$, the new equation will be a degenerate one.

A more sophisticated problem was considered by Y.Giga [12], the typical example is the following equation

$$\begin{aligned} u_t - \frac{u_{xx}}{(1+u_x^2)^\alpha} &= u(1+u_x^2)^{1/2} \quad \text{in } (a,b)\times\mathbf{R}_+ \\ u(x,0) &= u_0(x) \in C^{2,\mu}, \quad u(a,t) = u(b,t) = 0 \end{aligned} \tag{1.4}$$

where $\alpha > 1/2$. This type of nonuniformly, nonmonotone parabolic equation appears in some geometric problems with the mean curvature flow. In this case interior gradient blow up (with discontinuity of the solution after the blow up time) is due to the choice of special initial data and to the fact that $\alpha > 1/2$ implies superquadratic growth in u_x. Since the solution is equal to zero on the boundary, boundary gradient blow up is eliminated.

In all these examples (1.1)–(1.4) the gradient of the solution blows up in an interior point and after the blow up time the solution is a discontinuous function. Note that in (1.1)–(1.4) the equation is not monotone with respect to u and implicitly degenerate. On the contrary, if the equation is uniformly proper (see (2.3) below), i.e. uniformly monotone in u, viscosity solutions are known to be unique and continuous in the interior, see [9]. The inbetween case, that the equation is proper but not uniformly proper, can lead to blow up of the gradient or the second derivatives and to subsequent discontinuity on the boundary as in [19], or even to nonuniqueness of viscosity solutions.

Let us discuss uniqueness and nonuniqueness of viscosity solutions in more detail. A simple example of nonuniqueness of the viscosity solutions was considered by H.Ishii and N.Ramaswany [16] for first order Hamilton-Jacobi equations, namely

$$|Du(x)| = |x| \text{ in } B = \{x \in \mathbf{R}^n \ , \ |x| < 1\} \ , \ u = 0 \text{ on } \partial B. \qquad (1.5)$$

This problem has infinitely many continuous viscosity solutions

$$2u_c = \min\left\{1 - |x|^2, c + |x|^2\right\} \text{ for } -1 \le c \le 1 \ .$$

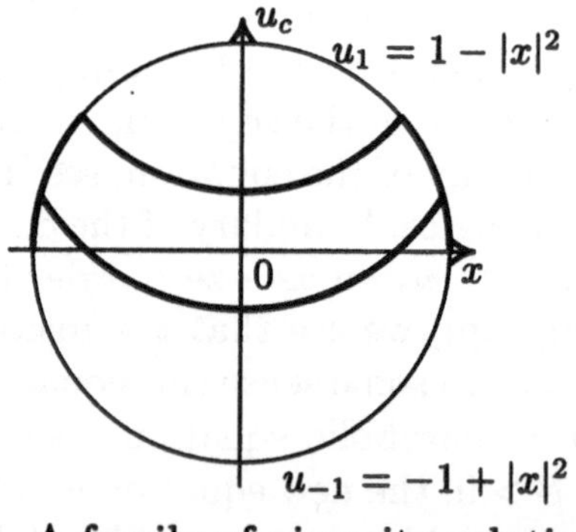

Figure 4. A family of viscosity solutions, vary c

The authors explain the nonuniqueness of the viscosity solutions with the degeneracy of the right-hand side of the equation. Indeed, considering the new equation $|Du(x)|\ |x|^{-1} = 1$ with singular coefficients, they "naturally" regularized it to $|Du(x)|\ (|x| + \varepsilon)^{-1} = 1$, so that the perturbed problem has a unique viscosity solution $u^\varepsilon(x)$ for every $\varepsilon > 0$ and $u^\varepsilon(x)$ tends to the maximal solution $u_1 = 1 - |x|^2$ when $\varepsilon \to 0$. However, as we will see in Theorems 1 and 2 below, the comparison principle and the uniqueness result for viscosity solutions of degenerate elliptic equations is true provided the equation is strictly proper, see (2.11). Note that (nonstrict) properness, as in (1.5) does not suffice.

2. DEFINITION AND COMPARISON RESULTS

Let us recall the definition of viscosity solutions. For simplicity we will illustrate the definitions only in the one dimensional case for the equation

$$F(u, u', u'') = 0 \quad \text{in } (a, b) \tag{2.1}$$

which is independent of x.

Definition 1. i) An upper-semicontinuous function $u(x)$ is a *viscosity subsolution of* (2.1) if for every $x_0 \in (a, b)$ and for every $\varphi \in C^2$ such that $\varphi(x) \geq u(x)$ and $\varphi(x_0) = u(x_0)$ the inequality $F(\varphi(x_0), \varphi'(x_0), \varphi''(x_0)) \leq 0$ holds.

ii) A lower-semicontinuous function $v(x)$ is a *viscosity supersolution of* (2.1) if for every $x_1 \in (a, b)$ and for every $\psi \in C^2$ such that $\psi(x) \leq v(x)$, $\psi(x_1) = v(x_1)$ the inequality $F(\psi(x_1), \psi'(x_1), \psi''(x_1)) \geq 0$ holds.

iii) A continuous function $u(x)$ is a *viscosity solution of* (2.1) if $u(x)$ is a viscosity sub- and supersolution

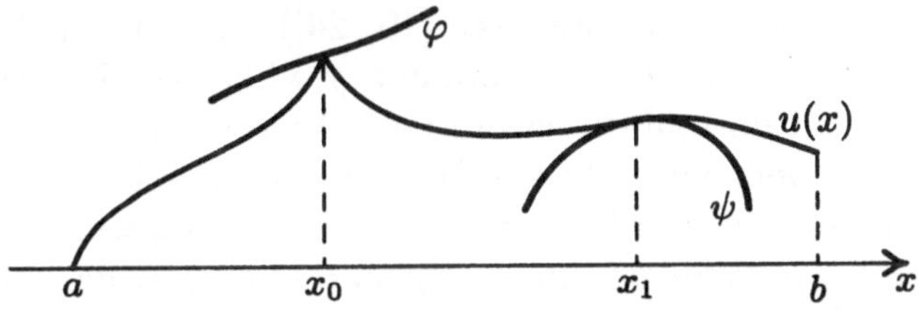

Figure 5. Illustration of Definition 1

The main assumptions under which the theory of the viscosity solutions works are

- Degenerate ellipticity

$$F(r, p, X) \leq F(r, p, Y) \quad \text{whenever} \quad Y \leq X, \text{ and} \tag{2.2}$$

- Uniform proper operator (uniform monotonicity w.r.t. $u(x)$)

$$\exists_{\gamma > 0} \text{ s.t.} \quad \gamma(r - s) \leq F(r, p, X) - F(s, p, X) \quad \text{whenever} \quad r \geq s. \tag{2.3}$$

Note that (2.2) is weaker than ellipticity,

$$F(r, p, X) < F(r, p, Y) \qquad \text{whenever} \qquad X > Y$$

in which the inequality signs are strict. Under the above conditions (2.2) and (2.3) the classical Dirichlet problem has a unique continuous viscosity solution. The uniqueness follows from the comparison principle for USC subsolutions and LSC supersolutions while the existence results are based on Perron's method.

However, in the case of degenerate elliptic equations in general the classical Dirichlet problem does not have a solution. To understand this phenomenon, recall that viscosity solutions are stable in the sense that viscosity solutions u^ε of regularized problems converge (in a suitable sense) to a viscosity solution of the limit problem. Sometimes this limit solution violates the Dirichlet condition in the classical sense. A typical example is the elliptic problem

$$\varepsilon u_{tt} + u_{xx} - u_t = f \qquad \text{in } (a,b)\times(0,T) \tag{2.4}$$

with Dirichlet data $u = g$ on the entire boundary. In the limit $\varepsilon \to 0$ we obtain a degenerate elliptic equation, whose solution satisfies $u_{xx} - u_t = f$ also for $t = T$, but not necessarily $u = g$ for $t = T$.

For linear degenerate elliptic equations of type

$$a^{ij}(x)u_{x_i,x_j} + b^i(x)u_{x_i} + c(x)u = f(x) \;\text{ in } \Omega, \;\; a^{ij}(x)\xi^i\xi^j \geq 0 \tag{2.5}$$

it is still possible to define the *right* boundary value problem by means of the so-called Fichera function $\beta(x) = \sum_{i=1}^{n}(b^i(x) - \sum_{j=1}^{n} a^{ij}_{x_j}(x))\nu^i$, where ν is the exterior unit normal to the boundary of Ω, such that the classical Dirichlet problem has a solution (see [10, 24]). The Dirichlet data are prescribed only on the part of the boundary where the elliptic equation is nondegenerate in normal direction, i.e. where $a^{ij}(x)\nu^i\nu^j > 0$, and on the characteristic one, i.e. where $a^{ij}(x)\nu^i\nu^j = 0$, only in those points where the Fichera function $\beta(x)$ is positive. Now, if we want to define the right Dirichlet problem for *nonlinear* degenerate elliptic equations and to find a solution of the problem we have to know a priori the value of the Fichera function which is impossible. In fact, $\beta(x)$ will depend on the unknown solution $u(x)$ by means of the coefficients of the equation and, in general, we will not know the sign of $\beta(x)$. A suitable definition which overcomes all these difficulties and is stable under small perturbations of the domain and data was suggested in [9]. In contrast to equation (2.1) we speak now about the Dirichlet problem (2.6):

$$F(u,u',u'') = 0 \;\text{ in } (a,b) \qquad u(a) = A, \quad u(b) = B. \tag{2.6}$$

Definition 2. i) An upper semicontinuous function $u(x)$ is a *viscosity subsolution of the Dirichlet problem* (2.6) if $u(x)$ is a viscosity subsolution of (2.1) in (a,b) and if it satisfies the inequalities

$$\begin{aligned} \min(F(\varphi(a), \varphi'(a), \varphi''(a)), u(a) - A) \le 0 \\ \min(F(\phi(b), \phi'(b), \phi''(b)), u(b) - B) \le 0 \end{aligned} \tag{2.7}$$

for every $\varphi, \phi \in C^2, \varphi(x) \ge u(x), \phi(x) \ge u(x)$, $\varphi(a) = u(a)$ and $\phi(b) = u(b)$.

ii) A lower semicontinuous function $v(x)$ is *a viscosity supersolution of the Dirichlet problem* (2.6) if $v(x)$ is a viscosity supersolution of (2.1) in (a, b) and if it satisfies the inequalities

$$\begin{aligned} \max(F(\psi(a), \psi'(a), \psi''(a)), v(a) - A) \ge 0 \\ \max(F(\Psi(b), \Psi'(b), \Psi''(b)), v(b) - B) \ge 0 \end{aligned} \tag{2.8}$$

for every $\psi, \Psi \in C^2$, $\psi(x) \le v(x)$, $\Psi(x) \le v(x)$, $\psi(a) = v(a)$ and $\Psi(b) = v(b)$.

iii) A continuous function $w(x) \in C([a, b])$ is *a viscosity solution of the Dirichlet problem* (2.6) if $w(x)$ is both a viscosity sub- and supersolution.

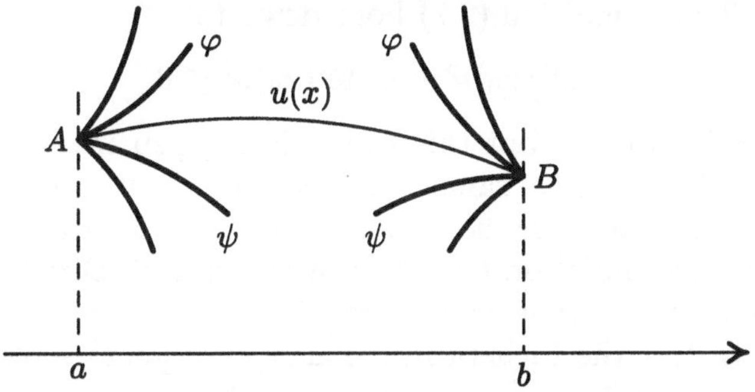

Figure 6. Illustration of Definition 2

Loosely speaking, Definition 2 says that either the solution satisfies the Dirichlet data in a classical sense, or it satisfies the equation on the boundary in the viscosity sense. Let us note, however, that if the solution $u(x)$ satisfies the equation in a classical sense at the boundary, then Definition 2 is not automatically satisfied. In fact, if $u \in C([a, b])$ is a viscosity solution of (2.6) and $u(a) > A$, then according to (2.7) the inequality

$$F(\varphi(a), \varphi'(a), \varphi''(a)) \le 0 \tag{2.9}$$

holds for every $\varphi \in C^2, \varphi(x) \ge u(x)$, $\varphi(a) = u(a)$. These are more admissible functions φ than described in Definition 1, because for interior points x_0 and $u \in C^1$ we have $u'(x_0) = \varphi'(x_0)$ as an additional constraint in Definition 1. If φ is an admissible test function at a, then so is $\varphi + \lambda(x - a) + \mu(x - a)^2/2$ for any $\lambda > 0$, $\mu \in \mathbf{R}$. Hence (2.9) implies

$$F(\varphi(a), \varphi'(a) + \lambda, \varphi''(a) + \mu) \le 0$$

for any $\lambda > 0$, μ, and if F is nondegenerate elliptic as in (2.11) we would reach a contradiction by varying μ. Hence F must be degenerate on the boundary, if u violates Dirichlet data in the classical sense.

Remark 1. A similar situations occurs in several space dimensions. In fact, if $u(x)$ is a smooth solution of the degenerate linear elliptic problem

$$\begin{aligned} a^{ij}(x)u_{x_i,x_j} + b^i(x)u_{x_i} + c(x)u = f(x) \quad &\text{in } \Omega, \\ u(x) = g(x) \quad &\text{on } \partial\Omega, \end{aligned} \tag{2.10}$$

if $u(x_0) > g(x_0)$ at some $x_0 \in \partial\Omega$, and if φ is a test function at x_0, then for any $\lambda > 0$ and $\mu \in \mathbf{R}$ the function

$$\varphi(x) - \lambda(\nu, x - x_0) + \mu(\nu \otimes \nu(x - x_0), x - x_0)/2$$

is also an admissible test function at x_0. Here ν is the exterior normal to $\partial\Omega$ at x_0. (Equivalently, if we work with the upper superjet, we have that when $(p, X) \in J^{2,+}u(x_0)$, then $(p - \lambda\nu, X + \mu\nu \otimes \nu) \in J^{2,+}u(x_0)$, too, see [20]. Thus condition (2.7) boils down to

$$-\mu a^{ij}(x_0)\nu^i\nu^j + \lambda b^i(x_0)\nu^i \leq 0$$

for every $\lambda > 0$, $\mu \in \mathbf{R}$, but this can only happen if $a^{ij}(x_0)\nu^i\nu^j = 0$ and $b^i(x_0)\nu^i \leq 0$. The same condition holds if we consider the corresponding inequality for the supersolutions. In this way we naturally obtain Fichera's formulation (see [10]) of the "right" Dirichlet problem assuming, so to speak, that some part of $\partial\Omega$ belongs to the interior of the domain. In fact the Fichera function $\beta(x) = [b^i(x) - a^{ij}_{x_j}(x)]\nu^i$ was originally introduced for the solvability of the Dirichlet problem in $L^p(\Omega)$ by means of the energy method and "naturally" appears if one rewrites equation (2.5) in divergence form

$$-(a^{ij}(x)u_{x_i})_{x_j} - [b^i(x) - a^{ij}_{x_j}(x)]u_{x_i} - c(x)u + f(x) = 0.$$

However, in the necessary condition of D.Gilbarg and N.Trudinger [13] for the classical solvability of (2.10) in $C^2(\Omega) \cap C(\overline{\Omega})$ the important function on the boundary instead of the Fichera function $\beta(x)$ is the function $b^i(x)\nu^i - \sum_{\ell=1}^{n-1} a^{ij}(x)k_\ell\lambda_i^\ell\lambda_j^\ell$, where λ^ℓ are the principal directions and k_ℓ are the principal curvatures of $\partial\Omega$ at the point $x \in \partial\Omega$. In general, the Fichera function and the Gilbarg-Trudinger function are different and coincide when the matrix $\{a^{ij}(x)\} \in C^2(\Omega')$ is a nonnegative one in a larger domain $\Omega' \supset \Omega$ and the set of the chracteristic points of the boundary, i.e. $\{x_0 \in \partial\Omega;\ a^{ij}(x_0)\nu^i\nu^j = 0\}$, has a nonempty interior.

Now, going back to our original problem (2.6), let us first understand why the notion of viscosity solution, i.e. Definition 2, is useful in the context of the Dirichlet problem. Since uniqueness and existence theorems are based on a comparison principle we will first focus on this question. In [9, Thm. 7.9] it was proved that a comparison principle holds for continuous viscosity sub- and supersolutions of problem (2.6), provided that the operator F is degenerate elliptic (2.2) and uniform proper (2.3) and that it satisfies an additional technical assumption, namely [9, (7.15)]. This condition (7.15) is, in general, hard to check for nonlinear equations. We will show in Theorem 1 that this condition is not necessary, at least when the function F is independent of x. In this case even the uniform monotonicity condition (2.3) can be weakened to the strict monotonicity condition (2.11)

$$0 < F(r,p,X) - F(s,p,X) \quad \text{whenever} \quad r > s. \tag{2.11}$$

However, if the strict monotonicity condition (2.11) is replaced by the weak monotonicity condition (2.12)

$$0 \leq F(r,p,X) - F(s,p,X) \quad \text{whenever} \quad r \geq s. \tag{2.12}$$

then even for uniformly elliptic equations the comparison principle fails. Uniform ellipticity is characterized by

$$\exists_{\gamma>0} \text{ s.t.} \qquad \gamma(X-Y) \leq F(r,p,Y) - F(r,p,X) \text{ whenever } X \geq Y, \tag{2.13}$$

while locally uniform ellipticity means

$$\exists_{\gamma_K>0} \text{ s.t.} \forall_{X \geq Y,\ |X|+|Y|+|p|<K} \quad \gamma_K(X-Y) \leq F(r,p,Y) - F(r,p,X). \tag{2.14}$$

In fact, we show in Theorem 2 that when the operator F satisfies (2.12) and (2.14), then the comparison principle for problem (2.6) is true, aside for some special cases of so-called extremal solutions. Those are characterized by a gradient blow up or a blow up of the second derivatives on the whole boundary of the domain, and then viscosity solutions are unique modulo additive constants.

Theorem 1. *Suppose $F(r,p,X) \in C(\mathbf{R} \times \mathbf{R} \times \mathbf{R})$ satisfies the degenerate ellipticity condition* (2.2) *and the strict monotonicity condition* (2.11). *If u and $v \in C([a,b])$ are viscosity sub- and supersolutions of problem* (2.6), *then $u \leq v$ in $[a,b]$.*

The proof of this and the following theorem are given in Section 4.

Theorem 2. *Suppose $F(r,p,X) \in C(\mathbf{R} \times \mathbf{R} \times \mathbf{R})$ satisfies the locally uniform ellipticity condition* (2.14) *and the weak monotonicity condition* (2.12). *Furthermore suppose that $F(r,p,X)$ is locally Lipschitz continuous in p. If $u, v \in C[a,b]$ are viscosity sub- and supersolutions of the Dirichlet problem* (2.6) *then either*

- *i)* $u \leq v$ *in* $[a,b]$, *or*
- *ii)* $u \equiv v + c$ *in* $[a,b]$ *for some positive constant* $c > 0$.

In fact, case ii) in Theorem 2 appears iff either the gradient of the solution blows up or the second derivatives of the solution blow up on the whole boundary. These types of solution were mentioned in passing only for the mean curvature equation as so-called extremal solutions, see [14]. They are the unique classical solutions of the equation without any boundary conditions. Depending on the mean curvature of $\partial\Omega$ they are unbounded or bounded with infinite gradient. Here we will not give the precise conditions which guarantee either gradient or Hessian blow up of the solution on the boundary. Let us just illustrate part ii) of Theorem 2 with the following two typical Examples 1 and 2.

Example 1. Consider first the problem

$$\begin{aligned} -u'' - (1 + u'^2)^{3/2} &= 0 \text{ in } (-1,1), \\ u(-1) = u(1) &= 0, \end{aligned} \tag{2.15}$$

or its multidimensional analogue for $n > 1$

$$-\text{div}\left(\frac{\nabla u}{\sqrt{1+|\nabla u|^2}}\right) - n = 0 \text{ in } B_1 = \{x \in \mathbf{R}^n,\ |x| < 1\},$$
$$u = 0 \text{ on } \partial B_1.$$

This Dirichlet problem (2.15) has infinitely many continuous viscosity solutions

$$u_c(x) = \sqrt{1 - |x|^2} + c \text{ for every } c \geq 0,$$

but only one of them, $u_0(x)$, is a classical solution of (2.15).

Since the solution $u_c(x)$ is $C^{1/2}$ Hölder continuous up to the boundary it is clear from Figure 7 that the set of the test functions from above at the boundary points ($|x| = 1$) is empty, so that according to (2.7) u_c is automatically a subsolution of (2.15). Moreover, $u_c(x)$ satisfies (2.8) and is a supersolution of (2.15) even in the classical sense. Therefore problem (2.15) has infinitely many continuous viscosity solutions $u_c(x)$.

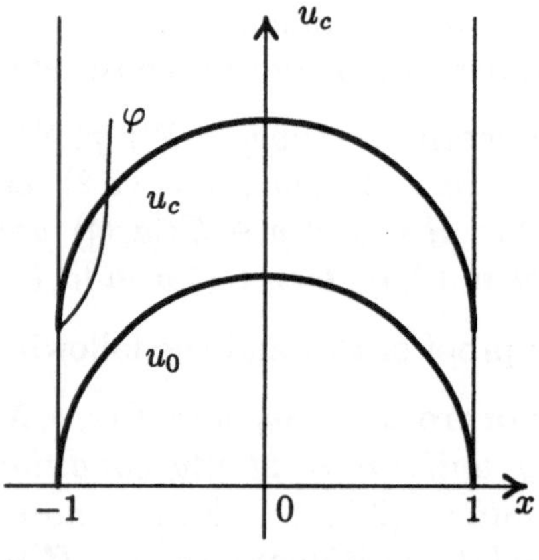

Figure 7. Two solutions of (2.15)

Example 2.

$$
\begin{gathered}
-\frac{u''}{\sqrt{1+u''^2}} + f(u') = 0 \quad \text{in } (0,1) \\
u(0) = u(1) = 0
\end{gathered}
\tag{2.16}
$$

where $f(p) = -\sqrt{(2-p^2)/(3-3p^2+p^4)}$ for $|p| \le 1$ and $f(p) \equiv -1$ for $|p| > 1$ is a $C^{1,1}$ function depicted in Figure 8.

Problem (2.16) has infinitely many viscosity solutions

$$
u_c(x) = -\frac{2}{3}\left[x^{3/2} + (1-x)^{3/2}\right] + \frac{2}{3} + c
$$

for every $c \ge 0$ and $u_c(x) \in C^{3/2}[0,1] \cap C^\infty(0,1)$, but only one of them, $u_0(x)$ is a classical one.

According to Definition 2, trivially $u_c(x)$ are viscosity supersolutions because they are classical supersolutions. As for the proof that $u_c(x)$ are viscosity subsolutions, it follows from $|u_c'(x_0)| = 1$ in the boundary points $x_0 = 0$ and $x_0 = 1$, that every test function $\varphi \in C^2$ from above satisfies the inequality $|\varphi'(x_0)| \ge 1$ (see Figure 9) and hence that the inequality

$$
-\frac{\varphi''(x_0)}{\sqrt{1+\varphi''^2(x_0)}} + f(\varphi'(x_0)) \le 0
$$

holds because $f(\varphi'(x_0)) = -1$.

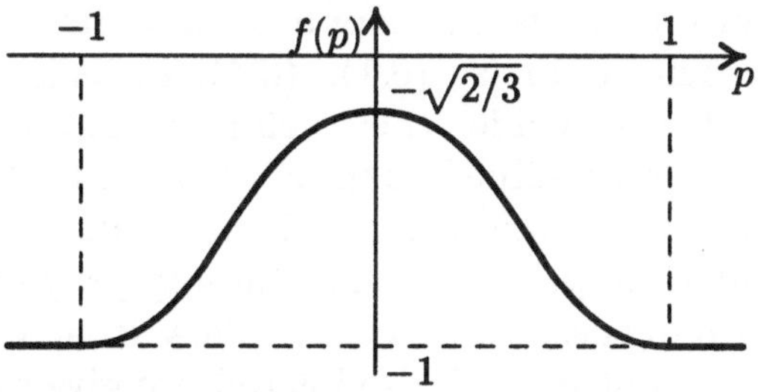

Figure 8. Shape of f from (2.16)

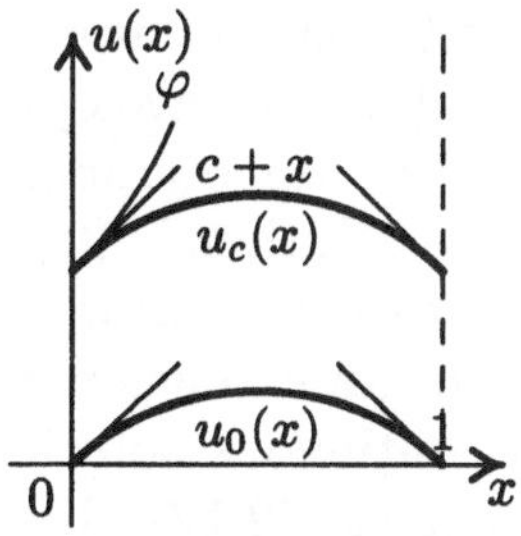

Figure 9. Two solutions of (2.16)

Let us comment that in some sense the results in Theorems 1 and 2 can be considered as an extension of the comparison result of R. Jensen in [17 , Thm. 3.1]. The result of R. Jensen is for viscosity solutions in $C(\overline{\Omega}) \cap W^{1,\infty}(\Omega)$ of equations satisfying (2.13), so that case ii) of Theorem 2 cannot occur.

Remark 2. In the multidimensional case the results in Theorems 1 and 2 hold under the additional C^1 smoothness of the solution on the "free boundary", i.e. the part of $\partial\Omega$ where the solution satisfies the equation but not the boundary data.

3. LIMIT OF VISCOSITY SOLUTIONS U^ε FOR NONMONOTONE EQUATIONS

The discontinuous solutions which appear in Section 1, namely (1.1)–(1.5), are not viscosity solutions in the sense of Definition 1 or 2 because of their lack of regularity. And the last two examples (2.15) and (2.16) are not compatible with the basic concept of viscosity solutions, namely that problem (2.6) should have a unique viscosity solutions in the class $C([a,b])$. In fact, the Dirichlet problem (2.6) in these examples has a unique classical solution, but also infinitely many continuous viscosity solutions. It is therefore clear that for equations satisfying (2.12) a major feature (uniqueness) from the theory of viscosity solutions breaks down and that in this case new concepts must be developed.

Motivated by the stability properties of viscosity solutions for monotone problems [8], we suggest selection criteria that may lead to a new concept of a solution. Like viscosity solutions, this vague concept of a new solution should be stable under small perturbations of the coefficients, in particular under vanishing viscosity. Roughly speaking, we regularize a possibly (explicitly or implicitly) degenerate equation by means of small viscosity depending on ε so that the new equation is no longer explicitly or implicitly degenerate. If the regularized problem has a unique viscosity solution $u^\varepsilon(x)$ for every small $\varepsilon > 0$, then the pointwise limit of $u^\varepsilon(x)$ as $\varepsilon \to 0$ may or may not exist, but its upper semicontinuous enveloppe u^* and its lower semicontinuous enveloppe u_* defined in [9, (6.1), (6.4)] do exist, as long as u^ε are bounded. If they coincide, they define a continuous viscosity solution. Otherwise the setvalued map $x \mapsto [u_*(x), u^*(x)]$ will be called a *limit solution* of the original problem. This notion allows for an extension of comparison principles and stability properties for discontinuous solutions. Uniqueness of u^ε could follow from weak monotonicity (2.12), for instance, but if (2.12) is violated, we give cases below where the u^ε are not unique, so that we have many limit solutions. If the limit solution is not single valued (i.e. discontinuous) only on the boundary, it may be modified to be continuous up to the boundary.

In some sense the concept of a limit solution can be considered as an equivalent of the existence of a global minimizer for variational problems, for example for the Cahn-Hilliard equation. However our concept has the potential to be applicable to nonvariational problems as well. As we shall see in Examples 4 and 5 below, our limit solutions have certain Perron properties, i.e. their minimal and maximal elements u_* and u^* are infima of continuous supersolutions or suprema of continuous subsolutions. This gives an idea how to prove existence for nonmonotone equations.

This intuitive notion of *limit solutions* is an extension of the concept of viscosity solutions to degenerate and nonmonotone equations. Indeed, if the classical Dirichlet problem has a unique viscosity solution, then as is well known, the limit solution will coincide with the unique viscosity solution. If problem (2.6), however, has a unique viscosity solution, then the limit solution obtained by means of a pointwise limit procedure will coincide with the viscosity solution, in general, only in the interior of the domain. On the boundary, there can be nonuniqueness, but loosely speaking by continuous continuation one can get uniqueness. In fact the special upper or lower star limits yield the maximal or minimal viscosity solution which coincide if the viscosity solution is unique. Otherwise one can identify a maximal and minimal viscosity solution.

In the light of these considerations recall that (1.1)–(1.5) have discontinuous solutions and the regularized problems have a unique solution u^ε, which becomes discontinuous in a special point as $\varepsilon \to 0$, see Figures 1,2 and 3. After the onset of discontinuity nothing seemed to be known for the time dependent problems (1.2) (1.3) (1.4).

It is interesting to note that in the example (1.5) of H. Ishii and N. Ramaswamy [16] the limit solution of the regularized equation

$$-\varepsilon\Delta u + |Du| - |x| = 0$$

under zero boundary darta in ± 1 is unique and coincides with the maximal solution $1-|x|^2$ which was obtained in [16] by regularization of the equation with singular coefficients to $|Du|/(|x|+\varepsilon) = 1$ and $\varepsilon \to 0$. Incidentally the limit solution for $-\varepsilon\Delta u - |Du| + |x| = 0$ is the minimal one $|x|^2 - 1$ of (1.5).

As for the nonuniqueness part ii) of Theorem 2, one can easily see (e.g. from Examples 1,2 and 3 below) that in the class of limit solutions the second statement ii) of Theorem 2 simply disappears. Thus Theorems 1 and 2 can be considered as a standard comparison principle for continuous limit solutions of (2.6).

Let us illustrate the above phenomena of degeneracy of the equation on the boundary with a more general example.

Example 3. Consider

$$F(x,u,Du,D^2u) = 0 \text{ in } \Omega, \quad u = 0 \text{ on } \partial\Omega, \tag{3.1}$$

where F is a smooth function which satisfies the uniform monotonicity condition (2.3), the equation is uniformly elliptic, F is a convex function w.r.t. the D^2u variables, and all "natural" structure conditions of N.Trudinger (see [26]) are fulfilled so that (3.1) has a unique classical

$C^2(\overline{\Omega})$ solution. If $\phi(x) \in C^\infty(\mathbf{R}^n)$ is a smooth function such that $\phi^2(x) > 0$ in $\mathbf{R}^n \setminus \partial\Omega$ and $\phi(x) = 0$ on $\partial\Omega$, then let us look at the modified problem

$$\phi^4(x)F(x,u,Du,D^2u) = 0 \text{ in } \Omega, \quad u = 0 \text{ on } \partial\Omega. \tag{3.2}$$

This equation is degenerate elliptic on the boundary with Fichera function $\beta(x) = 0$ on $\partial\Omega$. According to Fichera's notion (see line below (2.5)) of the *right* boundary value problem (at least for linear equations) the whole boundary should be free from boundary data. One can easily check that every classical solution of (3.1) with arbitrary Dirichlet data $u = g \in C^\infty$ on $\partial\Omega$, $g \neq 0$, will be a viscosity solution of (3.2). Thus (3.2) has infinitely many viscosity solutions. However, (3.2) has a unique limit solution $u(x)$ which is the unique classical solution of (3.1) with prescribed zero Dirichlet data. In fact, after the regularization of equation (3.2), i.e.,

$$\phi^4(x)F(x,u,Du,D^2u) - \varepsilon\Delta u = 0 \text{ in } \Omega, \quad u = 0 \text{ on } \partial\Omega, \tag{3.3}$$

the perturbed problem (3.3) has a unique classical solution $u^\varepsilon(x)$. One can easily prove global a priori estimates for $u^\varepsilon(x)$ in $\overline{\Omega}$, for instance

$$\sup_{\overline{\Omega}} |u^\varepsilon(x)| \leq \sup_{\overline{\Omega}} |F(x,0,0,0)|/\gamma,$$

and global boundary gradient estimates, such as

$$\pm u^\varepsilon(x) \leq N(2d(x)d_0^{-1} - d^2(x)d_0^{-2})$$

with $d(x) = \text{dist}(x, \partial\Omega)$, N sufficiently large and $0 < d(x) < d_0$, d_0 sufficiently small, which are uniform w.r.t. the small parameter $\varepsilon > 0$. Using the interior gradient and C^2 a priori estimates (see [26]) after the limit $\varepsilon \to 0$ we obtain the unique limit solution of (3.2). Note that problems (3.1) and (3.2) are not equivalent in the class of viscosity solutions but they are equivalent in the class of limit solutions.

The remaining examples illustrate the situation when the limit problem has more than one viscosity solution. This happens when the weak monotonicity condition (2.12), which was satisfied in Examples 1,2 and 3, fails.

Example 4. Cahn Hilliard

$$u^3 - u = 0 \text{ in } (0,b), \quad u(0) = A, \ u(b) = B, \tag{3.4}$$

$$-\varepsilon^2 u'' + u^3 - u = 0 \text{ in } (0,b), \quad u(0) = A, \ u(b) = B. \tag{3.4$_\varepsilon$}$$

Example 5. Quasilinear Cahn Hilliard

$$-(u')^2u'' + u^3 - u = 0 \text{ in } (0,b), \quad u(0) = A, \ u(b) = B, \qquad (3.5)$$
$$-(u'^2 + \varepsilon^2)u'' + u^3 - u = 0 \text{ in } (0,b), \quad u(0) = A, \ u(b) = B, \qquad (3.5)_\varepsilon$$
$$-|u'|^p u'' + u^3 - u = 0 \text{ in } (0,b), \quad u(0) = A, \ u(b) = B, \qquad (3.6)$$

for $p > 0$. Notice that (3.5) is a special case (p=2) of (3.6), and that (3.4) is a limiting case $p = 0$ of (3.6). For $p = n$ (space dimension) problem (3.6) was suggested in [5].

There is no difference in the qualitative properties of the solutions of (3.5) and (3.6), except the regularity of solutions so that we will consider only (3.5).

The Dirichlet problem for the simplest Cahn-Hillard equation (3.4) has three (even continuous) viscosity solutions $u_0 \equiv 0$, $u_{\pm 1} \equiv \pm 1$, independently of the choice of boundary data. Checking this is a simple exercise using Definition 2, and we omit it. Moreover, it has infinitely many discontinuous viscosity solutions in the sense of Definition 2, namely the set of functions $u(x)$ taking values in $\{-1, 0, 1\}$.

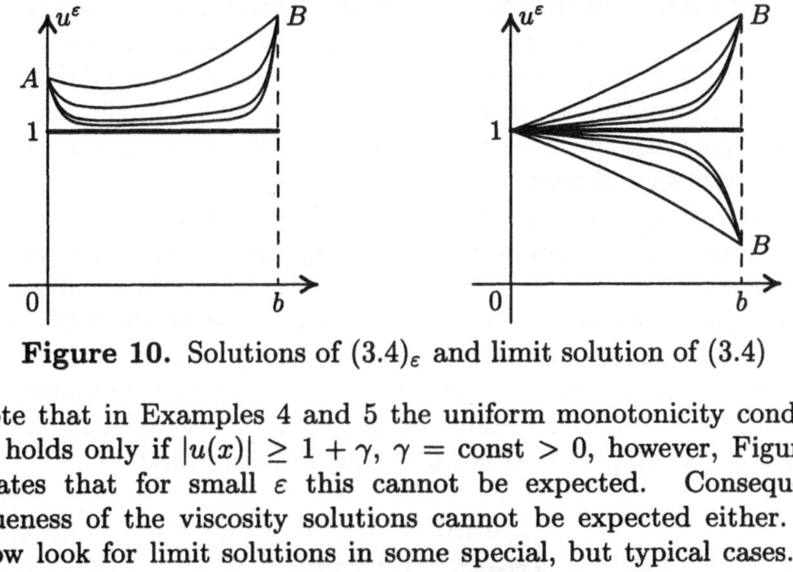

Figure 10. Solutions of $(3.4)_\varepsilon$ and limit solution of (3.4)

Note that in Examples 4 and 5 the uniform monotonicity condition (2.3) holds only if $|u(x)| \geq 1 + \gamma$, $\gamma = \text{const} > 0$, however, Figure 10 indicates that for small ε this cannot be expected. Consequently uniqueness of the viscosity solutions cannot be expected either. Let us now look for limit solutions in some special, but typical cases. For $A \geq 1$ and $B \geq 1$ problem (3.4) has a unique limit solution $u_1 \equiv 1$ (see left part of Figure 10).

In this case the pointwise and upper star limit $\varepsilon \to 0$ is $U(x) = 1$ for $x \in (0,b)$, $U(0) = A$, $U(b) = B$, but the lower star limit is $u_1 \equiv 1$ on $[0,b]$, and because it has maximal smoothness, we pick this one as limit solution. When $A \geq 1$, $-1 < B < 0$, for instance, the pointwise limit solution $U(x)$ is neither upper nor lower semicontinuos and the unique limit solution $U(x)$ is a discontinuous function on the boundary, $U(0) = A$, $U(b) = B$, $U(x) \equiv 1$ in (a,b) (see right half of Figure 11). In this case again $u_1 \equiv 1$ on $[0, b]$ has maximal continuity and will be the unique limit solution.

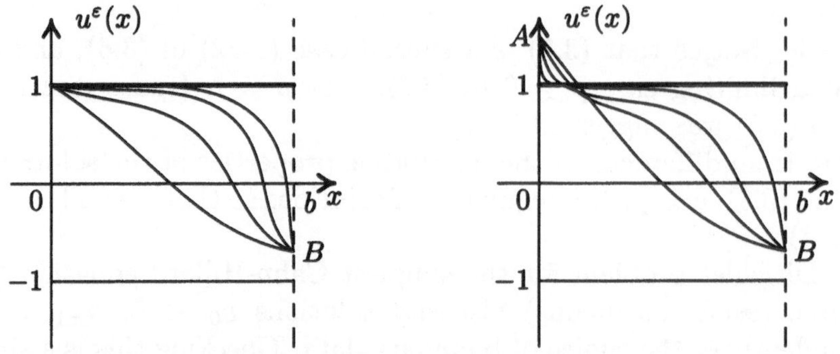

Figure 11. Solutions of $(3.4)_\varepsilon$ and limit solution of (3.4)

In contrast, when $A \geq 1$ and $B \leq -1$ we have an interior discontinuity of the limit solution at the center $b/2$ of the domain $(0, b)$ (see Figure 12).

If, for example, $A = 1$, $B = -1$, the pointwise limit solution $U(x) \equiv 1$ for $x \in [0, b/2)$, $U\,(b/2) = 0$, $U(x) \equiv -1$ for $x \in (b/2, b]$ is neither upper nor lower semicontinuos. In this case we shall speak of the upper and lower semicontious envelope of U as (two) limit solutions, because they have maximal smoothness.

It is interesting to point out that these limit solutions of (3.4) are not limit solutions on a subinterval any more, e.g. on the subinterval $(b/3, b)$ with boundary data $u(b/3) = 1$ and $u(b) = -1$. In fact limit solutions on $(b/3, b)$ are discontinuos in $2b/3$. This distinguishes them from classical solutions and is due to the degeneracy of the equation. In other words, the union of the unique limit solution on $(0, b/3)$ with boundary data $u(0) = u(b/3) = 1$ and of a limit solution on $(b/3, b)$ with boundary data $u(b/3) = 1$ and $u(b) = -1$ is not a limit solution of (3.4).

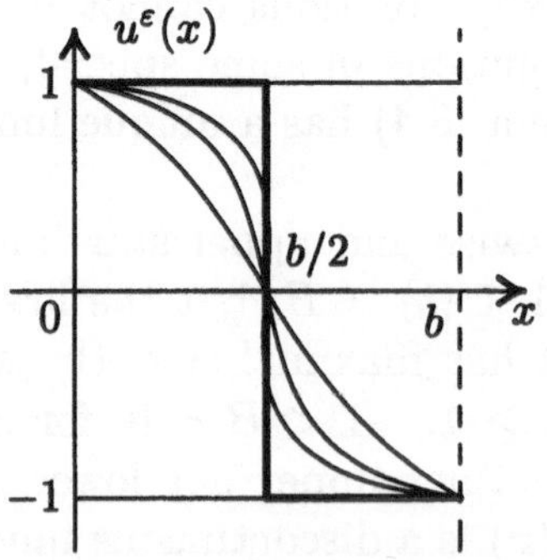

Figure 12. Solutions of $(3.4)_\varepsilon$ and discontinuous limit solution of (3.4)

To better understand this nonlocal phenomenon of the limit solutions let us consider problem (3.5) in the interval $(0, 2\pi)$ with data $A = 1$, $B = -1$. (3.5) has infinitely many continuous viscosity solutions.

The unique limit solution of problem (3.5) is the function $U \in C^{1,1}$ defined by $U(x) = 1$ in $[0, \pi/2]$, $U(x) = \sin x$ in $[\pi/2, 3\pi/2]$ and $U(x) = -1$ in $[3\pi/2, 2\pi]$. Any horizontal shift of $U(x)$ by at most of $\pi/2$ is again a viscosity solution of (3.5). Note that the limit solution is not the maximal viscosity solution $V \in C^{1,1}$ defined by $V(x) = 1$ in $[0, \pi]$ and $V(x) = -\cos x$ in $[\pi, 2\pi]$.

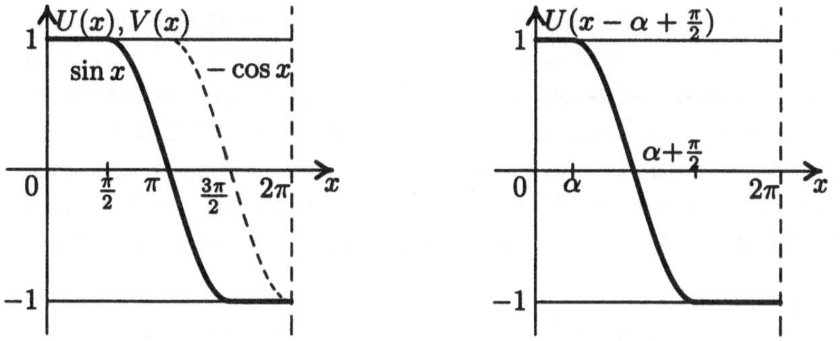

Figure 13. Viscosity solutions and unique limit solution of (3.5)

If we divide now the interval $[0, 2\pi]$ into two subintervals $I_1 = (0, c)$, $I_2 = (c, 2\pi)$ with $\pi/2 < c < 3\pi/2$ then the union of the two unique limit solutions in I_1 and I_2 coincides with the unique limit solution on the whole interval $[0, 2\pi]$. However, if $0 < c \leq \pi/2$ or $3\pi/2 \leq c < \pi$, then we have the same nonlocal effects as in (3.4), see right part of Figure 13. It is due to the fact that both boundary points of the subinterval are characteristic in the sense of Remark 1, iff $|u(c)| = 1$ when the nonlocal effect occurs. On the other hand, c is noncharacteristic when $u(c) \in (-1, 1)$, and then the nonlocal phenomenon does not occur.

We will finish these examples with the case $|A| < 1$, $|B| < 1$. Let us first consider the simplest subcase $A = B = 0$. Now solutions u^ε of $(3.4)_\varepsilon$ have three pointwise limits u_k $(k = 0, \pm 1)$ which take values $+1$, 0 or -1. Moreover, for $(3.5)_\varepsilon$ and they can have many pointwise limits u_k, $k \in \mathbf{N}$, which are periodic functions. To see this is left as an exercise to the reader. Note that the three limit solutions of (3.4) have Perron properties from above (infimum of supersolutions) or from below (supremum of subsolutions). The unique classical solution $u_0 \equiv 0$ of (3.4) has both Perron properties from below and from above in a trivial way, because both the set of supersolutions and the set of subsolutions contain only one function, namely the solution u_0 itself. This explains the unstable character of the trivial solution $u_0 \equiv 0$.

There are various selection criteria in the literature that lead to uniqueness results for nonmonotone problems. Some of them consider the unique positive solution, the unique convex solution or the unique maximal solution depending on the equations under consideration. We focus on solutions with maximal regularity. Our Examples show that those have Perron properties, and that they are sometimes convex, sometimes positive, sometimes maximal, but not always. In some sense Perron properties of the solutions correspond to being global minimizers of the variational problems.

It remains to discuss the typical cases $A \in [0,1)$, $B \in (0,1)$, or $A \in [0,1)$, $B \in (-1,0)$. These lead like in the case of zero boundary data to one or two limit functions u_k. If A and B have different sign, $-u_1$ can also be a limit solution, but $u_0 \equiv 0$ is not a limit solution, because it is not a pointwise limit a.e. of u^ϵ, see Figure 14. What happens with problem (3.5) when $A \in (0,1)$ and $B \in (-1,0)$? Also in this case there exist two limit solutions U and V with Perron properties, and many periodic limits in $C^{1,1}$ of u^ε without Perron properties (see Figure 15).

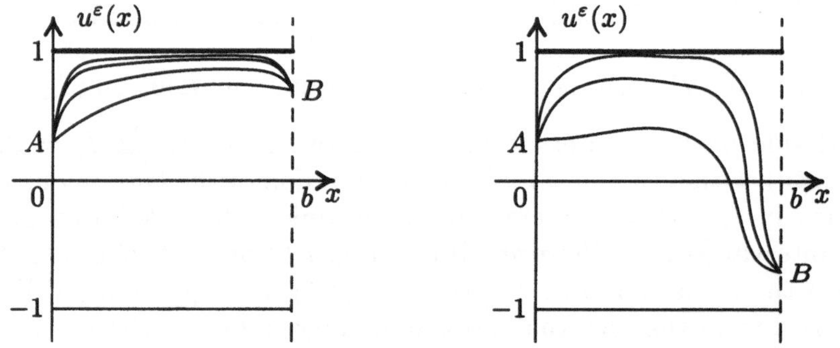

Figure 14. Structure of solutions to $(3.4)_\varepsilon$

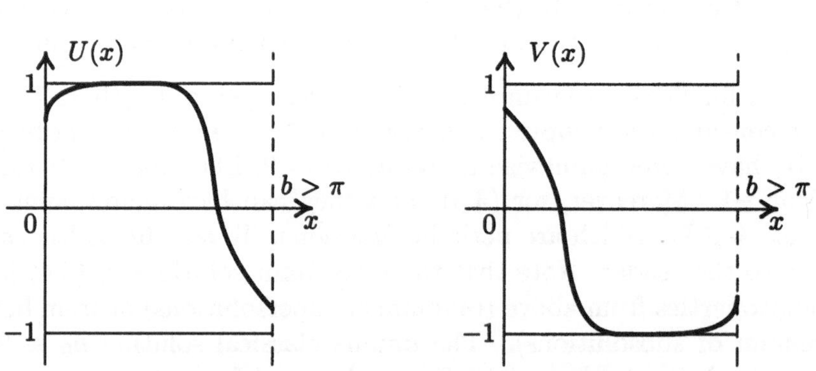

Figure 15. Limit solutions of (3.5)

The cause for such a difference in regularity of the solutions of (3.4) and (3.5) is the high degeneracy of equation (3.4) compared to (3.5). The limit solution $U(x)$ in Figure 15 is an infimum of viscosity supersolutions of (3.5) while the limit solution $V(x)$ in Figure 15 is a supremum of viscosity subsolutions of (3.5).

Let us note that a comparison principle holds for all limit solutions which have the same Perron properties as sub- or supersolutions. The comparison theorem fails for limit solutions with different Perron properties (see Figure 16 below). In fact, in Figure 16

$$U(0) = A_1 > A_2 = V(0), \; U(b) = B_1 > B_2 = V(b),$$

but it is not true that $U(x) \geq V(x)$ in $[0, b]$.

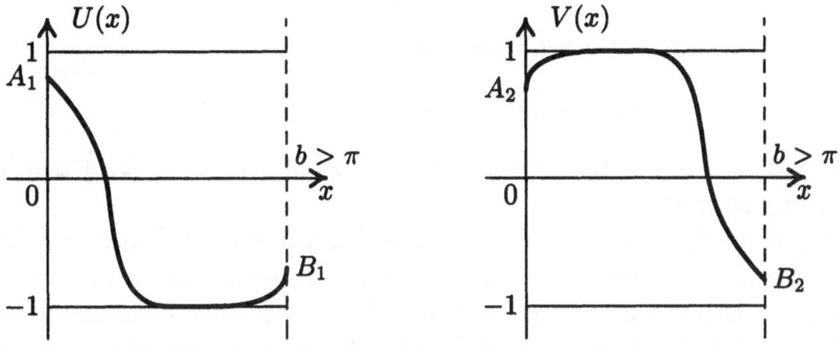

Figure 16. Comparison principle violated for limit solutions of (3.5)

A similar phenomenon of periodic structure and infinitely many solutions of the problem was noticed by St.Müller [23] in some variational problems, i.e.,

$$\begin{aligned} I^{\varepsilon}(u) &= \int_0^1 [\varepsilon^2 u_{xx}^2 + (u_x^2 - 1)^2 + u^2] dx \\ u(0) &= u(1) = 0 \end{aligned}$$

which appear in solid-solid phase transitions in crystals.

The results mentioned in examples 4 and 5 are true also for more general degenerate elliptic equations

$$F(u', u'') + u^3 - u = 0 \text{ in } (0, b)$$

for which $F(0, X) = 0$ for every X.

4. PROOF OF THEOREMS 1 AND 2

Proof of Theorem 1: If Theorem 1 is not true then $u(x) - v(x)$ has a positive maximum in $[a, b]$ at some point x_0 and $u(x) - v(x) \leq u(x_0) - v(x_0) = M > 0$. Let us first consider the case that $u - v \not\equiv$ const, i.e, there exists $z \in (a, b)$ such that $u(z) - v(z) < M$. Without loss of generality we will consider the interval $[z, b]$ and assume $x_0 \in (z, b]$. If $u(b) - v(b) = M$ then either $u(b) > B$ or $v(b) < B$ or both. For simplicity suppose that $u(b) > B$. Let $\varepsilon \in (0, \sqrt{2M})$ and consider $h(x, y) = u(x) - v(y) - |\alpha(x - y) - \varepsilon|^2/2$ in $[z, b] \times [z, b]$. (If $v(b) < B$, similar considerations apply to $g(x, y) = u(x) - v(y) - |\alpha(x - y) + \varepsilon|^2/2$.) Since $h(x_0, x_0) = M - \varepsilon^2/2 > 0$, $h(x, y)$ attains its positive maximum M_α at some points x_α, $y_\alpha \in [z, b]$. It is clear that $x_\alpha - y_\alpha \to 0$ as $\alpha \to \infty$. Moreover, the inequality $h(x_\alpha, y_\alpha) \geq h(x_0, x_0 - \varepsilon/\alpha)$ holds for all large α, i.e.,

$$|\alpha(x_\alpha - y_\alpha) - \varepsilon|^2 \leq 2u(x_\alpha) - 2v(y_\alpha) - 2u(x_0) + 2v(x_0 - \varepsilon/_\alpha).$$

From the continuity of v we conclude that

$$\alpha(x_\alpha - y_\alpha) - \varepsilon \to 0 \qquad \text{as } \alpha \to \infty, \tag{4.1}$$

so that $\lim_{\alpha\to\infty} M_\alpha = u(x_1) - v(x_1) = M$ whenever $x_1 \in [z, b]$ is a limit point of x_α as $\alpha \to \infty$.

a) If $x_1 \in (z, b)$, then x_α, y_α are interior points for large α. From maximizing h it follows [9, Thm. 3.2] that the set of test fuctions for u, v at x_α, y_α is not empty. More precisely, there exist constants X, Y, $X \leq Y$ and C^2–functions ϕ and ψ with the properties: $\phi(x_\alpha) = u(x_\alpha)$, $\phi(x) \geq u(x)$, $\psi(y_\alpha) = v(y_\alpha)$, $\psi(y) \leq v(y)$, $\phi'(x_\alpha) = \psi'(y_\alpha) = \alpha^2(x_\alpha - y_\alpha) - \varepsilon\alpha$, $\phi''(x_\alpha) = X$ and $\psi''(y_\alpha) = Y$. According to Definition 1 and (2.2), (2.11) we get the chain of inequalities

$$\begin{aligned} F(u(x_\alpha), \alpha^2(x_\alpha - y_\alpha) - \varepsilon\alpha, X) \leq 0 &\leq F(v(y_\alpha), \alpha^2(x_\alpha - y_\alpha) - \varepsilon\alpha, Y) \\ &\leq F(v(y_\alpha), \alpha^2(x_\alpha - y_\alpha) - \varepsilon\alpha, X) \\ &< F(u(x_\alpha), \alpha^2(x_\alpha - y_\alpha) - \varepsilon\alpha, X), \end{aligned}$$

an obvious contradiction.

b) Since by construction z cannot be an accumulation point of x_α, y_α, we have to check only the remaining case that $x_1 = b$. As long as x_α and y_α are interior points of (z, b), the same contradiction as in a) can be reached. Hence either $x_\alpha = b$ or $y_\alpha = b$ or both as α is large enough. Suppose that $y_\alpha = b$. Noting that $\alpha(x_\alpha - y_\alpha) - \varepsilon = \alpha(x_\alpha - b) - \varepsilon \to 0 \leq -\varepsilon < 0$ as $\alpha \to \infty$ we get a contradiction to (4.1). Finally, if $x_\alpha = b$ and $y_\alpha < b$, we can use Definition 2 and the fact that $u(b) > B$ to derive from (2.7): $F(u(b), \tilde{\phi}'(b), \tilde{\phi}''(b)) \leq 0$ for every test function $\tilde{\phi} \in C^2$ satisfying $\tilde{\phi}(x) \geq u(x)$ and $\tilde{\phi}(b) = u(b)$. Choosing $\tilde{\phi} = \phi$ from a) we can reach again a contradiction.

To complete the proof of Theorem 1, we still have to bring the assumption $u(x) - v(x) \equiv M > 0$ to a contradiction. In this case we will show that in a given small subintervall $[t, s] \subset (a, b)$, in which

$\text{osc}\,v(x) < M$, the function v is either convex (or concave) on $[t, s]$. For every $p, q \in [t, s]$ with $p < q$ we will show that $v(x), x \in (p, q)$, lies either entirely below (or above) the line segment $l(x)$ connecting $(p, v(p))$ with $(q, v(q))$. Otherwise there exist linear functions $l_1(x)$ and $l_2(x)$ parallel to $l(x)$ which are tangent to $v(x)$ from below at a point x_1 and to $u(x) \equiv v(x)+M$ at a point x_2 from above, see Figure 17. Without loss of generality (otherwise vary p, q) we may assume $x_1, x_2 \in (p, q)$.

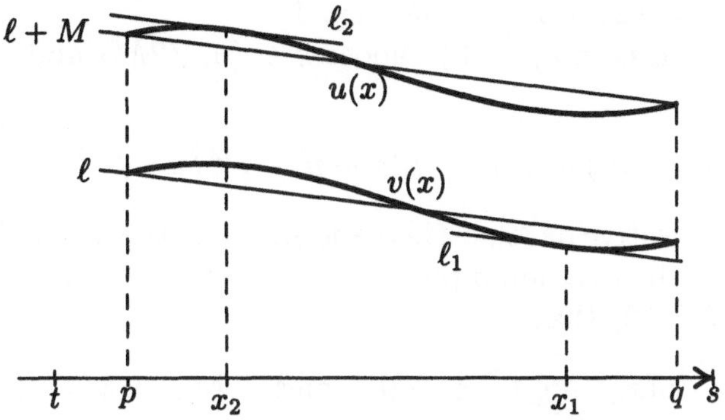

Figure 17. An impossible situation

Using $l_1(x)$ and $l_2(x)$ as test functions in Definition 1 and the strict monotonicity (2.11) of F with respect to u, as well as the fact that by choice of $[t, s]$ we have $\inf u(x) > \sup v(x)$ in $[t, s]$, we arrive at

$$\begin{aligned} F(v(x_1), l_1'(x_1), 0) \;&\geq 0 \geq F(u(x_2), l_2'(x_2), 0) \\ &= F(u(x_2), l_1'(x_1), 0) > F(v(x_1), l_1'(x_1), 0), \end{aligned}$$

another contradiction. Therefore v is either convex or concave in every subinterval $[t, s]$ such that $\text{osc}\{v(x) \mid x \in [s, t]\} < M$. But as is well known, see e.g. [9, Thm.A2], continuous convex (or concave) functions are almost everywhere twice differentiable.

Let $x_0 \in (t, s)$ be a point where u (and thus $v = u - M$) is twice differentiable. Then there exist constants r, X such that

$$u(x) = u(x_0) + r(x - x_0) + X(x - x_0)^2/2 + o(|x - x_0|^2) \qquad \text{as } x \to x_0.$$

Hence for every $\varepsilon > 0$ we have test functions in the sense of Definition 1

$$\begin{aligned} \psi(x) \;&:= v(x_0) + r(x - x_0) + (X - \varepsilon)(x - x_0)^2/2 \leq v(x), \\ \phi(x) \;&:= u(x_0) + r(x - x_0) + (X + \varepsilon)(x - x_0)^2/2 \geq u(x). \end{aligned}$$

Definition 1 implies now $F(u(x_0), r, X + \varepsilon) \leq 0 \leq F(v(x_0), r, X - \varepsilon)$ for every $\varepsilon > 0$, and by continuity of F also for $\varepsilon = 0$. Hence, using the strict monotonicity (2.11) we reach another contradiction, namely $F(u(x_0), r, X) \leq 0 \leq F(v(x_0), r, X) < F(u(x_0), r, X)$. This completes the proof of Theorem 1.

Proof of Theorem 2: Suppose that $\max\{u(x) - v(x) \mid x \in [a,b]\} = u(z) - v(z) = M > 0$. If $u - v \equiv M$ on $[a,b]$ then statement ii) of Theorem 2 holds and there is nothing to prove. Otherwise there exists a point $a_0 \in (a,b)$ such that $u(a_0) - v(a_0) < M$. Without loss of generality $z \in (a_0, b]$. The following two subcases can occur: a) $u(b) - v(b) < M$ or b) $u(b) - v(b) = M$. In case b) either $u(b) > B$ or $v(b) < B$ or both. We will cover only the case $u(b) > B$, since the other one is symmetric, see proof of Theorem 1.
Independent from case a) or b) choose $\varepsilon \in (0, \sqrt{2M})$ and consider the function

$$h(x,y) = u(x) - v(y) - |\alpha(x-y) - \varepsilon|^2/2 - \beta^{-2}\, e^{\beta(a-x)}$$

in $[a_0, b]$ for all positive parameters α and β. As in the proof of Theorem 1 we can show for the maximum points (x_α, y_α) of h on $[a_0,b] \times [a_0,b]$ (with $h(x_\alpha, y_\alpha) = M_\alpha > 0$) that

$$x_\alpha - y_\alpha \to 0, \quad \alpha(x_\alpha - y_\alpha) - \varepsilon \to 0 \quad \text{and} \quad M_\alpha \to M \quad \text{as } \alpha, \beta \to \infty. \tag{4.2}$$

From [9, Thm. 3.2] there exist test functions $\phi, \psi \in C^2$ such that $\phi(x) \geq u(x)$, $\phi(x_\alpha) = u(x_\alpha)$, $\psi(x) \leq v(x)$, $\psi(x_\alpha) = v(x_\alpha)$, $\phi'(x_\alpha) = \alpha^2(x_\alpha - y_\alpha) - \varepsilon\alpha - \beta^{-1}\, e^{\beta(a-x_\alpha)}$, $\psi'(y_\alpha) = \alpha^2(x_\alpha - y_\alpha) - \varepsilon\alpha$, $\phi''(x_\alpha) = X$ and $\psi''(y_\alpha) = Y$, where X, Y satisfy the inequalities [9, (3.8) with $\varepsilon = 1/\alpha$]

$$\begin{aligned} &-\left[2\alpha^2 - \frac{1}{2}e^{\beta(a-x_\alpha)} + \left(\alpha^4 + \frac{1}{4}e^{2\beta(a-x_\alpha)}\right)^{1/2}\right]\begin{pmatrix} 1 & 0 \\ 0 & 1 \end{pmatrix} \\ &\leq \begin{pmatrix} X & 0 \\ 0 & Y \end{pmatrix} \\ &\leq \begin{pmatrix} 3\alpha^2 - 3e^{\beta(a-x_\alpha)} + \alpha^{-2}e^{2\beta(a-x_\alpha)} & -3\alpha^2 + e^{\beta(a-x_\alpha)} \\ -3\alpha^2 + e^{\beta(a-x_\alpha)} & 3\alpha^2 \end{pmatrix} \end{aligned} \tag{4.3}$$

From the right inequality of (4.3) we have

$$X - Y \leq e^{\beta(a-x_\alpha)}[-1 + \alpha^{-2}e^{\beta(a-x_\alpha)}], \tag{4.4}$$

because we can multiply the matrix inequality (4.3) with the vector (1,1) from the right and from the left.
In case a) all accumulation points of x_α and y_α are in the open interval (a_0, b), while in case b) they can also accumulate in b. However, even in this case as in the proof of Theorem 1 it follows from (4.2) that either $x_\alpha, y_\alpha \in (a_0, b)$ or $x_\alpha = b$ for large α and $y_\alpha < b$. This means that in any case we can apply (2.7), (2.8) and, using (2.2), (2.13) and (4.4), arrive at the following absurd chain of inequalities

$$
\begin{aligned}
0 \le\ & F(v(y_\alpha), \alpha^2(x_\alpha - y_\alpha) - \varepsilon\alpha, Y) \\
& -F(u(x_\alpha), \alpha^2(x_\alpha - y_\alpha) - \varepsilon\alpha - \beta^{-1}e^{\beta(a-x_\alpha)}, X) \\
\le\ & F(u(x_\alpha), \alpha^2(x_\alpha - y_\alpha) - \varepsilon\alpha, X) \\
& -F(u(x_\alpha), \alpha^2(x_\alpha - y_\alpha) - \varepsilon\alpha - \beta^{-1}e^{\beta(a-x_\alpha)}, X) - \gamma_K(Y - X) \\
\le & \frac{e^{\beta(a-x_\alpha)}}{\beta} \sup \left| \int_0^1 F_p(u(x_\alpha), \alpha^2(x_\alpha - y_\alpha) - \varepsilon\alpha + \frac{1-t}{\beta} e^{\beta(a-x_\alpha)}, X)\, dt \right| \\
& +\gamma_K\ e^{\beta(a-x_\alpha)}[-1 + \alpha^{-2}e^{\beta(a-x_\alpha)}] \\
\le\ & e^{\beta(a-x_\alpha)} \left[C/\beta + \gamma_K e^{\beta(a-x_\alpha)} - \gamma_K \right] < 0,
\end{aligned}
$$

where

$$C = \sup \left| \int_0^1 F_p(u(x_\alpha), \alpha^2(x_\alpha - y_\alpha) - \varepsilon\alpha + \frac{1-t}{\beta} e^{\beta(a-x_\alpha)}, X)\, dt \right|,$$

and γ_K is the ellipticity constant from (2.14) corresponding to

$$K = \alpha^2 |x_\alpha - y_\alpha| + \varepsilon\alpha + \beta^{-1}e^{\beta(a-x_\alpha)} + |X| + |Y|.$$

In fact, to see the last strict inequality, note that for fixed and large α the numbers K, γ_K and C are independent of $\beta \ge \beta_0 >> 1$. Hence $\beta \to \infty$ provides the contradiction and completes the proof of Theorem 2.

Acknowledgments

This work was financially supported by the Deutsche Forschungsgemeinschaft (DFG).

References

[1] Alvarez, O., Lasry, J.M. and Lions, P.L. (1997) Convex viscosity solutions and state constraints. J. Math. Pures Appl. Ser. IX, 76:265–288.

[2] Angenent, S. and Fila, M. (1996). Interior gradient blow-up in a semilinear parabolic equation. Differential Integral Equation, 9:865–877.

[3] Barles, G., Diaz, G. and Diaz, J.I. (1992). Uniqueness and continuum of foliated solutions for a quasilinear elliptic equation with a non-lipschitz nonlinearity. Comm. Partial Differential Equation 17:1037–1050.

[4] M.Bertsch, R.dal Passo. Hyperbolic phenomena in a strongly degenerate parabolic equation. Arch. Rational Mech. Anal. **117** (1992), 349–387.

[5] Bethuel, F., Brezis, H. and Helein, F. (1994). *Ginzburg Landau vortices.* Birkhäuser, Boston.

[6] Capuzzo-Dolcetta, I. and Lions, P.L. (1990). Hamilton-Jacobi equations with state constraints. Trans. Amer. Math. Soc., 318:643–683.

[7] Crandall, M. and Houghdan Huan. (1992). On nonuniqueness of viscosity solutions. Differential Integral Equation, 5:1247–1265.

[8] Crandall, M. and Lions, P.L. (1983). Viscosity solutions of Hamilton-Jacobi equations. Trans. Amer. Math. Soc., 277:1–42.

[9] Crandall, M., Ishii, H. and Lions, P.L. (1992). User's Guide to viscosity solutions of second order partial differential equations. Bull. Amer. Math. Soc., 27:1–67.

[10] Fichera, G. (1956). Sulle equazioni differentiali lineari ellitico-paraboliche del secondo ordine. Atti Accad. Naz. Lincei. Mem. Cl. Sci. Fis. Mat. Nat. Ser.1, 8:1–30.

[11] Freidlin, M. (1985). *Functional investigation and Partial differential equations.* Princeton Univ. Press.

[12] Giga, Y. (1995). Interior derivative blow-up for quasilinear parabolic equations. Discrete Contin. Dynam. Systems, 1:449–461.

[13] Gilbarg, D. and Trudinger, N. *Elliptic partial differential equations of second order.* 2nd ed., Springer-Verlag, New York.

[14] Giusti, E. (1978). On the equations of surfaces of prescribed mean curvature. Invent. Math., 46:111–137.

[15] Ishii, H. (1989). On uniqueness and existence of viscosity solutions of fully nonlinear second order elliptic PDE's. Comm. Pure Appl. Math., 42:14–45.

[16] Ishii, H. and Ramaswamy, M. (1995). Uniqueness results for a class of Hamilton-Jacobi equations with singular coefficients. Comm. Partial Differential Equations, 20:2187–2213.

[17] Jensen, R. (1988). The maximum principle for viscosity solutions of fully nonlinear second order partial differential equations. Arch. Rational Mech. Anal., 101:1–27.

[18] Katsoulakis, J.M. (1994). Viscosity solutions of second order fully nonlinear elliptic equations with state constraints. Indiana Univ. Math. J., 43:493–519.

[19] Kawohl, B. and Kutev, N. (1995). Global behaviour of solutions to a parabolic mean curvature equation, Differential Integral Equations, 8:1923–1946.

[20] Kawohl, B. and Kutev, N. (1998). Strong maximum principle for semicontinuous viscosity solutions of nonlinear partial differential equations. Archiv der Mathematik, 70:470–479.

[21] Lions, P.L. (1983). Optimal control of diffusion processes and Hamilton-Jacobi-Bellman equations. Part II. Viscosity solutions and uniqueness. Comm. Partial Differential Equation, 8:1229–1276.

[22] Loretti, P. (1987). Some properties of constrained viscosity solutions of HJB equations. SIAM J. Control Optim., 25:1244–1252.

[23] Müller, St. (1990). Minimizing sequences for nonconvex functionals, phase transitions and singular perturbations. Springer Lecture Notes in Physics, 359:31–40.

[24] Oleinik, O.A. and Radkevič, E.V. (1973). *Second order equations with nonnegative characteristic form.* Amer. Math. Soc., Providence.

[25] Soner, H.M. (1986). Optimal control with state-space constraint I. SIAM J. Control. Optim., 24:552–562.

[26] Trudinger, N. (1983). Fully nonlinear, uniformly elliptic equations under natural structure conditions. Trans. Amer. Math. Soc., 278:751–769.

REMARKS ON COMPACTNESS IN THE FORMATION OF FINE STRUCTURES

Petr Klouček

Abstract: We investigate the interplay and intertwining of the amplitude and frequency of highly oscillatory sequences of bounded weakly differentiable maps forming laminated microstructures and fine structures.

Keywords: Microstructures, fine structures, weak convergence, strong convergence, finite elements.

1. INTRODUCTION

Processes involving phase-separation kinetics and dynamics of interfaces are often very complicated. For active materials, such as magnetostrictive, ferroelectric or shape memory alloys, the phase-separation kinetics and dynamics of coherent or semi-coherent phase boundaries depend on a domain microstructure. For these materials, it is crucial to understand how structures with a variety of tiny length scales arise with changes of temperature and stress in order to model and to understand their responsivness to external macroscopic stimuli.

Phase-separation kinetics involves a continuum description down to a scale slightly greater than atomic dimensions. Thus, a study of formation of microstructures has to deal with length scales which vary from nanometers to centimeters. Traditionally, one is not attempting to study details at the atomic scale but is trying to find a correlation between microscopic structure and large-scale material properties. This approach is less desirable for understanding of formation of microstructures in active materials because many such patterns are only meta-stable. Hence, such patterns are confined to compact subsets of some Sobolev spaces. Meta-stable microstructures with a wide variety of different scales determine fundamental properties of active materials through exchange of stability among competing microstructures. Laboratory experiments and recent theoretical work

Applied Nonlinear Analysis, edited by Sequeira *et al.*
Kluwer Academic / Plenum Publishers, New York, 1999.

[2], [9], [10], [13], indicate that spatial microstructural complexity depends on the rate of pattern formation. This issue has received a little attention so far compared to the analysis and computation of macroscopic equilibrium properties of constrained crystalline materials studied in [6], [15], [13], [19], [17] among others.

Alloys such as Nickel-Titanium or Indium-Thallium are typical examples of active materials. These alloys exhibit multiscale domain patterns, c.f. Figure 1, described in part by binomial microstructures [3], [18]. Such patterns can be modeled by gradients of weakly differentiable maps $u : \mathbb{R}^N \to \mathbb{R}^N$ that can come close to the solution of the following *differential inclusion*:

$$\begin{aligned}
&\nabla u \in \{F_1, F_2\}, \qquad \text{a.e. in } \Omega \subset \mathbb{R}^N, \quad N = 2,3,\\
&F_i \in M^{N\times N}, \quad i = 1,2,\\
&u(x) = (\lambda_1 F_1 + \lambda_2 F_2)\, x, \quad x \in \partial\Omega, \\
&\lambda_i > 0, \quad i = 1,2, \quad \lambda_1 + \lambda_2 = 1,\\
&F_2 = F_1 + a \otimes b, \qquad a, b \in \mathbb{R}^N, \quad (a \otimes b)_{ij} = a_i b_j.
\end{aligned} \tag{1.1}$$

The matrices F_i are assumed to be positive definite and linearly independent. Typical examples of these matrices are associated with crystallographic theories. For the face-centered cubic to face-centered tetragonal phase change, these matrices are given by

$$F_i = \nu_1 \mathrm{I} + (\nu_2 - \nu_1) e_i \otimes e_i, \qquad i = 1, \ldots, N, \tag{1.2}$$

where $\nu_i > 0$, $\nu_1 \neq \nu_2$, e_i are canonical basis vectors in $\mathbb{R}^N$. Another example is associated with the orthorhombic to monoclinic transformations. In this case

$$\begin{aligned}
&F_i = \left(\mathrm{I} + \eta(-1)^i e_2 \otimes e_1\right) \sum_{i=1}^{N} d_i e_i \otimes e_i, \quad d_i > 0, \quad \eta > 0,\\
&i = 1, \ldots, N.
\end{aligned} \tag{1.3}$$

It is possible to show [20], [13] that suitable use of self-similar scaling (homogenization) leads to approximation techniques for such differential inclusions. We note that there does not exist any functional representation for the solution of (1.1). This is because it is necessary to create oscillations in the gradients with unlimited frequency to meet the boundary condition.

Minimization of a stored energy with local minima at $SO(N)F_i$, where $SO(N)$ is the set of proper rotations in $\mathbb{R}^N$, is the direct counterpart of the differential inclusions such as (1.1). The lack of a functional representation manifests itself in the minimization framework as the lack of quasi-convexity. Thus the minimizing sequences do not carry, in the limit, any point-wise information about the pattern structure. It is remarkable that dynamical systems associated with such non-attainment problems seem to favor creation of patterns with finite scales (patterns with limited structural frequency) [9], [10], [14]. This means that the microstructural patterns obtained by dynamical processes represent local minima rather than the global minimum of the associated non-quasiconvex energy [2]. We introduce in this paper *comparison principles* that will help us to study the compactness of multiscale pattern formation. We anticipate that dynamics, as well as nonlinear optimization and approximation biases, could serve as "selection mechanisms" missing in the continuum equilibrium models of crystalline and polycrystalline materials.

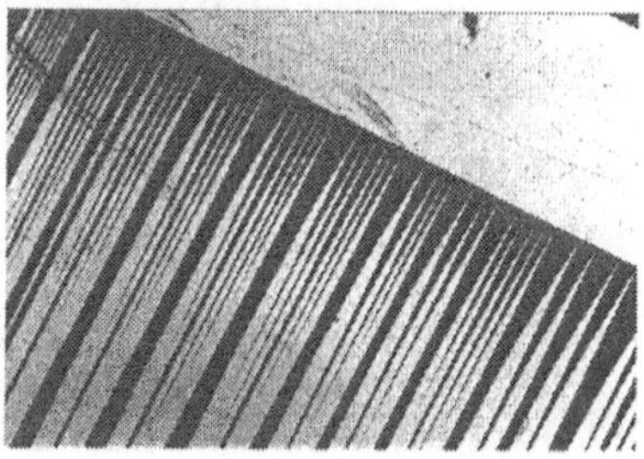

Figure 1 An image of an Austenitic-Finely Twinned Martensitic interface, [7]. The homogeneous (undeformed) part on the right is Austenite, and the laminated microstructure on the left is the Finely Twinned Martensite. The picture is 370 microns wide. One of the important considerations involved in obtaining a good shape memory effect is the high mobility of the Austenite-Finely Twinned Martensite Interface. Applications of the shape memory effect also include vibrations and energy absorbing dampers, bendable surgical tools or fatigue resistant wires, martensitic thin films, etc.

2. COMPARISON PRINCIPLES

Let $\Omega \subset \mathbb{R}^N$, $N = 2, 3$, be an open bounded domain with a Lipschitz (or "smooth" if necessary) boundary. We will understand ∇ to be a column vector of partial derivatives taken with respect to the untransformed domain Ω. We denote by $M^{N\times N}$ the space of $N \times N$ real matrices. Matrix multiplication in this space is understood in the sense $AB = A : B \overset{\text{def}}{=} \mathrm{Tr}(A^T B)$, where the matrix $A^T B$ is obtained by standard matrix multiplication. Consequently, the matrix norm is given by $\|A\| = \sqrt{A:A}$, which is the natural Euclidian norm on the space $\mathbb{R}^{N\times N}$. We will use $|.|$ to denote a norm on the space of $N-$dimensional vectors.

Lemma 2.1. *Let $\{u_m\}_{m=1}^{\infty} \subset W^{2,p}(\Omega,\mathbb{R}^1)$ and $2 \le p < \infty$. Then there exists a continuous function $\delta = \delta(\epsilon) : \mathbb{R}^+ \mapsto \mathbb{R}^+$, which converges to zero as $\epsilon \to 0$, and there exists a positive constant C, independent of m and n, such that for any $\epsilon > 0$ sufficiently small, we have*

$$\begin{aligned} &C\,\|\nabla u_n - \nabla u_m\|_{L^p_{Loc}(\Omega,\mathbb{R}^N)} \\ &\le \frac{1}{\epsilon^{N+1}}\|u_n - u_m\|_{L^p(\Omega,\mathbb{R}^1)} + \epsilon\left\|\nabla^2 u_m\right\|_{L^p(\Omega,\mathbb{R}^{N\times N})} + \delta(\epsilon), \end{aligned} \tag{2.1}$$

where $\nabla^2 = \nabla\nabla^T = \left(\partial^2_{x_i x_j}\right)_{i,j=1}^{N}$.

Proof. We shall consider any function defined on Ω to be extended outside of Ω by zero. Let R_ϵ be the Friedrich's mollifier. We write

$$\begin{aligned} &\nabla u_n(x) - \nabla u_m(x) \\ &= \nabla u_n(x) - (R_\epsilon \nabla u_n)(x) + (R_\epsilon(\nabla u_n - \nabla u_m))(x) \\ &\quad + (R_\epsilon \nabla u_m)(x) - \nabla u_m(x), \quad x \in \Omega. \end{aligned} \tag{2.2}$$

The $p-$mean continuity of L^p-functions yields a positive continuous function $\delta = \delta(\epsilon)$ such that

$$\|\nabla u_n - R_\epsilon \nabla u_n\|_{L^p(\Omega,\mathbb{R}^N)} \le \delta(\epsilon) \to 0 \quad \text{as } \epsilon \to 0_+. \tag{2.3}$$

Integration by parts and the Hölder inequality yield a positive constant $C > 0$, independent of m and n, such that for $p \in [2,\infty)$ and $q \in [2,\infty)$ conjugate we have

$$\begin{aligned} &\|R_\epsilon(\nabla u_n - \nabla u_m)\|^p_{L^p(\Omega,\mathbb{R}^N)} \\ &= \int_\Omega \epsilon^{-pN-p}\left|\int_\Omega \nabla\varphi\left(\frac{x-y}{\epsilon}\right)(u_n(y) - u_m(y))\,dy\right|^p dx \\ &\le \epsilon^{-pN-p}\,\text{meas}(\Omega)^{p/q+1}\,\|\nabla\varphi\|^p_{L^\infty(\Omega,\mathbb{R}^N)}\int_\Omega (u_n(x) - u_m(x))^p\,dx. \end{aligned} \tag{2.4}$$

Thus

$$\begin{aligned} &\|R_\epsilon(\nabla u_n - \nabla u_m\|_{L^p(\Omega,\mathbb{R}^N)} \\ &\le \epsilon^{-N-1}\,\text{meas}(\Omega)\,\|\nabla\varphi\|_{L^\infty(\Omega,\mathbb{R}^N)}\,\|u_n - u_m\|_{L^p(\Omega,\mathbb{R}^1)}. \end{aligned} \tag{2.5}$$

The Mean Value Theorem and definition of the Friedrich's mollifier yield, for any $x \in \Omega_1$, $\overline{\Omega}_1 \subset \Omega$, and $\epsilon > 0$ sufficiently small so that $x - \epsilon y \in \Omega$, a differentiable map $\eta : \mathbb{R}^2 \to \mathbb{R}^2$ such that

$$(R_\epsilon \nabla u_m)(x) - \nabla u_m(x) = \int_{B(0,1)} \varphi(y)\left(\nabla u_m(x - \epsilon y) - \nabla u_m(x)\right) dy$$

$$= \epsilon \int_{B(0,1)} \varphi(y) \nabla^2 u_m(\eta(x,y)) y \, dy, \quad \eta(x,y) \in (x - \epsilon y, x). \tag{2.6}$$

The integration by parts with respect to the y-variable yields

$$\begin{aligned} &\int_{\Omega_1} |(R_\epsilon \nabla u_m)(x) - \nabla u_m(x)|^p \, dx \\ &= \epsilon^p \int_{\Omega_1} \left| \int_{B(0,1)} \varphi(y) \nabla^2 u_m(\eta(x,y)) y \, dy \right|^p dx \\ &= \epsilon^p \int_{\Omega_1} \left| \int_{B(0,1)} \nabla(\varphi(y) y) \nabla u_m(\eta(x,y)) \, dy \right|^p dx. \end{aligned} \tag{2.7}$$

Using the Poincaré inequality for the unit ball, and the continuous imbedding of $W^{2,p}(\Omega)$ into the space $W^{1,p}(\Omega)$, we obtain a positive constant C, independent of m and n, such that

$$\begin{aligned} &\int_{\Omega_1} \left| \int_{B(0,1)} \nabla(\varphi(y) y) \nabla u_m(\eta(x,y)) \, dy \right|^p dx \\ &\le \int_{\Omega_1} |\nabla(\varphi(x) x) \nabla u_m(\eta(x,x))|^p \, dx \\ &+ \int_{\Omega_1} \left| \nabla(\varphi(x) x) \nabla u_m(\eta(x,x)) - \int_{B(0,1)} \nabla(\varphi(y) y) \nabla u_m(\eta(x,y)) \, dy \right|^p dx \\ &\le C \int_{\Omega_1} \left| \nabla^2 u_m(\eta(x,x)) \right|^p dx \le C \int_{\Omega_1} \left| \nabla^2 u_m(x,x) \right|^p dx. \end{aligned}$$

The proof follows from (2.3), (2.5) and the above estimates.

□

Theorem 2.2. *Let $p \in [2, \infty)$, and let $\{u_m\}_{m=1}^\infty \subset W_0^{2,\infty}(\Omega, \mathbb{R}^1)$ be a sequence such that*

$$u_m \rightharpoonup u, \quad \textit{weakly in } W^{1,p}(\Omega, \mathbb{R}^1). \tag{2.8}$$

Let us assume moreover that

$$\lim_{m \to 0} \|u - u_m\|_{L^p(\Omega, \mathbb{R}^1)}^{1-\gamma} \|\Delta u_m\|_{L^p(\Omega, \mathbb{R}^1)} = 0, \quad \textit{for some } \gamma \in (0,1). \tag{2.9}$$

Then

$$u_m \to u, \quad \textit{strongly in } W^{1,p}_{Loc}(\Omega, \mathbb{R}^1). \tag{2.10}$$

Proof. We take

$$\epsilon \stackrel{\text{def}}{=} \epsilon(m) = \|u - u_m\|_{L^p(\Omega,\mathbb{R}^1)}^{\frac{1-\gamma}{N+1}}, \quad \gamma \in (0,1) \tag{2.11}$$

in Lemma 2.1 and obtain a function $\delta(m) = \delta(\epsilon(m))$ such that $\delta(m) \to 0$ as $m \to \infty$ and

$$\|\nabla u - \nabla u_m\|_{L^p_{\text{Loc}}(\Omega,\mathbb{R}^N)} \leq \tag{2.12}$$

$$\leq \|u - u_m\|^{\gamma}_{L^p(\Omega,\mathbb{R}^1)} + \|u - u_m\|_{L^p(\Omega,\mathbb{R}^1)}^{1-\frac{N+\gamma}{N+1}} \left\|\nabla^2 u_m\right\|_{L^p(\Omega,\mathbb{R}^2)} + \delta(m).$$

We have for any smooth function v which vanishes on $\partial\Omega$ [8]

$$\int\limits_{\Omega} \left\|\nabla^2 v(x)\right\|^2 dx + \int_{\partial\Omega} \frac{1}{\kappa(s)} \left|\frac{\partial v}{\partial n}\right|^2 dS = \int\limits_{\Omega} |\Delta\, v(x)|^2\, dx, \tag{2.13}$$

where κ is the radius of curvature of $\partial\Omega$. The radius is positive if Ω is convex. Hence the locally strong convergence of the gradients follows from (2.12), (2.13), and assumption (2.9). □

Corollary 2.3 (The Uncertainty Principle). *Let $p \in [2,\infty)$, and let $\{u_m\}_{m=1}^{\infty}$ be a subset of $W^{2,p}(\Omega,\mathbb{R}^1)$. Let us assume that*

$$u_m \rightharpoonup u, \quad \textit{weakly in } W^{1,p}(\Omega, \mathbb{R}^N). \tag{2.14}$$

Then

$$u_m \not\to u, \quad \textit{strongly in } W^{1,p}_{Loc}(\Omega, \mathbb{R}^N) \tag{2.15}$$

if and only if

$$\lim_{m\to 0} \|u - u_m\|^{1-\gamma}_{L^p(\Omega,\mathbb{R}^1)} \|\Delta\, u_m\|_{L^p(\Omega,\mathbb{R}^1)} > 0, \quad \textit{for any } \gamma \in (0,1). \tag{2.16}$$

Proof. (i) Let us assume that (2.15) is true. Then Theorem 2.2 and (2.15) yield

$$0 < \|\nabla u - \nabla u_m\|_{L^p_{\text{Loc}}(\Omega,\mathbb{R}^N)} \leq \tag{2.17}$$

$$\leq \|u - u_m\|^{\gamma}_{L^p(\Omega,\mathbb{R}^1)} + \|u - u_m\|_{L^p(\Omega,\mathbb{R}^1)}^{1-\frac{N+\gamma}{N+1}} \left\|\nabla^2 u_m\right\|_{L^p(\Omega,\mathbb{R}^{N\times N})} + \delta(m),$$

for any $\gamma \in (0,1)$.

The continuous imbedding of $W^{1,p}(\Omega)$ into $L^p(\Omega)$ and (2.17), (2.13) give (2.16).

(ii) Let us assume that (2.16) is true. In order to deduce (2.15) we will prove that if (2.15) is false then (2.16) is false as well. The strong convergence of the sequence $\{u_m\}_{m=1}^{\infty}$ in $W^{1,p}(\Omega, \mathbb{R}^1)$ yields weak convergence of this sequence in $W^{2,p}(\Omega, \mathbb{R}^1)$. Reflexivity of the L^p-spaces for $p \in [2, \infty)$ gives in turn a uniform upper bound

$$\|\Delta u_m\|_{L^p(\Omega,\mathbb{R}^1)} \leq C. \tag{2.18}$$

Hence the proof follows. □

2.1. Piecewise affine functions

Theorem 2.4. *Let $p \in [2, \infty)$, and let $\{u_m\}_{m=1}^{\infty}$ be a sequence of continuous, piecewise affine functions such that*

$$u_m \rightharpoonup u, \qquad \textit{weakly in } W^{1,2}(\Omega, \mathbb{R}^1). \tag{2.19}$$

Then

$$u_m \to u, \qquad \textit{strongly in } W^{1,p}_{Loc}(\Omega, \mathbb{R}^1), \tag{2.20}$$

if and only if

$$\lim_{m\to 0} \|u - u_m\|_{L^\infty(\Omega,\mathbb{R}^1)} \|\Delta u_m\|_{W^{-1,2}(\Omega,\mathbb{R}^1)} = 0. \tag{2.21}$$

Proof. (i) Let us assume that (2.20) is true. Strong convergence of the sequence $\{u_m\}_{m=1}^{\infty}$ in $W^{1,p}(\Omega, \mathbb{R}^1)$ yields weak convergence of this sequence in $W^{2,p}(\Omega, \mathbb{R}^1)$. Reflexivity of the L^p-spaces for $p \in [2, \infty)$ gives, in turn, a uniform upper bound for $\|\Delta u_m\|_{L^p(\Omega,\mathbb{R}^1)}$. Hence (2.20) yields

$$\lim_{m\to\infty} \|u - u_m\|_{L^p(\Omega,\mathbb{R}^1)} \|\Delta u_m\|_{W^{-1,2}(\Omega),\mathbb{R}^1)} = 0 \tag{2.22}$$

We obtain from (2.22) a constant C, independent of m, such that, for any $s \geq p \geq 2$, and $\alpha \in (0, 1)$, we have

$$\begin{aligned} &\lim_{m\to\infty} \|u - u_m\|_{L^p(\Omega,\mathbb{R}^1)} \|\Delta u_m\|_{W^{-1,2}(\Omega),\mathbb{R}^1)} \\ &\leq \lim_{m\to\infty} \|u - u_m\|^{\alpha}_{L^2(\Omega,\mathbb{R}^1)} \|u - u_m\|^{1-\alpha}_{L^s(\Omega,\mathbb{R}^n)} \|\Delta u_m\|_{W^{-1,2}(\Omega),\mathbb{R}^1)} \\ &\leq C \lim_{m\to\infty} \|u - u_m\|^{\alpha}_{L^2(\Omega,\mathbb{R}^1)} \|\Delta u_m\|_{W^{-1,2}(\Omega),\mathbb{R}^1)} = 0 \end{aligned} \tag{2.23}$$

and (2.21) follows.

(ii) Let us assume that (2.21) is true. We have again for any $\phi \in C_0^1(\overline{\Omega}, \mathbb{R}^1)$ the relation

$$\begin{aligned} -\int_\Omega \phi(x)^2\,|\nabla u(x) - \nabla u_m(x)|^2\,dx \\ = 2\int_\Omega \phi(x)\nabla\phi(x)(u(x) - u_m(x))\nabla(u(x) - u_m(x))\,dx \\ + \langle\phi(u - u_m), \phi\,\Delta(u - u_m)\rangle. \end{aligned} \tag{2.24}$$

Rellich's Theorem, (2.24), and weak convergence (2.19) yield

$$-\lim_{m\to\infty}\int_\Omega \phi(x)^2\,|\nabla u(x) - \nabla u_m(x)|^2\,dx = \lim_{m\to\infty}\langle\phi(u - u_m), \phi\,\Delta\,u_m\rangle. \tag{2.25}$$

Let us denote by $\mathcal{E}_m$ the set of all maximal simply-connected subdomains E of Ω such that $\Delta\,u_m(s) = 0$, $s \in E$. Using the definition of duality pairing, the fact that

$$\Omega = \bigcup_{E\in\mathcal{E}_m} E \cup \mathcal{N}, \qquad \mathrm{meas}_N(\mathcal{N}) = 0, \tag{2.26}$$

integration by parts, and the trace theorem, we obtain, for any $\varphi \in W_0^{1,2}(\Omega, \mathbb{R}^1)$,

$$\langle\varphi, \Delta\,u_m\rangle = -\frac{1}{2}\sum_{E\in\mathcal{E}_m}\int_{\partial E}[\nabla u_m](s)n(s)\varphi(s)\,dS, \tag{2.27}$$

where

$$[\nabla u_m]\,(s_0) \stackrel{\mathrm{def}}{=} \lim_{\substack{s\to s_0\\ s\in E}} \nabla u_m(s) - \lim_{\substack{s\to s_0\\ s\in\Omega\setminus E}} \nabla u_m(s), \quad s_0 \in \partial E. \tag{2.28}$$

The vector n(s) is the unit outward normal to ∂E at $s \in \partial E$. Since the domain Ω is bounded, there exists a finite, positive constant C, independent of m, such that

$$\begin{aligned} &\|\Delta\,u_m\|_{W^{-1,2}(\Omega),\mathbb{R}^1)} \\ &= \sup\left\{\left|\frac{1}{2}\sum_{E\in\mathcal{E}_m}\int_{\partial E}[\nabla u_m](s)n(s)\varphi(s)\,dS,\right|, \|\varphi\|_{W_0^{1,2}(\Omega,\mathbb{R}^1} \le 1\right\} \\ &\ge C\,\mathrm{moh}(\mathcal{E}_m). \end{aligned} \tag{2.29}$$

Thus we obtain from (2.25), (2.27), the assumption (2.21), and (2.29)

$$\begin{aligned}
&-\lim_{m\to\infty}\int_{\Omega}\phi(x)^2\,|\nabla u(x)-\nabla u_m(x)|^2\,dx \\
&\le C\lim_{m\to 0}\|u-u_m\|_{L^\infty(\Omega,\mathbb{R}^1)}\,\mathrm{moh}(\mathcal{E}_m) \\
&\le C\lim_{m\to 0}\|u-u_m\|_{L^\infty(\Omega,\mathbb{R}^1)}\,\|\Delta\, u_m\|_{W^{-1,2}(\Omega,\mathbb{R}^1)}=0,
\end{aligned}\tag{2.30}$$

which concludes the proof. □

3. MICROSTRUCTURES

Let us return to the problem of approximating the differential inclusion (1.1). The condition $\mathrm{rank}(F_1 - F_2) = 1$ is required by the Hadamard jump condition in order to allow a continuous map $u : \mathbb{R}^N \to \mathbb{R}^N$ such that $\nabla u \in \{F_1, F_2\}$. The possible discontinuities in the gradient occur along planar interfaces with the normal b. The boundary condition excludes the solvability of (1.1), however.

Let us define a projection $\Pi : M^{N\times N} \mapsto SO(N)F_1 \cup SO(N)F_2$ by

$$\|A-\Pi A\| = \min_{M\in SO(N)F_1\cup SO(N)F_2}\|A-M\|\,. \tag{3.1}$$

This projection may not be invertible. If $\Pi_{1,2} : M^{N\times N} \mapsto \{F_1, F_2\}$ we can find for any nonsingular matrix A a matrix $B(A)$ such that

$$\Pi(A) = B(A)\Pi_{1,2}(A). \tag{3.2}$$

We observe that any sequence $\{u_h\}_{h>0} \subset V_h$, $V_h \subset C(\overline{\Omega})$, of finite dimensional weakly differentiable functions converging to the solution of (1.1) has to satisfy the following two conditions

$$\begin{aligned}
&u_h = Fx \text{ on } \partial\Omega, \qquad \text{and} \\
&\lim_{h\to 0}\|\nabla u_h - \Pi\nabla u_h\|_{L^p(\Omega,\mathbb{R}^{N\times N})} = 0, \qquad \text{for some } p\ge 2.
\end{aligned}\tag{3.3}$$

Let $r > 0$, and let D be an open measurable subset of Ω. We define

$$D^i_{r,h} \overset{\text{def}}{=} \{x\in D\,|\,\Pi_{1,2}\nabla u_h(x) = F_i,\ \|\Pi_{1,2}\nabla u_h(x)-\nabla u_h(x)\| < r\},$$
$$i = 1,2.$$

The *approximate Young's measure* [1] is given by

$$\mu_{x,R,r,\nabla u_h}(F_i) \overset{\text{def}}{=} \mathrm{meas}(D^i_{r,h})/\,\mathrm{meas}(D), \tag{3.4}$$

where $D \overset{\text{def}}{=} B_R(x)$ is an open ball with radius $R > 0$ contained in Ω.

We have the following Theorem [12]: *Let the sequence* $\{u_h\}_{h>0} \subset V_h$, $\dim V_h = \mathcal{O}(1/h)$, $V_h \subset W^{1,p}(\Omega, \mathbb{R}^n)$, $p \geq 2$, $N = 2, 3$, *satisfies* (3.3). *Then*

$$\begin{aligned} &u_h \rightharpoonup Fx, \qquad \text{weakly in } W^{1,2}(\Omega, \mathbb{R}^n), \qquad \text{and} \\ &\lim_{R\to 0_+} \lim_{r\to 0_+} \lim_{h\to 0_+} \mu_{x,R,r,\nabla u_h} \overset{*}{\rightharpoonup} \lambda_1 \delta_{F_1} + \lambda_2 \delta_{F_2}. \end{aligned} \tag{3.5}$$

The convergence is weak-$*$ in a sense of measure. The results (3.5) yield the following Theorem.

Theorem 3.1. *Let* $\{u_h\}_{h>0}$ *be a sequence of continuous, piecewise affine functions satisfying* (3.3). *Then*

$$\lim_{h\to 0} \|u_h - Fx\|_{L^\infty(\Omega,\mathbb{R}^1)} \|\Delta u_h\|_{W^{-1,2}(\Omega,\mathbb{R}^1)} > 0. \tag{3.6}$$

Thus, in particular, there does not exists a solution to the differential inclusion (1.1).

Proof. (i) We first show that there exists $h_0 > 0$ such that for any $0 < h < h_0$ we have

$$\|\nabla u_h(x) - F\| \geq \lambda_1 \lambda_2 \|F_1 - F_2\|, \qquad \text{a.e. in } \Omega. \tag{3.7}$$

Let D is an arbitrary open measurable subset of Ω. We have

$$\begin{aligned} &\int_D \|\nabla u_h(x) - F\| \, dx \\ &\geq \text{meas}(D^1_{r,h}) \|F_1 - F\| + \text{meas}(D^2_{r,h}) \|F_2 - F\| \\ &- \int_{D^1_{r,h}} \|F_1 - \nabla u_h(x)\| \, dx - \int_{D^2_{r,h}} \|F_2 - \nabla u_h(x)\| \, dx. \end{aligned} \tag{3.8}$$

It follows from the definition of D^i_r, $i = 1, 2$, that

$$\begin{aligned} &\|F_i - \nabla u_h(x)\| = \|F_i - \Pi_{1,2}(\nabla u_h(x)) + (\Pi_{1,2} \nabla u_h)(x) - \nabla u_h(x)\| \\ &= \|(\Pi_{1,2} \nabla u_h)(x) - \nabla u_h(x)\| < r, \qquad \text{for all } x \in D^i_r. \end{aligned} \tag{3.9}$$

Hence, we have from (3.8) and (3.9)

$$\begin{aligned} &\int_D \|\nabla u_h(x) - F\| \, dx \geq \text{meas}(D^1_{r,h}) \|F_1 - F\| + \text{meas}(D^2_{r,h}) \|F_2 - F\| \\ &- r \left(\text{meas}(D^1_{r,h}) + \text{meas}(D^2_{r,h}) \right). \end{aligned} \tag{3.10}$$

We can write the first two terms on the right-hand side of (3.10) as

$$\left(\lambda_2 \operatorname{meas}(D^1_{r,h}) + \lambda_1 \operatorname{meas}(D^2_{r,h})\right) \|F_1 - F_2\| . \tag{3.11}$$

Thus taking $r \leq \lambda_1\lambda_2 \|F_1 - F_2\|$ we obtain

$$\begin{aligned} &\left(\lambda_2^2 \operatorname{meas}(D^1_{r,h}) + \lambda_1^2 \operatorname{meas}(D^2_{r,h})\right) \\ &\leq \int_D \|\nabla u_h(x) - F\| \, dx \, \|F_1 - F_2\| . \end{aligned} \tag{3.12}$$

Weak$-*$ convergence in measure (3.5) and (3.3) yield

$$\lim_{r\to 0_+} \lim_{h\to 0_+} \left(\lambda_2^2 \operatorname{meas}(D^1_{r,h}) + \lambda_1^2 \operatorname{meas}(D^2_{r,h})\right) = \lambda_1\lambda_2 \operatorname{meas}(D). \tag{3.13}$$

Hence (3.7) follows from (3.13) and (3.12).

(ii) The Hadamard jump condition, the fact that $\operatorname{rank}(F_1 - F_2) = 1$ and the convergence in measure (3.5) yield existence of a continuous, piecewise affine function $w_{r,h}$ such that

$$\begin{aligned} &\nabla w_{r,h} \in \{F_1, F_2\}, \\ &\nabla w_{r,h}(x) = \Pi_{1,2} \nabla u_h(x), \qquad x \in \Omega^1_{r,h} \cup \Omega^2_{r,h}, \quad \text{where} \\ &\Omega^i_{r,h} \stackrel{\text{def}}{=} \{x \in \Omega,\ \Pi_{1,2}\nabla u_h(x) = F_i,\ \|\nabla u_h(x) - \Pi_{1,2}\nabla u_h(x)\| < r\}, \\ &i = 1, 2, \quad \varepsilon > 0, \quad \text{and} \\ &w_{r,h} \rightharpoonup F, \qquad \text{weakly as } r, h \to 0. \end{aligned} \tag{3.14}$$

The strong convergence of directional derivatives of u_h parallel to b guaranteed by the inequality [19], [12]

$$\begin{aligned} &\|(\nabla u_h - F)m\|_{L^p(\Omega,\mathbb{R}^N)} \leq \|\nabla u_h - \Pi\nabla u_h\|^{1/2}_{L^2(\Omega,\mathbb{R}^{N\times N})}, \\ &\text{for any } p \in [2, \infty], \quad m \in b^\perp, \end{aligned} \tag{3.15}$$

yields

$$\|\Delta w_{r,h} - \Delta u_h\|_{W^{-1,2}(\Omega,\mathbb{R}^n)} \to 0, \qquad \text{as } r, h \to 0. \tag{3.16}$$

Thus we have from (3.7), (3.14) and (3.16)

$$\begin{aligned} \lambda_1^2\lambda_2^2 \|F_1 - F_2\|^2 &\leq \int_\Omega \|\nabla u_h(x) - F\|^2 \, dx = \langle u_h - Fx, \Delta u_h \rangle \\ &= \langle u_h - Fx, \Delta w_{r,h} \rangle - \langle u_h - Fx, \Delta w_{r,h} - \Delta u_h \rangle. \end{aligned} \tag{3.17}$$

Since $w_{r,h}$ is a piecewise affine function we have

$$\begin{aligned} |\langle u_h - Fx, \Delta\, w_{r,h}\rangle| &\leq \sum_{E\in\mathcal{E}} \int_{\partial E} |[\nabla w_{r,h}]\,(s) n(s)(u_h(s) - Fs)|\; dS \\ &\leq C\; \|u_h - Fx\|_{L^\infty(\Omega,\mathbb{R}^3)}\, \|\Delta\, w_{r,h}\|_{W^{-1,2}(\Omega,\mathbb{R}^3)}\,, \end{aligned} \tag{3.18}$$

where we denote by $\mathcal{E}$ the set of all maximal simply connected subdomains E of Ω with Lipschitz boundary such that $\nabla w_{r,h}|_E$ is independent of x and $[\nabla w_{r,h}]\,(s) = \pm a \otimes b$. The inequality (3.6) now follows from (3.17) and (3.18) upon the replacement of $w_{r,h}$ by u_h. Nonexistence of a solution to (1.1) follows from (3.17) and Theorem 2.4. □

We can use the inequality (3.6) to derive a lower estimate for the convergence of the "macroscopic" quantities such as $u_h - Fx$.

Corollary 3.2. *Let* $\{u_h\}_{h>0}$ *be a sequence of continuous, piecewise affine functions satisfying* (3.3). *Then*

$$h^{1+n/p} \leq \|u_h - Fx\|_{L^p(\Omega,\mathbb{R}^n)}\,, \qquad p \in [1,\infty]. \tag{3.19}$$

Consequently,

$$\|\Delta\, w_{r,h}\|_{W^{-1,2}(\Omega,\mathbb{R}^n)} \geq\; const.\; h^{-1}. \tag{3.20}$$

Let us assume, moreover, that there exists a positive constant C such that

$$\|\nabla u_h - \Pi\nabla u_h\|_{L^2(\Omega,\mathbb{R}^{3\times 3})} \leq C\, h^{2(1-\gamma)}, \qquad \gamma \in [0,1). \tag{3.21}$$

Then there exist $r_0 > 0$ *and a positive constant* C, *independent of* h, *such that*

$$h^{\gamma-1} \leq C\; \|\Delta\, w_{r,h}\|_{W^{-1,2}(\Omega,\mathbb{R}^3)}\,, \qquad r \in (0, r_0). \tag{3.22}$$

The function $w_{r,h}$ *is defined by* (3.14).

Proof. (i) The discrete inverse inequality [4] implies that for any subset D of Ω such that $D = \cup_{Q_h \in \tau_h^0} Q_h$, where τ_h^0 is a subset of τ_h, we have

$$h^{1+n\max\left\{0,\frac{1}{q}-\frac{1}{p}\right\}}\; \|\nabla u_h - F\|_{L^p(D,\mathbb{R}^{N\times N})} \leq C\, \|u_h - Fx\|_{L^q(D,\mathbb{R}^n)}\,, \tag{3.23}$$

$p, q \in [1,\infty]$.

We note that (3.23) holds true because $Fx \in V_h$. The positive constant C in (3.23) depends on the uniformity and regularity of the partition τ_h

and also on p and q. This constant is uniformly bounded with respect to p and q, though. Taking $p = \infty$ in (3.23) and recalling (3.7), we obtain

$$h^{1+n/q} \leq C \|u_h - Fx\|_{L^q(D,\mathbb{R}^n)}, \quad \text{for any } q \in [1,\infty]. \tag{3.24}$$

Thus $q = \infty$ in (3.24) and (3.14), (3.15) of Theorem 3.1 yields (3.20)

(ii) We have for some $r \in (0,1)$ from (3.17)

$$\begin{aligned}(1-r)\lambda_1^2\lambda_2^2 \|F_1 - F_2\|^2 &\leq \|u_h - Fx\|_{L^\infty(\Omega,\mathbb{R}^n)} \|\Delta w_{r,h}\|_{W^{-1,2}(\Omega,\mathbb{R}^n)} \\ &\leq C_2 \|\nabla u_h - \Pi\nabla u_h\|^{1/2}_{L^2(\Omega,\mathbb{R}^{N\times N})} \|\Delta w_{r,h}\|_{W^{-1,2}(\Omega,\mathbb{R}^n)} \\ &\leq C_3 h^{1-\gamma} \|\Delta w_{r,h}\|_{W^{-1,2}(\Omega,\mathbb{R}^n)}.\end{aligned} \tag{3.25}$$

The upper estimate of $\|u_h - Fx\|_{L^\infty(\Omega,\mathbb{R}^n)}$ by $\|\nabla u_h - \Pi\nabla u_h\|^{1/2}_{L^2(\Omega,\mathbb{R}^{N\times N})}$ follows from [[12], Theorem 9.1]. □

Remark 3.3. We note that it does not seem at all possible to derive an upper bound for $\|\Delta w_{r,h}\|_{W^{-1,2}(\Omega,\mathbb{R}^n)}$ without any additional structural analysis.

Corollary 3.4. *Let* $\{u_h\}_{h>0}$ *be a sequence of continuous, piecewise affine functions satisfying* (3.3). *Let us assume, moreover, that*

$$\|\Delta w_{r,h}\|_{W^{-1,2}(\Omega,\mathbb{R}^n)} \leq h^{-\gamma}, \gamma \in [1/2,1], \quad r \in (0,1),\ 0 < h << 1. \tag{3.26}$$

Then

$$h^{\gamma+n/p} \leq \|u_h - Fx\|_{L^p(\Omega,\mathbb{R}^n)}, \qquad p \in [2,\infty]. \tag{3.27}$$

Proof. The proof follows from Theorem 3.1 and the discrete inverse inequalities (3.23) because $Fx \in V_h$ for any $h > 0$. □

Remark 3.5. It has been proven in [12] that

$$\|\nabla u_h - \Pi\nabla u_h\|_{L^p(\Omega,\mathbb{R}^{N\times N})} \leq h^{2(1-\gamma)} \implies \text{meas}(D) \geq h^\gamma, \gamma \in [1/2,1].$$

Here D is an arbitrary "laminate", i.e. the maximal simply connected subdomain of Ω such that $\Pi_{1,2}\nabla u_h|_D$ is independent of x. We can draw the following conclusion from this implication:

The faster the formation of a microstructure occurs, the coarser it will be.

□

4. FINE STRUCTURES

It has been proven in [20] that there exists a sequence of approximations u_ε of u such that

$$\begin{aligned} &\operatorname{dist}(\nabla u_\varepsilon(x), \{F_1, F_2\}) \le \varepsilon, \qquad \text{in } \Omega \\ &\sup_\Omega |u_\varepsilon(x) - (\lambda_1 F_1 + \lambda_2 F_2)x| \le \varepsilon, \qquad \text{and} \\ &u_\varepsilon(x) = (\lambda_1 F_1 + \lambda_2 F_2)x, \quad x \in \partial\Omega. \end{aligned} \tag{4.1}$$

A typical sequence of approximate functions u_ε that satisfy (4.1) is displayed in Figure 2.

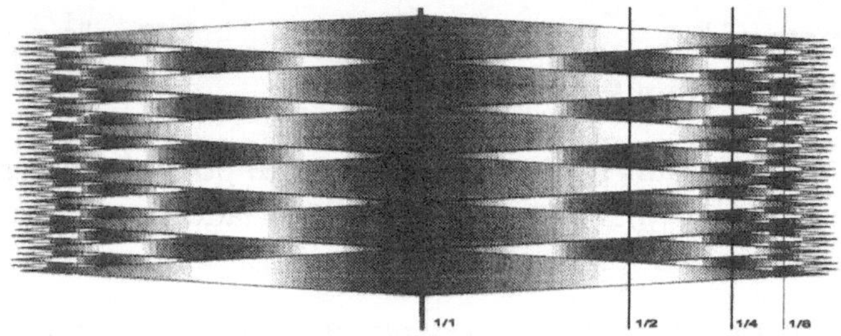

Figure 2 The four generations of fine structures u_ε satisfying (4.1). The "needle-like" structures near the boundary resemble the structures near the semi-coherent Austenitic-Finely Twinned Martensitic interface, c.f. Figure 1.

Crystallographic theory does not allow relaxation of the structure of the matrices yielding the equilibrium configurations. Theorem 3.1 shows that if we cannot reach an equilibrium with matrices other than those from $SO(N)F_1 \cup SO(N)F_2$, the only possibility left is to relax the boundary condition. In other words, we may seek a fine structure generated by compact or pre-compact sequences of weakly differentiable functions with gradients approaching the set $SO(N)F_1 \cup SO(N)F_2$ and such that their values on the boundary will be in some sense close to the desired condition $(\lambda_1 F_1 + \lambda_2 F_2)x$. This possibility is studied in [20] using convex integration [11].

Definition 4.1. *The laminated convex hull* $\mathcal{F}_{Lam}$ *of a set* $\mathcal{F} \subset M^{N\times N}$ *is the smallest set that contains with any* $A, B \in \mathcal{F}$, $\operatorname{rank}(A - B) = 1$, *also the convex combination* $\lambda A + (1-\lambda)B$, $\lambda \in (0,1)$.

It is proved in [20] that differential inclusions such as (1.1) are sometimes solvable by fine structures if the boundary condition is generated by a matrix that is in the interior of the laminated convex hull of $SO(N)F_1 \cup SO(N)F_2$. Following this hint we have the following conjecture.

Conjecture 4.2. *Let $\mathcal{F}_{Lam}$ be the laminated convex hull of the set $SO(N)F_1 \cup SO(N)F_2$, $N = 2$, and let $\{u_h\}_{h>0}$ be a sequence of continuous, piecewise affine functions such that*

$$\|\nabla u_h - \Pi \nabla u_h\|_{L^2(\Omega,\mathbb{R}^{N\times N})} \to 0, \qquad as\ h \to 0_+. \tag{4.2}$$

Let us assume, moreover, that

$$u_h(x) = F_\delta x, \quad x \in \partial\Omega, \tag{4.3}$$

where $F_\delta \in$ int $\mathcal{F}_{Lam}$, and dist$(F_\delta, \mathcal{F}_{Lam}) \le \delta$, for some $\delta > 0$.
Then

$$\lim_{h\to 0} \|u - u_h\|_{L^\infty(\Omega,\mathbb{R}^N)} \|\Delta u_h\|_{W^{-1,2}(\Omega,\mathbb{R}^N)} = 0. \tag{4.4}$$

Acknowledgments

The author was partially supported by the National Science Foundation, through the Center for Research on Parallel Computation, under Cooperative Agreement No. CCR-9120008.

References

[1] Ball, J.M. (1989). A version of the fundamental theorem for Young measures. (Partial Differential Equations and Continuum Models of Phase Transitions (M.Rascle, D.Serre and M.Slemrods, eds.), Lecture Notes in Physics) (344) (Springer-Verlag, New-York).

[2] Ball, J.M., Holmes, P.J., James, R.D., Pego, R.L. and Swart, P.J. (1971). On the dynamics of fine structure. *J. Nonlinear Sci.*, 1:17–70.

[3] Ball, J.M. and James, R.D. (1987). Fine phase mixtures as minimizers of energy. *Arch. Rational Mech. Anal.*, 100:13–52.

[4] Ciarlet, P.G. (1991). Basic Error Estimates for Elliptic Problems. *in: Handbook of Numerical Analysis*, P.G. Ciarlet, J.L. Lions Editors, North-Holland.

[5] Collins, C. and Luskin, M. (1991). Optimal order estimates for the finite element approximation of the solution of a non-convex variational problem. *Math.Comp.*, 621–637

[6] Collins, C., Luskin, M. and Kinderlehrer, D. (1991). Numerical approximation of the solution of a variational problem with a double well potential. *SIAM J. Numer. Anal.*, 28.

[7] Chu, C. (1993). Hysteresis and microstructures: a study of biaxial loading on compound twins of copper-aluminium-nickel single crystals. *Ph.D. dissertation* (University of Minnesota).

[8] Ericksson, K., Estep, D., Hansbo, P. and Johnson, C. (1996). Computational Differential Equations. *Cambridge University Press.*

[9] Friesecke, G. and McLeod, J.B. (1996). Dynamics as a mechanism preventing the formation of finer and finer microstructure. *Arch. Rat. Mech. Analysis*, 133.

[10] Friesecke, G. and McLeod, J.B. (1997). Dynamic stability of nonminimizing phase mixtures. *Preprint*.

[11] Gromov, M. (1986). Partial differential relations. *Springer*.

[12] Klouček, P. (1999). The Finite Element Approximations of Binomial Microstructures. *CAAM Technical report TR98-14, Rice University*, to appear in *SIAM J. Numer. Anal.*.

[13] Klouček, P. (1998). The Relaxation of Non-Quasiconvex Variational Integrals. *Num. Math.*, in press.

[14] Klouček, P. (1998). The computational modeling of nonequilibrium thermodynamics of the martensitic transformation. *J. Comp. Mech.*, 22(3).

[15] Klouček, P., Bo Li and Luskin, M. (1996). Nonconforming finite element approximation of the microstructure. *Math. Comp.*, 65(215):1111–1135.

[16] Kohn, R. and Müller, S. (1992). Branching of twins near an austenite/twinned-martensite interface. *Philosophical Magazine*, 66A:697–715.

[17] Li, B. and Luskin, M. (1996). Finite element analysis of microstructure for the cubic to tetragonal transformation. *To appear in SIAM J. Mumer. Math.*

[18] Luskin, M. (1996). Approximation of a laminated microstructure for a rotationally invariant, double well energy density. *Numer. Math.*

[19] Luskin, M. (1996). On the computation of Crystalline Microstructure. *Acta Numerica*.

[20] Müller, S. and Šverák, V. (1995). Attainment results for the two-well problem by convex integration. *Manuscript*.

FINITE ELEMENT ANALYSIS OF A NONLINEAR ELLIPTIC PROBLEM WITH A PURE RADIATION CONDITION

Michael Křížek, Liping Liu, Pekka Neittaanmäki

Abstract: We deal with a second order elliptic problem in a bounded plane domain with a pure Stefan-Boltzmann condition on the whole boundary. We prove that the finite element method converges in the H^1-norm without any additional smoothness assumptions on the true solution.

Keywords: Finite elements, nonlinear boundary conditions, error estimates.

1. INTRODUCTION

There is a great amount of monographs dealing with nonlinear elliptic problems (see, e.g., [2, 3, 5, 6] and the references therein). In [3, Chapt. 8], we propose a finite element analysis for solving a second order elliptic equation

$$-\mathrm{div}(A\,\mathrm{grad}\ u) = f \quad \text{in} \quad \Omega \tag{1.1}$$

with mixed boundary conditions: on a relatively open part Γ_1 of the boundary $\partial\Omega$, we prescribe nonhomogeneous Dirichlet boundary conditions, and on the remaining part Γ_2, we consider a nonlinear Stefan-Boltzmann radiation condition (compare also [1, 7])

$$\alpha u + n^\top A\,\mathrm{grad}\ u + \beta u^4 = g, \tag{1.2}$$

where $\Omega \subset R^2$ is a bounded domain with a Lipschitz-continuous boundary $\partial\Omega$, n is the outward unit normal to $\partial\Omega$, $A \in (L^\infty(\Omega))^{2\times2}$ is a symmetric uniformly positive definite matrix of heat conductivities, $\alpha \in L^\infty(\Omega)$ is a nonnegative heat transfer coefficient, $\beta = \sigma f_{\mathrm{em}}$, $\sigma = 5.669 \times 10^{-8}$ is the Stefan-Boltzmann constant, $0 \le f_{\mathrm{em}} \le 1$ is the relative emissivity function from $L^\infty(\partial\Omega)$, $f \in L^2(\Omega)$ is the density of volume heat sources and $g \in L^2(\partial\Omega)$ is the density of surface heat sources.

Applied Nonlinear Analysis, edited by Sequeira *et al.*
Kluwer Academic / Plenum Publishers, New York, 1999.

If $\Gamma_1 \neq \emptyset$ or α is positive on some part of Γ_2 with a positive measure, then we can use Friedrichs' inequality [4, p. 20] to prove that the problem is elliptic (cf. [3]).

The aim of this paper is to extend the above problem to the case $\alpha \equiv 0$ on the whole boundary $\partial\Omega$. In this case we are not able to use Friedrichs' inequality like in [3]. That is why we have to develop another approach to prove the existence and uniqueness of the weak and discrete solutions, etc. We also prove the convergence of finite element approximations to the weak solution u from the Sobolev space $H^1(\Omega)$ without any additional regularity assumptions upon u.

2. WEAK FORMULATION

Consider equation (1.1) with the pure radiation boundary condition

$$n^\top A \operatorname{grad} u + \beta u^4 = g \quad \text{on} \partial\Omega. \tag{2.1}$$

Remark 2.1. The case $\alpha \equiv 0$ on $\partial\Omega$ means that there is no heat transfer caused by a conduction through the boundary, i.e., the heat does not leave the body Ω by a conductive process. An example of such a situation can be a star, which is surrounded by vacuum, and all escaping heat is thus produced by radiation only. By the Kirchhoff law, energy losses are proportional to the fourth power of the surface temperature.

Set

$$\begin{aligned} V &= H^1(\Omega), \\ a(v,w) &= \int_\Omega (\operatorname{grad} v)^\top A \operatorname{grad} w \, dx, \quad v,\, w \in V, \\ F(v) &= \int_\Omega fv \, dx + \int_{\partial\Omega} gv \, ds, \quad v \in V, \end{aligned}$$

and

$$J(v) = \tfrac{1}{2} a(v,v) + \tfrac{1}{5} \textstyle\int_{\partial\Omega} \beta v^5 \, ds - F(v), \quad v \in V.$$

We see that the bilinear from $a(\cdot,\cdot)$ is not V-elliptic, which is the main source of all difficulties. From the heat conduction equation (1.1), Green's formula and the boundary condition (2.1) we get the problem: Find $u \in V$ such that

$$a(u,v) + \int_{\partial\Omega} \beta u^4 v \, ds = F(v) \quad \forall v \in V. \tag{2.2}$$

Throughout this and next section we shall assume that $\beta > 0$ on some nonempty relatively open subset Γ of $\partial\Omega$.

Remark 2.2. The case $\beta \equiv 0$ on $\partial\Omega$ means that the whole boundary is "isolated". This corresponds to the standard Neumann boundary conditions, which are well-studied in [3, p. 35].

Define the set of admissible temperatures

$$U = \{v \in V \mid v \geq 0 \quad \text{in} \quad \Omega\}.$$

We will use the following lemma instead of Friedrichs' inequality to prove that the energy functional J is coercive on U.

Lemma 2.3. *Let $\Omega \subset R^2$ be a bounded domain with a Lipschitz continuous boundary. Let $\beta > 0$ on $\Gamma \subset \partial\Omega$, where $\Gamma \neq \emptyset$ is a relatively open set in $\partial\Omega$. Then there exists a constant $C > 0$ such that*

$$|w|_{1,2}^2 + \int_{\partial\Omega} \beta w^5 \, ds \geq C\|w\|_{1,2}^2 \quad \forall w \in U, \quad \|w\|_{1,2} \geq 1.$$

Proof. First, we show that there exists a constant $C > 0$ such that for any $v \in U$, $\|v\|_{1,2} = 1$, we have

$$|v|_{1,2}^2 + \int_{\partial\Omega} \beta v^5 \, ds \geq C. \tag{2.3}$$

On the contrary, assume that (2.3) does not hold. Then there exists a sequence $\{v_k\}_{k=1}^{\infty} \subset U$, $\|v_k\|_{1,2} = 1$, such that

$$|v_k|_{1,2}^2 + \int_{\partial\Omega} \beta v_k^5 \, ds \leq \frac{1}{k}. \tag{2.4}$$

Since V is reflexive, we may choose, by the Eberlein-Schmulyan Theorem, a subsequence, still denoted by $\{v_k\}$ such that $v_k \rightharpoonup z$ for some $z \in V$. By the Rellich Theorem (see [4, p. 17]), the space $V = H^1(\Omega)$ is compactly imbedded into $L^2(\Omega)$. That is why $v_k \to z$ in $L^2(\Omega)$. Since v_k are nonnegative, their limit function z is also nonnegative, i.e., $z \in U$.

According to [4, p. 107], the trace operator from $H^1(\Omega)$ to $L^5(\partial\Omega)$ is compact. Therefore,

$$v_k|_{\partial\Omega} \to z|_{\partial\Omega} \quad \text{in} \quad L^5(\partial\Omega).$$

Obviously, the seminorm $|\cdot|_{1,2}$ is a convex and continuous functional in $H^1(\Omega)$. Hence, it is lower weakly semicontinuous functional and thus from (2.4), we have

$$|z|_{1,2}^2 + \int_{\partial\Omega} \beta z^5 \, ds = 0.$$

As $z \geq 0$ in Ω, the both terms are equal to zero. Therefore, since $|z|_{1,2} = 0$, we find that z is constant almost everywhere in Ω. The condition $\beta > 0$ on Γ guarantees that $z = 0$ on Γ, and thus $z \equiv 0$ in Ω.

We know that $v_k \to z$ in $L^2(\Omega)$, i.e.,

$$\lim_{k\to 0} \|v_k\|_{0,2} = 0.$$

From (2.4) we further see that

$$\lim_{k\to 0} |v_k|_{1,2} = 0.$$

But this is a contradiction, because

$$1 = \|v_k\|_{1,2}^2 = \|v_k\|_{0,2}^2 + |v_k|_{1,2}^2.$$

Hence, (2.3) holds.

Second, we choose an arbitrary $w \in U$, $w \neq 0$, and set

$$v = \frac{w}{\|w\|_{1,2}}.$$

Then by (2.3),

$$\frac{|w|_{1,2}^2}{\|w\|_{1,2}^2} + \frac{1}{\|w\|_{1,2}^5} \int_{\partial\Omega} \beta w^5 \, ds \geq C.$$

If $\|w\|_{1,2} \geq 1$ then clearly

$$|w|_{1,2}^2 + \int_{\partial\Omega} \beta w^5 \, ds \geq |w|_{1,2}^2 + \frac{1}{\|w\|_{1,2}^3} \int_{\partial\Omega} \beta w^5 \, ds \geq C\|w\|_{1,2}^2,$$

which proves the lemma. □

Lemma 2.3 will help us now to prove that the functional J is coercive on U. Indeed,

$$\begin{aligned} J(v) &\geq C|v|_{1,2}^2 + \frac{1}{5}\int_{\partial\Omega} \beta v^5 \, ds - F(v) \\ &\geq C_1\|v\|_{1,2}^2 - C_2\|v\|_{1,2} \quad \forall v \in U, \quad \|v\|_{1,2} \geq 1. \end{aligned} \tag{2.5}$$

The existence of a constant $C_1 > 0$ is guaranteed by Lemma 2.3 and from the continuity of the linear functional F we get a constant C_2.

Definition 2.4. A *weak solution* of the classical problem (1.1)+(2.1) is a function $u \in U$, which satisfies the variational inequality

$$a(u, v-u) + \int_{\partial\Omega} \beta u^4 (v-u) \, ds \geq F(v-u) \quad \forall v \in U. \tag{2.6}$$

If some $u \in U$ satisfies (2.2), then obviously it satisfies (2.6). Therefore, any classical solution is also a weak solution.

Let us consider the above variational inequality (2.6) in the form

$$\langle Au, v-u \rangle \geq 0 \quad \forall v \in U, \tag{2.7}$$

where $\langle \cdot, \cdot \rangle$ is the scalar product in the space $H^1(\Omega)$ and the operator $\mathcal{A} : V \to V$ is uniquely determined (due to the Riesz, compare [3, p. 136]) by the relation

$$\langle \mathcal{A}w, v \rangle = a(w, v) + \int_{\partial\Omega} \beta w^4 v \, ds - F(v), \quad w, v \in V, \tag{2.8}$$

since the right-hand side is a linear continuous functional for $v \in V$.

Lemma 2.5. *Let the assumptions of Lemma 2.3 hold. Then the operator $\mathcal{A}$ is strictly monotone on U.*

Proof. Let $w \neq v$ be two arbitrary elements from U. Then

$$\langle \mathcal{A}w - \mathcal{A}v, w - v \rangle = a(w - v, w - v) + \int_{\partial\Omega} \beta (w^4 - v^4)(w - v) \, ds.$$

We see that

$$(w^4 - v^4)(w - v) = (w^2 + v^2)(w + v)(w - v)^2 \geq 0 \quad \forall w, v \in U.$$

If there exists a subset of Γ with nonzero measure, where $w = v$, then by Friedrichs' inequality there exists a constant $C(w, v) > 0$ such that

$$\begin{aligned} \langle \mathcal{A}w - \mathcal{A}v, w - v \rangle &\geq a(w - v, w - v) \geq \overline{C} |w - v|_1^2 \\ &\geq C(w, v) \|w - v\|_{1,2}^2 > 0. \end{aligned}$$

If such a set does not exist, then

$$\beta (w^4 - v^4)(w - v) > 0$$

on some subset of Γ with nonzero measure. Hence, again

$$\langle \mathcal{A}w - \mathcal{A}v, w - v \rangle > 0$$

and the lemma is proved. □

Theorem 2.6. *Let the assumptions of Lemma 2.3 hold. Then there exists exactly one weak solution of the problem (1.1)+(2.1). This solution is the unique point of minimum of the functional J on U.*

Proof. The functional J is clearly convex on U, by (2.5) it is coercive and like in [3, p. 152] we find that it is continuous on V. Therefore, J attains minimum over U in some point u. Since J is Gâteaux differentiable, we get that

$$DJ(u; v - u) \geq 0 \quad \forall v \in U.$$

However, this inequality is equivalent to (2.6) (and also to (2.7)). The function $u \in U$ is thus the weak solution of the classical problem (1.1)+(2.1).

From Lemma 2.5 we know that $\mathcal{A}$ is strictly monotone. This implies the uniqueness of u (compare [3, p. 132]). □

3. FINITE ELEMENT ANALYSIS

Denote by X_h a finite dimensional space of continuous piecewise linear functions over a triangulation $\mathcal{T}_h$. By [3, p. 59],

$$X_h \subset H^1(\Omega).$$

Let $\pi_h : C(\overline{\Omega}) \to X_h$ be a standard linear interpolation operator and let

$$U_h = U \cap X_h.$$

This set is clearly closed, convex, and nonempty.

Lemma 3.1. *Let $\Omega \subset R^2$ be a bounded domain with a Lipschitz continuous boundary. Let $v, w \in V$, $\{v_k\}_{k=1}^{\infty} \subset V$, $\|v - v_k\|_{1,2} \to 0$ as $k \to \infty$ and let*

$$\widehat{v} = \max(v, w), \quad \widehat{v}_k = \max(v_k, w).$$

Then $\|\widehat{v} - \widehat{v}_k\|_{1,2} \to 0$ as $k \to \infty$.

Proof. We have

$$\begin{aligned} \widehat{v} &= \max(v, w) = \frac{1}{2}((v + w) + |v - w|), \\ \widehat{v}_k &= \max(v_k, w) = \frac{1}{2}((v_k + w) + |v_k - w|). \end{aligned}$$

Since $v_k - w \to v - w$ as $k \to \infty$ in the $H^1(\Omega)$-norm, then also $|v_k - w| \to |v - w|$ and thus $\|\widehat{v} - \widehat{v}_k\|_{1,2} \to 0$ as $k \to \infty$. □

For an arbitrary $\rho > 0$ denote by R_ρ the standard regularization operator, which assigns to any function $v \in L^1(\Omega)$ a function $R_\rho(v) \in C^\infty(\overline{\Omega})$ (see [4, p. 58]):

$$(R_\rho v)(x) = \int_{\|y\|<\rho} \overline{v}(x + y)\omega(y, \rho)\, dy,$$

where $\overline{v}(x) = v(x)$ for $x \in \Omega$, $\overline{v}(x) = 0$ for $x \in R^2 \setminus \Omega$,

$$\omega(y, \rho) = \begin{cases} \rho^2 \exp(\|y\|^2/(\|y\|^2 - \rho^2)) & \text{for } \|y\| < \rho, \\ 0 & \text{for } \|y\| \geq \rho. \end{cases}$$

By a simple calculation we get

$$\int_{\|y\|<\rho} \omega(y, \rho)\, dy = 1.$$

Recall (see [3, p. 67]) that a family $\mathcal{F}$ of triangulations is said to satisfy the *maximum angle condition* if there exists a constant γ_0 such that for any $\mathcal{T}_h \in \mathcal{F}$ and any $K \in \mathcal{T}_h$ we have

$$\gamma_K \leq \gamma_0 < \pi,$$

where γ_K is the maximum angle of the triangle K.

Theorem 3.2. *Let $\Omega \subset R^2$ be a bounded domain with a polygonal Lipschitz continuous boundary. Let the family $\mathcal{F} = \{\mathcal{T}_h\}$ of triangulations satisfy the maximum angle condition. Then for each $v \in U$ there exists $\{v_h\}$, $v_h \in U_h$, such that*

$$\lim_{h\to 0} \|v - v_h\|_{1,2} = 0.$$

Proof. Let $v \in U$ be given. We know from the density $\overline{C^\infty(\overline{\Omega})} = H^1(\Omega)$ that there exists a sequence $\{w_k\}_{k=1}^{\infty} \subset C^\infty(\overline{\Omega})$ such that $w_k \to v$ in the $H^1(\Omega)$-norm. Set

$$\widehat{w}_k = \max(0, w_k),$$

and extend $\widehat{w}_k$ by zero outside Ω. Then, $\widehat{w}_k \geq 0$ and $\widehat{w}_k \in H^1(\Omega)$. Since $v = \max(0, v)$, we find by Lemma 3.1 that

$$\|v - \widehat{w}_k\|_{1,2} \to 0 \quad \text{as} \quad k \to \infty. \tag{3.1}$$

Now let us fix k and perform the regularization of $\widehat{w}_k$. Choose $\rho > 0$. The functions $R_\rho \widehat{w}_k$ are from $C^\infty(\overline{\Omega})$. From the definition of R_ρ it is clear that

$$R_\rho \widehat{w}_k \geq 0 \quad \text{in} \quad \Omega.$$

By [4, p. 59],

$$\lim_{\rho\to 0} \|\widehat{w}_k - R_\rho \widehat{w}_k\|_{1,2} = 0. \tag{3.2}$$

Since $C^\infty(\overline{\Omega}) \subset H^2(\Omega)$, we know (see [3, p. 72]) due to the maximum angle condition that

$$\lim_{h\to 0} \|R_\rho \widehat{w}_k - \pi_h(R_\rho \widehat{w}_k)\|_{1,2} = 0. \tag{3.3}$$

Moreover, obviously $\pi_h(R_\rho \widehat{w}_k) \in U_h$, and from the triangle inequality

$$\|v - \pi_h(R_\rho \widehat{w}_k)\|_{1,2} \leq \|v - \widehat{w}_k\|_{1,2} + \|\widehat{w}_k - R_\rho \widehat{w}_k\|_{1,2} + \|R_\rho \widehat{w}_k - \pi_h(R_\rho \widehat{w}_k)\|_{1,2}.$$

For a given $\varepsilon > 0$ we find, by (3.1), (3.2) and (3.3), that there exist k and h_0 such that for all $h \in (0, h_0)$ all three terms in the right-hand side are less than $\varepsilon/3$. □

Since U_h is convex, closed and nonempty for any $h \in (0, h_0)$, we may show like in Theorem 2.6, that there exists exactly one *discrete solution* $u_h \in U_h$ of the problem

$$J(u_h) = \min_{v_h \in U_h} J(v_h), \tag{3.4}$$

and that (3.4) is equivalent to the problem: Find $u_h \in U_h$ such that

$$\langle A u_h, v_h - u_h \rangle \geq 0 \quad \forall v_h \in U_h, \tag{3.5}$$

where $\langle \mathcal{A}u_h, \cdot \rangle$ is a linear continuous functional defined on subspace X_h.

Lemma 3.3. *Let the assumptions of Lemma 2.3 hold and let $\mathcal{F}$ be a family of triangulations satisfying the maximum angle condition. Then the sequence $\{u_h\}$ of the discrete solutions of (3.4) weakly converges to the weak solution u.*

Proof. By (2.8) and the trace theorem the operator $\mathcal{A}$ is continuous. Since it is also monotone, the problem (3.5) is equivalent to (see [2, p. 321])

$$\langle \mathcal{A}v_h, v_h - u_h \rangle \geq 0 \quad \forall v_h \in U_h. \tag{3.6}$$

By Theorem 3.2, there exists a sequence $z_h \in U_h$ such that $z_h \to u$ in the $H^1(\Omega)$-norm. Then,

$$\limsup_{h\to 0} J(u_h) \leq \lim_{h\to 0} J(z_h) = J(u) < \infty.$$

We know that $u_h \in U_h \subset U$ and that J is coercive (see (2.5)). Consequently, the sequence $\{u_h\}$ is bounded in the $H^1(\Omega)$-norm.

From the reflexivity of the Sobolev space $H^1(\Omega)$ it follows that there exists a subsequence of $\{u_h\}$, which we denote again by $\{u_h\}$, which weakly converges to some element $\widehat{u} \in H^1(\Omega)$. Since $U_h \subset U$ for all $h \in (0, h_0)$, and U is a convex closed set, we have $\widehat{u} \in U$.

Now choose arbitrarily some function $w \in U$. Then, by Theorem 3.2, we can find a sequence $\{w_h\}$, $w_h \in U_h$ and $w_h \to w$ in $H^1(\Omega)$. According to (3.6), we have

$$\langle \mathcal{A}w_h, w_h - u_h \rangle \geq 0. \tag{3.7}$$

Obviously, $w_h - u_h \rightharpoonup w - \widehat{u}$ in $H^1(\Omega)$. Hence, from the continuity of $\mathcal{A}$, we get by (3.7) that

$$\langle \mathcal{A}w, w - \widehat{u} \rangle \geq 0 \quad \forall w \in U.$$

Moreover, since $\mathcal{A}$ is monotone, we get again by [2, p. 321] that

$$\langle \mathcal{A}\widehat{u}, w - \widehat{u} \rangle \geq 0 \quad \forall w \in U$$

and thus, by Theorem 2.6, $\widehat{u}$ is the weak solution of the problem (1.1)+(2.1). □

Lemma 3.4. *Let the assumptions of Lemma 3.3 hold. Then*

$$|u - u_h|_{1,2} \to 0 \quad \text{as} \quad h \to 0,$$

where u and u_h are weak and discrete solutions, respectively, of the radiation problem (1.1)+(2.1).

Proof. From the definition of $\mathcal{A}$ we have

$$\langle \mathcal{A}u_h - \mathcal{A}u, u_h - u \rangle = a(u_h - u, u_h - u) + \int_{\partial\Omega} \beta(u_h^4 - u^4)(u_h - u)\, ds.$$

Since

$$\beta(u_h^4 - u^4)(u_h - u) = \beta(u_h^2 + u^2)(u_h + u)(u_h - u)^2 \geq 0,$$

we have

$$\begin{aligned} \langle \mathcal{A}u_h - \mathcal{A}u, u_h - u \rangle &\geq a(u_h - u, u_h - u) \\ &= \int_\Omega \operatorname{grad}(u_h - u))^\top A \operatorname{grad}(u_h - u)\, dx \qquad (3.8) \\ &\geq C|u_h - u|_{1,2}^2, \end{aligned}$$

where C a positive constant.

Moreover, by (2.7), we find that

$$\begin{aligned} \langle \mathcal{A}u_h - \mathcal{A}u, u_h - u \rangle &= \langle \mathcal{A}u_h, u_h - u \rangle - \langle \mathcal{A}u, u_h - u \rangle \leq \qquad (3.9) \\ &\leq \langle \mathcal{A}u_h, u_h - u \rangle. \end{aligned}$$

According to Theorem 3.2, there exists a sequence $\{v_h\}$, $v_h \in U$, such that

$$\|u - v_h\|_{1,2} \to 0 \quad \text{as} \quad h \to 0. \qquad (3.10)$$

Clearly,

$$\langle \mathcal{A}u_h, u_h - u \rangle = \langle \mathcal{A}u_h, u_h - v_h \rangle + \langle \mathcal{A}u_h, v_h - u \rangle, \qquad (3.11)$$

and thus

$$\langle \mathcal{A}u_h, u_h - v_h \rangle = -\langle \mathcal{A}u_h, v_h - u_h \rangle \leq 0, \qquad (3.12)$$

since u_h solves the inequality

$$\langle \mathcal{A}u_h, w_h - u_h \rangle \geq 0 \quad \forall w_h \in U_h.$$

From (3.8), (3.9), (3.11) and (3.12), we get

$$C|u - u_h|_{1,2}^2 \leq |\langle \mathcal{A}u_h, v_h - u \rangle| \qquad (3.13)$$

By Lemma 3.3, $u_h \rightharpoonup u$ and thus the sequence $\{u_h\}$ is bounded. By (2.8) and the trace theorem the sequence $\{\mathcal{A}u_h\}$ is also bounded. Hence, by (3.10),

$$|\langle \mathcal{A}u_h, v_h - u \rangle| \leq C\|u - v_h\|_{1,2} \to 0 \quad \text{as} \quad h \to 0,$$

where $C > 0$ is a constant independent of h. From here and (3.13) the lemma follows. □

Theorem 3.5. *Let the assumptions of Lemma 3.3 hold. Then*

$$\|u - u_h\|_{1,2} \to 0 \quad \text{as} \quad h \to 0.$$

Proof. By Lemma 3.3, $u_h \rightharpoonup u$ in $H^1(\Omega)$. Due to the compact imbedding of $H^1(\Omega)$ into $L^2(\Omega)$ (the Rellich Theorem [4, p. 17]), we have

$$\|u - u_h\|_{0,2} \to 0 \quad \text{as} \quad h \to 0. \qquad (3.14)$$

Now Lemma 3.4, (3.14) and the equality yield

$$\|u - u_h\|_{1,2}^2 = \|u - u_h\|_{0,2}^2 + |u - u_h|_{1,2}^2,$$

which proves the theorem. □

Acknowledgments

This work was supported by the grant no. 201/98/1452 of the Grant Agency of the Czech Republic and the COMAS Graduate School at the University of Jyväskylä.

References

[1] Feistauer, M. and Najzar, K. (1998). *Finite element approximation of a problem with a nonlinear Newton boundary condition.* Numer. Math., 78:403–425.

[2] Fučík, S. and Kufner, A. (1980). *Nonlinear differential equations.* Elsevier, Amsterdam.

[3] Křížek, M. and Neittaanmäki, P. (1996). *Mathematical and numerical modelling in electrical engineering: theory and applications.* Kluwer, Dordrecht.

[4] Nečas, J. (1967). *Les méthodes directes en théorie des équations elliptiques.* Academia, Prague.

[5] Nečas, J. (1983). *Introduction to the theory of nonlinear elliptic equations.* Teubner, Leipzig.

[6] Nečas, J. and Hlaváček, I. (1981). *Mathematical theory of elastic and elasto-plastic bodies: an introduction.* Elsevier, Amsterdam, Oxford, New York.

[7] Tiihonen, T. (1997). *Stefan-Boltzamnn radiation on non-convex surfaces.* Math. Methods Appl. Sci., 20:47–57.

ESTIMATES OF THREE-DIMENSIONAL OSEEN KERNELS IN WEIGHTED L^P SPACES

Stanislav Kračmar, Antonín Novotný, Milan Pokorný

Abstract: We study convolutions with Oseen kernels (weakly singular and singular) in three-dimensional space. We give a detailed weighted L^p theory for $p \in (1; \infty]$ for anisotropic weights.

Keywords: Oseen fundamental solution, anisotropically weighted L^p spaces, singular integrals.

1. INTRODUCTION AND BASIC NOTATION

This paper concerns convolution integrals whose kernels are given by the Oseen fundamental tensor $\mathcal{O}(\cdot\,;\lambda)$ and its first or second gradients $\nabla\mathcal{O}(\cdot\,;\lambda)$, $\nabla^2\mathcal{O}(\cdot\,;\lambda)$ as well as by the $\nabla\mathcal{E}(\mathbf{x})$ and $\nabla^2\mathcal{E}(\mathbf{x})$, $\mathcal{E}(\mathbf{x})$ being the fundamental solution to the Laplace equation, which play the role of the fundamental pressure. We derive estimates of these weakly and strongly singular integral operators in anisotropically weighted L^p spaces. Such estimates can be applied to the investigation of qualitative properties of solutions of the stationary compressible Navier–Stokes equations in exterior domains using the method of decomposition, see e.g. [8]. They can also be applied with some modifications to the case of stationary flows of certain non–Newtonian fluids, see e.g. [9].

It is well known that the Oseen tensor exhibits various decay properties in various directions in $\mathbb{R}^N$, this is the mathematical reason for dealing with anisotropically weighted L^p spaces. Our work is based on the technique proposed by Farwig in [2], [3], where the volume potentials $\nabla^k\mathcal{O}{*}f$, $k = 0,1,2$ are studied in anisotropically weighted L^2 spaces with the weight function given by the formula $\eta^\alpha_\beta(\mathbf{x}) = (1+|\mathbf{x}|)^\alpha(1+s(\mathbf{x}))^\beta$, $s(\mathbf{x}) = |\mathbf{x}| - x_1$. We also used the results of Kurtz and Wheeden concerning singular integrals in L^p weight spaces, see [7]. The aim of

Applied Nonlinear Analysis, edited by Sequeira *et al.*
Kluwer Academic / Plenum Publishers, New York, 1999.

our work was to generalize Farwig's results on the case of L^p, $p \in (1;\infty]$ with the weight function η^α_β, and with the weight function $\nu^\alpha_\beta(\mathbf{x}) = |\mathbf{x}|^\alpha(1+s(\mathbf{x}))^\beta$. In order to study the dependence of estimates on Reynolds number Re, the weight functions $\nu^\alpha_\beta(\mathbf{x};\lambda) = |\mathbf{x}|^\alpha(1+s(\lambda\mathbf{x}))^\beta$, $\eta^\alpha_\beta(\mathbf{x};\lambda) = \eta^\alpha_\beta(\lambda\mathbf{x};1) = (1+\lambda|\mathbf{x}|)^\alpha(1+s(\lambda\mathbf{x}))^\beta$, $\lambda = 2Re$ are also used.

Our paper is organized as follows. We first introduce the fundamental Oseen solution and show its asymptotic properties. In Section 3 we calculate L^∞–weighted estimates of a certain convolution which plays an essential role in the next section where the L^∞–estimates of convolutions with Oseen kernels are studied. Applying this results we get, in Sections 5 and 6, the L^p–weighted theory of Oseen potentials (both weakly singular and singular kernels).

In this paper, we use the following notation

$c, c_0, c_1, \dots, C, C_0, C_1, \dots$ — positive constants

$s(\mathbf{x}) = |\mathbf{x}| - x_1$, $|\mathbf{x}| = (x_1 + x_2 + \ldots + x_N)^{1/2}$, $\mathbf{x} \in \mathbb{R}^N$

$\mathcal{E}(\mathbf{x})$ — fundamental solution of the Laplace equation

$(\mathcal{O}(\mathbf{x};\lambda), \mathcal{P}(\mathbf{x}))$— fundamental solution of the Oseen problem

$(\mathcal{S}(\mathbf{x}), \mathcal{P}(\mathbf{x}))$— fundamental solution of the Stokes problem

$\eta^\alpha_\beta(\mathbf{x}) = (1+|\mathbf{x}|)^\alpha(1+s(\mathbf{x}))^\beta$, $\sigma^\alpha_\beta(\mathbf{x}) = |\mathbf{x}|^\alpha s(\mathbf{x})^\beta$,
$\nu^\alpha_\beta(\mathbf{x}) = |\mathbf{x}|^\alpha(1+s(\mathbf{x}))^\beta$, $\mu^{\alpha,\gamma}_\beta(\mathbf{x}) = \eta^{\alpha-\gamma}_\beta(\mathbf{x})\nu^\gamma_0(\mathbf{x})$
$\nu^\alpha_\beta(\mathbf{x};\lambda) = |\mathbf{x}|^\alpha(1+s(\lambda\mathbf{x}))^\beta$, $\eta^\alpha_\beta(\mathbf{x};\lambda) = (1+|\lambda\mathbf{x}|)^\alpha(1+s(\lambda\mathbf{x}))^\beta$

$L^p(\Omega;w) = \left\{f; \|f\|^p_{p,(w),\Omega} = \int_\Omega |f|^p w\,\mathrm{d}\mathbf{x} < +\infty\right\}$,
$\mathcal{L}^p(\Omega;w) = \overline{C_0^\infty(\Omega)}^{\|\cdot\|_{p,(w),\Omega}}$, $p > 1$, $w > 0$ (usually $w = \eta^\alpha_\beta, \sigma^\alpha_\beta, \nu^\alpha_\beta$), $\Omega \subseteq \mathbb{R}^N$, $N = 3$

$B_r(a) = \{\mathbf{x} \in \mathbb{R}^N, |\mathbf{x} - \mathbf{a}| < r\}$, $B^r(a) = \{\mathbf{x} \in \mathbb{R}^N, |\mathbf{x} - \mathbf{a}| > r\}$, $\partial B_r(a) := \{\mathbf{x} \in \mathbb{R}^N, |\mathbf{x} - \mathbf{a}| = r\}$, $r \in \mathbb{R}^1$, $\mathbf{a} \in \mathbb{R}^N$, $|\cdot|$—norm in $\mathbb{R}^N$

2. OSEEN FUNDAMENTAL SOLUTION

In this section we recall some basic facts about the fundamental solution to the Oseen problem. Denote by $\mathcal{O}(\cdot;\lambda) = (\mathcal{O}_{ij}(\cdot;\lambda))$, $\mathcal{P} = (\mathcal{P}_i)$ its fundamental solution; it satisfies the identities

$$\begin{aligned} \partial_j \mathcal{O}_{ij} &= 0 \\ \Delta\mathcal{O}_{ij} + \partial_j\mathcal{P}_i - \lambda\partial_1\mathcal{O}_{ij} &= \delta_{ij}\delta \end{aligned} \tag{2.1}$$

in the sense of distributions, where δ_{ij} denotes the Kronecker delta, while δ denotes the Dirac delta–distribution. The latter is equivalent to

$$\eta_i(\mathbf{x}) = [\partial_k \mathcal{O}_{ij}(\cdot\,;\lambda) * \partial_k \eta_j + \mathcal{P}_i * \partial_j \eta_j - \lambda \mathcal{O}_{ij}(\cdot\,;\lambda) * \partial_1 \eta_j](\mathbf{x})$$
$$\forall \eta \in C_0^\infty(\mathbb{R}^N).$$

In particular, it holds,

$$\Delta \mathcal{O}_{ij}(\mathbf{x};\lambda) - \lambda \partial_1 \mathcal{O}_{ij}(\mathbf{x};\lambda) + \partial_j \mathcal{P}_i(\mathbf{x}) = 0$$

pointwise in $\mathbb{R}^N \setminus \{\mathbf{0}\}$.

In three space dimensions we can easily verify (see e.g. [6]) that the fundamental solution can be written as

$$\mathcal{P}_i(\mathbf{x}) = \partial_i \mathcal{E}(\mathbf{x}) = \frac{1}{4\pi} \frac{x_i}{|\mathbf{x}|^3} \tag{2.2}$$

$$\mathcal{O}_{ij}(\mathbf{x};\lambda) = (\delta_{ij}\Delta - \partial_i \partial_j) \varphi_{\mathcal{O}}(\mathbf{x};\lambda)\,, \tag{2.3}$$

where

$$\varphi_{\mathcal{O}}(\mathbf{x};\lambda) = \frac{-1}{4\pi\lambda} \psi\Big(\frac{\lambda s(\mathbf{x})}{2}\Big) \tag{2.4}$$

with

$$\psi(z) = \int_0^z \frac{1 - e^{-t}}{t} \mathrm{d}t = \sum_{i=1}^{\infty} \frac{(-1)^{i+1}}{i!i} z^i \tag{2.5}$$

and

$$s(\mathbf{x}) = |\mathbf{x}| - x_1. \tag{2.6}$$

The formulas (2.4)–(2.6) yield useful rescaling property

$$\lambda\, \mathcal{O}(\lambda \mathbf{x}; 1) = \mathcal{O}(\mathbf{x};\lambda). \tag{2.7}$$

The integral representation (2.5) implies

$$\psi'(t) = \tfrac{1-e^{-t}}{t}\,, \quad \psi''(t) = \tfrac{-1+e^{-t}+te^{-t}}{t^2}\,,$$
$$\psi'''(t) = \tfrac{2-2e^{-t}-2te^{-t}-t^2e^{-t}}{t^3}\,,$$
$$\psi^{(iv)}(t) = \tfrac{-6+6e^{-t}+6te^{-t}+3t^2e^{-t}+t^3e^{-t}}{t^4}\,.$$

The representation by the sum in (2.5) yields,

$$\psi^{(k)}(t) = \frac{(-1)^{k+1}}{k} + O(t) \quad \text{as } t \to 0\,, \quad k = 1, 2, \ldots. \tag{2.8}$$

When differentiating (2.6), we obtain

$$\frac{\partial s(\mathbf{x})}{\partial x_i} = \frac{x_i}{|\mathbf{x}|} - \delta_{1i}\,. \tag{2.9}$$

From here we get the estimates

$$\left|\frac{\partial s(\mathbf{x})}{\partial x_k}\right| \le \begin{cases} \frac{s(\mathbf{x})}{|\mathbf{x}|} & (k=1) \\ \sqrt{2}\sqrt{\frac{s(\mathbf{x})}{|\mathbf{x}|}} & (k \neq 1) \end{cases} \qquad |D^\alpha s(\mathbf{x})| \le \frac{c(\alpha)}{|\mathbf{x}|^{|\alpha|-1}}. \tag{2.10}$$

From (2.4)–(2.6) and (2.10) it is seen that $\mathcal{O}(\cdot\,;\cdot) \in C^\infty((\mathbb{R}^3\setminus\{0\})\times\mathbb{R})$ and for fixed $\mathbf{x} \neq \mathbf{0}$, $\mathcal{O}(\mathbf{x};\cdot)$ is an analytic function.

Now we calculate the derivatives of $\varphi_{\mathcal{O}}(\cdot\,;\lambda)$ in order to establish the asymptotic behaviour of $\mathcal{O}(\cdot\,;\lambda)$ and of its first and second derivatives near zero and at infinity.

$$\begin{aligned}
-\partial_i\varphi_{\mathcal{O}}(\mathbf{x};\lambda) &= \tfrac{\lambda}{8\pi}\psi'(\tfrac{\lambda s(\mathbf{x})}{2})\,\partial_i s(\mathbf{x}) \\
-\partial_r\partial_i\varphi_{\mathcal{O}}(\mathbf{x};\lambda) &= \tfrac{1}{16\pi}\psi''(\tfrac{\lambda s(\mathbf{x})}{2})\,\partial_r s(\mathbf{x})\,\partial_i s(\mathbf{x}) + \tfrac{1}{8\pi}\psi'(\tfrac{\lambda s(\mathbf{x})}{2})\,\partial_r\partial_i s(\mathbf{x}) \\
-\partial_k\partial_r\partial_i\varphi_{\mathcal{O}}(\mathbf{x};\lambda) &= \tfrac{\lambda^2}{32\pi}\psi'''(\tfrac{\lambda s(\mathbf{x})}{2})\,\partial_k s(\mathbf{x})\partial_r s(\mathbf{x})\partial_i s(\mathbf{x}) \\
&\quad+\tfrac{\lambda}{16\pi}\psi''(\tfrac{\lambda s(\mathbf{x})}{2})[\partial_k\partial_r s(\mathbf{x})\,\partial_i s(\mathbf{x}) + \partial_k\partial_i s(\mathbf{x})\,\partial_r s(\mathbf{x}) \\
&\quad+\partial_r\partial_i s(\mathbf{x})\,\partial_k s(\mathbf{x})] + \tfrac{1}{8\pi}\psi'\tfrac{\lambda s(\mathbf{x})}{2})\,\partial_k\partial_r\partial_i s(\mathbf{x}) \\
-\partial_l\partial_k\partial_r\partial_i\varphi_{\mathcal{O}}(\mathbf{x};\lambda) &= \tfrac{\lambda^3}{64\pi}\psi^{(iv)}(\tfrac{\lambda s(\mathbf{x})}{2})\,\partial_l s(\mathbf{x})\,\partial_k s(\mathbf{x})\,\partial_r s(\mathbf{x})\,\partial_i s(\mathbf{x}) \\
&\quad+\tfrac{\lambda^2}{32\pi}\psi'''(\tfrac{\lambda s(\mathbf{x})}{2})\,[\partial_l\partial_k s(\mathbf{x})\partial_r s(\mathbf{x})\partial_i s(\mathbf{x}) \\
&\quad+\partial_l\partial_r s(\mathbf{x})\partial_k s(\mathbf{x})\partial_i s(\mathbf{x}) + \partial_l\partial_i s(\mathbf{x})\partial_k s(\mathbf{x})\partial_r s(\mathbf{x}) \\
&\quad+\partial_k\partial_r s(\mathbf{x})\partial_i s(\mathbf{x})\partial_l s(\mathbf{x}) + \partial_k\partial_i s(\mathbf{x})\partial_r s(\mathbf{x})\partial_l s(\mathbf{x}) \\
&\quad+\partial_r\partial_i s(\mathbf{x})\partial_k s(\mathbf{x})\partial_l s(\mathbf{x})] + \tfrac{\lambda}{16\pi}\psi''(\tfrac{\lambda s(\mathbf{x})}{2}) \\
&\quad\cdot[\partial_l\partial_k\partial_r s(\mathbf{x})\,\partial_i s(\mathbf{x}) + \partial_l\partial_k\partial_i s(\mathbf{x})\,\partial_r s(\mathbf{x}) \\
&\quad+\partial_l\partial_r\partial_i s(\mathbf{x})\,\partial_k s(\mathbf{x}) + \partial_k\partial_r\partial_i s(\mathbf{x})\,\partial_l s(\mathbf{x}) \\
&\quad+\partial_i\partial_k s(\mathbf{x})\,\partial_r\partial_l s(\mathbf{x}) + \partial_r\partial_k s(\mathbf{x})\,\partial_i\partial_l s(\mathbf{x}) \\
&\quad+\partial_i\partial_r s(\mathbf{x})\,\partial_k\partial_l s(\mathbf{x})] + \tfrac{1}{8\pi}\psi'(\tfrac{\lambda s(\mathbf{x})}{2})\,\partial_l\partial_k\partial_r\partial_i s(\mathbf{x})
\end{aligned}$$

These formulas, together with (2.8), (2.10) and (2.3) yield

$$\begin{aligned}
\mathcal{O}(\mathbf{x};\lambda) &= \mathcal{S}(\mathbf{x}) + \lambda O(1) \quad \text{as } \lambda|\mathbf{x}| \to 0 \\
\nabla\mathcal{O}(\mathbf{x};\lambda) &= \nabla\mathcal{S}(\mathbf{x}) + \lambda^2 O(\tfrac{1}{\lambda|\mathbf{x}|}) \quad \text{as } \lambda|\mathbf{x}| \to 0 \\
\nabla^2\mathcal{O}(\mathbf{x};\lambda) &= \nabla^2\mathcal{S}(\mathbf{x}) + \lambda^3 O(\tfrac{1}{\lambda^2|\mathbf{x}|^2}) \quad \text{as } \lambda|\mathbf{x}| \to 0,
\end{aligned} \tag{2.11}$$

where $(\mathcal{S},\mathcal{P})$ is the Stokes fundamental solution (see e.g. [6]),

$$\mathcal{S}_{ij}(\mathbf{x}) = \frac{-1}{8\pi}\left[\frac{\delta_{ij}}{|\mathbf{x}|} + \frac{x_i x_j}{|\mathbf{x}|^3}\right]. \tag{2.12}$$

It can be shown (see e.g. [6] and also Section 6) that both the second derivative of $\mathcal{S}$ and $\nabla\mathcal{P}$ represent Calderón–Zygmund singular integral kernels.

In particular, for $\lambda \in (0;\lambda_0)$, $R > 0$ and $|\lambda \mathbf{x}| \le R$

$$\left|\nabla^k \mathcal{O}(\mathbf{x};\lambda)\right| \le \frac{c(R;\lambda_0,k)}{|\mathbf{x}|^{k+1}}. \tag{2.13}$$

For any $\mathbf{x} \neq \mathbf{0}$ formulas (2.3) and (2.10) together with the properties of the function $s(\mathbf{x})$ give

$$\begin{aligned}
|\mathcal{O}(\mathbf{x};\lambda)| &\le \frac{c}{\lambda}\frac{1-e^{-\frac{\lambda s(\mathbf{x})}{2}}}{s(\mathbf{x})|\mathbf{x}|}\\
|\nabla\mathcal{O}(\mathbf{x};\lambda)| &\le \frac{c}{\lambda}\Big[\frac{1-e^{-\frac{\lambda s(\mathbf{x})}{2}}}{s(\mathbf{x})|\mathbf{x}|^2} + \frac{1-e^{-\frac{\lambda s(\mathbf{x})}{2}} - \frac{\lambda s(\mathbf{x})}{2}e^{-\frac{\lambda s(\mathbf{x})}{2}}}{s^{3/2}(\mathbf{x})|\mathbf{x}|^{3/2}}\Big]\\
|\nabla^2\mathcal{O}(\mathbf{x};\lambda)| &\le \frac{c}{\lambda}\Big[\frac{1-e^{-\frac{\lambda s(\mathbf{x})}{2}}}{s(\mathbf{x})|\mathbf{x}|^3} + \frac{1-e^{-\frac{\lambda s(\mathbf{x})}{2}} - \frac{\lambda s(\mathbf{x})}{2}e^{-\frac{\lambda s(\mathbf{x})}{2}}}{s^{2}(\mathbf{x})|\mathbf{x}|^{2}}\Big].
\end{aligned} \tag{2.14}$$

This yields for $|\lambda\mathbf{x}| \ge R$ and any $\kappa > 0$

$$\begin{aligned}
|\mathcal{O}(\mathbf{x};\lambda)| &\le \frac{c(\kappa,R)}{|\mathbf{x}|(\kappa+s(\lambda\mathbf{x}))}\\
|\nabla\mathcal{O}(\mathbf{x};\lambda)| &\le \frac{c(\kappa,R)\lambda^{1/2}}{|\mathbf{x}|^{3/2}(\kappa+s(\lambda\mathbf{x}))^{3/2}}\\
|\nabla^2\mathcal{O}(\mathbf{x};\lambda)| &\le \frac{c(\kappa,R)\lambda}{|\mathbf{x}|^{2}(\kappa+s(\lambda\mathbf{x}))^{2}}.
\end{aligned} \tag{2.15}$$

Formulas (2.12) and (2.15) give us in particular that $\mathcal{O}$ and $\nabla\mathcal{O}$ are analogous to $\mathcal{P}$ weakly singular kernels while the second derivative of $\mathcal{O}$ can be written as a sum of a singular kernel ($\mathcal{S}$) and a weakly singular part.

With $(\mathcal{O},\mathcal{P})$ at hand, we can write explicitly a C^∞–solution of the problem

$$-\Delta u_i + \lambda\partial_1 u_i + \partial_i\Pi = f_i$$
$$\nabla\cdot\mathbf{u} = g$$

with $\mathbf{f}, g \in C_0^\infty(\mathbb{R}^3)$. Namely

$$u_i = -\mathcal{O}_{ij}(\cdot\,;\lambda) * f_j + \mathcal{P}_i * g$$
$$\Pi = \mathcal{P}_j * f_j + g - \lambda\partial_1(\mathcal{E} * g) = \mathcal{P}_j * f_j + g - \lambda\mathcal{P}_1 * g\,.$$

In the case of $\mathbf{f} = \mathrm{div}\mathcal{F}$ we have

$$\begin{aligned}
u_i &= -\partial_s\mathcal{O}_{ij} * \mathcal{F}_{js} + \mathcal{P}_i * g\\
\Pi &= \partial_s\mathcal{P}_j * \mathcal{F}_{js} + g - \lambda\mathcal{P}_1 * g + c_{js}\mathcal{F}_{js}\,,
\end{aligned}$$

where c_{js} is a constant.

Let us summarize the asymptotic behaviour of $\mathcal{O}$ and its derivatives using the weights ν_β^α, $\mu_\beta^{\alpha,\gamma}$ and η_β^α, introduced in Section 1. Moreover, we assume $\lambda = 1$ and for $\lambda \neq 1$ we may use the homogeneity property (2.7). Then we have for $\mathbf{x} \in \mathbb{R}^3 \setminus \{\mathbf{0}\}$

$$\begin{aligned} |\mathcal{O}(\mathbf{x};1)| &\leq C\nu_{-1}^{-1}(\mathbf{x}) \\ |\nabla \mathcal{O}(\mathbf{x};1)| &\leq C\mu_{-\frac{3}{2}}^{-\frac{3}{2},-2}(\mathbf{x}) \\ |\partial_1 \mathcal{O}(\mathbf{x};1)| &\leq C\nu_{-1}^{-2}(\mathbf{x}) \\ |\nabla^2 \mathcal{O}(\mathbf{x};1) - \nabla^2 \mathcal{S}(\mathbf{x})| &\leq C\nu_{-1}^{-2}(\mathbf{x}) \,. \end{aligned} \tag{2.16}$$

3. L^∞–ESTIMATES OF A CONVOLUTION IN $\mathbb{R}^N$

This section is devoted to the study of an auxiliary problem — the L^∞ estimates of certain convolution which will play a fundamental role in the following sections. Our aim is to give conditions on a, b, c, d, e, f such that

$$(\eta_{-b}^{-a} * \eta_{-d}^{-c})(\mathbf{x}) \leq K\eta_{-f}^{-e}(\mathbf{x}) \,, \quad \mathbf{x} \in \mathbb{R}^N \,. \tag{3.1}$$

We shall calculate the estimates for $N \geq 2$. Since we study the physically interesting case $N = 3$, the results will be summarized in Tables 1 and 2 only in this situation. Nevertheless, the calculations will be performed for general dimensions and the results can be easily read from the integrals $I_0 - I_{15}$.

Before calculating estimates of the type (3.1), we shall first study the asymptotic behaviour of the function

$$s(\mathbf{x}) = |\mathbf{x}| - x_1 \,.$$

Let us denote by $\mathbf{x}'$ the vector of the last $N-1$ components of $\mathbf{x}$, i.e. $\mathbf{x} = (x_1, \mathbf{x}')$. We have

Lemma 3.1. *If $x_1 > 0$ then $s(\mathbf{x}) \sim \frac{|\mathbf{x}'|^2}{|\mathbf{x}|}$; otherwise $s(\mathbf{x}) \sim |\mathbf{x}|$.*

Proof. Introducing the generalized spherical coordinates ($N \geq 3$)

$$\begin{aligned} x_1 &= R\cos\theta_1 \\ x_2 &= R\sin\theta_1 \cos\theta_2 \\ &\cdots \\ x_{N-1} &= R\sin\theta_1 \cdots \sin\theta_{N-2} \cos\theta_{N-1} \\ x_N &= R\sin\theta_1 \cdots \sin\theta_{N-2} \sin\theta_{N-1} \,, \end{aligned} \tag{3.2}$$

where $\theta_1, \cdots, \theta_{N-2} \in (0;\pi)$, $\theta_{N-1} \in (0;2\pi)$, we have

$$s = R(1 - \cos\theta_1) = 2\frac{(R\sin\theta_1)^2}{R}\Big(\frac{\sin(\theta_1/2)}{\sin\theta_1}\Big)^2 \,.$$

For $x_1 > 0$ we have

$$\theta_1 \in (0;\pi/2)\,, \text{ i.e. } 2\Big(\frac{\sin(\theta_1/2)}{\sin\theta_1}\Big)^2 \in (1/2;1)$$

which implies

$$\frac{1}{2}\frac{|\mathbf{x}'|^2}{R} \le s(\mathbf{x}) \le \frac{|\mathbf{x}'|^2}{R}\,.$$

Analogously we proceed for $x_1 < 0$ where $\theta_1 \in (\pi/2;\pi)$ and $|\mathbf{x}| \le s(\mathbf{x}) \le 2|\mathbf{x}|$. If $N = 2$ we use the polar coordinates and the only change consists in the fact that $\varphi = \theta_1 \in (-\pi/2;\pi/2)$ for $x_1 > 0$ and $\varphi \in [\pi/2; 3/2\pi]$ for $x_1 \le 0$. ■

Next we study the integral of $\eta_{-b}^{-a}(\mathbf{x})$ over the sphere for sufficiently large $R = |\mathbf{x}|$. In what follows, by $f(\mathbf{x}) \sim g(\mathbf{x})$ as $|\mathbf{x}| \to A$ we mean the following: there exist C_1, $C_2 > 0$ and $U(A)$, a neighbourhood of A, such that

$$C_1 f(\mathbf{x}) \le g(\mathbf{x}) \le C_2 f(\mathbf{x})$$

for all $\mathbf{x}$ such that $|\mathbf{x}|$ is from $U(A)$.

Lemma 3.2. *Let $N \ge 2$. Then for the exponents a, $b \in \mathbb{R}$ we have*

$$\int_{\partial B_R} \eta_{-b}^{-a}(\mathbf{x})dS \sim R^{N-1-a-\min(\frac{N-1}{2},b)} \cdot (\ln R \text{ if } b = (N-1)/2) \tag{3.3}$$

as $R \to \infty$.

Consequently, $\int_{\mathbb{R}^N} \eta_{-b}^{-a}(\mathbf{x})d\mathbf{x} < \infty \iff a + \min(\frac{N-1}{2}, b) > N$.

Proof. Using the generalized spherical coordinates (if $N \ge 3$ — see (3.2)) or the polar ones ($N = 2$) we get

$$\begin{aligned}\int_{\partial B_R} \eta_{-b}^{-a}(\mathbf{x})\mathrm{d}S &= C\int_0^\pi (1+R)^{-a}(1+s)^{-b}R^{N-1}(\sin\theta_1)^{N-2}\mathrm{d}\theta_1 \\ &= C\int_0^\pi (1+R)^{-a}(1+R(1-\cos\theta_1))^{-b}R^{N-1}\sin\theta_1^{N-2}\mathrm{d}\theta_1\,.\end{aligned}$$

Changing the variables $s = R(1-\cos\theta_1)$ we estimate the last integral by

$$C(1+R)^{1-a}\int_0^{2R}(1+s)^{-b}(\sqrt{2sR-s^2})^{N-3}\mathrm{d}s\,. \tag{3.4}$$

We estimate the integral (3.4) over three subintervals. Let us also note that for $N = 3$ it can be calculated explicitly. We have

$$\begin{aligned}\int_0^1 (1+s)^{-b}(2sR-s^2)^{\frac{N-3}{2}}\mathrm{d}s &\sim R^{\frac{N-3}{2}}\int_0^1 s^{\frac{N-3}{2}}\mathrm{d}s \sim R^{\frac{N-3}{2}}\,, \\ \int_1^R (1+s)^{-b}(2sR-s^2)^{\frac{N-3}{2}}\mathrm{d}s &\sim R^{\frac{N-3}{2}}\int_1^R s^{-b+\frac{N-3}{2}}\mathrm{d}s \\ &\sim R^{N-2-\min(b,\frac{N-1}{2})}\cdot(\ln R \text{ if } b = \tfrac{N-1}{2})\,, \\ \int_R^{2R} (1+s)^{-b}(2sR-s^2)^{\frac{N-3}{2}}\mathrm{d}s &\sim R^{N-2-b}\end{aligned}$$

which implies (3.3). As $\eta_{-b}^{-a}(\cdot) \in C(\mathbb{R}^N)$, the condition implying global integrability follows trivially. ∎

We can start to deal with the convolution (3.1). Recall that similar estimates were for the first time studied by Finn (see [4], [5]) in the three-dimensional case (but only for special values of c, d) and by Smith (see [10]) in the two-dimensional case. A generalization of their approach due to Farwig (see [2]) for the three-dimensional case was adapted to the two-dimensional case by Dutto (see [1]). We shall repeat their calculation in N dimensions where $N \geq 2$, arbitrary.

Finally note that the estimate (3.1) remains true if we replace the kernel $\eta_{-d}^{-c}(\mathbf{x}-\mathbf{y})$ by

$$K(\mathbf{z}) \sim \begin{cases} |\mathbf{z}|^{-\gamma} & \mathbf{z} \in B_1(\mathbf{0}), \gamma < N \\ \eta_{-d}^{-c}(\mathbf{z}) & \mathbf{z} \in B^1(\mathbf{0}) \end{cases} \tag{3.5}$$

(see also I_1 below).

In the sequel we shall use the following notation

$$\begin{array}{lll} \mathbf{x} = (x_1, \mathbf{x}') & \mathbf{y} = (y_1, \mathbf{y}') & \\ R = |\mathbf{x}| & r = |\mathbf{y}| & \tilde{r} = |\mathbf{x}-\mathbf{y}| \\ s = s(\mathbf{x}) & t = y_1 & \tilde{t} = x_1 - y_1 \\ & \varrho = |\mathbf{y}'| & \tilde{\varrho} = |\mathbf{x}' - \mathbf{y}'| . \end{array} \tag{3.6}$$

In order to capture the anisotropic structure of the function $\eta_{-b}^{-a}(\cdot)$ we shall study the convolution (3.1) in four different situations:

A) $R \leq R_0$

B) $x_1 > 0$, $|\mathbf{x}'| \leq \sqrt{x_1}$, $R > R_0$

C) $x_1 > 0$, $|\mathbf{x}'| = \frac{1}{2} R^{\frac{1}{2}+\sigma}$, $R > R_0$, $\sigma \in [0; \frac{1}{2}]$

D) $x_1 > 0$, $|\mathbf{x}'| \geq \frac{R}{2}$, $R > R_0$ or $x_1 < 0$, $R > R_0$.

Using Lemma 3.1 we easily verify that

$$\begin{gathered} \eta_{-b}^{-a}(\mathbf{y}) \sim \begin{cases} 1, & r \leq 1 \\ r^{-a}, & r > 1, \quad t > 0, \quad \varrho < \sqrt{t} \\ r^{-a+b}\varrho^{-2b}, & r > 1, \quad t > 0, \quad \varrho \geq \sqrt{t} \\ r^{-a-b}, & r > 1, \quad t \leq 0 \end{cases} \\ \eta_{-d}^{-c}(\mathbf{x}-\mathbf{y}) \sim \begin{cases} 1, & \tilde{r} \leq 1 \\ \tilde{r}^{-c}, & \tilde{r} > 1, \quad \tilde{t} > 0, \quad \tilde{\varrho} < \sqrt{\tilde{t}} \\ \tilde{r}^{-c+d}\tilde{\varrho}^{-2d}, & \tilde{r} > 1, \quad \tilde{t} > 0, \quad \tilde{\varrho} \geq \sqrt{\tilde{t}} \\ \tilde{r}^{-c-d}, & \tilde{r} > 1, \quad \tilde{t} \leq 0 . \end{cases} \end{gathered} \tag{3.7}$$

For notational convenience we denote $\nu = \sigma + \frac{1}{2}$ and $b^* = \min(\frac{N-1}{2}, b)$; analogously $d^* = \min(\frac{N-1}{2}, d)$.

We start with the case A). Applying Lemma 3.2 to the halfspaces $y_1 > 0$ and $y_1 < 0$ we find that the convolution is uniformly bounded if

$$a + b^* + c + d > N \qquad a + b + c + d^* > N\,. \tag{3.8}$$

Next we continue with the most complicated case C). We follow Farwig (see [2]) and divide $\mathbb{R}^N$ into 16 subdomains as shown in Figure 1. If $N = 2$ the subdomains are plane, otherwise they are cylindrical.

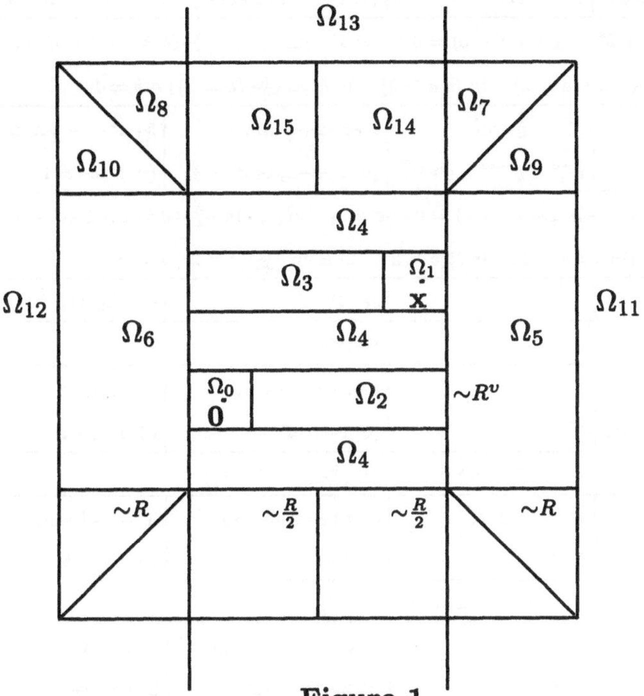

Figure 1

We calculate the convolutions separately on each subdomain. For the reader's convenience, the results are summarized in Tables 1 and 2 (for $N = 3$). We denote by I_k the corresponding part of the integral (3.1) over Ω_k, $k = 0, 1, \cdots, 15$. We shall get

$$I_i(\mathbf{x}) \leq KR^{-e_i - 2\sigma f_i} \sim \eta_{-f_i}^{-e_i}(\mathbf{x})\,.$$

Unfortunately in many cases additional logarithmic terms will appear which will cause some losses in the weighted estimates later on.

Remark 3.3. Let A be a positive function. We denote

$$\ln_+ A = \max(\ln A, 1)\,.$$

Dom.	$\eta^{-a}_{-b}(\mathbf{y})$	$\eta^{-c}_{-d}(\mathbf{x}-\mathbf{y})$	e	f
	logarithmic factors			
Ω_0	$r^{-a}(1+s(\mathbf{y}))^{-b}$	$R^{-c-2\sigma d}$	$c+\frac{1}{2}\min(0, a+b^*-3)$	$d+\frac{1}{2}\min(0, a+b^*-3)$
	$\ln R (b=1 \wedge a<2) \vee (b \neq 1 \wedge a+b^*=3) \quad \ln^2 R (b=1 \wedge a=2)$			
Ω_1	$R^{-a-2\sigma b}$	$\tilde{r}^{-c}(1+s(\mathbf{x}-\mathbf{y}))^{-d}$	$a+\frac{1}{2}\min(0, c+d^*-3)$	$b+\frac{1}{2}\min(0, c+d^*-3)$
	$\ln R (d=1 \wedge c<2) \vee (d \neq 1 \wedge c+d^*=3) \quad \ln^2 R (d=1 \wedge c=2)$			
Ω_2	$r^{-a} \quad \varrho<\sqrt{t}$ $r^{-a+b}\varrho^{-2b} \quad \varrho>\sqrt{t}$	$\tilde{r}^{-c+d}R^{-2d\upsilon}$	$a+c-2+\frac{1}{2}\min$ $(0, 1+b^*-a, 1+d-c)$	$b^*+d-1-\frac{1}{2}\min$ $(0, 1+b^*-a, 1+d-c)$
	$\ln \frac{R}{1+s}(\min(1+b^*-a, 1+d-c)=0 \wedge b \neq 1) \quad \ln_+ s \cdot \ln \frac{R}{1+s}(b=1 \wedge 1+d-c=0)$ $(\ln_+ s (a<2) \quad \ln R (a>2) \quad \ln R \ \ln \frac{R}{1+s}(a=2)) \wedge b=1$			
Ω_3	$r^{-a+b}R^{-2b\upsilon}$	$\tilde{r}^{-c} \quad \tilde{\varrho}<\sqrt{\tilde{t}}$ $\tilde{r}^{-c-d}\varrho^{-2d} \quad \varrho>\sqrt{\tilde{t}}$	$a+c-2+\frac{1}{2}\min$ $(0, 1+b-a, 1+d^*-c)$	$b+d^*-1-\frac{1}{2}\min$ $(0, 1+b-a, 1+d^*-c)$
	$\ln \frac{R}{1+s}(\min(1+b-a, 1+d^*-c)=0 \wedge d \neq 1) \quad \ln_+ s \cdot \ln \frac{R}{1+s}(d=1 \wedge 1+b-a=0)$ $(\ln_+ s (c<2) \quad \ln R (c>2) \quad \ln R \ \ln \frac{R}{1+s}(c=2)) \wedge d=1$			
Ω_4			see Ω_2, Ω_3	see Ω_2, Ω_3
	see Ω_2, Ω_3			
Ω_5	$R^{-a} \quad \varrho<\sqrt{t}$ $R^{-a+b}\varrho^{-2b} \quad \varrho>\sqrt{t}$	$\lvert\tilde{t}\rvert^{-c-d}$	$a+c+d-2+\frac{1}{2}\min$ $(0, 1-c-d)$	$b^*-1-\frac{1}{2}\min$ $(0, 1-c-d)$
	$(\ln_+ s (b=1)) \cdot (\ln \frac{R}{1+s}(c+d=1))$			
Ω_6	$\lvert t\rvert^{-c-d}$	$R^{-c} \quad \tilde{\varrho}<\sqrt{\tilde{t}}$ $R^{-c+d}\varrho^{-2d} \quad \tilde{\varrho}>\sqrt{\tilde{t}}$	$a+b+c-2+\frac{1}{2}\min$ $(0, 1-a-b)$	$d^*-1-\frac{1}{2}\min$ $(0, 1-a-b)$
	$(\ln_+ s (d=1)) \cdot (\ln \frac{R}{1+s}(a+b=1))$			
Ω_7	$R^{-a+b}\varrho^{-2b}$	ϱ^{-c-d}	$a+b+c+d-3+\frac{1}{2}\min$ $(0, 3-2b-c-d)$	$-\frac{1}{2}\min(0,$ $3-2b-c-d)$
	$\ln \frac{R}{1+s}(2b+c+d=3)$			
Ω_8	$\tilde{\varrho}^{-c-d}$	$R^{-a+b}\tilde{\varrho}^{-2b}$	$a+b+c+d-3+\frac{1}{2}\min$ $(0, 3-a-b-2d)$	$-\frac{1}{2}\min(0,$ $3-a-b-2d)$
	$\ln \frac{R}{1+s}(a+b+2d=3)$			

Table 1

Dom.	$\eta_{-b}^{-a}(\mathbf{y})$	$\eta_{-d}^{-c}(\mathbf{x}-\mathbf{y})$	e	f
	logarithmic factors			
Ω_9			$b>1$ see Ω_5	$b>1$ see Ω_5
			$b<1$ see Ω_7	$b<1$ see Ω_7
	$R^{-a+b}\varrho^{-2b}$	$\lvert\tilde{t}\rvert^{-c-d}$	$b=1$ see Ω_5	$b=1$ see Ω_5
	$(\ln\frac{R}{1+s}(c+d<1)\quad \ln^2\frac{R}{1+s}(c+d=1))\wedge b=1$			
Ω_{10}			$d>1$ see Ω_6	$d>1$ see Ω_6
			$d<1$ see Ω_8	$d<1$ see Ω_8
	$\lvert t\rvert^{-a-b}$	$R^{-c+d}\tilde{\varrho}^{-2d}$	$d=1$ see Ω_6	$d=1$ see Ω_6
	$(\ln\frac{R}{1+s}(a+b<1)\quad (\ln^2\frac{R}{1+s}(a+b=1))\wedge d=1$			
Ω_{11}	$r^{-a}(1+s(\mathbf{y}))^{-b}$	r^{-c-d}	$a+b^*+c+d-3>0$	0
	$\ln R(b=1)$			
Ω_{12}	$\tilde{r}^{-a-b}$	$\tilde{r}^{-c}(1+s(\mathbf{x}-\mathbf{y}))^{-d}$	$a+b+c+d^*-3>0$	0
	$\ln R(d=1)$			
Ω_{13}			see Ω_{11},Ω_{12}	see Ω_{11},Ω_{12}
	see Ω_{11},Ω_{12}			
Ω_{14}			$b+d>1\wedge 1+d-c>0$	see Ω_2,Ω_3
			$1+d-c<0$	see Ω_7
		$(\tilde{t}+\tilde{\varrho})^{-c+d}\,\tilde{\varrho}^{-2d}\ \tilde{t}>0$	$1+d-c=0\wedge b+d>1$	see Ω_2
	$R^{-a+b}\,\varrho^{-2b}$	$\tilde{\varrho}^{-c-d}\quad \tilde{t}<0$	$a+b+c+d-3$	0 otherwise
	$\ln\frac{R}{1+s}(b+d=1\wedge 1+d-c>0)\quad \ln^2\frac{R}{1+s}(b+d=1\wedge 1+d-c=0)$			
Ω_{15}			$b+d>1\wedge 1+b-a>0$	see Ω_2,Ω_3
			$1+b-a<0$	see Ω_8
	$(t+\varrho)^{-a+b}\,\varrho^{-2d}\ t>0$		$1+b-a=0\wedge b+d>1$	see Ω_3
	$\varrho^{-a-b}\quad t<0$	$R^{-c+d}\,\tilde{\varrho}^{-2d}$	$a+b+c+d-3$	0 otherwise
	$\ln\frac{R}{1+s}(b+d=1\wedge 1+b-a>0)\quad \ln^2\frac{R}{1+s}(b+d=1\wedge 1+b-a=0)$			

Table 2

We start by estimating the convolutions over the sixteen subdomains.

$\underline{\mathrm{I}_0}$ We have $\Omega_0 = \{\mathbf{y} \in \mathbb{R}^N : |t| \le \frac{1}{8}R^\nu; \varrho \le \frac{1}{8}R^\nu\}$ and therefore $\eta^{-a}_{-b}(\mathbf{y}) \sim r^{-a}(1+s(\mathbf{y}))^{-b}$, $\eta^{-c}_{-d}(\mathbf{x}-\mathbf{y}) \sim R^{-c-2d\sigma}$. Applying Lemma 3.2 we get

$$I_0 \sim R^{-c-2\sigma d}\int_0^{R^\nu}(1+r)^{N-1-a-b^*}\cdot(\ln_+ r\,,\ \text{if}\ b=\tfrac{N-1}{2})\mathrm{d}r$$
$$\sim R^{-(c+\frac{1}{2}\min(0,a+b^*-N))-2\sigma(d+\min(0,a+b^*-N))}$$
$$\cdot\begin{cases}\ln R & \text{if } a+b^*=N\,,\ b\ne\frac{N-1}{2}\\ & \text{or } a+b^*<N\,,\ b=\frac{N-1}{2}\\ \ln^2 R & \text{if } a+b^*=N\,,\ b=\frac{N-1}{2}\,.\end{cases}$$

The results are summarized in Tables 1 and 2.

$\underline{\mathrm{I}_1}$ The integral can be estimated in the same way by exchanging a, b for c, d. Assuming the kernel (3.5) instead of η^{-c}_{-d} we have

$$\tilde{I}_1(\mathbf{x}) = \int_{B_1(\mathbf{0})}K(\mathbf{x}-\mathbf{y})\eta^{-a}_{-b}(\mathbf{x})\mathrm{d}\mathbf{x} \sim R^{-a-2\sigma b}$$

since $\gamma < N$. Again, the summarized results can be found in Tables 1 and 2.

$\underline{\mathrm{I}_2}$ We have in Ω_2 that $r \sim t \in (R^\nu; R)$. So $\eta^{-a}_{-b}(\mathbf{y}) \sim r^{-a}$ for $\varrho < \sqrt{t}$ and $\eta^{-a}_{-b}(\mathbf{y}) \sim r^{-a+b}\varrho^{-2b}$ for $\varrho > \sqrt{t}$. Further $\tilde{\varrho} \sim R^\nu$ and $\tilde{r} \sim R + R^\nu - r$; therefore $\eta^{-c}_{-d}(\mathbf{x}-\mathbf{y}) \sim \tilde{r}^{-c+d}R^{-2\nu d}$. Thus

$$I_2 \sim R^{-2d\nu}\int_{R^\nu}^{R}\mathrm{d}r(R+R^\nu-r)^{d-c}\Big[r^{-a}\int_0^{\sqrt{r}}\varrho^{N-2}\mathrm{d}\varrho$$
$$+r^{b-a}\int_{\sqrt{r}}^{R^\nu}\varrho^{N-2-2b}\mathrm{d}\varrho\Big] = R^{-2\nu d}\int_{R^\nu}^{R}\mathrm{d}r(R+R^\nu-r)^{d-c}$$
$$\cdot\Big[r^{\frac{N}{2}-\frac{1}{2}-a}+r^{b-a}(R^{\nu(N-1-2b)}-r^{\frac{N}{2}-\frac{1}{2}-b})\Big]\cdot(\ln\tfrac{R^\nu}{\sqrt{r}}\ \text{if}\ b=\tfrac{N-1}{2})$$
$$\sim R^{-2\nu(b^*-\frac{N-1}{2}+d)}\int_{R^\nu}^{R}\mathrm{d}r(R+R^\nu-r)^{d-c}r^{b^*-a}$$
$$\cdot(\ln\tfrac{R^\nu}{\sqrt{r}}\ \text{if}\ b=\tfrac{N-1}{2}) \equiv J\,.$$

In order to verify the last equivalence it is enough to consider first $b \ge \frac{N-1}{2}$, i.e. $b^* = \frac{N-1}{2}$, and then $b < \frac{N-1}{2}$, i.e. $b^* = b$, and estimate the integrals.

Let us divide J into two parts — the integral over $(R^\nu; \frac{R}{2})$ and the integral over $(\frac{R}{2}; R)$. We estimate these two parts separately.

$$
\begin{aligned}
&R^{-c+d-2\nu(b^*-\frac{N-1}{2}+d)} \int_{R^\nu}^{\frac{R}{2}} \mathrm{d}r\, r^{b^*-a} \cdot (\ln \tfrac{R^\nu}{\sqrt{r}} \text{ if } b = \tfrac{N-1}{2}) \\
&\sim R^{-(c+a-\frac{N+1}{2})-2\sigma(b^*-\frac{N-1}{2}+d)+(\sigma-\frac{1}{2})\min(0,b^*-a+1)} \\
&\quad \cdot(\ln_+ \tfrac{R}{1+s} \text{ if } b \neq \tfrac{N-1}{2}\,,\ a = b^*+1) \\
&\quad \cdot\Big((\ln_+ s\,,\ a < \tfrac{N+1}{2}) \cdot (\ln R\,,\ a > \tfrac{N+1}{2}) \\
&\quad \cdot(\ln R \ln_+ \tfrac{R}{1+s}\,,\ a = \tfrac{N+1}{2}) \text{ if } b = \tfrac{N-1}{2}\Big)\,.
\end{aligned}
$$

We used the fact that $s(\mathbf{x}) \sim \frac{|\mathbf{x}'|^2}{|\mathbf{x}|} \sim R^{2\sigma}$ and $\ln_+ \frac{R}{R^\nu} = \frac{1}{2} \ln_+ \frac{R}{R^{2\sigma}} \sim \ln_+ \frac{R}{1+s}$. Analogously

$$
\begin{aligned}
&R^{-2\nu(b^*-\frac{N-1}{2}+d)+b^*-a} \int_{\frac{R}{2}}^{R} \mathrm{d}r (R+R^\nu-r)^{d-c} \cdot (\ln \tfrac{R^\nu}{\sqrt{r}} \text{ if } b = \tfrac{N-1}{2}) \\
&\sim R^{-(c+a-\frac{N+1}{2})-2\sigma(b^*-\frac{N-1}{2}+d)+(\sigma-\frac{1}{2})\min(0,d-c+1)} \\
&\quad \cdot(\ln_+ \tfrac{R}{1+s} \text{ if } c = d+1) \cdot (\ln_+ s \text{ if } b = \tfrac{N-1}{2})\,.
\end{aligned}
$$

The results may again be found in Tables 1 and 2.

$\underline{I_3}$ We proceed analogously as for I_2 exchanging a, b for c, d.

$\underline{I_4}$ Ω_4 can be considered as a subset of Ω_2 and Ω_3. Therefore I_4 can be estimated by I_2 and I_3.

$\underline{I_5}$ We have in Ω_5 $t \sim r \sim R$, so $\eta_{-b}^{-a}(\mathbf{y}) \sim R^{-a}$ $(\varrho < \sqrt{t})$ or $\eta_{-b}^{-a}(\mathbf{y}) \sim R^{-a+b}\varrho^{-2b}$ $(\varrho > \sqrt{t})$ where ϱ varies between 0 and R^ν. Further $\tilde{r} \sim |\tilde{t}| \in (R^\nu; R)$. As $\tilde{t} < 0$ we have $\eta_{-d}^{-c}(\mathbf{x}-\mathbf{y}) \sim |\tilde{t}|^{-c-d}$.

$$
\begin{aligned}
I_5 &\sim \int_{R^\nu}^{R} \mathrm{d}\tau\, \tau^{-c-d}\Big(R^{-a}\int_0^{\sqrt{R}} \varrho^{N-2}\mathrm{d}\varrho + R^{b-a}\int_{\sqrt{R}}^{R^\nu} \varrho^{N-2-2b}\mathrm{d}\varrho\Big) \\
&\sim R^{1-c-d+(\sigma-\frac{1}{2})\min(0,1-c-d)}[R^{\frac{N-1}{2}-a} + (R^{(N-2-2b)\nu} - R^{\frac{N-2-2b}{2}}) \\
&\quad \cdot R^{b-a}(\ln_+ s\,,\ \text{if } b = \tfrac{N-1}{2})] \cdot (\ln \tfrac{R}{R^\nu} \text{ if } c+d = 1) \\
&\sim R^{1-c-d+\frac{N-1}{2}-a+(\sigma-\frac{1}{2})\min(0,1-c-d)+2\sigma(\frac{N-1}{2}-b^*)} \\
&\quad \cdot(\ln_+ s \text{ if } b = \tfrac{N-1}{2}) \cdot (\ln_+ \tfrac{R}{1+s} \text{ if } c+d = 1)\,.
\end{aligned}
$$

$\underline{I_6}$ It is sufficient to interchange a, b by c, d and use the result for Ω_5.

$\underline{I_7}$ Denoting $\tau = |\tilde{t}| \in (R^\nu, R)$ we have in Ω_7 that $t \sim r \sim R$, $\varrho \sim \tilde{\varrho} \sim \tilde{r} \in (\tau; R)$. Therefore $\eta_{-b}^{-a}(\mathbf{y}) \sim R^{-a+b}\rho^{-2b}$, $\eta_{-d}^{-c}(\mathbf{x}-\mathbf{y}) \sim \rho^{-c-d}$

and

$$\begin{aligned}
I_7 &\sim R^{-a+b}\int_{R^\nu}^{R}\mathrm{d}\tau\Big(\int_{\tau}^{R}\varrho^{N-2-d-c-2b}\mathrm{d}\varrho\Big)\\
&\sim R^{b-a}\int_{R^\nu}^{R}\mathrm{d}\tau(R^{N-1-c-d-2b}-\tau^{N-1-c-d-2b})\\
&\quad\cdot(\ln\tfrac{R}{\tau}\ \text{ if } c+d+2b=N-1)\\
&\sim R^{N-a-b-c-d}+R^{N-a-b-c-d+(\sigma-\frac{1}{2})\min(0,N-c-d-2b)}\\
&\quad\cdot(\ln\tfrac{R}{1+s}\ \text{ if } \ c+d+2b=N)\\
&\sim R^{N-a-b-c-d+(\sigma-\frac{1}{2})\min(0,N-c-d-2b)}\cdot(\ln\tfrac{R}{1+s}\ \text{ if } \ c+d+2b=N)\,.
\end{aligned}$$

$\underline{I_8}$ We get the result by interchanging a, b by c, d and using the result for Ω_7.

$\underline{I_9}$ Analogously as in Ω_7 we have $t\sim r\sim R$, $\varrho\sim\tilde{\varrho}\in(R^\nu;\tau)$, $\tilde{r}\sim\tau=|\tilde{t}|\in(R^\nu,R)$; so $\eta_{-b}^{-a}(\mathbf{y})\sim R^{-a+b}\rho^{-2b}$, $\eta_{-d}^{-c}(\mathbf{x}-\mathbf{y})\sim\rho^{-c-d}$ and

$$I_9\sim R^{-a+b}\int_{R^\nu}^{R}\mathrm{d}\tau\,\tau^{-d-c}\Big(\int_{R^\nu}^{\tau}\varrho^{N-2-2b}\mathrm{d}\varrho\Big)\,.$$

If $b>\frac{N-1}{2}$ the significant term in the inner integral will be the lower bound and we can use I_5. If $b<\frac{N-1}{2}$ the significant term in the inner integral will be the upper bound and we can use I_7. If $b=\frac{N-1}{2}$ then

$$I_9\sim R^{-a+\frac{N-1}{2}}\int_{R^\nu}^{R}\tau^{-d-c}\ln\frac{\tau}{R^\nu}\mathrm{d}\tau\,.$$

In comparison with I_5 we get some additional logarithmic factors

$$b=\tfrac{N-1}{2}\ :(\ln_+\tfrac{R}{1+s}\ \text{ if } \ c+d<1)\cdot(\ln_+^2\tfrac{R}{1+s}\ \text{ if } \ c+d=1)\,.$$

$\underline{I_{10}}$ As in I_9, we may use I_6 for $d>\frac{N-1}{2}$, I_8 for $d>\frac{N-1}{2}$ and get some additional logarithmic factors to I_8 for $d=\frac{N-1}{2}$.

$\underline{I_{11}}$ The domain Ω_{11} is unbounded. We have $\tilde{r}\sim r\in(R;\infty)$. Therefore $\eta_{-d}^{-c}(\mathbf{x}-\mathbf{y})\sim r^{-c-d}$, $\eta_{-b}^{-a}(\mathbf{y})\sim r^{-a}(1+s(\mathbf{y}))^{-b}$ and applying Lemma 3.2 under the assumption $a+b^*+c+d>N$ we get

$$\begin{aligned}
I_{11}&\sim\int_R^\infty\mathrm{d}r\,r^{N-1-a-b^*-c-d}\cdot(\ln r\ \text{if}\ b=\tfrac{N-1}{2})\\
&\sim R^{N-a-b^*-c-d}\cdot(\ln R\ \text{if}\ b=\tfrac{N-1}{2})\,.
\end{aligned}$$

$\underline{I_{12}}$ We proceed as in the previous case and under the assumption $a+b+c+d^*>N$ we get

$$I_{12}\sim R^{N-a-b-c-d^*}\cdot(\ln R\ \text{if}\ d=\tfrac{N-1}{2})\,.$$

I$_{13}$ The domain Ω_{13} can be considered as a subset of Ω_{11} and Ω_{12}. Therefore I_{13} can be bounded by I_{11} and I_{12}.

I$_{14}$ In this subdomain we have $r \sim R$, $\varrho \sim \tilde{\varrho} \in (R^\nu; R)$. Moreover $\tilde{r} \sim |\tilde{t}| + \tilde{\varrho}$ where $\tilde{t} \in (-\frac{1}{8}R^\nu; \frac{R}{2})$. Then $\eta_{-b}^{-a}(\mathbf{y}) \sim R^{-a+b}\varrho^{-2b}$, $\eta_{-d}^{-c}(\mathbf{x}-\mathbf{y}) \sim (\tilde{t}+\tilde{\varrho})^{-c+d}\varrho^{-2d}$ if $\tilde{t} > 0$ and $\eta_{-d}^{-c}(\mathbf{x}-\mathbf{y}) \sim \tilde{\varrho}^{-c-d}$ if $\tilde{t} < 0$. Let us note that the strip $\tilde{t} \in (-\frac{1}{8}R; 0)$ has no influence on the asymptotic behaviour since $\tilde{\varrho} > |\tilde{t}|$ there.

$$I_{14} \sim R^{b-a} \int_{R^\nu}^{R} \mathrm{d}\varrho\, \varrho^{N-2-2b-2d} \int_0^{R/2} (\tilde{t}+\rho)^{d-c} \mathrm{d}\tilde{t}$$

$$\sim R^{b-a} \int_{R^\nu}^{R} \mathrm{d}\varrho\, \varrho^{N-2-2b-2d} \cdot \begin{cases} R^{1+d-c} & 1+d-c>0 \\ \varrho^{1+d-c} & 1+d-c<0 \\ \ln\frac{R}{\varrho} & 1+d-c=0. \end{cases}$$

Now we distinguish three cases.

a) $1+d-c>0$
If $b+d \le \frac{N-1}{2}$ then

$$I_{14} \sim R^{N-a-b-c-d} \cdot (\ln_+ \tfrac{R}{1+s} \text{ if } b+d = \tfrac{N-1}{2}),$$

while for $b+d > \frac{N-1}{2}$ we have

$$I_{14} \sim R^{-a-c+\frac{N+1}{2}+2\sigma(\frac{N-1}{2}-b-d)} \qquad (\text{see } I_2\,,\, I_3)\,.$$

b) $1+d-c<0$

$$I_{14} \sim R^{b-a} \int_{R^\nu}^{R} \varrho^{-2b-c-d+N-1} \mathrm{d}\varrho$$

and the integral can be estimated by I_7.

c) $1+d-c=0$

$$I_{14} \sim R^{b-a} \int_{R^\nu}^{R} \varrho^{N-2-2b-2d} \ln \tfrac{R}{\varrho} \mathrm{d}\varrho$$
$$= R^{N-a-b-c-d} \int_1^{R^{1-\nu}} z^{2b+2d-N} \ln z\, \mathrm{d}z\,.$$

Now for $b+d \le \frac{N-1}{2}$

$$I_{14} \sim R^{N-a-b-c-d} \cdot (\ln_+^2 \tfrac{R}{1+s} \text{ if } b+d = \tfrac{N-1}{2})$$

and for $b+d > \frac{N-1}{2}$

$$I_{14} \sim R^{\frac{N+1}{2}-a-c-2\sigma(b+d-\frac{N-1}{2})} \ln_+ \tfrac{R}{1+s}$$

which can be estimated by I_2.

I$_{15}$ Interchanging a, b by c, d we can use the results from I$_{14}$.

We have now completed investigation of the situation C). The results are summarized in Tables 1 and 2 ($N = 3$).

The situation D) is almost trivial since we are left with subdomains of the type Ω_1, Ω_2, Ω_{11}, Ω_{12} and Ω_{13}. The integrals can be estimated by the corresponding integrals in C) taking $\sigma = \frac{1}{2}$, i.e. $\nu = 1$.

Finally in the case B) we proceed as in case C) but the subdomains Ω_2, Ω_3 and Ω_4 coincide. The other integrals can again be estimated by the corresponding ones from the part C) taking $\sigma = 0$, i.e. $\nu = \frac{1}{2}$.

The study of the convolution (3.1) is therefore completed. In the next section, we shall apply the results in the study of L^∞–estimates of convolutions with the Oseen kernels.

4. L^∞–ESTIMATES FOR WEAKLY SINGULAR OSEEN KERNELS

Now we intend to get estimates for the functions ν_β^α, $\mu_\beta^{\alpha,\gamma}$ by analogy with the preceding section,

$$\nu_\beta^\alpha(\mathbf{x}) = \nu_\beta^\alpha(\mathbf{x};1) = |\mathbf{x}|^\alpha(1+s(\mathbf{x}))^\beta, \qquad \mu_\beta^{\alpha,\gamma}(\mathbf{x}) = \eta_\beta^{\alpha-\gamma}(\mathbf{x})\nu_0^\gamma(\mathbf{x}).$$

This is the aim of Lemma 4.1 and Lemma 4.2. We will formulate these lemmas in the case of $\mathbb{R}^N$, $N \in \mathbb{N}$, $N \geq 2$. Afterward, we apply the lemmas in the study of L^∞-estimates of convolutions with Oseen kernels. Let $I_{\alpha,\gamma}(\mathbf{x})$ denote the following integral

$$I_{\alpha,\gamma}^{(N)}(\mathbf{x}) = \int_{B_1(\mathbf{0})} \nu_0^{-\alpha}(\mathbf{y})\nu_0^{-\gamma}(\mathbf{x}-\mathbf{y})d\mathbf{y} = \int_{B_1(\mathbf{x})} \nu_0^{-\alpha}(\mathbf{x}-\mathbf{y}')\nu_0^{-\gamma}(\mathbf{y}')d\mathbf{y}',$$

where $\mathbf{x},\mathbf{y},\mathbf{y}' \in \mathbb{R}^N$. For the notational convenience we also denote $\ln_-|\mathbf{x}| := \max(1, -\ln|\mathbf{x}|)$, for $\mathbf{x} \neq \mathbf{0}$.

Lemma 4.1. *For $\alpha < N$, $\gamma < N$ there exists a positive constant C_1 such that for $\mathbf{x} \in B_2(\mathbf{0}) \setminus \{\mathbf{0}\} \subset \mathbb{R}^N$*

$$I_{\alpha,\gamma}^{(N)}(\mathbf{x}) \leq C_1 \begin{cases} \nu_0^{-(\alpha+\gamma-N)}(\mathbf{x}), & \text{if } \alpha+\gamma > N \\ \ln_-|\mathbf{x}|, & \text{if } \alpha+\gamma = N \\ 1, & \text{if } \alpha+\gamma < N. \end{cases}$$

Moreover, there exists a positive constant C_2, such that for $\mathbf{x} \in B^2(\mathbf{0}) \subset \mathbb{R}^N$

$$I_{\alpha,\gamma}^{(N)}(\mathbf{x}) \leq C_2\,\nu_0^{-\gamma}(\mathbf{x}).$$

Proof. We divide the proof into two parts:

a) First we assume $|\mathbf{x}| \leq 2$. We will estimate integrals over sets, whose union contains unit ball $B_1(\mathbf{0})$.

$B_1(\mathbf{0}) \subset B_{|\mathbf{x}|/2}(\mathbf{0}) \cup B_{|\mathbf{x}|/2}(\mathbf{x}) \cup \Big\{B_{2|\mathbf{x}|}(\mathbf{0}) \setminus \Big(B_{|\mathbf{x}|/2}(\mathbf{0}) \cup B_{|\mathbf{x}|/2}(\mathbf{x})\Big)\Big\} \cup$
$\Big\{B_4(\mathbf{0}) \setminus B_{2|\mathbf{x}|}(\mathbf{0})\Big\} \equiv M_1 \cup M_2 \cup M_3 \cup M_4$

$$\int_{M_1} \frac{1}{|\mathbf{y}|^\alpha} \frac{1}{|\mathbf{x}-\mathbf{y}|^\gamma} \, d\mathbf{y} \le \frac{c_1}{|\mathbf{x}|^\gamma} \int_0^{|\mathbf{x}|/2} \frac{1}{r^\alpha} r^{N-1} dr \le \frac{c_2}{|\mathbf{x}|^{\alpha+\gamma-N}}, \quad \alpha < N$$

$$\int_{M_2} \frac{1}{|\mathbf{y}|^\alpha} \frac{1}{|\mathbf{x}-\mathbf{y}|^\gamma} \, d\mathbf{y} \le \frac{c_3}{|\mathbf{x}|^\alpha} \int_0^{|\mathbf{x}|/2} \frac{1}{r^\gamma} r^{N-1} dr \le \frac{c_4}{|\mathbf{x}|^{\alpha+\gamma-N}}, \quad \gamma < N$$

$$\int_{M_3} \frac{1}{|\mathbf{y}|^\alpha} \frac{1}{|\mathbf{x}-\mathbf{y}|^\gamma} \, d\mathbf{y} \le \frac{c_5}{|\mathbf{x}|^{\alpha+\gamma}} \int_0^{2|\mathbf{x}|} r^{N-1} dr \le \frac{c_6}{|\mathbf{x}|^{\alpha+\gamma-N}}$$

$$\int_{M_4} \frac{1}{|\mathbf{y}|^\alpha} \frac{1}{|\mathbf{x}-\mathbf{y}|^\gamma} \, d\mathbf{y} \le c_7 \int_{2|\mathbf{x}|}^4 \frac{r^{-\alpha} r^{N-1}}{|r-|\mathbf{x}||^\gamma} dr$$

$$\le c_8 \int_{2|\mathbf{x}|}^4 \left(\frac{r}{|r-|\mathbf{x}||}\right)^\gamma r^{-\alpha-\gamma+N-1} dr \le c_9 \int_{2|\mathbf{x}|}^4 r^{-\alpha-\gamma+N-1} dr;$$

here we use inequality $\frac{r}{|r-|\mathbf{x}||} \le 2$.
The last integral can be estimated by $c_{10}|\mathbf{x}|^{-\alpha-\gamma+N}$ if $\alpha+\gamma > N$, by $c_{11} \ln_- |\mathbf{x}|$ if $\alpha+\gamma = N$ and by some constant if $\alpha+\gamma < N$.

b) Now we assume $|\mathbf{x}| \ge 2$.

$$\int_{B_1(\mathbf{0})} \frac{1}{|\mathbf{y}|^\alpha} \frac{1}{|\mathbf{x}-\mathbf{y}|^\gamma} \, d\mathbf{y} \le \frac{c_{11}}{|\mathbf{x}|^\gamma} \int_{B_1(\mathbf{0})} \frac{1}{|\mathbf{y}|^\alpha} \, d\mathbf{y} \le \frac{c_{12}}{|\mathbf{x}|^\gamma}.$$

The assertion of Lemma 4.1 follows from these five estimates of convolution integrals. ∎

We define for pairs of real numbers $[a,b] \le [c,d]$: $a \le c$ and $a+b \le c+d$. It is evident that $\eta_b^a(\mathbf{x}) \le 2^{c-a}\eta_d^c(\mathbf{x})$, $\mathbf{x} \in \mathbb{R}^N$ if $[a,b] \le [c,d]$.

In the formulation of Lemma 4.2 we use functions $\mu_\beta^{\alpha,\gamma}(\mathbf{x}) = \eta_\beta^{\alpha-\gamma}(\mathbf{x})\nu_0^\gamma(\mathbf{x})$. Let us note that $\mu_\beta^{\alpha,\alpha}(\mathbf{x}) = \eta_\beta^0(\mathbf{x})\nu_0^\alpha(\mathbf{x}) = \nu_\beta^\alpha(\mathbf{x})$, $\mu_\beta^{\alpha,\gamma}(\mathbf{x}) \sim \eta_\beta^\alpha(\mathbf{x})$, $\mathbf{x} \in B^1(\mathbf{0})$, $\mu_\beta^{\alpha,\gamma}(\mathbf{x}) \sim \nu_\beta^\gamma(\mathbf{x}) \sim \nu_0^\gamma(\mathbf{x})$, $\mathbf{x} \in B_1(\mathbf{0}) \setminus \{\mathbf{0}\}$.

Lemma 4.2. *Let $a,b,c,d,e,f \in \mathbb{R}$ and positive constant C be such that for all*
$\mathbf{x} \in \mathbb{R}^N$: $\int_{\mathbb{R}^N} \eta_{-d}^{-c}(\mathbf{x}-\mathbf{y})\,\eta_{-b}^{-a}(\mathbf{y})\,d\mathbf{y} \le C\eta_{-f}^{-e}(\mathbf{x})$, $N \in \mathbb{N}$, $N \ge 2$.

Let $g < N$, $h < N$, $[e,f] \le [a,b]$, $[e,f] \le [c,d]$. Then there exists a positive constant C' such that the following inequality is satisfied for $\mathbf{x} \in \mathbb{R}^N \setminus \{\mathbf{0}\}$:

$$\int_{\mathbb{R}^N} \mu_{-d}^{-c,-h}(\mathbf{x}-\mathbf{y})\,\mu_{-b}^{-a,-g}(\mathbf{y})\,d\mathbf{y} \le C' \begin{cases} \mu_{-f}^{-e,-g-h+N}(\mathbf{x}) & g+h > N \\ \mu_{-f}^{-e,-\delta}(\mathbf{x}),\ \delta > 0 & g+h = N \\ \mu_{-f}^{-e,0}(\mathbf{x}) \equiv \eta_{-f}^{-e}(\mathbf{x}) & g+h < N. \end{cases}$$

Proof. Evidently, for α, $\beta \in \mathbb{R}^1$ there exist positive constants c_1, c_2, c_3 and c_4 such that

a) $\quad c_1\,\eta_\beta^\alpha(\mathbf{x}) \le \nu_\beta^\alpha(\mathbf{x}) \le c_2\,\eta_\beta^\alpha(\mathbf{x})$ for all $\mathbf{x} \in B^1(\mathbf{0})$

b) $\quad c_3\,\eta_\beta^\alpha(\mathbf{x}) \le \eta_\beta^\alpha(\mathbf{y}) \le c_4\,\eta_\beta^\alpha(\mathbf{x})$ for all $\mathbf{x} \in \mathbb{R}^N$, $\mathbf{y} \in B_1(\mathbf{x})$

Now we will prove the assertion of the Lemma 4.2:

$$\mu_{-d}^{-c,-h} * \mu_{-b}^{-a,-g}(\mathbf{x}) = \int_{\mathbb{R}^N} \mu_{-d}^{-c,-h}(\mathbf{x}-\mathbf{y})\,\mu_{-b}^{-a,-g}(\mathbf{y})\mathrm{d}\mathbf{y}$$

$$\le C_1 \int_{B_1(\mathbf{x})} \nu_0^{-h}(\mathbf{x}-\mathbf{y})\mu_{-b}^{-a,-g}(\mathbf{y})\,\mathrm{d}\mathbf{y} + C_2 \int_{B^1(\mathbf{x})} \eta_{-d}^{-c}(\mathbf{x}-\mathbf{y})\mu_{-b}^{-a,-g}(\mathbf{y})\mathrm{d}\mathbf{y}.$$

We will study these two integrals separately. By using Lemma 4.1 we get the estimate of

$\int_{B_1(\mathbf{x})} \nu_0^{-h}(\mathbf{x}-\mathbf{y})\,\mu_{-b}^{-a,-g}(\mathbf{y})\,\mathrm{d}\mathbf{y} = \int_{B_1(\mathbf{0})} \mu_{-b}^{-a,-g}(\mathbf{x}-\mathbf{y})\,\nu_0^{-h}(\mathbf{y})\,\mathrm{d}\mathbf{y}$.

$$\int_{B_1(\mathbf{0})} \mu_{-b}^{-a,-g}(\mathbf{x}-\mathbf{y})\,\nu_0^{-h}(\mathbf{y})\,\mathrm{d}\mathbf{y}$$

$$\le \max_{\mathbf{y}\in B_1(\mathbf{x})} \eta_{-b}^{-a+g}(\mathbf{y}) \int_{B_1(\mathbf{0})} \nu_0^{-g}(\mathbf{x}-\mathbf{y})\nu_0^{-h}(\mathbf{y})\mathrm{d}\mathbf{y}$$

$$\le C_3\eta_{-b}^{-a+g}(\mathbf{x}) \begin{cases} \left.\begin{array}{ll} \nu_0^{-g-h+N}(\mathbf{x}), & g+h>N \\ \ln_-|\mathbf{x}|, & g+h=N \\ 1, & g+h<N \end{array}\right\} & \mathbf{x}\in B_1\setminus\{\mathbf{0}\} \\ \eta_0^{-g}(\mathbf{x}) & \mathbf{x}\in B^1 \end{cases}$$

$$\le C_5 \begin{cases} \mu_{-b}^{-a,-g-h+N}(\mathbf{x}), & g+h>N \\ \mu_{-b}^{-a,-\delta}(\mathbf{x}),\ \delta>0, & g+h=N \\ \mu_{-b}^{-a,0}(\mathbf{x}), & g+h<N \end{cases}$$

$$\le C_6 \begin{cases} \mu_{-f}^{-e,-g-h+N}(\mathbf{x}), & g+h>N \\ \mu_{-f}^{-e,-\delta}(\mathbf{x}),\ \delta>0, & g+h=N \\ \mu_{-f}^{-e,0}(\mathbf{x}) \equiv \eta_{-f}^{-e}(\mathbf{x}), & g+h<N\,. \end{cases}$$

In the second inequality we use Lemma 4.1 and the relation b). In the last inequality we take into account the assumption $[e,f] \le [a,b]$.

We estimate the remaining integral $\int_{B^1(\mathbf{x})} \eta_{-d}^{-c}(\mathbf{x}-\mathbf{y})\mu_{-b}^{-a,-g}(\mathbf{y})\mathrm{d}\mathbf{y}$ for $\mathbf{x}\in\mathbb{R}^N\setminus\{\mathbf{0}\}$ in the following way:

$$\int_{B^1(\mathbf{x})} \eta_{-d}^{-c}(\mathbf{x}-\mathbf{y})\mu_{-b}^{-a,-g}(\mathbf{y})\mathrm{d}\mathbf{y} \le \int_{\mathbb{R}^N} \eta_{-d}^{-c}(\mathbf{x}-\mathbf{y})\mu_{-b}^{-a,-g}(\mathbf{y})\mathrm{d}\mathbf{y}$$

$$= \int_{B_1} \eta_{-d}^{-c}(\mathbf{x}-\mathbf{y})\,\nu_0^{-g}(\mathbf{y})\mathrm{d}\mathbf{y} + \int_{B^1} \eta_{-d}^{-c}(\mathbf{x}-\mathbf{y})\eta_{-b}^{-a}(\mathbf{y})\mathrm{d}\mathbf{y}$$

$$\le C_5 \max_{\mathbf{y}\in B_1(\mathbf{x})} \eta_{-d}^{-c}(\mathbf{y}) \int_{B_1(\mathbf{0})} \nu_0^{-g}(\mathbf{y})\mathrm{d}\mathbf{y} + C_6\eta_{-f}^{-e}(\mathbf{x}) \le C_7\eta_{-f}^{-e}(\mathbf{x})\,.$$

The proof of Lemma 4.2 follows from these estimates. ∎

We now formulate main results of this section. We will use the following notation

$$\overline{\eta}_F^E(\mathbf{x};\lambda) = \eta_F^E(\mathbf{x};\lambda) \ \text{ if no logarithmic factor appears, and}$$

$$\overline{\eta}_F^E(\mathbf{x};\lambda) = \eta_F^E(\mathbf{x};\lambda) \cdot \begin{cases} P(\ln_+^{-1}|\lambda\mathbf{x}|) \\ P(\ln_+^{-1} s(\lambda\mathbf{x})) \end{cases} \text{ if there are logarithmic factors,}$$

where function $P(\cdot)$ is a polynomial of the first or the second order, see also Remark 4.4. Similarly we define $\overline{\nu}_F^E(\cdot\,;\lambda)$. Then we have

Theorem 4.3. *Let $A+B^* > 1$, $i,j = 1,2,3$. Let $f \in L^\infty(\mathbb{R}^3, \eta_B^A(\cdot\,;\lambda))$. Then $\mathcal{O}_{ij} * f \in L^\infty(\mathbb{R}^3, \overline{\eta}_F^E(\cdot\,;\lambda))$, where*

$$E = \begin{cases} A-1 & \text{for } A \le B^*+1 \\ \frac{A+B^*-1}{2} & \text{for } A \ge B+1,\ A+B \le 3 \\ 1 & \text{for } A+B^* \ge 3 \end{cases} \tag{i}$$

$$E+F = \begin{cases} A+B^*-1 & \text{for } A+B^* \le 3 \\ 2 & \text{for } A+B^* \ge 3 \end{cases} \tag{ii}$$

with logarithmic factors

$$\ln_+(\lambda|\mathbf{x}|) \ \text{ for } \begin{cases} A+B^* = 3 \\ A = B+1,\ 0 \le B \le 1 \end{cases} \tag{iii}$$

$$\ln_+(\lambda s(\mathbf{x})) \ \text{ for } \ A+B<3,\ B \le 1, \tag{iv}$$

(see Remark 4.4). Moreover we have

$$\|\mathcal{O}_{ij}(\cdot\,;\lambda) * f\|_{\infty,(\overline{\eta}_F^E(\cdot\,;\lambda)),\mathbb{R}^3} \le C\lambda^{-2} \|f\|_{\infty,(\eta_B^A(\cdot\,;\lambda)),\mathbb{R}^3}\,. \tag{4.1}$$

Let in addition for A, B the following conditions be satisfied

$$1 \le A < 3,\quad B > 0,\quad \text{or}\quad A \le B+5,\ 1 < A+B \le 3,\ B \le 0\,. \tag{v}$$

Then for $f \in L^\infty(\mathbb{R}^3, \nu_B^A(\cdot\,;\lambda))$ we have $\mathcal{O}_{ij}(\cdot\,;\lambda) * f \in L^\infty(\mathbb{R}^3, \overline{\nu}_F^E(\cdot\,;\lambda))$ and

$$\|\mathcal{O}_{ij}(\cdot\,;\lambda) * f\|_{\infty,(\overline{\nu}_F^E(\cdot\,;\lambda)),\mathbb{R}^3} \le C\lambda^{-2+A-E} \|f\|_{\infty,(\nu_B^A(\cdot\,;\lambda)),\mathbb{R}^3}\,. \tag{4.2}$$

Remark 4.4. The inequalities (4.1), (4.2) must be understood in the following sense. If no logarithmic terms appear then

$$\|\mathcal{O}_{ij}(\cdot\,;\lambda) * f\|_{\infty,(\eta_F^E(\cdot\,;\lambda)),\mathbb{R}^3} \le C\lambda^{-2} \|f\|_{\infty,(\eta_B^A(\cdot\,;\lambda)),\mathbb{R}^3}\,.$$

Analogously for the weight $\nu_F^E(\cdot\,;\lambda)$. But for $A+B^* = 3$ or $A = B+1$, $0 \le B \le 1$ we have

$$\|\mathcal{O}_{ij}(\cdot\,;\lambda) * f\|_{\infty,(\eta_F^E(\cdot\,;\lambda)\,P(\ln_+^{-1}(\lambda|\cdot|))),\mathbb{R}^3} \le C\lambda^{-2} \|f\|_{\infty,(\eta_B^A(\cdot\,;\lambda)),\mathbb{R}^3} \tag{4.3}$$

and for $A+B<3$, $B\le 1$

$$\|\mathcal{O}_{ij}(\cdot\,;\lambda)*f\|_{\infty,(\eta_F^E(\cdot\,;\lambda)\,P(\ln_+^{-1}(s(\lambda\cdot)))),\mathbb{R}^3}\le C\lambda^{-2}\|f\|_{\infty,(\eta_B^A(\cdot\,;\lambda)),\mathbb{R}^3}\,, \tag{4.4}$$

where $P(\cdot)$ is a polynomial. (The order of polynomial $P(\cdot)$ can be traced out from the proof of Theorem 4.3 using Tables 1 and 2.) Analogously for the weights $\nu_F^E(\cdot\,;\lambda)$. We can use instead of (4.3), (4.4) for $\varepsilon>0$

$$\|\mathcal{O}_{ij}(\cdot\,;\lambda)*f\|_{\infty,(\eta_F^{E-\varepsilon}(\cdot\,;\lambda)),\mathbb{R}^3}\le C\lambda^{-2}\|f\|_{\infty,(\eta_B^A(\cdot\,;\lambda)),\mathbb{R}^3}\,, \tag{4.3'}$$

$$\|\mathcal{O}_{ij}(\cdot\,;\lambda)*f\|_{\infty,(\eta_{F-\varepsilon}^{E}(\cdot\,;\lambda)),\mathbb{R}^3}\le C\lambda^{-2}\|f\|_{\infty,(\eta_B^A(\cdot\,;\lambda)),\mathbb{R}^3}\,, \tag{4.4'}$$

respectively.

Finally, in the case of $f=0$ in $B_{1/2}(\mathbf{0})$ (this is usually the case for $\Omega\subset\mathbb{R}^N$, exterior domain) we can get for the weight $\nu_B^A(\cdot\,;\lambda)$

$$\|\mathcal{O}_{ij}(\cdot\,;\lambda)*f\|_{\infty,(\nu_F^{E-\varepsilon}(\cdot\,;\lambda)),\mathbb{R}^3}\le C\lambda^{-2+A-E+\varepsilon}\|f\|_{\infty,(\nu_B^A(\cdot\,;\lambda)),\mathbb{R}^3}\,.$$

Proof of Theorem 4.3. Let $f\in L^\infty(\mathbb{R}^3,\eta_B^A(\cdot\,;1))$. Recalling that $|\mathcal{O}_{ij}(\mathbf{x}-\mathbf{y};1)|\le C_0\nu_{-1}^{-1}(\mathbf{x}-\mathbf{y};1)$ we have

$$\begin{aligned}|\mathcal{O}_{ij}(\cdot\,;1)*f(\mathbf{x})|&\le C_1\,\nu_{-1}^{-1}(\cdot\,;1)*\eta_{-B}^{-A}(\cdot\,;1)(\mathbf{x})\\&\le C_2\,\eta_{-1}^{-1}(\cdot\,;1)*\eta_{-B}^{-A}(\cdot\,;1)(\mathbf{x})\,.\end{aligned} \tag{4.5}$$

We have therefore to study the convolution (4.5); we apply Tables 1 and 2 with $c=d=1$, $a=A, b=B$ and we get, under the condition $A+B^*>1$, that (we skip the logarithmic factors, for the moment)

$$(\eta_{-1}^{-1}(\cdot\,;1)*\eta_{-B}^{-A}(\cdot\,;1))(\mathbf{x})\le C\eta_{-F}^{-E}(\mathbf{x};1)$$

with

$$E\le\min\left(1,\,\tfrac{A+B^*-1}{2},\,A-\tfrac{1}{2},\,A-1,\,A+B-1,\,\tfrac{A+B-1}{2},\,A+B^*-1\right)$$

$$=\min\left(1,\,\tfrac{A+B^*-1}{2},\,A-1\right) \tag{4.6}$$

$$E+F\le\min(2,\,A+B-1,\,A+B^*-1)=\min(2,A+B^*-1)\,.$$

We therefore easily get (i) and (ii), see Fig. 2 below:

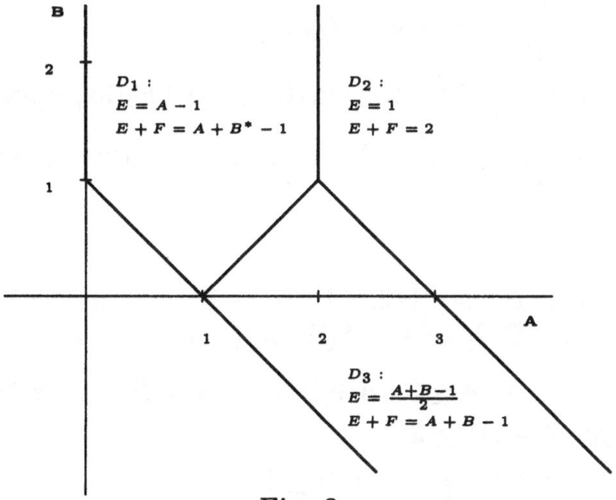

Fig. 2

Let us now regard the logarithmic factors. From Ω_0 we have $\ln_+(\lambda|\mathbf{x}|)$ whenever $B = 1$, $A \le 2$ or $A + B^* = 3$ and $e_0 = 1 + \frac{1}{2}\min(0,\, A + B^* - 3)$, $e_0 + f_0 = 2 + \min(0,\, A + B^* - 3)$. Therefore, if $A + B^* = 3$ the factor $\ln_+(\lambda|\mathbf{x}|)$ must be taken into account. But for $B = 1$, $0 < A < 2$ we have $e_0 = \frac{A}{2} > A - 1$, $e_0 + f_0 = A + B^* - 1 = A$; therefore, we can assume only $\ln_+(\lambda s(\mathbf{x}))$ here.

Next in Ω_1 we have $\ln_+(\lambda|\mathbf{x}|)$ due to $d = 1$. But $e_1 = A - \frac{1}{2} > \min\left(1, \frac{A+B^*-1}{2}, A - 1\right)$ and $e_1 + f_1 = A + B - 1$. So for $B > 1$ we easily see $A + B - 1 > A + B^* - 1$, but for $1 < A + B < 3$, $B \le 1$ we have $e_1 + f_1 = A + B^* - 1$; therefore, we must assume $\ln_+(\lambda s(\mathbf{x}))$ in this case.

Analogously we proceed in other sub-domains and get *(iii)* and *(iv)*. The estimate (4.1) for $\lambda = 1$ is therefore shown. In order to show (4.1) for $\lambda \ne 1$, let us recall the homogeneity property of $\mathcal{O}_{ij}(\mathbf{x} - \mathbf{y}; \lambda)$. Namely, for $N = 3$ we have $\mathcal{O}_{ij}(\mathbf{x} - \mathbf{y}; \lambda) = \lambda \mathcal{O}_{ij}(\lambda(\mathbf{x} - \mathbf{y}); 1)$ and therefore

$$\left|\int_{\mathbb{R}^3} \mathcal{O}_{ij}(\mathbf{x} - \mathbf{y}; \lambda)\, f(\mathbf{y})\, \mathrm{d}\mathbf{y}\right| = \lambda^{-2}\left|\int_{\mathbb{R}^3} \mathcal{O}_{ij}(\lambda\mathbf{x} - \mathbf{z}) f(\frac{\mathbf{z}}{\lambda})\, \mathrm{d}\mathbf{z}\right|$$
$$\le \lambda^{-2} \sup_{\mathbf{y}\in\mathbb{R}^3} \left|f(\mathbf{y})\, \eta^A_B(\lambda\mathbf{y}; 1)\right| \eta^{-E}_{-F}(\lambda\mathbf{x}; 1)\, P_1(\ln_+(\lambda|\mathbf{x}|))\, P_2(\ln_+ s(\lambda\mathbf{x}))$$

and so, as $\eta^A_B(\lambda\mathbf{x}; 1) = \eta^A_B(\mathbf{x}; \lambda)$, we have (4.1).

Let us study the weight $\nu^A_B(\mathbf{x}; \lambda)$. From Lemma 4.2 we have the following conditions $E \ge \max(0, A - 2)$ and $A < 3$ and therefore we get on D_1 that $A \ge 1$, on D_2 that $A < 3$, and on D_3 $\frac{A+B-1}{2} \ge A - 2$, i.e. $A \le B + 5$. Finally, to show (4.2) we proceed as in the case of the estimate (4.1). Evidently, (4.2) holds for $\lambda = 1$. Therefore

$$\begin{aligned}&\Big|\int_{\mathbb{R}^3} \mathcal{O}_{ij}(\mathbf{x}-\mathbf{y};\lambda) f(\mathbf{y})\mathrm{d}\mathbf{y}\Big| \\ &\le \lambda^{-2} \sup_{\mathbf{y}\in\mathbb{R}^3} \Big|f(\mathbf{y})\nu^A_B(\lambda\mathbf{y};1)\Big| \nu^{-E}_{-F}(\lambda\,\mathbf{x};1)\, P_1(\ln_+(\lambda|\mathbf{x}|))\, P_2(\ln_+ s(\lambda\mathbf{x})) \\ &= \lambda^{-2+A-E}\, \|f\|_{\infty,(\nu^A_B(\cdot\,;\lambda)),\mathbb{R}^3}\, \nu^{-E}_{-F}(\mathbf{x};\lambda)\, P_1(\ln_+(\lambda|\mathbf{x}|)) P_2(\ln_+ s(\lambda\mathbf{x}))\,.\end{aligned}$$

■

Theorem 4.5. *Let $A + B > 1/2$, $A > -1$. Then for $f \in L^\infty(\mathbb{R}^3, \eta^A_B(\cdot\,;\lambda))$ we have $|\nabla \mathcal{O}| * f \in L^\infty(\mathbb{R}^3, \overline{\eta}^E_F(\cdot\,;\lambda))$, where*

$$E = \begin{cases} A - \frac{1}{2} & \text{for } -1 < A \le 2,\ A+B > \frac{1}{2},\ A \le B+1,\ B \ge 0 \\ \frac{3}{2} & \text{for } A + B^* \ge 3 \\ A+B-\frac{1}{2} & \text{for } B < 0,\ \frac{1}{2} < A+B \le 1 \\ \frac{A+B}{2} & \text{for } B \le A-1,\ 1 \le A+B \le 3 \end{cases} \tag{i}$$

$$E + F = \begin{cases} A + B^* & \text{for } -1 < A \le 2,\ B \ge \frac{3}{2} \\ A+B-\frac{1}{2} & \text{for } \frac{1}{2} < A+B \le \frac{7}{2} \\ 3 & \text{for } A+B \ge \frac{7}{2},\ A \ge 2 \end{cases} \tag{ii}$$

with logarithmic factors

$$\ln_+(\lambda|\mathbf{x}|) \ \text{ for } \begin{cases} A + B^* = 3 \\ A = B+1,\ 0 \le B \le 1 \\ A+B = 1,\ B \le 0 \\ B = -\frac{1}{2},\ \frac{3}{4} \le A \le \frac{5}{4} \end{cases} \tag{iii}$$

$$\ln_+(\lambda s(\mathbf{x})) \ \text{ for } \begin{cases} A = B+1,\ 1 < B \le \frac{5}{4} \\ A+B = 1,\ 0 < B \le \frac{3}{2}. \end{cases} \tag{iv}$$

Moreover we have

$$\||\nabla \mathcal{O}(\cdot\,;\lambda)| * f\|_{\infty,(\overline{\eta}^E_F(\cdot\,;\lambda)),\mathbb{R}^3} \le C\lambda^{-1}\, \|f\|_{\infty,(\eta^A_B(\cdot\,;\lambda)),\mathbb{R}^3}\,. \tag{4.7}$$

Let in addition for A, B the following conditions be satisfied:

$$\frac{1}{2} \le A \le \frac{5}{2},\quad B \ge -\frac{1}{2},\quad A \le B+2. \tag{v}$$

Then for $f \in L^\infty(\mathbb{R}^3, \nu^A_B(\cdot\,;\lambda))$ we have $|\nabla \mathcal{O}(\cdot\,;\lambda)| * f \in L^\infty(\mathbb{R}^3, \overline{\nu}^E_F(\cdot\,;\lambda))$ and

$$\||\nabla \mathcal{O}(\cdot\,;\lambda)| * f\|_{\infty,(\overline{\nu}^E_F(\cdot\,;\lambda)),\mathbb{R}^3} \le C\lambda^{-1+A-E}\, \|f\|_{\infty,(\nu^A_B(\cdot\,;\lambda)),\mathbb{R}^3}\,.$$

The theorem (and the following theorems of this section) can be proved by analogy with Theorem 4.3.

Theorem 4.6. *Let $A + B^* > 0$, $i, j = 1,2,3$. Let $R = |\nabla^2 \mathcal{O} - \nabla^2 \mathcal{S}|$ or $R = \partial_1 \mathcal{O}_{ij}$. Then for $f \in L^\infty(\mathbb{R}^3, \eta^A_B(\cdot\,;\lambda))$ we have $R * f \in L^\infty(\mathbb{R}^3, \overline{\eta}^E_F(\cdot\,;\lambda))$, where*

$$E = \begin{cases} A & \text{for } -1 < A \le 2,\ A \le B+1,\ B \ge 0 \\ 2 & \text{for } A + B^* \ge 3 \\ A+B & \text{for } B \le 0,\ 0 < A+B < 1 \\ \frac{A+B+1}{2} & \text{for } B \le A-1,\ 1 \le A+B \le 3 \end{cases} \tag{i}$$

$$E + F = \begin{cases} A + B^* & \text{for } 0 < A + B^* \le 3 \\ 3 & \text{for } A + B^* \ge 3 \end{cases} \tag{ii}$$

with logarithmic factors

$$\ln_+(\lambda|\mathbf{x}|) \quad \text{for} \quad 0 < A + B^* \le 3. \tag{iii}$$

Moreover we have

$$\Big\| |\nabla^2 \mathcal{O}(\cdot\,;\lambda) - \nabla^2 \mathcal{S}(\cdot)| * f \Big\|_{\infty,(\overline{\eta}^E_F(\cdot\,;\lambda)),\mathbb{R}^3} \le C \, \|f\|_{\infty,(\eta^A_B(\cdot\,;\lambda)),\mathbb{R}^3}, \tag{4.8}$$

$$\|\partial_1 \mathcal{O}_{ij}(\cdot\,;\lambda) * f\|_{\infty,(\overline{\eta}^E_F(\cdot\,;\lambda)),\mathbb{R}^3} \le C\lambda^{-1} \, \|f\|_{\infty,(\eta^A_B(\cdot\,;\lambda)),\mathbb{R}^3}. \tag{4.9}$$

Let in addition for A, B the following conditions be satisfied

$$0 \le A < 3, \quad B \ge -1, \quad A < B + 3. \tag{iv}$$

Then for $f \in L^\infty(\mathbb{R}^3, \nu^A_B(\cdot\,;\lambda))$ we have $R * f \in L^\infty(\mathbb{R}^3, \overline{\nu}^E_F(\cdot\,;\lambda))$ and

$$\Big\| |\nabla^2 \mathcal{O}(\cdot\,;\lambda) - \nabla^2 \mathcal{S}(\cdot)| * f \Big\|_{\infty,(\overline{\nu}^E_F(\cdot\,;\lambda)),\mathbb{R}^3} \le C\lambda^{A-E} \, \|f\|_{\infty,(\nu^A_B(\cdot\,;\lambda)),\mathbb{R}^3}, \tag{4.10}$$

$$\|\partial_1 \mathcal{O}_{ij}(\cdot\,;\lambda) * f\|_{\infty,(\overline{\nu}^E_F(\cdot\,;\lambda)),\mathbb{R}^3} \le C\lambda^{-1+A-E} \, \|f\|_{\infty,(\nu^A_B(\cdot\,;\lambda)),\mathbb{R}^3}. \tag{4.11}$$

Theorem 4.7. *Let $A + B^* > 1$ and $i = 1,2,3$. Then for $f \in L^\infty(\mathbb{R}^3, \eta^A_B(\cdot\,;\lambda))$ we have $\mathcal{P}_i * f \in L^\infty(\mathbb{R}^3, \overline{\eta}^E_F(\cdot\,;\lambda))$, where*

$$E = \begin{cases} 2 & \text{for } A + B^* \ge 3,\ A \ge \frac{5}{2} \\ A - \frac{1}{2} & \text{for } A \le \frac{5}{2},\ B \ge \frac{1}{2} \\ A + B^* - 1 & \text{for } B \le \frac{1}{2},\ A + B^* \le 3 \end{cases} \tag{i}$$

$$E + F = \begin{cases} 2 & \text{for } A + B^* \ge 3 \\ A + B^* - 1 & \text{for } A + B^* \le 3 \end{cases} \tag{ii}$$

with logarithmic factors

$$\ln_+(\lambda|\mathbf{x}|) \quad \text{for} \quad \begin{cases} B = \frac{1}{2},\ \frac{1}{2} < A < \frac{5}{2} \\ B = 1,\ 2 \le A \le \frac{5}{2} \\ A + B^* = 3,\ A \ge \frac{5}{2} \end{cases} \tag{iii}$$

$$\ln_+(\lambda s(\mathbf{x})) \quad for \quad \begin{cases} A+B^*=3, \; 2\le A<\frac{5}{2} \\ B=1, \; 0<A<2. \end{cases} \tag{iv}$$

Moreover we have

$$\|\mathcal{P}_i * f\|_{\infty,(\overline{\eta}^E_F(\cdot\,;\lambda)),\mathbb{R}^3} \le C\lambda^{-1}\|f\|_{\infty,(\eta^A_B(\cdot\,;\lambda)),\mathbb{R}^3}\,. \tag{4.12}$$

Let in addition for A, B the following conditions be satisfied

$$\frac{1}{2}\le A<3, \quad B\ge 0. \tag{v}$$

Then for $f\in L^\infty(\mathbb{R}^3,\nu^A_B(\cdot\,;\lambda))$ we have $\mathcal{P}_i * f\in L^\infty(\mathbb{R}^3,\overline{\nu}^E_F(\cdot\,;\lambda))$ and

$$\|\mathcal{P}_i * f\|_{\infty,(\overline{\nu}^E_F(\cdot\,;\lambda)),\mathbb{R}^3} \le C\lambda^{-1+A-E}\|f\|_{\infty,(\nu^A_B(\cdot\,;\lambda)),\mathbb{R}^3}\,. \tag{4.13}$$

5. L^p–ESTIMATES FOR WEAKLY SINGULAR OSEEN KERNELS

This section is devoted to the L^p-estimates of convolutions with Oseen kernels. Here we shall use the results from the previous section, i.e. the L^∞-theory.

Theorem 5.1. *Let T be an integral operator with the kernel $\mathcal{O}_{ij}(\cdot\,;\lambda)$, $T: f\mapsto \mathcal{O}_{ij}(\cdot\,;\lambda)*f$, $i,j=1,2,3$ and let $1<p<\infty$. Then T is a well defined continuous operator:*

a)
$$L^p(\mathbb{R}^3;\eta_\beta^{\alpha+p/2}(\cdot\,;\lambda)) \longmapsto L^p(\mathbb{R}^3;\eta_\beta^{\alpha-p/2-\varepsilon}(\cdot\,;\lambda))$$

for $-\varepsilon(p-1)/p<\beta<p-1+\varepsilon(p-1)/p$, $p/2-1-\varepsilon/p<\alpha+\beta<5p/2-3+\varepsilon$, $\alpha-\beta<p/2-1+\varepsilon+\varepsilon(p-1)/p$, $-p/2-\varepsilon/p<\alpha<3p/2-2+\varepsilon$, $0<\varepsilon\le p/2$

b)
$$L^p(\mathbb{R}^3;\nu_\beta^{\alpha+p/2}(\cdot\,;\lambda)) \longmapsto L^p(\mathbb{R}^3;\nu_\beta^{\alpha-p/2-\varepsilon}(\cdot\,;\lambda))$$

for $-\varepsilon(p-1)/p<\beta<p-1+\varepsilon(p-1)/p$, $p/2-1-\varepsilon/p<\alpha+\beta<5p/2-3+\varepsilon$, $\alpha-\beta<p/2-1+\varepsilon+\varepsilon(p-1)/p$, $\max\{-p/2-\varepsilon/p, p/2-3+\varepsilon\}<\alpha<\min\{3p/2-2+\varepsilon, 5p/2-3\}$, $0<\varepsilon\le p/2$

Moreover we have for α, β specified in a) and b), respectively

ad a)

$$\|\mathcal{O}_{ij}(\cdot\,;\lambda)*f\|_{p,(\eta_\beta^{\alpha-p/2-\varepsilon}(\cdot\,;\lambda)),\mathbb{R}^3} \le C\lambda^{-2}\|f\|_{p,(\eta_\beta^{\alpha+p/2}(\cdot\,;\lambda)),\mathbb{R}^3}\,, \tag{5.1}$$

ad b)

$$\|\mathcal{O}_{ij}(\cdot\,;\lambda)*f\|_{p,(\nu_\beta^{\alpha-p/2-\varepsilon}(\cdot\,;\lambda)),\mathbb{R}^3} \le C\lambda^{\frac{\varepsilon}{p}-1}\|f\|_{p,(\nu_\beta^{\alpha+p/2}(\cdot\,;\lambda)),\mathbb{R}^3}\,. \tag{5.2}$$

Proof. We proceed similarly as in the case of L^∞–weighted estimates. Studying first $\eta^\alpha_\beta(\cdot\,;\lambda)$ weights we show (5.1) for $\lambda = 1$. Applying the homogeneity properties of $\mathcal{O}_{ij}(\cdot;\lambda)$ we get (5.1) in the general situation $\lambda \neq 1$. Next, using the results from a) together with Lemma 4.2 we show (5.2).

Let us denote

$$K(\mathbf{x},\mathbf{y}) = \mathcal{O}_{ij}(\mathbf{x}-\mathbf{y};1)\left(\eta_\beta^{\alpha-p/2-\varepsilon}(\mathbf{x})\right)^{1/p}\left(\eta_\beta^{\alpha+p/2}(\mathbf{y})\right)^{-1/p},$$
$$F(\mathbf{y}) = f(\mathbf{y})\left(\eta_\beta^{\alpha+p/2}(\mathbf{y})\right)^{1/p}$$

We easily observe that, in order to verify (5.1) with $\lambda = 1$, it is sufficient to show that there exists $C > 0$, independent of f, such that

$$\left\| \int_{\mathbb{R}^3} K(\cdot\,,\mathbf{y})\,F(\mathbf{y})\,\mathrm{d}\mathbf{y}\right\|_p \le C\|F\|_p. \tag{5.3}$$

Let $L(\cdot)$ and $M(\cdot)$ be non-negative functions defined on $\mathbb{R}^3$ such that for all $\mathbf{x}$, $\mathbf{y} \in \mathbb{R}^3$

$$J_0(\mathbf{x}) = \int_{\mathbb{R}^3} |K(\mathbf{x},\mathbf{y})|\; L(\mathbf{y})^q\,\mathrm{d}\mathbf{y} \le C^q\, M(\mathbf{x})^q, \tag{5.4}$$

$$J_1(\mathbf{y}) := \int_{\mathbb{R}^3} |K(\mathbf{x},\mathbf{y})|\; M(\mathbf{x})^p\,\mathrm{d}\mathbf{x} \le C^p\, L(\mathbf{y})^p, \tag{5.5}$$

where $C > 0$, $1 < p < \infty$ and $p^{-1} + q^{-1} = 1$. Then relation (5.3) is satisfied. Indeed,

$$\begin{aligned}
&\left\| \int_{\mathbb{R}^3} K(\cdot\,,\mathbf{y})\,F(\mathbf{y})\,\mathrm{d}\mathbf{y}\right\|_p^p \\
&\le \int_{\mathbb{R}^3}\Big\{\Big(\int_{\mathbb{R}^3}|K(\mathbf{x},\mathbf{y})|\,|F(\mathbf{y})|^p L(\mathbf{y})^{-p}\,\mathrm{d}\mathbf{y}\Big)^{\frac{1}{p}} J_0(\mathbf{x})^{\frac{1}{q}}\Big\}^p \mathrm{d}\mathbf{x} \\
&\le C^p \int_{\mathbb{R}^3} M(\mathbf{x})^p \int_{\mathbb{R}^3}|K(\mathbf{x},\mathbf{y})|\,|F(\mathbf{y})|^p L(\mathbf{y})^{-p}\mathrm{d}\mathbf{y}\,\mathrm{d}\mathbf{x} \\
&= C^p \int_{\mathbb{R}^3}|F(\mathbf{y})|^p J_1(\mathbf{y}) L(\mathbf{y})^{-p}\mathrm{d}\mathbf{y} \le C^{2p}\|F\|_p^p,
\end{aligned}$$

i.e. we get (5.3). We shall suppose the functions $L(\cdot)$, $M(\cdot)$ in the form $L(\mathbf{x}) = M(\mathbf{x}) = \eta^{-A}_{-\frac{\beta}{p^2}}(\mathbf{x})$, $A \in \mathbb{R}^1$. Denoting

$$\begin{aligned}
a_0 &= qA + \tfrac{\alpha}{p} + \tfrac{1}{2} & a_1 &= pA - \tfrac{\alpha}{p} + \tfrac{1}{2} + \tfrac{\varepsilon}{p} \\
b_0 &= \tfrac{\beta q}{p^2} + \tfrac{\beta}{p} & b_1 &= 0
\end{aligned} \tag{5.6}$$

we get that in order to verify (5.4) we have to find such a_0 b_0 that

$$\int_{\mathbb{R}^3} \mathcal{O}_{ij}(\mathbf{x}-\mathbf{y};1)\,\eta_{-b_0}^{-a_0}(\mathbf{y})\,\mathrm{d}\mathbf{y} \le C\,\eta_{-b_0}^{-a_0+1+\frac{\varepsilon}{p}}(\mathbf{x})$$

for all $\mathbf{x} \in \mathbb{R}^3$ and in order to verify (5.5) we have to find a_1 such that

$$\int_{\mathbb{R}^3} \mathcal{O}_{ij}(-\mathbf{x}-\mathbf{y};1)\,\eta_0^{-a_1}(-\mathbf{x})\,d\mathbf{x} \le C\,\eta_0^{-a_1+1+\frac{\varepsilon}{p}}(\mathbf{y})$$

for all $\mathbf{y} \in \mathbb{R}^3$.

Applying Theorem 4.3 with $f = \eta_{-b_0}^{-a_0}(\cdot)$ we get for the first inequality the following set of conditions

$$\begin{gathered} -\varepsilon/p < b_0 < 1+\varepsilon/p, \qquad a_0 < b_0 + 1 + 2\varepsilon/p, \qquad a_0 + b_0^* > 1, \\ a_0 + b_0 < 3 + \varepsilon/p, \qquad a_0 < 2 + \varepsilon/p, \end{gathered} \tag{5.7}$$

while for the second inequality we directly apply Tables 1 and 2 to get[1]

$$1 < a_1 < 1 + 2\varepsilon/p, \tag{5.8}$$

a_i b_i defined in (5.6). The conditions on a_i, b_i can be satisfied for some $A \in \mathbb{R}^1$ if we have for $0 < \varepsilon \le p/2$: $-\varepsilon(p-1)/p < \beta < p-1+\varepsilon(p-1)/p$, $p/2-1-\varepsilon/p < \alpha+\beta < 5p/2-3+\varepsilon$, $\alpha-\beta < p/2-1+\varepsilon+\varepsilon(p-1)/p$, $-p/2-\varepsilon/p < \alpha < 3p/2-2+\varepsilon$. Thus (5.1) is proved for $\lambda = 1$.

Next let $\lambda \ne 1$. As $\mathcal{O}_{ij}((\mathbf{x}-\mathbf{y});\lambda) = \lambda\mathcal{O}_{ij}(\lambda(\mathbf{x}-\mathbf{y});1)$ we easily have

$$\begin{aligned} &\int_{\mathbb{R}^3}\Big|\int_{\mathbb{R}^3}\mathcal{O}_{ij}(\mathbf{x}-\mathbf{y};\lambda)f(\mathbf{y})d\mathbf{y}\Big|^p \eta_\beta^{\alpha-p/2-\varepsilon}(\mathbf{x};\lambda)d\mathbf{x} \\ &= \lambda^{-2p}\int_{\mathbb{R}^3}\Big|\int_{\mathbb{R}^3}\mathcal{O}_{ij}(\lambda\mathbf{x}-\mathbf{z};1)f(\mathbf{z}/\lambda)d\mathbf{z}\Big|^p \eta_\beta^{\alpha-p/2-\varepsilon}(\lambda\mathbf{x};1)d\mathbf{x} \\ &\le C\lambda^{-2p-3}\int_{\mathbb{R}^3}\Big|f\Big(\frac{\mathbf{z}}{\lambda}\Big)\Big|^p \eta_\beta^{\alpha+p/2}(\mathbf{z};1)d\mathbf{z} \\ &= C\lambda^{-2p}\int_{\mathbb{R}^3}|f(\mathbf{y})|^p\eta_\beta^{\alpha+p/2}(\mathbf{y};\lambda)d\mathbf{y} \end{aligned}$$

and we have (5.1) with $\lambda \ne 1$.

In order to prove (5.2) we redefine functions $K(\cdot,\cdot)$ and $F(\cdot)$:

$$\begin{gathered} K(\mathbf{x},\mathbf{y}) = \mathcal{O}_{ij}(\mathbf{x}-\mathbf{y};1)\left(\nu_\beta^{\alpha-p/2-\varepsilon}(\mathbf{x})\right)^{1/p}\left(\nu_\beta^{\alpha+p/2}(\mathbf{y})\right)^{-1/p}, \\ F(\mathbf{y}) = f(\mathbf{y})\left(\nu_\beta^{\alpha+p/2}(\mathbf{y})\right)^{1/p}. \end{gathered}$$

We will now proceed as in the first part of the proof, but now we search the functions $L(\cdot)$, $M(\cdot)$ in the form $L(\mathbf{x}) = \mu_{-\frac{\beta}{p^2}}^{-A,-G}(\mathbf{x})$, $M(\mathbf{x}) = \mu_{-\frac{\beta}{p^2}}^{-A,-H}(\mathbf{x})$. Denoting

$$\begin{aligned} c_0 &= qG + \tfrac{\alpha}{p} + \tfrac{1}{2} \qquad & c_1 &= pH - \tfrac{\alpha}{p} + \tfrac{1}{2} + \tfrac{\varepsilon}{p} \\ d_0 &= qH + \tfrac{\alpha}{p} - \tfrac{1}{2} \qquad & d_1 &= pG - \tfrac{\alpha}{p} - \tfrac{1}{2} \end{aligned} \tag{5.9}$$

[1] The convolution on the left-hand side of (5.5) is again the same convolution as treated in Section 3, but at the point $-\mathbf{y}$.

we see that in order to verify (5.4) and (5.5) we have to find such a_i, b_i see (5.6), (5.7) and c_i d_i such that

$$\int_{\mathbb{R}^3} \mathcal{O}_{ij}(\mathbf{x}-\mathbf{y};1)\mu_{-b_0}^{-a_0,-c_0}(\mathbf{y})d\mathbf{y} \le C\mu_{-b_0}^{-a_0+1+\frac{\varepsilon}{p},-d_0}(\mathbf{x}),\ \mathbf{x}\in\mathbb{R}^3\setminus\{\mathbf{0}\}$$
$$\int_{\mathbb{R}^3} \mathcal{O}_{ij}(-\mathbf{x}-\mathbf{y};1)\mu_0^{-a_1,-c_1}(-\mathbf{x})d\mathbf{x} \le C\mu_0^{-a_1+1+\frac{\varepsilon}{p},-d_1}(\mathbf{y}),\ \mathbf{y}\in\mathbb{R}^3\setminus\{\mathbf{0}\}.$$

Recalling that $|\mathcal{O}_{ij}(\mathbf{x}-\mathbf{y};1)| \le C_0\nu_{-1}^{-1}(\mathbf{x}-\mathbf{y};1)$ we get from Lemma 4.2 the following two possible sets of conditions for c_i, d_i:

$$(i)\quad \begin{aligned} &c_i < 3\\ &c_i+1>3\\ &d_i \ge c_i-2 \end{aligned} \qquad\qquad (ii)\quad \begin{aligned} &c_i<2\\ &d_i\ge 0, \end{aligned}$$

where in both cases $i = 0,1$. Conditions for a_i, b_i are the same as in the first part of this proof. From the conditions (i) we get the following additional condition

$$p/2-3+\varepsilon < \alpha < 5p/2-3.$$

Case (ii) gives more restrictive conditions on α, no extension of the result. So, (5.2) is proved in the case $\lambda = 1$.

Finally to get (5.2) with $\lambda \ne 1$ we proceed as in the case of the weights $\eta_B^A(\cdot\,;\lambda)$. We have

$$\begin{aligned}
&\int_{\mathbb{R}^3}\Big|\int_{\mathbb{R}^3}\mathcal{O}_{ij}(\mathbf{x}-\mathbf{y};\lambda)f(\mathbf{y})d\mathbf{y}\Big|^p\nu_\beta^{\alpha-p/2-\varepsilon}(\mathbf{x};\lambda)d\mathbf{x}\\
&=\lambda^{-2p-\alpha+p/2+\varepsilon}\int_{\mathbb{R}^3}\Big|\int_{\mathbb{R}^3}\mathcal{O}_{ij}(\lambda\mathbf{x}-\mathbf{z};1)f(\mathbf{z}/\lambda)d\mathbf{z}\Big|^p\nu_\beta^{\alpha-p/2-\varepsilon}(\lambda\mathbf{x};1)d\mathbf{x}\\
&\le C\lambda^{-2p-\alpha+p/2+\varepsilon-3}\int_{\mathbb{R}^3}\big|f(\mathbf{z}/\lambda)\big|^p\nu_\beta^{\alpha+p/2}(\mathbf{z};1)d\mathbf{z}\\
&=C\,\lambda^{-p+\varepsilon}\int_{\mathbb{R}^3}|f(\mathbf{y})|^p\nu_\beta^{\alpha+p/2}(\mathbf{y};\lambda)d\mathbf{y}
\end{aligned}$$

and we have (5.2) with $\lambda \ne 1$. This completes our proof. ∎

Theorem 5.2. *Let T be an integral operator with the kernel $|\nabla\mathcal{O}|$, $T: f \mapsto |\nabla\mathcal{O}| * f$, and let $1 < p < \infty$. Then T is a well defined continuous operator:*

a) $$L^p(\mathbb{R}^3;\eta_\beta^{\alpha+p/2}(\cdot\,;\lambda)) \longmapsto L^p(\mathbb{R}^3;\eta_\beta^\alpha(\cdot\,;\lambda))$$

for $0<\beta<3p/2-3/2$, $-1<\alpha+\beta$, $\alpha<3p/2-2$, $\alpha-\beta<p/2-1$

b) $$L^p(\mathbb{R}^3;\nu_\beta^{\alpha+p/2}(\cdot\,;\lambda)) \longmapsto L^p(\mathbb{R}^3;\nu_\beta^\alpha(\cdot\,;\lambda))$$

for $0 < \beta < 3p/2 - 3/2$, $-1 < \alpha + \beta$, $-3 < \alpha < 3p/2 - 2$, $\alpha - \beta < p/2 - 1$

Moreover we have for α, β *specified in a) and b), respectively*

ad a)

$$\| |\nabla \mathcal{O}(\cdot\,;\lambda)| * f \|_{p,(\eta^{\alpha}_{\beta}(\cdot\,;\lambda)),\mathbb{R}^3} \le C, \lambda^{-1} \|f\|_{p,(\eta^{\alpha+p/2}_{\beta}(\cdot\,;\lambda)),\mathbb{R}^3}, \tag{5.10}$$

ad b)

$$\| |\nabla \mathcal{O}(\cdot\,;\lambda)| * f \|_{p,(\nu^{\alpha}_{\beta}(\cdot\,;\lambda)),\mathbb{R}^3} \le C \lambda^{-\frac{1}{2}} \|f\|_{p,(\nu^{\alpha+p/2}_{\beta}(\cdot\,;\lambda)),\mathbb{R}^3}. \tag{5.11}$$

The theorem (and the following theorems of this section) can be proved analogously as Theorem 5.1.

Theorem 5.3. *Let* $R = |\nabla^2 \mathcal{O} - \nabla^2 \mathcal{S}|$ *or* $R = \partial_1 \mathcal{O}_{ij}$, $i, j = 1, 2, 3$. *Let* T *be an integral operator with the kernel* R, $T : f \mapsto R * f$, *and let* $1 < p < \infty$. *Then* T *is a well defined continuous operator:*

a) $$L^p(\mathbb{R}^3; \eta^{\alpha+p/2}_{\beta}(\cdot\,;\lambda)) \longmapsto L^p(\mathbb{R}^3; \eta^{\alpha+p/2-\varepsilon}_{\beta}(\cdot\,;\lambda))$$

for $-\varepsilon(p-1)/p < \beta < p - 1 + \varepsilon(p-1)/p$, $-p/2 - 1 - \varepsilon/p < \alpha + \beta < 5p/2 - 3 + \varepsilon$, $\alpha - \beta < p/2 - 1 + \varepsilon + \varepsilon(p-1)/p$, $-3p/2 - \varepsilon/p < \alpha < 3p/2 - 2 + \varepsilon$, $0 < \varepsilon \le p/2$

b) $$L^p(\mathbb{R}^3; \nu^{\alpha+p/2}_{\beta}(\cdot\,;\lambda)) \longmapsto L^p(\mathbb{R}^3; \nu^{\alpha+p/2-\varepsilon}_{\beta}(\cdot\,;\lambda))$$

for $-\varepsilon(p-1)/p < \beta < p - 1 + \varepsilon(p-1)/p$, $-p/2 - 1 - \varepsilon/p < \alpha + \beta < 5p/2 - 3 + \varepsilon$, $\alpha - \beta < p/2 - 1 + \varepsilon + \varepsilon(p-1)/p$, $\max\{-3p/2 - \varepsilon/p, -p/2 - 3 + \varepsilon\} < \alpha < \min\{3p/2 - 2 + \varepsilon, 5p/2 - 3\}$, $0 < \varepsilon \le p/2$

Moreover we have for α, β *specified in a) and b) respectively*

ad a)

$$\left\| |\nabla^2 \mathcal{O}(\cdot\,;\lambda) - \nabla^2 \mathcal{S}(\cdot)| * f \right\|_{p,(\eta^{\alpha+p/2-\varepsilon}_{\beta}(\cdot\,;\lambda)),\mathbb{R}^3} \le C \|f\|_{p,(\eta^{\alpha+p/2}_{\beta}(\cdot\,;\lambda)),\mathbb{R}^3}, \tag{5.12}$$

$$\|\partial_1 \mathcal{O}_{ij}(\cdot\,;\lambda) * f\|_{p,(\eta^{\alpha+p/2-\varepsilon}_{\beta}(\cdot\,;\lambda)),\mathbb{R}^3} \le C \lambda^{-1} \|f\|_{p,(\eta^{\alpha+p/2}_{\beta}(\cdot\,;\lambda)),\mathbb{R}^3}, \tag{5.13}$$

ad b)

$$\left\| |\nabla^2 \mathcal{O}(\cdot\,;\lambda) - \nabla^2 \mathcal{S}(\cdot)| * f \right\|_{p,(\nu^{\alpha+p/2-\varepsilon}_{\beta}(\cdot\,;\lambda)),\mathbb{R}^3} \le C \lambda^{\frac{\varepsilon}{p}} \|f\|_{p,(\nu^{\alpha+p/2}_{\beta}(\cdot\,;\lambda)),\mathbb{R}^3} \tag{5.14}$$

$$\|\partial_1 \mathcal{O}_{ij}(\cdot\,;\lambda) * f\|_{p,(\nu^{\alpha+p/2-\varepsilon}_{\beta}(\cdot\,;\lambda)),\mathbb{R}^3} \le C \lambda^{\frac{\varepsilon}{p}-1} \|f\|_{p,(\nu^{\alpha+p/2}_{\beta}(\cdot\,;\lambda)),\mathbb{R}^3}. \tag{5.15}$$

Theorem 5.4. *Let T be an integral operator with the kernel $\mathcal{P}_i$, $T: f \mapsto \mathcal{P}_i * f$, $i = 1,2,3$, and let $1 < p < \infty$. Then T is a well defined continuous operator:*

a)
$$L^p(\mathbb{R}^3; \eta_\beta^{\alpha+p/2}(\cdot\,;\lambda)) \longmapsto L^p(\mathbb{R}^3; \eta_\beta^{\alpha-p/2}(\cdot\,;\lambda))$$

for $0 < \beta < p-1$, $p/2 - 3 < \alpha + \beta < 5p/2 - 3$

b)
$$L^p(\mathbb{R}^3; \nu_\beta^{\alpha+p/2}(\cdot\,;\lambda)) \longmapsto L^p(\mathbb{R}^3; \nu_\beta^{\alpha-p/2}(\cdot\,;\lambda))$$

for $0 < \beta < p-1$, $p/2 - 3 < \alpha + \beta < 5p/2 - 3$, $p/2 - 3 < \alpha < 5p/2 - 3$

Moreover we have for α, β specified in a) and b), respectively

ad a)
$$\|\mathcal{P}_i * f\|_{p,(\eta_\beta^{\alpha-p/2}(\cdot\,;\lambda)),\mathbb{R}^3} \le C\lambda^{-1} \|f\|_{p,(\eta_\beta^{\alpha+p/2}(\cdot\,;\lambda)),\mathbb{R}^3}, \tag{5.16}$$

ad b)
$$\|\mathcal{P}_i * f\|_{p,(\nu_\beta^{\alpha-p/2}(\cdot\,;\lambda)),\mathbb{R}^3} \le C \|f\|_{p,(\nu_\beta^{\alpha+p/2}(\cdot\,;\lambda)),\mathbb{R}^3}. \tag{5.17}$$

6. SINGULAR INTEGRALS

The aim of this section is to present some results concerning L^p-estimates of certain singular operators and finally to apply them on the convolutions (defined in the sense of principal value) of the type $\nabla\mathcal{P} * f$, $\nabla^2\mathcal{S} * f$ and $\nabla^2\mathcal{O} * f$.

We shall use the idea of Farwig (see [2]). Before formulating the result from [7] we need to define some notation.

Definition 6.1. *The weight w, $w \ge 0$ belongs to the Muckenhoupt class A_p, $1 \le p < +\infty$ if there is a constant C such that*

$$\begin{aligned} &\sup_Q \left[\left(\tfrac{1}{|Q|} \int_Q w(\mathbf{x})d\mathbf{x} \right) \left(\tfrac{1}{|Q|} \int_Q w(\mathbf{x})^{-\frac{1}{p-1}} d\mathbf{x} \right)^{p-1} \right] \le C < +\infty \\ &\sup_Q \tfrac{1}{|Q|} \int_Q w(\mathbf{x})d\mathbf{x} \le Cw(\mathbf{x}_0), \quad \forall \mathbf{x}_0 \in \mathbb{R}^N \end{aligned} \tag{6.1}$$

for $p \in (1;\infty)$ and $p = 1$, respectively. In the first case, the supremum is taken over all cubes Q in $\mathbb{R}^N$, in the second case only over those cubes which contain $\mathbf{x}_0$; $|Q|$ denotes the Lebesgue measure of Q. The constant does not depend on $\mathbf{x}_0$.

Remark 6.2.

a) For $p = 1$, the condition $(6.1)_2$ can be replaced by

$$Mw(\mathbf{x}) \le Cw(\mathbf{x}) \quad \text{for a.a. } \mathbf{x} \in \mathbb{R}^N \tag{6.2}$$

where $Mg(\mathbf{x})$ is the Hardy-Littlewood maximal function which is defined by the left hand side in $(6.1)_2$.

b) In (6.1) it is enough to take the supremum over all cubes with edges parallel to an arbitrary chosen Cartesian system X.

To show this, let X be a Cartesian system in $\mathbb{R}^N$ and X' another one arising from X by any rotation. Then we have

$$\frac{1}{N^{\frac{N}{2}}}\frac{1}{|Q_1|}\int_{Q_1} w(\mathbf{x})d\mathbf{x} \le \frac{1}{|Q'|}\int_{Q'} w(\mathbf{x})d\mathbf{x} \le \frac{N^{\frac{N}{2}}}{|Q_2|}\int_{Q_2} w(\mathbf{x})d\mathbf{x} \tag{6.3}$$

for any locally integrable function $w \ge 0$. In (6.3) Q' is a cube with edges parallel to the axes of X', Q_1 is the greatest cube with edges parallel to the axes of X such that $Q_1 \subset Q'$, and Q_2 is the smallest cube with edges parallel to the axes of X such that $Q' \subset Q_2$.

Next part is devoted to the investigation under what condition the weights defined in Section 1 belong to A_p for some $1 \le p < +\infty$. First we recall several general results:

Lemma 6.3.

a) *If* $w_1, w_2 \in A_1$ *then for any* $1 \le p < +\infty$ *the weight* $w \equiv w_1\, w_2^{1-p} \in A_p$.

b) *If* $w_1,\ w_2 \in A_p$ *for some* $1 \le p < +\infty$ *then for any* $\hbar \in [0;1]$, $w_\hbar \equiv w_1^\hbar w_2^{1-\hbar} \in A_p$.

c) *If* $w \in A_p$ *for some* $1 \le p < +\infty$ *then for any* $\hbar \in [0;1]$, $w^\hbar \in A_p$.

Proof. It follows directly from the definition of A_p and in case b) from the Hölder inequality. ∎

Definition 6.4. *Let* μ *be a non-negative Borel measure. We define the maximal function*

$$M\mu(x) = \sup_Q \frac{1}{|Q|}\int_Q d\mu(y) \tag{6.4}$$

where the supremum is taken over all cubes Q *such that* $x \in Q$. *Analogously we define* $Mf(x)$ *for* $f \in L^1_{loc}(\mathbb{R}^N)$, *replacing* $d\mu(y)$ *by* $|f|dy$. *(See also Remark 6.2 a) and b)).*

The proof of the following lemma can be found in [11] (Theorems IX.5.5 and IX.3.4).

Lemma 6.5.

a) *If* $M\mu$ *is finite for a.a.* $\mathbf{x} \in \mathbb{R}^N$ *then, for any* $\hbar \in [0;1)$, $(M\mu)^{\hbar} \in A_1$.

b) *Let* $w \in A_1$. *Then there exists a function* $f \in L^1_{loc}(\mathbb{R}^N)$ *such that* $w \sim (Mf)^{\hbar}$ *for some* $\hbar \in [0;1)$.

Using Lemma 6.5 we can easily show

Lemma 6.6. *The weights* $|\mathbf{x}|^{-a}$ *and* $(1+|\mathbf{x}|)^{-a}$ *satisfy the* A_1*-condition on* $\mathbb{R}^N$ *for each* $a \in [0;N)$.

Proof. We have that for $\mu = \delta_0$ the maximal function $M\mu(x) \sim |\mathbf{x}|^{-N}$ and so $|\mathbf{x}|^{-N\hbar} \in A_1$, $\forall \hbar \in [0;1)$. Further, if we define $\mu(A) = |A \cap B_1(\mathbf{0})|$, then $M\mu(x) \sim (1+|x|)^{-N}$ and again Lemma 6.5b) furnishes the result. ■

Lemma 6.7. *For* $b \in (-1;2]$ *and* $\hbar \in [0;1)$ *the function* $w_0^{(3)}(\mathbf{x}) = \left(\frac{|\mathbf{x}|^{b-1}}{s(\mathbf{x})}\right)^{\hbar}$, $\mathbf{x} \in \mathbb{R}^3$ *is a weight of the class* A_1 *in* $\mathbb{R}^3$.

Proof. For the proof see [2].

Using this result we can now show the following

Lemma 6.8. *For* $b \in (-1;1]$ *and* $\hbar \in [0;1)$ *the function* $w_1^{(3)} = \left(\nu_{-1}^{b-1}\right)^{\hbar}$ *is a weight of class* A_1 *in* $\mathbb{R}^3$.

Proof. We have to verify that $(Mw_1^{(3)})(\mathbf{x}) \leq Cw_1^{(3)}(\mathbf{x})$ a.e. in $\mathbb{R}^3$. Let Q_a denotes a closed cube with sides parallel to the axes and with the side length a; R will be a sufficiently large constant. We distinguish several cases:

A) $s(\mathbf{x}) \leq 1, \quad r = |\mathbf{x}| \geq R$

α) $a \leq \frac{1}{2} r^{\frac{1}{2}}$

Then for all $\mathbf{y} \in Q_a$ we have that $w_1^{(3)}(\mathbf{y}) \sim w_1^{(3)}(\mathbf{x})$ and

$$\int_{Q_a} w_1^{(3)}(\mathbf{y})\,\mathrm{d}\mathbf{y} \leq C|Q_a| w_1^{(3)}(\mathbf{x}) .$$

β) $a = \frac{1}{2} r^{\frac{1}{2}+\sigma}, \quad \sigma \in (0;\frac{1}{2}]$

Now $Q_a \subset \{\mathbf{y} \in \mathbb{R}^3;\ ||\mathbf{y}| - r| \leq cr^{\frac{1}{2}+\sigma};\ s(\mathbf{y}) \leq cr^{2\sigma}\}$ and proceeding analogously as in Lemma 3.2 we get

$$\int_{Q_a} w_1^{(3)}(\mathbf{y})\,\mathrm{d}\mathbf{y} \leq C \int_{r-cr^{\frac{1}{2}+\sigma}}^{r+cr^{\frac{1}{2}+\sigma}} \varrho^{1+(b-1)\hbar} \mathrm{d}\varrho \int_0^{Cr^{2\sigma}} \frac{\mathrm{d}s}{(1+s)^{\hbar}} \leq$$
$$\leq C|Q_a| r^{\frac{3}{2}+(b-1)\hbar+(3-2\hbar)\sigma} \leq C|Q_a| r^{(b-1)\hbar} \leq Cw_1^{(3)}(\mathbf{x})$$

as $r \geq R \gg 1$, $s(\mathbf{x}) \leq 1$.

γ) $a \geq \frac{r}{2}$
In this case all Q_a such that $\mathbf{x} \in Q_a$ are contained in the ball B_{4a} and therefore similarly as above

$$\int_{Q_a} w_1^{(3)}(\mathbf{y})\mathrm{d}\mathbf{y} \leq C \int_0^{4a} (1+\varrho)^{2-\hbar+(b-1)\hbar} \mathrm{d}\varrho \leq C|Q_a| a^{(b-2)\hbar}.$$

Evidently,

$$\int_{Q_a} w_1^{(3)}(\mathbf{y})\,\mathrm{d}\mathbf{y} \leq C|Q_a| r^{(b-2)\hbar} \leq C|Q_a| r^{(b-1)\hbar}.$$

B) $s(\mathbf{x}) \geq 1, \quad r \geq R$
Now, as $b \leq 2$ and $s \leq r$, we have

$$\left(1+\frac{1}{r}\right)^{b-1} \leq C\left(1+\frac{1}{r}\right) \leq C\left(1+\frac{1}{s}\right)$$

and so

$$w_1^{(3)}(\mathbf{x}) \leq \left(\frac{(1+r)^{b-1}}{(1+s)}\right)^{\hbar} \leq C\frac{r^{(b-1)\hbar}}{s^{\hbar}} = C\, w_0^{(3)}(\mathbf{x}).$$

Therefore for all $\mathbf{y} \in \mathbb{R}^3$ we have that $w_1^{(3)}(\mathbf{y}) \leq w_0^{(3)}(\mathbf{y})$. But then, as $w_0^{(3)} \in A_1$, we have

$$\int_{Q_a} w_1^{(3)}(\mathbf{y})\,\mathrm{d}\mathbf{y} \leq C \int_{Q_a} w_0^{(3)}(\mathbf{y})\,\mathrm{d}y \leq C\,|Q_a|\, w_0^{(3)}(\mathbf{x}).$$

As $|\mathbf{x}| \geq 1$, $s(\mathbf{x}) \geq 1$, we have $w_0^{(3)}(\mathbf{x}) \leq C w_1^{(3)}(\mathbf{x})$ and the required inequality follows.

C) $r \leq R$

α) If $\frac{a}{2} \leq R$, then trivially

$$\int_{Q_a} w_1^{(3)}(\mathbf{y})\,\mathrm{d}\mathbf{y} \leq C\,|Q_a|\, w_1^{(3)}(\mathbf{x}).$$

β) If $\frac{a}{2} > R$, then $Q_a \subset B_{3a}$ and analogously as in Aγ)

$$\int_{Q_a} w_1^{(3)}(\mathbf{y})\mathrm{d}\mathbf{y} \leq \int_{B_{3a}} w_1^{(3)}(\mathbf{y})\,\mathrm{d}\mathbf{y}$$

$$\leq C \int_0^{3a} (\varrho+1)^{2-\hbar+(b-1)\hbar} \mathrm{d}\varrho \leq C a^{3+(b-2)\hbar} \leq C_1 |Q_a| a^{(b-2)\hbar}.$$

As $b \leq 2$,

$$\int_{Q_a} w_1^{(3)}(\mathbf{y})\mathrm{d}\mathbf{y} \leq C|Q_a| \leq C_2 |Q_a| w_1^{(3)}(\mathbf{x}).$$

■

Lemma 6.9. *For $b \in (-1;2]$ and $\hbar \in [0;1)$ the function $w_2^{(3)} = \left(\eta_{-1}^{b-1}\right)^{\hbar}$ is a weight of class A_1 in $\mathbb{R}^3$.*

Proof. We proceed as in Lemma 6.8. Part A) remains the same. In part B) we use the fact that for some $C > 0$

$$\left(1+\frac{1}{r}\right)^{b-1} \leq \left(1+\frac{1}{r}\right) \leq C\left(1+\frac{1}{s}\right)$$

for all $\mathbf{y} \in \mathbb{R}^3$ such that $s(\mathbf{y}) \geq 1$, $r(\mathbf{y}) \geq R$, if $b \leq 2$. In part Cα) the condition $w_2^{(3)} \geq C > 0$ for $|\mathbf{x}| \leq R$ is satisfied without the necessity of any additional condition on b. In part Cβ) we proceed as in Lemma 6.8 but we use

$$\int_{S_\tau} w_2^{(3)} \mathrm{d}S \sim \tau^2(\tau+1)^{1+(b-2)\hbar}.$$

∎

Combining Lemmas 6.7, 6.8 and 6.9 with the properties of Muckenhoupt classes A_p we can show the following

Theorem 6.10. *Let $-1 < \beta < p-1$, $-3 < \alpha+\beta < 3(p-1)$. Then the weights $\eta_\beta^\alpha(\mathbf{x})$ and $\sigma_\beta^\alpha(\mathbf{x})$ are A_p-weights in $\mathbb{R}^3$ for $p \in (1;\infty)$.*

Let $-1 < \beta < p-1$, $-3 < \alpha+\beta < 3(p-1)$, $-3 < \alpha < 3(p-1)$. Then the weight $\nu_\beta^\alpha(\mathbf{x})$ is A_p-weight in $\mathbb{R}^3$ for $p \in (1;\infty)$.

In order to formulate the fundamental theorem used in this section we need to define the so called L^∞-Dini condition. We will use the following notation in the definition.

$B_1(\mathbf{0})$ - the unit ball with the center at the point $\mathbf{0}$ in $\mathbb{R}^N$,
ρ - rotation of $\partial B_1(\mathbf{0})$ with magnitude $|\rho| = \sup_{x\in\partial B_1(\mathbf{0})} |\rho x - x|$

Definition 6.11. *Let Ω be a function defined on $\partial B_1(\mathbf{0})$, $\Omega \in L^\infty(\partial B_1(\mathbf{0}))$. We say that function Ω satisfies the L^∞-Dini condition if*

$$\int_0^1 \theta_\infty(\delta)\frac{1}{\delta}d\delta < +\infty, \quad \text{where } \theta_\infty(\delta) = \sup_{|\rho|<\delta} \|\Omega(\rho x) - \Omega(x)\|_{L^\infty(\partial B_1(\mathbf{0}))}.$$

Notation: $\Omega \in L^\infty$-Dini.

For the proof of the following theorem see [7].

Theorem 6.12. *Let $N \in \mathbb{N}$, $N \geq 2$, $\Omega \in L^\infty(\partial B_1(\mathbf{0}))$, $\Omega \in L^\infty$-Dini, $\int_{\partial B_1(\mathbf{0})} \Omega dS = 0$, Ω is a positively homogeneous function of degree zero, $R(\mathbf{x}) := \Omega(\mathbf{x}')/|\mathbf{x}|^N$, $\mathbf{x}' = \mathbf{x}/|\mathbf{x}|$.*

Let T be an operator with the kernel R, i.e. $Tf(x) = (R*f)(x)$ in the principal–value sense and $w \in A_p$ in $\mathbb{R}^N$, $p > 1$. Then T is a continuous operator $\mathcal{L}^p(\mathbb{R}^N; w) \mapsto L^p(\mathbb{R}^N; w)$.

Remark 6.13. The fact, that $R(\cdot)$ satisfies the conditions formulated in Theorem 6.12 means that $R(\cdot)$ represents a Calderón-Zygmund singular integral kernel. We will write in this case $R(\cdot) \in CZ$. It is known that $\nabla^2 \mathcal{S}$, $\nabla \mathcal{P} \in CZ$.

Remark 6.14. If w is a weight such that $f \in L^p(\mathbb{R}^N; w) \Rightarrow f \in L^1_{loc}(\mathbb{R}^N)$ and $1 \le p < +\infty$ then it is an easy matter to see that $\mathcal{L}^p(\mathbb{R}^N; w)$ and $L^p(\mathbb{R}^N; w)$ coincide. The weights η^α_β have this property. For the other weights we will distinguish spaces $\mathcal{L}^p(\mathbb{R}^N; w)$ and $L^p(\mathbb{R}^N; w)$.

As the corollary of Theorem 6.12 we get

Corollary 6.15. *Let $N = 2, 3$, R be either $\partial_i\partial_j\mathcal{S}_{rs}$ or $\partial_i\mathcal{P}_r$, $i, j, r, s = 1, 2, ..., N$ and $f \in \mathcal{L}^p(\mathbb{R}^N; w)$, where w stands for η^α_β, σ^α_β or ν^α_β with $1 < p < \infty$ and α, β be such that the corresponding weights are A_p-weights. Then v.p.$(R * f) \in L^p(\mathbb{R}^N; w)$ and*

$$\|v.p.\,(R * f)\|_{p,(w),\mathbb{R}^N} \le C\|f\|_{p,(w),\mathbb{R}^N}.$$

We will now formulate the results for $\nabla\mathcal{P}$ which follow from Theorem 6.12.

Theorem 6.16. *$\nabla\mathcal{P} : f \mapsto \nabla\mathcal{P} * f$ defines continuous operators:*

a)
$$L^p(\mathbb{R}^3; \eta_\beta^{\alpha+p/2}) \longmapsto L^p(\mathbb{R}^3; \eta_\beta^{\alpha+p/2})$$

for $p > 1$, $-1 < \beta < p - 1$, $-3 - p/2 < \alpha + \beta < 5p/2 - 3$

b)
$$\mathcal{L}^p(\mathbb{R}^3; \nu_\beta^{\alpha+p/2}(\cdot\,; \lambda)) \longmapsto L^p(\mathbb{R}^3; \nu_\beta^{\alpha+p/2})$$

for $p > 1$, $-1 < \beta < p - 1$, $-3 - p/2 < \alpha < 5p/2 - 3$, $-3 - p/2 < \alpha + \beta < 5p/2 - 3$.

The last theorem follows from Theorem 5.3, Theorem 6.10 and Corollary 6.15.

Theorem 6.17. *Let T be an integral operator in the principal-value sense with the kernel $\partial_k\partial_l\mathcal{O}_{ij}(\cdot\,; \lambda)$, $i, j, k, l = 1, 2, 3$, $T : f \mapsto R * f$ and let $1 < p < \infty$. Then T is a well defined continuous operator:*

a) $$L^p(\mathbb{R}^3;\eta_\beta^{\alpha+p/2}(\cdot\,;\lambda)) \longmapsto L^p(\mathbb{R}^3;\eta_\beta^{\alpha+p/2-\varepsilon}(\cdot\,;\lambda))$$

for $\max\{-1,-\varepsilon(p-1)/p\} < \beta < p-1$, $-p/2-1-\varepsilon/p < \alpha+\beta < 5p/2-3$, $\alpha-\beta < p/2-1+\varepsilon+\varepsilon(p-1)/p$, $-3p/2-\varepsilon/p < \alpha < 3p/2-2+\varepsilon$, $0 < \varepsilon \le p/2$

b) $$L^p(\mathbb{R}^3;\nu_\beta^{\alpha+p/2}(\cdot\,;\lambda)) \longmapsto L^p(\mathbb{R}^3;\mu_\beta^{\alpha+p/2-\varepsilon,\alpha+p/2}(\cdot\,;\lambda))$$

$$L^p(\mathbb{R}^3\setminus\Omega;\nu_\beta^{\alpha+p/2}(\cdot\,;\lambda)) \longmapsto L^p(\mathbb{R}^3;\nu_\beta^{\alpha+p/2-\varepsilon}(\cdot\,;\lambda))$$

for $\max\{-1,-\varepsilon(p-1)/p\} < \beta < p-1$, $-p/2-1-\varepsilon/p < \alpha+\beta < 5p/2-3$, $\alpha-\beta < p/2-1+\varepsilon+\varepsilon(p-1)/p$, $\max\{-p/2-3,-3p/2-\varepsilon/p\} < \alpha < \min\{5p/2-3, 3p/2-2+\varepsilon\}$, $0 < \varepsilon \le p/2$, $\Omega \subset \mathbb{R}^3$ *- an arbitrary domain,* $0 \in \Omega$

Moreover, we have for α, β *specified in a) and b), respectively*

ad a)

$$\left\| v.p.\,(\partial_k\partial_l\mathcal{O}_{ij}(\cdot\,;\lambda) * f)\right\|_{p,(\eta_\beta^{\alpha+p/2-\varepsilon}(\cdot\,;\lambda)),\mathbb{R}^3} \le C\,\|f\|_{p,(\eta_\beta^{\alpha+p/2}(\cdot\,;\lambda)),\mathbb{R}^3}, \tag{6.5}$$

ad b)

$$\left\| v.p.\,(\partial_k\partial_l\mathcal{O}_{ij}(\cdot\,;\lambda) * f)\right\|_{p,(\nu_\beta^{\alpha+p/2-\varepsilon}(\cdot\,;\lambda)),\mathbb{R}^3} \le C\,\|f\|_{p,(\mu_\beta^{\alpha+p/2-\varepsilon,\alpha+p/2}(\cdot\,;\lambda)),\mathbb{R}^3} \tag{6.6}$$

$$\left\| v.p.\,(\partial_k\partial_l\mathcal{O}_{ij}(\cdot\,;\lambda) * f)\right\|_{p,(\nu_\beta^{\alpha+p/2-\varepsilon}(\cdot\,;\lambda)),\mathbb{R}^3\setminus\Omega} \le C\,\lambda^{\frac{\varepsilon}{p}}\,\|f\|_{p,(\nu_\beta^{\alpha+p/2}(\cdot\,;\lambda)),\mathbb{R}^3}. \tag{6.7}$$

Acknowledgments

The authors would like to thank to the referee for several suggestions which helped to improve the presentation of the paper. S. Kračmar was partially supported by the Grant Agency of the Czech Republic, grant No. 201/96/0313, M. Pokorný was partially supported by the Grant Agency of the Czech Republic, grant No. 201/96/0228.

References

[1] Dutto, P. (1998). Solutions physiquement raisonnables des équations de Navier–Stokes compressibles stationnaires dans un domain extérieur du plan. Ph.D. Thesis, University of Toulon.

[2] Farwig, R. (1992). The stationary exterior 3D-problem of Oseen and Navier–Stokes equations in anisotropically weighted Sobolev spaces. *Mathematische Zeitschrift*, 211:409–447.

[3] Farwig, R. (1990). Das stationäre Außenraumproblem der Navier-Stokes-Gleichungen bei nichtverschwindender Anströmgeschwindigkeit in anisotrop gewichteten Sobolevräumen. SFB 256 preprint no. **110** (Habilitationsschrift), University of Bonn.

[4] Finn, R. (1959). Estimates at infinity for stationary solution of Navier–Stokes equations. *Bult. Math. Soc. Sci. Math. Phys R.P. Roumaine*, Tome 3, 53(4):387–418.

[5] Finn, R. (1965). On the Exterior Stationary Problem and Associated Pertubation Problems for the N.S. Eq's. *Arch. Rational Mech. Anal.*, 19:363–406.

[6] Galdi, G.P. (1994). *An Introduction to the Mathematical Theory of Navier-Stokes Equations I.* Springer Verlag,.

[7] Kurtz, D.S. and Wheeden, R.L. (1979). Results on Weighted Norm Inequalities for Multipliers. *Trans. Amer. Math. Soc.*, 255:343–364.

[8] Novotný, A. and Padula, M. (1997). Physically reasonable solutions to steady compressible Navier–Stokes equations in 3D–exterior domains ($v_\infty \neq 0$). *Math. Ann.*, 370:439–489.

[9] Pokorný, M. (1999). Asymptotic behaviour of some equations describing the flow of fluids in unbounded domains. Ph.D. thesis, Charles University Prague & University of Toulon.

[10] Smith, D. (1965). Estimates at Infinity for Stationary Solutions of the N.S. Equations in Two Dimensions. *Arch. Rational Mech. Anal.*, 20:341–372.

[11] Torchinski, A. (1986). *Real Variable Methods in Harmonic Analysis*, Academic Press, Orlando.

HARDY'S INEQUALITY AND SPECTRAL PROBLEMS OF NONLINEAR OPERATORS

Alois Kufner

Abstract: It will be shown how the Hardy inequality is connected with the spectral properties of certain degenerated/singular nonlinear operators. A problem will be formulated.

Keywords: Hardy's inequality, spectral analysis, degenerated differential operators.

1. INTRODUCTION

1.1. The N-dimensional case. It is well known that the constant C in the Friedrichs (Poincaré) inequality

$$\int_\Omega |u(x)|^2 dx \le C \int_\Omega |\nabla u(x)|^2 dx, \quad \Omega \subset \mathbb{R}^N, \tag{1.1}$$

is closely connected with the value of the first eigenvalue for the Dirichlet (Neumann) problem for the Laplace operator, i.e., for the equation

$$-\Delta u = \lambda u \quad \text{in} \quad \Omega \tag{1.2}$$

(see, e.g., MICHLIN [7]).

Much less is known, if the Laplace operator Δ is replaced by the p-Laplace operator

$$\Delta_p u = \operatorname{div}\,(|\nabla u|^{p-2}\nabla u),$$

or, more generally, for the equation

$$-\Delta_p u = \lambda |u|^{q-2} u \quad \text{in} \quad \Omega \tag{1.3}$$

(see, e.g., LINQVIST [5]). Here, the Sobolev inequality

$$\Big(\int_\Omega |u(x)|^q dx\Big)^{1/q} \le C\Big(\int_\Omega |\nabla u(x)|^p dx\Big)^{1/p} \tag{1.4}$$

plays an important role.

Applied Nonlinear Analysis, edited by Sequeira *et al.*
Kluwer Academic / Plenum Publishers, New York, 1999.

In Section 2, we will shortly indicate how the N-dimensional Hardy inequality (= the weighted Friedrichs/Poincaré inequality)

$$\Big(\int_\Omega w_0(x)|u(x)|^q dx\Big)^{1/q} \le C\Big(\int_\Omega w_1(x)|\nabla u(x)|^p dx\Big)^{1/p} \tag{1.5}$$

with w_0, w_1 given weight functions (i.e., functions measurable and positive a.e. in Ω) is connected with the spectral problem

$$Au = \lambda w_0 |u|^{q-2} u \qquad \text{in} \quad \Omega \tag{1.6}$$

where A is the perturbed (degenerated and/or singular) p-Laplace operator:

$$(Au)(x) = -\text{div}\ (w_1(x)|\nabla u(x)|^{p-2}\nabla u(x)). \tag{1.7}$$

1.2. Remark. More about the connection between the Hardy inequality (1.5) and differential equations can be found in OPIC and KUFNER [9], Section 14.

1.3. The one dimensional case. The spectral theory of the Sturm-Liouville problem

$$-(w_1(x)u'(x))' + q(x)u(x) = \lambda w_0(x)u(x), \qquad x \in (a,b) \tag{1.8}$$

is well elaborated. For the nonlinear analogue of (1.8), i.e., for the equation

$$-(w_1(x)|u'(x)|^{p-2}u'(x))' + q(x)|u(x)|^{p-2}u(x) = \lambda w_0(x)|u(x)|^{p-2}u(x), \tag{1.9}$$

the structure of the spectrum was described by NEČAS [8] and numerous results are known for the special case $w_1 = w_0 \equiv 1$, $q \equiv 0$ (see, e.g., LINDQVIST [5]).

Recently, DRÁBEK and MANÁSEVICH [1] have given a complete description of the spectrum and a closed form representation of the corresponding eigenfunctions for the problem

$$-(|u'(x)|^{p-2}u'(x))' = \lambda|u(x)|^{q-2}u(x) \quad \text{in} \quad (0,T) \tag{1.10}$$

using sharp Poincaré (Sobolev) inequalities of the type

$$\Big(\int_0^T |u(x)|^q dx\Big)^{1/q} \le C\Big(\int_0^T |u'(x)|^p dx\Big)^{1/p}. \tag{1.11}$$

In Section 3, we will formulate a problem about the connection of the Hardy inequality

$$\Big(\int_a^b |u(x)|^q w_0(x) dx\Big)^{1/q} \le C\Big(\int_a^b |u'(x)|^p w_1(x) dx\Big)^{1/p} \tag{1.12}$$

(which is obviously a generalization of (1.11)) with the boundedness below and discreteness of the spectrum of the equation

$$-(w_1(x)|u'(x)|^{p-2}u'(x))' = \lambda w_0(x)|u(x)|^{q-2}u(x) \quad \text{in} \quad (a,b). \tag{1.13}$$

Moreover, we will formulate the problem for the higher order equation

$$(-1)^k(w_1(x)|u^{(k)}(x)|^{p-2}u^{(k)}(x))^{(k)} = \lambda w_0(x)|u(x)|^{q-2}u(x), \quad k \geq 1, \tag{1.14}$$

in connection with the k-th order Hardy inequality

$$\Big(\int_a^b |u(x)|^q w_0(x)dx\Big)^{1/q} \leq C\Big(\int_a^b |u^{(k)}(x)|^p w_1(x)dx\Big)^{1/p}. \tag{1.15}$$

2. THE MORE DIMENSIONAL CASE

2.1. Let us consider the spectral problem (1.6) for $p = q$, i.e.

$$-\text{div }(w_1(x)|\nabla u(x)|^{p-2}\nabla u(x)) = \lambda w_0(x)|u(x)|^{p-2}u(x) \quad \text{in} \quad \Omega, \tag{2.1}$$

$\Omega \subset \mathbb{R}^N$, $1 < p < \infty$. In its *weak formulation*, it means that the integral identity

$$\int_\Omega w_1(x)|\nabla u(x)|^{p-2}\nabla u(x)\nabla v(x)dx = \lambda \int_\Omega w_0(x)|u(x)|^{p-2}u(x)v(x)dx \tag{2.2}$$

holds for all $u, v \in V$ with an appropriate Banach space V, which is a subspace of the weighted Sobolev space $W^{1,p}(\Omega; w_1, w)$ normed by

$$\|u\|_W = \Big(\int_\Omega w_1(x)|\nabla u(x)|^p dx + \int_\Omega w(x)|u(x)|^p dx\Big)^{1/p}.$$

Here w_0, w_1 and w are (given) weight functions on Ω.

Assume that the N-dimensional Hardy inequality (1.5) holds for $u \in V$ with $p = q$, i.e., we have

$$\int_\Omega w_0(x)|u(x)|^p dx \leq C^p \int_\Omega w_1(x)|\nabla u(x)|^p dx, \quad u \in V. \tag{2.3}$$

Taking $v = u$ in (2.2) and using the inequality (2.3), we obtain that

$$\int_\Omega w_1(x)|\nabla u(x)|^p dx = \lambda \int_\Omega w_0(x)|u(x)|^p dx \leq \lambda C^p \int_\Omega w_1(x)|\nabla u(x)|^p dx,$$

i.e. $\lambda C^p \geq 1$ or

$$\lambda \geq C^{-p}. \tag{2.4}$$

Thus, *the spectrum is bounded from below* and one can expect that its infimum (the smallest eigenvalue) is C^{-p} with C the best constant in Hardy's inequality (2.3).

2.2. Remark. Inequality (2.3) tells in fact that the imbedding

$$V \subset L^p(\Omega; w_0) \tag{2.5}$$

[where $L^p(\Omega; w_0)$ is the weighted Lebesgue space normed by $\|u\|_L = (\int_\Omega |u(x)|^p w_0(x)dx)^{1/p}$] is *continuous.* Thus the foregoing result can be formulated also in the following form: *The continuity of the imbedding* (2.5) *implies the boundedness of the spectrum of the problem* (2.1) *from below.*

2.3. Problem. Does the inverse implication hold, too?

2.4. Remark. Moreover, as it will be seen in Section 3, one could expect also some information about the connection between the *compactness* of the imbedding (2.5) and the *discreteness* of the spectrum of the problem (2.1).

3. THE ONE DIMENSIONAL HIGHER ORDER CASE

3.1. The k-th order Hardy inequality. For simplicity, we will deal with the case $(a, b) = (0, \infty)$, i.e., we will consider the inequality

$$\Big(\int_0^\infty |u(x)|^q w_0(x)dx\Big)^{1/q} \le C\Big(\int_0^\infty |u^{(k)}(x)|^p w_1(x)dx\Big)^{1/p} \tag{3.1}$$

with $k \ge 1$ and $1 < p \le q < \infty$.

Furthermore, we introduce the functions

$$\begin{aligned} B_1(x) &= \Big(\textstyle\int_0^x (x-t)^{q(k-1)} w_0(t)dt\Big)^{1/q} \Big(\int_x^\infty w_1^{1-p'}(t)dt\Big)^{1/p'}, \\ B_2(x) &= \Big(\textstyle\int_0^x w_0(t)dt\Big)^{1/q} \Big(\int_x^\infty (t-x)^{p'(k-1)} w_1^{1-p'}(t)dt\Big)^{1/p'} \end{aligned} \tag{3.2}$$

with $p' = \frac{p}{p-1}$.

If we denote by $W_R^{k,p}(0, \infty; w_1)$ the weighted Sobolev space of functions $u \in AC^{k-1}(0, \infty)$ satisfying the "right endpoint conditions"

$$u(\infty) = u'(\infty) = \ldots = u^{(k-1)}(\infty) = 0 \tag{3.3}$$

and with finite norm

$$\|u\|_R = \Big(\int_0^\infty |u^{(k)}(x)|^p w_1(x)dx\Big)^{1/p},$$

then inequality (3.1) expresses the *continuity* of the imbedding

$$W_R^{k,p}(0, \infty; w_1) \hookrightarrow L^q(0, \infty; w_0). \tag{3.4}$$

Necessary and sufficient conditions for the continuity and, moreover, for the *compactness* of this imbedding are given by the following assertion (see, e.g., [9]]).

3.2. Proposition. *Let* $1 < p \le q < \infty$, $k \ge 1$. *Then inequality* (3.1) *holds for every function* u *satisfying* (3.3) *if and only if*

$$\sup_{0<x<\infty} B_i(x) = B_i < \infty, \quad i = 1, 2. \tag{3.5}$$

Moreover, the imbedding (3.4) *is compact if and only if*

$$\lim_{x\to 0} B_i(x) = \lim_{x\to\infty} B_i(x) = 0, \quad i = 1, 2. \tag{3.6}$$

3.3. A linear spectral problem. As it was shown in GLAZMAN [2], the spectrum of the operator

$$(Au)(x) = (-1)^k (w_1(x)u^{(k)}(x))^{(k)} \quad \text{in} \quad (0, \infty) \tag{3.7}$$

with a weight function w_1 such that $w^{-1} \in L^1(0,\infty)$, is *discrete and bounded from below* (= the operator A has the so-called property **BD**) if the following condition is satisfied:

$$\lim_{x\to\infty} x^{2k-1} \int_x^\infty \frac{1}{w_1(t)} dt = 0. \tag{3.8}$$

In 1973, KALYABIN [3], and in 1974, LEWIS [4] showed (independently) that condition (3.8) is also *necessary* for A to have property **BD**.

Now, the spectral problem

$$Au = \lambda u \quad \text{in} \quad (0, \infty)$$

with A given by (3.7) is exactly the problem (1.14) for $a = 0$, $b = \infty$, $p = q = 2$ and $w_0(x) \equiv 1$, and the corresponding k-th order Hardy inequality (1.15) has the form

$$\Big(\int_0^\infty |u(x)|^2 dx \Big)^{1/2} \le C \Big(\int_0^\infty |u^{(k)}(x)|^2 w_1(x) dx \Big)^{1/2}, \tag{3.9}$$

and the functions $B_1(x)$, $B_2(x)$ from (3.2) have the form

$$B_1(x) = \Big(\int_0^x (x-t)^{2(k-1)} dt \Big)^{1/2} \Big(\int_x^\infty w_1^{-1}(t) dt \Big)^{1/2},$$

$$B_2(x) = \Big(\int_0^x dt \Big)^{1/2} \Big(\int_x^\infty (t-x)^{2(k-1)} w_1^{-1}(t) dt \Big)^{1/2}.$$

Now, it is

$$B_1^2(x) = \frac{1}{2k-1} x^{2k-1} \int_x^\infty w^{-1}(t) dt,$$

and a comparison with condition (3.8) indicates, that the property **BD** of the operator A from (3.7) is connected with the condition (3.6) (for $i = 1$) which is necessary and sufficient for the compactness of the imbedding

$$W_R^{k,2}(0,\infty;w_1) \hookrightarrow L^2(0,\infty)$$

expressed by the Hardy inequality (3.9).

This gives rise to the following question:

3.4. Problem. Is it possible to obtain some information about the spectrum of the problem (1.14) via the Hardy inequality (1.15)? More precisely, does the validity of inequality (1.15) — i.e., the *continuity* of the corresponding imbedding

$$W_R^{k,p}(a,b;w_1) \hookrightarrow L^q(a,b;w_0) \text{ —}$$

guarantee the *boundedness of the spectrum from below*, and does the *compactness* of the imbedding guarantee the *discreteness* of the spectrum?

In particular, if we consider the case $k = 1$, i.e., a second order differential operator

$$(Au)(x) = -(w_1(x)|u'(x)|^{p-2}u'(x))',$$

is it possible to extend the results concerning the *linear* case and mentioned in Subsection 3.3, also to the nonlinear spectral problem

$$(Au)(x) = \lambda w_0(x)|u(x)|^{q-2}u(x) \quad \text{on} \quad (0,\infty)$$

(with $1 < p \le q < \infty$) and show that the condition

$$\lim_{x\to 0} B(x) = \lim_{x\to\infty} B(x) = 0$$

with

$$B(x) = \Big(\int_0^x w_0(t)dt\Big)^{1/q}\Big(\int_x^\infty w_1^{1-p'}(t)dt\Big)^{1/p'}$$

is necessary and sufficient for A to have the property **BD**?
(Notice that for $k = 1$ the functions B_1, B_2 from (3.2) coincide: $B_1(x) = B_2(x) = B(x)$.)

Acknowledgments

The work on this paper was supported by the Grant Agency of Czech Republic, project No. 201/97/0395.

References

[1] Drábek, P. and Manásevich, R. *On the solution to some p-Laplacian nonhomogeneous eigenvalue problems in closed form.* Submitted to Diff. and Integral Equations.

[2] Glazman, I.M. (1960). *Direct methods of qualitative spectral analysis of singular differential operators* (Russian). Fizmatgiz, Moscow.

[3] Kalyabin, G.A. (1973). *A necessary and sufficient condition for the spectrum of a homogeneous operator to be discrete in the matrix case* (Russian). Diff. Urav., 9:951–954.

[4] Lewis, R.T. (1974). *The discreteness of the spectrum of self-adjoint, even order, one term differential operator.* Proc. Amer. Math. Soc., 42:480–482.

[5] Lindqvist, P. (1990). *On the equation* $\mathrm{div}(|\nabla u|^{p-2}\nabla u) + \lambda |u|^{p-2}u = 0$. Proc. Amer. Math. Soc., 109:157–164.

[6] Lindqvist, P. (1993). *Note on a nonlinear eigenvalue problem.* Rocky Mountain J. Math., 23:281–288.

[7] Michlin, S.G. (1981). *Konstanten in einigen Ungleichungen der Analysis.* Teubner, Leipzig.

[8] Nečas, J. (1971). *About the discreteness of the spectrum of a nonlinear Sturm-Liouville equation* (Russian). Dokl. Akad. Nauk SSSR, 201:1045–1048.

[9] Opic, B. and Kufner, A. (1990). *Hardy-type inequalities.* Longman Scientific & Technical. Harlow.

REMARKS ON THE REGULARITY OF SOLUTIONS OF ELLIPTIC SYSTEMS

Salvatore Leonardi

Abstract: We obtain a result contained in [10] in a different manner.

An optimal sufficient condition is supplied in order to get the interior Hölder regularity of a weak H^1-solution to a linear elliptic system with measurable bounded coefficients. Explicit estimate of the norm of the gradient of a solution in a suitable weighted Morrey space is given.

The condition obtained is then applied to quasilinear and nonlinear elliptic systems.

Keywords: Linear elliptic systems, quasilinear elliptic systems, Hölder regularity.

1. INTRODUCTION

We will prove, in an easy way, an optimal sufficient condition which allows us to select a class of linear elliptic systems with measurable bounded coefficients having Hölder continuous weak solutions (see [10]). Such systems will be named De Giorgi systems.

We will also provide an explicit estimate of the norm of the gradient of a solution in a suitable weighted Campanato-Morrey space (see also [13]) and moreover we will apply the condition found to quasilinear and nonlinear systems with nonregular coefficients.

It is well known that the strong ellipticity condition is not generally sufficient to ensure the Hölder regularity in the interior of the domain for a solution to a linear elliptic system with measurable bounded coefficients in dimension $m \geq 3$.

In fact, while in 1957 De Giorgi [3] had proved that a weak solution $u \in H^1$ of the single equation ([1])

$$\frac{\partial}{\partial x_i}\left(a_{ij}(x)\frac{\partial u}{\partial x_j}\right) = \frac{\partial f_i(x)}{\partial x_i}, \tag{1.1}$$

$$a_{ij} \in L^\infty, \tag{1.2}$$

$$\Lambda_1 |\xi|^2 \leq a_{ij}\xi_i\xi_j \leq \Lambda_2 |\xi|^2, \tag{1.3}$$

$$f_i \text{ "smooth enough"}, \tag{1.4}$$

was of class $C^{0,\mu}_{loc}$ for any $m \geq 2$, in 1968 he showed that, when $m \geq 3$, the previous regularity result could not be extended to systems

$$\frac{\partial}{\partial x_i}\left(A^{rs}_{ij}(x)\frac{\partial u_s}{\partial x_j}\right) = \frac{\partial f^r_i(x)}{\partial x_i}, \tag{1.5}$$

$$A^{rs}_{ij} \in L^\infty, \tag{1.6}$$

$$\Lambda_2 |\eta|^2 \geq A^{rs}_{ij}\eta^r_i\eta^s_j \geq \Lambda_1 |\eta|^2, \tag{1.7}$$

$$f^r_i \text{ "smooth enough"}, \tag{1.8}$$

i.e. to systems with L^∞-coefficients satisfying only the strong ellipticity condition (1.7) (see also [13]); he presented the following counterexample

$$\frac{\partial}{\partial x_i}\left(A^{rs}_{ij}(x)\frac{\partial u_s}{\partial x_j}\right) = 0, \quad x \in B(0,1)\ ({}^2),$$

$$u_s = \frac{x_s}{|x|^\alpha} \text{ (solution)}, \quad 1 \leq \alpha < m/2.$$

If $m = 2$ it is known that a solution of the system (1.5)-(1.8) is Hölder continuous.

De Giorgi's counterexample was modified by Giusti-Miranda [7] to construct, for $m \geq 3$, a quasilinear system of the type

$$\frac{\partial}{\partial x_i}\left(A^{rs}_{ij}(u)\frac{\partial u_s}{\partial x_j}\right) = 0, \quad x \in B(0,1),$$
$$A^{rs}_{ij}(u) \text{ bounded real analytic},$$
$$u_s = \frac{x_s}{|x|} \text{ (solution)}$$

---

[1] The Einstein convention will be used.
[2] B(0,1) is the unit ball centered at the origin.

and also by Nečas and Stará [20] to provide a system of the type

$$\frac{\partial}{\partial x_i}\left(A_{ij}^{rs}(x,u)\frac{\partial u_s}{\partial x_j}\right) = 0, \quad x \in B(0,1),$$
$$A_{ij}^{rs}(x,u) \text{ bounded continuous},$$
$$u_s = \frac{x_s}{|x|^\gamma} \text{ (solution) } (^3), \quad 1 \le \gamma < m/2.$$

For what concerns the nonlinear case, Nečas, John and Stará [19] (see also [5]) constructed, for $m \geq 3$, a counterexample of the type

$$\frac{\partial}{\partial x_k}\left(a_k^{ij}(\nabla u)\right) = 0, \quad x \in B(0,1),$$
$$a_k^{ij}(\nabla u) \text{ bounded real analytic} \tag{1.9}$$
$$u_{ij} = \frac{x_{ij}}{|x|} - \frac{1}{m}\delta_{ij}|x| \text{ (solution)},$$

while for systems of the Euler-Lagrange type, again with bounded real analytic coefficients like (1.9), it is available for $m \geq 5$ a counterexample from [9] showing that such systems have Lipschitz but not C^1 - solution; the question of the regularity for $m = 3,4$ is still an open interesting problem.

The author is deeply indebted to Prof. Nečas for suggesting the idea of this work and for his constant guidance, help and encouragement throughout its writing and the years of studying at the Departments of Mathematics of Northern Illinois University and Prague.

2. NOTATIONS, FUNCTION SPACES, PRELIMINARY RESULTS

In R^m $(m \geq 3)$, with a generic point $x = (x_1, x_2, \ldots, x_m)$, we shall denote by Ω an open nonempty bounded set with diameter d_Ω and smooth boundary $\partial\Omega$.

Let $N \geq 1$ be a positive integer.

For $\rho > 0$ we define

$$B(x_0,\rho) = \{x \in R^m : |x - x_0| < \rho\}$$
$$\Omega(x_0,\rho) = \Omega \cap B(x_0,\rho).$$

If $u \in L^1(A)$, A being an open nonempty set of Ω, then we will set

$$u_A = \frac{1}{\text{meas}(A)}\int_A u(x)\,dx.$$

[3]With respect to the previous counterexample, in this case the solution can be unbounded.

Definition 2.1. (Morrey space) *Let $p \geq 1$ and $0 \leq \lambda \leq m$. By $L^{p,\lambda}(\Omega, R^N)$ we denote the linear space of functions $u \in L^p(\Omega, R^N)$ such that*

$$\|u\|_{L^{p,\lambda}(\Omega,R^N)} = \left\{ \sup_{x_0\in\Omega, 0<\rho\leq d_\Omega} \rho^{-\lambda} \int_{\Omega(x_o,\rho)} |u(x)|^p dx \right\}^{1/p} < +\infty. \quad (2.1)$$

$L^{p,\lambda}(\Omega, R^N)$ equipped with the norm (2.1) is a Banach space.

Definition 2.2. (Campanato space) *Let $p \geq 1$ and $0 \leq \lambda \leq m+p$. By $\mathcal{L}^{p,\lambda}(\Omega, R^N)$ we denote the linear space of functions $u \in L^p(\Omega, R^N)$ such that*

$$[u]_{\mathcal{L}^{p,\lambda}(\Omega,R^N)} = \left\{ \sup_{x_0\in\Omega, 0<\rho\leq d_\Omega} \rho^{-\lambda} \int_{\Omega(x_0,\rho)} |u - u_{\Omega(x_0,\rho)}|^p dx \right\}^{1/p} < +\infty. \quad (2.2)$$

Definition 2.3. (Weighted Morrey space) *Let $p \geq 1$ and $0 < \lambda < m$. By $N^{p,\lambda}(\Omega, R^N)$ we denote the linear space of functions $u \in L^p(\Omega, R^N)$ such that*

$$\begin{aligned} \|u\|_{N^{p,\lambda}(\Omega,R^N)} = & \left\{ \sup_{x_0\in\Omega} \int_0^{d_\Omega} \rho^{-\lambda-1} \left(\int_{\Omega(x_0,\rho)} |u|^p dx \right) d\rho \right. \\ & \left. + \frac{1}{\lambda} d_\Omega^{-\lambda} \int_\Omega |u|^p dx \right\}^{1/p} < +\infty. \end{aligned} \quad (2.3)$$

$N^{p,\lambda}(\Omega, R^N)$ equipped with the norm (2.3) is a Banach space.

Definition 2.4. (Weighted Campanato space) *Let $p \geq 1$ and $0 < \lambda < m+p$. By $\mathcal{N}^{p,\lambda}(\Omega, R^N)$ we denote the linear space of functions $u \in L^p(\Omega, R^N)$ such that*

$$\begin{aligned} [[u]]_{\mathcal{N}^{p,\lambda}(\Omega,R^N)} = & \left\{ \sup_{x_0\in\Omega} \int_0^{d_\Omega} \rho^{-\lambda-1} \left(\int_{\Omega(x_0,\rho)} |u - u_{\Omega(x_0,\rho)}|^p dx \right) d\rho \right. \\ & \left. + \frac{1}{\lambda} d_\Omega^{-\lambda} \int_\Omega |u - u_{\Omega(x_0,\rho)}|^p dx \right\}^{1/p} < +\infty. \end{aligned} \quad (2.4)$$

In the space $N^{p,\lambda}(\Omega, R^N)$ we can introduce also the following norm

$$\|u\|'_{N^{p,\lambda}(\Omega,R^N)} = \left\{ \sup_{x_0\in\Omega} \int_\Omega |x - x_0|^{-\lambda} |u|^p dx \right\}^{1/p}. \quad (2.5)$$

Proposition 2.1. *The norms (2.3) and (2.5) are equivalent.*

Proof. Let $u \in N^{p,\lambda}(\Omega, R^N)$ and extend it outside Ω with zero value. We also fix x_0 in R^m and without loss of generality (see [15] p. 214) we suppose it be a Lebesgue point of $|u|^p$.

Thus, integration by parts gives:

$$\begin{aligned} &\int_0^{d_\Omega} \rho^{-\lambda-1}\left(\int_{\Omega(x_0,\rho)} |u|^p dx\right) d\rho = \left[-\lambda^{-1}\rho^{-\lambda}\int_{\Omega(x_0,\rho)} |u|^p dx\right]_0^{d_\Omega} \\ + &\ \lambda^{-1}\int_0^{d_\Omega} \rho^{-\lambda}\frac{d}{d\rho}\left(\int_{\Omega(x_0,\rho)} |u|^p dx\right) d\rho = -\lambda^{-1} d_\Omega^{-\lambda}\int_\Omega |u|^p dx + \\ + &\ \lambda^{-1}\int_0^{d_\Omega} \rho^{-\lambda}\frac{d}{d\rho}\left(\int_0^\rho r^{m-1}\int_{\partial B(x_0,1)} |u|^p d\sigma dr\right) d\rho \\ = &\ -\lambda^{-1} d_\Omega^{-\lambda}\int_\Omega |u|^p dx + \lambda^{-1}\int_\Omega |x-x_0|^{-\lambda}|u|^p dx \end{aligned} \tag{2.6}$$

and taking the supremum for $x_0 \in R^m$ we get the thesis.

Proposition 2.2. *If* $0 < \lambda < m$ *then*

$$N^{p,\lambda}(\Omega, R^N) \subset L^{p,\lambda}(\Omega, R^N).$$

Proof. Let $u \in N^{p,\lambda}(\Omega, R^N)$. From (2.6) we have

$$\begin{aligned} &\int_0^{d_\Omega} \rho^{-\lambda-1}\left(\int_{\Omega(x_0,\rho)} |u|^p dx\right) d\rho + \lambda^{-1} d_\Omega^{-\lambda}\int_\Omega |u|^p dx \\ = &\ \lambda^{-1}\int_\Omega |x-x_0|^{-\lambda}|u|^p dx \end{aligned} \tag{2.7}$$

for a. a. $x_0 \in \Omega$.

On the other hand

$$\int_\Omega |x-x_0|^{-\lambda}|u|^p dx \geq \rho^{-\lambda}\int_{\Omega(x_0,\rho)} |u|^p dx \tag{2.8}$$

and thus from (2.7) and (2.8) we deduce

$$\|u\|_{L^{p,\lambda}(\Omega,R^N)} \leq \lambda \|u\|_{N^{p,\lambda}(\Omega,R^N)}.$$

Theorem 2.1. *If*

$$\frac{\lambda_2 - m}{p} \leq \frac{\lambda_1 - m}{q},$$

with $1 \leq p \leq q < \infty$, *then*

$$N^{q,\lambda_1}(\Omega, R^N) \subset N^{p,\lambda_2}(\Omega, R^N).$$

Proof. Fixed $x_0 \in R^m$, from Hölder inequality it turns out

$$\begin{aligned} &\int_{\Omega(x_0,\rho)} |u|^p dx \\ \leq\ & |\Omega(x_0,\rho)|^{1-p/q} \left(\int_{\Omega(x_0,\rho)} |u(x)|^q dx \right)^{p/q} \\ \leq\ & c\rho^{m(1-p/q)+(\lambda_1+1)p/q} \left(\rho^{-\lambda_1-1} \int_{\Omega(x_0,\rho)} |u(x)|^q dx \right)^{p/q}. \end{aligned} \tag{2.9}$$

Because

$$\lambda_2 + 1 \leq m\left(1 - \frac{p}{q}\right) + (\lambda_1+1)\frac{p}{q} + (1 - p/q)$$

one has

$$\left(\frac{\rho}{d_\Omega}\right)^{m(1-p/q)+(\lambda_1+1)p/q+(1-p/q)} \leq \left(\frac{\rho}{d_\Omega}\right)^{\lambda_2+1}, \quad 0 < \rho \leq d_\Omega$$

so that

$$\rho^{m(1-p/q)+(\lambda_1+1)p/q} \leq \rho^{\lambda_2+1} d_\Omega^{m(1-p/q)+\lambda_1 p/q-\lambda_2}, \quad 0 < \rho \leq d_\Omega.$$

From last inequality and formula (2.9) we infer

$$\rho^{-\lambda_2-1} \int_{\Omega(x_0,\rho)} |u|^p dx \leq c \left(\rho^{-\lambda_1-1} \int_{\Omega(x_0,\rho)} |u|^q dx \right)^{p/q}, \quad 0 < \rho \leq d_\Omega. \tag{2.10}$$

Integrating (2.10) on $]0, d_\Omega]$ and applying again Hölder inequality to the right side of (2.10) we obtain

$$\begin{aligned} &\int_0^{d_\Omega} \rho^{-\lambda_2-1} \left(\int_{\Omega(x_0,\rho)} |u|^p dx \right) d\rho \\ \leq\ & c \int_0^{d_\Omega} \left(\rho^{-\lambda_1-1} \int_{\Omega(x_0,\rho)} |u|^q dx \right)^{p/q} d\rho \\ \leq\ & c d_\Omega^{\frac{q-p}{q}} \int_0^{d_\Omega} \left(\rho^{-\lambda_1-1} \int_{\Omega(x_0,\rho)} |u|^q dx \right) d\rho. \end{aligned} \tag{2.11}$$

The thesis now follows readily from (2.11) and the embedding of L^q into L^p.

Remark 2.1. *Proposition 2.2 and Theorem 2.1 hold also for weighted Campanato spaces i.e.:*
1) $\mathcal{N}^{p,\lambda}(\Omega, R^N) \subset \mathcal{L}^{p,\lambda}(\Omega, R^N)$,
2) $\mathcal{N}^{q,\lambda_1}(\Omega, R^N) \subset \mathcal{N}^{p,\lambda_2}(\Omega, R^N)$ *if* $\frac{\lambda_2 - m}{p} \leq \frac{\lambda_1 - m}{q}$, $1 \leq p \leq q < \infty$.

Proposition 2.3. *If* $0 < \lambda < m$ *then*

$$N^{p,\lambda}(\Omega, R^N) \subset \mathcal{N}^{p,\lambda}(\Omega, R^N).$$

Proof. Let $u \in N^{p,\lambda}$. Thus

$$|u_{\Omega(x_0,\rho)}|^p \leq |\Omega(x_0,\rho)|^{-1} \int_{\Omega(x_0,\rho)} |u|^p dx \tag{2.12}$$

and so

$$\begin{aligned} &\int_0^{d_\Omega} \rho^{-\lambda-1} \left(\int_{\Omega(x_0,\rho)} |u_{\Omega(x_0,\rho)}|^p dx \right) d\rho \\ \leq &\int_0^{d_\Omega} \rho^{-\lambda-1} \left(\int_{\Omega(x_0,\rho)} |u|^p dx \right) d\rho. \end{aligned} \tag{2.13}$$

On the other hand

$$\begin{aligned} &\int_0^{d_\Omega} \rho^{-\lambda-1} \left(\int_{\Omega(x_0,\rho)} |u - u_{\Omega(x_0,\rho)}|^p dx \right) d\rho \\ \leq \; & 2^{p-1} \left[\int_0^{d_\Omega} \rho^{-\lambda-1} \left(\int_{\Omega(x_0,\rho)} |u|^p dx \right) d\rho \right. \\ + &\left. \int_0^{d_\Omega} \rho^{-\lambda-1} \left(\int_{\Omega(x_0,\rho)} |u_{\Omega(x_0,\rho)}|^p dx \right) d\rho \right] \end{aligned} \tag{2.14}$$

and thus from (2.12), (2.13) and (2.14) the thesis follows.

Let us now define Morrey and Campanato spaces when $\Omega = R^m$ [4].

Definition 2.5. *Let* $p \geq 1$ *and* $0 \leq \lambda \leq m$. *By* $L^{p,\lambda}(R^m, R^N)$ *we denote the linear space of functions* $u \in L^p(R^m, R^N)$ *such that*

$$\|u\|_{L^{p,\lambda}(R^m,R^N)} = \left\{ \sup_{x_0 \in R^m, \rho > 0} \rho^{-\lambda} \int_{B(x_0,\rho)} |u|^p dx \right\}^{1/p} < +\infty. \tag{2.15}$$

$L^{p,\lambda}(R^m, R^N)$ equipped with the norm (2.15) is a Banach space.

[4]We shall always assume $d_{R^m} = +\infty$.

Definition 2.6. *Let $p \geq 1$ and $0 \leq \lambda \leq m+p$. By $\mathcal{L}^{p,\lambda}(R^m, R^N)$ we denote the linear space of functions $u \in L^p(R^m, R^N)$ such that*

$$[u]_{\mathcal{L}^{p,\lambda}(R^m,R^N)} = \left\{ \sup_{x_0 \in R^m, \rho > 0} \rho^{-\lambda} \int_{B(x_0,\rho)} |u - u_{B(x_0,\rho)}|^p dx \right\}^{1/p} < +\infty. \tag{2.16}$$

Definition 2.7. (Weighted Morrey space) *Let $p \geq 1$ and $0 < \lambda < m$. By $N^{p,\lambda}(R^m, R^N)$ we denote the linear space of functions $u \in L^p(R^m, R^N)$ such that*

$$\|u\|_{N^{p,\lambda}(R^m,R^N)} = \left\{ \sup_{x_0 \in R^m} \int_0^\infty \rho^{-\lambda-1} \left(\int_{B(x_0,\rho)} |u|^p dx \right) d\rho \right\}^{1/p} < +\infty. \tag{2.17}$$

$N^{p,\lambda}(R^m, R^N)$ equipped with the norm (2.17) is a Banach space.

Definition 2.8. (Weighted Campanato space) *Let $p \geq 1$ and $0 < \lambda < m+p$. By $\mathcal{N}^{p,\lambda}(R^m, R^N)$ we denote the linear space of functions $u \in L^p(R^m, R^N)$ such that*

$$\begin{aligned} &[[u]]_{\mathcal{N}^{p,\lambda}(R^m,R^N)} = \\ &= \left\{ \sup_{x_0 \in R^m} \int_0^\infty \rho^{-\lambda-1} \left(\int_{B(x_0,\rho)} |u - u_{B(x_0,\rho)}|^p dx \right) d\rho \right\}^{1/p} < +\infty. \end{aligned} \tag{2.18}$$

Remark 2.2. *In the space $N^{p,\lambda}(R^m, R^N)$ we can introduce also the norm*

$$\|u\|'_{N^{p,\lambda}(R^m,R^N)} = \left\{ \sup_{x_0 \in R^m} \int_{R^m} |x - x_0|^{-\lambda} |u|^p dx \right\}^{1/p}. \tag{2.19}$$

Observe that propositions 2.1 and 2.2 continue to hold for $L^{p,\lambda}(R^m, R^N)$ and $N^{p,\lambda}(R^m, R^N)$.

The following Hardy type inequality (see also [21], Sec. 14.3) will be helpful in the sequel.

Theorem 2.2. *Let $m \geq 3$ and $0 \leq \lambda < m-2$. Let $u \in H^1(R^m)$ be such that*

$$\int_{R^m} |\nabla u|^2 |x - x_0|^{-\lambda} dx < +\infty, \quad \forall x_0 \in R^m.$$

Then

$$\int_{R^m} |u|^2 |x-x_0|^{-\lambda-2}\, dx \le \frac{4}{(m-\lambda-2)^2} \int_{R^m} |\nabla u|^2 |x-x_0|^{-\lambda}\, dx, \quad \forall x_0 \in R^m. \tag{2.20}$$

Proof. Fixed $x_0 \in R^m$, let us introduce radial and angular coordinates $\rho = |x - x_0|$, $\omega = \frac{x-x_0}{\rho}$ and let us write

$$u(x) = u(x_0 + \rho\omega) \equiv u(\omega, \rho) \ (^5).$$

For a.a. ω such that $|\omega| = 1$ we have:

$$\int_0^\infty \left| \frac{\partial u(\omega,\rho)}{\partial \rho} \right|^2 \rho^{m-1-\lambda} d\rho < +\infty \, . \tag{2.21}$$

Let us set

$$u_s(\omega, r) = \begin{cases} -\int_r^s \frac{\partial u(\omega,\rho)}{\partial \rho} d\rho & \text{for } r \le s \\ 0 & \text{elsewhere.} \end{cases} \tag{2.22}$$

Thus we deduce

$$\int_0^s (u_s(\omega,r))^2 r^{m-3-\lambda} dr = \left[\frac{1}{m-\lambda-2} r^{m-2-\lambda} (u_s(\omega,r))^2 \right]_0^s$$
$$- \frac{2}{m-\lambda-2} \int_0^s r^{m-2-\lambda} u_s(\omega,r) \frac{\partial u_s(\omega,r)}{\partial r} dr$$
$$\le \frac{2}{m-\lambda-2} \left(\int_0^s (u_s(\omega,r))^2 r^{m-3-\lambda} dr \right)^{1/2} \times$$
$$\times \left(\int_0^s \left(\frac{\partial u_s(\omega,r)}{\partial r} \right)^2 r^{m-1-\lambda} dr \right)^{1/2}.$$

On the other hand

$$u_s(\omega, r) = -u(\omega, s) + u(\omega, r) \tag{2.23}$$

and so, by (2.21), (2.22) and (2.23), letting $s \to +\infty$ we obtain

$$\int_0^\infty (u(\omega,r))^2 r^{m-3-\lambda} dr \le \frac{4}{(m-\lambda-2)^2} \int_0^\infty \left(\frac{\partial u(\omega,r)}{\partial r} \right)^2 r^{m-1-\lambda} dr. \tag{2.24}$$

Integrating (2.24) for $|\omega| = 1$ and taking the supremum for $x_0 \in R^m$ we achieve the thesis.

[5] In what follows we will omit the explicit dependence on x_0.

3. HÖLDER REGULARITY: THE PERTURBATION METHOD

In $\Omega \subset R^m$, $m \geq 3$, we will be concerned with the interior Hölder regularity of a weak solution $u \in H^1(\Omega, R^N)$ of the linear system

$$\frac{\partial}{\partial x_i}\left(A_{ij}^{rs}(x)\frac{\partial u_s}{\partial x_j}\right) = \frac{\partial f_i^r}{\partial x_i}, \tag{3.1}$$

$$i,j = 1,2,\ldots m;\ r,s = 1,2,\ldots N$$

under the conditions

$$A_{ij}^{rs}(x) \in L^\infty(\Omega), \tag{3.2}$$

$$A_{ij}^{rs}(x) = A_{ji}^{sr}(x), \tag{3.3}$$

$$\begin{gathered}\exists\ \Lambda_1,\ \Lambda_2 > 0 : \Lambda_1|\eta|^2 \leq A_{ij}^{rs}(x)\eta_i^r\eta_j^s \leq \Lambda_2|\eta|^2,\\ \forall \xi \in R^{mN},\ \text{a.e. } x \in \Omega,\end{gathered} \tag{3.4}$$

$$f = (f_i^r) \in N^{2,\lambda}(\Omega, R^{mN}),\ \ m-2 < \lambda < m. \tag{3.5}$$

As explained in the introduction, De Giorgi's counterexample shows that, when $m \geq 3$, systems of the above type satisfying only the strong ellipticity condition (3.4) do not generally have Hölder-continuous solutions even if $f \equiv 0$.

Definition 3.1. *Let conditions (3.2) through (3.5) be satisfied. We say that the system (3.1) is a De Giorgi system if any weak solution* $u \in H^1(\Omega, R^N)$ *belongs to* $C^{0,\mu}_{loc}(\Omega, R^N)$.

Our final goal will be to prove the following

Theorem 3.1. *Let the conditions (3.2) through (3.5) and the conditions*

$$\left\{\begin{array}{l} m-2 < \lambda < \dfrac{m-2+\left(m^2+4m-4\right)^{1/2}}{2} \\ \lambda \text{ sufficiently close to } m-2 \end{array}\right., \tag{3.6}$$

$$\frac{\Lambda_1}{\Lambda_2} > \frac{\left(1+\dfrac{(m-2)^2}{m-1}\right)^{1/2} - 1}{\left(1+\dfrac{(m-2)^2}{m-1}\right)^{1/2} + 1} \tag{3.7}$$

be satisfied and let $u \in H^1(\Omega, R^N)$ *be a solution of the system (3.1).*

Then there exists a constant $C(\lambda) > 0$ *such that*

$$\|\nabla u\|_{N^{2,\lambda}(\Omega)} \leq C(\lambda)\|f\|_{N^{2,\lambda}(\Omega)}, \tag{3.8}$$

i.e. u *is of class* $C^{0,\mu}_{loc}(\Omega)$ ([6]) *and (3.1) is a De Giorgi system.*

Remark 3.1. *It has been already observed in [10] that the condition (3.7) is optimal.*

We will deduce the interior regularity from the regularity in the whole space.

For this purpose, let us now consider in R^m the diagonal system

$$\frac{\partial}{\partial x_i}\left(D^{rs}_{ij}(x)\frac{\partial \omega_s}{\partial x_j}\right) = \frac{\partial g^r_i}{\partial x_i} \tag{3.9}$$

with

$$D^{rs}_{ij} \in L^\infty(R^m),\ D^{rs}_{ij} = 0 \text{ if } r \neq s,$$
$$g^r_i \in N^{2,\lambda}(R^m)$$

and let us suppose

$$\exists\, \nu_1, \nu_2 > 0 : \nu_2|\xi|^2 \geq D^{rs}_{ij}\xi^r_i\xi^s_j \geq \nu_1|\xi|^2$$
$$\forall \xi \in R^{mN},\ \text{a.e. } x \in R^m.$$

From the general theory of one single elliptic equation (see [13]) we know that for the solution $\omega = (\omega_1, \ldots, \omega_N) \in H^1(R^m, R^N)$ of the system (3.9) ([7]) the following estimates holds

$$\|\omega\|_{H^1} \leq 1/\nu_1\|g\|_{L^2}, \tag{3.10}$$

$$\|\nabla\omega\|_{N^{2,\lambda}} \leq C_1(\lambda)\|g\|_{N^{2,\lambda}}, \tag{3.11}$$

where $g \equiv (g^r_i)$ ($C_1(\lambda)$ has been explicitly calculated in [13]).

Thus, we can establish the following fundamental

Theorem 3.2. *Let* $u \in H^1(R^m, R^N)$ *be the solution* ([7]) *of the system*

$$\frac{\partial}{\partial x_i}\left(A^{rs}_{ij}(x)\frac{\partial u_s}{\partial x_j}\right) = \frac{\partial f^r_i}{\partial x_i} \text{ in } R^m \tag{3.12}$$

$$i, j = 1, 2, \ldots m;\ r, s = 1, 2, \ldots N$$

[6] Recall that $N^{2,\lambda}(\Omega) \subset L^{2,\lambda}(\Omega)$ and that, by virtue of known properties of the spaces $L^{2,\lambda}(\Omega)$ (see [2]), inequality (3.8) implies that $u \in C^{0,\mu}(B)$ where B is for example a ball in the interior of Ω.

[7] Existence and uniqueness of the solution in $H^1(R^m, R^N)$ are ensured by the Lax-Milgram theorem.

with the structural conditions

$$f = (f_i^r) \in N^{2,\lambda}(R^m, R^{mN}), \;\; m-2 < \lambda < m, \tag{3.13}$$

$$A_{ij}^{rs}(x) \in L^\infty(R^m), \tag{3.14}$$

$$A_{ij}^{rs}(x) = A_{ji}^{sr}(x), \tag{3.15}$$

$$\begin{gathered} \exists\, \Lambda_1,\, \Lambda_2 > 0 : \Lambda_1|\eta|^2 \le A_{ij}^{rs}(x)\eta_i^r\eta_j^s \le \Lambda_2|\eta|^2, \\ \forall \xi \in R^{mN}, \;\; a.e.\; x \in R^m, \end{gathered} \tag{3.16}$$

$$\exists\, M > 0 : \begin{cases} |(A_{ij}^{rs} - D_{ij}^{rs})\xi_i^r\eta_j^s| \le M|\xi|\,|\eta| \\ \max(1/\nu_1, C_1(\lambda))\, M < 1 \end{cases} \quad (^8). \tag{3.17}$$

Then it holds

$$\|\nabla u\|_{N^{2,\lambda}} \le \frac{\max(1/\nu_1, C_1(\lambda))}{1 - \max(1/\nu_1, C_1(\lambda))\, M} \|f\|_{N^{2,\lambda}}. \tag{3.18}$$

Proof. Fix $V \equiv (V_j^s) \in N^{2,\lambda}(R^m, R^{mN})$ and set

$$g_i^r \equiv (A_{ij}^{rs} - D_{ij}^{rs})V_j^s - f_i^r. \tag{3.19}$$

Fixed moreover $x_0 \in R^m$ and a ball $B(x_0, \rho)$, let us observe that, by virtue of $(3.17)_1$ and Hölder inequality, it turns out

$$\begin{aligned} \left| \int_{B(x_0,\rho)} (A_{ij}^{rs} - D_{ij}^{rs})V_j^s \Phi_i^r dx \right| \;&\le\; M \int_{B(x_0,\rho)} |V|\,|\Phi| dx \\ &\le M \left(\int_{B(x_0,\rho)} |V|^2 dx \right)^{1/2} \left(\int_{B(x_0,\rho)} |\Phi|^2 dx \right)^{1/2} \end{aligned}$$

for any $\Phi \equiv (\Phi_i^r) \in L^2(R^m, R^N)$.

Thus, from the above formula ([9]) we have

$$\|(A - D)V\|_{L^2(B(x_0,\rho))} \le M\|V\|_{L^2(B(x_0,\rho))} \tag{3.20}$$

[8] $1/\nu_1$ and $C_1(\lambda)$ are the constants from (3.10) and (3.11).

[9] Recall that

$$\|w\|_{L^2} = \sup_{\|\Phi\|_{L^2}=1} \left| \int w\Phi dx \right|$$

see Miranda [15].

where $A = (A_{ij}^{rs})$, $D = (D_{ij}^{rs})$; and so from (3.20), multiplying by $\rho^{-\lambda-1}$ and integrating in $]0, +\infty[$, it follows

$$\|(A-D)V\|_{N^{2,\lambda}} \leq M\,\|V\|_{N^{2,\lambda}}. \tag{3.21}$$

Analogously we deduce

$$\|(A-D)V\|_{L^2(R^m)} \leq M\|V\|_{L^2(R^m)}. \tag{3.22}$$

Fixed now a function $v \in H^1(R^m, R^N)$ such that $\nabla v \in N^{2,\lambda}(R^m, R^{mN})$, let us consider the problem of finding $\omega = (\omega_s) \in H^1(R^m, R^N)$ weak solution of the system

$$-\frac{\partial}{\partial x_i}\left(D_{ij}^{rs}\frac{\partial \omega_s}{\partial x_j}\right) = \frac{\partial}{\partial x_i}\left[\left(A_{ij}^{rs} - D_{ij}^{rs}\right)\frac{\partial v_s}{\partial x_j} - f_i^r\right] \quad \text{in } R^m. \tag{3.23}$$

By the Lax-Milgram theorem the problem (3.23) is uniquely solvable in $H^1(R^m, R^N)$; moreover, from the previous remarks, (3.10), (3.11) and (3.19) we get the following estimates

$$\|\omega\|_{H^1} \leq 1/\nu_1\left[M\|v\|_{H^1} + \|f\|_{L^2}\right], \tag{3.24}$$

$$\|\nabla\omega\|_{N^{2,\lambda}} \leq C_1(\lambda)\left[M\|\nabla v\|_{N^{2,\lambda}} + \|f\|_{N^{2,\lambda}}\right]. \tag{3.25}$$

Let

$$T : S \equiv \left\{v \in H^1(R^m) : \nabla v \in N^{2,\lambda}(R^m)\right\} \to S$$

be the operator, which maps $v \mapsto \omega$, defined through the law given by (3.23) and assume as a norm in S the following

$$\|v\|_S = \|v\|_{H^1} + \|\nabla v\|_{N^{2,\lambda}};$$

S equipped with the above norm is a Banach space.

Then, for every v^1, $v^2 \in S$, from (3.23), (3.24) and (3.25) we deduce

$$\begin{aligned} &\|T(v^1) - T(v^2)\|_S \equiv \|\omega^1 - \omega^2\|_{H^1} + \|\nabla(\omega^1 - \omega^2)\|_{N^{2,\lambda}} \\ \leq\; &\max(1/\nu_1, C_1(\lambda))\,M\left[\|v^1 - v^2\|_{H^1} + \|\nabla(v^1 - v^2)\|_{N^{2,\lambda}}\right] \\ \equiv\; &\max(1/\nu_1, C_1(\lambda))M\|v^1 - v^2\|_S, \end{aligned}$$

i.e., due to $(3.17)_2$, T is a contraction and so there exists a unique $\omega \in S$ such that

$$\int_{R^m} D_{ij}^{rs}\frac{\partial \omega_s}{\partial x_j}\frac{\partial \varphi_r}{\partial x_i}dx = -\int_{R^m}(A_{ij}^{rs} - D_{ij}^{rs})\frac{\partial \omega_s}{\partial x_j}\frac{\partial \varphi_r}{\partial x_i}dx + \int_{R^m} f_i^r\frac{\partial \varphi_r}{\partial x_i}dx.$$

But this means that ω is a solution of the system (3.12) and that it turns out

$$\|\nabla\omega\|_{N^{2,\lambda}} \leq \max(1/\nu_1, C_1(\lambda))\,[M\|\nabla\omega\|_{N^{2,\lambda}} + \|f\|_{N^{2,\lambda}}]. \tag{3.26}$$

Being necessarily $\omega = u$, the thesis follows from (3.26).

Remark 3.2. *If the hypotheses (3.13)-(3.17) hold then, by virtue of remark 2.2, (3.12) is a De Giorgi system.*

3.1. Proof of theorem 3.1

Let us premise the following two theorems.

Theorem 3.3. (Nečas [17]) *Let*

$$m-2 < \lambda < \frac{m-2+(m^2+4m-4)^{1/2}}{2}$$

and let $g = (g_i) \in N^{2,\lambda}(R^m)$ be compactly supported. Let $v \in H^1(R^m)$ be the solution of the problem

$$\Delta v = \frac{\partial g_i}{\partial x_i}.$$

Then it holds

$$\|\nabla v\|'_{N^{2,\lambda}(R^m)} \leq C_2(\lambda)\|g\|'_{N^{2,\lambda}(R^m)}, \tag{3.27}$$

being

$$C_2(\lambda) = \left[1 + \frac{(m-2)^2}{m-1}\right]^{1/2} (\alpha(\lambda))^{1/2}, \tag{3.28}$$

$$\alpha(\lambda) = \frac{1 - \dfrac{2y(\lambda)(m-2)}{m+(m-2)^2}}{\left(1 - \dfrac{2y^2(\lambda) - y(\lambda)(m-2)}{m-1}\right)^2}, \quad y(\lambda) = -1/2(\lambda - m + 2),$$

$$\lim_{\lambda \to m-2} \alpha(\lambda) = 1. \tag{3.29}$$

Theorem 3.4. *Let the conditions (3.13) through (3.16) and the conditions*

$$\begin{cases} m-2 < \lambda < \dfrac{m-2+(m^2+4m-4)^{1/2}}{2} \\ \lambda \text{ sufficiently close to } m-2 \end{cases}, \tag{3.30}$$

$$\frac{\Lambda_1}{\Lambda_2} > \frac{\left(1+\dfrac{(m-2)^2}{m-1}\right)^{1/2} - 1}{\left(1+\dfrac{(m-2)^2}{m-1}\right)^{1/2} + 1} \tag{3.31}$$

be satisfied and let $u \in H^1(R^m, R^N)$ be the solution of the system (3.12).
Then it holds

$$\|\nabla u\|_{N^{2,\lambda}} \le \frac{2\,C_2(\lambda)}{\Lambda_1+\Lambda_2 - C_2(\lambda)(\Lambda_2-\Lambda_1)} \|f\|_{N^{2,\lambda}} \ (^{10}). \tag{3.32}$$

Proof. Proceeding as in the proof of Theorem 3.2, let us perturb the given system by the operator

$$D_{ij}^{rs} \equiv \frac{\Lambda_1+\Lambda_2}{2}\delta_{ij}\,\delta_{rs}.$$

Then observe that

$$\nu_1 \equiv \frac{\Lambda_1+\Lambda_2}{2},$$

$$\begin{aligned} C_1(\lambda) &\equiv \left[1+\frac{(m-2)^2}{m-1}\right]^{1/2} (\alpha(\lambda))^{1/2}\,\frac{2}{\Lambda_1+\Lambda_2} \\ &\equiv C_2(\lambda)\frac{2}{\Lambda_1+\Lambda_2}, \end{aligned}$$

$$\begin{aligned} \left|\left(A_{ij}^{rs} - \frac{\Lambda_1+\Lambda_2}{2}\delta_{ij}\,\delta_{rs}\right)\xi_i^r\eta_j^s\right| &\le \frac{\Lambda_2-\Lambda_1}{2}|\xi||\eta| \\ &\equiv M|\xi||\eta|. \end{aligned}$$

Now, being λ sufficiently close to $m-2$, by virtue of (3.29) we deduce

$$\max(1/\nu_1, C_2(\lambda)) = C_2(\lambda);$$

so that, to verify $(3.17)_2$ from Theorem 3.2, by virtue of remark 2.2, it will be sufficient to show that

$$C_1(\lambda)M \equiv C_2(\lambda)\frac{\Lambda_2-\Lambda_1}{\Lambda_1+\Lambda_2} < 1.$$

But, since λ is sufficiently close to $m-2$, the above inequality follows from (3.29) and (3.31).

Proof of Theorem 3.1. The proof is based on an iterative process.

[10] $C_2(\lambda)$ is the constant from (3.28)

Let $u \in H^1(\Omega, R^N)$ be a solution of the system (3.1) satisfying, besides conditions (3.2) through (3.5), also the conditions (3.6) and (3.7).

Let $\varphi \in C_0^\infty(R^m)$ such that $0 \leq \varphi \leq 1$, $\varphi(x) \equiv 1$ in $\Omega' \subset\subset \Omega$ and $\psi \in H^1(R^m, R^N)$.

For $s = 1, 2, \ldots, N$ let us put

$$v_s(x) = u_s(x)\varphi(x). \tag{3.33}$$

Then one has

$$\begin{aligned}
\int_{R^m} A_{ij}^{rs} \frac{\partial \psi_r}{\partial x_i} \frac{\partial v_s}{\partial x_j} dx &= \int_{R^m} A_{ij}^{rs} \frac{\partial(\psi_r \varphi)}{\partial x_i} \frac{\partial u_s}{\partial x_j} dx - \int_{R^m} A_{ij}^{rs} \frac{\partial \varphi}{\partial x_i} \psi_r \frac{\partial u_s}{\partial x_j} dx \\
&\quad + \int_{R^m} A_{ij}^{rs} \frac{\partial \psi_r}{\partial x_i} \frac{\partial \varphi}{\partial x_j} u_s dx \\
&= \int_{R^m} \frac{\partial \psi_r}{\partial x_i} \left(A_{ij}^{rs} \frac{\partial \varphi}{\partial x_j} u_s + f_i^r \varphi \right) dx \\
&\quad + \int_{R^m} \psi_r \left(f_i^r \frac{\partial \varphi}{\partial x_i} - A_{ij}^{rs} \frac{\partial \varphi}{\partial x_i} \frac{\partial u_s}{\partial x_j} \right) dx.
\end{aligned}$$

That is $v = (v_s)$ is variational solution in R^m of the problem

$$\begin{aligned}
-\frac{\partial}{\partial x_i}\left(A_{ij}^{rs} \frac{\partial v_s}{\partial x_j} \right) &= -\frac{\partial}{\partial x_i}\left(A_{ij}^{rs} \frac{\partial \varphi}{\partial x_j} u_s + f_i^r \varphi \right) \\
&\quad + f_i^r \frac{\partial \varphi}{\partial x_i} - A_{ij}^{rs} \frac{\partial \varphi}{\partial x_i} \frac{\partial u_s}{\partial x_j}.
\end{aligned} \tag{3.34}$$

Let us put

$$F_i^r \equiv f_i^r \varphi + A_{ij}^{rs} \frac{\partial \varphi}{\partial x_j} u_s \tag{3.35}$$

and observe that $f_i^r \varphi \in N^{2,\lambda}(R^m)$ and that by virtue of Theorem 2.2 also

$$A_{ij}^{rs} \frac{\partial \varphi}{\partial x_j} u_s \in N^{2,2}(R^m).$$

Set now

$$G^r \equiv f_i^r \frac{\partial \varphi}{\partial x_i} - A_{ij}^{rs} \frac{\partial \varphi}{\partial x_i} \frac{\partial u_s}{\partial x_j} \; (\in L^2(R^m)),$$

taking into account that G^r has compact support, let us solve in $H^1(R^m)$ the problem

$$\begin{aligned} -\Delta \omega_r &= G^r \text{ in } R^m \\ r &= 1, 2, \ldots, N. \end{aligned} \tag{3.36}$$

Then it follows

$$-\Delta\left(\frac{\partial \omega_r}{\partial x_i}\right) = \frac{\partial G^r}{\partial x_i} \text{ in } R^m \tag{3.37}$$

and so by Theorem 2.2 we infer

$$\frac{\partial \omega_r}{\partial x_i} \in N^{2,2}(R^m). \tag{3.38}$$

Let us define

$$H_i^r \equiv F_i^r + \frac{\partial \omega_r}{\partial x_i}. \tag{3.39}$$

1) If $0 < \lambda < 2$, being $N^{2,2}(\Omega', R^N) \subset N^{2,\lambda}(\Omega', R^N)$, we deduce $H_i^r \in N^{2,\lambda}(R^m)$ and then from Theorem 3.4 it follows

$$\|\nabla v\|_{N^{2,\lambda}} \leq C\,\|H\|_{N^{2,\lambda}}, \;\; 0 < \lambda < m, \;\; m \geq 3. \tag{3.40}$$

Finally, from (3.40), (3.33), (3.35), (3.37) and (3.39) we infer

$$\|\nabla u\|_{N^{2,\lambda}(\Omega', R^{mN})} \leq C\,\|f\|_{N^{2,\lambda}(\Omega', R^{mN})} \tag{3.41}$$

so that, due to proposition 2.2 and the known properties of Morrey spaces, u is locally Hölder continuous i.e. (3.1) is a De Giorgi system.
2) If $2 \leq \lambda < m$, then $f_i^r \in N^{2,\lambda}(\Omega) \subset N^{2,2}(\Omega)$ and thus arguing analogously to the case *1)*, from Theorem 3.4, we get

$$\|\nabla u\|_{N^{2,2}(\Omega', R^{mN})} \leq C\,\|f\|_{N^{2,2}(\Omega', R^{mN})}. \tag{3.42}$$

From (3.42) and Poincaré inequality we deduce that

$$u \in \mathcal{N}^{2,4}(\Omega', R^N). \tag{3.43}$$

2i) Property (3.43), in the case $m = 3$, infers again the Hölder continuity of u (see remark 2.1).

2ii) If $m \geq 4$ $(2 \leq \lambda < 4)$ from (3.36), (3.37), (3.42) and from inequality (2.20) it follows $\frac{\partial \omega_r}{\partial x_i} \in N^{2,\lambda}(R^m)$ and $A_{ij}^{rs}\frac{\partial \varphi}{\partial x_j} u_s \in N^{2,\lambda}(R^m)$ from which we deduce again (3.41) and thus the Hölder continuity of u.

2iii) If $4 \leq \lambda < m$, as in case *2)* we have

$$\|\nabla u\|_{N^{2,4}(\Omega', R^{mN})} \leq C\,\|f\|_{N^{2,4}(\Omega', R^{mN})}. \tag{3.44}$$

2iv) From (3.44) we get $u \in \mathcal{N}^{2,6}$ from which, if $m = 5$, we obtain immediately the Hölder regularity of u.

2v) If $m \geq 6$ and $4 < \lambda < 6$, arguing as in the case *2ii)* it follows

$$\|\nabla u\|_{N^{2,\lambda}(\Omega', R^{mN})} \leq C\,\|f\|_{N^{2,\lambda}(\Omega', R^{mN})}$$

and from here we can go on as in *2iii)* iterating the process a finite number of times.

This completes the proof.

4. APPLICATIONS: QUASILINEAR AND NONLINEAR SYSTEMS

The results of the previous sections can be readily applied in any dimension ≥ 3 to a H^1-solution of the following quasilinear system

$$\frac{\partial}{\partial x_i}\left(A_{ij}^{rs}(x,u)\frac{\partial u_s}{\partial x_j}\right)=\frac{\partial f_i^r(x)}{\partial x_i} \quad \text{in } \Omega \tag{4.1}$$

$$i,j=1,2,\dots,m;\ r,s=1,2,\dots,N$$

with

$$A_{ij}^{rs}(x,u) \text{ bounded Carathéodory functions on } R^m\times R^N, \tag{4.2}$$

$$A_{ij}^{rs}=A_{ji}^{sr}, \tag{4.3}$$

$$\Lambda_1|\xi|^2\leq A_{ij}^{rs}(x,u)\xi_i^r\xi_j^s\leq\Lambda_2|\xi|^2, \tag{4.4}$$

$$\frac{\Lambda_1}{\Lambda_2}>\frac{\left(1+\dfrac{(m-2)^2}{m-1}\right)^{1/2}-1}{\left(1+\dfrac{(m-2)^2}{m-1}\right)^{1/2}+1}, \tag{4.5}$$

$$f_i^r\in N^{2,\lambda},\ m-2<\lambda<m. \tag{4.6}$$

Furthermore, in our framework we can consider the nonlinear system

$$\frac{\partial}{\partial x_i}\left(a_i^r(\nabla u)\right)=f^r(x),\ \text{ in } \Omega \tag{4.7}$$

$$i=1,2,\dots,m;\ r=1,2,\dots,N$$

with

$$a_i^r(p)\in C^1(R^{mN}), \tag{4.8}$$

$$|a_i^r(p)|\leq c|p| \tag{4.9}$$

$$\Lambda_1|\xi|^2\leq\frac{\partial a_i^r}{\partial p_j^s}\xi_i^r\xi_j^s\leq\Lambda_2|\xi|^2 \tag{4.10}$$

$$i,j=1,2,\dots,m;\ r,s=1,2,\dots,N$$

$$\frac{\Lambda_1}{\Lambda_2}>\frac{\left(1+\dfrac{(m-2)^2}{m-1}\right)^{1/2}-1}{\left(1+\dfrac{(m-2)^2}{m-1}\right)^{1/2}+1}, \tag{4.11}$$

$$A_{ij}^{rs} = A_{ji}^{sr} \tag{4.12}$$

$$f^r \in N^{2,\lambda}, \; m-2 < \lambda < m. \tag{4.13}$$

As it is well known (see [2, 5, 6]), without the condition (4.11) the only available result for such systems is the following

Theorem 4.1. *Let $u \in H^1(\Omega, R^N)$ be a solution of the system (4.7)-(4.10) then $u \in H^2_{loc}(\Omega, R^N)$ and the derivatives of u satisfy the quasilinear system*

$$\frac{\partial}{\partial x_i}\left(\delta_{lk} A_{ij}^{rs}(U) \frac{\partial U_k^s}{\partial x_j}\right) = \frac{\partial f^r}{\partial x_l} \tag{4.14}$$

where $U = (U_k^s) := \left(\frac{\partial u_s}{\partial x_k}\right)$ *and* $A_{ij}^{rs}(U) = \frac{\partial a_i^r}{\partial p_j^s}(U)$.

Indeed, a solution of the system (4.7)- (4.10), without (4.11), is Hölder continuous only for $2 \leq m \leq 4$ and its gradient is only partially Hölder continuous (see [2]).

With condition (4.11) we obtain the local Hölderianity of the gradient of a solution for any dimension.

References

[1] Barre Saint-Venant, A.J.C. (1855). *De la torsion des prismes.* Mem., Divers. Savants, Acad. Sci. Paris, 14.

[2] Campanato, S. (1980). *Sistemi ellittici in forma divergenza. Regolarità all'interno.* Quaderno Sc. Norm. Sup. Pisa, Pisa.

[3] De Giorgi, E. (1957). *Sulla differenziabilitá e l'analiticità delle estremali degli integrali multipli regolari.* Memorie dell'Accademia delle Scienze di Torino, Serie 3, 3(I).

[4] De Giorgi, E. (1968). *Un esempio di estremali discontinue per un problema variazionale di tipo ellittico.* Boll. U.M.I. 4.

[5] Giaquinta, M. (1983). *Multiple integrals in the calculus of variations and nonlinear elliptic systems.* Annals of Math. Studies, vol. 105 Princeton University Press, Princeton.

[6] Giaquinta, M. (1993). *Introduction to the regularity theory for nonlinear elliptic systems*, Birkhäuser Verlag.

[7] Giusti, E. and Miranda, M. (1968). *Un esempio di soluzioni discontinue per un problema di minimo relativo ad un integrale regolare del calcolo delle variazioni*, Boll. Un. Mat. Italiana 2.

[8] Hao, W., Leonardi, S. and Nečas, J. (1996). *An example of irregular solution to a nonlinear Euler-Lagrange elliptic system with real analytic coefficients.* Ann. Scuola Norm. Sup. Pisa (IV), XXIII (1).

[9] Hao, W., Leonardi, S. and Steinhauer, M. (1995). *Examples of discontinuous divergence-free solutions to elliptic variational problems.* Comm. Math. Univ. Carolinae 36(3).

[10] Koshelev, A.I. and Chelkak, S.I. (1985). *Regularity of solutions of quasilinear elliptic systems.* Teubner Texte zur Mathematik, Leipzig.

[11] Koshelev, A.I. (1995). *Regularity problem for quasilinear elliptic and parabolic systems.* Springer.

[12] Ladyzhenskaya, O.A. and Ural'tseva, N.N. (1968). *Linear and quasilinear elliptic equations.* English Translation: Academic Press.

[13] Leonardi, S. (1998). *On constants of some regularity theorems. De Giorgi's type counterexample.* Math. Nach., Vol. 190.

[14] Leonardi, S. (1996). *Doctoral Thesis.* Roma 1996.

[15] Miranda, C. (1978). *Istituzioni di analisi funzionale lineare.* U.M.I.

[16] Nečas, J. (1967). *Les methodes directes en théorie des equations elliptiques.* Masson et C. Paris, Academia Prague.

[17] Nečas, J. (1979). *On the regularity of weak solution to nonlinear elliptic systems of partial differential equations.* Lezioni Scuola Normale Superiore Pisa.

[18] Nečas, J. (1986). *Introduction to the theory of nonlinear elliptic equations.* Teubner Texte zur Mathematik, Leipzig 1983 o J. Wiley, Chichester.

[19] Nečas, J., John, O. and Stará, J. (1980). *Counterexample to the regularity of weak solution of elliptic systems.* Comm. Math. Univ. Carolinae 21(1).

[20] Nečas, J. and Stará, J. (1972). *Principio di massimo per i sistemi ellittici quasi lineari non diagonali.* Boll. Un. Mat. Italiana 6.

[21] Opic, B. and Kufner, A. (1990). *Hardy-type inequalities.* Pitman Res. Notes. in Math. Series, n. 219, Longman.

[22] Simader, C.G. and Sohr, H. (1996). *The Dirichlet problem for the Laplacian in bounded and unbounded domains,* Pitman Res. Notes. in Math. Series, n. 360, Longman.

SINGULAR PERTURBATIONS IN OPTIMAL CONTROL PROBLEM

Ján Lovíšek

Abstract: This paper concerns an optimal control problem of elliptic singular perturbations in variational inequalities (with controls appearing in coefficients, right hand sides an convex sets of states as well). The existence of an optimal control is verified.

Keywords: Optimal control problem, singular perturbations in variational inequalities, convex set.

1. INTRODUCTION

Some interesting optimal design problems of structural analysis can be solved within the range of the Optimal Control Theory of variational inequalities. The coefficients of the nonlinear operator, occuring in the inequality play the role of control variables. The aim of asymptotic methods on optimal control is to simplify the state inequality. The most classical approach is the use of asymptotic expansion in terms of small parameter that may enter the state inequality, i.e. the method of perturbations, in particular the method of singular perturbations. Singular perturbations play a special role as an adequate mathematical tool for describing several problems in the plasticity or elasticity theory.

We introduce an abstract framework for the theoretical study of an optimal control problem governed by the variational inequality for the operators: $v \to \varepsilon\mathscr{A}(e)v + \mathscr{B}(e)v, \varepsilon > 0$, e plays the role of control variables. We present a general theorem, yielding the existence of at least one optimal control for $\varepsilon > 0$ and $\varepsilon = 0$. We prove the convergence of solutions of optimal control problems when ε tends to zero.

Applied Nonlinear Analysis, edited by Sequeira *et al.*
Kluwer Academic / Plenum Publishers, New York, 1999.

2. EXISTENCE OF A SOLUTION TO THE OPTIMAL CONTROL PROBLEMS

Let the control space $U(\Omega)$ be a reflexive Banach space with a norm $\|\cdot\|_{U(\Omega)}$. Let $U_{ad}(\Omega) \subset U(\Omega)$ be the set of admissible controls in $U(\Omega)$. Let $X(\Omega)$ be a real Hilbert space with the inner product $\langle\cdot,\cdot\rangle_{X(\Omega)}$. Furthermore, let $V(\Omega)$ and $W(\Omega)$ be two Banach spaces with norms $\|\cdot\|_{V(\Omega)}$, $\|\cdot\|_{W(\Omega)}$, being compactly imbedded into $X(\Omega)$ by imbeddings $\mathcal{J}_{V(\Omega)}$, $\mathcal{J}_{W(\Omega)}$, respectively, such that the ranges $\mathcal{J}_H(H)$ are dense $X(\Omega)$ for $H = V(\Omega)$, $W(\Omega)$. Let us denote by $V^*(\Omega)$ and $W^*(\Omega)$ the dual spaces of $V(\Omega)$ and $W(\Omega)$ and by $\|\cdot\|_{V^*(\Omega)}$, $\|\cdot\|_{W^*(\Omega)}$ their norms with respect to given duality pairings $\langle\cdot\rangle_{V(\Omega)}$, $\langle\cdot\rangle_{W(\Omega)}$, where, by convention, $\langle x,y\rangle_H \equiv \langle \mathcal{J}_H^{*^{-1}} x, \mathcal{J}_H y\rangle_{X(\Omega)}$ for $H = V(\Omega), W(\Omega)$, $y \in H$ and $x \in \mathcal{J}_H^*(X(\Omega))$ which is dense in H^* and will be identified with $X(\Omega)$. For a Banach space $\mathcal{H}$ and two nonnegative constants λ, Λ we denote by $\mathcal{E}_{\mathcal{H}}(\lambda,\Lambda)$ the set of all operators D from $\mathcal{H}$ into $\mathcal{H}^*$ for which the inequalities

$$\left.\begin{array}{l} \lambda\|v - w\|^2_{\mathcal{H}} \le \langle Dv - Dw, v - w\rangle_{\mathcal{H}} \quad \text{and} \\ \|Dv - Dw\|_{\mathcal{H}^*} \le \Lambda\|v - w\|_{\mathcal{H}} \quad \text{for all} \quad v, w \in \mathcal{H} \end{array}\right\} \tag{2.1}$$

hold. We assume

$$\left.\begin{array}{l} V(\Omega) \subset W(\Omega),\ V(\Omega) \text{ is dense in } W(\Omega) \text{ and} \\ U_{ad}(\Omega) \subset U(\Omega) \text{ is compact in } U(\Omega). \end{array}\right\} \tag{N0}$$

We introduce the system $\{\mathcal{K}(e,\Omega)\}_{e\in U_{ad}(\Omega)}$, $\{\mathcal{O}(e,\Omega)\}_{e\in U_{ad}(\Omega)}$ of nonempty closed convex sets $\mathcal{K}(e,\Omega) \subset V(\Omega)$, $\mathcal{O}(e,\Omega) \subset W(\Omega)$, $e \in U_{ad}(\Omega)$, and the family of operators $\{\mathcal{A}(e); e \in U_{ad}(\Omega)\}$ acting from $V(\Omega)$ into $V^*(\Omega)$ and $\{\mathcal{B}(e); e \in U_{ad}(\Omega)\}$ acting from $W(\Omega)$ into $W^*(\Omega)$ satisfying the following assumptions

$$\left.\begin{array}{l} 1^0\ e_n \to e_0 \text{ strongly in } U(\Omega) \Rightarrow \\ \qquad\qquad \mathcal{K}(e_0,\Omega) = \mathrm{Lim}_{n\to+\infty}\mathcal{K}(e_n,\Omega) \\ \quad 0 \in \prod_{n\in N}\mathcal{K}(e_n,\Omega), \\ 2^0\ \{\mathcal{A}(e); e \in U_{ad}(\Omega)\} \subset \mathcal{E}_{V(\Omega)}(0, M_{\mathcal{A}}), \mathcal{A}(e)0 = 0, \\ 3^0\ e_n \to e_0 \text{ strongly in } U(\Omega) \Rightarrow \mathcal{A}(e_n)v \to \mathcal{A}(e_0)v \\ \quad \text{strongly in } V^*(\Omega) \text{ for all } v \in V(\Omega), \\ 4^0 \text{ there is } \alpha_{\mathcal{A}} > 0 \text{ such that for all } e \in U_{ad}(\Omega) \\ \quad \text{and all } v, w \in V(\Omega) \text{ the inequality} \\ \quad \langle \mathcal{A}(e)v - \mathcal{A}(e)w, v - w\rangle_{V(\Omega)} + \|v - w\|^2_{W(\Omega)} \ge \\ \qquad\qquad \alpha_{\mathcal{A}}\|v - w\|^2_{V(\Omega)} \text{ holds.} \end{array}\right\} \tag{H$\mathcal{A}$}$$

and

$$\left.\begin{array}{l} 1^0\ cl\mathscr{K}(e,\Omega)=\mathcal{O}(e,\Omega), e\in U_{ad}(\Omega)(\text{closure in } W(\Omega)),\\ 2^0\ e_n\to e_0 \text{ strongly in } U(\Omega)\Rightarrow \\ \qquad\qquad \mathcal{O}(e_0,\Omega)=\mathrm{Lim}_{n\to\infty}\mathcal{O}(e_n,\Omega),\\ 3^0\ \{\mathscr{B}(e); e\in U_{ad}(\Omega)\}\subset \mathscr{E}_{W(\Omega)}(\alpha_{\mathscr{B}},M_{\mathscr{B}})\\ \quad \text{with } \alpha_{\mathscr{B}}>0, \mathscr{B}(e)0=0,\\ 4^0\ e_n\to e_0 \text{ in } U(\Omega)\Rightarrow \mathscr{B}(e_n)v\to\mathscr{B}(e_0)v\\ \quad \text{strongly in } W^*(\Omega) \text{ for all } v\in W(\Omega). \end{array}\right\} \qquad (H\mathscr{B})$$

Note that $W^*(\Omega)\hookrightarrow V^*(\Omega)$ continuously, and one has the transposition formula

$$\langle F,v\rangle_{V(\Omega)}=\langle F,v\rangle_{W(\Omega)} \text{ for any } v\in V(\Omega) \text{ and for any } F\in W^*(\Omega).$$

We assume, moreover, that

$$f\in W^*(\Omega) \text{ and } B: U(\Omega)\to W^*(\Omega) \text{ is a continuous operator.} \qquad (B0)$$

For every $0<\varepsilon<\alpha_{\mathscr{B}}$ and for every $e\in U_{ad}(\Omega)$ there exists a unique state function $u_\varepsilon(e)\in\mathscr{K}(e,\Omega)$ such that

$$\langle\varepsilon\mathscr{A}(e)u_\varepsilon(e)+\mathscr{B}(e)u_\varepsilon(e),v-u_\varepsilon(e)\rangle_{V(\Omega)}\geq\langle f+Be,v-u_\varepsilon(e)\rangle_{W(\Omega)}, \qquad (2.2)$$

for all $v\in\mathscr{K}(e,\Omega)$.

Indeed, thanks to the general theory of variational inequalities enough to prove that there is $c_\varepsilon>0$ such that

$$\begin{aligned}\langle\varepsilon(\mathscr{A}(e)v-\mathscr{A}(e)w),v-w\rangle_{V(\Omega)}+\langle\mathscr{B}(e)v-\mathscr{B}(e)w,v-w\rangle_{W(\Omega)}\\ \geq c_\varepsilon\|v-w\|^2_{V(\Omega)}, v,w\in V(\Omega)\end{aligned} \qquad (2.3)$$

and this immediately follows from $((H\mathscr{A}),2^0,4^0)$, $((H\mathscr{B}),3^0)$ (e.g. by contradiction).

Thanks to $((H\mathscr{B}),3^0)$, for any $e\in U_{ad}(\Omega)$

$$\begin{aligned}&\text{There exists } u(e)\in\mathcal{O}(e,\Omega) \text{ such that for any } v\in\mathcal{O}(e,\Omega),\\ &\langle\mathscr{B}(e)u(e),v-u(e)\rangle_{W(\Omega)}\geq\langle f+Be,v-u(e)\rangle_{W(\Omega)}.\end{aligned} \qquad (2.4)$$

Let us consider a functional $\mathscr{L}: U(\Omega)\times W(\Omega)\to R^+\equiv\{a\in R: a\geq 0\}$ for which the following condition holds

$$\left.\begin{array}{l} 1^0\ \{v_n\}_{n\in N}\subset V(\Omega),\ v\in W(\Omega)\\ \quad v_n\to v \text{ strongly in } W(\Omega)\Rightarrow\mathscr{L}(e,v)=\lim_{n\to+\infty}\mathscr{L}(e,v_n),\\ 2^0\ \{v_n\}_{n\in N}\subset W(\Omega),\ v\in W(\Omega),\ \{e_n\}_{n\in N}\subset U_{ad}(\Omega),\\ \quad e\in U_{ad}(\Omega), e_n\to e \text{ strongly in } U(\Omega),\\ \quad v_n\to v \text{ weakly in } W(\Omega)\Rightarrow\mathscr{L}(e,v)\leq\liminf_{n\to\infty}\mathscr{L}(e_n,v_n). \end{array}\right\} \qquad (E0)$$

We introduce the functional J_ε by

$$J_\varepsilon(e) = \mathscr{L}(e, u_\varepsilon(e)), e \in U_{ad}(\Omega), \tag{2.5}$$

where $u_\varepsilon(e)$ is the uniquely determined solution of (2.2), $e \in U_{ad}(\Omega)$. We shall solve the following optimization problem $(\mathscr{P}_\varepsilon)$.

Find a control $e_\varepsilon \in U_{ad}(\Omega)$ such that

$$J_\varepsilon(e_\varepsilon) = \inf_{e \in U_{ad}(\Omega)} J_\varepsilon(e). \qquad (\mathscr{P}_\varepsilon)$$

We say that e_ε is an optimal control of the problem $(\mathscr{P}_\varepsilon)$.

Theorem 2.1. *Let the assumptions* $(N0)$, $(H\mathscr{A})$, $(H\mathscr{B})$, $(B0)$ *and* $(E0)$ *be satisfied. Then there exists at least one solution to* $(\mathscr{P}_\varepsilon)$ *for any* $0 < \varepsilon < \alpha_{\mathscr{B}}$.

Proof. Due to the compactness of $U_{ad}(\Omega)$ in $U(\Omega)$, there exists a sequence $\{e_{\varepsilon\langle n\rangle}\}_{n\in N} \subset U_{ad}(\Omega)$ such that

$$\left.\begin{array}{l} \lim_{n\to+\infty} e_{\varepsilon\langle n\rangle} = e_{\varepsilon\langle 0\rangle} \text{ in } U(\Omega), e_{\varepsilon\langle n\rangle} \in U_{ad}(\Omega), \\ \lim_{n\to+\infty} J_\varepsilon(e_{\varepsilon\langle n\rangle}) = \inf_{e\in U_{ad}(\Omega)} J_\varepsilon(e). \end{array}\right\} \tag{2.6}$$

For any $e_{\varepsilon\langle n\rangle} \in U_{ad}(\Omega)$ consider the following variational inequality: $u_\varepsilon(e_{\varepsilon\langle n\rangle}) \in \mathscr{K}(e_{\varepsilon\langle n\rangle}, \Omega)$

$$\begin{aligned} &\langle \varepsilon\mathscr{A}(e_{\varepsilon\langle n\rangle})u_\varepsilon(e_{\varepsilon\langle n\rangle}) + \mathscr{B}(e_{\varepsilon\langle n\rangle})u_\varepsilon(e_{\varepsilon\langle n\rangle}), v - u_\varepsilon(e_{\varepsilon\langle n\rangle})\rangle_{V(\Omega)} \\ &\qquad \geq \langle f + Be_{\varepsilon\langle n\rangle}, v - u_\varepsilon(e_{\varepsilon\langle n\rangle})\rangle_{W(\Omega)} \end{aligned} \tag{2.7}$$

for all $v \in \mathscr{K}(e_{\varepsilon\langle n\rangle}, \Omega)$.

We take an arbitrary $v_0 \in \mathscr{K}(e_{\varepsilon\langle 0\rangle}, \Omega)$ and (by $(H\mathscr{A}), 1^0$) a sequence $\{v_n\}_{n\in N} \in \prod_{n\in N} \mathscr{K}(e_{\varepsilon\langle n\rangle}, \Omega)$ such that $v_n \to v_0$ strongly in $V(\Omega)$. Putting $v = v_n$ in (2.7), $\varepsilon\langle \mathscr{A}(e_{\varepsilon\langle n\rangle})v_n, u_\varepsilon(e_{\varepsilon\langle n\rangle}) - v_n\rangle_{V(\Omega)} + \langle \mathscr{B}(e_{\varepsilon\langle n\rangle})v_n, u_\varepsilon(e_{\varepsilon\langle n\rangle}) - v_n\rangle_{W(\Omega)}$ adding to its both sides and multiplying the resulting inequality by minus one, we obtain

$$\begin{aligned} &\langle \varepsilon(\mathscr{A}(e_{\varepsilon\langle n\rangle})u_\varepsilon(e_{\varepsilon\langle n\rangle}) - \mathscr{A}(e_{\varepsilon\langle n\rangle})v_n), u_\varepsilon(e_{\varepsilon\langle n\rangle}) - v_n\rangle_{V(\Omega)} \\ &+ \langle \mathscr{B}(e_{\varepsilon\langle n\rangle})u_\varepsilon(e_{\varepsilon\langle n\rangle}) - \mathscr{B}(e_{\varepsilon\langle n\rangle})v_n, u_\varepsilon(e_{\varepsilon\langle n\rangle}) - v_n\rangle_{W(\Omega)} \\ &\leq \langle \varepsilon\mathscr{A}(e_{\varepsilon\langle n\rangle})v_n, v_n - u_\varepsilon(e_{\varepsilon\langle n\rangle})\rangle_{V(\Omega)} \\ &+ \langle \mathscr{B}(e_{\varepsilon\langle n\rangle})v_n, v_n - u_\varepsilon(e_{\varepsilon\langle n\rangle})\rangle_{W(\Omega)} \\ &+ \langle f + Be_{\varepsilon\langle n\rangle}, u_\varepsilon(e_{\varepsilon\langle n\rangle}) - v_n\rangle_{W(\Omega)}, \end{aligned} \tag{2.8}$$

for $n \in N$. From (2.3), (2.8), $((H\mathscr{A}), 2^0, 3^0)$, $((H\mathscr{B}), 4^0)$ and $(B0)$ it follows that

$$\|u_\varepsilon(e_{\varepsilon\langle n\rangle})\|_{V(\Omega)} \leq \text{constant}(\varepsilon),\ n \in N \text{ for fixed } \varepsilon > 0. \tag{2.9}$$

This yields the existence of a subsequence $\{u_\varepsilon(e_{\varepsilon\langle n_k\rangle})\}_{k\in N}$ and of an element $u_{\varepsilon\langle 0\rangle} \in V(\Omega)$ such that

$$u_\varepsilon(e_{\varepsilon\langle n_k\rangle}) \to u_{\varepsilon\langle 0\rangle} \text{ weakly in } V(\Omega). \tag{2.10}$$

As $u_\varepsilon(e_{\varepsilon\langle n\rangle}) \in \mathscr{K}(e_{\varepsilon\langle n\rangle}, \Omega)$, the assumption $((H\mathscr{A}), 1^0)$ yields

$$u_{\varepsilon\langle 0\rangle} \in \mathscr{K}(e_{\varepsilon\langle 0\rangle}, \Omega). \tag{2.11}$$

By $((H\mathscr{A}), 1^0)$ there exists a sequence $\{\theta_k\}_{k\in N}$, $\theta_k \in \mathscr{K}(e_{\varepsilon\langle n_k\rangle}, \Omega)$, such that $\theta_k \to u_\varepsilon(e_{\varepsilon\langle 0\rangle})$ strongly in $V(\Omega)$. Inserting $v := \theta_k$ in (2.7), adding

$$\langle \varepsilon\mathscr{A}(e_{\varepsilon\langle n_k\rangle})\theta_k + \mathscr{B}(e_{\varepsilon\langle n_k\rangle})\theta_k, u_\varepsilon(e_{\varepsilon\langle n_k\rangle}) - \theta_k\rangle_{V(\Omega)},$$

to its both sides and multiplying the resulting inequality by minus one, we obtain

$$\begin{aligned}
&\limsup_{k\to+\infty}\langle(\varepsilon\mathscr{A}(e_{\varepsilon\langle n_k\rangle}) + \mathscr{B}(e_{\varepsilon\langle n_k\rangle}))\\
&\qquad (u_\varepsilon(e_{\varepsilon\langle n_k\rangle}) - \theta_k), u_\varepsilon(e_{\varepsilon\langle n_k\rangle}) - \theta_k\rangle_{V(\Omega)}\\
&\limsup_{k\to+\infty}\left|\langle\varepsilon\mathscr{A}(e_{\varepsilon\langle n_k\rangle})\theta_k, \theta_k - u_\varepsilon(e_{\varepsilon\langle n_k\rangle})\rangle_{V(\Omega)}\right|\\
&+ \limsup_{k\to+\infty}\left|\langle\mathscr{B}(e_{\varepsilon\langle n_k\rangle})\theta_k, \theta_k - u_\varepsilon(e_{\varepsilon\langle n_k\rangle})\rangle_{W(\Omega)}\right|\\
&+ \limsup_{k\to+\infty}\left|\langle f + Be_{\varepsilon\langle n_k\rangle}, u_\varepsilon(e_{\varepsilon\langle n_k\rangle}) - \theta_k\rangle_{W(\Omega)}\right| = 0.
\end{aligned} \tag{2.12}$$

The last inequality follows from $(B0)$ and from the facts

$$\begin{aligned}
&e_n \to e \text{ strongly in } U(\Omega), v_n \to v \text{ strongly in } V(\Omega)\\
&\quad \text{for } n \to +\infty \Rightarrow \|\mathscr{A}(e_n)v_n - \mathscr{A}(e)v\|_{V^*(\Omega)}\\
&\leq c_{\mathscr{A}}\|v_n - v\|_{V(\Omega)} + \|\mathscr{A}(e_n)v - \mathscr{A}(e)v\|_{V^*(\Omega)} \to 0\\
&\qquad\qquad \text{for } n \to +\infty,
\end{aligned} \tag{2.13}$$

$$\begin{aligned}
&e_n \to e \text{ strongly in } U(\Omega), w_n \to w \text{ strongly in } W(\Omega)\\
&\quad \text{for } n \to +\infty \Rightarrow \|\mathscr{B}(e_n)w_n - \mathscr{B}(e)w\|_{W^*(\Omega)}\\
&\leq c_{\mathscr{A}}\|w_n - w\|_{W(\Omega)} + \|\mathscr{B}(e_n)w - \mathscr{B}(e)w\|_{W^*(\Omega)} \to 0\\
&\qquad\qquad \text{for } n \to +\infty,
\end{aligned} \tag{2.14}$$

which are consequence of $((H\mathscr{A}), 2^0, 3^0)$, $((H\mathscr{B}), 3^0, 4^0)$, respectively. Due to the uniform monotonicity of $[\varepsilon\mathscr{A}(e_{\varepsilon\langle n_k\rangle}) + \mathscr{B}(e_{\varepsilon\langle n_k\rangle})]$ (cf. (2.3)) we obtain the strong convergence

$$u_\varepsilon(e_{\varepsilon\langle n_k\rangle}) \to u_{\varepsilon\langle 0\rangle} \text{ strongly in } V(\Omega) \text{ for } k \to +\infty. \tag{2.15}$$

Moreover, (2.15) together with (2.13) and (2.14) yield

$$\begin{aligned} &\mathscr{A}(e_{\varepsilon\langle n_k\rangle})u_\varepsilon(e_{\varepsilon\langle n_k\rangle}) \to \mathscr{A}(e_{\varepsilon\langle 0\rangle})u_{\varepsilon\langle 0\rangle} \text{ strongly in } V^*(\Omega), \\ &\mathscr{B}(e_{\varepsilon\langle n_k\rangle})u_\varepsilon(e_{\varepsilon\langle n_k\rangle}) \to \mathscr{B}(e_{\varepsilon\langle 0\rangle})u_{\varepsilon\langle 0\rangle} \text{ strongly in } W^*(\Omega), \end{aligned} \tag{2.16}$$

for $k \to +\infty$. Given a $v \in \mathscr{K}(e_{\varepsilon\langle 0\rangle}, \Omega)$, by the assumption $((H\mathscr{A}), 1^0)$ there exists a sequence $\{v_k\}_{k\in N}, v_k \in \mathscr{K}(e_{\varepsilon\langle n_k\rangle}, \Omega), v_k \to v$ strongly in $V(\Omega)$. Limiting (2.7) with $v = v_k$, we have

$$\begin{aligned} \langle \varepsilon\mathscr{A}(e_{\varepsilon\langle 0\rangle})u_{\varepsilon\langle 0\rangle}, v - u_{\varepsilon\langle 0\rangle})\rangle_{V(\Omega)} + \langle \mathscr{B}(e_{\varepsilon\langle 0\rangle})u_{\varepsilon\langle 0\rangle}, v - u_{\varepsilon\langle 0\rangle}\rangle_{W(\Omega)} \\ \geq \langle f + Be_{\varepsilon\langle 0\rangle}, v - u_{\varepsilon\langle 0\rangle}\rangle_{W(\Omega)} \end{aligned} \tag{2.17}$$

and, as $v \in \mathscr{K}(e_{\varepsilon\langle 0\rangle}, \Omega)$ is chosen arbitrarily, we get

$$\left.\begin{aligned} &u_{\varepsilon\langle 0\rangle} \equiv u_\varepsilon(e_{\varepsilon\langle 0\rangle})(\text{is the unique solution of (2.2) for } e = e_{\varepsilon\langle 0\rangle}) \\ &\text{and} \\ &u_\varepsilon(e_{\varepsilon\langle n\rangle}) \to u_\varepsilon(e_{\varepsilon\langle 0\rangle}) \text{ strongly in } V(\Omega). \end{aligned}\right\} \tag{2.18}$$

Then $((E0), 2^0)$ and (2.18) yield

$$\begin{aligned} \mathscr{L}(e_{\varepsilon\langle 0\rangle}, u_\varepsilon(e_{\varepsilon\langle 0\rangle})) &\leq \liminf_{n\to+\infty} \mathscr{L}(e_{\varepsilon\langle n\rangle}, u_\varepsilon(e_{\varepsilon\langle n\rangle})) \\ &= \inf_{e\in U_{ad}(\Omega)} \mathscr{L}(e, u_\varepsilon(e)). \end{aligned} \tag{2.19}$$

Hence $\mathscr{L}(e_{\varepsilon\langle 0\rangle}, u_\varepsilon(e_{\varepsilon\langle 0\rangle})) = \inf\{\mathscr{L}(e, u_\varepsilon(e)); e \in U_{ad}(\Omega)\}$, which completes the proof.

Limit state function and limit cost function. We define the limit state function for any $e \in U_{ad}(\Omega)$ by the variational inequality

$$\left.\begin{aligned} &\text{Find } u_0(e) \in \mathcal{O}(e, \Omega) \text{ such that} \\ &\langle \mathscr{B}(e)u_0(e), v - u_0(e)\rangle_{W(\Omega)} \geq \langle f + Be, v - u_0(e)\rangle_{W(\Omega)} \\ &\text{for any } v \in \mathcal{O}(e, \Omega) \end{aligned}\right\} \tag{2.20}$$

and the limit cost function

$$J_0(e) = \mathscr{L}(e, u_0(e)). \tag{2.21}$$

In this case one has the limit control problem $(\mathscr{P}_0)$ defined as follows

$$\text{Find } e_0 \in \operatorname{Arg}\inf\{J_0(e); e \in U_{ad}(\Omega)\}. \tag{$\mathscr{P}_0$}$$

Theorem 2.2. *Let the assumptions* $(N0)$, $(H\mathscr{B})$, $(B0)$ *and* $(E0)$ *be satisfied. Then there exists at least one solution to* $(\mathscr{P}_0)$.

The proof is analogous to that of Theorem 2.1.

There arises a natural question concerning the type of relation between solutions to $(\mathscr{P}_0)$ and $(\mathscr{P}_\varepsilon)$ if $\varepsilon \to 0_+$.

We prove the following theorem.

Theorem 2.3. *Let the assumption* $(N0)$, $(H\mathscr{A})$, $(H\mathscr{B})$, $(B0)$ *and* $(E0)$ *be satisfied. Let* e_{ε_n}, e_0 *be solutions of the problem* $(\mathscr{P}_{\varepsilon_n})$, $(\mathscr{P}_0)$ *respectively,* $\varepsilon_n \to 0_+$. *Then there exists a sequence* $\{\varepsilon_n\}_{n\in N}$, $\varepsilon_n \to 0_+$ *for* $n \to +\infty$, *such that*

$$\left.\begin{array}{l} e_{\varepsilon_n} \to e_0 \text{ strongly in } U(\Omega), \\ u_{\varepsilon_n}(e_{\varepsilon_n}) \to u_0(e_0) \text{ strongly in } W(\Omega), \\ J_{\varepsilon_n}(e_{\varepsilon_n}) = \inf_{e\in U_{ad}} J_{\varepsilon_n}(e) \to J_0(e_0) = \inf_{e\in U_{ad}(\Omega)} J_0(e). \end{array}\right\} \tag{2.22}$$

Proof. Due to compactness of $U_{ad}(\Omega)$ there exists $\{e_{\varepsilon_n}\}_{n\in N} \subset U_{ad}(\Omega)$ such that $e_{\varepsilon_n} \to e_0$ strongly in $U(\Omega)$. The "state function" $u_{\varepsilon_n}(e_{\varepsilon_n}) \in \mathscr{K}(e_{\varepsilon_n}, \Omega)$ is a solution of the state variational inequality

$$\left.\begin{array}{l} \langle \varepsilon_n \mathscr{A}(e_{\varepsilon_n})u_{\varepsilon_n}(e_{\varepsilon_n}) + \mathscr{B}(e_{\varepsilon_n})u_{\varepsilon_n}(e_{\varepsilon_n}), v - u_{\varepsilon_n}(e_{\varepsilon_n})\rangle_{V(\Omega)} \\ \geq \langle f + Be_{\varepsilon_n}, v - u_{\varepsilon_n}(e_{\varepsilon_n})\rangle_{W(\Omega)}, \\ \text{for any } v \in \mathscr{K}(e_0, \Omega), \text{ for given } e_{\varepsilon_n} \in U_{ad}(\Omega), \varepsilon_n > 0, n \in N. \end{array}\right\} \tag{2.23}$$

We take an arbitrary $v_0 \in \mathscr{K}(e_0, \Omega)$ and sequence $\{v_n\}_{n\in N} \in \prod_{n\in N} \mathscr{K}(e_{\varepsilon_n}, \Omega)$ such that $v_n \to v_0$ strongly in $V(\Omega)$. In the equality (2.23) we take the fixed $v = v_n$, add $\varepsilon_n\langle \mathscr{A}(e_{\varepsilon_n})v_n, u_{\varepsilon_n}(e_{\varepsilon_n}) - v_n\rangle_{V(\Omega)} + \langle \mathscr{B}(e_{\varepsilon_n})v_n, u_{\varepsilon_n}(e_{\varepsilon_n}) - v_n\rangle_{W(\Omega)}$ to both sides of (2.23), multiply the resulting inequality by minus one and use $((H\mathscr{A}), 2^0)$ and $((H\mathscr{B}), 3^0)$. It follows that

$$\left.\begin{array}{l} \varepsilon_n(\langle \mathscr{A}(e_{\varepsilon_n})u_{\varepsilon_n}(e_{\varepsilon_n}) - \mathscr{A}(e_{\varepsilon_n})v_n, u_{\varepsilon_n}(e_{\varepsilon_n}) - v_n\rangle_{V(\Omega)} \\ \quad + \|u_{\varepsilon_n}(e_{\varepsilon_n}) - v_n\|^2_{W(\Omega)}) + (\alpha_{\mathscr{B}} - \varepsilon_n)\|u_{\varepsilon_n}(u_{\varepsilon_n}) - v_n\|^2_{W(\Omega)} \\ \leq \langle f + Be_{\varepsilon_n}, u_{\varepsilon_n}(e_{\varepsilon_n}) - v_n\rangle_{W(\Omega)} \\ \quad + \langle(\varepsilon_n\mathscr{A}(e_{\varepsilon_n}) + \mathscr{B}(e_{\varepsilon_n}))v_n, v_n - u_{\varepsilon_n}(e_{\varepsilon_n})\rangle_{V(\Omega)}. \end{array}\right\}$$

Setting $\varepsilon_n \leq \alpha_{\mathscr{B}}/2$ applying $((H\mathscr{A}), (H\mathscr{B}))$, we get

$$\left.\begin{array}{l} (\varepsilon_n\alpha_{\mathscr{A}})\|u_{\varepsilon_n}(e_{\varepsilon_n}) - v_n\|^2_{V(\Omega)} + (\alpha_{\mathscr{B}}/2)\|u_{\varepsilon_n}(e_{\varepsilon_n}) - v_n\|^2_{W(\Omega)} \\ \leq \text{constant}_{\langle 1\rangle}\|f + Be_{\varepsilon_n}\|_{W^*(\Omega)}\|u_{\varepsilon_n}(e_{\varepsilon_n}) - v_n\|_{W(\Omega)} \\ + \text{constant}_{\langle 2\rangle}\varepsilon_n\|u_{\varepsilon_n}(e_{\varepsilon_n}) - v_n\|_{V(\Omega)}\|v_n\|_{V(\Omega)} \\ + \text{constant}_{\langle 3\rangle}\|u_{\varepsilon_n}(e_{\varepsilon_n}) - v_n\|_{W(\Omega)}\|v_n\|_{W(\Omega)}, \end{array}\right\}$$

where $\text{constant}_{\langle i\rangle}$, $i = 1, 2, 3$ are constants which do not depend on n.

Hence we conclude that

$$\left.\begin{array}{l} \|u_{\varepsilon_n}(e_{\varepsilon_n})\|_{W(\Omega)} \leq \text{constant}, \\ \sqrt{\varepsilon_n}\|u_{\varepsilon_n}(e_{\varepsilon_n})\|_{V(\Omega)} \leq \text{constant} \;\Rightarrow \\ \sqrt{\varepsilon_n}\|\mathscr{A}(e_{\varepsilon_n})u_{\varepsilon_n}(e_{\varepsilon_n})\|_{V^*(\Omega)} \leq \text{constant} \end{array}\right\} \tag{2.24}$$

for some constants independent of n.

We can therefore extract a subsequence $\{u_{\varepsilon_{n_k}}(e_{\varepsilon_{n_k}})\}_{k\in N}$ such that

$$\begin{aligned} &u_{\varepsilon_{n_k}}(e_{\varepsilon_{n_k}}) \to w_O \text{ weakly in } W(\Omega) \text{ for } k \to +\infty, w_O \in \mathcal{O}(e_0, \Omega), \\ &\sqrt{\varepsilon_n} u_{\varepsilon_{n_k}}(e_{\varepsilon_{n_k}}) \to v_O \text{ weakly in } V(\Omega) \text{ for } k \to +\infty \end{aligned} \tag{2.25}$$

exploiting the assumption $((H\mathscr{B}), 2^0)$, too. Moreover, there is a sequence $\{w_{Ok}\}_{k\in N}$ such that $w_{Ok} \in \mathscr{K}(e_{\varepsilon_{n_k}}, \Omega)$ and $w_{Ok} \to w_O$ strongly in $W(\Omega)$.

We put $v = w_{Ok}$ into (2.23) formulated for the index n_k, add

$$\langle \varepsilon_{n_k} \mathscr{A}(e_{\varepsilon_{n_k}}) w_{Ok} + \mathscr{B}(e_{\varepsilon_{n_k}}) w_{Ok}, u_{\varepsilon_{n_k}}(e_{\varepsilon_{n_k}}) - w_{Ok} \rangle_{V(\Omega)},$$

to its both sides, multiply the resulting inequality by minus one and employ $((H\mathscr{A}), 2^0)$, $((H\mathscr{B}), 3^0)$, $(B0)$, (2.13) and (2.14) again. As the right hand side of the resulting inequality tends to zero (cf. (2.12)), we obtain

$$\begin{aligned} &\limsup_{k\to+\infty} \alpha_{\mathscr{B}} \| w_{Ok} - u_{\varepsilon_{n_k}}(e_{\varepsilon_{n_k}}) \|^2_{W(\Omega)} = 0 \\ &\Rightarrow u_{\varepsilon_{n_k}}(e_{\varepsilon_{n_k}}) \to w_O, \text{ strongly in } W(\Omega), \limsup_{k\to+\infty} \\ &\langle \varepsilon_{n_k} \mathscr{A}(e_{\varepsilon_{n_k}})(u_{\varepsilon_{n_k}}(e_{\varepsilon_{n_k}}) - w_{Ok}), u_{\varepsilon_{n_k}}(e_{\varepsilon_{n_k}}) - w_{Ok} \rangle_{V(\Omega)} \\ &= \limsup_{k\to+\infty} \langle \varepsilon_{n_k} \mathscr{A}(e_{\varepsilon_{n_k}})(u_{\varepsilon_{n_k}}(e_{\varepsilon_{n_k}}), u_{\varepsilon_{n_k}}(e_{\varepsilon_{n_k}})) \rangle_{V(\Omega)} = 0. \end{aligned} \tag{2.26}$$

We take $w \in \mathcal{O}(e_0, \Omega)$ arbitrary. We can find $\{w_k\}_{k\in N} \subset V(\Omega)$ such that $w_k \in \mathscr{K}(e_{\varepsilon_{n_k}}, \Omega)$ and $w_k \to w$ strongly in $W(\Omega)$. To prove

$$w_O = u_0(e_0), \tag{2.27}$$

we return to (2.23) for the index n_k and put $v = w_k$ there. Due to $(B0)$, (2.25) and (2.26) it is easy to see that for $k \to +\infty$ we obtain

$$\langle \mathscr{B}(e_0) u_0(e_0), w - u_0(e_0) \rangle_{W(\Omega)} \geq \langle f + Be_0, v - u_0(e_0) \rangle_{W(\Omega)}, \tag{2.28}$$

and (2.27) is valid. Moreover, the method of the proof shows that the convergence

$$u_\varepsilon(e) \to u_0(e) \text{ strongly in } W(\Omega) \text{ for any } e \in U_{ad}(\Omega), \tag{2.29}$$

holds. Indeed, if it were not true, there would be a sequence $\varepsilon_k \to 0$ and a constant $\tilde{\mathcal{O}} > 0$ independent of k such that

$$\| u_{\varepsilon_k}(e) - u_0(e) \|_{W(\Omega)} \geq \tilde{\mathcal{O}} \text{ for any } k \in \mathcal{N}. \tag{2.30}$$

Putting an arbitrary fixed $v \in \mathscr{K}(e, \Omega)$ into the appropriate variational inequalities, we arrive at

$$\begin{aligned} &\| u_{\varepsilon_k}(e) \|_{W(\Omega)} \leq \text{constant}, \\ &\sqrt{\varepsilon_k} \| u_{\varepsilon_k}(e) \|_{V(\Omega)} \leq \text{constant} \\ &\Rightarrow \sqrt{\varepsilon_k} \| \mathscr{A}(e) u_{\varepsilon_k}(e) \|_{V^*(\Omega)} \leq \text{constant}, \end{aligned} \tag{2.31}$$

where constants do not depend on k. The existing $W(\Omega)$ - weak limit of a suitable subsequence $\{u_{\varepsilon_{k_n}}(e)\}_{n\in N} \subset \{u_{\varepsilon_k}(e)\}_{k\in N}$ must be $u_0(e)$ due to $((H\mathscr{B}), 1^0)$ and due to quite analogous arguments to those used in deriving (2.27) through (2.26) and (2.28). This is a contradiction to (2.30).

Now, from (2.29), from the fact that $J_{\varepsilon_{n_k}}(e_{\varepsilon_{n_k}}) \leq J_{\varepsilon_{n_k}}(e)$ for all $e \in U_{ad}(\Omega)$ and all k, and from $((E0), 1^0)$, we get

$$\begin{array}{l} \limsup_{k\to+\infty} J_{\varepsilon_{n_k}}(e_{\varepsilon_{n_k}}) \leq J_0(e) \text{ for any } e \in U_{ad}(\Omega) \\ \Rightarrow \limsup_{k\to+\infty} J_{\varepsilon_{n_k}}(e_{\varepsilon_{n_k}}) \leq \inf_{e\in U_{ad}(\Omega)} J_0(e) \leq J_0(e_0). \end{array} \tag{2.32}$$

Furthermore, we observe that $((E0), 2^0)$, (2.26) and (2.27) imply $\liminf_{k\to+\infty} J_{\varepsilon_{n_k}}(e_{\varepsilon_{n_k}}) \geq \mathscr{L}(e_0, u_0(e_0)) = J_0(e_0)$. Comparing this result with (2.32) we see that necessarily

$$\inf_{e\in U_{ad}(\Omega)} J_0(e) = J_0(e_0). \tag{2.33}$$

Theorem 2.3 is proved.

References

[1] Barbu, V. (1987). *Optimal Control of Variational Inequalities.* Pitman Advanced Publishing Program, Boston-London.

[2] Huet, D. (1973). Singular Perturbations of Elliptic Problems. *Ann. Mat. Pura Appl.*, 95:77–114.

[3] Khludnev, A.M. and Sokolowski, J. (1997). *Modelling and Control in Solid Mechanics.* Birkhäuser Verlag, Basel-Boston-Berlin.

[4] Lions, J.L. (1973). *Perturbations singuliéres dans les problémes aux limites et en contrôle optimal.* Lect Notes in Math. 232, Springer-Verlag, Berlin.

[5] Nečas, J. (1967). *Les Méthodes Directes an Théorie des Équations Elliptiques.* Masson, Paris/Academia, Prague.

OPTIMIZATION OF STEADY FLOWS FOR INCOMPRESSIBLE VISCOUS FLUIDS

Josef Málek, Tomáš Roubíček

Abstract: An optimal-control problem for the stationary Navier-Stokes system are investigated. The maximum principle is derived by a suitable relaxation. Its sufficiency is shown provided data involved in the control problem are small enough (depending on the Reynolds number). Regularity of the Navier-Stokes system and its adjoint problem is used.

Keywords: Navier-Stokes equations, regularity, optimal control, existence, relaxation, maximum principle, sufficiency.

Introduction

In this paper we deal with optimization of steady two- and three-dimensional fluid flows governed by the Navier-Stokes system. Analysis of the various problems of optimal control of viscous flows enjoys recently significant attention within mathematical comunity. Optimal control problem of this sort was already studied in [1, 2, 3, 7, 18, 19, 20] and [25, Section III.11]. For an optimal shape design problem see [21]. Besides, optimal control of evolutionary Navier-Stokes system was treated in [4, 5, 8, 9, 10, 11, 12, 13, 14, 22, 29, 32, 33, 34, 35] and also in [25, Section I.18].

Our main goal is to adapt the relaxation method by convex compactification [30] for Navier-Stokes equations and to exploit (quite standard) regularity results for the stationary (linearized) Navier-Stokes system to derive nontrivial results concerning sufficiency of the maximum principle.

The scheme of the paper is the following. In Section 1, we specify an optimal-control problem $(\mathcal{P})$ we will deal with, and in Section 2 we pose the relaxed problem to $(\mathcal{P})$ and show its correctness under the assumption that a driving force is sufficiently small so that the state response is uniquely defined. In Section 3, we confine ourselves

Applied Nonlinear Analysis, edited by Sequeira *et al.*
Kluwer Academic / Plenum Publishers, New York, 1999.

to a special form of the data and derive the corresponding maximum principle (i.e. necessary condition of optimality), and show that this maximum principle forms a sufficient condition provided that the desired velocity profile and the driving force are small enough (depending on the Reynolds number), which ensures that the relaxed cost functional is "enough" uniformly convex with respect to the state; cf. Remark 4 below. For this purpose, L^q-regularity results for a dual (adjoint) equation to the linearized Navier-Stokes system are exploited.

We wish to remark that the result presented here can be extended to the power-law-like fluids in two dimensions (at least for $p > \frac{3}{2}$, where p denotes the power-law exponent, see [26] for more details). While the regularity results applied here to the Navier-Stokes system are well traced in the literature, their extension to the power-law fluids is possible because of recent nontrivial $C^{1,\alpha}$-regularity results for this class of fluids performed in [24] (see also contribution of the same authors in this volume); the method is based on the approach introduced in [27] and [28] but will not be presented in this paper.

1. AN OPTIMAL-CONTROL PROBLEM

We will confine ourselves to steady flows of an incompressible fluid in a two- or three-dimensional bounded domain $\Omega \subset \mathbb{R}^n$ (i.e. $n = 2$ or 3) with no-slip (i.e. homogeneous Dirichlet) boundary condition.

We will first deal with the following optimal control problem for flows governed by the Navier-Stokes system:

$$(\mathcal{P}) \qquad \begin{cases} \text{Minimize} & J(z,u) := \int_\Omega h(x, u(x), z(x))\,\mathrm{d}x \\ \text{subject to} & (u\cdot\nabla)u - \mu\Delta u + \nabla p = f(\cdot, u, z) \qquad \text{on } \Omega, \\ & \operatorname{div} u = 0 \qquad \text{on } \Omega, \\ & z(x) \in S(x) \qquad \text{for a.a. } x \in \Omega, \\ & u \in W_0^{1,2}(\Omega;\mathbb{R}^n), \quad p \in L_0^2(\Omega), \quad z \in L^q(\Omega;\mathbb{R}^m). \end{cases}$$

Here, z denotes the control, u represents the velocity field and p is the pressure. Not completely rigorous but frequently used notation $(u\cdot\nabla)u$ means $\sum_{k=1}^n u_k \frac{\partial u}{\partial x_k}$. By $\mu > 0$ we denote a fluid viscosity which is indirectly proportional to the Reynolds number. Further, $h : \Omega \times \mathbb{R}^n \times \mathbb{R}^m \to \mathbb{R}$ and $f : \Omega \times \mathbb{R}^n \times \mathbb{R}^m \to \mathbb{R}^n$ are given Carathéodory functions. Finally, $S : \Omega \rightrightarrows \mathbb{R}^m$ is a given multi-valued function forming the control constraints.

We use standard notation of function spaces: $C(\cdot)$ for the spaces of bounded continuous functions; $C_0(\mathbb{R}^m)$ for the space of continuous functions on $\mathbb{R}^m$ vanishing at infinity; $L^q(\Omega)$, $q \in [1,\infty]$, for the

Lebesgue spaces and $W_0^{1,q}(\Omega)$ for the Sobolev spaces having zero trace at the boundary $\partial\Omega$. The corresponding vector-valued spaces are denoted by $L^q(\Omega;\mathbb{R}^n)$ and $W_0^{1,q}(\Omega;\mathbb{R}^n)$, respectively. By (g,f) we mean $\int_\Omega g(x)\cdot f(x)\mathrm{d}x$. Finally we use the shorthand notation $L_0^q(\Omega)$ and $W_{0,\mathrm{DIV}}^{1,q}(\Omega;\mathbb{R}^n)$ for subspaces of zero-mean-value functions in $L^q(\Omega)$ and divergence-free functions in $W_0^{1,q}(\Omega;\mathbb{R}^n)$, respectively, i.e.

$$L_0^q(\Omega) := \{p\in L^q(\Omega);\ \int_\Omega p\,\mathrm{d}x = 0\}\,, \tag{1.1a}$$

$$W_{0,\mathrm{DIV}}^{1,q}(\Omega;\mathbb{R}^n) := \{u\in W_0^{1,q}(\Omega;\mathbb{R}^n);\mathrm{div}\,u=0\}\,. \tag{1.1b}$$

Of course, the solution (u,p) to the Navier-Stokes system in $(\mathcal{P})$ is understood in the weak sense, which means that $u\in W_{0,\mathrm{DIV}}^{1,2}(\Omega;\mathbb{R}^n)$ and, for a given z,

$$((u\cdot\nabla)u,v)+\mu(\nabla u,\nabla v) = (f(u,z),v)\quad \forall v\in W_{0,\mathrm{DIV}}^{1,2}(\Omega;\mathbb{R}^n)\,. \tag{1.2}$$

The basic data qualification we will need are the following:

$$|h(x,r,s)|\le a_1(x)+\beta(|r|)+c|s|^q, \tag{1.3a}$$

$$|h(x,r_1,s)-h(x,r_2,s)| \le (\tilde a_1(x)+\beta(\max(|r_1|,|r_2|))+c|s|^q)|r_1-r_2|, \tag{1.3b}$$

$$h(x,r,s)\ge c_0|s|^q, \tag{1.3c}$$

$$|f(x,r,s)|\le a_2(x), \tag{1.3d}$$

$$|f(x,r_1,s)-f(x,r_2,s)|\le(\tilde a_2(x)+\beta(\max(|r_1|,|r_2|)))|r_1-r_2|, \tag{1.3e}$$

S admits a measurable q-integrable selection, (1.3f)

where $\beta:\mathbb{R}^+\to\mathbb{R}^+ := [0,\infty)$ is a continuous increasing function with $\beta(0)=0$, $a_1,\tilde a_1\in L^1(\Omega)$, $a_2,\tilde a_2\in L^2(\Omega)$, $c\in\mathbb{R}^+$, and $c_0>0$.

Remark 1. Let us recall that, having a solution $u\in W_{0,\mathrm{DIV}}^{1,2}(\Omega;\mathbb{R}^n)$ satisfying (1.2) and assuming that f fulfills (1.3d), it is standard (see [35]) to construct the corresponding pressure $p\in L_0^2(\Omega)$ such that

$$((u\cdot\nabla)u,v)+\mu(\nabla u,\nabla v)-(p,\mathrm{div}\,v) = (f(u,z),v) \tag{1.4}$$

for all $v\in W_0^{1,2}(\Omega;\mathbb{R}^n)$. We will involve the pressure only in the formulations of the theorems and lemmas but not in the proofs, because p can always be reconstructed uniquely if one knows that $u\in W_{0,\mathrm{DIV}}^{1,2}(\Omega;\mathbb{R}^n)$ satisfies (1.2),

Remark 2. Note that (1.3d) leads, just by taking $v:=u$ in (1.2), to the energy estimate

$$\|\nabla u\|_{L^2(\Omega;\mathbb{R}^{n\times n})}\le\frac{k_0}{\mu}\|a_2\|_{L^2(\Omega)}\ , \tag{1.5}$$

where the constant k_0 comes from the Poincaré inequality.

2. RELAXED PROBLEM

We will extend continuously the problem $(\mathcal{P})$ on a suitable convex locally (sequentially) compact envelope of the set of admissible controls

$$Z_{\rm ad} := \{z \in L^q(\Omega;\mathbb{R}^m);\ z(x) \in S(x) \text{ for a.a. } x \in \Omega\}; \tag{2.1}$$

note that (1.3f) just means that $Z_{\rm ad} \neq \emptyset$. To do this, we take a suitable linear space of Carathéodory integrands containing all possible nonlinearities occurring in the problem $(\mathcal{P})$, e.g.

$$H := \mathrm{span}\Big\{ g_1 \cdot (h \circ u) + g_2 \cdot (f \circ u);\ g_1 \in C(\bar\Omega),\ g_2 \in L^2(\Omega;\mathbb{R}^n),\ u \in W^{1,2}(\Omega;\mathbb{R}^n)\Big\}, \tag{2.2}$$

where $[g_1 \cdot (h \circ u)](x,s) := g_1(x)h(x,u(x),s)$, and similarly $[g_2 \cdot (f \circ u)](x,s) := g_2(x) \cdot f(x,u(x),s)$. It is natural to equip H by

$$\|h\|_H := \inf_{\substack{\{a \in L^1(\Omega),\, c \in \mathbb{R};\ \forall x \in \Omega,\ s \in \mathbb{R}^m \\ |h(x,s)| \le a(x) + c|s|^q\ \}}} \|a\|_{L^1(\Omega)} + c\,, \tag{2.3}$$

which is a norm (see [30, Example 3.4.13]) making H separable (see [31, Lemma 1]). Then we imbed $L^q(\Omega;\mathbb{R}^m)$ (norm,weak*)-continuously into H^* by

$$i : z \mapsto \left(h \mapsto \int_\Omega h(x,z(x))\,\mathrm{d}x \right) \tag{2.4}$$

and define the set of the so-called generalized Young functionals by $Y_H^q(\Omega;\mathbb{R}^m) := \text{w*-cl}\, i(L^q(\Omega;\mathbb{R}^m))$. It is known (cf. [30]) that, as a consequence of (2.2) with (1.3a–e), $Y_H^q(\Omega;\mathbb{R}^m)$ is a convex locally (sequentially) compact envelope of $L^q(\Omega;\mathbb{R}^m)$. The set of admissible relaxed controls is then defined by

$$\bar Z_{\rm ad} := \text{w*-cl}\, i(Z_{\rm ad}). \tag{2.5}$$

Thanks to the special form (2.1), also the set $\bar Z_{\rm ad}$ is convex and locally compact in H^* if the weak* topology on H^* is considered.

We will need a continuous extension of the Nemytskiĭ mapping $z \mapsto f_0(x,z(x)) : L^q(\Omega;\mathbb{R}^m) \to L^1(\Omega;\mathbb{R}^{m_1})$ with some $f_0 : \Omega \times \mathbb{R}^m \to \mathbb{R}^{m_1}$ satisfying $|f_0(x,s)| \le a(x) + c|s|^q$ for some $a \in L^1(\Omega)$ and $c \in \mathbb{R}$. This extension is defined by

$$f_0 \bullet \eta \in \mathrm{rca}(\bar\Omega;\mathbb{R}^{m_1}) \cong C(\bar\Omega;\mathbb{R}^{m_1})^* : \tag{2.6}$$

$$\int_{\bar\Omega} g(x)[f_0 \bullet \eta](\mathrm{d}x) \equiv \langle f_0 \bullet \eta, g\rangle = \langle \eta, g \cdot f_0\rangle$$

for any $g \in C(\bar\Omega;\mathbb{R}^{m_1})$, where $[g \cdot f_0](x,s) := \sum_{k=1}^{m_1} g_k(x) f_{0k}(x,s)$. Note that, due to (2.2), $g \mapsto g \cdot f_0 : C(\bar\Omega;\mathbb{R}^{m_1}) \to H$ is continuous because $\|g \cdot f_0\|_H \le \|g\|_{C(\bar\Omega;\mathbb{R}^{m_1})}\|f_0\|_H$ and $\eta \mapsto f_0 \bullet \eta$ is linear; cf. [30, Example 3.6.3]. Obviously, $f_0 \bullet i(z) = f_0(z)$. We will use this extension for $f_0 := h \circ u$ (and $m_1 := 1$) and also for $f_0 := f \circ u$ (and $m_1 := n$). In the later case, we always have $(f \circ u) \bullet \eta \in L^2(\Omega)$ due to (1.3d).

Then the continuous extension of the original problem $(\mathcal{P})$ looks naturally as follows:

$$(\mathcal{RP}) \quad \begin{cases} \text{Minimize} & \bar J(\eta, u) := \int_{\bar\Omega} [(h \circ u) \bullet \eta](\mathrm{d}x) \\ \text{subject to} & (u \cdot \nabla)u - \mu\Delta u + \nabla p = (f \circ u) \bullet \eta\,, \\ & \operatorname{div} u = 0, \\ & u \in W_0^{1,2}(\Omega;\mathbb{R}^n),\ p \in L_0^2(\Omega),\ \eta \in \bar Z_{\mathrm{ad}} \subset Y_H^q(\Omega;\mathbb{R}^m). \end{cases}$$

Again, by (u,p) we understand a weak solution, which means analogously to (1.2) that $u \in W_{0,\mathrm{DIV}}^{1,2}(\Omega;\mathbb{R}^n)$, $p \in L_0^2(\Omega)$, and, for a given $\eta \in Y_H^q(\Omega;\mathbb{R}^m)$, the following identity holds:

$$((u \cdot \nabla)u, v) + \mu(\nabla u, \nabla v) = \langle \eta, v \cdot (f \circ u)\rangle \quad \forall v \in W_{0,\mathrm{DIV}}^{1,2}(\Omega;\mathbb{R}^n), \tag{2.7}$$

or, in accord with Remark 1, for all $v \in W_{0,\mathrm{DIV}}^{1,2}(\Omega;\mathbb{R}^n)$,

$$((u \cdot \nabla)u, v) + \mu(\nabla u, \nabla v) - (p, \operatorname{div} v) = \langle \eta, v \cdot (f \circ u)\rangle. \tag{2.8}$$

Next lemma shows that, in fact, the solution to (2.7) is regular and the Navier-Stokes system in $(\mathcal{RP})$ holds almost everywhere. Moreover, assuming certain condition on the smallness of a_2, $\tilde a_2$ and β occurring at (1.3d,e) we obtain the continuous dependence of u on η. Note that $W^{2,2}$-regularity of the velocity will be used in assumption (1.3b,e) (because β may have an arbitrary growth) and in other places, too.

Lemma 1. *Let Ω be a C^2-domain and (1.3d) hold. Let $u \in W_{0,\mathrm{DIV}}^{1,2}(\Omega;\mathbb{R}^n)$ denote the solution to the Navier-Stokes system with the relaxed control $\eta \in \bar Z_{\mathrm{ad}}$. Then*

$$\forall \eta \in \bar Z_{\mathrm{ad}}: \ \|u\|_{L^\infty(\Omega;\mathbb{R}^n)} \le c\,\|u\|_{W^{2,2}(\Omega;\mathbb{R}^n)} \le C \equiv C(\Omega, \|a_2\|_{L^2(\Omega)}). \tag{2.9}$$

Moreover, let also (1.3e) hold and u^1 and $u^2 \in W_{0,\mathrm{DIV}}^{1,2}(\Omega;\mathbb{R}^n)$ be two solutions to the Navier-Stokes system with relaxed controls η^1 and $\eta^2 \in \bar Z_{\mathrm{ad}}$, respectively. Then

$$\|u^1 - u^2\|_{W^{1,2}(\Omega;\mathbb{R}^n)} \le C_0\|\eta^1 - \eta^2\|_{H^*} \tag{2.10}$$

provided that a_2, β and c occurring at (1.3d,e) satisfy

$$k_1\|\tilde a_2\|_{L^2(\Omega)} + k_0^2\beta(C(\|a_2\|_{L^2(\Omega)})) + \frac{k_0k_1}{\mu}\|a_2\|_{L^2(\Omega)} < \mu\,, \tag{2.11}$$

where k_0 comes from the Poincaré inequality (cf. Remark 2) and k_1 comes from the inequality

$$\|\omega\|_{L^4(\Omega;\mathbb{R}^n)} \leq \sqrt{k_1}\|\nabla\omega\|_{L^2(\Omega;\mathbb{R}^{n\times n})}. \tag{2.12}$$

Proof. Due to (1.3d), the right-hand side $(f\circ y)\bullet\eta$ of the relaxed Navier-Stokes equation (2.7) is bounded in $L^2(\Omega,\mathbb{R}^n)$ if η ranges $Y_H^q(\Omega;\mathbb{R}^m)$. Then we can directly use nowdays standard regularity approach to the stationary Navier-Stokes equations (cf. [16], or [6]) to obtain (2.9).

Next, let u^1 and u^2 solve the identity (2.7) with $\eta := \eta^1$ and $\eta := \eta^2$, respectively. Subtracting these identities and putting $v := u^1 - u^2$ gives

$$\begin{aligned}\mu\|\nabla u^1-\nabla u^2\|^2_{L^2(\Omega;\mathbb{R}^{n\times n})} &= \langle\eta^1,(u^1-u^2)\cdot(f\circ u^1)\rangle-\langle\eta^2,(u^1-u^2)\cdot(f\circ u^2)\rangle\\ &\quad+((u^2\cdot\nabla)u^2,u^1-u^2)-((u^1\cdot\nabla)u^1,u^1-u^2)\\ &= \langle\eta^1,(u^1-u^2)\cdot[(f\circ u^1)-(f\circ u^2)]\rangle+\langle\eta^1-\eta^2,(u^1-u^2)\cdot(f\circ u^2)\rangle\\ &\quad+((u^2-u^1)\cdot\nabla)u^2,u^1-u^2)+((u^1\cdot\nabla)(u^2-u^1),u^1-u^2).\end{aligned}$$

Due to divergence-free constraint, the last term vanishes. The other terms are estimated by means of (1.3d,e). Thus, we obtain

$$\begin{aligned}\mu\|\nabla u^1-\nabla u^2\|^2_{L^2(\Omega;\mathbb{R}^{n\times n})} &\leq \|\tilde a_2\|_{L^2(\Omega)}\|u^1-u^2\|^2_{L^4(\Omega;\mathbb{R}^n)}\\ &\quad+\beta(\max(\|u^1\|_{L^\infty(\Omega;\mathbb{R}^n)},\|u^2\|_{L^\infty(\Omega;\mathbb{R}^n)}))\|u^1-u^2\|^2_{L^2(\Omega;\mathbb{R}^n)}\\ &\quad+\|\eta^1-\eta^2\|_{H^*}\|u^1-u^2\|_{L^2(\Omega;\mathbb{R}^n)}\|a_2\|_{L^2(\Omega)}\\ &\quad+\|\nabla u^2\|_{L^2(\Omega;\mathbb{R}^{n\times n})}\|u^1-u^2\|^2_{L^4(\Omega;\mathbb{R}^n)},\end{aligned}$$

where we used also the estimate

$$\begin{aligned}\left\|(u^1-u^2)\cdot(f\circ u^2)\right\|_H &\leq \left\||u^1-u^2|\,a_2\right\|_{L^1(\Omega)}\\ &\leq \|u^1-u^2\|_{L^2(\Omega;\mathbb{R}^n)}\|a_2\|_{L^2(\Omega)},\end{aligned}$$

which follows by the Hölder inequality from (1.3d) and (2.3). By (2.9), $\|u^i\|_{L^\infty(\Omega,\mathbb{R}^n)} \leq C = C(\|a_2\|_{L^2(\Omega)})$ for $i = 1,2$. As β is increasing we see that $\beta(\max(\|u^1\|_{L^\infty(\Omega;\mathbb{R}^n)},\|u^2\|_{L^\infty(\Omega;\mathbb{R}^n)})) \leq \beta(C)$. Using (2.12), Poincaré inequality (see (1.5)) and Young inequality, we have

$$\begin{aligned}\mu\|\nabla u^1-\nabla u^2\|^2_{L^2(\Omega;\mathbb{R}^{n\times n})} &\leq (k_1\|\tilde a_2\|_{L^2(\Omega)}\\ &\quad+k_0^2\beta(C))\|\nabla u^1-\nabla u^2\|^2_{L^2(\Omega;\mathbb{R}^{n\times n})}+\delta\|\nabla u^1-\nabla u^2\|^2_{L^2(\Omega;\mathbb{R}^{n\times n})}\\ &\quad+\frac{k_0^2\|a_2\|^2_{L^2(\Omega)}}{4\delta}\|\eta^1-\eta^2\|^2_{H^*}+\frac{k_0k_1}{\mu}\|a_2\|_{L^2(\Omega)}\|\nabla u^1-\nabla u^2\|^2_{L^2(\Omega;\mathbb{R}^{n\times n})}\end{aligned}$$

for arbitrary $\delta > 0$, which implies (2.10) if (2.11) holds. ∎

In particular, (2.11) ensures a unique response $u = u(\eta)$ for $(\mathcal{RP})$ to a given generalized control η. Hencefore, we can then put

$$\Phi(z) := J(z, u(z)) \quad \& \quad \bar{\Phi}(\eta) := \bar{J}(\eta, u(\eta)). \tag{2.13}$$

Although the unique response is desirable, some results, as e.g. Proposition 1 below, hold even without this assumption.

Now, we can state the existence of a solution to the relaxed problem and relations between this problem and the original one, see also [30] for such a kind of results.

Proposition 1. *Let Ω be a C^2-domain, and let (1.3) and (2.10) be satisfied. Then*

1. *$(\mathcal{RP})$ possesses at least one optimal control.*
2. *Moreover, $\inf(\mathcal{P}) = \min(\mathcal{RP})$.*
3. *For any optimal solution $(\eta, u, p) \in Y_H^q(\Omega; \mathbb{R}^m) \times W_0^{1,2}(\Omega; \mathbb{R}^n) \times L_0^2(\Omega)$ to $(\mathcal{RP})$ and any sequence $\{(z^k, u^k, p^k)\}_{k\in\mathbb{N}}$ such that $z^k \in Z_{\text{ad}}$ and (u^k, p^k) solves the Navier-Stokes system in $(\mathcal{P})$ with $z := z^k$ and $i(z^k) \to \eta$ weakly* in H^*, it holds $\Phi(z^k) \to \inf(\mathcal{P})$, so that this sequence is minimizing for $(\mathcal{P})$.*
4. *Conversely, having a minimizing sequence $\{(z^k, u^k, p^k)\}_{k\in\mathbb{N}}$ for $(\mathcal{P})$, there exists a subsequence of $\{(i(z^k), u^k, p^k)\}_{k\in\mathbb{N}}$ converging weakly* in $H^* \times W^{1,2}(\Omega; \mathbb{R}^n) \times L^2(\Omega)$ and the limit of any such subsequence solves $(\mathcal{RP})$.*

Sketch of the proof. In accord with Remark 1, we will omit p's in this proof. Always, there is a minimizing sequence $\{(z^k, u^k)\}_{k\in\mathbb{N}}$ for $(\mathcal{P})$. In view of (1.2), it holds

$$((u^k \cdot \nabla)u^k, v) + \mu(\nabla u^k, \nabla v) = (f(u^k, z^k), v) \tag{2.14}$$

for all $v \in W_{0,\text{DIV}}^{1,2}(\Omega; \mathbb{R}^n)$. By (1.3a,d,f), $\inf(\mathcal{P}) < +\infty$. Then, by (1.3c), the following apriori estimate holds:

$$\limsup_{k\to\infty} \int_\Omega c_0 |z^k|^q \mathrm{d}x \le \lim_{k\to\infty} \int_\Omega h(x, u^k(x), z^k(x)) \mathrm{d}x = \inf(\mathcal{P}) < +\infty. \tag{2.15}$$

This implies $\{z^k\}_{k\in\mathbb{N}}$ bounded in $L^q(\Omega; \mathbb{R}^m)$. Then $i(z^k)$ converges weakly* to some $\eta \in \bar{Z}_{\text{ad}}$, if a suitable subsequence is selected. Due to (1.5) and the Poincaré inequality, $\{u^k\}_{k\in\mathbb{N}}$ is bounded in $W^{1,2}(\Omega; \mathbb{R}^n)$. Thus, taking another subsequence if necessary, we obtain that $u^k \to$

u weakly in $W_0^{1,2}(\Omega;\mathbb{R}^n)$, which implies that $u^k \to u$ strongly in $L^4(\Omega;\mathbb{R}^n)$. Thus, for any $v \in W^{1,2}_{0,\mathrm{DIV}}(\Omega;\mathbb{R}^n)$,

$$((u^k\cdot\nabla)u^k, v) = -((u^k\cdot\nabla)v, u^k) \to -((u\cdot\nabla)v, u) = ((u\cdot\nabla)u, v)\,.$$

Moreover, by (1.3d), $\{f(u^k,z^k)\}_{k\in\mathbb{N}}$ is bounded in $L^2(\Omega;\mathbb{R}^n)$ and, by using also (1.3e), $f(u^k,z^k)\to(f\circ u)\bullet\eta$ weakly in $L^2(\Omega;\mathbb{R}^n)$; cf. [30, Lemma 3.6.7]. Similarly, by (1.3a,b) we can see that $\{h(u^k,z^k)\}_{k\in\mathbb{N}}$ is bounded in $L^1(\Omega)$ and converges to $(h\circ u)\bullet\eta$ weakly* in $\mathrm{rca}(\bar\Omega)$.

Altogether, it enables us to pass to the limit in the integral identity (2.14), which gives just (2.7). Thus u satisfies (2.7), i.e. (η,u) is admissible for $(\mathcal{RP})$.

Moreover,

$$\begin{aligned}\lim_{k\to\infty} J(z^k,u^k) &= \lim_{k\to\infty}\int_\Omega h(x,u^k(x),z^k(x))\,\mathrm{d}x\\ &= \int_{\bar\Omega}[(h\circ u)\bullet\eta](\mathrm{d}x) = \bar J(\eta,u).\end{aligned} \tag{2.16}$$

As $J(z^k,u^k)\to\inf(\mathcal{P})$, we showed $\bar J(\eta,u) = \inf(\mathcal{P})$ so that certainly $\inf(\mathcal{RP})\le\inf(\mathcal{P})$.

Taking a minimizing sequence $\{(\eta^k,u^k)\}_{k\in\mathbb{N}}$ for $(\mathcal{RP})$, we can prove similarly as above that $\{\eta^k\}_{k\in\mathbb{N}}$ converges (after taking possibly a subsequence) weakly* in H^* and the limit solves $(\mathcal{RP})$, as claimed in 1.

Taking (η,u) a solution to $(\mathcal{RP})$, there is a sequence $\{z^k\}_{k\in\mathbb{N}}\subset Z_{\mathrm{ad}}$ bounded in $L^q(\Omega;\mathbb{R}^m)$ such that w*-$\lim_{k\to\infty} i(z^k) = \eta$. Then one can prove similarly as above that $\Phi(z^k) = \bar\Phi(i(z^k)) \to \bar\Phi(\eta) = \min(\mathcal{RP})$, so that $\min(\mathcal{RP})\ge\inf(\mathcal{P})$.

Thus 2 was proved, justifying also the points 3–4 as a side effect. ∎

Remark 3. If Ω were only a Lipschitz domain, we do not know whether (2.9) holds; then the growth of β in (1.3) would have to be specified appropriately.

Remark 4. As H is separable and $h(x,r,\cdot)$ has a q-growth while $f(x,r,\cdot)$ is bounded (see (1.3c,d)), by using [30, Lemmas 4.2.3–4] one can see that any optimal relaxed control $\eta\in Y_H^q(\Omega;\mathbb{R}^m)$ is q-nonconcetrating in the sense that there is a sequence of controls $\{z^k\}_{k\in\mathbb{N}}$ such that w*-$\lim_{k\to\infty} i(z^k) = \eta$ and the set $\{|z^k|^q;\ k\in\mathbb{N}\}$ is relatively weakly compact in $L^1(\Omega)$. Every such η has a so-called L^q-Young-measure representation $\nu\in\mathcal{Y}^q(\Omega;\mathbb{R}^m)$ (possibly not determined uniquely) satisfying

$$\forall h\in H: \qquad \langle\eta,h\rangle = \int_\Omega\int_{\mathbb{R}^m} h(x,s)\nu_x(\mathrm{d}s)\,\mathrm{d}x, \tag{2.17}$$

where $\mathcal{Y}^q(\Omega;\mathbb{R}^m)$ denotes the set of all L^q-Young measures, i.e. weakly measurable families $\nu := \{\nu_x\}_{x\in\Omega}$ of probability Radon measures on $\mathbb{R}^m$ satisfying $\int_\Omega \int_{\mathbb{R}^m} |s|^q \nu_x(\mathrm{d}s)\,\mathrm{d}x < +\infty$; the adjective "weakly measurable" means that for any $v \in C_0(\mathbb{R}^m)$ the mapping $\Omega \to \mathbb{R}$: $x \mapsto \langle \nu_x, v\rangle := \int_{\mathbb{R}^m} v(s)\nu_x(\mathrm{d}s)$ is measurable in the usual sense.

If S is measurable and closed-valued, the relaxed problem $(\mathcal{RP})$ can be rewritten in terms of L^q-Young measure into the following form:

$$
(\mathcal{RP}') \quad \left\{\begin{array}{ll}
\text{Minimize} & \bar{J}(\eta, u) := \displaystyle\int_\Omega \int_{\mathbb{R}^m} h(x, u(x), s)\nu_x(\mathrm{d}s)(\mathrm{d}x) \\
\text{subject to} & (u \cdot \nabla)u - \mu\Delta u + \nabla p = \displaystyle\int_{\mathbb{R}^m} f(x, u(x), s)\nu_x(\mathrm{d}s) , \\
 & \operatorname{div} u = 0, \\
 & \operatorname{supp}(\nu_x) \subset S(x) \quad \text{for a.a. } x \in \Omega, \\
 & u \in W_0^{1,2}(\Omega; \mathbb{R}^n), \quad p \in L_0^2(\Omega), \quad \nu \in \mathcal{Y}^q(\Omega; \mathbb{R}^m).
\end{array}\right.
$$

For an extension in terms of classical relaxed controls (i.e. L^∞-Young measures) we refer also [8, 12, 13, 33, 34].

An example for usage of $(\mathcal{RP}')$ is the following existence result.

Proposition 2. *Let Ω be a C^2-domain, let (1.3) and (2.11) hold, let S be measurable and closed-valued. Denote by $h \times f$ the mapping of $\Omega \times \mathbb{R}^n \times \mathbb{R}^n$ onto $\mathbb{R} \times \mathbb{R}^n$ such that $[h \times f](x, r, s) = (h(x, r, s), f(x, r, s))$. Assume that for all $r \in \mathbb{R}^n$ and a.a. $x \in \Omega$*

$$\overline{\mathrm{co}}\,[h \times f](x, r, S(x)) \subset Q(x, r), \tag{2.18}$$

where the "orientor field" Q is defined by

$$Q(x, r) := \{(a, y) \in \mathbb{R} \times \mathbb{R}^n;\ \ a \geq h(x, r, s),\ \ y = f(x, r, s),\ \ s \in S(x)\}. \tag{2.19}$$

Then $(\mathcal{P})$ has a solution.

Sketch of the proof. (For more details see [31, Lemma 2].) Take a solution η which does exist by Proposition 1(i). By Remark 4, η is q-nonconcentrating and (every) its L^q-Young-measure representation ν solves $(\mathcal{RP}')$. For any $x \in \Omega$ for which $\int_{\mathbb{R}^m} |s|^q \nu_x(\mathrm{d}s) < \infty$ we have

$$\int_{\mathbb{R}^m} [h \times f](x, u(x), s)\nu_x(\mathrm{d}s) \in \overline{\mathrm{co}}\,[h \times f](x, u(x), S(x)) \subset Q(x, u(x)) , \tag{2.20}$$

where we used also (2.18). Let us put

$$R(x) := \Big\{ s \in S(x);\ h(x, u(x), s) \leq \int_{\mathbb{R}^m} h(x, u(x), \sigma)\nu_x(\mathrm{d}\sigma), \quad f(x, u(x), s) = \int_{\mathbb{R}^m} f(x, u(x), \sigma)\nu_x(\mathrm{d}\sigma) \Big\} . \tag{2.21}$$

By (2.19), for any $(a,y) \in Q(x,u(x))$ there is $s \in S(x)$ such that $a \geq h(x,u(x),s)$ and $y = f(x,u(x),s)$. Hence, for the particular choice

$$(a,y) = (a(x),y(x)) := \int_{\mathbb{R}^m} [h \times f](x,u(x),s)\nu_x(\mathrm{d}s), \tag{2.22}$$

the inclusion (2.20) implies that $a(x) \geq h(x,u(x),s)$ and $y(x) = f(x,u(x),s)$ for some $s \in S(x)$, hence $R(x) \neq \emptyset$. Besides, the multi-valued mapping $R : \Omega \rightrightarrows \mathbb{R}^m$ defined by (2.21) is measurable and closed-valued, thus it possesses a measurable selection $z(x) \in R(x)$. In particular, $z(x) \in S(x)$. Moreover, in view of (3.1) with (3.3),

$$f(x,u(x),z(x)) = y(x) = \int_{\mathbb{R}^m} f(x,u(x),s)\nu_x(\mathrm{d}s) \tag{2.23}$$

for a.a. $x \in \Omega$, so that z and ν give the same response u, i.e. $u(z) = u(\nu) := u(\eta)$ with η given by (2.17). Hence the pair (z,u) is admissible for $(\mathcal{P})$. Moreover, by using also Proposition 1(ii), we get $\int_\Omega h(x,u(x),z(x))\,\mathrm{d}x \leq \int_\Omega a(x)\,\mathrm{d}x = \int_\Omega \int_{\mathbb{R}^m} h(x,u(x),s)\nu_x(\mathrm{d}s)\mathrm{d}x = \min(\mathcal{RP}') = \min(\mathcal{RP}) = \inf(\mathcal{P})$. In particular, the coercivity (1.3c) implies $c_0 \int_\Omega |z(x)|^q\,\mathrm{d}x \leq \int_\Omega h(x,u(x),z(x))\,\mathrm{d}x \leq \inf(\mathcal{P}) < +\infty$; note that (1.3a,d,f) makes $\inf(\mathcal{P})$ indeed finite. Therefore, $z \in L^q(\Omega;\mathbb{R}^m)$, which completes the proof that z solves $(\mathcal{P})$. ∎

Remark 5. Note that (2.18) is fulfilled if for example $Q(x,r)$ is convex and compact. This is ensured if $S(x)$ is compact for a.a. x (as h, f are Carathéodory functions) and $Q(x,r)$ is convex, which is a slightly generalized variant of the Filippov–Roxin condition. A very special case that can be however handled by a direct method occurs if $S(x)$ is convex, $f(x,r,\cdot)$ is affine and $h(x,r,\cdot)$ is convex on $S(x)$ for a.a. $x \in \Omega$; cf. e.g. [19] for such a type of existence result.

3. MAXIMUM PRINCIPLE

In this section we formulate first-order necessary optimality conditions for $(\mathcal{RP})$ in terms of a maximum principle. For maximum principle for Navier-Stokes optimal control problems, we refer also to [4, 13, 33] or for other type of first-order optimality conditions also to [3, 5, 7, 19, 20, 22, 25, 29, 35]. To give as simple proofs as possible, we confine ourselves to the special case

$$h(x,r,s) := \frac{1}{2}|r - u_{\mathrm{d}}(x)|^2 + \hat{h}(x,s)\,, \qquad f(x,r,s) := \hat{f}(x,s)\,, \tag{3.1}$$

where $u_{\mathrm{d}} \in L^{q_0}(\Omega;\mathbb{R}^n)$ is a desired (given) velocity profile, and

$$q_0 > n. \tag{3.2}$$

The first term in (3.1) realizes the so-called flow tracking often used in literature, cf. [5, 2, 18, 19, 22, 23, 25, 29].

To formulate the maximum principle we will need the so-called adjoint state $w \in W^{1,2}_{0,\mathrm{DIV}}(\Omega;\mathbb{R}^n)$ satisfying the integral identity

$$\mu(\nabla w,\nabla v)-((u\cdot\nabla)w,v)+(w,(v\cdot\nabla)u) = (u_{\mathrm{d}}-u,v) \quad \forall v\in W^{1,2}_{0,\mathrm{DIV}}(\Omega;\mathbb{R}^n)\,. \tag{3.3}$$

It is worth mentioning that w in (3.3) is a weak solution of the adjoint system to the linearized Navier-Stokes equations, i.e.

$$-\mu\Delta w_i+\frac{\partial\pi}{\partial x_i} = (u_{\mathrm{d}}-u)_i-\sum_{k=1}^{n}\Big(w_k\frac{\partial u_k}{\partial x_i}-u_k\frac{\partial w_k}{\partial x_i}\Big)\,, \quad i=1,\dots,n, \tag{3.4a}$$

$$\operatorname{div} w = 0\,, \tag{3.4b}$$

where $\pi \in L^2_0(\Omega)$, cf. Remark 1. The following regularity of the adjoint state, higher than e.g. in [19, Theorem 3.2], will be essential for (3.18) below. Let us remark that, in context of fluid control, condition (3.5) was already used by Bilič [1].

Lemma 2. *Let Ω be a C^2-domain, let (1.3) with (3.1) with $u_{\mathrm{d}} \in L^{q_0}(\Omega;\mathbb{R}^n)$ hold, and let a_2 from (1.3d) satisfy*

$$\frac{k_0k_1}{\mu^2}\|a_2\|_{L^2(\Omega)} < 1. \tag{3.5}$$

Then there is C_1 depending on Ω, μ and $\|a_2\|_{L^2(\Omega)}$ such that for arbitrary small ϵ

$$\forall \eta \in \bar{Z}_{\mathrm{ad}}: \quad \|\nabla w(\eta)\|_{L^\infty(\Omega;\mathbb{R}^{n\times n})} \le C_1\|u-u_{\mathrm{d}}\|_{L^{n+\epsilon}(\Omega;\mathbb{R}^n)}\,, \tag{3.6}$$

where $w = w(\eta) \in W^{1,2}_{0,\mathrm{DIV}}(\Omega;\mathbb{R}^n)$ solves (3.3) with $u = u(\eta)$.

Proof. Let us first observe that (3.5) implies the existence of C depending on the above mentioned quantities such that

$$\|\nabla w\|_{L^2(\Omega;\mathbb{R}^{n\times n})} \le C\,. \tag{3.7}$$

Indeed, testing in (3.3) by $v := w$, and using the Hölder inequality and (1.5) we obtain (notice that $((u\cdot\nabla)w,w)=0$)

$$\begin{aligned}\mu\|\nabla w\|^2_{L^2(\Omega;\mathbb{R}^{n\times n})} &= ((u\cdot\nabla)w,w)-(w,(w\cdot\nabla)u)+(u_{\mathrm{d}}-u,w)\\ &\le \|w\|^2_{L^4(\Omega;\mathbb{R}^n)}\|\nabla u\|_{L^2(\Omega;\mathbb{R}^{n\times n})}+\|u-u_{\mathrm{d}}\|_{L^2(\Omega;\mathbb{R}^n)}\|w\|_{L^2(\Omega;\mathbb{R}^n)}\\ &\le \frac{k_0k_1}{\mu}\|a_2\|_{L^2(\Omega)}\|\nabla w\|^2_{L^2(\Omega;\mathbb{R}^n)}+k_0\|u-u_{\mathrm{d}}\|_{L^2(\Omega;\mathbb{R}^n)}\|\nabla w\|_{L^2(\Omega;\mathbb{R}^{n\times n})}.\end{aligned}$$

Therefore

$$\left(\mu - \frac{k_0 k_1}{\mu}\|a_2\|_{L^2(\Omega)}\right)\|\nabla w\|_{L^2(\Omega;\mathbb{R}^{n\times n})} \leq k_0\|u - u_{\rm d}\|_{L^2(\Omega;\mathbb{R}^n)}\,, \qquad (3.8)$$

and (3.7) follows due to (3.5).

Now, using the facts that $u \in W^{2,2}(\Omega;\mathbb{R}^n)$ and $w \in W^{1,2}(\Omega;\mathbb{R}^n)$, we can view (3.4) as the Stokes system with the right-hand side belonging at least to $L^2(\Omega;\mathbb{R}^n)$ (the restriction comes from the term $\sum_{k=1}^n u_k\frac{\partial w_k}{\partial x_i}$). Then applying standard L^2-regularity result for the Stokes system one obtains $w \in W^{2,2}(\Omega;\mathbb{R}^n)$ with

$$\|w\|_{W^{2,2}(\Omega;\mathbb{R}^n)} \leq c\|u - u_{\rm d}\|_{L^2(\Omega;\mathbb{R}^n)}\,.$$

However, using this we easily observe that the right-hand side of (3.4a) belongs now to $L^{n+\epsilon}(\Omega;\mathbb{R}^n)$, $\epsilon > 0$, $\epsilon \leq \min\left(q_0, \frac{2n}{n-2}\right)$. The L^q-regularity theory for the Stokes system (cf. [16] for example) then implies

$$\|w\|_{W^{2,n+\epsilon}(\Omega;\mathbb{R}^n)} \leq \tilde{C}_1(\Omega,\mu,\|a_2\|_{L^2(\Omega)})\,\|u - u_{\rm d}\|_{L^{n+\epsilon}(\Omega;\mathbb{R}^n)}\,. \qquad (3.9)$$

The assertion then follows from the imbedding $W^{2,n+\epsilon}(\Omega)$ into $W^{1,\infty}(\Omega)$. Then C_1 is $\tilde{C}_1$ multiplied by the norm of the imbedding $W^{2,n+\epsilon}(\Omega) \subset W^{1,\infty}(\Omega)$. ∎

Lemma 3. *Defining the so-called Hamiltonian* $\mathcal{H}_w : \Omega \times \mathbb{R}^m \to \mathbb{R}$ *by*

$$\mathcal{H}_w(x,s) := w(x)\cdot \hat{f}(x,s) - \hat{h}(x,s)\,, \qquad (3.10)$$

the following increment formula holds

$$\bar{\Phi}(\tilde{\eta}) - \bar{\Phi}(\eta) + \int_{\bar{\Omega}} \mathcal{H}_w \bullet (\tilde{\eta}-\eta)\,{\rm d}x = \int_\Omega \frac{1}{2}|\tilde{u}-u|^2{\rm d}x - \Big(((\tilde{u}-u)\cdot\nabla)w, \tilde{u}-u\Big) \qquad (3.11)$$

provided $\eta, \tilde{\eta} \in Y_H^q(\Omega;\mathbb{R}^m)$, $u = u(\eta)$, $\tilde{u} = u(\tilde{\eta})$, *and the adjoint state* $w \in W^{1,2}_{0,\rm DIV}(\Omega;\mathbb{R}^n)$ *solves (3.3).*

Proof. We use successively the formula for the Hamiltonian (3.10), the weak formulation (2.7) both for $u = u(\eta)$ and for $\tilde{u} = u(\tilde{\eta})$ with $v := w$, the adjoint equation (3.3) with $v := \tilde{u} - u$, the algebraic identity $\frac{1}{2}|\tilde{u} - u_{\rm d}|^2 - \frac{1}{2}|u - u_{\rm d}|^2 - (u - u_{\rm d})\cdot(\tilde{u} - u) = \frac{1}{2}|\tilde{u} - u|^2$, and the Green theorem. Thus we can obtain:

$$\bar{\Phi}(\tilde{\eta}) - \bar{\Phi}(\eta) + \int_{\bar{\Omega}} \mathcal{H}_w \bullet (\tilde{\eta}-\eta){\rm d}x = \frac{1}{2}\int_\Omega |\tilde{u}-u_{\rm d}|^2 - |u-u_{\rm d}|^2{\rm d}x + \langle \tilde{\eta}-\eta, w\cdot\hat{f}\rangle$$

$$= \frac{1}{2}\int_\Omega |\tilde{u}-u_{\rm d}|^2 - |u-u_{\rm d}|^2{\rm d}x + \mu(\nabla\tilde{u}-\nabla u, \nabla w) - ((u\cdot\nabla)u, w) + ((\tilde{u}\cdot\nabla)\tilde{u}, w)$$

$$= \int_\Omega \frac{1}{2}|\tilde{u}-u_{\rm d}|^2 - \frac{1}{2}|u-u_{\rm d}|^2 + (u_{\rm d}-u)\cdot(\tilde{u}-u)\,{\rm d}x - ((u\cdot\nabla)u, w)$$

$$+((\tilde{u}\cdot\nabla)\tilde{u},w)+((u\cdot\nabla)w,\tilde{u}-u)+(w,((u-\tilde{u})\cdot\nabla)u)$$
$$=\int_\Omega\frac{1}{2}|\tilde{u}-u|^2\mathrm{d}x+((\tilde{u}\cdot\nabla)\tilde{u},w)-(w,(\tilde{u}\cdot\nabla)u)+((u\cdot\nabla)w,\tilde{u}-u)$$
$$=\int_\Omega\frac{1}{2}|\tilde{u}-u|^2\mathrm{d}x-\Big(((\tilde{u}-u)\cdot\nabla)w,\tilde{u}-u\Big)\,. \qquad \blacksquare$$

As a simple consequence we can now get the integral maximum principle for the relaxed problem as the first-order necessary optimality condition.

Proposition 3. *Let the assumptions of Lemma 2 hold, and let* $(\eta,u)\in Y^q_H(\Omega;\mathbb{R}^m)\times W^{1,2}_0(\Omega;\mathbb{R}^n)$ *be an optimal solution for* $(\mathcal{RP})$. *Then there is* $w\in W^{1,2}_{0,\mathrm{DIV}}(\Omega;\mathbb{R}^n)$ *solving (3.3) such that, for the Hamiltonian* $\mathcal{H}_w$ *from (3.10), the following maximum principle holds:*

$$\int_{\bar\Omega}\mathcal{H}_w\bullet\eta\,\mathrm{d}x=\sup_{z\in Z_{\mathrm{ad}}}\int_\Omega\mathcal{H}_w(x,z(x))\,\mathrm{d}x\,. \tag{3.12}$$

Sketch of the proof. Let us calculate the directional derivative of $\bar\Phi$, which is by definition:

$$\mathrm{D}\bar\Phi(\eta,\tilde\eta-\eta):=\lim_{\varepsilon\searrow0}\frac{\bar\Phi(\eta+\varepsilon(\tilde\eta-\eta))-\bar\Phi(\eta)}{\varepsilon}=\int_{\bar\Omega}(\hat{h}-w\cdot\hat{f})\bullet(\tilde\eta-\eta)\mathrm{d}x$$
$$+\lim_{\varepsilon\searrow0}\frac{1}{\varepsilon}\int_\Omega\frac{1}{2}|u_\varepsilon-u|^2-(((u_\varepsilon-u)\cdot\nabla)w)\cdot(u_\varepsilon-u)\,\mathrm{d}x$$

where we used also (3.11) with $\tilde{u}:=u_\varepsilon$ denoting the solution of the relaxed Navier-Stokes equation (2.7) but with $\eta_\varepsilon:=\eta+\varepsilon(\tilde\eta-\eta)$ in place of η. Let us agree to consider only $\tilde\eta\in\bar Z_{\mathrm{ad}}$ and $0<\varepsilon\le1$, which will be sufficient for usage in (3.14) and which will guarantee $\eta_\varepsilon\in\bar Z_{\mathrm{ad}}$.

Note that (3.5) now implies (2.11) because (1.3e) now holds with with $\tilde{a}_2=0$ and $\beta=0$. Hencefore we have Lemma 1 at our disposal, so that (2.10) gives

$$\|u_\varepsilon-u\|_{L^2(\Omega;\mathbb{R}^n)}\le\|u_\varepsilon-u\|_{W^{1,2}(\Omega;\mathbb{R}^n)}\le C_0\|\eta_\varepsilon-\eta\|_{H^*}=\varepsilon C_0\|\tilde\eta-\eta\|_{H^*}\,.$$

This yields $\int_\Omega|u_\varepsilon-u|^2\mathrm{d}x=\mathcal{O}(\varepsilon^2)$. Similarly, the term

$$\left|\int_\Omega(((u_\varepsilon-u)\cdot\nabla)w)\cdot(u_\varepsilon-u)\,\mathrm{d}x\right|\le\|\nabla w\|_{L^\infty(\Omega;\mathbb{R}^{n\times n})}\|u_\varepsilon-u\|^2_{L^2(\Omega;\mathbb{R}^n)}$$

is $\mathcal{O}(\varepsilon^2)$ because ∇w is bounded in $L^\infty(\Omega;\mathbb{R}^{n\times n})$ due to (3.6) with (1.5). Altogether, we have proved the expression for the directional derivative, which apparently depends linearly and continuously on the direction as soon as $\tilde{\eta} \in \bar{Z}_{\mathrm{ad}}$. Thus $\bar{\Phi}$ has a Gâteaux differential $\nabla\bar{\Phi}$ given by

$$\langle \nabla\bar{\Phi}(\eta), \tilde{\eta} - \eta \rangle = \langle \eta - \tilde{\eta}, \mathcal{H}_w \rangle, \quad w \text{ solves (3.3) with } u = u(\eta). \tag{3.13}$$

Then (η, u) solves $(\mathcal{P})$, which means that η minimizes $\bar{\Phi}$ on $\bar{Z}_{\mathrm{ad}}$, implies that $-\nabla\bar{\Phi}(\eta)$ belongs to the normal cone to the convex set $\bar{Z}_{\mathrm{ad}}$ at η, which is just equivalent to

$$\forall \tilde{\eta} \in \bar{Z}_{\mathrm{ad}}: \qquad \langle \nabla\bar{\Phi}(\eta), \tilde{\eta} - \eta \rangle \le 0\,. \tag{3.14}$$

This means precisely

$$\begin{aligned} \int_{\bar{\Omega}} \mathcal{H}_w \bullet \eta \,\mathrm{d}x &= \langle -\nabla\bar{\Phi}(\eta), \eta \rangle = \max_{\tilde{\eta} \in \bar{Z}_{\mathrm{ad}}} \langle -\nabla\bar{\Phi}(\eta), \tilde{\eta} \rangle \\ &= \max_{\tilde{\eta} \in \bar{Z}_{\mathrm{ad}}} \int_{\bar{\Omega}} \mathcal{H}_w \bullet \tilde{\eta} \,\mathrm{d}x = \sup_{z \in Z_{\mathrm{ad}}} \int_{\Omega} \mathcal{H}_w(x, z(x)) \,\mathrm{d}x\,. \end{aligned} \tag{3.15}$$

■

As in [30, Theorem 4.2.2], one can modify the integral maximum principle (3.12) to a pointwise (sometimes called Pontryagin's) maximum principle (cf. also [4, 13, 33]):

Corollary 1. *Let the assumptions of Lemma 2 hold, and let S be measurable and closed-valued. Then for any solution (η, u) to $(\mathcal{RP})$ it holds*

$$[\mathcal{H}_w \bullet \eta](x) = \max_{s \in S(x)} \mathcal{H}_w(x, s) \qquad \text{for a.a. } x \in \Omega \tag{3.16}$$

with the Hamiltonian $\mathcal{H}_w$ from (3.10) with $w \in W^{1,2}_{0,\mathrm{DIV}}(\Omega; \mathbb{R}^n)$ solving (3.3).

Having an L^q-Young-measure representation ν of an optimal relaxed control η, (3.16) says that ν_x is supported on the set where $\mathcal{H}_w(x, \cdot)$ attains its maximum. By Lemma 2, we have in particular an L^∞-regularity of the multiplier w, which then gives the following assertion:

Corollary 2. *If $\hat{h}$ and $\hat{f}$ are independent of $x \in \Omega$, then any optimal relaxed control η for $(\mathcal{RP})$ has an L^∞-Young measure representation ν, i.e. ν_x is compactly supported independently of $x \in \Omega$.*

The following assertion states an important global property of $\bar{\Phi}$ if the Reynolds number is small, see Remark 6 below.

Lemma 4. *Let the assumptions of Lemma 2 hold, and let a_2 from (1.3d) satisfy (3.5) and also*

$$C_{n,q_0} \frac{k_0}{\mu} \|a_2\|_{L^2(\Omega)} + c\|u_{\mathrm{d}}\|_{L^{q_0}(\Omega;\mathbb{R}^n)} \le \frac{1}{2C_1} \tag{3.17}$$

with $C_1 = C_1(\Omega, \mu, \|a_2\|_{L^2(\Omega)})$ from Lemma 2 and C_{n,q_0} denoting the norm of the imbedding $W^{1,2}(\Omega) \subset L^{q_0}(\Omega)$. Then the extended cost functional $\bar{\Phi} : \bar{Z}_{\rm ad} \to \mathbb{R}$ is convex with respect to the geometry of the space H^*.

Proof. Using (1.5), Lemma 2 and (3.17), the second-order term in (3.11) in nonnegative because of the following estimate:

$$\begin{aligned}
&\int_\Omega \frac{1}{2}|\tilde{u} - u|^2 \mathrm{d}x - \Big(((\tilde{u} - u) \cdot \nabla)w, \tilde{u} - u\Big) \qquad (3.18)\\
&\geq \left(\frac{1}{2} - \|\nabla w\|_{L^\infty(\Omega;\mathbb{R}^{n\times n})}\right) \|\tilde{u} - u\|^2_{L^2(\Omega;\mathbb{R}^n)}\\
&\geq \left(\frac{1}{2} - C_1\|u - u_{\rm d}\|_{L^{n+\epsilon}(\Omega;\mathbb{R}^n)}\right) \|\tilde{u} - u\|^2_{L^2(\Omega;\mathbb{R}^n)}\\
&\geq \left(\frac{1}{2} - C_1\|u\|_{L^{n+\epsilon}(\Omega;\mathbb{R}^n)} - C_1\|u_{\rm d}\|_{L^{n+\epsilon}(\Omega;\mathbb{R}^n)}\right) \|\tilde{u} - u\|^2_{L^2(\Omega;\mathbb{R}^n)}\\
&\geq \left(\frac{1}{2} - C_1 C_{n,q_0}\frac{k_0}{\mu}\|a_2\|_{L^2(\Omega)} - C_1\ c\|u_{\rm d}\|_{L^{q_0}(\Omega;\mathbb{R}^n)}\right) \|\tilde{u} - u\|^2_{L^2(\Omega;\mathbb{R}^n)}.
\end{aligned}$$

By (3.17) and the proof of Proposition 3, we have just obtained $\bar{\Phi}(\tilde{\eta}) - \bar{\Phi}(\eta) - [\nabla\bar{\Phi}(\eta)](\tilde{\eta} - \eta) \geq 0$ and, replacing the roles of η and $\tilde{\eta}$, also $\bar{\Phi}(\eta) - \bar{\Phi}(\tilde{\eta}) - [\nabla\bar{\Phi}(\tilde{\eta})](\eta - \tilde{\eta}) \geq 0$. Therefore, by addition, we obtain $[\nabla\bar{\Phi}(\eta) - \nabla\bar{\Phi}(\tilde{\eta})](\tilde{\eta} - \eta) \geq 0$, which just says that $\nabla\bar{\Phi}$ is monotone, from which the convexity of $\bar{\Phi}$ follows by well-known arguments. ∎

We are now ready to state also the sufficiency of the maximum principle (3.12).

Proposition 4. *Let condition (3.17) be satisfy. Then the maximum principle consisting of (3.3), (3.10), and (3.12) is sufficient in the sense that, having a triple $(\eta, u, w) \in \bar{Z}_{\rm ad} \times W^{1,2}_{0,\rm DIV}(\Omega; \mathbb{R}^n)^2$ such that u solves the Navier-Stokes system (2.7), and w solves the adjoint problem to the linearized Navier-Stokes system (3.3), and the maximum principle (3.12) holds, then (η, u) is the optimal solution to $(\mathcal{RP})$.*

Proof. By Lemma 4, $\bar{\Phi}$ is convex, so that (3.14) is also a sufficient optimality condition. Yet, (3.14) is equivalent with (3.15). ∎

Remark 6. The constant C_1 from (3.6) depends on μ as $\mathcal{O}(\mu^{-1})$. Then, for given $u_{\rm d}$ and a_2, the condition (3.17) requires μ sufficiently large. Hencefore, (3.17) needs a sufficiently small Reynolds number. As the fluid (and its viscosity μ) is usually given, we rather need a sufficiently small driving force and desired velocity profile, as expressed in (3.17), indeed.

Acknowledgments

This research has been partly covered by the grant 201/96/0228 (Grant agency of the Czech Republic). Moreover, the second author has been partly supported also by the grant A 107 5707 (Grant agency of the Academy of Sciences of the Czech Republic).

References

[1] BILIČ, N. (1985). Approximation of optimal distributed control problem for Navier-Stokes equations. In: *Numerical Methods and Approx. Th.*, Univ. Novi Sad, Novi Sad, 177–185.

[2] BURKARD, J. AND PETERSON, J. (1995). Control of steady incompressible 2D channel flow. In: *Flow Control* (M.D.Gunzburger, ed.), IMA Vol. Math. Appl. **68**, Springer, New York, 111–126.

[3] CASAS, E. (1995). Optimality conditions for some control problems of turbulent flow. In: *Flow Control* (M.D.Gunzburger, ed.), IMA Vol. Math. Appl. **68**, Springer, New York, 127–147.

[4] CHEBOTAREV, A.YU. (1993). Principle of maximum in the problem of boundary control of flow of viscous fluid (in Russian). *Sib. Mat. J.*, 34:189–197.

[5] CHOI, H., HINZE, M AND KUNISH, K. Instantaneous control of backward-facing-step flows. (Preprint No.571/1997, TU Berlin.) To appear in: *Theoret. Comput. Fluid Mechanics.*

[6] CONSTANTIN, P. AND FOIAS, P. (1989). *Navier-Stokes equations.* The University of Chicago Press.

[7] DESAI, M.C. AND ITO, K. (1994). Optimal control of Navier-Stokes equations. *SIAM J. Control Optim.*, 32:1428–1446.

[8] FATTORINI, H.O. (1995). Optimal chattering control for viscous flow. *Nonlinear Anal., Th. Meth. Appl.*, 25:763–797.

[9] FATTORINI, H.O. (1997). Robustness and convergence of suboptimal controls in distributed parameter systems. *Proc. Roy. Soc. Edinburgh*, 127A:1153–1179.

[10] FATTORINI, H.O. AND SRITHARAN, S.S. (1992). Existence of optimal controls for viscous flow problems. *Proc. Roy. Soc. London*, 439A:81–102.

[11] FATTORINI, H.O. AND SRITHARAN, S.S. Optimal chattering control for viscous flow. *Nonlinear Anal., Th. Meth. Appl.*, 25:763–797.

[12] FATTORINI, H.O. AND SRITHARAN, S.S. (1995). Relaxation in semilinear infinite dimensional systems modelling fluid flow control problems. In: *Control and Opt. Design of Dist. Parameter Systems.* (Ed. J.E.Lagnese.) Springer, New York, 93–111.

[13] FATTORINI, H.O. AND SRITHARAN, S.S. (1994). Necessary and sufficient conditions for optimal control in viscous flow problems. *Proc. Roy. Soc. Edinburgh Sect. A*, 124:211–251.

[14] FURSIKOV, A.V., IMANUVILOV O.YU. (1996). On exact boundary zero controllability of two-dimensional Navier-Stokes equations. (In Russian) *Mat. Sb.*, 187:103–138.

[15] GABASOV, R. AND KIRILLOVA, F. (1971). *Qualitative Theory of Optimal Processes.* Nauka, Moscow.

[16] GALDI, P.G. (1994). *An introduction to the Navier-Stokes Equations.* Springer-Verlag.

[17] GAMKRELIDZE, R.V. (1978). *Principles of Optimal Control Theory.* (In Russian.) Tbilisi Univ. Press, Tbilisi, 1977. Engl. transl.: Plenum Press, New York.

[18] GUNZBURGER, M.D. (1995). A prehistory of flow control and optimization. In: *Flow Control* (M.D.Gunzburger, ed.), IMA Vol. Math. Appl. **68**, Springer, New York, 185–195.

[19] GUNZBURGER, M.D., HOU, L. AND SVOBODNY, T.P. (1991). Analysis and finite element approximation of optimal control problems for stationary Navier-Stokes equations with distributed and Neumann controls. *Math. Comp.*, 57:123–151.

[20] GUNZBURGER, M.D., HOU, L. AND SVOBODNY, T.P. (1992). Boundary velocity control of incompressible flow with an application to viscous drag reduction. *SIAM J. Control Optim.*, 30:167–181.

[21] GUNZBURGER, M.D. AND KIM, H. (1998). Existence of an optimal solution of a shape control problem for the stationary Navier-Stokes equations. *SIAM J. Control Optim.*, 36:895–909.

[22] HINZE, M. AND KUNISH, K. Control strategies for fluid flows – optimal versus suboptimal control. *ENUMATH 97* (Eds. H.G.Bock et al.), World Scientific, Singapore, 351–358.

[23] IVANOVIČ, L.D. AND SÜLI, E.E. (1986). Approximation and regularization of an optimal control problem for linearized Navier-Stokes equations. *Z. Angew. Math. Mech.*, 66:306–308.

[24] KAPLICKÝ, P., MÁLEK, J. AND STARÁ, J. $C^{1,\alpha}$-solutions to a class of nonlinear fluids in two dimensions – stationary Dirichlet problem. *Math. Ann.* (submitted).

[25] LIONS, J.L. (1985). *Contrôle des systémes distribués singuliers.* Bordas, Paris, 1983. Engl. transl.: *Control of Distributed Singular Systems.* Gauthier-Villars.

[26] MÁLEK, J., NEČAS, J., ROKYTA, M. AND RŮŽIČKA, M. (1996). *Weak and measure-valued solutions to the evolutionary PDE's.* Chapman & Hall, London.

[27] NEČAS, J. (1968). Sur la régularitè des solutions faibles des équations elliptiques non linéaires. *Comment. Math. Univ. Carolin.*, 9(3):365-413.

[28] NEČAS, J. (1967). Sur la régularité des solutions variationnelles des équations elliptiques non-linéaires d'ordre 2k en deux dimensions. *Ann. Scuola Norm. Sup. Pisa* XXI, Fasc. III, 427–457.

[29] OU, Y.-R. (1995). Mathematical modelling and numerical simulation in external flow control. In: *Flow Control* (M.D.Gunzburger, ed.), IMA Vol. Math. Appl. **68**, Springer, New York, 219–255.

[30] ROUBÍČEK, T. (1997). *Relaxation in Optimization Theory and Variational Calculus.* W. de Gruyter, Berlin.

[31] ROUBÍČEK, T. Convex locally compact extensions of Lebesgue spaces and their applications. In: Proc. Conf. *Calculus of Variations and Related Topics.* Haifa, March 1998. Longmann, to appear.

[32] SRITHARAN, S.S. (1992). An optimal control problem in exterior hydrodynamic. *Proc. Roy. Soc. Edinburgh*, 121A:5–32.

[33] SRITHARAN, S.S. (1995). Optimal feedback control of hydrodynamics: a progress report. In: *Flow Control* (M.D.Gunzburger, ed.), IMA Vol. Math. Appl. **68**, Springer, New York, 257–274.

[34] SRITHARAN, S.S. Deterministic and stochastic control of Navier-Stokes equation with linear, monotone and hyper viscosities. A preprint.

[35] TEMAM, R. (1984). *Navier-Stokes equtions, Theory and Numerical Analysis.* 3rd ed., North-Holland, Amsterdam-New York.

[36] TEMAM, R. (1995). Remarks on the control of turbulent flows. In: *Flow Control* (M.D.Gunzburger, ed.), IMA Vol. Math. Appl. **68**, Springer, New York, 357–381.

[37] WARGA, J. (1972). *Optimal Control of Differential and Functional Equations.* Academic Press, New York.

[38] YOUNG, L.C. (1937). Generalized curves and the existence of an attained absolute minimum in the calculus of variations. *Comptes Rendus de la Société des Sciences et des Lettres de Varsovie*, Classe III, 30:212–234.

ASYMPTOTIC BEHAVIOUR OF COMPRESSIBLE MAXWELL FLUIDS IN EXTERIOR DOMAINS

Šárka Matušů-Nečasová, Adélia Sequeira, Juha Hans Videman

Abstract: This article is concerned with the steady motion of a compressible viscoelastic non-Newtonian fluid of Maxwell type around a three-dimensional rigid body. Results on existence, uniqueness and asymptotic behaviour of the solution are obtained for small data.

The method of proof is based on an appropriate decomposition of the original nonlinear set of equations into auxiliary problems (Neumann problem for the Laplacian, Stokes problem and two transport equations) and on a suitable fixed point argument. The asymptotic decay of the solution, as regards the velocity and pressure, is defined by the linearized part, i.e. by the asymptotic behaviour of the fundamental solution of the Stokes system.

Keywords: Viscoelastic fluid, compressible fluid, Maxwell fluid, exterior domain, asymptotic behaviour.

1. INTRODUCTION

Over the past 50 years, the motion of viscoelastic non-Newtonian fluids has been modelled by several authors, cf. e.g. [16, 28, 22, 18] and all the references cited therein. A significant number of mathematical results has been obtained, as it comes to existence, uniqueness and asymptotic behaviour of solutions to the equations governing the motion of viscoelastic fluids, see *e.g.* [1, 20, 21, 7, 4, 29].

All the above mentioned articles focus on incompressible viscoelastic fluids. However, since viscoelastic fluids have certain properties similar to compressible gases, it is natural to consider also compressible viscoelastic fluids. One of the first works towards this direction can be found in [19], where the authors consider compressible viscoelastic fluids of integral type. As the fluids of the Maxwell type are concerned, even a small degree of compressibility changes the character

Applied Nonlinear Analysis, edited by Sequeira *et al.*
Kluwer Academic / Plenum Publishers, New York, 1999.

of the governing equations from hyperbolic-parabolic type to pure hyperbolic type, see [2]. From a numerical point of view, this allows the application of numerical methods designed specifically for solving hyperbolic problems, [17].

As it comes to the mathematical analysis, the equations governing the steady motion of slightly compressible viscoelastic fluids and of (fully) compressible viscoelastic fluids of White-Metzner type have been shown to admit a unique classical solution under suitable smallness conditions on the data, see Talhouk [27] and Sy [26].

More recently, Nečasová, Sequeira and Videman [8, 9] have investigated the existence of classical solutions for the compressible viscoelastic fluids of the Oldroyd type in an exterior domain both for zero and nonzero velocity at infinity.

In this paper, we shall consider the steady flow of compressible viscoelastic fluids of the Maxwell type around a three-dimensional obstacle $\mathcal{B}$. The main idea behind our proof is a suitable decomposition of the problem into a coupled system consisting of an elliptic equation for the velocity field and of two hyperbolic equations, one for the density and the other one for the elastic part of the stress tensor. Consequently, by linearizing the nonlinear coupled system, we obtain four linear problems; the Neumann problem for the Laplacian, the Stokes problem and two transport equations. After proving existence to the linearized problem, an appropriate contraction argument provides, for sufficiently small data, the existence of a unique solution to the original problem in a functional framework that readily yields the asymptotic behaviour of the solution.

The decomposition procedure and the functional setting is similar to the one used in [12, 13] for the compressible Navier-Stokes equations. The equations governing the steady flow of incompressible viscoelastic fluids past an obstacle were studied in [5, 14, 15].

The paper is organized as follows. After formulating the problem in Section 2, we introduce our notations and recall some auxiliary results in Section 3. Finally, in Section 4 we prove two main theorems; the first one is concerned with the linearized decomposed problem and the second one yields existence, uniqueness and asymptotic behaviour for a nonlinear system of equations corresponding to the original problem.

2. FORMULATION OF THE PROBLEM

We consider the following system of equations modeling the steady motion of a compressible non-Newtonian fluid of Maxwell type in an exterior domain Ω of $\mathbb{R}^3$, cf. [17, 2]

$$\left.\begin{array}{ll} \rho\mathbf{v}\cdot\nabla\mathbf{v}+\nabla p=\rho\mathbf{f}+\nabla\cdot\mathbf{T} & \\ \nabla\cdot(\rho\mathbf{v})=0 & \text{in } \Omega. \\ \mathbf{T}+\lambda_1(\mathbf{v}\cdot\nabla)\mathbf{T}-\lambda_1\mathbf{g}(\mathbf{T},\nabla\mathbf{v})=2\mu_e\mathbf{D}(\mathbf{v}) & \end{array}\right\} \tag{2.1}$$

Here, $\rho > 0$ denotes the fluid density, $\mathbf{v}$ the velocity field, $p = p(\rho)$ the pressure, $\mathbf{T}$ the elastic part of the stress tensor and $\mathbf{f}$ the external force. Moreover, $\mathbf{D}(\mathbf{v}) = \frac{1}{2}(\nabla\mathbf{v}+(\nabla\mathbf{v})^T)$ denotes the rate of deformation tensor and $\mathbf{g}(\cdot,\cdot)$ is a quadratic function given by

$$\mathbf{g}(\mathbf{T},\nabla\mathbf{v}) = \mathbf{T}(\nabla\mathbf{v})^T + (\nabla\mathbf{v})\mathbf{T}.$$

Finally, $\mu_e > 0$ stands for the elastic viscosity coefficient and $\lambda_1 > 0$ denotes the relaxation time[1]. We supplement equations (2.1) with the boundary condition

$$\mathbf{v}=0 \quad \text{on} \quad \partial\Omega \tag{2.2}$$

and with the conditions on the asymptotic behaviour

$$\mathbf{v}\to 0,\ \ \rho\to 1, \quad \text{as} \quad |x|\to\infty. \tag{2.3}$$

Taking (formally) the divergence of $(2.1)_3$, substituting into $(2.1)_1$, and assuming that the fluid motion is isothermal, i.e. $p = k\rho$, with k a positive constant, ($k = 1$, for simplicity), we rewrite the equations (2.1)-(2.3) for the perturbation $(\sigma, \mathbf{v})$ with $\rho = 1 + \sigma$, obtaining (see [20] for similar splitting of the problem in the case of an incompressible Maxwell fluid)

$$\left.\begin{array}{rcll} -\mu_e\Delta\mathbf{v}-(\mu_e+\lambda_1)\nabla(\nabla\cdot\mathbf{v})+\nabla\sigma &=& \mathbf{P}(\mathbf{v},\sigma,\mathbf{f},\mathbf{T}) & \\ \nabla\cdot\mathbf{v}+\nabla\cdot(\sigma\mathbf{v}) &=& 0 & \\ \mathbf{T}+\lambda_1\mathbf{v}\cdot\nabla\mathbf{T}-\lambda_1\mathbf{g}(\mathbf{T},\nabla\mathbf{v}) &=& 2\mu_e\mathbf{D}(\mathbf{v}) & \text{in } \Omega \\ \mathbf{v}_{|\partial\Omega} &=& 0 & \\ \mathbf{v}(x),\sigma(x) &\to& 0 \ \text{ as } \ |x|\to\infty & \end{array}\right\} \tag{2.4}$$

where

$$\mathbf{P}(\mathbf{v},\sigma,\mathbf{f},\mathbf{T}) = (1+\sigma)\mathbf{f}+\nabla\cdot\mathbf{F}(\mathbf{v},\sigma,\mathbf{f})+\nabla\cdot(\lambda_1(\nabla\cdot\mathbf{v})\mathbf{T}+\lambda_1(\nabla\mathbf{v})\mathbf{T})$$

with

$$\begin{aligned} \mathbf{F}(\mathbf{v},\sigma,\mathbf{f}) = {} & \lambda_1\sigma(\nabla\mathbf{v})^T-\lambda_1((1+\sigma)\mathbf{v}\cdot\nabla\mathbf{v})\otimes\mathbf{v}+ \\ & +\lambda_1(1+\sigma)\mathbf{f}\otimes\mathbf{v}-(1+\sigma)\mathbf{v}\otimes\mathbf{v}. \end{aligned}$$

[1] If $\lambda_1 = 0$, system (2.1) reduces to the compressible Navier-Stokes equations.

Let us start by linearizing system (2.4). This yields the set of equations

$$\left.\begin{aligned} -\mu_e\Delta\mathbf{v} - (\mu_e + \lambda_1)\nabla(\nabla\cdot\mathbf{v}) + \nabla\sigma &= \mathbf{P}(\mathbf{z},\tau,\mathbf{f},\mathbf{S}) \\ \nabla\cdot\mathbf{v} + \nabla\cdot(\sigma\mathbf{z}) &= 0 \\ \mathbf{v}_{|\partial\Omega} &= 0 \\ \mathbf{v}(x),\sigma(x) &\to 0 \quad \text{as} \quad |x|\to+\infty \\ \mathbf{T} + \lambda_1\mathbf{z}\cdot\nabla\mathbf{T} &= 2\mu_e\mathbf{D}(\mathbf{v}) + \lambda_1\mathbf{g}(\mathbf{S},\nabla\mathbf{z}). \end{aligned}\right\} \tag{2.5}$$

Assuming that $\mathbf{z},\tau$ and $\mathbf{S}$ are given with $\mathbf{z}|_{\partial\Omega} = 0, \mathbf{z}(x), \tau(x) \to 0$ as $|x| \to +\infty$, we analyse the solvability of (2.5). Observe that (2.5) consists of two separate problems: compressible Stokes equations $(2.5)_{1,2,3,4}$ and linear transport equation $(2.5)_5$ for $\mathbf{T}$. To investigate the system $(2.5)_{1,2,3,4}$ we shall decouple its elliptic and hyperbolic parts. Towards this end, recall that in view of the Helmholtz decomposition we can uniquely write $\mathbf{v}$ in the form

$$\mathbf{v} = \mathbf{u} + \nabla\varphi, \qquad \nabla\cdot\mathbf{u} = 0, \qquad \mathbf{u}\cdot\mathbf{n}|_{\partial\Omega} = 0, \qquad \nabla\varphi\cdot\mathbf{n}|_{\partial\Omega} = 0.$$

This splits $(2.5)_{1,2,3,4}$ into a coupled system for $(\mathbf{u},\pi)$, σ and φ that can be studied by considering a mapping

$$\mathcal{L} : \xi \to \varphi \tag{2.6}$$

defined in the following way:

Assuming that $\mathbf{P} = \mathbf{P}(\mathbf{z},\tau,\mathbf{f},\mathbf{S})$ and ξ are given, we solve the Stokes problem for $(\mathbf{u},\pi)$

$$\left.\begin{aligned} -\mu_e\Delta\mathbf{u} + \nabla\pi = \mathbf{P} \\ \nabla\cdot\mathbf{u} = 0 \\ \mathbf{u}|_{\partial\Omega} = -\nabla\xi|_{\partial\Omega} \\ \mathbf{u}(x)\to 0, \pi(x)\to 0 \quad \text{as} \quad |x|\to+\infty. \end{aligned}\right\} \tag{2.7}$$

After $(\mathbf{u},\pi)$ is obtained from (2.7) and given $\mathbf{z}$, we look for σ as a solution to the transport equation

$$\left.\begin{aligned} \sigma + (2\mu_e + \lambda_1)\nabla\cdot(\sigma\mathbf{z}) = \pi \\ \sigma(x)\to 0, \quad \text{as} \quad |x|\to+\infty. \end{aligned}\right\} \tag{2.8}$$

Finally, φ is obtained as the solution of the Neumann problem for the Laplacian

$$\left.\begin{aligned} \Delta\varphi = -\nabla\cdot(\sigma\mathbf{z}) \\ \frac{\partial\varphi}{\partial n}|_{\partial\Omega} = 0 \\ \nabla\varphi(x)\to 0 \quad \text{as} \quad |x|\to\infty. \end{aligned}\right\} \tag{2.9}$$

Provided the mapping $\mathcal{L}$ admits a unique fixed point φ, it is easy to check that the pair $(\mathbf{v}, \sigma)$ where $\mathbf{v} = \mathbf{u} + \nabla\varphi$ solves uniquely $(2.5)_{1,2,3,4}$. After obtaining $\mathbf{v}$, one may consider the solvability of $(2.5)_5$ for given $\mathbf{g} = \mathbf{g}(\mathbf{S}, \nabla\mathbf{z})$ and $\mathbf{z}$, and conclude that the linearized system (2.5) admits a unique solution $(\mathbf{v}, \sigma, \mathbf{T})$. Finally, we look for a solution $(\mathbf{v}, \sigma, \mathbf{T})$ to the nonlinear system (2.4) as a fixed point of the map

$$\mathcal{N} : (\mathbf{z}, \tau, \mathbf{S}) \to (\mathbf{v}, \sigma, \mathbf{T}) \tag{2.10}$$

defined in an obvious way by (2.5).

3. MATHEMATICAL PRELIMINARIES. NOTATIONS AND AUXILIARY LEMMAS.

First, we introduce our notations. By $\Omega \subset \mathbb{R}^3$ we denote a domain exterior to a simply-connected compact set $\mathcal{B}$. For simplicity, we assume that the origin of coordinates is located in $\mathcal{B}$. By $\partial\Omega$ we denote the boundary of Ω and $x = (x_1, x_2, x_3)$ stands for a generic point in $\mathbb{R}^3$. We set

$$B_R(x) := \{y \in \mathbb{R}^3 : |x - y| < R\}$$

$$\Omega_R = \Omega \cap B_R(0), \qquad \Omega^R = \Omega - \overline{\Omega}_R.$$

The regularity of the domain Ω is defined by the regularity of its boundary $\partial\Omega$, i.e. we say that Ω is of class C^m if $\partial\Omega$ is of class C^m, with $m \geq 0$ an integer, see [10] for more details.

The Lebesgue spaces are denoted by $L^p(\Omega), 1 \leq p \leq \infty$, and equipped with the norms $\|\cdot\|_{0,p}$. By $W^{k,p}(\Omega)$, $k \geq 0$ an integer, $1 \leq p \leq \infty$, we denote the usual Sobolev spaces with the norms

$$\|\cdot\|_{k,p} = \Big(\sum_{|\alpha|=0}^{k} \|D^\alpha \cdot\|_{0,p}^p\Big)^{1/p},$$

where $\alpha = (\alpha_1, \alpha_2, \alpha_3)$ is the standard multi-index. We set $H^k(\Omega) = W^{k,2}(\Omega)$ and $\|\cdot\|_k = \|\cdot\|_{k,2}$. Further, for $1 < q < 3$ we define the Banach space

$$\mathcal{H}_0^{1,q}(\Omega) = \overline{C_0^\infty(\Omega)}^{|\cdot|_{1,q}}$$

equipped with the norm $|\cdot|_{1,q} = \|\nabla \cdot\|_{0,q}$. The dual space of $\mathcal{H}_0^{1,q}(\Omega)$ is denoted by $\mathcal{H}^{-1,q}(\Omega)$, its norm by $|\cdot|_{-1,q}$ and $\mathcal{H}^{-1}(\Omega) = \mathcal{H}^{-1,2}(\Omega)$ with the norm $|\cdot|_{-1}$. Moreover, the Banach spaces $\mathcal{H}^{k,q}(\Omega)$, are obtained as a completion of $C_0^\infty(\overline{\Omega})$-functions in the norm $\|\nabla \cdot\|_{k-1,q}$. One has the following characterization for the space $\mathcal{H}^{k,2}(\Omega) = \mathcal{H}^k(\Omega)$

$$\mathcal{H}^k(\Omega) = \left\{u \in L^6(\Omega) : \nabla u \in \mathbf{H}^{k-1}(\Omega)\right\}.$$

In particular, for all $u \in \mathcal{H}^1(\Omega)$ the Sobolev inequality $\|u\|_{0,6} \le c\|\nabla u\|_{0,2}$ is valid and $\|\frac{u}{|x|}\|_{0,2} \le c\|\nabla u\|_0$. All the corresponding spaces for vector- or tensor-valued functions are denoted by the boldface letter.

For $k \ge 1$, an integer, and for $r \in (3,\infty)$, $q_1 \in (1,3/2)$, $q_2 \in (3/2,3)$, we define the following function spaces

$$\mathbf{U_k} = \Big\{\mathbf{u} \in \mathcal{H}^{k+2}(\Omega) : |x|\mathbf{u}, |x|^2\nabla\mathbf{u}, |x|\nabla^2\mathbf{u} \in \mathbf{L}^\infty(\Omega)\},$$

$$G_k = \Big\{\varphi \in \mathcal{H}^{k+3} : |x|\nabla\varphi, |x|\nabla^3\varphi \in L^\infty(\Omega), |x|^2\nabla^2\varphi \in W^{1,r}(\Omega)\Big\},$$

$$\mathbf{V_k} = \Big\{\mathbf{v} = \mathbf{u} + \nabla\varphi : \mathbf{u} \in \mathbf{U_k}, \varphi \in G_k\Big\},$$

$$S_k = \Big\{\pi \in H^{k+1}(\Omega) : |x|^2\pi, |x|\nabla\pi \in L^\infty(\Omega), |x|^2\nabla\pi \in L^r(\Omega)\},$$

$$\mathbf{Y_k} = \Big\{\mathbf{S} \in \mathbf{H}^{k+1}(\Omega) : |x|^2\mathbf{S}, |x|\nabla\mathbf{S} \in \mathbf{L}^\infty(\Omega)\Big\},$$

$$\mathbf{Z_k} = \Big\{\mathbf{f} \in \mathbf{H}^k(\Omega) : |x|\mathbf{f} \in L^{q_1}(\Omega) \cap L^{q_2}(\Omega), \quad |x|^{3+\epsilon}\mathbf{f},\ |x|^2\nabla\mathbf{f},\ |x|\nabla^2\mathbf{f} \in \mathbf{L}^\infty(\Omega)\Big\},$$

equipped with the corresponding product norms $\|\cdot\|_{\mathbf{U}_k}$, $\|\cdot\|_{G_k}$, etc.

Finally, we remark that throughout the paper c denotes a generic positive constant that may depend on several parameters and may take different values even within the same calculation, but does not depend on the data.

We shall recall existence results in a three-dimensional exterior domain for the following linear problems: the Neumann problem for the Laplacian, the Stokes problem and two transport equations.

First, let us consider the Neumann problem

$$\left.\begin{array}{lll} \Delta\varphi = g & \text{in } \Omega & \\ \dfrac{\partial\varphi}{\partial n}|_{\partial\Omega} = \psi, & \nabla\varphi(x) \to 0 \quad \text{as} & |x| \to \infty. \end{array}\right\} \tag{3.1}$$

One has the following result, cf. [23].

Lemma 3.1. *Let $k \ge 1$, let $\Omega \subset \mathbb{R}^3$ be an exterior domain of class C^{k+2} and suppose that $g \in H^k(\Omega) \cap \mathcal{H}^{-1}(\Omega)$ and $\psi \in H^{k+1-1/2}(\partial\Omega)$. Then problem (3.1) admits a unique solution $\varphi \in \mathcal{H}^{k+2}(\Omega)$ satisfying the estimate*

$$\|\nabla\varphi\|_{k+1} \le c(\|g\|_k + |g|_{-1} + \|\psi\|_{k+1-1/2,\partial\Omega}). \tag{3.2}$$

Next, let us consider the Stokes problem

$$\left.\begin{aligned} -\mu\Delta\mathbf{u} + \nabla\pi &= \mathbf{f} \\ \nabla\cdot\mathbf{u} &= h \\ \mathbf{u}|_{\partial\Omega} &= \boldsymbol{\psi} \\ \mathbf{u}(x)\to 0,\quad \pi(x)\to 0 \quad &\text{as} \quad |x|\to\infty, \end{aligned}\right\} \tag{3.3}$$

for which we recall the following lemma, cf. [6, 3].

Lemma 3.2. *Let $\Omega \in C^{k+2}$ with $k \geq 0$ an integer, be an exterior domain in $\mathbb{R}^3$, assume that $\mathbf{f} \in \mathbf{H}^k(\Omega) \cap \mathcal{H}^{-1}(\Omega)$, $\boldsymbol{\psi} \in \mathbf{H}^{k+2-1/2}(\Omega)$ and $h \in \mathcal{H}^{-1}(\Omega) \cap H^{k+1}(\Omega)$. There exists a unique solution to problem (3.3) such that $\mathbf{u} \in \mathcal{H}^{k+1}(\Omega)$, $\pi \in H^{k+1}(\Omega)$. Moreover, this solution satisfies the estimate*

$$\|\nabla\mathbf{u}\|_{k+1} + \|\pi\|_{k+1} \leq c\Big(|\mathbf{f}|_{-1} + \|\mathbf{f}\|_k + |h|_{-1} + \|h\|_{k+1} + \|\boldsymbol{\psi}\|_{k+2-1/2,\partial\Omega}\Big).$$

Moreover, for the transport equation

$$w + \nabla\cdot(\mathbf{v}w) = g, \tag{3.4}$$

one has the following result, cf. [12, 11].

Lemma 3.3. *Let $\Omega \in C^{k+1}$ with $k \geq 1$ an integer be an exterior domain in $\mathbb{R}^3$. Moreover, let $g \in H^{k+1}(\Omega)$ and let $\mathbf{v} \in \mathcal{H}^{k+2}(\Omega)$ be such that $\mathbf{v}\cdot\mathbf{n} = 0$ at $\partial\Omega$. There exists a constant $\delta > 0$ such that if $\|\nabla\mathbf{v}\|_{k+1} < \delta$, with δ sufficiently small, problem (3.4) has a unique solution $w \in H^{k+1}(\Omega)$ such that $\nabla\cdot(\mathbf{v}w) \in H^{k+1}(\Omega)$ and $\Delta w \in H^{k-1}(\Omega)$. Moreover, this solution satisfies the estimates*

$$\|w\|_{k+1} + \|\nabla\cdot(\mathbf{v}w)\|_{k+1} \leq c\|g\|_{k+1}, \tag{3.5}$$

$$\|\Delta w\|_{k-1} \leq c\Big(\|\Delta g\|_{k-1} + \|\nabla\mathbf{v}\|_{k+1}\|w\|_{k+1}\Big) \tag{3.6}$$

where c denotes a positive constant depending in particular on δ.

The asymptotic behaviour of the solution of the transport problem (3.4) is characterized by the following lemma, see [15] for similar considerations.

Lemma 3.4. *Let $\Omega \in C^k$ be an exterior domain in R^3, $k \geq 3$ an integer. Let $g \in S_k$, $\mathbf{v} \in \mathcal{H}^{k+2}(\Omega)$ and $\mathbf{v}\cdot\mathbf{n} = 0$ at $\partial\Omega$. Further, assume that $|x|\mathbf{v} \in L^\infty(\Omega)$. There exists a constant $\delta > 0$ such that if*

$$\|\nabla\mathbf{v}\|_{k+1} + \||x|\mathbf{v}\|_{0,\infty} < \delta$$

then the unique solution of the problem (3.4) *belongs to* S_k *and satisfies the estimate*

$$\|w\|_{S_k} + \|\nabla \cdot (\mathbf{v}w)\|_{S_k} \le c\|g\|_{S_k}. \tag{3.7}$$

The following lemmas are concerned with the decay at infinity of weakly singular integrals of the form

$$I_j(g)(x) = \int_\Omega K_j(|x-y|)\, g(y)\, dy, \qquad j = 1,2,$$

where $K_j(x)$, $j = 1,2$, is a smooth function in $\mathbb{R} \setminus \{0\}$ having the decay property $K_j(x) = O(|x|^{-j})$, $j = 1,2$. The behaviour of these type of integrals has been studied in full generality, e.g., in [24].

Lemma 3.5. *Let* $|x|g \in L^{q_1}(\Omega) \cap L^{q_2}(\Omega)$, *with* $1 \le q_1 < 3/2 < q_2 < 3$. *Then* $|x|I_1(g) \in L^\infty(\Omega)$ *and*

$$\||x|I_1(g)\|_{0,\infty} \le c(\||x|g\|_{0,q_1} + \||x|g\|_{0,q_2}).$$

Lemma 3.6. *i) Let* $1 < k < 3$ *and* $|x|^k g \in L^\infty(\Omega)$. *Then* $|x|^{k-1} I_2(g) \in L^\infty(\Omega)$ *and it holds*

$$\||x|^{k-1} I_2(g)\|_{0,\infty} \le c\||x|^k g\|_{0,\infty}.$$

ii) Let $k > 3$ *and* $|x|^k g \in L^\infty(\Omega)$. *Then* $|x|^2 I_2(g) \in L^\infty(\Omega)$ *and it holds*

$$\||x|^2 I_2(g)\|_{0,\infty} \le c\||x|^k g\|_{0,\infty}.$$

iii) Let $|x|g \in L^q(\Omega) \cap L^p(\Omega)$, *with* $3/2 < q < 3 < p$. *Then* $|x|I_2(g) \in L^\infty(\Omega)$ *and*

$$\||x|I_2(g)\|_{0,\infty} \le c(\||x|g\|_{0,q} + \||x|g\|_{0,p}).$$

iv) Assume further that $\nabla K_2(x) = O(|x|^{-3})$ *and that* $g = \nabla \cdot \mathbf{h}$, *with* $\mathbf{h} \in \mathbf{W}^{1,\infty}$, $|x|^2\mathbf{h} \in \mathbf{L}^\infty(\Omega)$ *and* $|x|^3 \nabla \cdot \mathbf{h} \in \mathbf{L}^\infty(\Omega)$. *Then* $|x|^2 I_2(g) \in L^\infty(\Omega)$ *and it holds*

$$\||x|^2 I_2(g)\|_{0,\infty} \le c(\||x|^2\mathbf{h}\|_{0,\infty} + \||x|^3 \nabla \cdot \mathbf{h}\|_{0,\infty}).$$

Finally, we present a lemma about the decay of singular integrals of the Calderon-Zygmund type, i.e. we consider the integral

$$I_3(g)(x) = \int_\Omega \frac{1}{|x-y|^3} \mathcal{W}\Big(\frac{x-y}{|x-y|}\Big)\, g(y)\, dy$$

where $\mathcal{W}(x)$ is Hölder continuous on the unit sphere S_1 and such that $\int_{S_1} \mathcal{W}\, dS = 0$. The following lemma is due to Stein [25].

Lemma 3.7. *Let* $1 < t < \infty$, $-\frac{3}{t} < \alpha < \frac{3}{t'}$, $\left(t' = \frac{t}{t-1}\right)$. *Let* $|x|^\alpha g \in L^t(\Omega)$. *Then* $|x|^\alpha I_3(g) \in L^t(\Omega)$ *and*

$$\||x|^\alpha I_3(g)\|_{0,t} \leq c\||x|^\alpha g\|_{0,t}.$$

4. MAIN RESULTS

Here, we shall present two theorems containing the main results. The first one deals with the solvability of the linearized system (2.5), and the second theorem provides the existence of a unique solution to the nonlinear system (2.1).

Theorem 4.1. *Let* $k \geq 1$ *and let* $\Omega \subset \mathbb{R}^3$ *be an exterior domain of class* C^{k+2}. *Suppose that* $\mathbf{z}, \mathbf{P}, \mathbf{g}$ *are given and such that*

$$\mathbf{z} \in \mathcal{H}^{k+2}(\Omega), \quad \mathbf{P} \in \mathcal{H}^{-1}(\Omega) \cap \mathbf{H}^k(\Omega), \quad \mathbf{g} \in \mathbf{H}^{k+1}(\Omega),$$

and $\mathbf{z}|_{\partial\Omega} = 0$. *There exist constant* $\delta > 0$ *such that if*

$$\|\nabla \mathbf{z}\|_{k+1} < \delta$$

then problem (2.5) *admits a unique solution*

$$(\mathbf{v}, \sigma, \mathbf{T}) \in \mathcal{H}^{k+2}(\Omega) \times H^{k+1}(\Omega) \times \mathbf{H}^{k+1}(\Omega)$$

satisfying the estimate

$$\|\nabla \mathbf{v}\|_{k+1} + \|\sigma\|_{k+1} + \|\mathbf{T}\|_{k+1} \leq c(\|\mathbf{P}\|_k + |\mathbf{P}|_{-1} + \|\mathbf{g}\|_{k+1}), \qquad (4.1)$$

with $c = c(k, \Omega, \delta) > 0$.

Proof. By the results of Novotný and Padula [12] problem $(2.5)_{1,2,3,4}$ has a unique solution $(\mathbf{v}, \sigma) \in \mathcal{H}^{k+2}(\Omega) \times H^{k+1}(\Omega)$ satisfying the estimate

$$\|\nabla \mathbf{v}\|_{k+1} + \|\sigma\|_{k+1} \leq c(\|\mathbf{P}\|_k + |\mathbf{P}|_{-1}), \qquad (4.2)$$

provided δ is chosen sufficiently small.

Now consider the transport equation $(2.5)_5$, for given $\mathbf{z}, \mathbf{g}$ and $\mathbf{v}$. In view of Lemma 3.3, one sees that, if $\|\nabla \mathbf{z}\|_{k+1}$ is chosen small enough, $(2.5)_5$ admits a unique solution $\mathbf{T} \in \mathbf{H}^{k+1}(\Omega)$ satisfying the estimate

$$\|\mathbf{T}\|_{k+1} \leq c\Big(\|\nabla \mathbf{v}\|_{k+1} + \|\mathbf{g}\|_{k+1}\Big). \qquad (4.3)$$

Now, joining estimates (4.2) and (4.3), one obtains (4.1). ∎

Theorem 4.2. *Let $k \geq 4$ be an integer and let $\Omega \in C^{k+2}$ be an exterior domain of $\mathbb{R}^3$. Given*

$$\mathbf{f} \in \mathbf{Z}_k$$

there exists a constant β, depending on $k, \partial\Omega$ and λ_1, such that if

$$\|\mathbf{f}\|_{\mathbf{Z}_k} \leq \beta$$

then problem (2.1) admits a unique solution $(\mathbf{v}, p, \rho, \mathbf{T})$ with $p = \rho$ and $\rho = 1 + \sigma$, such that

$$\mathbf{v} \in \mathbf{V}_k, \quad \sigma \in S_k, \quad \mathbf{T} \in \mathbf{Y}_k.$$

Moreover, this solution satisfies the estimate

$$\|\nabla \mathbf{v}\|_{\mathbf{V}_k} + \|\sigma\|_{S_k} + \|\mathbf{T}\|_{\mathbf{Y_k}} \leq c\|\mathbf{f}\|_{\mathbf{Z}_k}, \tag{4.4}$$

with some constant $c = c(k, \partial\Omega, \beta, \lambda_1) > 0$.

Proof. Let us define the set

$$\mathbf{D}_\delta = \Big\{(\mathbf{z}, \tau, \mathbf{S}) \in \mathbf{V}_k \times S_k \times \mathbf{Y}_k \Big| \quad \|(\mathbf{z}, \tau, \mathbf{S})\|_k \leq \delta, \ \mathbf{z}|_{\partial\Omega} = 0\Big\},$$

where, for all $k \geq 1$, an integer,

$$\|(\mathbf{z}, \tau, \mathbf{S})\|_k = \|\mathbf{z}\|_{\mathbf{V}_k} + \|\tau\|_{S_k} + \|\mathbf{S}\|_{\mathbf{Y}_k}.$$

Note that $\mathbf{D}_\delta$ is a closed, convex and bounded subset of $\mathbf{V}_k \times S_k \times \mathbf{Y}_k$.

First, let us show that the map $\mathcal{N}$ introduced in (2.10) is well defined from $\mathbf{D}_\delta$ into $\mathbf{V}_k \times S_k \times \mathbf{Y}_k$ provided β and δ are chosen sufficiently small. In fact, it suffices to show that the nonlinear terms $\mathbf{P} = \mathbf{P}(\mathbf{z}, \tau, \mathbf{f}, \mathbf{S})$ and $\mathbf{g} = \mathbf{g}(\mathbf{S}, \nabla \mathbf{z})$ meet the regularity requirements of Theorem 4.1 and that $\mathbf{v}, \sigma$ and $\mathbf{T}$ have the same decay properties as $\mathbf{z}, \tau$ and $\mathbf{S}$, respectively.

By interpolation, one gets

$$\|\mathbf{f}\|_{0,\frac{6}{5}} \leq \|\mathbf{f}\|_{0,1} + \|\mathbf{f}\|_{0,\frac{3}{2}}.$$

On the other hand, for $q_1 \in (1, 3/2)$ and $q_2 \in (3/2, 3)$ it holds

$$\|\mathbf{f}\|_{0,1} \leq \||x|\mathbf{f}\|_{0,q_1} \||x|^{-1}\|_{0,q_1'} \leq c\||x|\mathbf{f}\|_{0,q_1},$$

$$\|\mathbf{f}\|_{0,\frac{3}{2}} \leq \||x|\mathbf{f}\|_{0,q_2} \||x|^{-1}\|_{0,\frac{3q_2}{2q_2-3}} \leq c\||x|\mathbf{f}\|_{0,q_2},$$

where $q_1' = q_1/(q_1 - 1)$. Therefore

$$\|\mathbf{f}\|_{0,\frac{6}{5}} \leq c(\||x|\mathbf{f}\|_{0,q_1} + \||x|\mathbf{f}\|_{0,q_2}).$$

Furthermore, since $\|\mathbf{z} \otimes \mathbf{z}\|_0 \leq \||x|\mathbf{z}\|_{0,\infty}\|\nabla \mathbf{z}\|_0$, $\|\mathbf{z}\|_\infty \leq c\|\nabla \mathbf{z}\|_1$ and $|\mathbf{f}|_{-1} \leq c\|\mathbf{f}\|_{0,6/5}$, we obtain by Sobolev imbeddings

$$\begin{aligned} |\mathbf{P}|_{-1} \leq \ & c\Big((1 + \|\tau\|_{k+1})\Big(\||x|\mathbf{f}\|_{0,q_1} + \||x|\mathbf{f}\|_{0,q_2} + \\ & + \||x|\mathbf{z}\|_{0,\infty}\|\nabla \mathbf{z}\|_0 + \|\nabla \mathbf{z}\|_1^3 + \||x|\mathbf{f}\|_{0,\infty}\|\nabla \mathbf{z}\|_0\Big) + \\ & + \|\tau\|_{k+1}\|\nabla \mathbf{z}\|_{k+1} + \|\nabla \mathbf{z}\|_{k+1}\|\mathbf{S}\|_{k+1}\Big), \end{aligned} \tag{4.5}$$

$$\begin{aligned}\|\mathbf{P}\|_k &\le c\Big((1+\|\tau\|_{k+1})\Big(\|\mathbf{f}\|_k+\|\nabla\mathbf{z}\|_{k+1}^2+\\ &+\|\nabla\mathbf{z}\|_{k+1}^3+\|\mathbf{f}\|_{k+1}\|\nabla\mathbf{z}\|_{k+1}\Big)+\\ &+\|\nabla\mathbf{z}\|_{k+1}\|\mathbf{S}\|_{k+1}+\|\tau\|_{k+1}\|\nabla\mathbf{z}\|_{k+1}\Big),\end{aligned} \tag{4.6}$$

On the other hand, one readily gets

$$\|\mathbf{g}\|_{k+1}\le c\,\|\nabla\mathbf{z}\|_{k+1}\|\mathbf{S}\|_{k+1}. \tag{4.7}$$

In order to get the asymptotic behaviour of $\mathbf{v}$ and σ we consider the splitting (2.7) – (2.9) of system $(2.5)_{1,2,3,4}$ into a Stokes system, transport equation and the Neumann problem. It is straightforward to show, cf. [12], that under the hypotheses of Theorem 4.1 there exists a unique solution $(\mathbf{u},\pi,\sigma,\varphi)$ to the coupled problem

$$\left.\begin{aligned}-\mu_e\Delta\mathbf{u}+\nabla\pi=\mathbf{P}(\mathbf{z},\tau,\mathbf{f},\mathbf{S})\\ \nabla\cdot\mathbf{u}=0\\ \mathbf{u}|_{\partial\Omega}=-\nabla\varphi|_{\partial\Omega}\\ \mathbf{u}(x)\to 0,\pi(x)\to 0\quad\text{as}\quad |x|\to+\infty.\end{aligned}\right\} \tag{4.8}$$

$$\left.\begin{aligned}\sigma+(2\mu_e+\lambda_1)\nabla\cdot(\sigma\mathbf{z})=\pi\\ \sigma(x)\to 0,\quad\text{as}\quad |x|\to+\infty.\end{aligned}\right\} \tag{4.9}$$

$$\left.\begin{aligned}\Delta\varphi=-\nabla\cdot(\sigma\mathbf{z})\\ \frac{\partial\varphi}{\partial n}|_{\partial\Omega}=0\\ \nabla\varphi(x)\to 0\quad\text{as}\quad |x|\to\infty.\end{aligned}\right\} \tag{4.10}$$

Now, let us consider the Stokes problem (4.8). We may derive the following representation for $\mathbf{u}$ and π, see [3]

$$\begin{aligned}\mathbf{u}(x) &= -\tfrac{1}{\mu_e}\textstyle\int_\Omega \mathcal{U}(x-y)\mathbf{P}(y)dy\\ &+\textstyle\int_{\partial\Omega}[-\nabla_y\mathbf{u}(y)\mathcal{U}(x-y)+\nabla_y\mathcal{U}(x-y)\mathbf{u}(y)+\tfrac{1}{\mu_e}\pi(y)\mathcal{U}(x-y)]\mathbf{n}(y)]dS_y,\\ \pi(x) &= \textstyle\int_\Omega q(x-y)\mathbf{P}(y)dy-\int_{\partial\Omega}q(x-y)\pi(y)\\ &\quad+\textstyle\int_{\partial\Omega}[q(x-y)\nabla_y\mathbf{u}+2\mathbf{u}(y)\nabla_y q(x-y)]\mathbf{n}(y)dS_y.\end{aligned} \tag{4.11}$$

Here $\mathcal{U}=(\mathcal{U}_{ij})$, $q=(q_i)$, i,j=1,2,3 is the fundamental solution of the Stokes operator, i.e.

$$\mathcal{U}_{ij}(x)=-\frac{1}{8\pi}(\delta_{ij}|x|^{-1}+x_ix_j|x|^{-3}),\qquad q_i=\frac{1}{4\pi}x_i|x|^{-3}$$

and the following estimates hold true

$$|\mathcal{U}(x)| \leq c|x|^{-1}, \quad |\nabla \mathcal{U}(x)| \leq c|x|^{-2}, \quad |q(x)| \leq c|x|^{-2} \tag{4.12}$$

The decay estimates for $\mathbf{u}$ and π can be easily obtained by fixing $R > 1$ in such a way that $2|x-y| \geq |x|$, for all $x \in \Omega^R$ and $y \in \partial\Omega$. In Ω_R one gets

$$\||x|^l D^\alpha \mathbf{u}\|_{\infty,\Omega_R} \leq R^l \|D^\alpha \mathbf{u}\|_{\infty,\Omega}, \quad l = 0,1,2, \quad |\alpha| = 0,1,2, \tag{4.13}$$

$$\||x|^l D^\alpha \pi\|_{\infty,\Omega_R} \leq R^l \|D^\alpha \pi\|_{\infty,\Omega}, \quad l = 0,1,2, \quad |\alpha| = 0,1. \tag{4.14}$$

Next, let us define a tensor field $\hat{\mathbf{F}}$ in such a way that $\mathbf{P} - (1+\tau)\mathbf{f} = \nabla \cdot \hat{\mathbf{F}}$. Taking into account the decay properties (4.12) and using Lemmas 3.5 and 3.6, one gets from the representation formulas (4.11) the following estimates

$$\left.\begin{aligned}
\||x|\mathbf{u}(x)\|_{\infty,\Omega^R} \leq\ & c\Big((1+\|\tau\|_{k+1})(\||x|\mathbf{f}\|_{0,q_1} + \||x|\mathbf{f}\|_{0,q_2}) + \\
& + \||x|^2\hat{\mathbf{F}}\|_{0,\infty} + \|\nabla \mathbf{u}\|_{0,\infty} + \|\pi\|_{0,\infty} + \|\hat{\mathbf{F}}\|_{0,\infty} + \|\nabla\varphi\|_{0,\infty}\Big). \\
\||x|^2\pi\|_{\infty,\Omega^R} \leq\ & c\Big((1+\|\tau\|_{k+1})\||x|^{3+\epsilon}\mathbf{f}\|_{0,\infty} + \||x|^3\nabla\cdot\hat{\mathbf{F}}\|_{0,\infty} + \\
& + \||x|^2\hat{\mathbf{F}}\|_{0,\infty} + \|\nabla \mathbf{u}\|_{0,\infty} + \|\pi\|_{0,\infty} + \|\nabla\varphi\|_{0,\infty}\Big).
\end{aligned}\right\} \tag{4.15}$$

Differentiating the formulas (4.11) with respect to x and using integration by part when necessary, one obtains

$$\left.\begin{aligned}
\||x|^2\nabla\mathbf{u}(x)\|_{\infty,\Omega^R} \leq\ & c\Big((1+\|\tau\|_{k+1})\||x|^{3+\epsilon}\mathbf{f}\|_{0,\infty} + \||x|^2\hat{\mathbf{F}}\|_{0,\infty} + \\
& + \||x|^3\nabla\cdot\hat{\mathbf{F}}\|_{0,\infty} + \|\pi\|_{0,\infty} + \|\nabla \mathbf{u}\|_{0,\infty} + \|\nabla\varphi\|_{0,\infty}\Big), \\
\||x|\nabla^2\mathbf{u}\|_{\infty,\Omega^R} \leq\ & c\Big((1+\|\tau\|_{k+1})\||x|^2\nabla\mathbf{f}\|_{0,\infty} + \\
& + \||x|^2\nabla\cdot(\nabla\cdot\hat{\mathbf{F}})\|_{0,\infty} + \|\nabla\varphi\|_{0,\infty} + \|\nabla \mathbf{u}\|_{0,\infty} + \|\pi\|_{0,\infty}\Big),
\end{aligned}\right\} \tag{4.16}$$

$$\left.\begin{aligned} \||x|\nabla\pi(x)\|_{\infty,\Omega^R} &\leq c\Big((1+\|\tau\|_{k+1})\||x|^2\nabla\mathbf{f}\|_{0,\infty}+ \\ &\quad +\||x|^2\nabla\cdot(\nabla\cdot\hat{\mathbf{F}})\|_{0,\infty}+\|\nabla\mathbf{u}\|_{0,\infty}+\|\pi\|_{0,\infty}+\|\nabla\varphi\|_{0,\infty}\Big), \\ \||x|^2\nabla\pi\|_{0,r,\Omega^R} &\leq c\Big((1+\|\tau\|_{k+1})\||x|^{3+\epsilon}\mathbf{f}\|_{0,\infty}+\||x|^2\nabla\cdot\hat{\mathbf{F}}\|_{0,r}+ \\ &\quad +\|\nabla\mathbf{u}\|_{0,\infty}+\|\pi\|_{0,\infty}+\|\nabla\varphi\|_{0,\infty}\Big). \end{aligned}\right\} \tag{4.17}$$

On the other hand, it is easy to check that

$$\begin{aligned} &\||x|^2\hat{\mathbf{F}}\|_{0,\infty}+\||x|^3\nabla\cdot\hat{\mathbf{F}}\|_{0,\infty}+\||x|^2\nabla\nabla\cdot\hat{\mathbf{F}}\|_{0,\infty}+\||x|^2\nabla\cdot\hat{\mathbf{F}}\|_{0,r}\leq \\ &\leq c(1+\|(\mathbf{z},\tau,\mathbf{S})\|_k)\Big(\|\mathbf{f}\|_{Z_k}\|(\mathbf{z},\tau,\mathbf{S})\|_k+ \\ &+\|(\mathbf{z},\tau,\mathbf{S})\|_k^2+\|(\mathbf{z},\tau,\mathbf{S})\|_k^3\Big). \end{aligned} \tag{4.18}$$

Using estimates (4.5) and (4.6) together with Theorem 4.1, and collecting estimates (4.13), (4.14), (4.15), (4.16), (4.17) and (4.18), we conclude that there exists a unique solution problem (4.8) satisfying the estimate

$$\begin{aligned} &\|\mathbf{u}\|_{\mathbf{U}_k}+\|\pi\|_{S_k}\leq c(1+\|(\mathbf{z},\tau,\mathbf{S})\|_k)\Big(\|\mathbf{f}\|_{Z_k}\|+ \\ &+\|\mathbf{f}\|_{Z_k}\|(\mathbf{z},\tau,\mathbf{S})\|_k+\|(\mathbf{z},\tau,\mathbf{S})\|_k^2+\|(\mathbf{z},\tau,\mathbf{S})\|_k^3\Big). \end{aligned} \tag{4.19}$$

Now, concerning the asymptotic behaviour of the transport equation (4.9), it follows from estimate (3.7) of Lemma 3.4 that, for $\delta > 0$ small enough, the unique solution σ of problem (4.9) satisfies the estimate

$$\|\sigma\|_{S_k}+\|\nabla\cdot(\sigma\mathbf{z})\|_{S_k}\leq c\,\|\pi\|_{S_k}. \tag{4.20}$$

Next, let us investigate the Neumann problem (4.10). The representation formula for the solution φ is given by, cf. *e.g.* [3]

$$\begin{aligned} \varphi(x) &= -\textstyle\int_\Omega \mathcal{E}(x-y)\eta(y)dy+ \\ &+ \textstyle\int_{\partial\Omega}[-\nabla_y\varphi(y)\mathcal{E}(x-y)+\nabla_y\mathcal{E}(x-y)\varphi(y)]\mathbf{n}(y)dS_y, \end{aligned} \tag{4.21}$$

where

$$\mathcal{E}(x) = -\frac{1}{4\pi|x|}$$

and

$$\eta = \nabla\cdot(\sigma\mathbf{z}).$$

Dividing Ω into Ω_R and Ω^R as in above while getting the estimates for the solution of the Stokes problem and using Lemmas 3.6 and 3.7 in order to estimate the volume integrals in (4.21), one obtains[2]

$$\| |x| \nabla \varphi \|_{0,\infty} \leq c\Big(\| |x|^2 \nabla \cdot (\sigma \mathbf{z}) \|_{0,\infty} + \| \nabla \varphi \|_{0,\infty} + \| \varphi \|_{0,\infty} \Big), \tag{4.22}$$

$$\| |x|^2 \nabla^2 \varphi \|_{1,r} \leq c\Big(\| |x|^2 \nabla \cdot (\sigma \mathbf{z}) \|_{1,r} + \| \nabla \varphi \|_{0,\infty} + \| \varphi \|_{0,\infty} \Big), \tag{4.23}$$

$$\begin{aligned} \| |x| \nabla^3 \varphi \|_{0,\infty} \leq c(& \| |x| \nabla^2 \nabla \cdot (\sigma \mathbf{z}) \|_{0,2} + \| |x| \nabla^2 \nabla \cdot (\sigma \mathbf{z}) \|_{0,6} + \\ & + \| \nabla \varphi \|_{0,\infty} + \| \varphi \|_{0,\infty}). \end{aligned} \tag{4.24}$$

Since $W^{1,r}(\Omega) \subset C^0(\bar{\Omega})$, it follows from (4.23) in particular that

$$\| |x|^2 \nabla^2 \varphi \|_{0,\infty} \leq c(\| |x|^2 \nabla \cdot (\sigma \mathbf{z}) \|_{0,\infty} + \| \nabla \varphi \|_{0,\infty} + \| \varphi \|_{0,\infty}). \tag{4.25}$$

Recalling Lemma 3.1 and using estimates (4.22)–(4.24), one easily deduces that

$$\| \varphi \|_{G_k} \leq c\Big(\| \sigma \|_{S_k} \| \nabla \mathbf{z} \|_{\mathbf{V}_k} + \| |x|^2 \nabla (\nabla \cdot (\sigma \mathbf{z})) \|_{0,r} \Big). \tag{4.26}$$

Finally, let us study the asymptotic behaviour of the solution $\mathbf{T}$ of the transport equation $(2.5)_5$. Provided $\delta > 0$ is chosen sufficiently small, one readily obtains by direct estimations

$$\left. \begin{aligned} \| |x|^2 \mathbf{T} \|_{0,\infty} \; & + \; \| |x| \nabla \mathbf{T} \|_{0,\infty} \leq c\Big(\| |x|^2 \nabla \mathbf{u} \|_{0,\infty} + \\ & + \; \| |x|^2 \nabla^2 \varphi \|_{0,\infty} + \| |x| \nabla^2 \mathbf{u} \|_{0,\infty} + \| |x| \nabla^3 \varphi \|_{0,\infty} + \\ & + \; \| |x|^2 \nabla \mathbf{g} \|_{0,\infty} + \| |x| \nabla^2 \mathbf{g} \|_{0,\infty} + \delta \| \mathbf{T} \|_{k+1} \Big) \end{aligned} \right\} \tag{4.27}$$

Hence, from Lemma 3.3 and estimates (4.27), (4.25), (4.7) one concludes that

$$\| \mathbf{T} \|_{\mathbf{Y}_k} \leq c\left(\| \mathbf{u} \|_{\mathbf{U}_k} + \| \varphi \|_{G_k} + \| \sigma \|_{S_k} \| \nabla \mathbf{z} \|_{\mathbf{V}_k} + \| |x|^2 \nabla (\nabla \cdot (\sigma \mathbf{z})) \|_{0,r} \right). \tag{4.28}$$

Collecting the estimates (4.19), (4.20), (4.26) and (4.28) and observing in particular that $\mathbf{v} = (\mathbf{u} + \nabla \varphi) \in \mathbf{V}_k$, one finally obtains

$$\begin{aligned} \| (\mathbf{v}, \sigma, \mathbf{T}) \|_k \leq C(1 + \| (\mathbf{z}, \tau, \mathbf{S}) \|_k) \Big(& \| \mathbf{f} \|_{\mathbf{Z}_k} + \\ + \| \mathbf{f} \|_{\mathbf{Z}_k} \| (\mathbf{z}, \tau, \mathbf{S}) \|_k + \| (\mathbf{z}, \tau, \mathbf{S}) \|_k^2 + \| (\mathbf{z}, \tau, \mathbf{S}) \|_k^3 \Big). \end{aligned}$$

[2] $\nabla^2 \mathcal{E} = \nabla q(x) = \frac{\sum(\frac{x}{|x|})}{|x|^3}$ with $\sum \in L^\infty(S_1)$, $\int_{S_1} \sum dS = 0$ where S_1 is a unit sphere , is a singular kernel of the Calderon-Zygmund type, see [25]

From this estimate it follows that the operator $\mathcal{N}$ maps the set $\mathbf{D}_\delta$ into itself and if β and δ are chosen sufficiently small, it holds

$$\|(\mathbf{v},\sigma,\mathbf{T})\|_k \le \delta.$$

Hence $\mathcal{N}(\mathbf{D}_\delta) \subset \mathbf{D}_\delta$.

Now, it remains to prove that $\mathcal{N}$ is a contraction in the topology of

$$\mathcal{H}^{k+1} \times H^k \times \mathbf{H}^{\mathbf{k}}.$$

Let us take $(\mathbf{z}_1,\tau_1,\mathbf{S}_1),(\mathbf{z}_2,\tau_2,\mathbf{S}_2) \in \mathbf{D}_\delta$ and denote by $(\mathbf{v}_1,\sigma_1,\mathbf{T}_1)$ and $(\mathbf{v}_2,\sigma_2,\mathbf{T}_2)$ their corresponding images via the mapping $\mathcal{N}$. Further, let us set

$$\mathbf{v}=\mathbf{v}_1-\mathbf{v}_2,\ \ \sigma=\sigma_1-\sigma_2,\ \ \mathbf{T}=\mathbf{T}_1-\mathbf{T}_2,\ \ \mathbf{z}=\mathbf{z}_1-\mathbf{z}_2,\ \ \tau=\tau_1-\tau_2,$$

$$\mathbf{S}=\mathbf{S}_1-\mathbf{S}_2,\ \mathbf{P}_1=\mathbf{P}(\mathbf{z}_1,\tau_1,\mathbf{f},\mathbf{S}_1),\ \mathbf{P}_2=\mathbf{P}(\mathbf{z}_2,\tau_2,\mathbf{f},\mathbf{S}_2),\ \mathbf{P}=\mathbf{P}_1-\mathbf{P}_2.$$

One obtains the following equations for the triple $(\mathbf{v},\sigma,\mathbf{T})$

$$\left.\begin{aligned}
-\mu_e\Delta\mathbf{v}-(\mu_e+\lambda_1)\nabla(\nabla\cdot\mathbf{v})+\nabla\sigma&=\mathbf{P}\\
\nabla\cdot\mathbf{v}&=-\nabla\cdot(\sigma\mathbf{z}_1)-\nabla\cdot(\sigma_2\mathbf{z})\\
\mathbf{T}+\lambda_1\mathbf{z}_1\cdot\nabla\mathbf{T}&=2\mu_e\mathbf{D}(\mathbf{v})-\lambda_1\mathbf{g}(\mathbf{S},\nabla\mathbf{z}_1)-\\
&\qquad-\lambda_1\mathbf{g}(\mathbf{S}_2,\nabla\mathbf{z})-\lambda_1\mathbf{z}\cdot\nabla\mathbf{T}_2\\
\mathbf{v}|_{\partial\Omega}&=0\\
\mathbf{v}(x)\to0,\ \sigma(x)\to0\quad&\text{as}\quad|x|\to\infty.
\end{aligned}\right\}\qquad(4.29)$$

Considering $(4.29)_{1,2,4,5}$ as a (compressible) Stokes system, we get for sufficiently small δ

$$\|\nabla\mathbf{v}\|_k+\|\sigma\|_k\le c\Big(\|\mathbf{P}\|_{k-1}+|\mathbf{P}|_{-1}+\|\sigma_2\mathbf{z}\|_{k+1}\Big).\qquad(4.30)$$

For the transport equation $(4.29)_3$, one obtains, for sufficiently small δ

$$\begin{aligned}\|\mathbf{T}\|_k\ \le\ &c\Big(\|\nabla\mathbf{v}\|_k+\|\mathbf{S}\|_k\|\nabla\mathbf{z}_1\|_{k+1}+\\
&+\|\mathbf{S}_2\|_{k+1}\|\nabla\mathbf{z}\|_k+\|\nabla\mathbf{z}\|_k\|\mathbf{T}_2\|_{k+1}).\end{aligned}\qquad(4.31)$$

Now, a straightforward calculation implies that

$$\left.\begin{aligned}&|\mathbf{P}|_{-1}+\|\mathbf{P}\|_{k-1}\le c\,(1+\|(\mathbf{z}_1,\tau_1,\mathbf{S}_1)\|_k+\|(\mathbf{z}_2,\tau_2,\mathbf{S}_2)\|_k)\times\\
&\times\Big(\|(\mathbf{z}_1,\tau_1,\mathbf{S}_1)\|_k+\|(\mathbf{z}_2,\tau_2,\mathbf{S}_2)\|_k+\|\mathbf{f}\|_{\mathbf{Z}_k}+\\
&+(\|(\mathbf{z}_1,\tau_1,\mathbf{S}_1)\|_k+\|(\mathbf{z}_2,\tau_2,\mathbf{S}_2)\|_k)^2\Big)\Big(\|\nabla\mathbf{z}\|_k+\|\tau\|_k+\|\mathbf{S}\|_k\Big).\end{aligned}\right\}\qquad(4.32)$$

Hence, one concludes from (4.30), (4.31) and (4.32) that

$$\|\nabla \mathbf{v}\|_k + \|\sigma\|_k + \|\mathbf{T}\|_k \leq$$
$$\leq (1+2\delta)(2\delta + 4\delta^2 + \beta)\Big(|\nabla \mathbf{z}\|_k + \|\tau\|_k + \|\mathbf{S}\|_k\Big)$$

Therefore, if δ and β are chosen sufficiently small, $\mathcal{N}$ is a contraction in $\mathcal{H}^{k+1} \times H^k \times \mathbf{H}^k$ and consequently the mapping $\mathcal{N}$ has a unique fixed point in the ball $\mathbf{D}_\delta$. Hence, we have found a unique solution to the system (2.4) and returning to the original variables we obtain the statement of the theorem. ■

Acknowledgments

Adélia Sequeira and Juha Videman are grateful to Research Centre CMA/IST (Centro de Matemática Aplicada) and to European Union FEDER/PRAXIS (Project Nr.2/2.1/MAT/380/94) for their financial support.

Šárka Matušů-Nečasová would like to express her gratitude to the Grant Agency of Czech Republic Nr. 201/98/1450. This work was concluded during her stay in Portugal funded by the grants from CMA/IST (Lisbon) and from CIM/Fundação Calouste Gulbenkian (Coimbra) – Thematic Term on Theoretical and Computational Fluid Dynamics. The portuguese financial support is deeply acknowledged.

References

[1] Cioranescu, D. and Ouazar, E.H. (1984). Existence and uniqueness for fluids of second grade. *Collège de France Seminars*, Pitman Research Notes in Mathematics Pitman, Boston, 109:178–197.

[2] Edwards, B.J. and Beris, A.N. (1990). Remarks concerning compressible viscoelastic fluid models. *J. Non-Newtonian Fluid Mech.*, 36:411–417.

[3] Galdi, G.P. (1994). *An Introduction to the Mathematical Theory of the Navier-Stokes Equations.* Springer Tracts in Natural Philosophy, Vols. 38 and 39, Springer, New York.

[4] Galdi, G.P. (1995). Mathematical theory of second-grade fluids. G.P. Galdi (ed.) *Stability and Wave Propagation in Fluids*, CISM Courses and Lectures Springer, New York, 344:66-103.

[5] Galdi, G.P., Sequeira, A. and Videman, J.H. (1997). Steady motions of a second-grade fluid in an exterior domain, *Adv. Math. Sci. Appl.*, 7:977–995.

[6] Galdi, G.P. and Simader, Ch. (1990). Existence, uniqueness and L^q estimates for the Stokes problem in an exterior domain. *Arch. Rational Mech. Anal.*, 112:291–318.

[7] Guillopé, C. and Saut, J.-C. (1990). Existence results for the flow of viscoelastic fluids with a differential constitutive law. *Nonlinear Analysis, Theory, Methods & Applications*, 15:849–869.

[8] Matušů-Nečasová, Š, Sequeira A. and Videman, J.H. (1999). Existence of classical solutions for compressible viscoelastic fluids of Oldroyd type past an obstacle. *Mathematical Methods in the Applied Sciences*, 22:449-460.

[9] Matušů-Nečasová, Š. Sequeira A. and Videman, J.H. Steady motion of compressible viscoelastic fluids of Oldroyd type in exterior domain. In preparation.

[10] Nečas, J. (1967). Les Méthodes Directes en la Théorie des Equations Elliptiques. Masson Paris.

[11] Novotný, A. (1998). On the steady transport equation. Málek, J., Nečas, J. and Rokyta, M. (eds.) *Advanced Topics in Theoretical Fluid Mechanics.* Pitman Research Notes in Mathematics Series 392, Longman, Essex, 118–146.

[12] Novotný, A. and Padula, M. (1994). L^p -approach to steady flows of viscous compressible fluids in exterior domains. *Arch. Rational Mech. Anal.*, 126:243–297.

[13] Novotný, A. and Padula, M. (1996). Physically reasonable solutions to steady compressible Navier-Stokes equations in 3D-exterior domains ($v_\infty = 0$). *J. Math. Kyoto Univ.*, 36(2):389–422.

[14] Novotný, A., Sequeira, A. and Videman, J.H. (1997). Existence of threedimensional flows of second-grade fluids past an obstacle. *Nonlinear Analysis, Theory, Methods & Applications*, 30:3051–3058.

[15] Novotný, A., Sequeira, A. and Videman, J.H. Steady Motions of Viscoelastic Fluids in 3-D Exterior Domains–Existence, Uniqueness and Asymptotic Behaviour. *Arch. Rational Mech. Anal.* In press.

[16] Oldroyd, J.G. (1950). On the formulation of rheological equations of state. *Proc. Roy. Soc. London*, A200:523–541.

[17] Phelan, F.R., Malone, M.F. and Winter, H.H. (1989). A purely hyperbolic model for unsteady viscoelastic flow. *J. Non-Newtonian Fluid Mech.*, 32:197–224.

[18] Rajagopal, K.R. (1993). Mechanics of non-Newtonian fluids. Galdi, G.P. and Nečas, J. (eds.) *Recent Developments in Theoretical Fluid Mechanics.* Pitman Research Notes in Mathematics, 291:129–162.

[19] Renardy, M., Hrusa, W.J. and Nohel, J.A. (1987). *Mathematical Problems in Viscoelasticity.* Longman, New York.

[20] Renardy, M. (1985). Existence of Slow Steady Flows of Viscoelastic Fluids with Differential Constitutive Equations, *ZAMM*, 65:449-451.

[21] Renardy, M. (1988). Recent advances in the mathematical theory of steady flow of viscoelastic fluids. *J. Non-Newtonian Fluid Mech.*, 29:11–24.

[22] Schowalter, W.R. (1978). *Mechanics of Non-Newtonian Fluids.* Pergamon Press, New York.

[23] Simader, C. (1990). *The weak Dirichlet and Neumann problem for the Laplacian in L^q for bounded and exterior domains.* Applications in Nonlinear analysis, function spaces and applications (editors: Krbec, Kufner, Opic, Rákosník) Leipzig, Teubner, 4:180–223.

[24] Smirnov, A. (1964). *A course of higher mathematics.* Pergamon Press, Addison - Wesley.

[25] Stein, E. (1957). Note on singular integrals. *Proc. Am. Math. Soc.*, 8:250–254.

[26] Sy, M.H. (1996). *Contributions à l'étude mathématique des problèmes issues dè la mécanique des fluides viscoélastiques. Lois de comportement de type intégral ou différentiel.* Thèse d'Université de Paris-Sud, Orsay.

[27] Talhouk, R. (1995). Écoulements stationnaires de fluides viscoélastiques faiblement compressibles. *C. R. Acad. Sci. Paris*, 320:1025–1030.

[28] Truesdell, C. and Noll, W. (1992). *The Nonlinear Field Theories of Mechanics*. 2nd edition, Springer, Berlin.

[29] Videman, J.H. (1997). *Mathematical Analysis of Viscoelastic Non-Newtonian Fluids*. PhD Thesis, Instituto Superior Técnico, Lisbon.

REGULARITY OF A SUITABLE WEAK SOLUTION TO THE NAVIER–STOKES EQUATIONS AS A CONSEQUENCE OF REGULARITY OF ONE VELOCITY COMPONENT

Jiří Neustupa, Patrick Penel

Abstract: We show that if $(v;p)$ is a suitable weak solution to the Navier–Stokes equations (in the sense of L.Caffarelli, R.Kohn & L.Nirenberg – see [1]) such that v_3 (the third component of v) is essentially bounded in a sub-domain D of a time–space cylinder Q_T then v has no singular points in D.

Keywords: Navier–Stokes equations, weak solutions, regularity.

1. INTRODUCTION

Let Ω be either $\mathbb{R}^3$ or a bounded domain in $\mathbb{R}^3$ with the boundary $\partial\Omega$ of the class $C^{2+\mu}$ for some $\mu > 0$ and let T be a positive number. Denote $Q_T = \Omega\times\,]0,T[$. We will deal with the Navier–Stokes initial–boundary value problem for viscous incompressible fluids which is defined by the following equations and conditions

$$\frac{\partial v}{\partial t} + v\cdot\nabla v = f - \nabla p + \nu\,\Delta v \qquad \text{in } Q_T, \tag{1.1}$$

$$\operatorname{div} v = 0 \qquad \text{in } Q_T, \tag{1.2}$$

$$v = 0 \qquad \text{on } \partial\Omega\times\,]0,T[, \tag{1.3}$$

$$v|_{t=0} = v_0 \tag{1.4}$$

where $v = (v_1, v_2, v_3)$ and p denote the unknown velocity and pressure, f is the external body force and $\nu > 0$ is the viscosity coefficient. We shall assume, for simplicity, that $f = 0$.

Qualitative properties of the problem (1.1)–(1.4) are discussed in many books (see, e.g., O.A.Ladyzhenskaya [12], R.Temam [19] or G.P.Galdi [3] and [4]) and at least hundreds of articles. It is well known that a weak solution v of (1.1)–(1.4) belongs to $L^\infty(0,T;\ L^2(\Omega)^3)$ and $\nabla v \in L^2(Q_T)^9$. The detailed information about the most of the known results on the existence and regularity of weak solutions can be found e.g. in the recent works of H.Kozono [11] and G.P.Galdi [4]. The global in time existence of the weak solutions is known already for a long time (see J.Leray [13] and E.Hopf [7]). The uniqueness is known only in the class $L^r(0,T;\ L^s(\Omega)^3)$, where $r \in [2,+\infty]$, $s \in [3,+\infty]$ and $2/r+3/s \leq 1$ (see e.g. G.Prodi [15], H.Sohr & W.von Wahl [17], H.Kozono & H.Sohr [10], H.Kozono [11], G.P.Galdi [4]). Furthermore, it is known that if v_0 and f are "smooth" then the weak solution v of the problem (1.1)–(1.4) is "smooth" locally in time (K.K.Kiselev & O.A.Ladyzhenskaya [9], V.A.Solonnikov [18], G.P.Galdi [4]). If, moreover, v_0 and f are "small enough" then v remains "smooth" globally in time (O.A.Ladyzhenskaya [12], J.G.Heywood [6], G.P.Galdi [4]). The regularity was also proved for the weak solutions which belong to the class $L^r(0,T;\ L^s(\Omega)^3)$ where $r \in [2,+\infty[$, $s \in]3,+\infty]$ and $2/r+3/s < 1$ (S.Kaniel & M.Shinbrot [8], Y.Giga [5], G.P.Galdi [4]). The regularity of weak solutions in the class $L^\infty(0,T;\ L^3(\Omega)^3)$ still remains an open problem.

J.Leray [13] proposed a possible counter–example to the global in time regularity of weak solutions. Nevertheless, this type of blow–up was later excluded by J.Nečas, M.Růžička & V.Šverák [14]. Thus, in spite of an enormous effort of many mathematicians, the question of the global in time uniqueness and regularity of weak solutions of the problem (1.1)–(1.4) (with v_0 and f "smooth enough") still remains, in its whole generality, open.

When trying to construct an example of a blow up of solutions of the problem (1.1)–(1.4) (i.e. an irregularity of a weak solution), it seems interesting to know whether the components of each weak solution v are coupled. A question is how some regularity of one of them already implies the regularity of all components. This paper gives the answer to this question in the case when v is a suitable weak solution.

$C_0^\infty(\Omega)^3$ denotes the set of all infinitely differentiable vector–functions defined in Ω, with a compact support in Ω. $C_{0,\sigma}^\infty(\Omega)^3$ is a set of all divergence–free vector functions which belong to $C_0^\infty(\Omega)^3$. $L^2_\sigma(\Omega)^3$ is the closure of $C_{0,\sigma}^\infty(\Omega)^3$ in $L^2(\Omega)^3$.

We shall denote by U open balls in $\mathbb{R}^3 \times \mathbb{R}$ and by B open balls in $\mathbb{R}^3$. (Thus, e.g. $B_\epsilon(x)$ will be an open ball in $\mathbb{R}^3$ with the center x and radius ϵ.)

A pair $(v;p)$ of measurable functions on Q_T is called a *suitable weak solution* of the problem (1.1)–(1.4) (with $f = 0$) if

1. $v \in L^2(0,T;\, W_0^{1,2}(\Omega)^3) \cap L^\infty(0,T;\, L^2_\sigma(\Omega)^3)$, $p \in L^{5/4}(Q_T)$,

2. v is a weak solution of the problem (1.1)–(1.4) and p is an associated pressure,

3. $(v;p)$ satisfies the so called *generalized energy inequality*

$$\int_0^T \int_\Omega |\nabla v|^2\, \phi\, dx\, dt \le \int_0^T \int_\Omega \Big[|v|^2 \Big(\frac{\partial \phi}{\partial t} + \Delta\phi\Big) + (|v|^2 + 2p)\, v \cdot \nabla\phi \Big]\, dx\, dt$$

for every infinitely differentiable function ϕ on Q_T with a compact support in Q_T.

L.Caffarelli, R.Kohn & L.Nirenberg [1] proved the existence of a suitable weak solution of the problem (1.1)–(1.4) under the assumption that the initial data v_0 is in $L^2_\sigma(\Omega)^3$ and moreover, under the additional assumption that $v_0 \in W^{2/5,5/4}(\Omega)^3$ (the space of vector functions whose components have fractional derivatives up to the order $\frac{2}{5}$ in $L^{5/4}(\Omega)$) in the case when Ω is bounded. (In fact, L.Caffarelli, R.Kohn & L.Nirenberg work with the external force f which need not be equal to zero and they also require some smoothness of f, but we wish not to go into details in this point because our f equals identically zero.)

A point $(x,t) \in Q_T$ is called a *regular point* of the weak solution v if there exists a neighbourhood U of (x,t) in Q_T such that $v \in L^\infty(U)^3$. Points of Q_T which are not regular are called *singular*. Let us denote by $S(v)$ the set of all singular points of v. It is obvious that $S(v)$ is closed in Q_T.

A further important result of the paper [1] says that if $(v;p)$ is a suitable weak solution of the problem (1.1)–(1.4) then its singular set $S(v)$ has a so called 1–dimensional parabolic measure equal to zero. Since the parabolic measure dominates the Hausdorff measure, this result implies that the 1–dimensional Hausdorff measure of $S(v)$ equals zero.

The main result of this paper is the following:

Theorem 1. *Let $(v;p)$ (where $v = (v_1, v_2, v_3)$) be a suitable weak solution to the problem (1.1)–(1.4) (with $f = 0$). Suppose that there exists a sub–domain D of Q_T such that v_3 is essentially bounded in D. Then v has no singular points in D (i.e. the set $S(v) \cap D$ is empty).*

2. AUXILIARY RESULTS

Suppose in the following that $(v;p)$ is a suitable weak solution to the problem (1.1)–(1.4) (with $f = 0$) which satisfies the assumptions of Theorem 1.

It is known that the interval $]0,T[$ can be expressed as $\bigcup_{\gamma\in\Gamma}]a_\gamma, b_\gamma[\cup G$ where set Γ is at most countable, $]a_\gamma, b_\gamma[$ (for $\gamma \in \Gamma$) are disjoint open intervals in $]0,T[$, the $\frac{1}{2}$–dimensional Hausdorff measure of G is finite and v is of the class C^∞ on the set $\Omega \times \mathcal{I}$ for every interval $\mathcal{I}$ whose closure is contained in some of the intervals $]a_\gamma, b_\gamma[$. (This follows e.g. from the results of J.Heywood [6], C.Foias & R.Temam [2] and G.P.Galdi [4].) Functions v and p satisfy equations (1.1) and (1.2) in a strong sense on each of the time intervals $]a_\gamma, b_\gamma[$. In accordance with G.P.Galdi [4], the time instants b_γ can be called *epochs of possible irregularity*.

Suppose further that D' is a domain in D such that $D' \subset \overline{D'} \subset D$ and $\partial D' \cap S(v) = \emptyset$. Then $D' \cap S(v)$ is closed. The orthogonal projection of $D' \cap S(v)$ onto the time axis t is a closed subset of G. Let us denote this subset by G'. (Thus, G' is a set of times at which solution v has irregular points in D'.) The time interval $]0,T[$ can now be written as $\bigcup_{\gamma\in\Gamma'}]a'_\gamma, b'_\gamma[\cup G'$ where set Γ' is at most countable and $]a'_\gamma, b'_\gamma[$ (for $\gamma \in \Gamma'$) are disjoint open intervals in $]0,T[$. Moreover, v is of the class C^∞ on every set $\{(x,t) \in D';\ t \in \mathcal{I}\}$ where $\mathcal{I}$ is an arbitrary interval whose closure is contained in some of the intervals $]a'_\gamma, b'_\gamma[$. We shall call the time instants b'_γ (for $\gamma \in \Gamma'$) *D'–epochs of irregularity.*

Lemma 1. *Suppose that t_0 is a D'–epoch of irregularity (i.e. $t_0 = b'_\gamma$ for some $\gamma \in \Gamma'$) and $(x_0, t_0) \in D'$. Then there exist positive numbers τ, ϵ_1, ϵ_2 such that $\epsilon_1 < \epsilon_2$ and*

1. *τ is so small that $a'_\gamma < b'_\gamma - \tau = t_0 - \tau$,*

2. $\overline{B_{\epsilon_2}(x_0)} \times [t_0 - \tau, t_0 + \tau] \subset D'$,

3. $\left(\overline{B_{\epsilon_2}(x_0)} - B_{\epsilon_1}(x_0)\right) \times [t_0 - \tau, t_0 + \tau] \cap S(v) = \emptyset$,

4. *v and p have all space derivatives continuous in $\left(\overline{B_{\epsilon_2}(x_0)} - B_{\epsilon_1}(x_0)\right) \times [t_0 - \tau, t_0 + \tau]$.*

Proof. Since D' is open, there exists $\epsilon_0 > 0$ and $\tau > 0$ such that $\overline{B_{\epsilon_0}(x_0)} \times [t_0 - \tau, t_0 + \tau] \subset D'$. τ can surely be chosen so small that $a'_\gamma < b'_\gamma - \tau = t_0 - \tau$.

Further, we claim that there exists $\epsilon_1 \in]0, \epsilon_0[$ such that the set

$$A(\epsilon_1) = \{(x,t) \in D';\ |x - x_0| = \epsilon_1,\ t \in [t_0 - \tau, t_0 + \tau]\}$$

has an empty intersection with $S(v)$. Suppose the opposite. Then to each $\epsilon \in]0, \epsilon_0[$ there exists a point $(x_\epsilon, t_\epsilon) \in S(v) \cap A(\epsilon)$. Since the 1–dimensional Hausdorff measure of $S(v) \cap \left(\cup_{\epsilon \in]0,\epsilon_0[} A(\epsilon)\right)$ is zero, it can be covered by m 4–dimensional balls $U^1, \ldots, U^m$ in $\mathbb{R}^3 \times \mathbb{R}$ with radii $\rho^1, \ldots, \rho^m$ such that

$$\rho^1 + \ldots + \rho^m < \tfrac{1}{4}\epsilon_0 . \tag{2.1}$$

The balls $U^1, \ldots, U^m$ can be chosen so that their centers $(x^1, t^1), \ldots, (x^m, t^m)$ successively belong to $A(\epsilon^1), \ldots, A(\epsilon^m)$ for some $\epsilon^1, \ldots \epsilon^m \in]0, \epsilon_0[$. Then the intervals $]\epsilon^1 - \rho^1, \epsilon^1 + \rho^1[, \ldots,]\epsilon^m - \rho^m, \epsilon^m + \rho^m[$ form a covering of the interval $]0, \epsilon_0[$. (Indeed, if $\epsilon \in]0, \epsilon_0[$ and $(x_\epsilon, t_\epsilon) \in A(\epsilon) \cap S(v)$ then $(x_\epsilon, t_\epsilon) \in U^i$ for some $i \in \{1; \ldots; m\}$. Since $|\epsilon - \epsilon^i| \leq \mathit{dist}\,((x_\epsilon, t_\epsilon); (x^i, t^i)) < \rho^i$, we also have $\epsilon \in]\epsilon^i - \rho^i, \epsilon^i + \rho^i[$.) This and (2.1) is the contradiction with the fact that the length of the interval $]0, \epsilon_0[$ is ϵ_0.

The existence of $\epsilon_2 \in]\epsilon_1, \epsilon_0]$ such that statement 3 of the lemma is true now follows from the compactness of the set $A(\epsilon_1)$ and the closedness of $D' \cap S(v)$.

The statement in item 4 is a consequence of the interior regularity results, due essentially to J.Serrin [16]. (See also G.P.Galdi [4].) □

Let us further denote for simplicity $B_1 = B_{\epsilon_1}(x_0)$ and $B_2 = B_{\epsilon_2}(x_0)$. Put $\epsilon_3 = (2\epsilon_1 + \epsilon_2)/3$, $\epsilon_4 = (\epsilon_1 + 2\epsilon_2)/3$ and $B_3 = B_{\epsilon_3}(x_0)$, $B_4 = B_{\epsilon_4}(x_0)$. Suppose that η is an infinitely differentiable function on $\mathbb{R}^3$ such that its values are in the interval $[0, 1]$, $\eta = 0$ on $\mathbb{R}^3 - B_4$ and $\eta = 1$ on B_3.

From now $\omega = (\omega_1, \omega_2, \omega_3)$ will denote $\operatorname{curl} v$.

Lemma 2. *Suppose that t_0 is a D'–epoch of irregularity and $(x_0, t_0) \in D'$. Let ϵ_1 and ϵ_2 be positive numbers given by Lemma 1. Then $\omega_3 \in L^\infty(t_0 - \tau, t_0;\, L^2(B_2)) \cap L^2(t_0 - \tau, t_0;\, W^{1,2}(B_2))$.*

Proof. Applying operator curl to equation (1.1), we obtain a vector equation for ω. Its third component is

$$\frac{\partial \omega_3}{\partial t} + v \cdot \nabla \omega_3 = \omega \cdot \nabla v_3 + \nu\, \Delta \omega_3 . \tag{2.2}$$

Multiplying equation (2.2) by $\eta^2 \omega_3$, integrating on B_2, applying the integration by parts and using the essential boundedness of v_3, we can obtain the estimate

$$\frac{d}{dt} \int_{B_2} (\eta\omega_3)^2\, dx + \frac{\nu}{2} \int_{B_2} |\nabla(\eta\omega_3)|^2\, dx \leq c_1 \int_{B_2} |\omega|^2\, dx + c_1 . \tag{2.3}$$

(Of course, the constant c_1 does not depend on t.) The integrability of the right hand side of inequality (2.3) on the time interval $]t_0 - \tau, t_0[$

and the form of function η imply that $\omega_3 \in L^\infty(t_0-\tau, t_0; L^2(B_1)) \cap L^2(t_0-\tau, t_0; W^{1,2}(B_1))$. Using the regularity of v and ω on $(\overline{B_2} - B_1) \times [t_0-\tau, t_0]$ (see statement 4 of Lemma 1), we obtain that $\omega_3 \in L^\infty(t_0-\tau, t_0; L^2(B_2)) \cap L^2(t_0-\tau, t_0; W^{1,2}(B_2))$. $\square$

Lemma 3. *Let $\mathcal{D}$ be a bounded Lipschitz domain in $\mathbb{R}^3$. Let further $1 < r < +\infty$ and $m \in (\mathbb{N} \cup \{0\})$. Then there exists a linear operator R from $W_0^{m,r}(\mathcal{D})$ into $W_0^{m+1,r}(\mathcal{D})^3$ with the properties:*

1. $\operatorname{div} Rf = f$ *for all* $f \in W_0^{m,r}(\mathcal{D})$ *with* $\int_{\mathcal{D}} f\, dx = 0$,
2. $\exists\ c_2 > 0 : \ \|\nabla^{m+1} Rf\|_{L^r(\mathcal{D})^{3m+2}} \le c_2 \|\nabla^m f\|_{L^r(\mathcal{D})^{3m}}$ *for all* $f \in W_0^{m,r}(\mathcal{D})$.

Lemma 3 immediately follows from G.P.Galdi [3, Theorem 3.2, Chap. III.3].

3. PROOF OF THEOREM 1

Let D' be a domain in D such that $D' \subset \overline{D'} \subset D$ and $\partial D' \cap S(v) = \emptyset$. Let the sets G', Γ' and $]a'_\gamma, b'_\gamma[$ (for $\gamma \in \Gamma'$) have the same meaning as in the previous section.

If set G' is empty then v has no singular points in D'.

Suppose that G' is nonempty, t_0 is a D'–epoch of irregularity and $(x_0, t_0) \in D'$. Suppose further that τ, ϵ_1, ϵ_2 are the numbers given by Lemma 1, B_1, ..., B_4 are the balls defined in Section 2 and η is the function which was also defined in Section 2.

We set $V(.,t) = R(\nabla\eta \cdot v(.,t))$ where R is the operator from Lemma 3 (with $\mathcal{D} = B_2$). It follows from the proof of Lemma 3 (see G.P.Galdi [3, Chap. III.3]) that since $\nabla\eta$ has a compact support in B_2, V also has a compact support in B_2. We have

$$\int_{B_2} \nabla\eta \cdot v\, dx = \int_{B_2} \operatorname{div}(\eta v)\, dx = \int_{\partial B_2} \eta v \cdot n\, dS = 0$$

where n is the outer normal vector to ∂B_2. Thus, $\operatorname{div} V = \nabla\eta \cdot v$ in $B_2 \times]t_0-\tau, t_0+\tau[$. V and ∇V are essentially bounded in $B_2 \times]t_0-\tau, t_0+\tau[$ and moreover,

$$V \in L^2(t_0-\tau, t_0+\tau; W^{2,2}(B_2)^3), \tag{3.1}$$

$$\frac{\partial V}{\partial t} \in L^2(B_2 \times]t_0-\tau, t_0+\tau[)^3. \tag{3.2}$$

The Navier–Stokes equation (1.1) (with $f = 0$) can be written in the form

$$\frac{\partial v}{\partial t} + \omega \times v = -\nabla(p + \tfrac{1}{2}|v|^2) + \nu\Delta v. \tag{3.3}$$

We put $u = \eta v - V$. It can be verified that u is a strong solution of the following problem:

$$\frac{\partial u}{\partial t} + \omega \times u = g - \nabla(\eta p + \tfrac{1}{2}\eta|v|^2) + \nu\Delta u \quad \text{in } B_2\times\,]t_0 - \tau, t_0[\tag{3.4}$$

$$\operatorname{div} u = 0 \quad \text{in } B_2\times\,]t_0 - \tau, t_0[\tag{3.5}$$

$$u = 0 \quad \text{on } \partial B_2\times\,]t_0 - \tau, t_0[\tag{3.6}$$

where

$$g = -\frac{\partial V}{\partial t} - \omega \times V + (p + \tfrac{1}{2}|v|^2)\,\nabla\eta + \nu\Delta V.$$

The third component u_3 of vector function u is obviously bounded at least on $(B_2 - B_1)\times\,]t_0 - \tau, t_0[$.

We shall now suppose that t is a fixed time from the interval $]t_0-\tau, t_0[$. We shall multiply equation (3.4) by $-\Delta u$ and integrate over B_2. Since u and η have a compact support in B_2, we have

$$\int_{B_2} \nabla(\eta p + \tfrac{1}{2}\eta|v|^2)\cdot\Delta u\,dx = \int_{B_2}(\eta p + \tfrac{1}{2}\eta|v|^2)\,\Delta\operatorname{div} u\,dx = 0.$$

Thus, we obtain

$$\frac{d}{dt}\,\frac{1}{2}\int_{B_2}|\nabla u|^2\,dx + \nu\int_{B_2}|\Delta u|^2\,dx = -\int_{B_2} g\cdot\Delta u\,dx + \int_{B_2}(\omega\times u)\cdot\Delta u\,dx. \tag{3.7}$$

In the following estimates, c_3 and c_4 will be generic constants, i.e. constants whose values may change from one line to the next. While c_3 will depend only on domain B_2, function η and $\sup\operatorname{ess}_{t\in(0,T)}\|v(\,.\,,t)\|_{L^2(\Omega)^3}$, c_4 will also depend on a certain number δ. They will not depend on t.

Using the properties of functions V and η and the boundedness of p and v on $\operatorname{supp}\nabla\eta\times\,]t_0 - \tau, t_0[$, we can estimate the first integral on the right hand side of (3.7):

$$\int_{B_2} g\cdot\Delta u\,dx \le \frac{\nu}{3}\int_{B_2}|\Delta u|^2\,dx + \frac{3}{4\nu}\int_{B_2} g^2\,dx \le$$

$$\le \frac{\nu}{3}\int_{B_2}|\Delta u|^2\,dx + c_3\int_{B_2}|\omega|^2\,dx + c_3. \tag{3.8}$$

The second integral on the right hand side of (3.7) can be written as

$$\int_{B_2}(\omega\times u)\cdot\Delta u\,dx =$$

$$= \int_{B_2}(\omega_2 u_3 - \omega_3 u_2)\,\Delta u_1\,dx + \int_{B_2}(\omega_3 u_1 - \omega_1 u_3)\,\Delta u_2\,dx +$$

$$+ \int_{B_2}(v_{3,2} - v_{2,3})\,u_2\,\Delta u_3\,dx - \int_{B_2}(v_{1,3} - v_{3,1})\,u_1\,\Delta u_3\,dx. \tag{3.9}$$

(For example, $v_{3,2}$ denotes the partial derivative of v_3 with respect to x_2. Analogously, $v_{3,22}$ will later denote the second partial derivative of v_3 with respect to x_2, etc.) We shall further use the compact imbedding of $W^{2,2}(B_2)$ into $L^\infty(B_2)$ and the estimate

$$\|\zeta\|_{L^\infty(B_2)} \leq \delta\, \|\Delta\zeta\|_{L^2(B_2)} + c_4(\delta)\, \|\zeta\|_{L^2(B_2)} \tag{3.10}$$

(for every $\delta > 0$ and $\zeta \in W^{2,2}(B_2) \cap W_0^{1,2}(B_2)$) which follows from this imbedding. We shall work with $\delta \in]0,1]$ and so we shall have $\delta^2 \leq \delta$. Using also the essential boundedness of v_3 and u_3 on $B_2\times]t_0-\tau, t_0[$, the boundedness of $\int_{B_2} |u|^2\, dx$ on the time interval $]t_0-\tau, t_0[$ and the properties of ω_3 following from Lemma 2, we can estimate the terms on the right hand side of (3.9) in this way:

$$\left|\int_{B_2} \omega_2\, u_3\, \Delta u_1\, dx\right| \leq \delta \int_{B_2} |\Delta u_1|^2\, dx + c_4(\delta) \int_{B_2} \omega_2^2\, dx, \tag{3.11}$$

$$\begin{aligned}
\left|\int_{B_2} \omega_3\, u_2\, \Delta u_1\, dx\right| &\leq \|u_2\|_{L^\infty(B_2)} \int_{B_2} |\omega_3|\, |\Delta u_1|\, dx \leq \\
&\leq \left[\delta\Big(\int_{B_2} |\Delta u_2|^2\, dx\Big)^{1/2} + c_4(\delta)\Big(\int_{B_2} |u_2|^2\, dx\Big)^{1/2}\right] \int_{B_2} |\omega_3|\, |\Delta u_1|\, dx \leq \\
&\leq \left[\delta\Big(\int_{B_2} |\Delta u_2|^2\, dx\Big)^{1/2} + c_4(\delta)\right] \Big(\int_{B_2} \omega_3^2\, dx\Big)^{1/2} \Big(\int_{B_2} |\Delta u_1|^2\, dx\Big)^{1/2} \leq \\
&\leq \delta\, c_3 \int_{B_2} |\Delta u|^2\, dx + c_4(\delta),
\end{aligned} \tag{3.12}$$

The integrals of $\omega_3\, u_1\, \Delta u_2$ and $\omega_1\, u_3\, \Delta u_2$ on B_2 can be estimated quite analogously. The integral of $v_{2,3}\, u_2\, \Delta u_3$ can be estimated in this way:

$$\begin{aligned}
\left|\int_{B_2} v_{2,3}\, u_2\, \Delta u_3\, dx\right| &= \left|\int_{B_2} [u_{2,3} + (1-\eta) v_{2,3} + V_{2,3}]\, u_2\, \Delta u_3\, dx\right| \leq \\
&\leq \left|\tfrac{1}{2} \int_{B_2} u_2^2\, \Delta u_{3,3}\, dx\right| + c_3 \int_{B_2} |u_2\, \Delta u_3|\, dx \leq \left|\int_{B_2} u_{2,j}\, u_{2,j}\, u_{3,3}\, dx\right| + \\
&+ \left|\int_{B_2} u_2\, \Delta u_2\, u_{3,3}\, dx\right| + \delta \int_{B_2} |\Delta u_3|^2\, dx + c_4(\delta) \int_{B_2} u_2^2\, dx \leq \\
&\leq \left|2 \int_{B_2} u_{2,j3}\, u_{2,j}\, u_3\, dx\right| + \delta \int_{B_2} |\Delta u_2|^2\, dx + c_4(\delta) \int_{B_2} u_2^2\, u_{3,3}^2\, dx + \\
&\quad + \delta \int_{B_2} |\Delta u_3|^2\, dx + c_4(\delta).
\end{aligned}$$

Using the boundedness of u_3 in the first integral and applying twice the integration by parts to the third integral on the right hand side, we obtain:

$$\left|\int_{B_2} v_{2,3}\, u_2\, \Delta u_3\, dx\right| \le 3\delta \int_{B_2} |\Delta u|^2\, dx + c_4(\delta) \int_{B_2} |\nabla u_2|^2\, dx + $$
$$+ c_4(\delta) \int_{B_2} u_{2,3}^2\, u_3^2\, dx + c_4(\delta) \int_{B_2} |u_2|\, |u_{2,33}|\, u_3^2\, dx + $$
$$+ c_4(\delta) \int_{B_2} u_2^2\, |u_{3,33}|\, |u_3|\, dx + c_4(\delta) \le$$
$$\le 3\delta \int_{B_2} |\Delta u|^2\, dx + c_4(\delta) \int_{B_2} |\nabla u|^2\, dx + c_4(\delta) \int_{B_2} |u_2|\, |u_{2,33}|\, dx + $$
$$+ \delta \int_{B_2} u_{3,33}^2\, dx + c_4(\delta) \int_{B_2} u_2^4\, dx + c_4(\delta) \le 5\delta \int_{B_2} |\Delta u|^2\, dx + $$
$$+ c_4(\delta) \int_{B_2} |\nabla u|^2\, dx + c_4(\delta) \int_{B_2} u_2^2\, dx + c_4(\delta)\, \|u_2\|_{L^\infty(B_2)}^2 \int_{B_2} u_2^2\, dx + c_4(\delta).$$

Applying now estimate (3.10) to $\|u_2\|_{L^\infty(B_2)}^2$, we get:

$$\left|\int_{B_2} v_{2,3}\, u_2\, \Delta u_3\, dx\right| \le 6\delta \int_{B_2} |\Delta u|^2\, dx + c_4(\delta) \int_{B_2} |\nabla u|^2\, dx + c_4(\delta). \tag{3.13}$$

Further, we have:

$$\left|\int_{B_2} v_{3,2}\, u_2\, \Delta u_3\, dx\right| \le \delta \int_{B_2} |\Delta u_3|^2\, dx + c_4(\delta) \int_{B_2} v_{3,2}^2\, u_2^2\, dx.$$

Integrating again twice by parts in the second integral, we obtain:

$$\left|\int_{B_2} v_{3,2}\, u_2\, \Delta u_3\, dx\right| \le \delta \int_{B_2} |\Delta u_3|^2\, dx + \delta \int_{B_2} v_{3,22}^2\, dx + $$
$$+ c_4(\delta) \int_{B_2} u_2^4\, dx + c_4(\delta) \int_{B_2} v_3^2\, u_{2,2}^2\, dx + c_4(\delta) \int_{B_2} v_3^2\, |u_2|\, |u_{2,22}|\, dx \le$$
$$\le \delta \int_{B_2} |\Delta u_3|^2\, dx + \delta \int_{B_2} v_{3,22}^2\, dx + c_4(\delta)\, \|u_2\|_{L^\infty(B_2)}^2 \int_{B_2} u_2^2\, dx + $$
$$+ c_4(\delta) \int_{B_2} u_{2,2}^2\, dx + \delta \int_{B_2} u_{2,22}^2\, dx + c_4(\delta) \int_{B_2} u_2^2\, dx \le$$
$$\le \delta \int_{B_2} |\Delta u_3|^2\, dx + \delta \int_{B_2} v_{3,22}^2\, dx + \left[\delta \Big(\int_{B_2} |\Delta u_2|^2\, dx\Big)^{1/2} + \right.$$
$$\left. + c_4(\delta) \Big(\int_{B_2} u_2^2\, dx\Big)^{1/2}\right]^2 \int_{B_2} u_2^2\, dx + c_4(\delta) \int_{B_2} u_{2,2}^2\, dx + $$
$$+ \delta \int_{B_2} u_{2,22}^2\, dx + c_4(\delta) \le$$

$$\leq 3\delta \int_{B_2} |\Delta u|^2 \, dx + \delta \int_{B_2} |\Delta v|^2 \, dx + c_4(\delta) \int_{B_2} u_{2,2}^2 \, dx + c_4(\delta). \quad (3.14)$$

All remaining terms on the right hand side of (3.9) can be estimated in a similar way. Thus, estimate (3.7) now gives:

$$\frac{d}{dt}\frac{1}{2}\int_{B_2} |\nabla u|^2 \, dx + \nu \int_{B_2} |\Delta u|^2 \, dx \leq \frac{\nu}{3}\int_{B_2} |\Delta u|^2 \, dx + \delta c_5 \int_{B_2} |\Delta u|^2 \, dx +$$
$$+ \delta c_6 \int_{B_2} |\Delta v|^2 \, dx + c_4(\delta) \int_{B_2} |\omega|^2 \, dx + c_4(\delta) \int_{B_2} |\nabla u|^2 \, dx + c_4(\delta).$$

Using the equality $u = v - V$ on B_3 and the regularity of v on $B_2 - B_3$, we obtain

$$\int_{B_2} |\Delta v|^2 \, dx \leq \int_{B_3} |\Delta u|^2 \, dx + \int_{B_3} |\Delta V|^2 \, dx + \int_{B_2 - B_3} |\Delta v|^2 \, dx \leq$$
$$\leq \int_{B_2} |\Delta u|^2 \, dx + c_3 \, .$$

Choosing now δ so small that $\delta c_5 + \delta c_6 \leq \frac{1}{3}\nu$, we get

$$\frac{d}{dt}\frac{1}{2}\int_{B_2} |\nabla u|^2 \, dx + \frac{\nu}{3}\int_{B_2} |\Delta u|^2 \, dx \leq$$
$$\leq c_7 \int_{B_2} |\omega|^2 \, dx + c_8 \int_{B_2} |\nabla u|^2 \, dx + c_9 \, . \quad (3.15)$$

The integrability of the right hand side of (3.15) on the time interval $]t_0 - \tau, t_0[$ implies that $\nabla u \in L^\infty(t_0 - \tau, t_0; L^2(B_2)^9)$ and $\Delta u \in L^2(B_2 \times]t_0 - \tau, t_0[)^3$. Since $u = v - V$ on $B_3 \times]t_0 - \tau, t_0[$, we have:

$$\nabla v \in L^\infty(t_0 - \tau, t_0; L^2(B_3)^9) \quad \text{and} \quad \Delta v \in L^2(B_3 \times]t_0 - \tau, t_0[)^3. \quad (3.16)$$

It is known that every weak solution to the problem (1.1)–(1.4) can be redefined on a set of measure zero in Q_T so that it becomes a weakly continuous mapping from the interval $]0, T[$ to $L^2(\Omega)^3$. (See e.g. G.P.Galdi [4].) Suppose that the weak solution v that we work with has already been redefined in this way. Then it has a sense to speak about its value at time t_0 – it is a function $v(.,t_0)$ from $L^2(\Omega)^3$. We claim that the function $v^0 = v(.,t_0)|_{B_3}$ (the restriction of $v(.,t_0)$ to B_3) belongs to $W^{1,2}(B_3)^3$. If $\{t_n\}$ is a sequence of time instants which converges to t_0 from the left then $\{v(.,t_n)|_{B_3}\}$ contains a subsequence which is weakly convergent in $W^{1,2}(B_3)^3$. (This follows from (3.16) and from the reflexivity of the space $W^{1,2}(B_3)^3$.) The weak limit of the subsequence, which is in $W^{1,2}(B_3)^3$, coincides with v^0.

Put $h(x,t) = v(x,t)$ for $(x,t) \in \partial B_3 \times [t_0 - \tau, t_0 + \tau]$. It follows from the regularity of v in $(\overline{B_2} - B_1)) \times [t_0 - \tau, t_0 + \tau]$ (see statement 4 of Lemma 1) that h is of the class C^∞ on $\partial B_3 \times [t_0 - \tau, t_0 + \tau]$.

Function v, restricted in space variables to B_3 and in time to the interval $]t_0, t_0 + \tau[$, satisfies in a weak sense the problem given by the equations

$$\frac{\partial v}{\partial t} + v \cdot \nabla v = -\nabla p + \nu \Delta v \quad \text{in } B_3 \times]t_0, t_0 + \tau[, \tag{3.17}$$

$$\operatorname{div} v = 0 \quad \text{in } B_3 \times]t_0, t_0 + \tau[, \tag{3.18}$$

by the boundary condition

$$v = h \qquad \text{on } \partial B_3 \times]t_0, t_0 + \tau[\tag{3.19}$$

and by the initial condition

$$v\,|_{t=t_0} = v^0 \qquad \text{on } B_3\,. \tag{3.20}$$

Due to the higher smoothness of functions h and v^0, there exists $\tau_1 \in]0, \tau[$ such that the problem (3.17)–(3.20) has a strong solution on the time interval $]t_0, t_0 + \tau_1[$. (This follows e.g. from the results of K.K.Kiselev & O.A.Ladyzhenskaya [9] and V.A.Solonnikov [18].) This solution is unique even in the class of weak solutions and so it coincides with our suitable weak solution v on $B_3 \times]t_0, t_0 + \tau_1[$. It satisfies

$$v \in L^\infty(t_0, t_0 + \tau_1;\, W^{1,2}(B_3)^3).$$

However, this fact and (3.16) imply that (x_0, t_0) is a regular point of solution v. (We can obtain this conclusion e.g. if we apply Theorem 7 from H.Kozono [11].) Since x_0 was an arbitrary point in Ω such that $(x_0, t_0) \in D'$, there cannot exist t_0 which is a D'–epoch of irregularity. This means that solution v has no singular points in set D'.

We can now show that v cannot have a singular point (x_0, t_0) even in D: If (x_0, t_0) is a singular point of v in D then there surely exists $r > 0$ and a 4–dimensional ball U_r with the radius r and the center (x_0, t_0) such that $U_r \subset \overline{U_r} \subset D$ and $\partial U_r \cap S(v) = \emptyset$. (Otherwise we could easily derive a contradiction with the fact that the 1–dimensional Hausdorff measure of $S(v)$ is zero.) Putting $D' = U_r$ and using the result saying that $D' \cap S(v) = \emptyset$, we obtain the contradiction with the assumption that (x_0, t_0) is a singular point of v in D. This completes the proof of Theorem 1. □

Acknowledgments

The research was supported by the Université de Toulon et du Var, France, and by the Grant Agency of the Czech Republic (grant No. 201/96/0313).

References

[1] Caffarelli, L., Kohn, R. and Nirenberg, L. (1982). Partial regularity of suitable weak solutions of the Navier–Stokes equations. *Comm. Pure Apl. Math.*, 35:771–831.

[2] Foias, C. and Temam, R. (1979). Some analytic and geometric properties of the solutions of the evolution Navier–Stokes equations. *J. Math. Pures Appl.*, 58:339–368.

[3] Galdi, G.P. (1994). *An Introduction to the Mathematical Theory of the Navier–Stokes Equations. Vol I: Linearized Steady Problems, Vol II: Nonlinear Steady Problems*, Springer Tracts in Natural Philosophy, Vol. 38, 39, Springer–Verlag, New York–Berlin–Heidelberg.

[4] Galdi, G.P. *An Introduction to the Navier–Stokes Initial–Boundary Value Problem.* To be published.

[5] Giga, Y. (1986). Solutions for semilinear parabolic equations in L^p and regularity of weak solutions of the Navier–Stokes equations. *J. Differential Equations*, 61:186–212.

[6] Heywood, J.G. (1980). The Navier–Stokes equations: On the existence, uniqueness and decay of solutions. *Indiana Univ. Math. J.*, 29:639–681.

[7] Hopf, E. (1950). Über die Anfangwertaufgabe für die Hydrodynamirmhen Grundgleichungen. *Math. Nachr.*, 4:213–231.

[8] Kaniel, S. and Shinbrot, M. (1967). Smoothness of weak solutions of the Navier–Stokes equations. *Arch. Rational Mech. Anal.*, 24:302–324.

[9] Kiselev, K.K. and Ladyzhenskaya, O.A. (1957). On existence and uniqueness of the solutions of the nonstationary problem for a viscous incompressible fluid (in Russian). *Izv. Akad. Nauk SSSR*, 21:655–680.

[10] Kozono, H. and Sohr, H. (1996). Remark on uniqueness of weak solutions to the Navier–Stokes equations. *Analysis*, 16:255–271.

[11] Kozono, H. (1998). Uniqueness and regularity of weak solutions to the Navier–Stokes equations. *Lecture Notes in Num. and Appl. Anal.*, 16:161–208.

[12] Ladyzhenskaya, O.A. (1967). Uniqueness and smoothness of generalized solutions of the Navier–Stokes equations (in Russian). *Zap. Nauch. Sem. LOMI*, 5:169–185.

[13] Leray, J. (1934). Sur le mouvements d'un liquide visqueux emplissant l'espace. *Acta Math.*, 63:193–248.

[14] Nečas, J., Růžička, M. and Šverák V. (1996). On Leray's self–similar solutions of the Navier–Stokes equations. *Acta Math.*, 176:283–294.

[15] Prodi, G. (1959). Un teorema di unicità per el equazioni di Navier–Stokes. *Ann. Mat. Pura Appl.*, 48:173–182.

[16] Serrin, J. On the interior regularity of weak solutions of the Navier–Stokes equations. *Arch. Rational Mech. Anal.*, 9:187–195.

[17] Sohr, H. and von Wahl, W. (1984). On the singular set and the uniqueness of weak solutions of the Navier–Stokes equations. *Manuscripta Math.*, 49:27–59.

[18] Solonnikov, V.A. (1973). Estimates of solutions of a non–stationary Navier–Stokes system (in Russian). *Zap. Nauch. Sem. LOMI*, 38:153–231.

[19] Temam, R. (1977). *Navier–Stokes Equations.* North–Holland, Amsterdam–New York–Oxford.

ON A CLASS OF HIGH RESOLUTION METHODS FOR SOLVING HYPERBOLIC CONSERVATION LAWS WITH SOURCE TERMS

Paula de Oliveira, José Santos

Abstract: A class of conservative numerical methods for solving hyperbolic nonhomogeneous scalar conservation laws is presented. Convergence and stability properties are studied. Particular attention is devoted to time depending point sources. Several numerical examples are presented.

Keywords: Conservation laws, point sources, finite volume methods, conservative numerical methods.

1. INTRODUCTION

Pollution problems can be modeled by convection-reaction equations in a case that diffusion can be neglected. These equations are nonhomogeneous scalar conservation laws of the type

$$u_t + (f(u))_x = (q(x,t))_x\,, \quad -\infty < x < \infty \quad \text{and} \quad t > 0. \tag{1.1}$$

Here f is a smooth function, depending in general nonlinearly on u, and q is a bounded, piecewise smooth function.

In a large class of such convection-reaction problems the reaction term is represented by time depending point sources. In order to represent a point source localized at $x = 0$, we consider for instance

$$q(x) = \begin{cases} b_l\,, & x < 0 \\ b_r\,, & x > 0 \end{cases}\,,$$

thus obtaining in the distributional sense, $q_x(x) = (b_l - b_r)\delta(x)$, where δ is the Dirac delta function.

For this class of problems, standard numerical methods obtained by a direct discretization of the differential form fail to converge even in the

Applied Nonlinear Analysis, edited by Sequeira *et al.*
Kluwer Academic / Plenum Publishers, New York, 1999.

linear case. Let us consider for example

$$\begin{cases} u_t + u_x = \sum_{i=1}^{3} c_i \delta(x - a_i) \ , & x \in \mathbb{R} \ , \quad t > 0 \\ u(x,0) = 0 \ , & x \in \mathbb{R} \end{cases}$$

with $c_1 = 1$, $c_2 = 3$, $c_3 = 2$, $a_1 = 0.1$, $a_2 = 0.3$, $a_3 = 0.4$. If we solve this problem using the upwind finite difference scheme

$$\frac{u_j^{n+1} - u_j^n}{k} + \frac{u_j^n - u_{j-1}^n}{h} = \sum_{i=1}^{3} c_i \delta(jh - a_i), \ n = 0, 1, \dots \ , \ j = 0, \pm 1, \dots \ .$$

with $k = h$, which means that we are integrating along the characteristic lines, we observe the vanishing of the numerical solution as h goes to zero. The numerical solution does not converge to the weak solution. We can see this fact in Figure 1, where is represented the numerical solution obtained at $t = 0.5$ computed, respectively, with $h = \frac{1}{10}$, $h = \frac{1}{20}$, and $h = \frac{1}{40}$. The exact solution is represented with a continuous line.

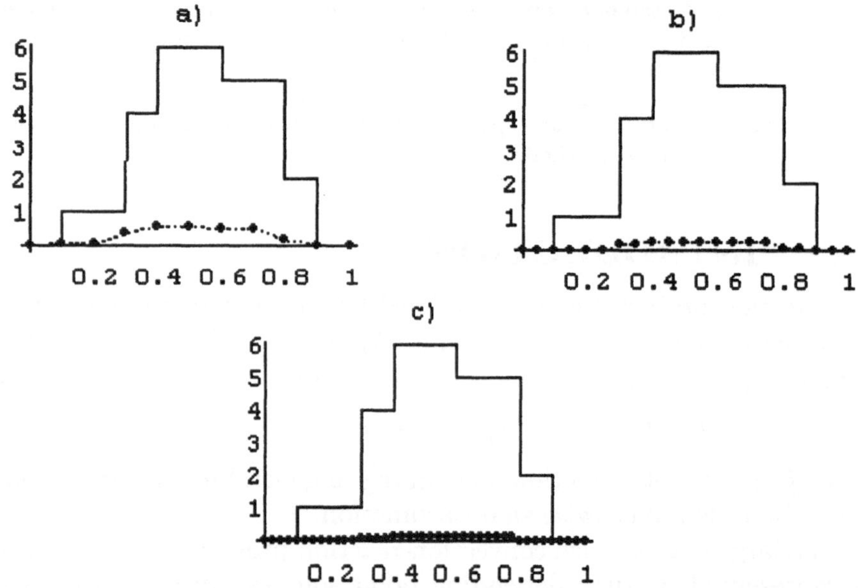

Figure 1. Numerical solution at $t = 0.5$ computed with a) $k = h = 1/10$, b) $k = h = 1/20$, c) $k = h = 1/40$.

Hyperbolic conservation laws with source terms have recently been analysed by several authors (see e.g., [1], [3], [4], [7], [10]). In particular in [3] a conservation law of type (1.1) with a steady source term is studied. L^∞ stable Godunov-type difference schemes, which have a similar equilibrium structure to the continuous case, are constructed.

Convergence problems are not considered. Koren in [4] suggests that by taking a steady source term as the derivative of a certain function, more accurate results can be achieved. In citeSantos, the authors present a general convergence result for a class of conservative numerical methods. Stability and entropy conditions are also established. Following the previous results, in the present paper, we study the convergence properties of a class of high resolution methods which are at least second order accurate on smooth solutions and give well resolved nonoscillatory discontinuities.

The paper is organized as follows. In Section 2 we recall briefly some convergence results established in [9]. In Section 3 we construct a class of high resolution methods and prove its convergence properties by using the concept of total variation diminishing (TVD). Finally, in Section 4 some numerical simulations are presented.

2. BASIC CONCEPTS AND CONVERGENCE RESULTS

In this section we present some preliminary concepts used in the study of convergence. We also present some of the convergence results established in [9] under a slightly modified version.

Let us consider the nonhomogeneous scalar conservation law (1.1) with the initial data

$$u(x,0) = u_0(x) \quad \text{for} \quad x \in \mathbb{R}. \tag{2.1}$$

A weak solution $u(x,t)$, of (1.1) and (2.1), satisfies the integral form

$$\begin{aligned} \int_{x_{j-1/2}}^{x_{j+1/2}} u(x,t_{n+1})dx &= \int_{x_{j-1/2}}^{x_{j+1/2}} u(x,t_n)dx - \\ -\Bigg\{\left[\int_{t_n}^{t_{n+1}} f\left(u\left(x_{j+1/2},t\right)\right)dt - \int_{t_n}^{t_{n+1}} f\left(u\left(x_{j-1/2},t\right)\right)dt\right] \\ -\left[\int_{t_n}^{t_{n+1}} q\left(x_{j+1/2},t\right)dt - \int_{t_n}^{t_{n+1}} q\left(x_{j-1/2},t\right)dt\right]\Bigg\}. \end{aligned} \tag{2.2}$$

In what follows we consider *conservative numerical methods* in the general form

$$u_j^{n+1} = u_j^n - \frac{k}{h}\left\{\left[F\left(u^n;j\right) - F\left(u^n;j-1\right)\right] - \left[S\left(q^n;j\right) - S\left(q^n;j-1\right)\right]\right\}, \tag{2.3}$$

where $F(u^n;j) = F\left(u_{j-p}^n,\dots,u_{j+l}^n\right)$, $S(q^n;j) = S\left(q_{j-p}^n,\dots,q_{j+l}^n\right)$ and $q_{j+b}^n = q\left((j+b)h,nk\right)$, $b = -(p+1),\dots,0,\dots,l$. F and S are functions of $p+l+1$ arguments. F is called the numerical homogeneous flux function and S is called the numerical source flux function. $F-S$ can be considered as an extended numerical flux function.

Method (2.3) is *consistent* with the nonhomogeneous conservation law (1.1) if the numerical homogeneous flux function reduces to the true homogeneous flux for the case of a constant flow and the numerical source flux satisfies $S(q,\dots,q)=q\ ,\ \forall q\in \mathbb{R}$.

In order to extend the grid function u_j^n, we define for $n\geq 0$ a piecewise constant function for all x and t,

$$u_m(x,t) := u_j^n \quad \text{for } (x,t)\in [(j-1/2)h_m,(j+1/2)h_m[\times[nk_m,(n+1)k_m[, \tag{2.4}$$

where $\frac{k_m}{h_m}=\lambda$ for a fixed λ.

We define the total variation of v, TV(v), for $v:\mathbb{R}\to\mathbb{R}$, by5

$$TV(v) := \sup_{\substack{-\infty=x_0<x_1<\dots<x_n=\infty\\ n\in\mathbb{N}}} \sum_{j=1} |v(x_j)-v(x_{j-1})|\,. \tag{2.5}$$

Theorem 2.1. *Let a sequence of grids indexed by $m=1,2,\dots,$ with mesh parameters $h_m,k_m\to 0$ as $m\to\infty$, be given. Let $(u_m)_m$ be a sequence of numerical solutions as defined in (2.3) and (2.4) with respect to h_m, k_m and the initial values u_j^0.*

Assume that

(i) the method is consistent with the nonhomogeneous conservation law (1.1);

(ii) the numerical homogeneous flux function F and the numerical source flux function S are Lipschitz continuous functions;

(iii) for each T there exist R_1 and R_2 such that

$$TV(u_m(\cdot,t))\leq R_1 \text{ for all } 0\leq t\leq T,\ m=1,2,\dots\ , \tag{2.6}$$

$$TV(q(\cdot,t))\leq R_2 \text{ for all } 0\leq t\leq T\ , \tag{2.7}$$

where, for each t, TV denotes the total variation defined in (2.5);

(iv) over every bounded set $\Omega=[a,b]\times[0,T]$ in the $x-t$ space, $\|u_m-u\|_{1,\Omega}\to 0$ as $m\to\infty$, where $\|\cdot\|_{1,\Omega}$ is the L_1 norm over the set Ω.

Then $u(x,t)$ is a weak solution of the conservation law (1.1).

When nonlinear problems are considered, we need some form of nonlinear stability to prove convergence results.

We will say that a numerical method is *total variation stable* [6], or simply *TV-stable*, if the approximations u_m for $k_m < k_0$ lie in some fixed set of the form

$$\Gamma = \{u \in L_{1,T} : TV_T(u) \leq R \text{ and } Supp\,(u\,(.,t)) \subset [-M, M] \ \forall t \in [0,T]\}, \tag{2.8}$$

where $L_{1,T}$ consists of all functions of x and t for which the norm

$$\|u\|_{1,T} = \int_0^T \int_{-\infty}^{\infty} |u\,(x,t)|\, dxdt \tag{2.9}$$

is finite, and

$$\begin{aligned} TV_T\,(u) = {} & \lim_{\epsilon \to 0} \sup \frac{1}{\epsilon} \int_0^T \int_{-\infty}^{\infty} |u\,(x+\epsilon,t) - u\,(x,t)|\, dxdt \\ & + \lim_{\epsilon \to 0} \sup \frac{1}{\epsilon} \int_0^T \int_{-\infty}^{\infty} |u\,(x,t+\epsilon) - u\,(x,t)|\, dxdt, \end{aligned}$$

represents the total variation of u over $[0, T]$.

Let u^n represents the vector of approximations u_j^n at time $t = nk$. Since the functions $u_m(x,t)$ are piecewise constant, the definition of TV_T reduces simply to

$$TV_T\,(u^n) = \sum_{n=0}^{T/k} \sum_{j=-\infty}^{\infty} \left[k \left|u_{j+1}^n - u_j^n\right| + h \left|u_j^{n+1} - u_j^n\right|\right],$$

or, in terms of the one-dimensional total variation and the discrete L_1 norm,

$$TV_T\,(u^n) = \sum_{n=0}^{T/k} \left[kTV\,(u^n) + \left\|u^{n+1} - u^n\right\|_1\right]. \tag{2.10}$$

In the definition of the set Γ, R and M may depend on the initial data u_0, the flux function $f(u)$ and the source term q_x, but not on m. We note that Γ is a compact set in $L_{1,T}$ [6].

Theorem 2.2. *Let us consider a conservative numerical method in the general form (2.3), where the numerical homogeneous flux F and the numerical source flux S are Lipschitz continuous functions. Suppose that for each initial data u_0, with compact support, there exist $k_0, R_1, R_2 > 0$ such that*

$$TV(u^n) \leq R_1 \text{ and } TV(q^n) \leq R_2 \ \forall n,k \text{ with } k < k_0,\ nk \leq T. \tag{2.11}$$

Then the method is TV-stable. Moreover, if the method is consistent with the scalar nonhomogeneous conservation law (1.1), then the method is convergent to a weak solution of (1.1).

Proof. We prove that there exists $\alpha > 0$ such that $\left\|u^{n+1} - u^n\right\|_1 \leq \alpha k$, $\forall n, k$ with $k < k_0$, $nk < T$. Let us consider a method in the conservation form (2.3). Then, we have

$$h \sum_{j=-\infty}^{\infty} \left|u_{j+1}^n - u_j^n\right| = \tag{2.12}$$

$$= k \sum_{j=-\infty}^{\infty} \left|\left[F(u^n;j) - F(u^n;j-1)\right] - \left[S(q^n;j) - S(q^n;j-1)\right]\right|,$$

that is,

$$\begin{aligned} \left\|u^{n+1} - u^n\right\|_1 \;&\leq\; k \sum_{j=-\infty}^{\infty} \left|F(u^n;j) - F(u^n;j-1)\right| + \\ &+\; k \sum_{j=-\infty}^{\infty} \left|S(q^n;j) - S(q^n;j-1)\right|. \end{aligned} \tag{2.13}$$

Using the bounds $TV(q^n) \leq R_2$ and $TV(u^n) \leq R_1$ together with the fact that $u_m(\cdot, nk)$ has compact support, we easily establish that $\left|u_j^n\right| \leq \frac{R_1}{2}$ and $\left|q_j^n\right| \leq R_2$ $\forall j, n$, with $nk \leq T$ (note that $q(\cdot, t_n)$ approach constant values as $x \to \pm\infty$ for each $t_n = nk$ with $t_n \leq T$, because $TV(q^n)$ is finite). Considering these uniform bounds and the continuity of $F(u^n;j)$ and $S(q^n;j)$, we have the following estimates

$$\left|F(u^n;j) - F(u^n;j-1)\right| \leq K_1 \sum_{i=-p}^{l} \left|u_{j+i}^n - u_{j+i-1}^n\right|, \tag{2.14}$$

$$\left|S(q^n;j) - S(q^n;j-1)\right| \leq K_2 \sum_{i=-p}^{l} \left|q_{j+i}^n - q_{j+i-1}^n\right| \tag{2.15}$$

and

$$\left\|u^{n+1} - u^n\right\|_1 \leq kK_1 \sum_{i=-p}^{l} TV(u^n) + kK_2 \sum_{i=-p}^{l} TV(q^n),$$

that is, $$\left\|u^{n+1} - u^n\right\|_1 \leq k\,(p + l + 1)\,(K_1 R_1 + K_2 R_2).$$

Let $\alpha = (p + l + 1)\,(K_1 R_1 + K_2 R_2)$. From (2.10) we have

$$TV_T(u^n) \leq (\alpha + R_1)\,T,$$

for all $k < k_0$. This shows that $TV_T(u^n)$ is uniformly bounded as $k \to 0$. The other requirement for TV-stability is $Supp(u)$ to be uniformly bounded over $[0, T]$, but this is always satisfied for any explicit method if $\frac{k}{h}$ is constant as $k \to 0$. Then all u^n lie in a set of the form (2.8) for all $k < k_0$ and the method is TV-stable.

Finally, we want to prove that if the method is TV-stable and consistent with the scalar nonhomogeneous conservation law (1.1) then the method is convergent. Let

$$W = \{u : u(x,t) \text{ is a weak solution of the conservation law (1.1)}\}.$$

If the global error, defined by $dist(u_m, W) = \inf_{u \in W} \|u_m - u\|_{1,T}$, does not converge to zero, then there exist some $\varepsilon > 0$ and some sequence of approximations u_{m_j} such that

$$k_j \to 0 \text{ as } j \to \infty \text{ while } dist\left(u_{m_j}, W\right) > \varepsilon, \text{ for all } j. \tag{2.16}$$

Since $u_{m_j} \in \Gamma$ for all j, there exists a subsequence converging to some function $v \in \Gamma$. Moreover, since the u_m are generated by a conservative and consistent method, it follows from Theorem 2.1, that the limit v must be a weak solution of the conservation law, i.e., $v \in W$, which contradicts (2.16). Hence, a sequence satisfying (2.16) cannot exist and we conclude that $dist(u_m, W) \to 0$ as $k \to 0$. ■

Sufficient conditions that guarantee the convergence of the numerical solution to the entropy solution have been established in [9].

3. A CLASS OF HIGH RESOLUTION METHODS

In this section we construct a class of numerical methods that are convergent, in the sense of Theorem 2.2, and are at least second order accurate on smooth solutions giving well resolved nonoscillatory discontinuities.

We consider the general initial value problem

$$\begin{cases} u_t + (f(u))_x = (q(x,t))_x, & -\infty < x < \infty \text{ and } t > 0 \\ u(x,0) = u_0(x), & x \in \mathbb{R} \end{cases}, \tag{3.1}$$

where q is a bounded piecewise smooth function and u_0 has compact support.

We will use conservative methods in the general form (2.2), that is

$$u_j^{n+1} = \underbrace{u_j^n - \frac{k}{h} F(u^n; j) - F(u^n; j-1)}_{\overline{u}_j^{n+1}} + \underbrace{\frac{k}{h}\left[S(q^n; j) - S(q^n; j-1)\right]}_{\varepsilon_j^{n+1}} \tag{3.2}$$

where

$$\overline{u}_j^{n+1} = u_j^n - \frac{k}{h}\left[F\left(u^n; j\right) - F\left(u^n; j-1\right)\right] \tag{3.3}$$

and

$$\varepsilon_j^{n+1} = \frac{k}{h}\left[S\left(q^n; j\right) - S\left(q^n; j-1\right)\right] . \tag{3.4}$$

We suppose that, the numerical homogeneous flux F and the numerical source flux S are Lipshitz continuous functions, $S\left(q^n; j\right) = S\left(q_{j-p}^n, \cdots, q_{j+l}^n\right)$ is a function of $(p+l+1)$ variables and $TV\left(q^n\right) \leq R_2\ \forall n, k$ with $nk \leq T$. This last bound together with the continuity of $S\left(q^n; j\right)$ allows us to derive a bound of the form (2.15).

From (3.2) we have that $TV\left(u^{n+1}\right) \leq TV\left(\overline{u}^{n+1}\right) + TV\left(\varepsilon^{n+1}\right)$.

We begin by establishing an estimate for $TV\left(\varepsilon^{n+1}\right)$. From (3.4), we have

$$\sum_{j=-\infty}^{\infty} \left|\varepsilon_{j+1}^{n+1} - \varepsilon_j^{n+1}\right| =$$

$$= \frac{k}{h} \sum_{j=-\infty}^{\infty} \left|\left[S\left(q^n; j+1\right) - S\left(q^n; j\right)\right] - \left[S\left(q^n; j\right) - S\left(q^n; j-1\right)\right]\right|,$$

and then by (2.15)

$$\sum_{j=-\infty}^{\infty} \left|\varepsilon_{j+1}^{n+1} - \varepsilon_j^{n+1}\right| \leq$$

$$\leq \frac{k}{h}\left(K_1 \sum_{i=-p}^{l} \sum_{j=-\infty}^{\infty} \left|q_{j+i}^n - q_{j+i-1}^n\right| + K_2 \sum_{i=-p}^{l} \sum_{j=-\infty}^{\infty} \left|q_{j+i+1}^n - q_{j+i}^n\right|\right) .$$

The last inequality can take the form

$$TV\left(\varepsilon^{n+1}\right) \leq \frac{k}{h}\left[K_1 \sum_{i=-p}^{l} TV\left(q^n\right) + K_2 \sum_{i=-p}^{l} TV\left(q^n\right)\right],$$

that is

$$TV\left(\varepsilon^{n+1}\right) \leq \frac{k}{h} R_2^{\varepsilon}, \tag{3.5}$$

where $R_2^{\varepsilon} = (p+l+1)\, R_2\left(K_1 + K_2\right)$.

Now, let us analyze $TV\left(\overline{u}^{n+1}\right)$. We note that if a total variation diminishing (TVD) method is used to solve (3.3) for each time level $t_{n+1} = (n+1)k$, then we have

$$TV\left(\overline{u}^{n+1}\right) \leq TV\left(u^n\right), \tag{3.6}$$

for all grid functions u^n, with $nk \leq T$. Thus, considering that $TV\left(u^0\right) \leq L$, it is a straightforward task, to show that

$$TV\left(u^n\right) \leq R_1, \tag{3.7}$$

where $R_1 = L + \frac{T}{h}R_2^\varepsilon$.

From the previous considerations we can conclude, using Theorem 2.2, that the methods of class (3.2) produce convergent numerical solutions.

Let us construct now a family of methods of class (3.2). For the homogeneous part of the problem we use a one-step non-linear TVD version of the Lax-Wendroff scheme [2]. If we consider $a(u) = f'(u)$, this method can be written in the conservative form (3.3), with

$$\begin{aligned} F\left(u^n; j\right) = &\frac{1}{2}\left[f\left(u_j^n\right) + f\left(u_{j+1}^n\right)\right] - \frac{1}{2}\frac{k}{h}a_{j+1/2}\left[f\left(u_{j+1}^n\right) - f\left(u_j^n\right)\right]\Psi\left(\theta_j\right) \\ &- \frac{1}{2}\sigma\left[1 - \Psi\left(\theta_j\right)\right]\left[f\left(u_{j+1}^n\right) - f\left(u_j^n\right)\right] \end{aligned} \tag{3.8}$$

where $a_{j+1/2}$ can be evaluated by the following formula

$$a_{j+1/2} = \begin{cases} \frac{f(u_{j+1}^n) - f(u_j^n)}{u_{j+1}^n - u_j^n}, & \text{if } u_j^n \neq u_{j+1}^n \\ a(u_j^n), & \text{if } u_j^n = u_{j+1}^n \end{cases}.$$

In (3.8) $\sigma = sgn\left(a_{j+1/2}\right)$, $\Psi\left(\theta_j\right)$ represents a flux-limiter, and

$$\theta_j = \frac{f\left(u_{j+1-\sigma}^n\right) - f\left(u_{j-\sigma}^n\right)}{f\left(u_{j+1}^n\right) - f\left(u_j^n\right)}.$$

At each grid point we define a local Courant number by $\nu_j = \frac{k}{h}a_{j+1/2}$.

In order that (3.3) is, locally, a TVD method and considering Theorem 16.3 (Harten) [6], it is sufficient that conditions

$$\begin{cases} |\nu_j| < \frac{1}{2}, \ \forall_j \\ 0 \leq \Psi(\theta)/\theta \leq 2, \ \forall_\theta \\ 0 \leq \Psi(\theta) \leq 2, \ \forall_\theta \end{cases},$$

are satisfied. We note that for linear problems the first stability condition can be replaced by $|\nu| \leq 1$.

In the numerical examples of Section 4 the so-called "superbee" limiter of Roe [8], will be used. This limiter, well adapted for sharp discontinuities, is defined by

$$\psi(r) = \begin{cases} \min(2, r), & r > 1 \\ \min(2r, 1), & 0 < r \leq 1 \\ 0 = 0, & r \leq 0 \end{cases} \tag{3.9}$$

To solve the nonhomogeneous problem we use an extended Lax-Wendroff method with a flux-limiter procedure, that is, we consider

$$S(q^n;j) = q_j^n + \frac{1}{2}(1-\nu_j)\left(q_{j+1}^n - q_j^n\right)\Psi(\theta_j), \tag{3.10}$$

where Ψ represents the flux-limiter (3.9) and

$$\theta_j = \frac{q_j^n - q_{j-1}^n}{q_{j+1}^n - q_j^n}.$$

4. NUMERICAL EXAMPLES

In this section we solve problems using the family of high resolution methods previously constructed. The computations have been performed using the program package *Mathematica*, a powerful computer algebra system.

Example 1. We consider the initial boundary value problem

$$\begin{aligned} u_t + u_x &= [\exp(t) - 1]\,\delta(x-0.1), && x \in (0,1),\ t>0, \\ u(x,0) &= 0, && x \in (0,1), \\ u(0,t) &= 0, && t \geq 0. \end{aligned}$$

The exact solution is

$$u(x,t) = \begin{cases} 0\,, & x < 0.1 \ \vee \ x \geq 0.1 + t, \\ \exp(0.1 - x + t) - 1, & 0.1 \leq x < 0.1 + t. \end{cases}$$

In this case as $(q(x,t))_x = [\exp(t) - 1]\,\delta(x-0.1)$ *we* have $q(x,t) = [\exp(t) - 1]\,H(x-0.1)$. Numerically $q_j^n = [\exp(nk) - 1]\,H(jh-0.1)$. As in this case we have a linear problem, expressions are simplified and, of course, we have only one Courant number $\nu = \frac{k}{h}$.

To outline the robustness of our approach we assume, for example, $k = \frac{h}{2}$, that is, we do not integrate along the characteristic lines. We analyse the behaviour of the numerical solution for two values of the spatial stepsize, $h = \frac{1}{20}$ and $h = \frac{1}{40}$, in three moments: $t = 0.25$, $t = 0.5$, $t = 0.75$ (Figure 2). The numerical solution is represented with a dashed line and the exact solution is represented with a continuous line.

If $k = h$ had been used, similar results would have been achieved.

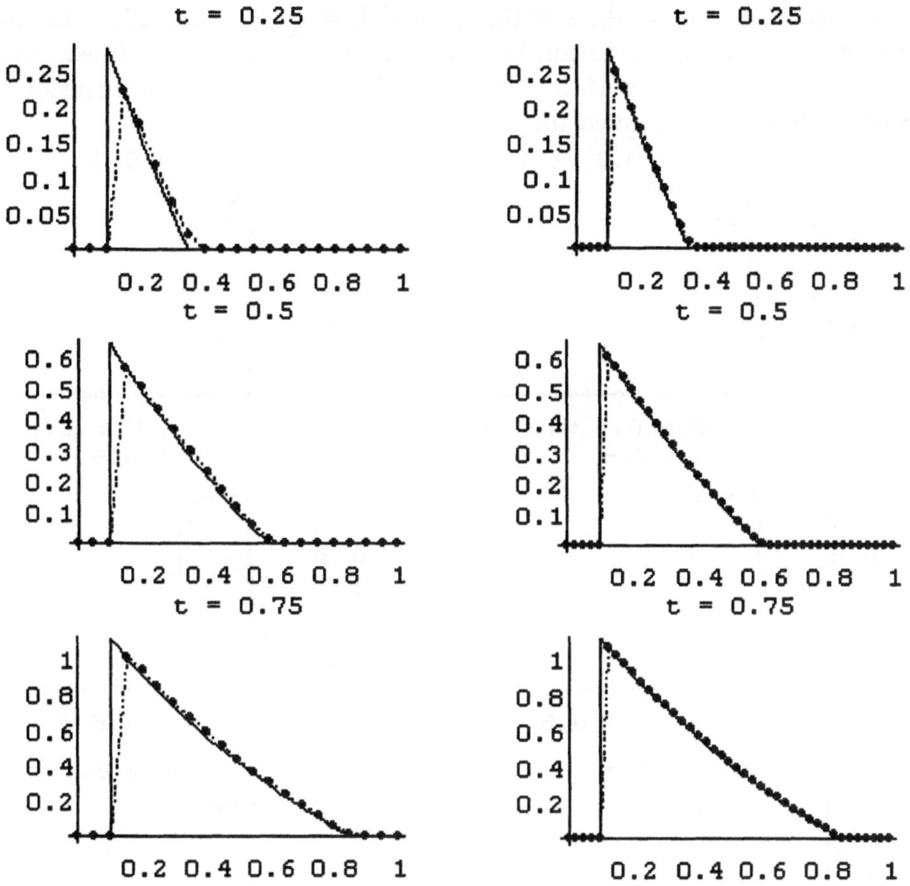

Figure 2. Numerical solutions at $t = 0.25$, $t = 0.5$ and $t = 0.75$ for $h = \frac{1}{20}$, $k = 0.025$ (in the left side) and for $h = \frac{1}{40}$, $k = 0.0125$ (in the right side).

Example 2. We consider the inviscid Burger's equation with a point source,

$$\begin{aligned} u_t + \left(\frac{u^2}{2}\right)_x &= \left(\frac{1+t}{2}\right)\delta(x-0.2), && x \in (0,1)\,, \quad t > 0, \\ u(x,0) &= 0, && x \in (0,1)\,, \\ u(0,t) &= 0, && t \geq 0. \end{aligned}$$

In this problem $(q(x,t))_x = \left(\frac{1+t}{2}\right)\delta(x-0.2)$ and then, $q(x,t) = \left(\frac{1+t}{2}\right)H(x-0.2)$. We consider $q_j^n = \left(\frac{1+nk}{2}\right)H(jh-0.2)$. We solve it numerically with the method defined above.

In Figure 3 we present the numerical solutions obtained at three distinct moments, $t = 0.2$, $t = 0.4$, $t = 0.8$, with two different pairs of values of k, h ($h = \frac{1}{20}$, $k = 0.0125$ and $h = \frac{1}{40}$, $k = 0.00625$). In this problem the "exact" solution has been obtained with a fine fixed mesh and it is represented with a continuous line. The numerical solution is represented with a dashed line.

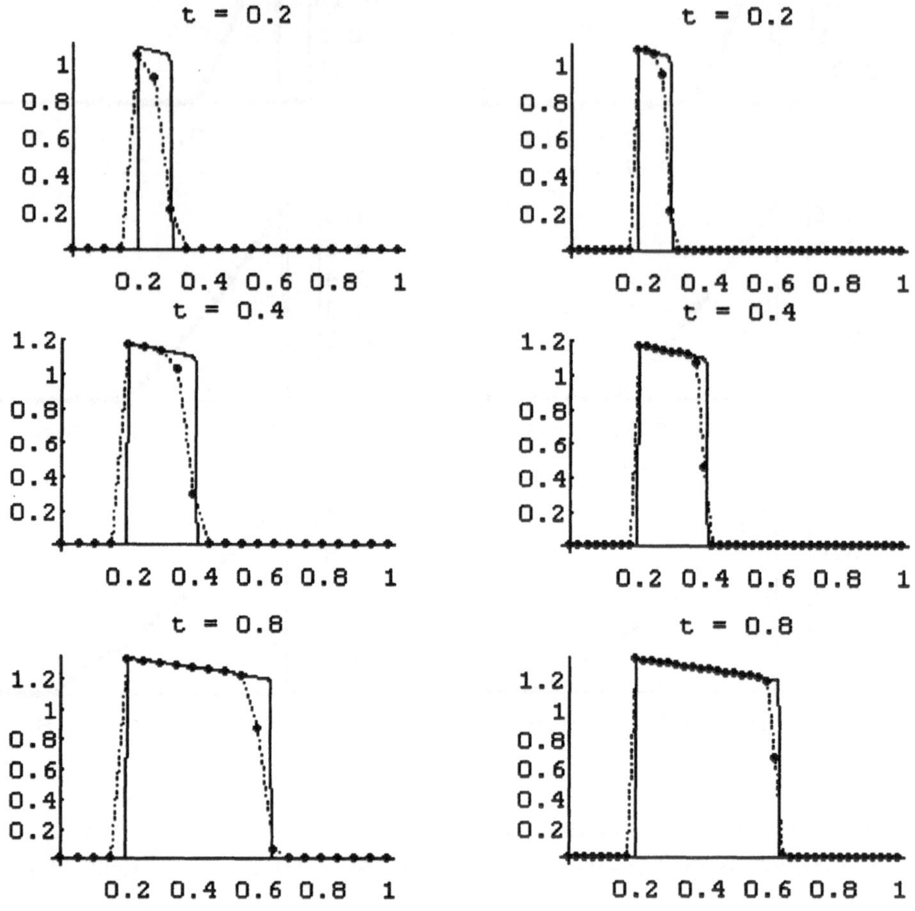

Figure 3. Numerical solutions at $t = 0.2$, $t = 0.4$ and $t = 0.8$ for $h = \frac{1}{20}, k = 0.0125$ (in the left side) and for $h = \frac{1}{40}, k = 0.00625$ (in the right side).

Example 3. We consider the initial value problem with a point source,

$$\begin{aligned} u_t + ((x+1)\,u)_x &= \delta(x - 0.1), && x \in (0,1)\,, \quad t > 0, \\ u(x,0) &= 0, && x \in (0,1)\,, \\ u(0,t) &= 0, && t \geq 0. \end{aligned}$$

The exact solution is

$$u(x,t) = \frac{1}{1+x}\left[H\left(x-0.1\right) - H\left(\left(0.9+x\right)e^{-t}-1\right)\right].$$

In Figure 4 we present the numerical solutions obtained at three distinct moments, $t = 0.2$, $t = 0.4$, $t = 0.5$, with two different couple of values of k, h ($h = \frac{1}{20}$, $k = 0.0125$ and $h = \frac{1}{40}$, $k = 0.00625$).

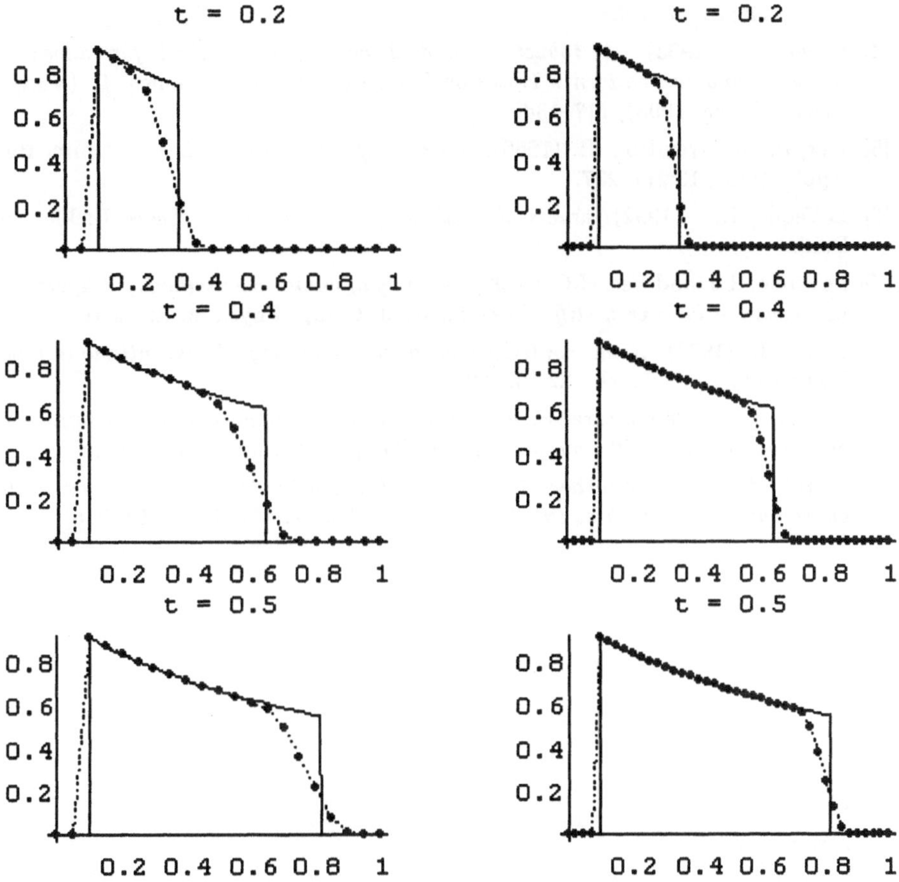

Figure 4. Numerical solutions at $t = 0.2$, $t = 0.4$ and $t = 0.5$ for $h = \frac{1}{20}, k = 0.0125$ (in the left side) and for $h = \frac{1}{40}, k = 0.00625$ (in the right side).

References

[1] Chalabey, A. (1997). *On convergence of numerical schemes for hyperbolic conservation laws with stiff source terms.* Math. Comp., 66(218):527–545.

[2] Fletcher, C.A.J. (1998). Computational Techniques for Fluid Dynamics. Vol. II, Spriger-Verlag.

[3] Greenberg, J.M., Leroux, A.Y., Baraille, R. and Noussair, A. (1997). *Analysis and approximation of conservative laws with source terms.* SIAM J. Numer. Anal., 34(5):1980–2007.

[4] Koren, B. (1993). *A robust upwind discretization method for advection, diffusion and source terms.* Notes on Numerical Fluid Mechanics, 45 (Vieweg, Braunschweig, 1993), 117–138.

[5] Lax, P. and Wendroff, B. (1960). *Systems of Conservation Laws.* Comm. Pure Appl. Math., 13:217–237.

[6] LeVeque, R.J. (1992)*Numerical Methods for Conservation Laws.* Birkhauser, Berlin.

[7] LeVeque, R.J. and Yee, H.C. (1990). *A study of numerical methods for hyperbolic conservation laws with stiff source terms.* J. Comp. Phys., 86:187–210.

[8] Roe, P.L. (1985). *Some contributions to the modeling of discontinuous flows.* Lectures in Appl. Math., 22:163–193.

[9] Santos, J. and de Oliveira, P. *A converging finite volume scheme for hyperbolic conservation laws with source terms.* J. Comp. Appl. Math. To appear.

[10] Schroll, H.J. and Winther, R. (1996). *Finite-difference schemes for scalar conservation laws with source terms.* IMA J. Numer. Anal., 16:201–215.

ON THE DECAY TO ZERO OF THE L^2-NORMS OF PERTURBATIONS TO A VISCOUS COMPRESSIBLE FLUID MOTION EXTERIOR TO A COMPACT OBSTACLE

Mariarosaria Padula

Abstract: We prove that the rest state of a viscous isothermal fluid filling a region exterior to a compact rigid obstacle, is stable with respect a class of sufficiently weak perturbations σ, $\mathbf{u}$ to the density and velocity fields (provided they exist globally in time). Under hypothesis of summability f or a weighted norm of perturbations at initial time we also prove the decay to z ero for L^2-norms of perturbations along infinitely many sequences of times.

Keywords: Nonlinear stability, energy methods, Navier-Stokes equations, compressible fluids, exterior domains, qualitative methods.

1. INTRODUCTION

It is known that a *homogeneous basic rest state* is unconditionally (for arbitrary large initial data) stable with respect to all values of the parameters in the class of incompressible fluid perturbations. The type of stability achieved is exponential for flows occurring in bounded domains, furthermore, while for flows occurring in exterior domains is only asymptotical, say the kinetic energy decays to zero.

For compressible fluids a *nonhomogeneous rest state* continues to be stable with respect to all values of the parameters, however, the stability has been proven only for small initial data, in bounded domains [8]. Recently, it has been proved unconditional exponential stability for a non homogeneous rest state in the class of viscous compressible fluid flows, corresponding to large potential forces, and for a steady flow corresponing to small external forces, in bounded rigid fixed domains

Applied Nonlinear Analysis, edited by Sequeira *et al.*
Kluwer Academic / Plenum Publishers, New York, 1999.

[11], [13], [14]. In this note, we are not interested to the problem of existence of solutions, only we make some qualitative considerations on the stability problem.

Aim of this paper is the proof of the unconditional decay to zero, along a sequence of instants going to infinity, of the total energy (summ of the kinetic energy and the enthalpy) of perturbations of a isothermal viscous fluid flow in domains exterior to a fixed obstacle Ω, in corrispondence of zero external forces.

As known, the motions of a viscous compressible fluid, in Ω, are governed by the Poisson-Stokes equations

$$\frac{\partial \rho}{\partial t} + \nabla \cdot (\rho \mathbf{v}) = 0,$$

$$\rho[\frac{\partial \mathbf{v}}{\partial t} + \mathbf{v} \cdot \nabla \mathbf{v}] - \mu \Delta \mathbf{v} - (\lambda + \mu) \nabla \nabla \cdot \mathbf{v} = -\nabla p,$$

$$\mathbf{v}(x,t)|_{\partial\Omega} = 0, \tag{1.1}$$

$$\lim_{|x|\to\infty} \rho(x,t) = m, \qquad \lim_{|x|\to\infty} \mathbf{v}(x,t) = 0,$$

$$\rho(0,x) = \rho_0(x), \quad \mathbf{v}(0,x) = \mathbf{v}_0(x),$$

where ρ, $\mathbf{v}$, $\mathbf{f}$ denote the density, the velocity and the external force, at the point x, at the instant t. Moreover, λ, μ are the constant viscosity coefficients, verifying $n\lambda+2\mu \geq 0$, where n is the dimension of the space. Furthermore, for isothermal fluids the pressure $p = p(\rho)$ is supposed a linear function of the density, $p = k\rho$, with $k = R\theta_0$, where R is the universal constant of the gas and θ_0 is a basic uniform temperature.

System (1.1) admits the rest solution

$$\rho(t,x) = \rho_b, \quad \mathbf{v}(t,x) = 0, \tag{1.2}$$

with $\rho_b = const.$, in correspondence of the initial data $\rho_0 = \rho_b$, and $\mathbf{v}_0 = 0$. Decay to the rest of nonsteady solutions corresponding to large initial data constitutes an interesting problem, to our knowledge, first results in this direction have been furnished in [2], [3], furthermore, for the linearized equations a local energy decay is given in [5], [6]. In [12] it was proposed a sketch of proof of stability in unbounded domains based on the energy method, here we develop the ideas of [12] and furnish a complete proof of unconditional stability of the rest in a exterior domain, moreover, we prove the decay to zero for the L^2-norms of the perturbations.

In this note, we prove that the rest state is stable in the mean, in a class of suitably regular perturbations.

In a work in preparation we give the proof of the decay to zero of perturbations to a steady flow corresponding to nonzero external forces, whatever large are taken the potential forces and the initial data (unconditional). We remind that in [9], [10] it has been proved existence of such steady flows.

In order to state our theorem we introduce the following regularity class

$$\begin{aligned}\mathcal{V} := \{(\rho, \mathbf{u}) \in C^1(0,T);\\ C^0(\Omega)) \cap C^0(0,T; C^1(\Omega)) \times C^1(0,T; C_0^0(\Omega)) \cap C^0(0,T; C^2(\Omega)) :\\ 0 < m_1 \le \rho(x,t) \le m_2, \quad |\mathbf{u}| = o(1/r),\\ |\nabla \rho| \le (c/r^2) \quad \rho = \rho_b + \sigma, \quad |\sigma| \le c/(r \ln^3 r)\}.\end{aligned}$$

and its subset

$$\begin{aligned}\mathcal{V}^e := \{(\rho, \mathbf{u}) \in \mathcal{V} : 0 < m_1 \le \rho(x,t) \le m_2, \quad |\mathbf{u}| \le c/r,\\ |\nabla \rho| \le c/(r \ln^2 r), \quad |\nabla \cdot \mathbf{u}| \le 1/\ln^2 r\}.\end{aligned}$$

In $\mathcal{V}$ we prove the stability result, in $\mathcal{V}^e$ the decay to zero for the L^2-norms of the perturbations. In the class $\mathcal{V}$ equations (1.1) can be written pointwise. Of course, such assumptions can be fairly weakened, however this is not in the aims of our note.

The result we prove is the following:

Theorem 1. *Assume there exist solutions to (1.1) $\tilde{\rho}(t,x), \mathbf{u}(t,x)) \in \mathcal{V}$, corresponding to the initial data*

$$\rho(0,x) = \rho_b + \sigma_0(x), \quad \mathbf{u}(0,x) = \mathbf{u}_0(x),$$

with ρ_b constant. Then, the L^2-norm of the perturbations in the class $\mathcal{V}^e$ decays to zero along infinitely many sequences of times $t_n \to \infty$.

2. PRELIMINARY LEMMAS

In this section we furnish some preliminary lemmas which will be used in the proof of the non linear stability given in section 3. Recall that Ω denotes the region of the space, exterior to a compact, fixed, rigid obstacle $\mathcal{B}$. In the sequel, O indicates a fixed point in $\mathcal{B}$ and (r, θ, φ), $r = |x|$, denote a spherical coordinate system with the origin in O. With such choice it will be $r > a > 0$, in the sequel we also assume $a > 1$.

The Lemma below, proved in Lemma 3.1 p.36, [4], furnishes a weighted Poincare' inequality.

Lemma 2.1. Weighted Poincare' inequality - *Let $\psi = \psi(r)$ be a positive function, non decreasing in r. In the class S of functions u such that*

$$\left\| \sqrt{\psi} \frac{u}{r} \right\|^2 + \|\nabla u\|^2 \leq \infty, \qquad u|_{\partial\Omega} = 0,$$

the following inequality holds

$$\left\| \sqrt{\psi} \frac{u}{r} \right\|^2 \leq 4 \|\sqrt{\psi} \nabla u\|^2. \tag{2.1}$$

The above Lemma is a generalization of a inequality first derived by Leray in [7], it will be used below with $\psi = \ln r$[1]. The interest of this Lemma lies in the value of the constant which has a universal value.

Lemma 2.2. *Denote by Ω_R, the intersection of Ω with a ball of radius R centered at O, and by Ω^R, its complementary in Ω, say $\Omega^R := \Omega - \Omega_R$. Then, for any $\eta > 0$ there exists a R, such that, for all sufficiently regular fields σ, $\mathbf{u}$, the following estimates hold true*

$$\begin{aligned} \|\sigma\|^2 &\leq c_1 R^2 \left\| \frac{\sigma}{r} \right\|^2_{\Omega_R} + \eta \|\sqrt{\ln r}\, \sigma\|^2_{\Omega^R} \\ \|\mathbf{u}\|^2 &\leq c_1 R^2 \|\nabla \mathbf{u}\|^2_{\Omega_R} + \eta \|\sqrt{\ln r}\, \mathbf{u}\|^2_{\Omega^R}. \end{aligned} \tag{2.2}$$

Proof. The proof follows easily by taking $\eta = \dfrac{1}{\ln R}$, and using the Poincare' inequality in Ω_R for $\mathbf{u}$.

Lemma 2.3. Weighted Poincare' inequality - *Given the vector field $\mathbf{u} \in C^2(\Omega)$, with $|x|\nabla \mathbf{u} \in L^2(\Omega)$, then the following inequality holds*

$$\|\mathbf{u}\| \leq c_2 \| |x| \nabla \mathbf{u} \|.$$

Lemma 2.4. Divergence problem - *Given the scalar fields $\sigma \in C^2(\Omega)$, there exists a unique vector field z with $\nabla z \in L^2(\Omega)$, solution of the following problem*

$$\nabla \cdot z = \frac{\sigma}{|x|^2} z|_{\partial\Omega} = 0,$$

$$\lim_{|x| \to \infty} |z| = 0.$$

Moreover, there exists a constant c_3, function of ρ_b, and Ω, such that the following estimate holds true

$$\|\nabla z\| \leq c_3 \left\| \frac{\sigma}{|x|^2} \right\|. \tag{2.3}$$

---

[1] If $\Omega = \mathbb{R}^3$ the proof continues to hold provided we take with $\psi = \ln r$

3. PROOF OF NON LINEAR STABILITY

In this section we give the proof of non linear stability in mean. We begin by deriving the following energy identity for perturbations

Theorem 2. — Energy identity. — *Let* $\mathbf{u}$, $\rho := \rho_b + \sigma$ *solve* (1.1), *with* $\mathbf{u} \in L^2(0,\infty; W_0^{1,2}(\Omega)) \cap L^\infty(0,\infty; L^2_\rho(\Omega))$, $\sigma \in L^\infty(0,\infty; L^2_{1/\overline{\rho}}(\Omega))$. *Then, the following identity holds*

$$\frac{d[K(t) + E_\sigma(t)])}{dt} + D(t) = 0, \tag{3.1}$$

where

$$K(t) := \frac{1}{2}\int_\Omega \rho v^2; \qquad E_\sigma(t) := \frac{R}{2M^2}\int_\Omega \frac{\sigma^2}{\overline{\rho}}$$
$$D(t) := \|\nabla \mathbf{u}\|^2 + (l+1)\|\nabla \cdot \mathbf{u}\|^2,$$

and $\overline{\rho}$ *is a value between* ρ_b, *and* ρ.

Proof. The pair σ, $\mathbf{u}$ represents the perturbation to the basic rest state $\mathbf{v}(t,x) = 0$, $\rho(t,x) = \rho_b$, it will satisfy the following system

$$\begin{aligned}
&\frac{\partial \sigma}{\partial t} + \nabla \cdot ((\rho_b + \sigma)\mathbf{u}) = 0,\\
&\rho[\frac{\partial \mathbf{u}}{\partial t} + \mathbf{u}\cdot\nabla\mathbf{u}] - \mu\Delta\mathbf{u} - (\lambda+\mu)\nabla\nabla\cdot\mathbf{u} = -k\nabla\rho, \qquad (3.2)\\
&\mathbf{u}(t,x)|_{\partial\Omega} = 0, \quad \lim_{|x|\to\infty}|\sigma| = 0, \quad \lim_{|x|\to\infty}|\mathbf{u}| = 0,\\
&\sigma(0,x) = \sigma_0(x), \quad \mathbf{u}(0,x) = \mathbf{u}_0(x).
\end{aligned}$$

Notice that it holds

$$\begin{aligned}
-\frac{d}{dt}E_\rho \;&:= -k\textstyle\int_\Omega \frac{\partial}{\partial t}[\rho(\ln\rho - 1)] + k\frac{d}{dt}\int_\Omega \rho\ln\rho_b =\\
&= \textstyle\int_\Omega \mathbf{u}\cdot(-k\nabla(\rho_b+\sigma) + \rho\nabla U),
\end{aligned}$$

where the integration element dx is omitted, and

$$E_\rho := k\int_\Omega \rho\{\ln\rho - 1 - \ln\rho_b\}.$$

Developing E_ρ around $\sigma = 0$ it delivers

$$E_\sigma := \frac{k}{2}\int_\Omega \frac{1}{2\overline{\rho}}\sigma^2,$$

where $\overline{\rho}$ is between ρ_b and ρ.

Multiply (3.1) by $\mathbf{u}$ and integrate over Ω, we find

$$\frac{d}{dt}[K(t) + E_\sigma(t)] + \mu\|\nabla\mathbf{u}\|^2 + (\lambda+\mu)\|\nabla\cdot\mathbf{u}\|^2 = 0, \tag{3.3}$$

where

$$K(t) := \frac{1}{2}\int_\Omega \rho u^2.$$

The regularity class $\mathcal{V}$ ensures that $K(t)$, $E_\sigma(t)$ are equivalent to the L^2 norms of $\mathbf{u}$ and σ, respectively.

We are, now, in position to prove the main theorem.

Proof of Theorem 1.

Multiply $(3.2)_2$ by $\frac{\partial \mathbf{u}}{\partial t}$, integrate over Ω to get

$$\frac{d}{dt}[\mu\|\nabla \mathbf{u}\|^2 + (\lambda+\mu)\|\nabla\cdot\mathbf{u}\|^2] + \left\|\sqrt{\rho}\frac{\partial \mathbf{u}}{\partial t}\right\|^2$$

$$= -k\left(\rho_b\frac{\partial \mathbf{u}}{\partial t}, \nabla\left(\frac{\sigma}{\rho_b}\right)\right) - \left(\rho\mathbf{u}\cdot\nabla\mathbf{u}, \frac{\partial \mathbf{u}}{\partial t}\right). \tag{3.4}$$

Devide $(3.2)_1$ by ρ_b take the ∇ on both sides, and multiply by $\rho_b\mathbf{u}$, integrating over Ω we obtain

$$\frac{d}{dt}\left(\rho_b\mathbf{u}, \nabla\left(\frac{\sigma}{\rho_b}\right)\right) = \left(\rho_b\frac{\partial \mathbf{u}}{\partial t}, \nabla\left(\frac{\sigma}{\rho_b}\right)\right) + \left(\nabla\cdot(\rho_b\mathbf{u}), \frac{\nabla\cdot(\rho\mathbf{u})}{\rho_b}\right) \tag{3.5}$$

Multiplying (3.5) times k, and adding it to (3.4) we deduce

$$\frac{d}{dt}\left(\nabla\cdot(\rho_b\mathbf{u}), \frac{\sigma}{\rho_b}\right) k\frac{d}{dt}\left(\rho_b\mathbf{u}, \nabla\left(\frac{\sigma}{\rho_b}\right)\right) +$$

$$+\frac{d}{dt}\left[\mu\|\nabla\times\mathbf{u}\|^2 + (\lambda+2\mu)\|\nabla\cdot\mathbf{u}\|^2\right] + \left\|\sqrt{\rho}\frac{\partial \mathbf{u}}{\partial t}\right\|^2 = g(\sigma,\mathbf{u}),$$

where using the regularity properties of the class $\mathcal{V}$, Lemma 2.1 with $\psi = 1$, and Cauchy inequality, we have

$$g(\sigma,\mathbf{u}) := +k\left(\nabla\cdot\mathbf{u}, \nabla\cdot(\rho\mathbf{u})\right) - \left(\rho\mathbf{u}\cdot\nabla\mathbf{u}, \frac{\partial \mathbf{u}}{\partial t}\right)$$

$$\leq c\|\nabla\mathbf{u}\|^2 + \frac{1}{2}\left\|\sqrt{\rho}\frac{\partial \mathbf{u}}{\partial t}\right\|^2. \tag{3.6}$$

In order to obtain the decay for the L^2-norms of the perturbations, we need a dissipative term in the L^2-norm of σ. This will be achieved by using the auxiliary lemmas given in section 2.

To this end, we rewrite equation $(3.2)_2$ in the equivalent form:

$$\rho\left(\frac{\partial \mathbf{u}}{\partial t} + \mathbf{u}\cdot\nabla\mathbf{u}\right) + \mu\nabla\nabla\times\mathbf{u} - (\lambda+\mu)\nabla\nabla\cdot\mathbf{u} = -k\nabla(\sigma). \tag{3.7}$$

Multiply (3.7) by $\mathbf{V}$, with $\mathbf{V}$ given in Lemma 2.4, integrating by parts, we get

$$k\left\|\frac{\sigma}{|x|}\right\|^2 = h, \tag{3.8}$$

where

$$h := -(\rho\mathbf{u}\cdot\nabla\mathbf{u},\mathbf{V}) + \left(\rho\frac{\partial\mathbf{u}}{\partial t},\mathbf{V}\right) + \mu(\nabla\mathbf{u},\nabla\mathbf{V}) + (\lambda+\mu)(\nabla\cdot\mathbf{u},\nabla\cdot\mathbf{V})$$

Applying Lemma 2.3, we deduce

$$h \le \frac{k}{2}\left\|\frac{\sigma}{|x|}\right\|^2 + c\left(\|\nabla\mathbf{u}\|^2 + \left\|\sqrt{\rho}\frac{\partial\mathbf{u}}{\partial t}\right\|^2\right),$$

where c is function of k, ρ, and $\mathbf{u}$. We add (3.1) to (3.6) multiplied by α, plus (3.8) multiplied by β it delivers

$$\frac{d\mathcal{E}}{dt} + \mathcal{D} = \mathcal{F}, \tag{3.9}$$

where

$$\mathcal{E} := E(t) + \frac{1}{2}\alpha\mu\|\nabla\mathbf{u}\|^2 + \alpha(\lambda+\mu)\|\nabla\cdot\mathbf{u}\|^2 - k\alpha\left(\sigma,\nabla\cdot(\mathbf{u})\right)$$

$$\mathcal{D} := \left\|\sqrt{\rho}\frac{\partial\mathbf{u}}{\partial t}\right\|^2 + \alpha(\mu\|\nabla\mathbf{u}\|^2 + (\lambda+\mu)\|\nabla\cdot\mathbf{u}\|^2)\beta k\left\|\frac{\sigma}{|x|}\right\|^2$$

and

$$\begin{aligned}\mathcal{F} &:= \alpha g + \beta h \le \\ &\le \frac{1}{2}\left(\left\|\sqrt{\rho}\frac{\partial\mathbf{u}}{\partial t}\right\|^2 + \beta\left\|\frac{\sigma}{r}\right\|^2\right) + c\left[\beta\left\|\sqrt{\rho}\frac{\partial\mathbf{u}}{\partial t}\right\|^2 + \alpha\|\nabla\mathbf{u}\|^2\right] \le \delta\mathcal{D}.\end{aligned}$$

Suppose we take

$$\alpha = c\ \max\left\{\frac{1}{4\mu},\frac{\rho_m k^2}{\lambda+\mu}\right\} \quad \beta \le \frac{1}{4c},$$

then, it is

$$\mathcal{E} \ge \gamma\left[\frac{1}{2}(\|\mathbf{u}\|^2 + \|\sigma\|^2 + \mu\|\nabla\mathbf{u}\|^2 + (\lambda+\mu)\|\nabla\cdot\mathbf{u}\|^2)\right]$$

$$\left\|\sqrt{\rho}\frac{\partial\mathbf{u}}{\partial t}\right\|^2 + \alpha\mu\|\nabla\mathbf{u}\|^2 + \beta k\left\|\frac{\sigma}{|x|}\right\|^2 \le \gamma_1\mathcal{D}$$

From (3.9) it follows

$$\frac{d\mathcal{E}}{dt} + \frac{1}{2}\mathcal{D} \le 0. \tag{3.10}$$

In particular, since

$$\mathcal{E} \simeq \|\mathbf{u}\|^2 + \|\sigma\|^2 + \|\nabla \mathbf{u}\|^2,$$

we deduce the boundedness of the L^2 norm of the perturbations in terms of the initial data. The stability of the rest with respect to arbitrary perturbations is so completely proved.

4. DECAY TO ZERO OF THE L^2-NORM OF THE PERTURBATIONS

The lack of validity of the Poincare' inequality in unbounded domains does not allow us to deduce the decay of the norm of perturbations from (3.10). In this section we wish to prove the decay to zero of the sum of the kinetic energy plus the enthalpy of the perturbations σ, $\mathbf{u}$ which is equivalent to their L^2 norms. To this end, we employ again the lemmas of section 2 and energy methods.

First of all we write the continuity equation in the following form

$$\frac{1}{\rho}\frac{\partial \sigma}{\partial t} = -\nabla \cdot \mathbf{u} - \frac{\mathbf{u} \cdot \nabla \rho}{\rho}. \tag{4.1}$$

Multiply (4.1), and $(3.1)_2$ respectively by $k \ln^2 r\sigma/\rho$, $\ln^2 r\mathbf{u}$, with $r = |x|$, integrating over Ω after several integrations by parts and using the transport theorem, it furnishes

$$\begin{aligned}
\frac{k}{2}\frac{d}{dt}\left\|\frac{\ln r\sigma}{\sqrt{\rho}}\right\|^2 &= -k\left(\ln^2 r\sigma, \nabla \cdot \mathbf{u}\right) \\
&-k\left(\ln^2 r\frac{\sigma^2}{\rho^2}, \mathbf{u} \cdot \nabla\rho\right) + k\left(\ln^2 r\frac{\sigma^2}{2\rho^2}, \nabla \cdot \mathbf{u}\right) \\
\frac{1}{2}\frac{d}{dt}\|\sqrt{\rho}\ln r\mathbf{u}\|^2 &+ \mu\|\ln r\nabla \times \mathbf{u}\|^2 + (\lambda + 2\mu)\|\ln r\nabla \cdot \mathbf{u}\|^2 = \\
&= -k(\ln^2 r\mathbf{u}, \nabla\sigma) - \mu(\nabla(\ln^2 r) \times \mathbf{u}, \nabla \times \mathbf{u}) - \\
&-(\lambda + 2\mu)(\mathbf{u} \cdot \nabla(\ln^2 r), \nabla \cdot \mathbf{u}) + (\ln^2 r\sigma\mathbf{u}, \nabla U).
\end{aligned} \tag{4.2}$$

Add the two resulting equations, in such a way the pressure term at the right hand side vanishes, and we obtain

$$\begin{aligned}
&\frac{1}{2}\frac{d}{dt}\left[\left\|\frac{\ln r\sigma}{\sqrt{\rho}}\right\|^2 + \|\sqrt{\rho}\ln r\mathbf{u}\|^2\right] \\
&+\mu\|\ln r\nabla \times \mathbf{u}\|^2 + (\lambda + 2\mu)\|\ln r\nabla \cdot \mathbf{u}\|^2 \le F(t),
\end{aligned} \tag{4.3}$$

where

$$F(t) := -\mu(\nabla(\ln^2 r) \times \mathbf{u}, \nabla \times \mathbf{u}) - (\lambda + 2\mu)(\mathbf{u} \cdot \nabla(\ln^2 r), \nabla \cdot \mathbf{u})$$
$$+k(\sigma\mathbf{u}, \nabla(\ln^2 r)) + (\ln^2 r\sigma\mathbf{u}, \nabla U)-$$
$$-k\left(\ln^2 r\frac{\sigma^2}{\rho^2}, \mathbf{u} \cdot \nabla\rho\right) + k\left(\ln^2 r\frac{\sigma^2}{\rho}, \nabla \cdot \mathbf{u}\right).$$

By Lemma 2.1, the properties of class $\mathcal{V}$ we check that

$$F(t) \le c(\|\nabla\mathbf{u}\|^2 + \|\sigma\|\|\nabla\mathbf{u}\|),$$

which claims that $F(t) \in L^2(0,\infty)$. Moreover, from (4.3) we infer that

$$\ln r(|u| + |\sigma|) \in L^\infty(0,\infty; L^2(\Omega)),$$

that is equivalent to

$$\|\ln r\ u\| + \|\ln r\ \sigma\| < c_1.$$

By use of Lemma 2.2 we know that

$$\|\mathbf{u}\|^2 + \|\sigma\|^2 \le cR^2\left(\left\|\frac{\sigma}{r}\right\|^2_{\Omega_R} + \|\nabla\mathbf{u}\|^2_{\Omega_R}\right)$$
$$+\eta(\|\sqrt{\ln r}\sigma\|^2_{\Omega R} + \|\sqrt{\ln r}\mathbf{u}\|^2_{\Omega R}) \le R^2\left(\left\|\frac{\sigma}{r}\right\|^2_{\Omega_R} + \|\nabla\mathbf{u}\|^2_{\Omega_R}\right) + c_1\eta. \tag{4.4}$$

The decay to zero of $\mathcal{E}$ along a sequence of times $t_n \to \infty$ it is now almost achieved. To this end, we use the summability of $\mathcal{D}$ which states the existence of a sequence $t_n \to \infty$ along which $\mathcal{D}(t_n) \to 0$. Therefore, for all $\delta > 0$, there exists a N such that for all $n \ge N$ it is

$$\mathcal{D}(t_n) < \delta.$$

Moreover, for all $\delta/2$ there exists a R such that

$$\mathcal{D}(t_n) = \left(\left\|\frac{\sigma}{r}\right\|^2_{\Omega_R} + \|\nabla\mathbf{u}\|^2_{\Omega_R}\right) + \frac{\delta}{2}$$

which delivers

$$\left(\left\|\frac{\sigma}{r}\right\|^2_{\Omega_R} + \|\nabla\mathbf{u}\|^2_{\Omega_R}\right) \le \frac{\delta}{2}.$$

Then, for all $\epsilon > 0$ we fix $\delta = \dfrac{\epsilon}{cR^2}$, $\eta = \dfrac{\epsilon}{2c_1}$. Substituting these values in (4.4) we deduce that for all ϵ there exisists a N such that for all $n > N$ it holds

$$\|\mathbf{u}(t_n)\|^2 + \|\sigma(t_n)\|^2 \le \epsilon,$$

and the decay of the L^2-norm of the perturbation is proved.

Acknowledgments

The author thanks the 60% contract MURST, also she acknowledges the GNFM of italian CNR for financial supports.

References

[1] Beirao da Veiga, H. (1987). An L^p-theory for the n-dimensional, stationary compressible Navier-Stokes equations and the incompressible limit for compressible fluids. The equilibrium solutions. *Comm. Math. Phys.*, 109:229–248.

[2] Deuring, P. (1992). Decay estimates for the compressible Navier-Stokes equations in unbounded domain. *Math. Z.*, 209:115–130.

[3] Deuring, P. (1993). L^2-decay for the compressible Navier-Stokes equations in unbounded domains. *Comm. Partial Differential Equations*, 18:1445–1476.

[4] Galdi, G.P. and Rionero, S. (1976). Weighted energy methods in fluid dynamics and elasticity. *Lecture Notes in Math.*, 1134.

[5] Kobayashi, T. (1997). On a local energy decay of solutions for the equations of motion of compressible viscous and heat-conductive gases in an exterior domain in R^3. *Tsukuba J. Math.*, 21:629–670.

[6] Kobayashi, T. (1997). On the local energy decay of higher derivatives of solutions for the equations of motion of compressible viscous and heat-conductive gases in an exterior domain in R^3. *Tsukuba J. Math.*, 21:629–670.

[7] Leray, J. (1933). Etude de diverses equations integrales non lineaires et de quelques problemes que pose l'hydridynamique. *J. Math. Pures Appl.*, 12:1–82.

[8] Matsumura, A. and Padula, M. (1992). Stability of the stationary soluitions of compressible viscous fluids with large external potential forces. *Stab. Appl. Anal. Cont. Media*, 2:183–202.

[9] Novotny, A. and Padula, M. (1993). Existence and uniqueness of - stationary solutions for viscous compressible heat conductive fluid with large potential and small nonpotential external forces. *Siberian Math. J.*, 34:120–146.

[10] Novotny, A. and Pileckas, K. (1998). Steady compressible Navier-Stokes equations with large potential forces via a method of decomposition. *Math. Methods Appl. Sci.*, 21:665–648.

[11] Novotny, A. and Straskraba, I. Convergence to equilibria for compressible Navier-Stokes equations with large data. Preprint of Lab. d'Analise non lineaire applique n.98/03.

[12] Padula, M. (1992). Stability properties of regular flows of heat-conducting compressible fluids. *J. Math. Kyoto Univ.*, 32:401–442.

[13] Padula, M. (1998). On the exponential decay to the rest state for a viscous isothermal fluid. *J. Fluid Mech. and Anal.*, 1.

[14] Padula, M. (1998). On the exponential stability of the rest of a viscous compressible fluid. Proc. of *Society of Trends in Application of Mathematics to Mechanics*, Nice 25–29 May 1998.

GLOBAL BEHAVIOR OF COMPRESSIBLE FLUID WITH A FREE BOUNDARY AND LARGE DATA

Patrick Penel, Ivan Straškraba

Abstract: The global behavior of the solutions to one-dimensional Navier-Stokes system with a free boundary is investigated for large data. It is shown that the solutions stabilize to equilibrium in general on subsequences, and completely, if the body force is such that the corresponding equilibrium is unique. Mild condition on state equation is imposed which is satisfied in important physical situations.

Keywords: Navier-Stokes equations, global behavior, free boundary, large data, compressible fluids, one-dimensional case.

1. INTRODUCTION

We consider the problem

$$\rho(u_t + uu_x) + p(\rho)_x - \mu u_{xx} = \rho f, \tag{1.1}$$

$$\rho_t + (\rho u)_x = 0, \quad x \in (0, y(t)), \quad t > 0, \tag{1.2}$$

$$u(0,t) = 0, \quad p(\rho(y(t),t)) - \mu u_x(y(t),t) = p_0, \tag{1.3}$$

$$y'(t) = u(y(t),t), \quad t > 0, \tag{1.4}$$

$$y(0) = \ell, \ u(x,0) = u_0(x), \ \rho(x,0) = \rho_0(x), \ x \in [0,\ell], \tag{1.5}$$

which is known [1] as the Navier-Stokes system for 1-D compressible isentropic flow with a free boundary. The unknown quantities are the density ρ, velocity u and the trajectory $x = y(t)$ of the free boundary. The given data are the functional dependance $p = p(\rho)$ between the density ρ and the pressure p, viscosity constant μ, density of external forces $f = f(x)$, the pressure p_0 outside the free boundary, the initial position ℓ of the free boundary and initial distribution u_0 and ρ_0 of the velocity and density, respectively. Along with the problem (1.1) - (1.5)

Applied Nonlinear Analysis, edited by Sequeira *et al.*
Kluwer Academic / Plenum Publishers, New York, 1999.

consider the problem

$$p(\widetilde{\rho})_x = \widetilde{\rho} f, \quad x \in (0, \widetilde{y}), \tag{1.6}$$

$$p(\widetilde{\rho}(\widetilde{y})) = p_0, \quad \int_0^{\widetilde{y}} \widetilde{\rho}\, dx =: M_0, \quad \widetilde{\rho} \geq 0. \tag{1.7}$$

Our intention is to prove that if $p(\cdot)$ satisfies certain natural assumptions then

$$\lim_{t\to\infty} \int_0^{y(t)} \rho(x,t) u(x,t)^2\, dx = 0, \tag{1.8}$$

$$\lim_{t\to\infty} \int_0^{\widetilde{y}} |\rho(x,t) - \widetilde{\rho}(x)|^q\, dx = 0. \tag{1.9}$$

where ρ is suitably extended if $y(t) > \widetilde{y}$, and $q \in [1, \infty)$ is arbitrary. A similar result has been proved in [10], when the equations (1.1) - (1.5) are written in Lagrangian mass coordinates, and in [7] in the Euler coordinates and for Dirichlet boundary conditions but under the restrictive condition $f' \leq 0$. Recently, this assumption has been removed in [8] and we are now going to do the same for the case of boundary conditions (1.3). Let us note that in threedimensional case, recently a similar result has been proved for space periodic [6] and Dirichlet [2] boundary conditions, with global weak solutions assumed (but still in the class of existence).

The Euler coordinates approach makes it possible to include also the cases when, for some x, $\rho(x,t)$ tends to zero as $t \to \infty$. Since in our case the global existence of solutions for arbitrarily large data is known [9] we need only to find sufficient uniform estimates which would lead (similarly as in [7]) to (1.8), (1.9). The crucial role plays the estimate

$$\sup\{\rho(x,t); 0 \leq x \leq y(t),\ t \geq 0\} < \infty, \tag{1.10}$$

which will be derived in Section 3. As a consequence of (1.10) and energy equality (3.2) below we obtain (1.8). Then, in Section 3, with the help of a comparative "quasistationary" problem we find a function $\bar{\rho}(x,t)$ which is for $t \to \infty$ equivalent to $\rho(x,t)$. The result then follows from compactness of the solution operator of the "quasistationary" problem in question.

It should be mentioned that [3] considered the free boundary problem with zero external body force but nonlinear viscosity and nonmonotone equation of state, and [5] examined it taking into account the energy equation. The spherically symmetric case is investigated in [4].

We adopt the usual notation, namely, $C^k(\cdot)$ for spaces of k-times continuously differentiable functions, $W_q^k(\cdot)$ for the Sobolev spaces of k-th order and power q, $L^q(\cdot)$ for Lebesgue spaces with power q. The norm in L^q will be denoted by $|\cdot|_q$ and in W_q^k by $\|\cdot\|_{k,q}$. Besides, given $g \in L^1(0,a), a \in (0,\infty]$ we denote $(Ig)(x) := \int_0^x g(\xi)\,d\xi$ for $x \in [0,a]$.

2. STATIONARY PROBLEM

In this section the stationary problem to the evolutionary problem (1.1) - (1.5) will be studied. First, let us define the solution of the general problem we intend to work with.

Definition 2.1. *A triple (u,ρ,y) is called the solution of (1.1) - (1.5) on the interval $[0,T]$, $(T>0)$ if $y \in W_1^1(0,T;R^+), u \in W_2^{2,1}(Q), \rho \in W_2^1(Q) \cap L^\infty(Q)$ with $Q = \{(x,t);\quad 0 < x < y(t),\quad t \in (0,T)\}$, the equations (1.1), (1.2) are satisfied in the sense of $L^2(Q)$, the equation (1.4) in the sense of $L^2(0,T)$ and (1.3), (1.5) hold in the sense of traces in the spaces $W_2^{2,1}(Q)$ and $W_2^1(0,l)$, respectively.*

Next, let us make the following fundamental assumptions:

$$p(\cdot) \in C^1_{\mathrm{loc}}((0,\infty)),\ p(0+) = p(0) = 0,\ p'(r) > 0 \ \text{ for } \ r > 0, \qquad \liminf_{r\to\infty} p(r) = \infty; \tag{2.1}$$

$$f = f(x), f \in L^\infty_{\mathrm{loc}}([0,\infty)) \cap L^1(0,\infty); \tag{2.2}$$

$$u_0 \in H^1(0,\ell),\ u_0(0) = 0,\ \rho_0 \in H^1(0,\ell),\ 0 < \alpha_0 \le \rho_0(x) \le \beta_0 < \infty \quad \text{for } x \in [0,\ell]. \tag{2.3}$$

The stationary problem corresponding to (1.1) - (1.5) is given by the equations

$$\rho u u_x + p(\rho)_x - \mu u_{xx} = \rho f, \tag{2.4}$$

$$(\rho u)_x = 0, \quad x \in (0,y_0) \tag{2.5}$$

$$u(0) = u(y_0) = 0, \quad p(\rho(y_0)) - \mu u_x(y_0) = p_0, \tag{2.6}$$

$$\int_0^{y_0} \rho\,dx = M_0, \quad \rho \ge 0. \tag{2.7}$$

Definition 2.2. *A triple (u,ρ,y_0) is called the solution of (2.4) - (2.7) if $y_0 > 0$, $u \in W_2^2(0,y_0) \cap \mathring{W}_2^1(0,y_0)$, $\rho \in W_2^1(0,y_0)$, the equations (2.4), (2.5) hold a.e. in $(0,y_0)$ and $(2.6)_2$, (2.7) hold.*

Theorem 2.3. *Let (u,ρ,y_0) be a solution of (2.4) - (2.7). Then $u \equiv 0$ and $\tilde\rho = \rho, \tilde y = y_0$ satisfy (1.6), (1.7).*

Proof. Let $\varphi \in C_0^\infty(0,y_0)$. If (u,ρ,y_0) satisfy (2.4) - (2.7) then by integration by parts we get

$$\int_0^{y_0} \rho f \varphi\,dx = \int_0^{y_0} (\rho u u_x + p(\rho)_x - \mu u_{xx})\varphi\,dx = -\int_0^{y_0} (\rho u^2 \varphi' + p(\rho)\varphi' + \mu u \varphi'')\,dx. \tag{2.8}$$

Define $\Omega^+ = \{x \in (0,y_0); \rho(x) > 0\}, \Omega^- = (0,y_0) \setminus \Omega^+$. Then, by continuity of ρ, Ω^+ is open and hence there exists $S \subset N$ and $0 < a_j < b_j < y_0$, $j \in S$ such that $\Omega^+ = \bigcup_{j\in S}(a_j, b_j)$. Since $\rho(a_j) = \rho(b_j) = 0$, $\rho > 0$ on (a_j, b_j) and $\rho u = \text{const}$ in the same interval, we have $u = 0$ on (a_j, b_j). This and (2.5) implies $u(a_j) = u(b_j) = u'(a_j+) = u'(b_j-) = 0$. So we have

$$\begin{aligned}\int_{a_j}^{b_j} \rho f \varphi \, dx &= \int_{a_j}^{b_j} ((\rho u^2)_x + p(\rho)_x - \mu u_{xx})\varphi \, dx \\ &= -\int_{a_j}^{b_j} (\rho u^2 \varphi' + p(\rho)\varphi' + \mu u \varphi'') \, dx,\end{aligned}$$

which, by summation, yields

$$\int_{\Omega^+} \rho f \varphi \, dx = -\int_{\Omega^+} (\rho u^2 \varphi' + p(\rho)\varphi' + \mu u \varphi'') \, dx. \tag{2.9}$$

Then by (2.8), (2.9) we find

$$\begin{aligned}0 &= \int_{\Omega^-} \rho f \varphi \, dx = \int_0^{y_0} \rho f \varphi \, dx - \int_{\Omega^+} \rho f \varphi \, dx \\ &= -\int_0^{y_0} (\rho u^2 \varphi' + p(\rho)\varphi' + \mu u \varphi'') \, dx + \int_{\Omega^+} (\rho u^2 \varphi' + p(\rho)\varphi' + \mu u \varphi'') \, dx \\ &= -\int_{\Omega^-} (\rho u^2 \varphi' + p(\rho)\varphi' + \mu u \varphi'') \, dx = -\mu \int_{\Omega^-} u \varphi'' \, dx.\end{aligned}$$

Since $u = 0$ on Ω^+, the last identity implies

$$\int_0^{y_0} u \varphi'' \, dx = 0 \qquad \text{for all} \qquad \varphi \in C_0^\infty((0, y_0)).$$

This yields $u(x) = c_1 + c_2 x$ for $x \in (0, y_0)$ and $c_1 = c_2 = 0$ follows from $u(0) = u(y_0) = 0$. Now the assertion easily follows. □

Later, in Section 4 we shall prove existence of the equilibrium state as a limit of the evolutionary solution, but uniqueness will not follow from this process. Here we give one sufficient condition for the uniqueness of the rest state.

Theorem 2.4. *If $(y_j, \rho_j), j = 1,2$ are two rest states, say with $y_2 < y_1$, and $f \leq 0$, then $y_1 = y_2$ and $\rho_1 \equiv \rho_2$.*

Proof. Assume we have two couples $(y_j, \rho_j), j = 1,2$ such that

$$\begin{aligned}p(\rho_j)_x &= \rho_j f, \quad x \in (0, y_j), \\ \int_0^{y_j} \rho_j \, dx &= M_0, \\ p(\rho_j(y_j)) &= p_0, \quad j = 1,2, \ y_1 > y_2, \ \inf_{x\in(0,y_j)} \rho_j \geq 0.\end{aligned} \tag{2.10}$$

First, it is clear that $\inf\{\rho_j(x); x \in [0, y_j], j = 1,2\} > 0$. Indeed, by assumption and $(2.10)_1$, ρ_j are nonincreasing and so, if there is an $x_j \in [0, y_j]$ such that $\rho_j(x_j) = 0$ then necessarily $0 < p_0 = p(\rho_j(y_j)) = 0$ which is a contradiction. Define $\pi(r) := \int_1^r \frac{p'(s)}{s}\, ds$. Then clearly

$$\pi(\rho_j(x)) = \pi(p^{-1}(p_0)) + \int_{y_j}^{x} f\, d\xi \quad \text{for} \quad x \in (0, y_j). \tag{2.11}$$

If there is an $x_0 \in [0, y_2]$ such that $\rho_1(x_0) = \rho_2(x_0)$ then clearly, $\rho_1 = \rho_2$ on $(0, y_2)$ and consequently $y_1 = y_2$ otherwise ρ_1 must be zero on (y_2, y_1) which is a contradiction. So let $\rho_1(x) \neq \rho_2(x)$ for all $x \in [0, y_2]$. Then necessarily $\rho_1(x) < \rho_2(x)$ in $[0, y_2]$ since otherwise we had $M_0 = \int_0^{y_2} \rho_2(x)\, dx < \int_0^{y_2} \rho_1(x)\, dx < \int_0^{y_1} \rho_1(x)\, dx = M_0$. But by (2.11) we have $0 < \pi(\rho_2(x)) - \pi(\rho_1(x)) = \int_{y_2}^{y_1} f\, dx \leq 0, x \in (0, y_2)$, which is a contradiction. □

3. ENERGY EQUALITY AND GLOBAL BOUNDEDNESS OF DENSITY

In this section we derive energy equality and the estimate (1.10).

Proposition 3.1. *Let* (u, ρ, y) *be a solution of* (1.1) - (1.5) *and* $P(r)$ *a function satisfying*

$$P(r) = P(1)r + r \int_1^r \frac{p(s)}{s^2}\, ds, \quad r > 0. \tag{3.1}$$

Then the following energy equality holds:

$$\frac{d}{dt} \int_0^{y(t)} (\frac{1}{2}\rho u^2 + P(\rho) + p_0 - \rho I f)\, dx + \mu \int_0^{y(t)} u_x^2\, dx = 0, \quad t \geq 0. \tag{3.2}$$

Proof. Multiply (1.2) by u and integrate $\int_0^{y(t)} dx$. Since $\rho u(u_t + u u_x) = \frac{1}{2}(\rho u^2)_t + \frac{1}{2}(\rho u^3)_x$ we have

$$\int_0^{y(t)} \rho u(u_t + u u_x)\, dx = \frac{1}{2} \int_0^{y(t)} (\rho u^2)_t\, dx + \frac{1}{2} \rho y'(t)^3 = \frac{1}{2} \frac{d}{dt} \int_0^{y(t)} \rho u^2\, dx.$$

Further, by (1.3), (1.4) we find

$$\int_0^{y(t)} (p(\rho) - \mu u_x)_x u\, dx = \mu \int_0^{y(t)} u_x^2\, dx - \int_0^{y(t)} p(\rho) u_x\, dx + p_0 y'(t).$$

The choice of P in (3.1) and (1.2) gives

$$- \int_0^{y(t)} p(\rho) u_x\, dx = \frac{d}{dt} \int_0^{y(t)} P(\rho)\, dx.$$

Finally, with the help of (1.2) we get

$$\int_0^{y(t)} f\rho u\, dx$$
$$= (If)(y(t))\rho(y(t),t)y'(t) - \int_0^{y(t)} If(\rho u)_x\, dx = \frac{d}{dt}\int_0^{y(t)} If\rho\, dx,$$

and (3.2) easily follows. □

Lemma 3.2. *Let the assumptions* (2.1) - (2.3) *be satisfied and let*

$$p_0 \geq 0. \tag{3.3}$$

Then $P(1)$ *in* (3.1) *can be chosen in such a way that*

$$E(t) := \int_0^{y(t)} (\frac{1}{2}\rho u^2 + P(\rho) + p_0 - \rho If)\, dx \geq 0 \quad \textit{for all } t \geq 0. \tag{3.4}$$

Proof. Since $\int_0^{y(t)} \rho\, dx = M_0$ and P is convex, by (3.1) and Jensen inequality we have

$$\int_0^{y(t)} P(\rho)\, dx =$$
$$P(1)M_0 + \int_0^{y(t)} \rho \int_1^{\rho} \frac{p(s)}{s^2}\, ds\, dx \geq P(1)M_0 + M_0 \int_1^{M_0} \frac{p(s)}{s^2}\, ds.$$

The result immediately follows. □

Lemma 3.3. *Let the assumptions of Lemma 3.2 be satisfied. Then*

$$\sup_{t\geq 0} y(t) < \infty, \quad \int_0^{\infty}\int_0^{y(t)} u_x^2\, dx\, dt < \infty, \tag{3.5}$$

in particular

$$\lim_{s,t\to\infty} \int_s^t \int_0^{y(\tau)} u_x^2\, dx\, d\tau. \tag{3.6}$$

Proof. The integration $\int_0^t ds$ of (3.2) yields

$$E(t) + \mu \int_0^t \int_0^{y(s)} u_x\, dx\, ds = E(0).$$

Since $E(t) \geq 0$ by (3.4), the relations (3.5) follow immediately. □

Let us now prove the estimate (1.10).

Lemma 3.4. *Let besides the assumptions* (2.1) - (2.3), (3.3) *the function P given by* (3.1) *satisfy*

$$p(r) \leq \text{ const } (1 + P(r)) \quad \text{for } r > 0. \tag{3.7}$$

Then (1.10) *holds.*

Proof. Denote $D_t = \frac{\partial}{\partial t} + u \cdot \frac{\partial}{\partial x}$. By integration $\int_x^{y(t)} d\xi$ the equation (1.1) we get $p(\rho) - \mu u_x = \int_x^{y(t)} (\rho D_t u - \rho f)\, dx + p_0$. Hence by (1.2)

$$\mu D_t \ln \rho = -\mu u_x = \int_x^{y(t)} (\rho D_t u - \rho f)\, d\xi - p(\rho) + p_0. \tag{3.8}$$

Since $\rho u_t = D_t(\rho u)$ and

$$D_t\Big(\int_x^{y(t)} \rho u\, dx\Big) =$$
$$\int_x^{y(t)} (\rho u)_t\, d\xi - \rho u^2 + y'(t)\rho(y(t),t)u(y(t),t),$$

we have

$$\begin{aligned}
\int_x^{y(t)} \rho D_t u\, d\xi &= \int_x^{y(t)} D_t(\rho u)\, d\xi + \int_x^{y(t)} \rho u u_x\, d\xi \\
&= \int_x^{y(t)} (\rho u)_t\, d\xi + \int_x^{y(t)} u(\rho u)_x\, d\xi + \int_x^{y(t)} \rho u u_x\, d\xi \\
&= \int_x^{y(t)} (\rho u)_t\, d\xi + \rho u^2(y(t),t) - \rho u^2(x,t) \\
&= D_t\Big(\int_x^{y(t)} \rho u\, d\xi\Big).
\end{aligned}$$

This, together with (3.8) yields

$$\mu D_t \ln \rho = D_t\Big(\int_x^{y(t)} \rho u\, d\xi\Big) - \int_x^{y(t)} \rho f\, d\xi + p_0 - p(\rho),$$

which can also be written as

$$D_t(\mu \ln \rho - r) = R - p(\rho), \tag{3.9}$$

where $r(x,t) = \int_x^{y(t)} \rho u\, d\xi$, $R(x,t) = p_0 - \int_x^{y(t)} \rho f\, d\xi$. By integration of (3.2) we get

$$\sup_{t \geq 0} y(t) < \infty, \quad \sup\Big\{\int_x^{y(t)} \rho u^2\, d\xi;\ 0 \leq x \leq y(t),\ t \geq 0\Big\} < \infty. \tag{3.10}$$

So we have

$$
\begin{aligned}
|r(x,t)| &\le \sup_{t\ge 0}\Big(\int_x^{y(t)} \rho\, d\xi\Big)^{1/2}\Big(\int_x^{y(t)} \rho u^2\, d\xi\Big)^{1/2} =: c_0 < \infty \\
|R(x,t)| &\le |p_0| + |f|_\infty \int_0^{y(t)} \rho\, d\xi =: C_0 < \infty \qquad (3.11)
\end{aligned}
$$

for $0 \le x \le y(t)$, $t \ge 0$. Now, let $x : 0 \le x < \sup_{s\ge 0} y(t)$ be arbitrary but fixed. If $p(\rho(x,t)) \le C_0$ for all $t \ge 0$ such that $x \le y(t)$ then we have $\rho(x,t) \le p^{-1}(C_0)$ for this particular x and all $t \ge 0 : y(t) \ge x$. So let us assume that there is a $t_2 > 0$ such that $x \le y(t_2)$ and $p(\rho(x,t_2)) \ge C_0$. Denote by $\xi := \xi(x,t;\tau)$ a function satisfying the equations $\xi_\tau = u(\xi,\tau)$, $\xi(x,t;t) = x$, where $0 \le \tau \le t, x \le y(t)$. By the regularity of the solution (u,ρ,y) such a function exists and $\xi(x,t;\cdot) \in H^1(0,t)$ for any $t > 0$ such that $y(t) \ge x$. Clearly, for z smooth enough we have $(d/d\tau)z(\xi(x,t;\tau),\tau) = z_t(\xi(x,t;\tau),\tau) + u(\xi(x,t;\tau),\tau)\cdot z_x(\xi(x,t;\tau),\tau)$. Let $t_1 \in [0,t_2)$ be a minimal number such that $p(\rho(\xi(x,t_2;\tau),\tau) \ge C_0$ for all $\tau \in [t_1,t_2]$. Then either $t_1 > 0$ and $p(\rho(\xi(x,t_2;t_1),t_1)) = C_0$ or $t_1 = 0$ and by (2.3) we have $p(\rho(\xi(x,t_2;t_1),t_1)) = p(\rho_0(\xi(x,t_2;0))) \le p(\beta_0) < \infty$. Consider (3.9) along the curve $x := \xi(x,t_2;\tau), \tau \in [t_1,t_2]$. Then by integration $\int_{t_1}^{t_2} d\tau$ we get

$$
\begin{aligned}
\mu \ln \rho(x,t_2) - r(x,t_2) &= \mu \ln \rho(\xi(x,t_2;t_1),t_1) - r(\xi(x,t_2;t_1),t_1) \\
&+ \int_{t_1}^{t_2} (R - p(\rho))(\xi(x,t_2;\tau),\tau)\, d\tau. \qquad (3.12)
\end{aligned}
$$

Since the last term on the right-hand side of (3.12) is negative, with the help of (3.11) we find

$$
\mu \ln \rho(x,t_2) \le \max\{\ln \beta_0, \ln p^{-1}(C_0)\} + 2c_0.
$$

Since x was arbitrary and the constants β_0, c_0, C_0 depend on the data only, (1.10) readily follows. □

4. STABILIZATION OF SOLUTIONS

Lemmas 3.3 and 3.4 allow us to show that the solution stabilizes to equilibrium, more precisely, we are able to prove the following

Theorem 4.1. *Let in addition to the assumptions* (2.1) - (2.3) *we have* $p_0 > 0$, *and* (3.7) *with* P *defined by* (3.1) *with* $P(1)$ *sufficiently large (see Lemma 3.2). Then*

$$
\lim_{t\to\infty} \int_0^{y(t)} (\rho u^2)(x,t)\, dx = 0, \qquad (4.1)
$$

and if $q \in [1, \infty)$ *and* $t_n \to \infty$ *is arbitrary then there exists a subsequence* $s_n \to \infty$ *of* $\{t_n\}$ *and* $\widetilde{y} \in R, \widetilde{\rho} \in L^\infty(0, \widetilde{y})$ *such that*

$$\lim_{n\to\infty} y(s_n) = \widetilde{y}, \tag{4.2}$$

$$\lim_{n\to\infty} \int_0^{\widetilde{y}} |\rho(s_n) - \widetilde{\rho}|^q \, dx = 0, \tag{4.3}$$

$$\begin{aligned} p(\widetilde{\rho})_x &= \widetilde{\rho} f, \quad x \in (0, \widetilde{y}), \\ \int_0^{\widetilde{y}} \widetilde{\rho} \, dx &= \int_0^{\ell} \rho_0 \, dx, \\ p(\widetilde{\rho}(\widetilde{y})) &= p_0 \end{aligned} \tag{4.4}$$

Moreover, if $\widetilde{\rho}$ *is determined by* (4.3) *uniquely as it is the case for example when* $f \leq 0$ *(see Theorem 2.4), then the above convergence holds for all* $t \to \infty$.

Proof. The energy equality (3.2) can be written in the form

$$\frac{d}{d\tau} \int_0^{y(\tau)} (\frac{1}{2}\rho u^2 + p_0) \, dx = \int_0^{y(\tau)} (\rho f u + p(\rho) u_x - \mu u_x^2) \, dx. \tag{4.5}$$

Integrate (4.5) $\int_s^t d\tau, 0 \leq s < t$:

$$\int_0^{y(t)} (\frac{1}{2}\rho(t)u(t)^2 + p_0) \, dx = \int_0^{y(s)} (\frac{1}{2}\rho(s)u(s)^2 + p_0) \, dx \tag{4.6}$$
$$+ \int_s^t \int_0^{y(\tau)} (\rho f u + p(\rho) u_x - \mu u_x^2) \, dx \, d\tau.$$

Now, integrate (4.6) $\int_{t-1}^t ds$ with $t \geq 1$:

$$\frac{1}{2} \int_0^{y(t)} \rho u(t)^2 \, dx = \frac{1}{2} \int_{t-1}^t \int_0^{y(s)} \rho(s) u(s)^2 \, dx \, ds \tag{4.7}$$
$$+ p_0 \Big(\int_{t-1}^t y(s) \, ds - y(t) \Big) + \int_{t-1}^t \int_s^t \int_0^{y(\tau)} (\rho f u + p(\rho) u_x - \mu u_x^2) \, dx \, d\tau \, ds.$$

Show that the right-hand side of (4.7) tends to zero as $t \to \infty$. First, putting $\sigma(t) := (\int_{t-1}^t \int_0^{y(\tau)} u_x^2 \, dx \, d\tau)^{1/2}$, by (3.6) we have

$$\int_{t-1}^t \int_s^t \int_0^{y(\tau)} u_x^2 \, dx \, d\tau \, ds \leq \sigma(t)^2 \to 0 \text{ as } t \to \infty.$$

Second, by (1.10), (3.10) and Poincaré inequality

$$\int_{t-1}^t \int_0^{y(s)} \rho(s) u(s)^2 \, dx \, ds \leq |\rho|_\infty (\sup_{s \geq 0} y(s)) \sigma(t)^2 \to 0 \text{ as } t \to \infty,$$

$$\Big|\int_{t-1}^{t}\int_{s}^{t}\int_{0}^{y(\tau)}\rho(\tau)fu(\tau)\,dx\,d\tau\,ds\Big|\le$$
$$\le|f|_\infty\Big(\int_{t-1}^{t}\int_{0}^{y(\tau)}\rho(\tau)\,dx\,d\tau\Big)^{1/2}\Big(\int_{t-1}^{t}\int_{0}^{y(\tau)}\rho(\tau)u(\tau)^2\,dx\,d\tau\Big)^{1/2}$$
$$\le|f|_\infty M_0^{1/2}\int_{t-1}^{t}\int_{0}^{y(\tau)}\rho(\tau)u(\tau)^2\,dx\,d\tau\to0\ \text{ as }\ t\to\infty,$$

$$\Big|\int_{t-1}^{t}\int_{s}^{t}\int_{0}^{y(\tau)}p(\rho)u_x\,dx\,d\tau\,ds\Big|$$
$$\le\ |p(\rho)|_\infty(\sup_{\tau\ge0}y(\tau))^{1/2}\Big(\int_{t-1}^{t}\int_{0}^{y(\tau)}u_x^2\,dx\,d\tau\Big)^{1/2}\to0\ \text{ as }\ t\to\infty.$$

Finally,

$$\Big|y(t)-\int_{t-1}^{t}y(s)\,ds\Big|\le\int_{t-1}^{t}|u(y(s),s)|\,ds=$$
$$\int_{t-1}^{t}\Big|\int_{0}^{y(\tau)}u_x(x,\tau)\,dx\Big|\,d\tau\le\Big(\int_{t-1}^{t}y(\tau)\int_{0}^{y(\tau)}u_x(x,\tau)^2\,dx\,d\tau\Big)^{1/2}$$
$$\le(\sup_{\tau\ge0}y(\tau))^{1/2}\sigma(t)\to0\ \text{ as }\ t\to\infty.$$

Now, from (4.7) and the above relations the first relation in (4.1) follows immediately.

It remains to prove the convergence of y and ρ on some subsequence of a given sequence $t_n\to\infty$. To this purpose put

$$\begin{aligned}w(x,t)&:=\int_{x}^{y(t)}(\rho f)(\xi,t)\,d\xi\\ m(t)&:=y(t)^{-1}\int_{0}^{y(t)}\Big(p(\rho)+\int_{x}^{y(t)}\rho f\,d\xi\Big)\,dx\\ \bar\rho(x,t)&:=p^{-1}\Big(m(t)-\int_{x}^{y(t)}\rho f\,d\xi\Big).\end{aligned}\tag{4.8}$$

Later, we will see that $w,m,\bar\rho$ and their derivatives may be well globally estimated. First, it is easy to see that

$$\begin{aligned}\int_{0}^{y(t)}p(\bar\rho)\,dx&=m(t)y(t)-\int_{0}^{y(t)}\int_{x}^{y(t)}\rho f\,d\xi\,dx\\ &=\int_{0}^{y(t)}\Big(p(\rho)+\int_{x}^{y(t)}\rho f\,d\xi\Big)\,dx-\int_{0}^{y(t)}\int_{x}^{y(t)}\rho f\,d\xi\,dx\\ &=\int_{0}^{y(t)}p(\rho)\,dx.\end{aligned}$$

Consequently, putting

$$\psi(x,t) := \int_0^x (p(\rho) - p(\bar{\rho}))\, dx. \tag{4.9}$$

we get $\psi(0,t) = \psi(y(t),t) = 0$. So we have

$$Q(t) := \int_0^{y(t)} (p(\rho) - p(\bar{\rho}))^2\, dx = \int_0^{y(t)} (p(\rho) - p(\bar{\rho}))\psi_x\, dx \tag{4.10}$$

$$= -\int_0^{y(t)} (p(\rho) - p(\bar{\rho}))_x \psi\, dx = \int_0^{y(t)} \Big((\rho u)_t + (\rho u^2)_x - \mu u_{xx}\Big)\psi\, dx.$$

Show that

$$\lim_{t\to\infty} \int_{t-1}^t Q(s)\, ds = 0. \tag{4.11}$$

In what follows we heavily rely on the estimate (1.10). First, if we assume (3.7) then it is easy to show that

$$\sup_{t\geq 0} |\psi(t)|_\infty < \infty.$$

By (3.2) we get $\sup_{t\geq 0} y(t) < \infty$. So we have

$$\Big|\int_{t-1}^t \int_0^{y(s)} (\rho u^2)_x \psi\, dx\, ds\Big| = \Big|\int_{t-1}^t \int_0^{y(s)} \rho u^2 (p(\rho) - p(\bar{\rho}))\, dx\, ds\Big|$$

$$\leq \text{ const } \sup_{s\geq 0} |\rho(p(\rho) - p(\bar{\rho}))(s)|_\infty \int_{t-1}^t \int_0^{y(s)} u^2\, dx\, ds$$

$$\leq \text{ const } \sigma(t)^2 \to 0 \quad \text{as} \quad t\to\infty.$$

Using (1.10) again, we find also

$$\Big|\int_{t-1}^t \int_0^{y(s)} u_x \psi_x\, dx\, ds\Big| \leq \text{ const } \sigma(t) \to 0 \quad as \quad t\to\infty.$$

It remains to estimate

$$\int_{t-1}^t \int_0^{y(s)} (\rho u)_s \psi\, dx\, ds = \int_0^{y(t)} \rho u\psi(t)\, dx - \int_0^{y(t-1)} \rho u\psi(t-1)\, dx$$

$$- \int_{t-1}^t \int_0^{y(s)} \rho u\psi_s\, dx\, ds - \int_{t-1}^t (\rho u\psi)(y(s),s) y'(s)\, ds.$$

First of all, by global boundedness of ψ and ρ we have

$$\Big|\int_0^{y(s)} \rho u\psi(s)\, dx\Big|$$

$$\leq \text{ const} \Big(\int_0^{y(s)} \rho u^2(s)\, dx\Big)^{1/2} \leq \text{ const } \sigma(s) \to 0 \text{ as } s\to\infty.$$

Further, since $p(\rho)_t + \rho p'(\rho)u_x + up(\rho)_x = 0$, we have

$$\begin{aligned}\psi_t &= \frac{\partial}{\partial t}\int_0^x (p(\rho) - p(\bar\rho))\,dx = \int_0^x (p(\rho)_t - p(\bar\rho)_t)\,dx \\ &= \int_0^x (p(\rho) - \rho p'(\rho))u_x\,d\xi - up(\rho) - m'(t)x + \int_0^x w_t\,d\xi.\end{aligned} \tag{4.12}$$

To estimate ψ_t show that

$$|m'(t)| \leq \text{ const } \Big(|y'(t)| + (\int_0^{y(t)} |u_x|^2\,dx)^{1/2}\Big) \tag{4.13}$$

holds true. By differentiation in $(4.7)_2$ and using (1.2) and its consequence for $p(\rho)_t$, after some evident cancellations we find

$$\begin{aligned}m'(t) &= \frac{y'(t)}{y(t)}m(t) - \frac{1}{y(t)}\int_0^{y(t)} (\rho p'(\rho) - p(\rho))u_x\,dx \\ &+ \frac{1}{y(t)}\int_0^{y(t)} \rho u f\,dx + \frac{1}{y(t)}\int_0^{y(t)}\int_x^{y(t)} \rho u f'\,d\xi\,dx.\end{aligned}$$

First of all, it follows from (1.10) that $\sup_{t\geq 0}|m(t)| < \infty$. Secondly, it is $\inf_{t\geq 0} y(t) > 0$. Indeed, by (3.2), $y(t)$ is globally bounded and $\int_0^{y(t)} \rho(t)\,dx = M_0$ for all $t \geq 0$. If there were $t_n \geq 0$ such that $y(t_n) \to 0$ then by (1.10) $M_0 = \int_0^{y(t_n)} \rho(t_n)\,dx \leq y(t_n)\sup_{x,t}\rho \to 0$ as $n \to \infty$ which is a contradiction. So we have $\inf_{t\geq 0} y(t) > 0$. Then, again by (1.10), we find (4.13) true. Further, by (1.2), (1.4) we obviously have

$$w_t = \rho u f + \int_x^{y(t)} \rho u f'\,d\xi$$

and consequently

$$|w_t(s)| \leq \text{ const } \Big(\int_0^{y(s)} |u_x(\xi, s)|^2\,d\xi\Big)^{1/2}.$$

As we have clearly

$$\int_{t-1}^t \int_0^{y(s)} \rho u \psi_t\,dx\,ds \leq \text{ const } \sigma(t)\Big(\int_{t-1}^t \int_0^{y(s)} |\psi_t|^2\,dx\,ds\Big)^{1/2},$$

it remains to show that

$$\sup_{t\geq 1}\int_{t-1}^t \int_0^{y(s)} |\psi_t|^2\,dx\,ds < \infty.$$

Since for the first two terms in the expression on the right of (4.12) for ψ_t it is the same job as above we need only to show that

$$\int_{t-1}^t |y'(s)|^2\,ds \leq \text{ const } < \infty \quad \text{for all} \quad t \geq 0.$$

But we have

$$|y'(s)|^2 = \Big|\int_0^{y(s)} u_x(\xi,s)\,d\xi\Big|^2 \le (\sup_{s\ge 0} y(s)) \int_0^{y(s)} |u_x(\xi,s)|^2\,d\xi$$

which yields

$$\int_{t-1}^t |y'(s)|^2 \le \text{ const } \sigma(t)^2.$$

This completes the proof of (4.11). Now, show that

$$\lim_{t\to\infty} \int_{t-1}^t |Q'(s)|\,ds = 0 \tag{4.14}$$

By (4.10) we have

$$\begin{aligned} Q'(s) &= 2\int_0^{y(s)} (p(\rho)-p(\bar\rho))(p(\rho)_t - p(\bar\rho)_t)\,dx \\ &+ y'(s)(p(\rho(y(s),s)) - p(\bar\rho(y(s),s))). \end{aligned}$$

Consequently, by boundedness of ρ we find

$$|Q'(s)| \le \text{ const } \Big(\int_0^{y(s)} |p(\rho)_t - p(\bar\rho)_t|\,dx + |y'(s)|\Big).$$

With the integral we can proceed as with estimating ψ_t above to obtain

$$\int_0^{y(s)} |p(\rho)_t - p(\bar\rho)_t|\,dx \le \text{ const } \Big(\int_0^{y(s)} |u_x|\,dx + |y'(s)|\Big).$$

So we find

$$\begin{aligned} &\int_{t-1}^t |Q'(s)|\,ds \le \text{ const } \int_{t-1}^t |y'(s)|\Big(\sqrt{y}(s)\Big(\int_0^{y(s)} |u_x|^2\,dx\Big)^{1/2} \\ &+|y'(s)|+1\Big)\,ds \le \text{ const } \Big(\int_{t-1}^t |y'(s)|^2\,ds\Big)^{1/2}\Big(1+\sigma(t) \\ &+\Big(\int_{t-1}^t |y'(s)|^2\,ds\Big)^{1/2}\Big) \le \text{ const } \sigma(t) \to 0 \quad \text{as} \quad t\to\infty. \end{aligned}$$

Since

$$Q(t) = \int_{t-1}^t Q(s)\,ds + \int_{t-1}^t \int_s^t Q'(\sigma)\,d\sigma\,ds,$$

(4.11) and (4.14) imply $\lim_{t\to\infty} Q(t) = 0$, or equivalently

$$\lim_{t\to\infty} \int_0^{y(t)} (p(\rho(t)) - p(\bar\rho(t)))^2\,dx = 0. \tag{4.15}$$

Next we show that $\lim_{t\to\infty} p(\rho(y(t),t)) = p_0$. To this purpose, let us introduce $r(t) := \mu\log\rho(y(t),t)$ and $g(r) := p(\exp(\mu^{-1}r))$. Then by (1.2), (1.3), (1.4) we have

$$\begin{aligned} r'(t) + g(r(t)) &= p_0, \\ r(0) &= \mu\log\rho_0(\ell). \end{aligned} \tag{4.16}$$

By (1.10) we apriori know that r is bounded from above on $[0,\infty)$. Hence by (4.15), $|r'|$ is globally bounded as well. So given $\tau_n \to \infty$ we can select $\sigma_n \to \infty$ such that $r(\sigma_n) \to r_\infty$, $r'(\sigma_n) \to p_0 - g(r_\infty)$ with some $r_\infty \in [-\infty,\infty)$. If $p_0 \neq g(r_\infty)$ then there are $n_0, \delta > 0$ such that $|r'(\sigma_n)| \geq \delta$ for all $n \geq n_0$. Then given such σ_n there exists a maximal interval $(\sigma_n, \bar{\sigma})$ where $|r'| > 0$. If $\bar{\sigma} = \infty$ then by monotonicity of r on (σ_n, ∞) we have $\lim_{\sigma\to\infty} r(\sigma) = r_\infty$. But then $|r'(t)| \geq \alpha > 0$ in some neighbourhood of ∞ and r would be unbounded. In the case $r'(t) \geq \alpha > 0$ this is a contradiction to the boundedness from above of r. If $r'(t) \leq -\alpha < 0$ then $r_\infty = -\infty$ and consequently $\lim_{t\to\infty} r'(t) = p_0 > 0$ which is impossible. Hence necessarily $\bar{\sigma} < \infty$. Then $r'(\bar{\sigma}) = 0$ and $r_1(\sigma) := r(\sigma)$ for $\sigma \in [0,\bar{\sigma}]$ and $r_1(\sigma) = r(\bar{\sigma})$ for $\sigma > \bar{\sigma}$ is a solution of (4.15). By uniqueness, $r_1 = r$ in $[0,\infty)$ and $r(\bar{\sigma}) = r_\infty = g^{-1}(p_0)$. Going back to p and ρ we have proved $\lim_{t\to\infty} p(\rho(y(t),t)) = p_0$. Now, let $t_n \to \infty, q \in [1,\infty)$ be arbitrary. Then we can select $\{s_n\} \subset \{t_n\}$ such that

$$y(s_n) \to y_\infty \text{ and } \rho(s_n) \to \rho_\infty \text{ weak in } L^q(0,y_\infty). \tag{4.17}$$

Note, that if eventually for some sufficiently large n we have $y(s_n) < y_\infty$ then we can prolong ρ by

$$\rho(x,s_n) = \pi^{-1}\Big(\pi(\rho(y(s_n),s_n)) + \int_{y(s_n)}^{x} f(\xi)\,d\xi\Big),\ x \in (y(s_n), y_\infty),$$

where $\pi(r) := \int_1^r \frac{p'(s)}{s}\,ds$. Indeed, the function under π^{-1} is close to $\pi(p^{-1}(p_0))$ and so it belongs to the definition region of π^{-1}. But this yields (modulo further selection)

$$p(\bar{\rho}(s_n)) \equiv m(s_n) - \int_x^{y(s_n)} \rho(s_n) f\,d\xi \to m_\infty - \int_x^{y_\infty} \rho_\infty f\,d\xi \tag{4.18}$$

strongly in $L^q_{\mathrm{loc}}([0,\infty))$ with some constant m_∞. Further, by (4.15) and (4.17) we have

$$\begin{aligned}
&\limsup_{n\to\infty} \int_0^{y_\infty} \big(p(\rho(s_n)) - p(\bar{\rho}(s_n))\big)^2\,dx \leq \\
&\limsup_{n\to\infty} \int_0^{y(s_n)} \big(p(\rho(s_n)) - p(\bar{\rho}(s_n))\big)^2\,dx \\
&+ \limsup_{n\to\infty} \Big| \int_{y(s_n)}^{y_\infty} \big(p(\rho(s_n)) - p(\bar{\rho}(s_n))\big)^2\,dx \Big| = 0.
\end{aligned}$$

This together with (4.18) yields $p(\rho(x,s_n)) \to m_\infty - \int_x^{y_\infty} \rho_\infty f\,d\xi$ a.e. in $(0,y_\infty)$ and in $L^q(0,y_\infty)$ (since $|\rho(s_n)|_\infty \leq C < \infty$). As $\rho(s_n) \to \rho_\infty$ weakly and, $\rho(s_n) \to p^{-1}(m_\infty - \int_x^{y_\infty} \rho_\infty f\,d\xi)$ strongly, we have $p(\rho_\infty(x)) = m_\infty - \int_x^{y_\infty} \rho_\infty f\, d\xi$ for a.e. $x \in (0, y_\infty)$, which yields $p(\rho_\infty)_x = \rho_\infty f$ in $(0, y_\infty)$. Since for large n,

$$p(\rho(x, s_n)) = p(\rho(y(s_n), s_n) + \int_{y(s_n)}^{x} \rho(\xi, s_n) f(\xi)\, d\xi$$

in a neighbourhood of y_∞, we have $m_\infty = p(\rho_\infty(y_\infty)) = p_0$. Finally, since $\int_0^{y(s_n)} \rho(s_n)\, dx = M_0$, we have also $\int_0^{y_\infty} \rho_\infty\, dx = M_0$. Now, we put $\widetilde{y} := y_\infty, \widetilde{\rho} := \rho_\infty$ and the proof is complete. Of course, by standard argument we get that in the case of uniqueness of equilibrium (see for example Theorem 2.4) $\rho(t) \to \rho_\infty$ in $L^q(0, y_\infty)$. $\square$

Acknowledgments

This work was written during the stay of the second author at U.T.V. in 1998 and partially supported by the Grant Agency of the Czech Republic, Grant No. 201/98/1450.

The authors are indebted to prof. A.A. Zlotnik for several remarks that helped to correct the argument.

References

[1] Antoncev, S.N., Kažichov, A.V. and Monachov, V.N. (1993). *Boundary value problems of mechanics of inhomogeneous fluids* (Russian). Nauka, Novosibirsk.

[2] Feireisl, E. and Petzeltová, H. (1997). On the long-time behaviour of solutions to the Navier-Stokes equations of compressible flow. Submitted to *Arch. Rat. Mech. Anal.*

[3] Kuttler, K. (1990). Initial boundary value problems for the displacement in an isothermal, viscous gas. Nonlinear Anal., 15(7):601–623.

[4] Matušů–Nečasová, Š., Okada, M. and Makino, T. (1997). Free boundary problem for the equation of spherically symmetric motion of viscous gas (III). Japan. J. Indust. Appl. Math., 14(2):199–213.

[5] Nagasawa, T. (1989). On the one-dimensional free boundary problem for the heat conductive compressible viscous gas. Lecture Notes in Num. Appl. Anal., 10:83–99.

[6] Novotný, A. and Straškraba, I. Convergence to equilibria for compressible Navier-Stokes equations with large data. Publications du Laboratoire d'Analyse Non Linéaire Appliquée, Publication no. **98/03,** UFR Science, Université de Toulon et du Var, BP 132, La Garde Cedex (France).

[7] Straškraba, I. (1996). Asymptotic development of vacuums for 1-D Navier-Stokes equations of compressible flow. To appear in *Nonlinear World*, 3:519–535.

[8] Straškraba, I. (1998). Large time behavior of solutions to compressible Navier-Stokes equations (Ed. R. Salvi). *Pitman Res. Notes in Math. Series*, 388:125–138.

[9] Zlotnik, A.A., and Amosov, A.A. (1998). Generalized solutions in the large of equations of one-dimensional motion of a viscous barotropic gas (Russian). *Doklady AN SSSR*, 299(6):1303–1307.

[10] Zlotnik, A.A. and Bao, N.Z. Properties and asymptotic behavior of solutions of a problem for one-dimensional motion of viscous barotropic gas (Russian). *Mat. Zametki*, 55(5):51–68.

A GEOMETRIC APPROACH TO DYNAMICAL SYSTEMS IN $\mathbb{R}^N$

Reimund Rautmann

Abstract: In 2-dimensional dynamical systems defined by some differential equation $\dot{x} = f(x)$, global existence and asymptotics of the solutions follow from a geometric condition concerning the characteristics, on which one component of the vector function f is vanishing. By extending this geometric condition to n dimensions we find 2 classes of differential equations which have global solutions for all positive times. Additional monotonicity of the characteristics implies the existence of a unique stationary point which is asymptotically stable and globally attractive.

Keywords: Dynamical systems, characteristics condition, flow invariant rectangles, global existence, attractors.

Introduction

Global existence and asymptotic behaviour of dynamical systems which are defined by a differential equation in the n-dimensional Euclidean space $\mathbb{R}^n$ belong to the classical problems in mathematics. For the study of autonomous systems given by $\frac{d}{dt}x = f(x)$, with a function $f : \Omega \to \mathbb{R}^n$ being defined on some open subset $\Omega \subset \mathbb{R}^n$, some important tools are well known: (i) asymptotical growth conditions for $|f(x)|$, [27], (ii) conditions for ∇f in case $\Omega = \mathbb{R}^n$, [5, 7, 15], (iii) monotonicity conditions, [8, 9, 11, 23, 24], (iv) Ljapunov functions in $\Omega \subset \mathbb{R}^n$, [13, 14, 22], (v) direction conditions for f on $\partial\Omega, \Omega \subsetneq \mathbb{R}^n$, or on the boundaries of n-dimensional rectangles contained in Ω, respectively, and resulting flow invariance [3, 4, 12, 17, 21, 25]. A survey of these methods, most of which apply to the non-autonomous case, too, can be found from the many citations in [1, 9, 12, 13, 24].

Working with one of the latter two approaches above we have to overcome the difficulty how to find a Ljapunov function or how to verify the required direction conditions for f in any concrete case. Precisely this is the starting point of the following studies: We will see that a geometric condition for the characteristics of the differential equation $\frac{d}{dt}x = f(x)$ implies the existence of flow invariant rectangles, the union of which is the whole of the domain Ω of f. In section 1 we will demonstrate the geometric condition above in two-dimensional model problems which (in section 2) will lead us to the results established in $\mathbb{R}^n$. – By means of the well known comparison theorems for quasimonotone weakly coupled parabolic systems, [26 p. 259, p. 267] from the results of section 2 there follow asymptotic estimates for solutions of reaction-diffusion problems, [16, 19], which will be studied elsewhere. Since – looking towards these applications – we are mainly interested in non-negative solutions, we will consider differential equations in the positive cone $\mathbb{R}^n_+$.

1. CHARACTERISTICS OF A DIFFERENTIAL EQUATION IN $\mathbb{R}^N$. MODEL PROBLEMS IN $\mathbb{R}^2$.

By $\mathbb{R}^n_+ = \{x \in \mathbb{R}^n | \ 0 < x_i \text{ for all } i = 1, \dots, n\}$ we will denote the open positive cone in $\mathbb{R}^n, n \geq 1$, by E_i the positive x_i-axis.

Definition 1.1. *Let $f = (f_1, \dots, f_n) : \mathbb{R}^n_+ \to \mathbb{R}^n$ denote a continuous map. For $i = 1, \dots, n$ we consider the (eventually void) sets*

$$C_i = \{x \in \mathbb{R}^n_+ | \ f_i(x) = 0\}$$

which represent the characteristics of the differential equation

$$\frac{d}{dt}x = f(x), \tag{1.1}$$

and we will write

$$V_i^{\pm} = \{x \in \mathbb{R}^n_+ | \ \pm f_i(x) > 0\}$$

for the point sets of constant sign of f_i in $\mathbb{R}^n_+$. (The notation $\pm$ here and below means either "+" or "-" on both sides of the equation.)

We will always consider t as the time variable and use the point notation for the velocity vector $\dot{x} = \frac{d}{dt}x$. In the following, a subset S of the domain of the direction field f will be called flow invariant for (1.1), if any solution $x(t)$ of (1.1) starting at $x(t_0) \in S$ remains in S for all t of its maximum right-hand interval $[t_0, t^+)$ of existence. Since we will only suppose continuity of f, the initial value problem of (1.1) may have several local solutions.

In $\mathbb{R}^2_+$, for differential equations having the general size

$$\begin{aligned} \dot{x}_1 &= (x_2 - \varphi_1(x_1)) \cdot g_1(x) \\ \dot{x}_2 &= (x_1 - \varphi_2(x_2)) \cdot g_2(x) \end{aligned} \tag{1.2}$$

with $g_i(x) > 0$, figure 1 demonstrates the case of oscillating characteristics, figure 2 below the case of monotone increasing characteristics.

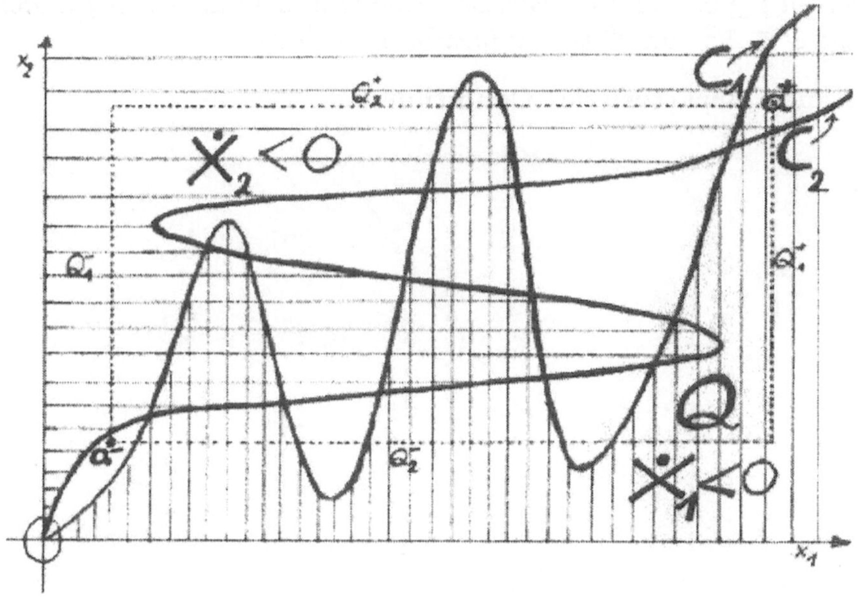

Figure 1

In figure 1 we find the
Characteristics Condition

$$(CC) \begin{cases} \dot{x}_1 = 0 & \Leftrightarrow \quad (x_1, x_2) \in C_1 = \{(y_1, y_2) \in \mathbb{R}^2_+ | y_2 = \varphi_1(y_1))\}, \\ \dot{x}_2 = 0 & \Leftrightarrow \quad (x_1, x_2) \in C_2 = \{(y_1, y_2) \in \mathbb{R}^2_+ | y_1 = \varphi_2(y_2))\}, \\ \dot{x}_i < 0 & \Leftrightarrow \quad (x_1, x_2) \text{ between } C_i \text{ and } E_i, \\ \varphi_i : \mathbb{R}^1_+ & \rightarrow \quad \mathbb{R}^1_+ \text{ being continuous.} \end{cases}$$

Because of (CC) we can draw axis-parallel rectangles $Q = [a^-, a^+] = \{x \in \mathbb{R}^2_+ | \; a_i^- \leq x_i \leq a_i^+\}$ which are flow invariant for (1.2), since on their boundaries the direction field in (1.2) is pointing strictly inwards. We note

Observation 1.2. **[19]:** *Assume*

(1) $f : \mathbb{R}^2_+ \rightarrow \mathbb{R}^2$ *denotes a continuous map,*

(2) (CC) holds,

(3) in any neighborhood of ∞ *in* $\mathbb{R}^2_+$ *there is a point* (a_1^+, a_2^+) *with* $f(a_1^+, a_2^+) < 0$,

(4) in any neighborhood of 0 *in* $\mathbb{R}^2_+$ *there is a point* (a_1^-, a_2^-) *with* $0 < f(a_1^-, a_2^-)$.

Then all solutions $x(t)$ *of (1.1)*

$$\dot{x} = f(x) \quad \text{with } x(0) \in \mathbb{R}^2_+$$

exist for all $t \geq 0$, *and each solution remains in a suitable axis-parallel rectangle* $Q \subset \mathbb{R}^2_+$, *which is flow invariant for (1.1).*

Monotone Characteristics Condition

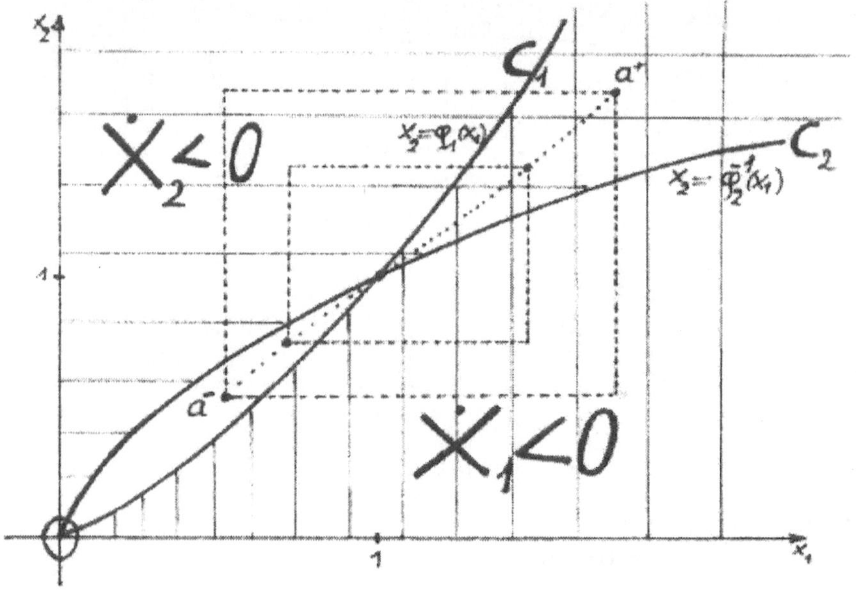

Figure 2

As sketched in figure 2, the characteristics condition (CC) holds there with the additional

Monotonicity Condition:

The functions $\varphi_i : \mathbb{R}^1_+ \to \mathbb{R}^1_+$ are strictly monotone and fulfil

$$(M) \begin{cases} s \uparrow \infty \Rightarrow \varphi_i(s) \uparrow \infty, & s \downarrow 0 \Rightarrow \varphi_i(s) \downarrow 0, \\ s \in (0,1) \Leftrightarrow \varphi_1(s) < \varphi_2^{-1}(s), & s \in (1,\infty) \Leftrightarrow \varphi_2^{-1}(s) < \varphi_1(s). \end{cases}$$

Due to (M) and (CC) we can draw a sequence of axis-parallel rectangles like $[a^-, a^+]$ which are contracting to $\{(1,1)\}$. We note

Observation 1.3. 18: *Assume*

(1) $f : I\!R^2_+ \to I\!R^2$ continuous,

(2) (CC),(M) hold.

Then all solutions $x(t)$ of (1.1)

$$\dot{x} = f(x) \quad \text{with } x(0) \in I\!R^2_+$$

exist for all $t \geq 0$, each solution enters each member of a suitable monotone family of axis parallel rectangles, which are flow invariant and contracting to the unique limit set $\{(1,1)\}$ of all solutions of (1.1) in $I\!R^2_+$.

In quite another approach to models for cooperating species, monotone characteristics have been used in [23 ,p. 372].

2. SOME GLOBAL RESULTS IN $I\!R^N_+$, $N \geq 2$.

Notation: In order to translate our observations above to $I\!R^n_+$, we consider n-dimensional axis-parallel rectangles

$$Q = [a^-, a^+] = \{x \in I\!R^n | a_i^- \leq x_i \leq a_i^+ \quad \text{for all } i = 1, \dots, n\}$$

containing all points between the lowest or uppest corner $a_- = (a_i^-)$ or $a_+ = (a_i^+)$, respectively, $0 < a_i^- < a_i^+$ for all i. The ith upper or lower (n-1)-dimensional face of Q is

$$Q_i^\pm = \{x \in Q | \ x_i = a_i^\pm\}, \quad \text{respectively.}$$

By

$$\begin{aligned} Q^\varepsilon &= \{x \in I\!R^n | a_i^- - \varepsilon \leq x_i \leq a_i^+ + \varepsilon \quad \text{for all } i = 1, \dots, n\} \quad \text{or} \\ (Q_i^\pm)^\varepsilon &= \{x \in Q^\varepsilon | a_i^\pm - \varepsilon \leq x_i \leq a_i^\pm + \varepsilon\} \end{aligned}$$

we will denote the closed ε-neighborhood of Q or $Q_i^\pm$, respectively, $\varepsilon > 0$. Finally for arbitrary $\delta \in \left(0, \min_i \frac{1}{2}\{a_i^+ - a_i^-\}\right)$, we define the $\delta-$retract $Q^{-\delta}$ of Q by

$$Q^{-\delta} = \{x \in I\!R^n | \ a_i^- + \delta \leq x_i \leq a_i^+ - \delta \text{ for all } i\}.$$

Definition 2.1. *We will say that the map $f : Q^\varepsilon \to I\!R^n$ fulfills the direction condition on ∂Q, if*

$$\pm f_{i|_{Q_i^\pm}} < 0 \tag{2.1}$$

holds for all i. Here $f_{i|_{Q_i^\pm}}$ stands for the restriction of f_i to $Q_i^\pm$.

Remark 2.2. *By definition 1.1 of $V_i^{\pm}$, (2.1) is equivalent to*

$$Q_i^{\pm} \subset V_i^{\mp}. \tag{2.2}$$

In order to extend the relation "x between C_i and E_i" to $I\!R^n$, we assume V_i^- to be an open neighborhood of E_i in the closed positive cone $\overline{I\!R^n_+}$. Then in case $C_i = \partial V_i^-$, the relation "x between C_i and E_i" clearly means "$x \in V_i^-$".

Lemma 2.3. *Let the continuous map $f : Q^\varepsilon \to I\!R^n$ obey the direction condition (2.1). Then*

(a) Q is flow invariant for (1.1), and

(b) each solution $x(t)$ of (1.1) starting at $t = 0$ at any point $x(0) \in Q$ exists for all $t \geq 0$ in Q.

Proof of (a) by contradiction: Assume for $t \in [0, t_1]$ that there is a solution $x(t)$ of (1.1), $x(0) \in Q$, $x(t_1) \notin Q$, thus for at least one index i we must have $x_i(0) \in [a_i^-, a_i^+]$, $x_i(t_1) \notin [a_i^-, a_i^+]$. If we take $t^* = \sup\{t \in [0, t_1) |\ x_i(t) \in [a_i^-, a_i^+]\}$, either $x_i(t_*) = a_i^+$ or $x_i(t_*) = a_i^-$ holds by continuity of $x(t)$. Integrating (1.1) along $x(t)$ gives

$$x_i(t) - a_i^{\pm} = \int_{t^*}^{t} f_i(x(s))\, ds.$$

But this contradicts the definition of t^* e.g. in case $x_i(t^*) = a_i^+$: Then $x_i(t) > a_i^+$ for all $t \in (t^*, t_1]$, but $f_i(x(s)) < 0$ holds for all s in a neighborhood of t^* because of the direction condition (2.1) and the uniform continuity of f on the compact set Q^ε. In case $x_i(t^*) = a_i^-$, we conclude similarly.

Proof of (b): Since f does not depend on t, we can consider f as a continuous map $\tilde{f} : I\!R \times Q^\varepsilon \to I\!R^n$, $\tilde{f}(t, x) \equiv f(x)$. Denoting by $[0, t^+)$ the right maximum interval of existence of a solution $x(t)$ of (1.1) starting at $x(0) \in Q$, we see $(t, x(t)) \in [0, t^+] \times Q$ from (a) if $t \to t^+ < \infty$, but this contradicts the well known fact that $(t, x(t))$ cannot remain in any compact subset of the open set $I\!R \times \mathring{Q}^\varepsilon$ where $\tilde{f}$ is continuous [6, p. 13]. Thus $t^+ = \infty$ must hold. ∎

Lemma 2.4. *Let the continuous map $f : Q^\varepsilon \to I\!R^n$ fulfil the direction condition (2.1). Then there are $\delta \in (0, \varepsilon]$ and $\tau > 0$, such that $Q^{-\delta}$ is attractor for (1.1) in Q^δ, and each solution $x(t)$ of (1.1) starting at $x(t_0) \in Q^\delta$ fulfills $x(t_0 + \tau) \in Q^{-\delta}$, i.e. $x(t)$ has entered $Q^{-\delta}$ at the latest at $t = t_0 + \tau$.*

Proof: Because of the direction condition, on (some of) the 2^{n-1} compact sets Q_i^+ some of the continuous functions f_i take on the negative maximum γ^+ and on (some of) the 2^{n-1} compact sets Q_i^- the positive minimum γ^-, respectively. Thus with $2\gamma = \min\{-\gamma^+, \gamma^-\} > 0$ we have

$$\pm f_i|_{Q_i^\pm} \leq -2\gamma.$$

From this, recalling the uniform continuity of f on the compact set Q^ε we get the direction condition for f even on suitable neighborhoods $\left(Q_i^\pm\right)^{\varepsilon'}$: Namely, we can find $\varepsilon' \in (0, \varepsilon]$, $\varepsilon' < \frac{1}{2}(a_i^+ - a_i^-)$, such that

$$\pm f_i|_{(Q_i^\pm)^{\varepsilon'}} \leq -\gamma \tag{2.3}$$

holds for all i for the negative upper or positive lower bound of the velocity component f_i on $(Q_i^+)^{\varepsilon'}$ or $(Q_i^-)^{\varepsilon'}$, respectively. Therefore by Lemma 2.3, the rectangles $Q^{\varepsilon'}$ and

$$Q' = Q^\delta \cap \{x \in \mathbb{R}^n | a_i^- + \delta \leq x_i \leq a_i^+ - \delta \text{ for any fixed } i\} \text{ for all } \delta \in [0, \varepsilon']$$

are flow invariant for (1.1), too, and any solution $x(t)$ of(1.1) starting at $x(t_0) \in Q^{\varepsilon'}$ or $x(t_0) \in Q'$ exists inside of $Q^{\varepsilon'}$ or Q', respectively, for all $t \geq t_0$. Thus for each solution $x(t)$ of (1.1) starting at $x(t_0) \in Q^\delta$, for which

$$a_i^- + \delta \leq x_i(t) \leq a_i^+ - \delta \tag{2.4}$$

holds initially at $t = t_0$, this inequality remains valid for all $t \geq t_0$. We will prove that after a suitable time τ, (2.4) must hold for all components of $x(t_0 + \tau)$, if $x(t_0) \in Q^\delta$:

We write

$$M = \max\{|f_i(x)| \mid x \in Q^{\varepsilon'},\ i = 1, \ldots, n\}$$

for the maximum absolute value of the velocity in any x_i-direction. Thus the estimate $\gamma \leq M$ follows. Taking any time length $\tau > 0$ with

$$\tau \cdot M < \frac{\varepsilon'}{2},$$

we find that $x(t_0) \in Q^{\frac{\varepsilon'}{2}}$ implies $x(t) \in Q^{\varepsilon'}$ for all $t \in [t_0, t_0 + \tau]$ and any solution $x(t)$ of (1.1). Now we take some distance $\delta \in (0, \frac{\varepsilon'}{2})$ fulfilling

$$2\delta \leq \gamma\tau.$$

Then if for the ith component of any solution $x(t)$ of (1.1) starting at any $x(t_0) \in Q^\delta$, one of the inequalities

$$a_i^\pm - \delta \leq x_i(t_0) \leq a_i^\pm + \delta$$

holds, from the flow invariance of Q^δ and (2.3) we conclude

$$\pm x_i(t_0+\tau) \le \pm x_i(t_0) - \gamma\tau \le \pm a_i^\pm - \delta,$$

which means

$$x_i(t_0+\tau) \in [a_i^- + \delta, a_i^+ - \delta]$$

by the definition of δ and ε'. This proves the statement of Lemma 2.4 ∎

Definition 2.5. *By $a^\pm : [1,\infty) \to I\!R_+^n$ we will denote two continuous, in the component-wise order of $I\!R^n$ strictly monotone increasing or decreasing maps, respectively. We will require*

$$\lim_{s\to\infty} a_i^+(s) = \infty, \quad \lim_{s\to\infty} a_i^-(s) = 0, \quad a_i^\pm(1) = 1$$

for all $i = 1,\ldots,n$. Then a monotone family (Q_s) of rectangles is given by

$$Q_s = [a^-(s), a^+(s)] = \{x \in I\!R^n \mid a_i^-(s) \le x_i \le a_i^+(s) \text{ for all } i\}, \quad 1 < s.$$

Proposition 2.6. *Let $(Q_s)_{1<s}$ denote a monotone family of rectangles. Then we have*

(a) $I\!R_+^n = \bigcup_{s>1} Q_s$,

(b) $\{E\} = \bigcap_{s>1} Q_s$, where $E = (1,\ldots,1)$,

and for any $s > 1$ and $\varepsilon \in [0, \min_i |a_i^\pm(s) - 1|)$ there exists $\delta_j \in (0, s-1)$ such that

(c) $Q_{s-\delta_1} \subset \overset{\circ}{Q_s^{-\varepsilon}}$,

(d) $Q_s^{-\varepsilon} \subset \overset{\circ}{Q}_{s-\delta_2}$.

Statements (a) and (b) of Proposition 2.6 result immediately from our requirements concerning monotonicity and asymptotics of the functions $a^\pm$. To get (c) from these requirements, for given $\varepsilon \in (0, \min_i |a_i^\pm(s) - 1|)$ take $\delta_1 \in (0, s-1)$ which fulfills

$$\begin{aligned} a_i^-(s-\delta_1) &\in (a_i^-(s)+\varepsilon, 1) \text{ and} \\ a_i^+(s-\delta_1) &\in (1, a_i^+(s)-\varepsilon) \text{ for all } i. \end{aligned}$$

For proving (d) take $\delta_2 \in (0, s-1)$ which fulfills

$$\begin{aligned} a_i^-(s-\delta_2) &\in (a_i^-(s), a_i^-(s)+\varepsilon) \text{ and} \\ a_i^+(s-\delta_2) &\in (a_i^+(s)-\varepsilon, a_i^+(s)) \text{ for all } i. \end{aligned}$$

Evidently this is always possible because of the continuity and strict monotonicity of $a^{\pm}$, and $a_i^{\pm}(1) = 1$. ∎

Lemma 2.7. **[20]**: *Let*

(1.) $f = (f_1, \ldots, f_n) : I\!R^n_+ \to I\!R^n$ *denote a continuous map,*

(2.) $(Q_s)_{1<s}$ *a monotone family of rectangles, such that the direction condition (2.1):*

$$\pm f_i|_{Q^{\pm}_{s,i}} < 0$$

holds for all $s > 1, \quad i = 1, \ldots, n.$

Then, for all $x_0 \in I\!R^n_+$*, each solution* $x(t)$ *of (1.1) with* $x(0) = x_0$ *exists for all* $t \geq 0$*. Its limit set is precisely* $\{E\}$.

Note: *under the conditions above:*

- *Each rectangle* Q_s *is flow invariant,*
- f *has precisely the one critical point* $E = (1, \ldots, 1)$,
- E *is asymptotically stable,* $\{E\}$ *being a global attractor of (1.1) in* $I\!R^n_+$.

Proof of Lemma 2.7. Because of $I\!R^n_+ = \bigcup_{1<s} Q_s, \quad \{E\} = \bigcap_{1<s} Q_s$ and Lemma 2.3, we only have to prove that each solution $x(t)$ of (1.1) starting at $t = 0$ in $x(0) \in Q_{s_0}$ for some $s_0 > 1$, will enter Q_s at finite time t_s for all $s \in (1, s_0)$. The entrance time t_s may depend on s and the solution $x(t)$ under consideration.

Let $x(t)$ denote a fixed solution of (1.1), $x(0) \in Q_{s_0}$, $1 < s_0$. In order to prove by contradiction, we assume

$$1 < s_* = \inf\{s \in (1, s_0] | \text{ there exists } t > 0, x(t) \in Q_s\}.$$

By our assumption concerning $\{Q_s\}$, the direction condition (2.1) holds on ∂Q_{s_*}. Thus by Lemma 2.4 we can find $\delta > 0$, $\tau > 0$ such that $Q^{-\delta}_{s_*}$ is attractor in $Q^{\delta}_{s_*}$, and $x(t_0) \in Q^{\delta}_{s_*}$ implies $x(t_0 + \tau) \in Q^{-\delta}_{s_*}$. By definition of s_* there is a sequence $(s_k) \downarrow s_*, s_k \in (s_*, s_0]$ for which some $t_k > 0$ exists with $x(t_k) \in Q_{s_k}$ for all $k \in I\!N$.

Recalling the monotonicity of $\{Q_s\}$ and $(s_k) \downarrow s_*$, we see $Q_{s_*} = \bigcap_{k \in I\!N} Q_{s_k}$. Therefore for sufficiently large $k_* \in I\!N$ we have $Q_{s_k} \subset Q^{\delta}_{s_*}$ for all $k \geq k_*$, thus $x(t_k) \in Q^{\delta}_{s_*}$. However, than $x(t_k + \tau) \in Q^{-\delta}_{s_*}$ follows from Lemma 2.4, and we have $Q^{-\delta}_{s_*} \subset Q^{-\delta'}_{s_*} \subset Q_{s_*-\varepsilon}$ for sufficiently small $\delta' \in (0, \delta)$ and suitable $\varepsilon > 0$ by Proposition 2.6, thus $x(t_k + \tau) \in Q_{s_*-\varepsilon}$ in contradiction to the definition of s_*. ∎

Notation: For any $x = (x_1, \dots, x_n) \in \mathbb{R}^n_+$, we define $x_{(i)} = (x_1, \dots, x_{i-1}, x_{i+1}, \dots, x_n) \in \mathbb{R}^{n-1}_+$. Its norm $|x_{(i)}| = \left(\sum_{j\neq i} x_j^2\right)^{1/2}$ equals the distance from x to the positive x_i-axis E_i.

As we will see below, under the following conditions, which altogether concern the geometry of the characteristics in (1.1), Observation 1.2 will extend to $\mathbb{R}^n_+$:

Theorem 2.8. *For all $i = 1, \dots, n$, assume*

$$f_i(x) = \psi_i\left(|x_{(i)}| - \varphi_i(x_i)\right) \cdot g_i(x), \text{ and}$$

(1) $\psi_i : \mathbb{R} \to \mathbb{R}$ continuous, odd, strictly monotone increasing, thus $\psi_i(s) = 0 \Leftrightarrow s = 0$,

(2) $\varphi_i : \mathbb{R}^1_+ \to \mathbb{R}^1_+$ continuous, $s \to \infty \Rightarrow \varphi_i(s) \to s^{\gamma_i} \cdot c$, $s \to 0 \Rightarrow \varphi_i(s) \to s^{\gamma_i} \cdot c$ where $c = \sqrt{n-1}$,

(3) $g_i : \mathbb{R}^n_+ \to \mathbb{R}^1_+$ continuous.

Then, if there exists some point $(A_i) \in \mathbb{R}^n_+$ for which

(4) $\frac{\max_{j\neq i}(A_j)}{A_i} < \gamma_i$ holds for all i,

all solutions $x(t)$ of (1.1)

$$\dot{x} = f(x) \text{ with } x(0) \in \mathbb{R}^n_+$$

exist for all $t \geq 0$, and each solution remains in a suitable rectangle $Q_s = \left\{x \in \mathbb{R}^n_+ |\ s^{-A_i} \leq x_i \leq s^{A_i}\right\}$.

Proof of Theorem 2.8. Recalling Lemma 2.3, in order to prove Theorem 2.8 we only have to show that each starting point $x(0) \in \mathbb{R}^n_+$ of a solution of (1.1) is contained in a rectangle Q with boundary ∂Q, on which the direction condition (2.1) holds.

Consider the family of rectangles $Q_s = \left\{x |\ s^{-A_i} \leq x_i \leq s^{A_i}\right\} = [a^-, a^+]$ for $s > 1$ with a vector $(A_i) > 0$ from condition (4) in Theorem 2.8We find

$$|a^+_{(i)}| = \left(\sum_{j\neq i} s^{2A_j}\right)^{1/2} \leq cs^{\max_{j\neq i} A_j} < c \cdot s^{\gamma_i A_i - \delta} \leq \varphi(s^{A_i}) \qquad (2.5)$$

from (4) for sufficiently small $\delta > 0$ and from (2) in Theorem 2.8 for sufficiently large s and all i. Similarly we get

$$|a^-_{(i)}| > \varphi(s^{-A_i}) \qquad (2.6)$$

for sufficiently large s and all i, which shows the existence of points $a^{\pm} = a^{\pm}(s) \in V_i^{\mp}$ for all $i = 1, \ldots, n$.

From (2.5), (2.6) recalling $|a^-_{(i)}| \le |x_{(i)}| \le |a^+_{(i)}|$ for $x \in [a^-, a^+]$ and any i, and from $x_i = s^{\pm A_i}$ for any $x \in Q^{\pm}_{s,i}$, we see $Q^{\pm}_{s,i} \subset V_i^{\mp}$, which proves the direction condition (2.1) on ∂Q_s. Finally because of Proposition 2.6(a), any initial value $x(0) \in \mathbb{R}^n_+$ is contained in some flow-invariant rectangle Q_s. ∎

In order to extend Observation 1.3 to $\mathbb{R}^n_+$ additionally we have to assume monotonicity of the characteristics in (1.1):

Corollary 2.9. *For all i, assume $f_i(x) = \psi_i\left(|(x_{(i)})| - cx_i^{\gamma_i}\right) \cdot g_i(x)$, $c = \sqrt{n-1}$ with ψ_i and g_i from Theorem 2.1. Then, if condition (4) of Theorem 2.8 holds, all solutions $x(t)$ of (1.1) exist for all $t \ge 0$, each solution remains in the smallest $Q_s = \left\{x \in \mathbb{R}^n_+ | s^{-A_i} \le x_i \le s^{A_i}\right\}$ containing $x(0)$, and has the limit set $\{(1, \ldots 1)\}$.*

Proof of Corollary 2.9. In the case $\varphi_i(s) = cs^{\gamma_i}$, (2.5) and (2.6) hold for all $s > 1$, thus the direction condition (2.1) holds on the boundary ∂Q_s of each rectangle Q_s. Since the family (Q_s) is monotone, the statement of Corollary 2.9 follows from Lemma 2.7 ∎

In $\mathbb{R}^2_+$, (1.1) with f from Corollary 2.9 reads

$$\begin{aligned} \dot{x}_1 &= \psi_1(x_2 - x_1^{\gamma_1}) \cdot g_1(x) \\ \dot{x}_2 &= \psi_2(x_1 - x_2^{\gamma_2}) \cdot g_2(x), \text{ and we have} \end{aligned}$$

$$(4) \qquad \left.\begin{array}{l} \frac{A_2}{A_1} < \gamma_1 \\ \frac{A_1}{A_2} < \gamma_2 \end{array}\right\} \Rightarrow (*) \quad 1 < \gamma_1 \cdot \gamma_2.$$

Inequality $(*)$ is the global stability condition for (1.1) in $\mathbb{R}^2_+$ with $\psi_i(s) \equiv s$, [18].

For more general questions of global existence it would be desirable to have all sets V_i^- as small as possible, because they restrict the admissible asymptotic growth of the direction field f. The sets V_i^- in the next Theorem 2.10 (roughly spoken) can be made smaller than the V_i^- in Theorem 2.8

Theorem 2.10. *For all $i = 1, \ldots, n$, assume*

$$f_i(x) = \psi_i\left(\prod_{j \ne i} x_j^{\frac{\alpha_{ij}}{m_j} m_i} - \varphi_i(x_i)\right) \cdot g_i(x), \text{ and}$$

(1) $\psi_i : \mathbb{R} \to \mathbb{R}$ continuous, odd, strictly monotone increasing, thus $\psi_i(s) = 0 \Leftrightarrow s = 0$,

(2) $\varphi_i : \mathbb{R}^1_+ \to \mathbb{R}^1_+$ *continuous,* $s \to \infty \Rightarrow \varphi_i(s) \to s^{\gamma_i}$,
$s \to 0 \Rightarrow \varphi_i(s) \to s^{\gamma_i}, \quad \varphi_i(1) = 1$

(3) $g_i : \mathbb{R}^n_+ \to \mathbb{R}^1_+$ *continuous,*

(4) $0 \le \alpha_{ij}, \quad \alpha_{ii} = 0, \quad 0 < m_j.$

Then, if in addition we assume

(5) there is a point $(A_j) \in \mathbb{R}^n_+, \quad (\delta_{ij} - \alpha_{ij})(A_j) \in \mathbb{R}^n_+$, *and*

(6) $1 \le \gamma_i$ *holds for all* i,

then all solutions $x(t)$ *of (1.1)*

$$\dot{x} = f(x) \qquad \text{with } x(0) \in \mathbb{R}^n_+$$

exist for all $t \ge 0$, *and each solution remains in a suitable rectangle* Q_s.

Note: *Assumption (5) holds if and only if the matrix* $(\delta_{ij} - \alpha_{ij})$ *belongs to the class of M-matrices [2, p.136].*

Proof of Theorem 2.10. For the proof of Theorem 2.10 recalling again Lemma 2.3 we will show that on the boundary of the rectangle

$$Q_s = \left\{ x |\ s^{-A_i m_i} \le x_i \le s^{A_i m_i} = [a^-, a^+] \right\}$$

the direction condition (2.1) holds, if s is large enough. From (4) in Theorem 2.10, for any $x \in Q^+_{s,i}$ we see

$$\prod_{j \ne i} x_j^{\frac{\alpha_{ij}}{m_j} m_i} \le s^{\sum_{j \ne i} \alpha_{ij} A_j m_i} < s^{A_i m_i - \varepsilon} \tag{2.7}$$

because of (5), $\varepsilon > 0$ being sufficiently small. Thus $f_i(x) < 0$ follows from

$$s^{A_i m_i - \varepsilon} \le \varphi_i(x_i) = \varphi_i(s^{A_i m_i}), \tag{2.8}$$

which holds by (2) in Theorem 2.10 for sufficiently large s. Similarly for any $x \in Q^-_{s,i}$ we get $f_i(x) > 0$ for s large enough. Thus for s sufficiently large, the direction condition (2.1) holds on ∂Q_s, and evidently with $s \to \infty$ any initial value $x(0) \in \mathbb{R}^n_+$ is contained in some flow invariant rectangle Q_s. ∎

Corollary 2.11. *For all* i, *assume* $f_i(x) = \psi_i \left(\prod_{j \ne i} x_j^{\frac{\alpha_{ij}}{m_j} m_i} - x_i^{\gamma_i} \right) \cdot g_i(x)$,
with $\psi_i, g_i, \alpha_{ij}, m_i$ *from Theorem 2.10,* $\gamma_i \ge 1$. *Then, if also (5) holds,*

all solutions $x(t)$ of (1.1) starting at any $x(0) \in \mathbb{R}^n_+$ exist for all $t \geq 0$, each solution remains in the smallest rectangle

$$Q_s = \left\{ x \in \mathbb{R}^n_+ | s^{-m_i A_i} \leq x_i \leq s^{m_i A_i} \right\}$$

containing $x(0)$, and has the limit set $\{(1, \dots 1)\}$.

Proof of Corollary 2.11. In the case $\varphi_i(s) \equiv s^{\gamma_i}$, the inequalities (2.7), (2.8) and therefore $Q^+_{s,i} \subset V^-_i$ as well as $Q^-_{s,i} \subset V^+_i$ hold for all $s > 1$. Thus the direction condition (2.1) holds on the boundary of each rectangle Q_s. The family (Q_s) being monotone, the statement of Corollary 2.11 follows from Lemma 2.7 ∎

Acknowledgments

Supported by Deutsche Forschungsgemeinschaft.

References

[1] Amann, H. (1990). Ordinary differential equations, de Gruyter, Berlin.

[2] Berman, A. and Plemmons, R.J. (1979). Nonnegative matrices in the mathematical sciences, Academic Press New York.

[3] Bony, J.M. (1969). *Principe du maximum, inégalité de Harnack et unicité du problème de Cauchy pour les opérators elliptiques dégénérés*, Ann. Inst. Fourier, 19:277–304.

[4] Brezis, H. (1970). *On a characterisation of flow-invariant sets*, Comm. Pure Appl. Math., 23:261–263.

[5] Gasull, A., Llibre, J. and Satomayor, J. (1991). *Global asymptotic stability of differential equations in the plane*, J. Differential Equations, 91:327–335.

[6] Hartman, Ph. (1964). Ordinary differential equations, Wiley New York.

[7] Hartman, Ph. and Olech, C. (1962). *On global asymptotic stability of solutions of differential equations*, Trans. Amer. Math. Soc., 104:154–178.

[8] Hirsch, M. (1982). *Systems of differential equations which are competitive or cooperative. I: Limit sets*, SIAM J. Math. Anal., 13:167–179.

[9] Hirsch, M. (1988b). *Stability and convergence in strongly monotone dynamical systems*, J. Reine Angew. Math., 383:1–53.

[10] Hirsch, M. and Smale, St. (1974). Differential equations, dynamical systems, and linear algebra, Academic Press New York.

[11] Jiang, J.F. (1998). *The complete classification of asymptotic behaviour for bounded cooperative Lotka-Volterra systems with assumption (SM)*, Quart. Appl. Math., LVI:37–53.

[12] Ladde, G.S. and Lakshmikantham, V. (1974). *On flow-invariant sets*, Pacific J. Math., 51:215–220.

[13] Lakshmikantham, V., Leela, S. and Martynyuk, A.A. (1989). Stability analysis of nonlinear systems, Decker New York.

[14] Lu, Zh. and Takeuchi, Y. (1994). *Qualitative stability and global stability for Lotka-Volterra systems*, J. Math. Anal. Appl., 182:260–268.

[15] Meisters, G. and Olech, C. (1988). *Solution of the global asymptotic stability Jacobian conjecture for the polynomial case*, in: Analyse Mathématique et Applications, Gauthier-Villars Paris, 373–381.

[16] Pao, C.V. (1992). Nonlinear parabolic and elliptic equations, Plenum Press New York.

[17] Rautmann, R. (1972). *A criterion for global existence in case of ordinary differential equations*, Applicable Anal., 2:187–194.

[18] Rautmann, R. (1985). *Eine Klasse asymptotisch stabiler autonomer Systeme und Normschranken für spezielle Reaktions-Diffusions-Prozesse*, Z. Angew. Math. Mech., 65:207–217.

[19] Rautmann, R. (1987). *Global bounds for some reaction diffusion processes*, Methoden Verfahren Math. Phys., 33:87–98.

[20] Rautmann, R. (1991). *On tests for stability*, Methoden Verfahren Math. Phys., 37:201–212.

[21] Redheffer, R.M. (1972). *The theorems of Bony and Brezis on flow-invariant sets*, Amer. Math. Monthly, 79:740–747.

[22] Redheffer, R.M. and Walter, W. (1984). *Solution of the stability problem for a class of generalized Volterra prey-predator systems*, J. Differential Equations, 52:245–263.

[23] Smith, H.L. (1986). *On the asymptotic behaviour of a class of deterministic models of cooperative species*, SIAM J. Appl. Math., 3:368–375.

[24] Smith, H.L. (1995). Monotone dynamical systems, AMS Mathematical surveys and monographs 41, Providence R.I.

[25] Volkmann, P. (1975). *Über die Invarianz-Sätze von Bony und Brezis in normierten Räumen*, Arch. Math., XXVI:89–93.

[26] Walter, W. (1970). Differential and integral inequalities, Springer Berlin.

[27] Wintner, A. (1945). *The non-local existence problem of ordinary differential equations*, Amer. J. Math., 67:277–284.

ON A THREE-DIMENSIONAL CONVECTIVE STEFAN PROBLEM FOR A NON-NEWTONIAN FLUID

José Francisco Rodrigues, José Miguel Urbano

Abstract: The coupling of diffusion and convection phenomena ina material under a change of phase is presented. The Stefan problem is adopt with convection only in the liquid phase. The existence result for weak solutions is proved.

Keywords: Non-Newtonian fluids, convective Stefan problem.

1. INTRODUCTION

In this work we consider the coupling of diffusion and convection phenomena in a material under a change of phase. As a model we adopt the Stefan problem, with convection only in the liquid phase. The main mathematical difficulty lies in the interesting and delicate question of defining the liquid zone and the corresponding formulation of the motion equations. The liquid zone can be defined as the set $\{\theta > 0\}$, where θ is the renormalized temperature. A natural requirement is the continuity of the temperature (or at least the lower semicontinuity) that leads to an open set $\{\theta > 0\}$ where the equations for the velocity $\mathbf{v}$ can be suitably formulated in a weak sense. To deal with this main issue we consider a non-Newtonian fluid of dilatant type for which $\nabla \mathbf{v} \in L^q$ with $q > N = 3$. This restriction is sufficient to assure enough integrability of the convective term in the Stefan problem to obtain a continuous solution.

Other conduction–convection problems, similar to this one from the point of view of the mathematical analysis, have been considered previously. We mention in particular the works [CDK1], where the Stefan problem is coupled with the Navier–Stokes equations in a stationary setting, the extension to the evolutionary problem in

Applied Nonlinear Analysis, edited by Sequeira *et al.*
Kluwer Academic / Plenum Publishers, New York, 1999.

[CDK2], that was successful only for (the Stokes system and) two spatial variables, and the partial regularity obtained in [DO1] and [DO2] in the three-dimensional case, respectively for the Stokes and Navier–Stokes equations. Related contributions are given in [R1,2] and [RU], where the stationary problem is studied in a very general setting. In [RU] the Stefan problem for the p-Laplacian operator ($1 < p < \infty$) is coupled with a variational inequality modeling a non-Newtonian flow with $\mathbf{v} \in W^{1,q}$, and a continuous temperature is obtained provided $pq > N$, N being the spatial dimension, thus showing the existence of a weak solution.

For the problem considered here we obtain an existence result for weak solutions via an approximation and penalization procedure and the consequent passage to the limit using appropriate *a priori* estimates. We combine monotonicity methods with a local compactness argument needed to show the convergence of the nonlinear convective term in the flow equations. The continuity of the temperature is obtained using the techniques developed by DiBenedetto (see [D1,2]). The proof lies on energy and logarithmic estimates, that are possible due to the integrability of the velocity that follows from our essential assumption that $q > N = 3$.

2. THE MATHEMATICAL FORMULATION AND MAIN RESULT

We consider a material occupying a bounded regular domain $\Omega \subset \mathbb{R}^3$ and coexisting in two phases, a solid phase, corresponding to a region $\mathcal{S}$, and a liquid phase, corresponding to a region $\mathcal{L}$. The two regions are separated by a surface Φ, through which the change of phase occurs, that constitutes a free boundary and is one of the unknowns of the problem.

With the aid of the usual transformation of Kirchoff we work with the renormalized temperature θ and make the assumption that the phase change occurs at the fixed temperature $\theta = 0$. Then we can write

$$\mathcal{S} = \{\theta < 0\}\ , \qquad \mathcal{L} = \{\theta > 0\}\ , \qquad \Phi = \{\theta = 0\}\ .$$

The strong formulation of the Stefan problem with convection reads (see [R3], [M])

$$(\partial_t + \mathbf{v}\cdot\nabla)\,\theta - \Delta\theta = 0 \quad \text{in } \{\theta < 0\} \cup \{\theta > 0\}\ , \tag{1}$$

$$[\nabla\theta]_-^+ \cdot \mathbf{n}_x = -\lambda\,\mathbf{w}\cdot\mathbf{n}_x = \lambda\, n_t \quad \text{in } \{\theta = 0\}\ , \tag{2}$$

where $(\mathbf{n}_x, n_t)$ is the normal vector to Φ in the space-time cylinder $Q = \Omega \times (0,T)$, $\mathbf{w}$ the velocity of Φ and $\lambda > 0$ the latent heat. The Stefan condition (2) measures, roughly speaking, the amount of heat used in the phase transition. To (1) and (2) we must add boundary conditions at the fixed boundary of Ω, $\partial\Omega$, and also an initial condition. We take, to simplify, $\theta = 0$ on $\partial\Omega \times (0,T)$, but non-homogeneous boundary conditions or the Neumann type conditions can equally be treated.

As is physically natural we suppose that $\mathbf{v} = 0$ in the solid region $\{\theta < 0\}$. In the liquid bulk, convection is ruled by the following system

$$\left.\begin{aligned} &\nabla \cdot \mathbf{v} = 0, && (3)\\ &\partial_t \mathbf{v} + (\mathbf{v}\cdot\nabla)\,\mathbf{v} - \nabla\cdot\mathbf{S} + \nabla p = \mathbf{f}(\theta) && (4)\end{aligned}\right\} \quad \text{in } \{\theta > 0\},$$

where $\mathbf{S} = (S_{ij})$ is the viscous stress tensor, p is the pressure and $\mathbf{f}$ a density of forces, that may depend continuously on the temperature θ. The constitutive assumption we choose is given by the relation

$$S_{ij} = \nu(\theta)\,[D_{II}(\mathbf{v})]^{\frac{q-2}{2}}\,D_{ij}(\mathbf{v})\,, \qquad q > 3\,, \tag{5}$$

where $\nu(\theta)$ is a (temperature dependent) viscosity coefficient, $D_{II}(\mathbf{v}) = \frac{1}{2}D_{ij}D_{ij}$ is the second scalar invariant with

$$D_{ij} = \frac{1}{2}(v_{i,j} + v_{j,i})\,, \qquad v_{i,j} = \frac{\partial v_i}{\partial x_j}\,.$$

We use the usual summation convention throughout.

A fluid governed by (3), (4) and (5) with $q > 2$ is called a dilatant fluid (see [C], for instance).

We also assume that the fluid adheres to the solid boundary so that

$$\mathbf{v} = 0 \quad \text{on } \partial\{\theta \geq 0\} \cup \{\theta < 0\} \tag{6}$$

and prescribe an initial condition $\mathbf{v}(x,0) = \mathbf{v}_0(x)$.

We next formulate the problem in a weak form following the standard procedure of multiplication by test functions and formal integration by parts, assuming, in addition, that Φ is a smooth surface.

Let ξ be a smooth test function and multiply (1) by ξ separately in $\{\theta < 0\}$ and $\{\theta > 0\}$. Upon addition and the use of the boundary conditions we arrive at

$$-\int_Q \theta\,\partial_t\xi + \int_Q \nabla\theta\cdot\nabla\xi - \int_Q \theta\,\mathbf{v}\cdot\nabla\xi = \int_\Phi \left([\nabla\theta]_-^+\cdot\mathbf{n}_x\right)\xi\,. \tag{7}$$

We introduce the maximal monotone graph

$$H(s) = \begin{cases} \lambda & ;\ s > 0\\ [0,\lambda] & ;\ s = 0\\ 0 & ;\ s < 0\end{cases}$$

and using (2) and with Φ a surface, we get

$$\int_Q H(\theta)\,\partial_t\xi = \int_{\{\theta>0\}} \lambda\,\partial_t\xi = \lambda\int_\Phi \xi\,n_t = -\lambda\int_\Phi (\mathbf{w}\cdot\mathbf{n}_x)\,\xi = \int_\Phi \left([\nabla\theta]_-^+\cdot\mathbf{n}_x\right)\xi\,.$$

Introducing the enthalpy $\eta \in \theta + H(\theta)$, we rewrite (7) in the form

$$-\int_Q \eta\, \partial_t \xi + \int_Q \nabla\theta \cdot \nabla\xi - \int_Q \theta \mathbf{v} \cdot \nabla\xi = 0\,, \quad \forall \xi \in C_0^\infty(Q)\,, \tag{8}$$

which means that

$$\partial_t \eta - \Delta\theta + \mathbf{v} \cdot \nabla\theta = 0 \quad \text{in } \mathcal{D}'(Q)\,. \tag{9}$$

Concerning the flow equations we need to introduce the spaces of solenoidal vectorial functions

$$\mathcal{V} = \left\{ \mathbf{v} = (v_i) : \; v_i \in \mathcal{D}(\Omega),\; \nabla \cdot \mathbf{v} = 0 \right\},$$

$$H = \text{closure of } \mathcal{V} \text{ in } [L^2(\Omega)]^3\,,$$

$$V^q = \text{closure of } \mathcal{V} \text{ in the norm } \|\cdot\|_{V^q}\,,$$

where

$$\|\mathbf{v}\|_{V^q} = \|D_{II}^{1/2}(\mathbf{v})\|_{L^q(\Omega)}\,, \qquad q > 3\,.$$

Assuming that Ω is a bounded domain with a Lipschitz boundary, Korn's inequality is valid and we can characterize

$$V^q = \left\{ \mathbf{v} \in [W_0^{1,q}(\Omega)]^3 : \; \nabla \cdot \mathbf{v} = 0 \; \text{ in } \Omega \right\}.$$

The integration by parts of the flow equation (4), with solenoidal smooth test functions, holds only in the liquid zone, so the test function must be supported in $\{\theta > 0\}$, which will be required to be an open subset of Q.

We are now in conditions of presenting the main result on the existence of a weak solution. The spaces of test functions are respectively

$$\mathcal{T}_1 = \left\{ \xi \in H^1(Q) : \; \xi(T) = 0 \right\},$$

$$\mathcal{T}(\theta) = \left\{ \mathbf{\Psi} \in L^q(0,T;V^q) \cap [H_0^1(Q)]^3 : \; \operatorname{supp} \mathbf{\Psi} \subset \{\theta > 0\} \right\},$$

$$\mathcal{T}_{\theta_0} = \left\{ \mathbf{\Pi} \in \mathcal{V}(\Omega) : \; \operatorname{supp} \mathbf{\Pi} \subset \{\theta_0 > 0\} \right\},$$

when $\theta = \theta(x,t)$ and $\theta_0 = \theta_0(x)$ are given continuous functions in Q and Ω respectively.

In order to obtain the existence result we suppose the following assumptions hold:

$$\eta_0 \in \theta_0 + H(\theta_0) \quad \text{and} \quad \exists M > 0 : \; \|\theta_0\|_{L^\infty(\Omega)} \le M\,, \tag{A1}$$

$$\nu \in C^0(\mathbb{R}) \quad \text{and} \quad \nu \ge \nu_* > 0\,, \tag{A2}$$

$$\mathbf{f} : \mathbb{R} \to \mathbb{R}^3 \quad \text{is continuous and } \mathbf{f}(0) = \mathbf{0}\,, \tag{A3}$$

$$\mathbf{v}_0 \in [L^2(\Omega)]^3\,, \nabla \cdot \mathbf{v} = 0\,. \tag{A4}$$

Theorem. *There exists at least one weak solution $(\theta, \eta, \mathbf{v})$ of the conduction–convection problem with change of phase for a dilatant fluid with $q > N = 3$, such that*

$$\theta \in L^2(0,T; H_0^1(\Omega)) \cap L^\infty(Q) \cap C^0(Q) , \tag{10}$$

$$\eta \in L^\infty(Q) ; \quad \eta \in \theta + H(\theta) \text{ a.e. in } Q , \tag{11}$$

$$\mathbf{v} \in L^\infty(0,T;H) \cap L^q(0,T;V^q) ; \quad \mathbf{v} = 0 \text{ a.e. in } \{\theta < 0\} , \tag{12}$$

$$-\int_Q \eta\, \partial_t \xi + \int_Q \nabla\theta \cdot \nabla\xi - \int_Q \theta\, \mathbf{v} \cdot \nabla\xi = \int_\Omega \eta_0\, \xi(0) , \quad \forall \xi \in \mathcal{T}_1 , \tag{13}$$

$$\begin{aligned} &-\int_{\{\theta>0\}} \mathbf{v} \cdot \partial_t \mathbf{\Psi} - \int_{\{\theta>0\}} \mathbf{v} \cdot (\mathbf{v} \cdot \nabla \mathbf{\Psi}) + \\ &+ \int_{\{\theta>0\}} \nu(\theta)\, [D_{II}(\mathbf{v})]^{\frac{q-2}{2}}\, D\mathbf{v} : D\mathbf{\Psi} = \int_{\{\theta>0\}} \mathbf{f}(\theta) \cdot \mathbf{\Psi} , \quad \forall \mathbf{\Psi} \in \mathcal{T}(\theta) , \end{aligned} \tag{14}$$

$$\int_\Omega \mathbf{v}(x,t) \cdot \mathbf{\Pi}(x) \underset{t \downarrow 0+}{\longrightarrow} \int_\Omega \mathbf{v}_0(x) \cdot \mathbf{\Pi}(x) , \quad \forall \mathbf{\Pi} \in \mathcal{T}_{\theta_0} . \tag{15}$$

Remark 1. In (13) we have added a term corresponding to an initial condition that has to be prescribed for the enthalpy and not for the temperature:

$$\eta(x,0) = \eta_0(x) ; \quad \eta_0 \in \theta_0 + H(\theta_0) .$$

In fact η_0 carries some additional information concerning the initial state with respect to the one given by the initial temperature θ_0.

Remark 2. The continuity of θ given by (10) assures the subset $\{\theta > 0\} \equiv \{(x,t) \in Q : \ \theta(x,t) > 0\}$, representing the liquid zone, is an open subset, and therefore (14) is well defined.

3. PROOF OF THE THEOREM

The proof consists in the study of an approximated problem penalized in the solid region and the consequent passage to the limit based on *a priori* estimates, monotonicity methods and a technical lemma.

Given $\varepsilon > 0$, define the continuous functions

$$\chi_\varepsilon(s) = \begin{cases} 1 & ; \ s \le -2\varepsilon \\ -s/\varepsilon - 1 & ; \ -2\varepsilon < s < -\varepsilon \\ 0 & ; \ s \ge -\varepsilon \end{cases} \tag{16}$$

and let

$$\gamma_\varepsilon(s) \quad \text{be a } C^\infty \text{ approximation of } \ s + H(s) .$$

Consider the regularized and penalized problem

($\mathbf{P}_\varepsilon$) Find $(\theta_\varepsilon, \mathbf{v}_\varepsilon)$ such that

$$\theta_\varepsilon \in H^1(0,T;L^2(\Omega)) \cap L^\infty(0,T;H_0^1(\Omega)) \cap L^\infty(Q) \cap C^0(Q) \tag{17}$$

$$\mathbf{v}_\varepsilon \in L^\infty(0,T;H) \cap L^q(0,T;V^q) \tag{18}$$

$$-\int_Q \gamma_\varepsilon(\theta_\varepsilon)\,\partial_t\xi + \int_Q \nabla\theta_\varepsilon \cdot \nabla\xi - \int_Q \theta_\varepsilon\,\mathbf{v}_\varepsilon^* \cdot \nabla\xi = \int_\Omega \eta_{0\varepsilon}\,\xi(0)\ , \quad \forall \xi \in \mathcal{T}_1\ , \tag{19}$$

$$-\int_Q \mathbf{v}_\varepsilon \cdot \partial_t \mathbf{\Psi} - \int_Q \mathbf{v}_\varepsilon \cdot (\mathbf{v}_\varepsilon \cdot \nabla \mathbf{\Psi}) + \int_Q \nu(\theta_\varepsilon)\,[D_{II}(\mathbf{v}_\varepsilon)]^{\frac{q-2}{2}}\,D\mathbf{v}_\varepsilon : D\mathbf{\Psi} +$$
$$+\frac{1}{\varepsilon}\int_Q \chi_\varepsilon(\theta_\varepsilon)\,\mathbf{v}_\varepsilon \cdot \mathbf{\Psi} = \int_Q \mathbf{f}(\theta_\varepsilon) \cdot \mathbf{\Psi}\ , \quad \forall \mathbf{\Psi} \in L^q(0,T;V^q) \cap [H_0^1(Q)]^3\ , \tag{20}$$

$$\int_\Omega \mathbf{v}_\varepsilon(x,t) \cdot \mathbf{\Pi}(x) \underset{t\downarrow 0^+}{\longrightarrow} \int_\Omega \mathbf{v}_0(x) \cdot \mathbf{\Pi}(x)\ , \quad \forall \mathbf{\Pi} \in \mathcal{V}\ , \tag{21}$$

where $\eta_{0\varepsilon}$ is an approximation of η_0 such that

$$\eta_{0\varepsilon} = \gamma_\varepsilon(\theta_{0\varepsilon})\ , \quad \theta_{0\varepsilon} \in H^1(\Omega)\ , \quad \|\theta_{0\varepsilon}\|_{L^\infty(\Omega)} \le M\ ,$$
$$\theta_{0\varepsilon} \to \theta_0 \ \text{ and } \ \eta_{0\varepsilon} \to \eta_0 \quad \text{in } L^2(\Omega)\ ;$$

and $\mathbf{v}_\varepsilon^*$ is a smooth mollification of $\mathbf{v}_\varepsilon$ that preserves its solenoidal property.

This approximated problem is solved with the use of Leray–Schauder's fixed point method. Given $\tau \in L^2(Q)$ solve (20)-(21) with $\theta_\varepsilon = \tau$ using a Galerkin procedure as for instance in [L], where the uniqueness is also established (cf. page 84). Obtaining

$$\mathbf{v}_\tau \in L^\infty(0,T;H) \cap L^q(0,T;V^q)\ ,$$

with this $\mathbf{v}_\tau$ we solve (19) with $\mathbf{v}_\varepsilon^* = \mathbf{v}_\tau^*$ using the results of [R3], that yields

$$\eta = \gamma_\varepsilon(\theta) \in W \equiv H^1(0,T;L^2(\Omega)) \cap L^\infty(0,T;H_0^1(\Omega)) \cap L^\infty(Q)\ .$$

Define an application $\mathcal{F}_\varepsilon$ putting

$$\mathcal{F}_\varepsilon(\tau) = \gamma_\varepsilon^{-1}(\eta)$$

in a ball of $L^2(Q)$ of the form

$$B = \left\{\tau \in L^2(Q) :\ \|\tau\|_{L^2(Q)} \le \rho\right\}\ ,$$

where ρ is a constant that is obtained using a standard estimate such that

$$\mathcal{F}_\varepsilon(B) \subset B\ .$$

The application can be seen to be continuous and it is also compact due to the compactness of the injection of

$$W \hookrightarrow L^2(Q) \, .$$

We can also show that the following *a priori* properties hold (see next section for the proof of (27))

$$\|\theta_\varepsilon\|_{L^\infty(Q)} \le M \, ; \quad \|\gamma_\varepsilon(\theta_\varepsilon)\|_{L^\infty(Q)} \le M' \, , \tag{22}$$

$$\|\nabla\theta_\varepsilon\|_{L^2(Q)} \le C \, , \tag{23}$$

$$\int_0^{T-h} h^{-1} \, \|\theta_\varepsilon(t+h) - \theta_\varepsilon(t)\|^2_{L^2(\Omega)} \, dt \le C \, , \quad \forall 0 < h < T \, , \tag{24}$$

$$\operatorname*{ess\,sup}_{0<t<T} \|\mathbf{v}_\varepsilon(\cdot,t)\|_{L^2(\Omega)} + \|\nabla\mathbf{v}_\varepsilon\|_{L^q(Q)} \le C \, , \tag{25}$$

$$\frac{1}{\varepsilon}\int_Q \chi_\varepsilon(\theta_\varepsilon)\,|\mathbf{v}_\varepsilon|^2 \le C \, , \tag{26}$$

$$\theta_\varepsilon \ \text{ is equicontinuous in any compact subset } \ K \subset\subset Q \, . \tag{27}$$

From these results we can extract subsequences $\varepsilon \to 0$ such that

$$\theta_\varepsilon \to \theta \quad \text{strongly in } \ L^2(Q) \ \text{ and uniformly in } K \subset\subset Q \, , \tag{28}$$

$$\nabla\theta_\varepsilon \rightharpoonup \nabla\theta \quad \text{weakly in } \ L^2(Q) \, , \tag{29}$$

$$\gamma_\varepsilon(\theta_\varepsilon) \rightharpoonup \eta \quad \text{weakly in } \ L^2(Q) \, , \tag{30}$$

$$\mathbf{v}_\varepsilon, \mathbf{v}_\varepsilon^* \rightharpoonup \mathbf{v} \quad \text{weakly* in } \ L^q(0,T;L^\infty(\Omega)) \cap L^\infty(0,T;L^2(\Omega)) \, , \tag{31}$$

$$\nabla\mathbf{v}_\varepsilon \rightharpoonup \nabla\mathbf{v} \quad \text{weakly in } \ L^q(Q) \, . \tag{32}$$

These convergences are enough to pass to the limit in the temperature equation and we find that θ solves (19) in the definition of weak solution. We can show $\eta \in \theta + H(\theta)$ using a monotonicity argument.

Passing to the limit in the flow equation (20) has a delicate point which is the nonlinear convective term. We use the strong convergence for the $\mathbf{v}_\varepsilon$ in the interior of the liquid zone $\{\theta > 0\}$ obtained with the limit function θ from (28).

Indeed, the uniform convergence (28) implies that, for any compact subset $K \subset\subset \{\theta > 0\}$, there exists an $\varepsilon_K > 0$, such that, due to the definition (16),

$$\chi_\varepsilon(\theta_\varepsilon) = 0 \ \text{ in } K \subset\subset \{\theta > 0\}, \quad \text{for all } \ 0 < \varepsilon < \varepsilon_K \, . \tag{33}$$

Hence, in the distributional sense in any open subset $\mathcal{O} \subset K$, from (20) we have

$$\int_{\mathcal{O}} \partial_t \mathbf{v}_\varepsilon \cdot \boldsymbol{\Psi} = \int_{\mathcal{O}} \Big\{\mathbf{v}_\varepsilon \otimes \mathbf{v}_\varepsilon - \nu(\theta_\varepsilon)\,[D_{II}(\mathbf{v}_\varepsilon)]^{\frac{q-2}{2}}\, D\mathbf{v}_\varepsilon\Big\} : D\boldsymbol{\Psi} + \int_{\mathcal{O}} \mathbf{f}(\theta_\varepsilon)\cdot\boldsymbol{\Psi} \, , \tag{34}$$

for any $\Psi \in \mathcal{V}$, such that supp $\Psi \subset \mathcal{O}$, which means, in particular, that

$$\partial_t \mathbf{v}_\varepsilon \text{ belongs to a bounded set of } L^{q'}(t_1,t_2;W^{-1,q'}(\omega))$$

for any open subset $\mathcal{O} = \omega \times]t_1,t_2[\subset K$, $q' = q/(q-1)$. Since the embedding $W_0^{1,q}(\Omega) \subset L^\infty(\Omega)$ is compact for $q > N = 3$, by well-known compactness results (see [S], for instance), we may conclude that

$$\mathbf{v}_\varepsilon \to \mathbf{v} \quad \text{in } L^q_{\text{loc}}(\{\theta > 0\}) \ . \tag{35}$$

Consequently, we may now let $\varepsilon \to 0$ in (34), obtaining the condition (14) for the equation of motion.

The conclusion of the proof follows now the same lines as in [CDK2] or in [RU].

Remark 3. In [DO1], since the equicontinuity in the case $q = 2$ was shown only up to a possible singular closed subset $\Sigma \subset Q$ of Hausdorff dimension at most 5/3 and Hausdorff measure arbitrarily small, the flow Stokes equations hold only in $\{\theta > 0\} \cap \{Q \backslash \Sigma\}$. The extension of this case $q = 2$ to the Navier–Stokes equations done in [DO2] is similar but also requires an additional Lemma, based in the Galerkin approximation, to show the strong convergence of the velocity field in the interior of the liquid zone. Our local compactness argument described above seems simpler than [DO2] and independent of the Galerkin approximation to the regularized–penalized problem (17)–(21).

4. THE EQUICONTINUITY OF THE TEMPERATURE

In this section we show that the equibounded sequence $(\theta_\varepsilon)_\varepsilon$ is also equicontinuous thus proving the uniform convergence $\theta_\varepsilon \to \theta$ and the continuity of θ. The idea is to obtain an implicit modulus of continuity for each θ_ε that is independent of ε, using the refinement of De Giorgi's iteration technique developed by DiBenedetto (see [D1] and [D2]). At the basis of this reasoning lie energy and logarithmic estimates that we derive in the next two lemmas.

We start with some notation and in order to show the dependence on the space dimension we work in $\mathbb{R}^N$, $N \geq 2$. Given a point $x_0 \in \mathbb{R}^N$, let $K_\rho(x_0)$ denote the N-dimensional cube with center at x_0 and wedge 2ρ:

$$K_\rho(x_0) := \left\{ x \in \mathbb{R}^N : \max_{1 \leq i \leq N} |x_i - x_{0i}| < \rho \right\} ;$$

given a point $(x_0,t_0) \in \mathbb{R}^{N+1}$, denote the cylinder of radius ρ and height $\tau > 0$ by

$$(x_0,t_0) + Q(\tau,\rho) := K_\rho(x_0) \times (t_0 - \tau, t_0) \ .$$

We'll write the inequalities in the following two lemmas for cylinders centered at the origin, the changes being obvious for a cylinder centered at a generic point (x_0, t_0).

Given a cylinder $Q(\rho, \tau) \subset Q$, consider a piecewise smooth cut off function $0 \le \zeta \le 1$ such that

$$|\nabla \zeta| < \infty \quad \text{and} \quad \zeta(x,t) = 0, \quad x \in K_\rho(0) \equiv K_\rho \,. \tag{36}$$

Lemma 1. *Let θ_ε be a solution of (P_ε) and $k \in \mathbb{R}$. There exists a constant $C > 0$, independent of ε, such that*

$$\sup_{-\tau<t<0} \int_{K_\rho \times \{t\}} (\theta_\varepsilon - k)_-^2 \, \zeta^2 + \int_{-\tau}^0 \int_{K_\rho} \left| \nabla (\theta_\varepsilon - k)_- \zeta \right|^2 \le$$

$$\le C \int_{-\tau}^0 \int_{K_\rho} (\theta_\varepsilon - k)_-^2 \, |\nabla \zeta|^2 + C \int_{K_\rho \times \{-\tau\}} (\theta_\varepsilon - k)_- \, \zeta^2 +$$

$$+ C \int_{-\tau}^0 \int_{K_\rho} (\theta_\varepsilon - k)_- \, \zeta \, \partial_t \zeta + C \left\{ \int_{-\tau}^0 \int_{K_\rho} \chi_{\{(\theta_\varepsilon - k)_- > 0\}} \right\}^\alpha$$

with

$$\alpha = \frac{(N+2)\, q - 2N}{(N+2)\, q} \,.$$

Proof. We take $\xi = -(\theta_\varepsilon - k)_- \, \zeta^2$ as a test function for the approximate equation

$$\partial_t [\gamma_\varepsilon(\theta_\varepsilon)] - \Delta \theta_\varepsilon + \mathbf{v}_\varepsilon^* \cdot \nabla \theta_\varepsilon = 0$$

and integrate in time in $(-\tau, t)$ with $t \in (-\tau, 0)$. The first term gives

$$- \int_{-\tau}^t \int_{K_\rho} \partial_t [\gamma_\varepsilon(\theta_\varepsilon)] \left((\theta_\varepsilon - k)_- \, \zeta^2 \right) = \int_{K_\rho} \int_{-\tau}^t \partial_t \Big(\int_0^{(\theta_\varepsilon - k)_-} \gamma_\varepsilon'(k-s) \, s \, ds \Big) \, \zeta^2 \ge$$

$$\ge \frac{1}{2} \int_{K_\rho \times \{t\}} (\theta_\varepsilon - k)_-^2 \, \zeta^2 - C \int_{K_\rho \times \{-\tau\}} (\theta_\varepsilon - k)_- \, \zeta^2 - C \int_{-\tau}^t \int_{K_\rho} (\theta_\varepsilon - k)_- \, \zeta \, \partial_t \zeta$$

since $\gamma_\varepsilon' \ge 1$. From the second term we obtain

$$\int_{-\tau}^t \int_{K_\rho} \nabla \theta_\varepsilon \cdot \nabla [-(\theta_\varepsilon - k)_- \, \zeta^2] \ge$$

$$\ge \int_{-\tau}^t \int_{K_\rho} |\nabla (\theta_\varepsilon - k)_- \, \zeta|^2 - C \int_{-\tau}^t \int_{K_\rho} (\theta_\varepsilon - k)_-^2 \, |\nabla \zeta|^2$$

by a simple use of Young's inequality. Finally, the convective term produces

$$- \int_{-\tau}^t \int_{K_\rho} \theta_\varepsilon \, \mathbf{v}_\varepsilon^* \cdot \nabla \big[-(\theta_\varepsilon - k)_- \, \zeta^2 \big] \ge -\frac{1}{2} \int_{-\tau}^t \int_{K_\rho} \left| \nabla (\theta_\varepsilon - k)_- \, \zeta \right|^2 -$$

$$-C\int_{-\tau}^{t}\int_{K_\rho}(\theta_\varepsilon-k)_-^2\,|\nabla\zeta|^2-C\int_{-\tau}^{t}\int_{K_\rho}|\mathbf{v}_\varepsilon^*|^2\,\chi_{\{(\theta_\varepsilon-k)_->0\}}$$

using again Young's inequalities and the fact that $\|\theta_\varepsilon\|_{L^\infty}\le M$.

To conclude, we observe that due to the estimate

$$\operatorname*{ess\,sup}_{0<t<T}\|\mathbf{v}_\varepsilon(\cdot,t)\|_{L^2(\Omega)}+\|\nabla\mathbf{v}_\varepsilon\|_{L^q(Q)}\le C$$

(that obviously still holds for $\mathbf{v}_\varepsilon^*$) and a well known embedding theorem (see [LSU, pp. 74–78]), we have

$$\|\mathbf{v}_\varepsilon^*\|_{L^{q\frac{N+2}{N}}(Q)}\le C$$

and so

$$\int_{-\tau}^{t}\int_{K_\rho}|\mathbf{v}_\varepsilon^*|^2\,\chi_{\{(\theta_\varepsilon-k)_->0\}}\le\|\mathbf{v}_\varepsilon^*\|^2_{L^{q\frac{N+2}{N}}(Q)}\left\{\int_{-\tau}^{t}\int_{K_\rho}\chi_{\{(\theta_\varepsilon-k)_->0\}}\right\}^{1-\frac{2N}{(N+2)q}}$$

$$\le C\left\{\int_{-\tau}^{t}\int_{K_\rho}\chi_{\{(\theta_\varepsilon-k)_->0\}}\right\}^{\alpha}.$$

Putting all the terms together we have the desired conclusion. ∎

Remark 4. A similar estimate holds with $(\theta_\varepsilon-k)_+$ but we will not reproduce it here. The difference is that in the right-hand side, in all the terms involving $(\theta_\varepsilon-k)_+$, this quantity is raised to the power 2. That is due to the fact that for $\varepsilon<k$ we are above the singularity of γ', so $\gamma'_\varepsilon\equiv 1$. For the details see [D1] or [U].

We turn to the logarithmic estimate. Consider the now standard function

$$\Psi^+=\left(\ln\left\{\frac{H_k^+}{H_k^+-(\theta_\varepsilon-k)_++b}\right\}\right)_+$$

where $H_k^+=\operatorname{ess\,sup}_{Q(\tau,\rho)}|(\theta_\varepsilon-k)_+|$, $0<b<H_k^+$ and $k>\varepsilon$. As before, consider a cutoff function satisfying (36), but independent of $t\in(-\tau,0)$.

Lemma 2. *Let θ_ε be a solution of (P_ε) and $k>\varepsilon$. There exists a constant $C>0$, independent of ε, such that*

$$\sup_{-\tau<t<0}\int_{K_\rho\times\{t\}}[\Psi^+]^2\,\zeta^2\le\int_{K_\rho\times\{-\tau\}}[\Psi^+]^2\,\zeta^2+C\int_{-\tau}^{0}\int_{K_\rho}\Psi^+\,|\nabla\zeta|^2$$

$$+\frac{C}{b^2}\left(1+\ln\frac{H_k^+}{b}\right)\left\{\int_{-\tau}^{0}\int_{K_\rho}\chi_{\{(\theta_\varepsilon-k)_+>0\}}\right\}^{\alpha}.$$

For the proof see [D1], [D2] or [U]; the last two references treat the more general degenerate problem.

Remark 5. We don't need a logarithmic estimate for $(\theta_\varepsilon - k)_-$.

The uniform convergence $\theta_\varepsilon \to \theta$ on the compacts follows easily from the theorem of Ascoli–Arzelá and the

Proposition. *The sequence $(\theta_\varepsilon)_\varepsilon$ is locally equicontinuous, i.e., there exists an interior modulus of continuity for each θ_ε, which is independent of ε.*

Proof. Given the inequalities established in the lemmas we can use the technique developed by DiBenedetto to obtain the result. See [D1] and [U] for further details.

We only remark that the exponent α in the term that, in the inequalities, comes from the lower order convective term of the equation is sufficiently large. Observe that, for example, the energy inequality corresponds to inequality (27) in [D1] for the choices

$$\widehat{q} = \widehat{r} = \frac{1}{1-\alpha} \, .$$

The condition needed for our proof to work is

$$\frac{1}{\widehat{r}} + \frac{N}{2\,\widehat{q}} < 1$$

which in this case reads

$$(1-\alpha) + \frac{N}{2}\,(1-\alpha) < 1 \iff 1-\alpha < \frac{2}{N+2} \iff$$

$$\iff \frac{2N}{(N+2)\,q} < \frac{2}{N+2} \iff q > N$$

and is precisely the assumption we made with $N = 3$. ∎

Remark 6. An interesting open problem is the case in which the convective term depends on the enthalpy. The proof of estimates that are independent of ε becomes a much more delicate matter.

Acknowledgments

J.F. Rodrigues was partially supported by FCT, Praxis/2/2.1/MAT/125/94 and Praxis XXI and J.M. Urbano by CMUC/FCT, Praxis/2/2.1/MAT/458/94 and Praxis XXI.

References

[C] Cioranescu, D. (1997). Appl. Math. Optim., 3:263–282.

[CDK1] Cannon, J.R., DiBenedetto, E. and Knightly, G.K. (1980). Arch. Rational Mech. Anal., 73:79–97.

[CDK2] Cannon, J.R., DiBenedetto, E. and Knightly, G.K., (1983). Comm. Partial Differential Equations, 8:1549–1604.

[D1] DiBenedetto, E. (1982). Ann. Mat. Pura Appl. (IV), 130:131–177.

[D2] DiBenedetto, E. (1993). Degenerate Parabolic Equations. Springer-Verlag, Berlin.

[DO1] DiBenedetto, E. and O'Leary, M. (1993). Arch. Rational Mech. Anal., 123:99–116.

[DO2] DiBenedetto, E. & O'Leary, M. (1993). Conduction–Convection with Change of Phase. Proc. Japan–Korean Conf. PDE, Seoul, Korea, Feb. 1993.

[L] Lions, J.L. (1969). Quelques Méthodes de Résolution des Problèmes aux Limites non Linéaires , Dunod, Gauthier-Villars, Paris.

[LSU] Ladyzenskaya, O.A., Solonnikov, V.A. and Ural'tseva, N.N. (1968). Linear and quasilinear equations of parabolic type. Amer. Math. Soc. Transl., Providence, 1968.

[M] Meirmanov, A.M. (1992). The Stefan Problem. Walter de Gruyter, Berlin.

[R1] Rodrigues, J.F. (1986). Ann. Mat. Pura Appl. (IV), 144:203–218.

[R2] Rodrigues, J.F. (1992). Weak solutions for thermoconvective flows of Boussinesq–Stefan type. In "Mathematical Topics in Fluid Mechanics" (edited by J.F. Rodrigues and A. Sequeira), pp. 93–116. Pitman Research Notes in Mathematics Series, 274.

[R3] Rodrigues, J.F. (1994). Variational methods in the Stefan problem. In "Lect. Notes in Math.s", #1584, Springer-Verlag, Berlin, 147–212.

[RU] Rodrigues, J.F. and Urbano, J.M. (1998). Int. J. Non-Linear Mech., 33:555–566.

[S] Simon, J. (1987). Ann. Mat. Pura Appl. (IV), 146:65–96.

[U] Urbano, J.M. Ann. Mat. Pura Appl. (IV). To appear

REPLACING H BY H^2

Mirko Rokyta

Abstract: We study higher order MUSCL type finite volume scheme applied to linear convection dominated diffusion problem in a bounded convex domain $\Omega \subset \mathbb{R}^2$. Although the original problem is linear, the numerical problem becomes non-linear, due to MUSCL type reconstruction/limiter technique. For second order schemes, an a-priori estimate of order h^2 for the discrete L^2 norm is obtained. Moreover, we discuss the solvability of the corresponding nonlinear discrete problem.

Keywords: Convection dominated diffusion equation in 2D, upwind finite volume scheme, higher order finite volume scheme, a-priori error estimates, MUSCL type reconstruction/limiter.

1. FOREWORD

One of the privileges which I have obtained from the life was undoubtedly the privilege to attend lectures delivered by Professor Jindřich Nečas. During the years of my undergraduate and graduate study at Faculty of Mathematics and Physics of the Charles University in Prague, I was often a member of the audience on Professor Nečas' courses. I remember that listening to his enthusiastic way of delivering lectures I learned successively that the life of a mathematician is not only the life framed by the formulas, but also a life full of fun and happiness, if one wants to. Besides the first class mathematics, we were presented a lot of stories and jokes, and a great passion and love for science.

When it became more or less obvious that my mathematical career is bound towards the numerical analysis, I occasionally mentioned this fact in front of Professor Nečas. I remember him making a little pause to clear his throat, and then stating to me, with smile in his eyes: "You know, numerical people always work hard to replace h by h^2." As usual, he was both brilliant and true in his observation.

I want to dedicate this contribution on replacing h by h^2 to Professor Jindřich Nečas on the occasion of his jubilee.

Applied Nonlinear Analysis, edited by Sequeira *et al.*
Kluwer Academic / Plenum Publishers, New York, 1999.

2. INTRODUCTION

In the theory of numerical algorithms for solving partial differential equations, independently of whether they are based on the concept of finite difference, finite element or finite volume, the basic role plays the concept of accuracy of a numerical approximation. In other words, if v denotes the exact solution of a continuous problem and u_h the approximative numerical solution corresponding to the parameter h of approximation, we are interested not only in the *convergence* result (in the suitable topology)

$$u_h \to v \quad \text{as } h \to 0+,$$

but also in the *accuracy of approximation*, i.e., closeness of u_h to v for small h. A useful information of this kind can be given by *a-priori* or *a-posteriori* error estimates, i.e. the estimates of the form

$$\|v - u_h\| \le c\,h^\alpha \|v\| \qquad \text{or} \qquad \|v - u_h\| \le c\,h^\alpha \|u_h\|, \tag{2.1}$$

respectively, with suitably chosen norms. While a-priori error estimates are more suitable for theoretical analysis of the problem, a-posteriori estimates are used in the context of practical computations, since the right-hand side of the estimate can be often explicitly computed, after the approximate solution u_h has been numerically evaluated.

In estimates (2.1), the $\alpha > 0$ denotes the order (or rate) of convergence, and we speak about schemes of first and higher order for $\alpha = 1$ and $\alpha > 1$, respectively. In the context of finite volume methods, theoretically justified distinction between first and higher order schemes is still not quite satisfactorily developed. In the literature, one can find a-priori error estimates of the type (2.1), e.g. for first order finite volume schemes on unstructured grids applied to nonlinear problems, cf. [1], [2], with $\alpha = \frac{1}{4}$, cf. [15]. Also, one has estimates for the streamline diffusion shock capturing method applied to the linear transport equation with $\alpha = \frac{3}{2}$, cf. [10], and for the streamline diffusion finite element method applied to the quasilinear convection-diffusion equation with $\alpha = \frac{3}{2}$. For diffusion problems there are results for first order schemes (cf. [9] for stationary case, [6] for combined finite volume–finite element method in non-stationary case) indicating $\alpha = 1$ for first order schemes. However, numerical experiments for upwind finite volume methods of second order [16] indicate that, at least for smooth solutions, one should expect $\alpha = 2$.

In this contribution we will be dealing with upwind finite volume schemes applied to linear convection dominated diffusion equation in 2D (cf. [14]). We use the so-called Engquist-Osher upwind numerical flux in the approximation of the convective term. In order to obtain a scheme which is hoped to be of higher order, we use the so-called MUSCL-type reconstruction (cf. [5]) in the evaluation of the convective term. For the resulting scheme, we get an *a-priori* error estimate of the type (2.1) with $\alpha = 2$. Thus, the convergence rate for higher order MUSCL type schemes is theoretically justified. To obtain this result, we work on regular triangular grids using linear reconstruction operators modified on those triangles where the discrete solution has a local extremum (see Definition 3.3, cf. [5]).

It is worth noting that, even if the continuous problem is linear, the discrete problem becomes highly non-linear due to the reconstruction process. Moreover, the resulting discrete operator fails to be continuous, since the reconstruction process is not so (see Example 5.2). Therefore, a question of solvability of the discrete problem is in order. We will discuss this question in the last section of this contribution.

The text of this contribution was presented in the form of a lecture on the International Conference on Applied Analysis, Lisbon, Portugal, February 26–March 1, 1997. The conference was dedicated to Professor Jindřich Nečas. The results of this contribution were partly obtained in cooperation with Dietmar Kröner (Freiburg University, Germany). Namely, the higher order error estimation part of this contribution is accepted for publication as an independent paper (cf. [13]).

3. THE PROBLEM AND ITS APPROXIMATION

Consider the following boundary value problem

$$Lv := -\varepsilon\Delta v + \operatorname{div}(bv) + cv = f \quad \text{in } \Omega\,, \tag{3.1}$$

$$v = 0 \quad \text{on } \partial\Omega\,, \tag{3.2}$$

where Ω is a convex polygonal domain in $\mathbb{R}^2$, $0 < \varepsilon < 1$, and $b(x)$, $c(x)$, $f(x)$ are functions which are sufficiently smooth on $\overline{\Omega}$ and such that $0 < c_0 \le c(x) \le c_1$, $\operatorname{div} b = 0$.

Notation 3.1. *a) We denote by $|T_j|$ and x_j the volume and the center of gravity of a triangle $T_j \in \mathcal{T}_h$, respectively. Denoting N_j the set of the numbers of the neighbouring triangles to T_j, we denote by S_{jl}, $l \in N_j$, the joint edge of T_j and T_l. We further denote by x_{jl} the mid-point of S_{jl}, and by n_{jl} the outward unit normal to T_j in the direction of T_l. Moreover we put: $T_{jl} := T_j \cup T_l$, $|d_{jl}| := |x_l - x_j|$, and $\gamma_{jl} := |S_{jl}|/|d_{jl}|$, $j \ne l$. By $F_h(x)$ we denote the piecewise constant approximant of $F \in L^2(\Omega)$, defined by*

$$F_h \equiv I_h F\,, \quad (I_h F)_j := \frac{1}{|T_j|}\int_{T_j} F\,. \tag{3.3}$$

b) If V_{jl} is a quantity assigned to an ordered pair of indices (j,l), $l \in N_j$, with $V_{lj} = 0$ for $S_{jl} \subset \partial\Omega$, we have

$$\sum_j \sum_{l \in N_j} V_{jl} = \sum_{\text{edges}} (V_{jl} + V_{lj}) \,, \tag{3.4}$$

where $\sum_j$ and $\sum_{\text{edges}}$ stand for the sum over all triangles of the triangulation and over all edges of the triangulation, respectively.

Assumption 3.2. *Throughout this contribution we will assume that $\overline{\Omega} = \bigcup_j \overline{T}_j$, where $T_j \in \mathcal{T}_h$ are equalsided triangles with side length h. This rather restrictive assumption is due to the technique used—see Section 4. In view of this assumption we have*

$$|S_{jl}| = h \,, \quad |d_{jl}| = \frac{h}{\sqrt{3}} \,, \quad \gamma := \gamma_{jl} = \sqrt{3} \,. \tag{3.5}$$

Let $u_h(x) = u_j$ for $x \in T_j$. We will give a-priori error estimates of the type (2.1) in case that u_h is a solution of the discrete problem

$$L_h u_h = f_h \,, \tag{3.6}$$

$$u_l = 0 \qquad \text{if } S_{jl} \in \partial\Omega \,, \tag{3.7}$$

with the discrete operator L_h given by

$$(L_h u_h)_j := -\frac{\varepsilon\gamma}{|T_j|} \sum_{l \in N_j} (u_l - u_j) + \frac{1}{|T_j|} \sum_{l \in N_j} g_{jl}\Big(L_j^u(x_{jl}), L_l^u(x_{jl})\Big) + c_j u_j \,. \tag{3.8}$$

The first term in (3.8) approximates the diffusion term $-\varepsilon\Delta v$, while the convective term $\operatorname{div}(bv)$ is approximated by $\sum_{l \in N_j} g_{jl}\big(L_j^u(x_{jl}), L_l^u(x_{jl})\big)$. Here, g_{jl} stands for an upwind finite volume flux. The easiest choice of its arguments,

$$L_j^u(x_{jl}) := u_j \,, \qquad L_l^u(x_{jl}) := u_l \,, \tag{3.9}$$

converts the scheme (3.8) into a *first order* numerical scheme. We will show that the more sophisticated definition of $L_j^u(x_{jl})$ and $L_l^u(x_{jl})$, namely using the MUSCL type reconstruction (see Definition 3.3) will turn (3.8) into a *second order* numerical scheme. The higher order of the scheme will be theoretically justified.

The upwind finite volume flux g_{jl} is, in general, any locally Lipschitz continuous function, satisfying the following three basic properties:

$$g_{jl}(u,u) = u \int_{S_{jl}} b n_{jl} \, ds \,, \tag{3.10}$$

$$g_{jl}(u,v) = -g_{lj}(v,u) \,, \tag{3.11}$$

$$\frac{\partial}{\partial u} g_{jl}(u,v) \geq 0 \geq \frac{\partial}{\partial v} g_{jl}(u,v) \,, \tag{3.12}$$

which are referred to as *consistency*, *conservativity*, and *monotonicity* of the numerical flux g_{jl}, respectively. (See [11] or [12] for more discussion on general upwind finite volume numerical fluxes.) Moreover, due to (3.10) and $\operatorname{div} b = 0$, we have that

$$\sum_{l \in N_j} g_{jl}(u_j, u_j) = 0 \,. \tag{3.13}$$

It is quite natural to use upwind numerical fluxes to approximate convective terms in the context of hyperbolic conservation laws. In fact, the amount of numerical viscosity introduced by upwinding (which is, from the point of view of a mathematician, encrypted in the monotonicity condition (3.12)), corresponds to small artificial viscosity, added to the original continuous problem. In view of this, solving hyperbolic conservation laws by upwind finite volume method, means solving the viscous parabolic perturbation of the original hyperbolic problem with small viscosity parameter. It is therefore not surprising that the convergence of upwind schemes to the weak *entropy* solution of conservation law was proven in the beginning of 1990's. For more details see [3], [10], [12].

Since in our case we still are dealing with equation containing just a small diffusion parameter $\varepsilon > 0$, i.e., the equation in which convection is dominating the diffusion, we are using the same idea of improving the properties of numerical scheme by handling the convective term the same way as it is natural in the case of conservation laws. Namely, we suggest to use *upwind* finite volume discretization of the convective term. In such a way, we therefore introduce an artificial viscosity, corresponding to the parameter h of the grid, into the scheme. Of course, at least from the intuitive point of view, since there is already viscous term in our equation, with diffusion parameter ε, the interplay between h and ε will be important in our analysis. We will discuss this question more carefully at the end of this section (see Remark 3.5 a)).

Throughout this contribution, we will be using a particular numerical flux, namely the Engquist-Osher type upwind finite volume flux g_{jl} defined by

$$g_{jl}(u_j, u_l) := b_{jl}^{+} u_j + b_{jl}^{-} u_l \,, \qquad b_{jl}^{\pm} := \int_{S_{jl}} (b n_{jl})^{\pm} \, ds \,. \tag{3.14}$$

It can be easily shown that (3.10)–(3.13) is satisfied by this particular numerical flux.

In the following definition we will describe the MUSCL type reconstruction function $L_j^u(x)$, which introduces the higher order approximation of the convective term into our considerations.

Definition 3.3. *Let T_k, T_l, T_m be all neighbouring triangles to T_j with centers of gravity x_k, x_l, x_m, x_j, respectively. Let $w \in L^\infty(\Omega)$ with $w|_{T_j} \in C^0(T_j)$ and $w_i := w(x_i)$ for $i = k, l, m, j$, respectively. Let (cf. [5])*

$$\begin{aligned} &R_k^w \text{ be a plane passing through } (x_l, w_l), (x_m, w_m), (x_j, w_j)\,, \\ &R_l^w \text{ be a plane passing through } (x_k, w_k), (x_m, w_m), (x_j, w_j)\,, \\ &R_m^w \text{ be a plane passing through } (x_k, w_k), (x_l, w_l), (x_j, w_j)\,. \end{aligned}$$

Define an index i by

$$|\nabla R_i^w| = \min\{|\nabla R_k^w|, |\nabla R_l^w|, |\nabla R_m^w|\} \tag{3.15}$$

and put

$$G_j^w := \nabla R_i^w\,. \tag{3.16}$$

If $w_j \geq \max\{w_k, w_l, w_m\}$ or $w_j \leq \min\{w_k, w_l, w_m\}$, we say that w_j is a local extremum. Let the coefficients $\alpha_j = \alpha_j^w \in \{0, 1\}$ be such that

$$\alpha_j^w = \begin{cases} 0 & \text{if } w_j \text{ is the local extremum,} \\ 1 & \text{otherwise.} \end{cases} \tag{3.17}$$

Then define

$$L_j^w(x) := w_j + \alpha_j^w G_j^w (x - x_j)\,. \tag{3.18}$$

In what follows we state main result of this section as a theorem. It concerns a-priori error estimate for the higher order scheme (3.6), (3.7), (3.8), (3.18). The difficulties in proofs arise due to the nonlinearity of the higher order part of the scheme (cf. (3.15)).

Theorem 3.4. *Let $f \in W^{1,2}(\Omega)$, $\Omega \subset \mathbb{R}^2$. Let $v \in W^{3,2}(\Omega)$ be the solution of (3.1), (3.2). Let u_h be the solution of (3.6), (3.7), with the discrete operator given by (3.8), (3.18). Denote*

$$z_h := v_h - u_h = I_h v - u_h\,.$$

Then there exists a constant $c > 0$ independent of ε and h such that

$$\begin{aligned} \|z_h\|_\varepsilon^2 &:= \varepsilon\gamma \sum_{\text{edges}} (z_j - z_l)^2 + c_0 \sum_j z_j^2 |T_j| \qquad (3.19) \\ &\leq c h^4 \varepsilon \|v\|_{W^{3,2}(\Omega)}^2 + c\frac{h^4}{\varepsilon} \|v\|_{W^{2,2}(\Omega)}^2 + c\frac{h^4}{\varepsilon^4} \|f\|_{L^2(\Omega)}^2\,. \end{aligned}$$

Remark 3.5. a) For the terms $\|v\|^2_{W^{2,2}(\Omega)}$, and $\|v\|^2_{W^{3,2}(\Omega)}$ we have the following estimates (cf. [13]):

$$\|v\|^2_{W^{3,2}(\Omega)} \le \frac{c}{\varepsilon^5}\|f\|^2_{L^2(\Omega)}, \quad \|v\|^2_{W^{2,2}(\Omega)} \le \frac{c}{\varepsilon^3}\|f\|^2_{L^2(\Omega)}. \tag{3.20}$$

Employing these estimates, we obtain finally the desired $O(h^2)$ estimate:

$$\|z_h\|_\varepsilon \le c\frac{h^2}{\varepsilon^2}\|f\|_{L^2(\Omega)}. \tag{3.21}$$

We see from (3.21) the interplay between h and ε. The final estimate is not uniform with respect to ε, and therefore cannot be used to study what happens when $\varepsilon \to 0+$. However, we see that for any $\varepsilon > 0$, arbitrarily small but fixed, the convergence rate is of order $O(h^2)$ in $\|\cdot\|_\varepsilon$ (or $\|\cdot\|_{L^2(\Omega)}$—see (3.19)) norm, as $h \to 0+$.

b) We need the exact solution v to be more regular, namely we assume $v \in W^{3,2}(\Omega)$. Conditions under which $v \in W^{3,2}(\Omega)$, if Ω is a convex domain, are given e.g. in [7]. The assumption $f \in W^{1,2}(\Omega)$ implies $v \in W^{3,2}(\Omega)$ for instance if all angles in the corners of Ω are less than $\pi/2$. This together with Assumption 3.2 forces Ω to be an equalsided triangle.

4. THE SKETCH OF PROOF OF THE MAIN THEOREM

To prove Theorem 3.4, we estimate the term

$$\left(L_h(I_hv) - I_hLv, z_h\right) := \sum_j \left(L_h(I_hv) - I_hLv\right)_j |T_j| z_j$$

both from above and from below. Combining these two estimates, we obtain the statement of Theorem 3.4. From the technical point of view, we use the Bramble-Hilbert lemma (see for example [8]), which needs all triangles to be equalsided. Furthermore, we employ the particular form of both the Engquist-Osher numerical flux g_{jl} (cf. (3.14)) and the reconstruction functional L^u_j (cf. (3.18)). The latter information allows us, among other, to develop the estimates

$$|L^u_j(x_l) - u_j| \le |u_j - u_l|, \tag{4.1}$$
$$(u_l - L^u_j(x_l))(u_j - u_l) \le 0, \tag{4.2}$$

holding for all $l \in N_j$, which control both size and the sign of the reconstructed value $L^u_j(x_l)$. The idea of "switching back to first order in local extrema" (i.e., using (3.17)) plays an important role in the above presented estimates. We refer the reader to [13] for the details on these estimates.

We proceed by formulating two basic lemmas and sketching their proofs. Combining these two lemmas, one immediately gets the assertion of Theorem 3.4. All main steps to follow the suggested strategy and the technical details necessary to prove all statements of this particular section can be found in [13].

Lemma 4.1. (Estimates from below) *Under the assumptions of Theorem 3.4 there exists a constant $c > 0$ independent of ε and h such that*

$$\begin{aligned}\Big(L_h(I_h v) - I_h L v, z_h\Big) \;\geq\; & \frac{\varepsilon\gamma}{2}\sum_{\text{edges}}(z_l - z_j)^2 + c_0\sum_j z_j^2|T_j| \\ & -c\frac{h^4}{\varepsilon}\|v\|^2_{W^{2,2}(\Omega)} - c\frac{h^4}{\varepsilon^4}\|f\|^2_{L^2(\Omega)}\,.\end{aligned}$$

Lemma 4.2. (Estimates from above) *Under the assumptions of Theorem 3.4 there exists a constant $c > 0$ independent of ε and h such that*

$$\begin{aligned}\Big(L_h(I_h v) - I_h L v, z_h\Big) \;\leq\; & \frac{\varepsilon\gamma}{4}\sum_{\text{edges}}(z_j - z_l)^2 + \frac{c_0}{8}\sum_j z_j^2|T_j| \\ & +c\,h^4\varepsilon\|v\|^2_{W^{3,2}(\Omega)} + c\frac{h^4}{\varepsilon}\|v\|^2_{W^{2,2}(\Omega)}\,.\end{aligned}$$

Proof of Theorem 3.4: The assertion of the theorem follows immediately from the assertions of Lemma 4.1 and Lemma 4.2.

An outline of the proof of Lemma 4.1: Using the fact that $I_h L v = I_h f = f_h = L_h u_h$ and the definition of L_h, one can write

$$\sum_j \Big(L_h(I_h v) - I_h L v\Big)_j |T_j| z_j = A_1 + A_2 + A_3$$

where

$$\begin{aligned}A_1 &= \sum_j \Big(-\varepsilon\sum_{l\in N_j}(v_l - v_j)\gamma + \varepsilon\sum_{l\in N_j}(u_l - u_j)\gamma\Big) z_j\,,\\ A_2 &= \sum_j\sum_{l\in N_j}\Big(g_{jl}(L_j^v(x_{jl}), L_l^v(x_{jl})) - g_{jl}(L_j^u(x_{jl}), L_l^u(x_{jl}))\Big) z_j\,,\\ A_3 &= \sum_j c_j(v_j - u_j) z_j |T_j|.\end{aligned}$$

We have straightforwardly that $A_1 \geq \varepsilon\gamma \sum_{\text{edges}}(z_l - z_j)^2$, and $A_3 \geq c_0 \sum_j z_j^2 |T_j|$. Therefore it remains to estimate A_2. Using the definition (3.14) of g_{jl}, we can write

$$A_2 = \sum_j \sum_{l \in N_j} \Big(b_{jl}^+ (L_j^v(x_{jl}) - L_j^u(x_{jl})) + b_{jl}^- (L_l^v(x_{jl}) - L_l^u(x_{jl})) \Big) z_j \,.$$

Then, using (3.18), we have

$$\begin{aligned} L_j^v(x_{jl}) - L_j^u(x_{jl}) &= \Big(L_j^v(x_{jl}) - v_j \Big) + (v_j - u_j) + \Big(u_j - L_j^u(x_{jl}) \Big) \\ &= \frac{1}{2}(z_j + z_l) + \frac{1}{2}\Big(L_j^v(x_l) - v_l \Big) + \frac{1}{2}\Big(u_l - L_j^u(x_l) \Big) \,, \end{aligned}$$

and similarly for $L_l^v(x_{jl}) - L_l^u(x_{jl})$. Therefore,

$$\begin{aligned} A_2 &= \frac{1}{2} \sum_j \sum_{l \in N_j} \Big(b_{jl}^+ (z_j + z_l) z_j + b_{jl}^- (z_j + z_l) z_j \Big) \\ &\quad + \frac{1}{2} \sum_{\text{edges}} \Big(b_{jl}^+ (L_j^v(x_l) - v_l) + b_{jl}^- (L_l^v(x_j) - v_j) \Big) (z_j - z_l) \\ &\quad + \frac{1}{2} \sum_{\text{edges}} \Big(b_{jl}^+ (u_l - L_j^u(x_l)) + b_{jl}^- (u_j - L_l^u(x_j)) \Big) (z_j - z_l) \\ &=: W_1 + W_2 + W_3 \,. \end{aligned} \tag{4.3}$$

Firstly, we have $W_1 = 0$ due to $\sum_{l \in N_j} b_{jl} = 0$ and $b_{jl} = -b_{lj}$. In order to estimate terms W_2 and W_3 we use the definition (3.18) to see that $u_l - L_j^u(x_l) = u_l - u_l - \alpha_j^u G_j^u(x_l - x_j)$, and we proceed by discussing the three possible cases: If $\alpha_j^u = 1$ and stencil of G_j^u contains x_l, then $u_l - L_j^u(x_l) = 0$. If $\alpha_j^u = 1$ and stencil of G_j^u does not contain x_l, we have

$$\begin{aligned} u_l - L_j^u(x_l) &= u_l - u_j - G_j^u(x_l - x_j) \\ &= u_l - u_j + G_j^u(x_m - x_j + x_k - x_j) \\ &= u_l - u_j + u_m - u_j + u_k - u_j \\ &= \left(\frac{1}{|T_j|} \sum_{l \in N_j} (u_l - u_j) \right) |T_j| = R_j |T_j|, \end{aligned}$$

with

$$R_j := \frac{1}{|T_j|} \sum_{l \in N_j} (u_l - u_j) \,.$$

Finally, if $\alpha_j^u = 0$, then $u_l - L_j^u(x_l) = u_l - u_j$. However, $\alpha_j^u = 0$ implies that there is local extremum in u_j. Therefore, we have also in this case

$$|u_l - u_j| \leq \Big| \sum_{l \in N_j} (u_l - u_j) \Big| = |R_j| |T_j| \,.$$

We conclude that in any case,

$$|u_l - L_j^u(x_l)| \le |R_j||T_j|\,. \tag{4.4}$$

By similar considerations we get that

$$|v_l - L_j^v(x_l)| \le \Big|\sum_{l\in N_j} (v_l - v_j)\Big|\,. \tag{4.5}$$

This particular part of the proof shows the way we employ the information that "switching back to first order" (i.e., choosing $\alpha_j^u = 0$) takes place only in local extrema.

In order to estimate (4.5), we use the Bramble-Hilbert technique. We define

$$E(v_j) := \sum_{l\in N_j} (v_l - v_j) = (v_l - v_j) + (v_k - v_j) + (v_m - v_j)\,. \tag{4.6}$$

Using then the technique of the reference triangles and Bramble-Hilbert lemma, we eventually obtain

$$|v_l - L_j^v(x_l)| \le c\,h\,|v|_{W^{2,2}(T)} \tag{4.7}$$

with $c > 0$ independent both of h and of the domain $T := T_j \cup T_l \cup T_k \cup T_m$. The same estimate holds when replacing j by l and vice versa. Using now (4.7) and Young inequality, we obtain

$$W_2 \ge -c\frac{h^4}{\varepsilon}\sum_{\text{edges}} |v|^2_{W^{2,2}(T)} - \frac{\varepsilon\gamma}{4}\sum_{\text{edges}} |z_j - z_l|^2\,,$$

while (4.4) implies, in a similar way, that

$$W_3 \ge -c\frac{h^4}{\varepsilon}\sum_j |R_j|^2|T_j| - \frac{\varepsilon\gamma}{4}\sum_{\text{edges}} |z_j - z_l|^2\,.$$

Moreover, it can be shown that (cf. [13]) there is a constant c independent of ε such that

$$\sum_j |R_j|^2|T_j| \le \frac{c}{\varepsilon^3}\,\|f\|^2_{L^2(\Omega)}\,.$$

This together with (4.2) and the fact that $W_1 = 0$ finishes the proof.

An outline of the proof of Lemma 4.2: We obtain, using the convention (3.4), that

$$\sum_j \Big(L_h(I_h v) \quad - \quad I_h L v\Big)_j |T_j| z_j$$

$$= -\varepsilon \sum_{\text{edges}} \Big((v_l - v_j)\gamma - \int_{S_{jl}} \partial_n v\, ds \Big)(z_j - z_l)$$
$$+ \sum_{\text{edges}} \Big(g_{jl}(L_j^v(x_{jl}), L_l^v(x_{jl})) - \int_{S_{jl}} n_{jl} bv\, ds \Big)(z_j - z_l)$$
$$+ \sum_j \int_{T_j} (c_j v_j - cv) z_j .$$

It can be shown that there is a constant c independent of both h and the domain $T_{jl} := T_j \cup T_l$ such that (see [13] for the details)

$$\Big| (v_l - v_j)\gamma - \int_{S_{jl}} \partial_n v\, ds \Big| \le c h^2 \|v\|_{W^{3,2}(T_{jl})}, \quad (4.8)$$
$$\Big| g_{jl}(L_j^v(x_{jl}), L_l^v(x_{jl})) - \int_{S_{jl}} n_{jl} bv\, ds \Big| \le c h^2 \|v\|_{W^{2,2}(T_{jl})}, \quad (4.9)$$
$$\Big| \int_{T_j} (c_j v_j - cv) \Big| \le c h^3 \|v\|_{W^{2,2}(T_{jl})} . \quad (4.10)$$

With these estimates, the proof can be straightforwardly finished using the Young inequality.

We refer the reader to [13] for the remaining technical details.

5. SOLVABILITY OF THE DISCRETE PROBLEM

Even if the original continuous problem is linear, the discrete problem (3.6), (3.7), (3.8) becomes nonlinear, since the MUSCL type reconstruction (3.18) is used in the approximation of the convective term. In this section we will be interested in questions of *solvability* of the resulting nonlinear discrete equation (3.6).

One of the possibilities to attack the solvability question could be the following result due to Deimling (cf. [4], §9, Theorem 2): if A is a continuous mapping $\mathbb{R}^n \to \mathbb{R}^n$ such that

$$\frac{(A(x), x)}{\|x\|} \to \infty \quad \text{as} \quad \|x\| \to \infty, \quad (5.1)$$

then A maps $\mathbb{R}^n$ *onto* $\mathbb{R}^n$.

In the following lemma we will show that the discrete operator L_h satisfies the coercivity condition (5.1). Therefore it is tempting to proclaim the problem to be solved, using the above Deimling result.

Unfortunately, the result of Deimling cannot be used, since the discrete operator L_h is *not continuous*. We will show this fact by constructing an explicit Example 5.2 and discuss the difficulties connected to this discontinuity later on. But first, for the sake of completeness, let us formulate and prove Lemma 5.1.

Lemma 5.1. *Let N be the number of triangles of $T_j \in \mathcal{T}_h$ and $u_h = (u_1, \dots, u_N)$ an element of $\mathbb{R}^n$. Defining the scalar product in $\mathbb{R}^n$ as $(u_h, v_h) := \sum_{j=1}^N u_j v_j |T_j|$, we have*

$$(L_h u_h, u_h)_N \geq \varepsilon\gamma \sum_{\text{edges}} (u_l - u_j)^2 + c_0 \sum_j u_j^2 |T_j| \geq c_0 \|u_h\|_N^2 .$$

Therefore, $L_h : \mathbb{R}^n \to \mathbb{R}^n$ satisfies (5.1), namely,

$$\frac{(L_h u_h, u_h)_N}{\|u_h\|_N} \to \infty \quad as \quad \|u_h\|_N \to \infty .$$

Proof: We obtain from (3.8) that

$$\begin{aligned}(L_h u_h, u_h)_N \;\geq\; & \varepsilon\gamma \sum_{\text{edges}} (u_l - u_j)^2 + \sum_j \sum_{l \in N_j} g_{jl}(L_j^u(x_{jl}), L_l^u(x_{jl}))u_j \\ & + c_0 \sum_j u_j^2 |T_j| ,\end{aligned} \tag{5.2}$$

and therefore it remains to show that

$$\sum_j \sum_{l \in N_j} g_{jl}(L_j^u(x_{jl}), L_l^u(x_{jl}))u_j \geq 0 .$$

We get that $\sum_j \sum_{l \in N_j} g_{jl}(L_j^u(x_{jl}), L_l^u(x_{jl}))u_j$ is equal to

$$\begin{aligned}& \sum_j \sum_{l \in N_j} \Big(b_{jl}^+(u_j + \alpha_j^u G_j^u(x_{jl} - x_j)) + b_{jl}^-(u_l + \alpha_l^u G_l^u(x_{jl} - x_l))\Big)u_j \\ &= \frac{1}{2} \sum_{\text{edges}} \Big(b_{jl}^+(u_j - u_l)^2 - b_{jl}^-(u_j - u_l)^2\Big) \\ &\quad + \sum_{\text{edges}} \Big((b_{jl}^+ \alpha_j^u G_j^u(x_{jl} - x_j) + b_{jl}^- \alpha_l^u G_l^u(x_{jl} - x_l))u_j \\ &\quad + (b_{lj}^+ \alpha_l^u G_l^u(x_{lj} - x_l) + b_{lj}^- \alpha_j^u G_j^u(x_{lj} - x_j))u_l\Big) \\ &\geq \frac{1}{2} \sum_{\text{edges}} \Big(b_{jl}^+(u_j - u_l)^2 - b_{jl}^-(u_j - u_l)^2\Big) \\ &\quad - \sum_{\text{edges}} \Big(b_{jl}^+ \alpha_j^u |G_j^u(x_{jl} - x_j)||u_j - u_l| \\ &\qquad + |b_{jl}^-| \alpha_l^u |G_l^u(x_{jl} - x_l)||u_j - u_l|\Big) .\end{aligned}$$

Now since

$$\alpha_j^u |G_j^u(x_{jl} - x_j)| = \frac{1}{2}\alpha_j^u |G_j^u(x_l - x_j)| = \frac{1}{2}|(L_j^u(x_l) - u_j)| \le \frac{1}{2}|u_l - u_j|$$

(see (4.1)), we can continue

$$\begin{aligned}\sum_j \sum_{l\in N_j} & g_{jl}(L_j^u(x_{jl}), L_l^u(x_{jl}))u_j \\ & \ge \frac{1}{2}\sum_{\text{edges}} \left(b_{jl}^+(u_j - u_l)^2 - b_{jl}^-(u_j - u_l)^2\right) \\ & \qquad -\frac{1}{2}\sum_{\text{edges}} \left(b_{jl}^+|u_j - u_l|^2 + |b_{jl}^-||u_j - u_l|^2\right) \\ & \ge 0,\end{aligned}$$

which proves the lemma.

Example 5.2. The higher order discrete operator L_h is not continuous. We will demonstrate this fact by constructing an explicit example. Let $u_h = (u_1, \dots, u_N)$ be a discrete function defined on the regular triangulation $\mathcal{T}_h$. All the values of u_j are equal to zero, except the three values $u_k = 1$, $u_{k+1} = 1$, $u_{k+2} = u_{k+2}(a) = a$, where $a \in (1-\delta, 1+\delta)$. Hence, $u_h = u_h(a)$ depends on the real parameter a.

The values of u_h are chosen in such a way that the triangle T_{k+1}, lying inside of Ω, has one of its sides parallel to the x-axis while T_k is its neighbour "to the left" and T_{k+2} is its neighbour "to the right". We define the function $\mathcal{L}(a)$ of one real variable by

$$\mathcal{L}(a) := \sum_j \sum_{l\in N_j} g_{jl}(L_j^u(x_{jl}), L_l^u(x_{jl})),$$

using $u_h(a)$ to evaluate the numerical flux on the right-hand side. We put, for simplicity, $h = 1$ for the purpose of these calculations. Then it can be explicitly, though tediously, computed that

$$\lim_{a\to 1-} \mathcal{L}(a) = -\frac{5+\sqrt{3}}{2} \approx -3.366\dots, \qquad \lim_{a\to 1+} \mathcal{L}(a) = -2.$$

This shows that the operator L_h is not continuous (cf. (3.8)).

The discontinuity of the operator L_h implies that there is no hope to prove the solvability of the discrete problem, at least in the general case. Discontinuity produces "jumps" and therefore "the holes" in the range of the operator L_h can occur, implying that there can be right-hand sides $f_h \in \mathbb{R}^N$ for which the equation $L_h u_h = f_h$ is not solvable.

Fortunately, at least partial result in this direction can be proven. Namely, we can "estimate the size of holes in the range" showing that to every right-hand side $f_h \in \mathbb{R}^n$ there is another function $\widetilde{f}_h$ which is "reasonably close" (in terms of h) to f_h, and for which the problem $L_h u_h = \widetilde{f}_h$ is solvable.

Theorem 5.3. *Let $f \in L^2(\Omega)$. Consider the convection-diffusion problem (3.1), (3.2). Let $f_h = I_h f \in \mathbb{R}^N$ be the discrete right-hand side of the discrete equation $L_h u_h = f_h$. Then there exist $\widetilde{f}_h \in \mathbb{R}^N$ such that*

$$\widetilde{f}_h \in Range(L_h) \qquad and \qquad \|\widetilde{f}_h - f_h\|_{L^1(\Omega)} \le c \frac{h}{\varepsilon^{3/2}} \|f\|_{L^2(\Omega)},$$

where $c = c(\Omega)$ is independent of h and ε.

Proof: Since $Lv = f$, we have $f_h = I_h(f) = I_h(Lv)$. Define

$$\widetilde{f}_h := L_h(I_h v).$$

Then clearly $\widetilde{f}_h \in \text{Range}(L_h)$. Moreover we have

$$\begin{aligned}
\|\widetilde{f}_h - f_h\|_1 &= \|L_h(I_h v) - I_h L v\|_1 = \sum_j \left(L_h(I_h v) - I_h L v\right)_j |T_j| \\
&\le \varepsilon \sum_j \sum_{l \in N_j} \left|(v_l - v_j)\gamma - \int_{S_{jl}} \partial_n v \, ds\right| \\
&\quad + \sum_j \sum_{l \in N_j} \left|g_{jl}(L_j^v(x_{jl}), L_l^v(x_{jl})) - \int_{S_{jl}} n_{jl} b v \, ds\right| \\
&\quad + \sum_j \left|\int_{T_j} (c_j v_j - cv)\right|.
\end{aligned}$$

Using then the estimates (4.8)–(4.10), we see that there is a constant c independent of both h and the domain $T_{jl} := T_j \cup T_l$ such that

$$\begin{aligned}
\|\widetilde{f}_h - f_h\|_1 &\le c\varepsilon h^2 \sum_j \sum_{l \in N_j} \|v\|_{W^{3,2}(T_{jl})} + c\, h^2 \sum_j \sum_{l \in N_j} \|v\|_{W^{2,2}(T_{jl})} \\
&\quad + c\, h^3 \sum_j \|v\|_{W^{2,2}(T_j)}.
\end{aligned}$$

By the Cauchy-Schwartz inequality,

$$\begin{aligned}
h \sum_j \sum_{l \in N_j} \|v\|_{W^{2,2}(T_{jl})} &\le \Big(\sum_j \sum_{l \in N_j} h^2\Big)^{1/2} \Big(\sum_j \sum_{l \in N_j} \|v\|^2_{W^{2,2}(T_{jl})}\Big)^{1/2} \\
&\le c(\Omega)\, \|v\|_{W^{2,2}(\Omega)}, \qquad (5.3)
\end{aligned}$$

and therefore, with $c = c(\Omega)$,

$$\begin{aligned} \|\tilde{f}_h - f_h\|_1 &\le c\varepsilon h\|v\|_{W^{3,2}(\Omega)} + c\, h\|v\|_{W^{2,2}(\Omega)} + c\, h^2\|v\|_{W^{2,2}(\Omega)} \\ &\le c\frac{h}{\varepsilon^{3/2}}\|f\|_{L^2(\Omega)} + c\frac{h}{\varepsilon^{3/2}}\|f\|_{L^2(\Omega)} + c\frac{h^2}{\varepsilon^{3/2}}\|f\|_{L^2(\Omega)} \end{aligned}$$

by (3.20), and the proof follows.

Acknowledgments

The results of this contribution were partly obtained in cooperation with Dietmar Kröner (Freiburg University, Germany). Namely, the higher order error estimation part of this contribution is accepted for publication as an independent paper (cf. [13]). The author was supported in part by grants GAČR 201/96/0313, GAČR 201/96/0228, and by the Ministry of Education of Czech Republic program Kontakt (Project ME 050/1998).

References

[1] Cockburn, B., Coquel, F. and LeFloch, P. (1994). *An error estimate for finite volume methods for multidimensional conservation laws.* Math. Comp., 63(207):77–103.

[2] Cockburn, B. and Gremaud, P.-A. (1996). *A priori error estimates for numerical methods for scalar conservation laws. Part I: The general approach.* Math. Comp., 65(214):533–573.

[3] Cockburn, B. and Shu, C.W. (1989). *TVB Runge-Kutta projection discontinuous Galerkin finite element method for conservation laws. II: General framework.* Math. Comp., 52:411–435.

[4] Deimling, K. (1974). *Nichtlineare Gleichungen und Abbildungsgrade.* Springer-Verlag.

[5] Durlofsky, L.J., Engquist, B. and Osher, S. (1992). *Triangle based adaptive stencils for the solution of hyperbolic conservation laws.* J. Comput. Phys., 98:64–73.

[6] Feistauer, M., Felcman, J. and Lukáčová-Medviďová, M. (1997). *On the convergence of a combined finite volume – finite element method for nonlinear convection – diffusion problems.* Numer. Methods Partial Differential Equations, 13:1–28.

[7] Grisvard, P. (1986). *Problèmes aux limites dans les polygones: mode d'emploi*, EDF Bull. Direction Etudes Rech. Ser., Série C, 1:21–59.

[8] Heinrich, B. (1987). *Finite difference methods on irregular networks.* Birkhäuser, Basel.

[9] Herbin, R. (1995). *An error estimate for a finite volume scheme for a diffusion – convection problem on triangular mesh.* Numer. Methods Partial Differential Equations, 11:165–173.

[10] Johnson, C. and Szepessy, A. (1987). *Convergence of a finite element method for a nonlinear hyperbolic conservation law.* Math. Comp., 49:427–444.

[11] Kröner, D., Noelle, S. and Rokyta, M. (1995). *Convergence of higher order upwind finite volume schemes on unstructured grids for scalar conservation laws in several space dimensions.* Numer. Math., 71(4):527–560.

[12] Kröner, D. and Rokyta, M. (1994). *Convergence of upwind finite volume schemes for scalar conservation laws in 2D.* SIAM J. Numer. Anal., 31(2):324–343.

[13] Kröner, D. and Rokyta, M. (1996). *A-priori error estimates for upwind finite volume schemes in several space dimensions.* Preprint No. 96-37, Univ. Freiburg. (SIAM J. Numer. Anal., accepted)

[14] Roos, H.-G., Stynes M. and Tobiska, L. (1996). *Numerical methods for singularly perturbed differential equations. Convection-diffusion and flow problems.* Springer Series in Computational Mathematics, 24, Springer-Verlag, Berlin.

[15] Vila, J.-P. (1994). *Convergence and error estimates in finite volume schemes for general multi-dimensional scalar conservation laws. I: Explicit monotone schemes.* RAIRO Model. Math. Anal. Numer., 28(3):267–295.

[16] Wierse, M. (1994). *Higher order upwind schemes on unstructured grids for the compressible Euler equations in time dependent geometries in 3D.* PhD Thesis, Universität Freiburg.

FLOW OF SHEAR DEPENDENT ELECTRORHEOLOGICAL FLUIDS: UNSTEADY SPACE PERIODIC CASE

Michael Růžička

Abstract: We study the existence of weak and strong solutions to the unsteady and steady system of partial differential equations with non-standard growth conditions describing the flow of shear dependent electrorheological fluids in the case of space periodic boundary conditions.

Keywords: Non-Newtonian fluid, weak and strong solutions, nonlinear elliptic and parabolic system with non-standard growth.

1. FORMULATION OF THE PROBLEM AND RESULTS

Electrorheological fluids are special viscous fluids, which are characterized by the ability to change dramatically their mechanical properties in dependence on an applied electric field. This behaviour can be exploited in many technological applications, as clutches, actuators, medical rehabilitation equipment and valves, to name a few.

Recently, Rajagopal, Růžička [9], [10] (see also Růžička [11]) have developed a model which treats the electric field not as a given constant but as a variable, which has to be determined. Thus this approach takes into account the complicated interactions between the electromagnetic fields and the moving fluid, which have been neglected in previous investigations. The starting point are the general balance laws of thermodynamics and electrodynamics, which are simplified by incorporating the physical properties of electrorheological fluids and accomplishing a dimensional analysis and a subsequent approximation.

The resulting system, which has to be completed by initial and boundary conditions, reads

$$\begin{aligned} \operatorname{div} \mathbf{E} &= 0\,, \\ \operatorname{curl} \mathbf{E} &= 0\,, \end{aligned} \tag{1.1}$$

$$\begin{aligned} \frac{\partial \mathbf{v}}{\partial t} - \operatorname{div} \mathbf{S} + [\nabla \mathbf{v}]\mathbf{v} + \nabla \pi &= \mathbf{f} + \chi^E [\nabla \mathbf{E}]\mathbf{E}\,, \\ \operatorname{div} \mathbf{v} &= 0\,, \end{aligned} \tag{1.2}$$

where[1] $\mathbf{E}$ is the electric field, $\mathbf{v}$ the velocity, $\mathbf{S}$ the extra stress tensor, π the pressure, $\mathbf{f}$ the mechanical force and χ^E the constant dielectric suspectibility. The material properties are modeled through the form of the extra stress, which is assumed to be[2]

$$\begin{aligned} \mathbf{S} = \alpha_{21}\big((1+|\mathbf{D}|^2)^{\frac{p-1}{2}} - 1\big)\mathbf{E}\otimes\mathbf{E} + (\alpha_{31} + \alpha_{33}|\mathbf{E}|^2)(1+|\mathbf{D}|^2)^{\frac{p-2}{2}}\mathbf{D} \\ + \alpha_{51}(1+|\mathbf{D}|^2)^{\frac{p-2}{2}}(\mathbf{D}\mathbf{E}\otimes\mathbf{E} + \mathbf{E}\otimes\mathbf{D}\mathbf{E})\,, \end{aligned} \tag{1.3}$$

where $p = p(|\mathbf{E}|^2)$ is a C^1-function such that

$$1 < p_\infty \le p(|\mathbf{E}|^2) \le p_0 < \infty\,. \tag{1.4}$$

We require that the coefficients α_{ij} and the function p are such that the operator induced by $-\operatorname{div} \mathbf{S}(\mathbf{D},\mathbf{E})$ is coercive, i.e.

$$\mathbf{S}(\mathbf{D},\mathbf{E})\cdot\mathbf{D} \ge c_0(1+|\mathbf{E}|^2)(1+|\mathbf{D}|^2)^{\frac{p(|\mathbf{E}|^2)-2}{2}}|\mathbf{D}|^2 \tag{1.5}$$

holds for all $\mathbf{D} \in X := \{\mathbf{D} \in \mathbb{R}^{3\times3}_{\text{sym}},\ \operatorname{tr}\mathbf{D} = 0\}$, and uniformly monotone, i.e.

$$\frac{\partial S_{ij}(\mathbf{D},\mathbf{E})}{\partial D_{kl}} B_{ij}B_{kl} \ge c_1(1+|\mathbf{E}|^2)(1+|\mathbf{D}|^2)^{\frac{p(|\mathbf{E}|^2)-2}{2}}|\mathbf{B}|^2 \tag{1.6}$$

is satisfied for all $\mathbf{B},\mathbf{D} \in X$.[3]

The quasi-static Maxwell's equations (1.1) are widely studied in the literature and thus we will investigate in this paper the system (1.2) only, in which $\mathbf{E}$ is assumed to be any given vector field, having certain regularity properties. System (1.2) is completed by space periodic boundary conditions and an initial condition $\mathbf{v}_0$.

[1]Here and in the following we use the notation $[\nabla\mathbf{u}]\mathbf{w} = (w_j \frac{\partial u_i}{\partial x_j})_{i=1,2,3}$, where the summation convention over repeated indices is used. Moreover, we have divided equation $(1.2)_1$ by the constant density ρ_0 and adapted the notation appropriately.
[2]This form of the extra stress is a prototype for a class of models, which is capable to explain many experimental observations (cf. Rajagopal, Růžička [9] or Růžička [11] for details).
[3]Conditions for α_{ij} and p that ensure the validity of (1.5) and (1.6) can be found in Růžička [11] Chapter 1.

Before we state our results, we shall introduce some notation. Let $\Omega = (0, L)^3$ be a cube of given length L, and $T > 0$ a given length of the time interval $I = (0, T)$. We denote by $(L^q(\Omega), \|.\|_q)$ and $(W^{k,q}(\Omega), \|.\|_{k,q})$, $q \in [1, \infty], k \in \mathbb{N}$, the usual Lebesgue and Sobolev spaces of periodic functions with mean value zero. The space of divergence free smooth functions is denoted by $\mathcal{V}$. The closure of $\mathcal{V}$ in the $\|.\|_2$-norm and the $\|\nabla.\|_q$-norm, resp. is labeled H and V_q, resp. We use the notation $L^q(I, X(\Omega))$ for Bochner spaces with values in some function space over Ω. We also need Lebesgue and Sobolev spaces with variable exponents, which are denoted by $L^{p(x)}(\Omega)$ and $W^{k,p(x)}(\Omega)$, respectively. For given $p(x) \in L^\infty(\Omega), 1 < p_\infty \le p(x) \le p_0 < \infty$, we define the modular

$$|f|_{p(x)} \equiv \int_\Omega |f(x)|^{p(x)}\, dx\,,$$

which can be used to define a norm on the generalized Lebesgue space

$$L^{p(x)}(\Omega) = \{f \in L^1(\Omega)\,; \quad |\lambda f|_{p(x)} < \infty \text{ for some } \lambda > 0\}\,.$$

Generalized Sobolev spaces are defined analogously. We refer to Kováčik, Rákosník [3] for a detailed treatment of these spaces.

For given $\mathbf{E} \in L^\infty(I, W^{1,\infty}(\Omega))$ we consider the system (1.2), where $\mathbf{S}$ is given by (1.3), (1.4) and satisfies (1.5), (1.6), on the time-space cylinder $Q_T = I \times \Omega$ together with an initial condition

$$\mathbf{v}(0) = \mathbf{v}_0 \qquad \text{in } \Omega\,. \tag{1.7}$$

Then we have

Theorem 1.1. *Let $\Omega = (0, L)^3$ be a given cube and assume that $T > 0, \mathbf{v}_0 \in V_2$, $\mathbf{E} \in L^\infty(I, W^{1,\infty}(\Omega))$, and $\mathbf{f} \in L^r(Q_T)$, $r = \max(p'_\infty, 2)$, are given.*

(i) *Whenever*

$$9/5 < p_\infty \le p(|\mathbf{E}|^2) \le p_0 < p_\infty + 1 \tag{1.8}$$

there exists a solution $\mathbf{v}$ of the problem (1.2), (1.7) such that

$$\begin{aligned} \mathbf{v} &\in L^\infty(I, H) \cap L^{p_\infty}(I, V_{p_\infty})\,, \\ \mathbf{D}(\mathbf{v}) &\in L^{p(|\mathbf{E}|^2)}(Q_T)\,, \end{aligned} \tag{1.9}$$

which satisfies (1.2) in the weak sense, i.e. for almost all $t \in I$ and all $\varphi \in \mathcal{V}$ we have

$$\begin{aligned} &\langle \frac{\partial \mathbf{v}}{\partial t}(t), \varphi\rangle_{W^{3,2}\cap V_2} + \int_\Omega \mathbf{S}(\mathbf{D}(\mathbf{v}(t)), \mathbf{E}(t)) \cdot \mathbf{D}(\varphi)\, dx \\ &+ \int_\Omega [\nabla \mathbf{v}(t)]\mathbf{v}(t) \cdot \varphi\, dx = \int_\Omega \mathbf{f}(t) \cdot \varphi\, dx - \chi^E \int_\Omega \mathbf{E}(t) \otimes \mathbf{E}(t) \cdot \mathbf{D}(\varphi)\, dx\,. \end{aligned} \tag{1.10}$$

(ii) *Moreover, if*

$$11/5 < p_\infty \le p(|\mathbf{E}|^2) \le p_0 < p_\infty + 4/3 \tag{1.11}$$

there exists a unique solution of the problem (1.2), (1.7) with the additional property.

$$\mathbf{v} \in L^\infty(I, V_2) \cap L^2(I, W^{2,2}(\Omega)) \cap L^{p_\infty}(I, V_{3p_\infty}) . \tag{1.12}$$

The main problem in the proof of the previous theorem consists in the identification of the limit

$$\lim_{N\to\infty} \int_0^T \int_\Omega \mathbf{S}(\mathbf{D}(\mathbf{v}^N), \mathbf{E}) \cdot \mathbf{D}(\varphi)\, dx\, dt , \tag{1.13}$$

where $\mathbf{v}^N$ is some approximate solution of (1.2). The method used here is based on Vitali's convergence theorem and the almost everywhere convergence of $\mathbf{D}(\mathbf{v}^N)$. This basic idea was initiated by Nečas [8] and developed in Málek, Nečas, Růžička [5], [6], Bellout, Bloom, Nečas [1], Málek, Nečas, Rokyta, Růžička [4] to handle situations, when monotonicity methods fail to identify the above limit. Theorem 1.1 contains the results in [5] as a special case (put $p = \text{const.}, \mathbf{E} \equiv 0$) and thus shows that the basic idea is widely applicable. It is worth noticing that unsteady problems for electrorheological fluids cannot be treated with the help of monotonicity methods even for large p_∞ due to the non-standard growth of the governing system. Besides the results of the author [11] it seems that Theorem 1.1 is the only result for parabolic systems with non-standard growth and a nonlinear right-hand side.

The next section is devoted to the proof of Theorem 1.1. In the course of this we can easily derive the following result for the steady system.

Theorem 1.2. *Let* $\Omega = (0, L)^3$ *be a given cube and assume that* $\mathbf{E} \in W^{1,\infty}(\Omega)$ *and* $\mathbf{f} \in L^{q'}(\Omega)$, $q = \min(p_\infty, 2)$, *are given. Then there exists a solution* $\mathbf{v}$ *of the steady problem (1.2) such that*

$$\begin{aligned} \mathbf{v} &\in W^{2,q}(\Omega) \cap V_{3p_\infty} , \\ \mathbf{D}(\mathbf{v}) &\in V_{p(|\mathbf{E}|^2)} , \end{aligned} \tag{1.14}$$

which satisfies the steady version of (1.10), *whenever*

$$9/5 < p_\infty \le p(|\mathbf{E}|^2) \le p_0 < 3p_\infty + 1 . \tag{1.15}$$

2. PROOF OF THEOREM 1.1

First we show the existence of solutions to the Galerkin approximation of the system (1.2) and then we derive apriori estimates, which enable the limiting process in the weak formulation (1.10) in all terms except the elliptic nonlinearity. In order to identify the limit also for this term we derive an additional apriori estimate, which ensures the almost everywhere convergence of $\mathbf{D}(\mathbf{v}^N)$ and we can apply Vitali's convergence theorem.

(i) Galerkin approximation and apriori estimates

Let $\omega_r, r \in \mathbb{N}$, be the eigenfunctions of the Stokes operator $A := -P\Delta$, i.e.,

$$A\omega_r = \lambda_r \omega_r\,, \qquad r = 1, 2, \ldots \tag{2.1}$$

The set $\{\omega_r\}_{r\in\mathbb{N}}$ forms a basis[4] in the space V_2, which is orthonormal in V_2. Let us denote by P_N the orthogonal continuous projector of H onto the linear hull of the first N eigenvectors $\omega^r, r = 1, \ldots, N$. We define

$$\mathbf{v}^N(t,x) := \sum_{r=1}^{N} c_r^N(t)\omega^r(x)\,, \tag{2.2}$$

where the coefficients $c_r^N(t)$ solve the so-called Galerkin system, for $r = 1, \ldots, N$,

$$\begin{aligned}&\frac{d}{dt}\int_\Omega \mathbf{v}^N \cdot \omega^r\,dx + \int_\Omega \mathbf{S}(\mathbf{D}(\mathbf{v}^N), \mathbf{E}) \cdot \mathbf{D}(\omega^r)\,dx + \int_\Omega [\nabla \mathbf{v}^N]\mathbf{v}^N \cdot \omega^r\,dx \\ &= \int_\Omega \mathbf{f} \cdot \omega^r\,dx - \chi^E \int_\Omega \mathbf{E} \otimes \mathbf{E} \cdot \mathbf{D}(\omega^r)\,dx\,,\end{aligned} \tag{2.3}$$

which initial condition $\mathbf{v}^N(0) = P_N(\mathbf{v}_0)$. One easily checks that the Galerkin system is solvable locally in time. The global solvability follows from the following apriori estimate

$$\sup_{t\in I} \|\mathbf{v}^N(t)\|_2^2 + \int_0^T \int_\Omega |\mathbf{D}(\mathbf{v}^N)|^{p(|\mathbf{E}|^2)} + |\nabla \mathbf{v}^N|^{p_\infty}\,dx\,dt \le c\,, \tag{2.4}$$

where the constant c depends on $\mathbf{f}, \mathbf{v}_0, \mathbf{E}$. To show this inequality we multiply the r-th equation in (2.3) by c_r^N and sum up, to obtain

$$\begin{aligned}&\frac{1}{2}\frac{d}{dt}\|\mathbf{v}^N\|_2^2 + \int_\Omega \mathbf{S}(\mathbf{D}(\mathbf{v}^N), \mathbf{E}) \cdot \mathbf{D}(\mathbf{v}^N)\,dx \\ &= \int_\Omega \mathbf{f} \cdot \mathbf{v}^N\,dx - \chi^E \int_\Omega \mathbf{E} \otimes \mathbf{E} \cdot \mathbf{D}(\mathbf{v}^N)\,dx\,.\end{aligned} \tag{2.5}$$

[4]We refer to Constantin, Foias [2] for a discussion of the properties of the Stokes operator.

From (1.5) and the pointwise inequalities

$$(1+y^2)^{\frac{q-2}{2}}y^2 \ge c(q)(y^q-1)\,, \qquad (1+y^2)^{\frac{p(|\mathbf{E}|^2)-2}{2}} \ge (1+y^2)^{\frac{p_\infty-2}{2}}$$

we deduce that the second term on the left-hand side of (2.5) is bounded from below by

$$c_2\int_\Omega (1+|\mathbf{E}|^2)\big(|\mathbf{D}(\mathbf{v}^N)|^{p(|\mathbf{E}|^2)} + |\mathbf{D}(\mathbf{v}^N)|^{p_\infty}\big)\,dx - c\int_\Omega 1+|\mathbf{E}|^2\,dx\,.$$

The terms on the right-hand side of (2.5) are bounded from above by

$$\frac{c_2}{2}\int_\Omega (1+|\mathbf{E}|^2)|\mathbf{D}(\mathbf{v}^N)|^{p_\infty}\,dx + c\,\|\mathbf{E}\|_2^2\,dx + c\,\|\mathbf{f}\|_2^2 + c\,.$$

Estimate (2.4) follows from (2.5) and the previous estimates if we use Korn's inequality and integrate over $(0,T)$. Estimate (2.4) implies that

$$\begin{aligned} \mathbf{v}^N &\text{ is bounded in } L^\infty(I,H)\cap L^{p_\infty}(I,V_{p_\infty})\,,\\ \mathbf{D}(\mathbf{v}^N) &\text{ is bounded in } L^{p(|\mathbf{E}|^2)}(Q_T)\,. \end{aligned} \tag{2.6}$$

This information together with an appropriate estimate of $\frac{\partial \mathbf{v}^N}{\partial t}$, which will be proved later on, is sufficient to pass to the limit as $N\to\infty$ in all terms of (2.3) except the elliptic nonlinearity. In order to identify the limiting element also for this term (cf. (1.13)) we need some additional information, which we shall derive next.

(ii) Additional apriori estimates

Multiplying the r-th equation of (2.3) by $\lambda_r c_r^N(t)$, where λ_r are the eigenvalues of the Stokes operator, we easily obtain

$$\begin{aligned} &\frac{1}{2}\frac{d}{dt}\|\nabla\mathbf{v}^N\|_2^2 + \int_\Omega \mathbf{S}(\mathbf{D}(\mathbf{v}^N),\mathbf{E})\cdot\mathbf{D}(A\mathbf{v}^N)\,dx \qquad\qquad (2.7)\\ &= -\int_\Omega [\nabla\mathbf{v}^N]\mathbf{v}^N\cdot A\mathbf{v}^N\,dx + \int_\Omega \mathbf{f}\cdot A\mathbf{v}^N\,dx - \chi^{\mathbf{E}}\int_\Omega \mathbf{E}\otimes\mathbf{E}\cdot\mathbf{D}(A\mathbf{v}^N)\,dx\,. \end{aligned}$$

Due to the space periodic boundary conditions we have that

$$A\mathbf{v}^N = -\Delta\mathbf{v}^N$$

and thus we compute

$$\begin{aligned} \int_\Omega \mathbf{S}(\mathbf{D}(\mathbf{v}^N),\mathbf{E})\cdot\mathbf{D}(A\mathbf{v}^N)\,dx &= \int_\Omega \frac{\partial S_{ij}(\mathbf{D}(\mathbf{v}^N),\mathbf{E})}{\partial E_k}\nabla E_k D_{ij}(\nabla\mathbf{v}^N)\,dx\\ &+ \int_\Omega \frac{\partial S_{ij}(\mathbf{D}(\mathbf{v}^N),\mathbf{E})}{\partial D_{kl}} D_{kl}(\nabla\mathbf{v}^N)D_{ij}(\nabla\mathbf{v}^N)\,dx\,, \end{aligned} \tag{2.8}$$

$$\int_\Omega [\nabla\mathbf{v}^N]\mathbf{v}^N\cdot A\mathbf{v}^N\,dx = \int_\Omega \frac{\partial v_j^N}{\partial x_k}\frac{\partial v_i^N}{\partial x_j}\frac{\partial v_i^N}{\partial x_k}\,dx\,, \tag{2.9}$$

$$-\chi^{\mathbf{E}}\int_\Omega \mathbf{E}\otimes\mathbf{E}\cdot\mathbf{D}(A\mathbf{v}^N)\,dx = 2\,\chi^{E}\int_\Omega E_i\frac{\partial E_j}{\partial x_k}D_{ij}\Big(\frac{\partial \mathbf{v}^N}{\partial x_k}\Big)\,dx\,. \tag{2.10}$$

From (1.6) follows that the second term on the right-hand side of (2.8) is bounded from below by

$$c_1\int_\Omega (1+|\mathbf{D}(\mathbf{v}^N(t))|^2)^{\frac{p(|\mathbf{E}(t)|^2)-2}{2}}|\mathbf{D}(\nabla\mathbf{v}^N(t))|^2\,dx =: c_1 J_p(\mathbf{v}^N(t))\,.$$

The term $J_p(\mathbf{v}(t))$ plays an important role and thus we establish various lower bounds.

Lemma 2.1. *There are constants depending only on* Ω *and* p *such that*

$$1+\|\nabla^2\mathbf{v}\|^{p_\infty}_{\frac{3p_\infty}{p_\infty+1}} \le c\,(1+J_p(\mathbf{v})) \qquad \text{if } p_\infty\in(1,2)\,, \tag{2.11}$$

$$\|\nabla^2\mathbf{v}\|_2^2 \le c\,J_p(\mathbf{v}) \qquad \text{if } p_\infty\ge 2\,. \tag{2.12}$$

Moreover, for $q\in[1,2]$ *and* $c=c(\Omega,p,\mathbf{E},q)$ *we have*

$$\|(1+|\mathbf{D}(\mathbf{v})|^2)^{\frac{p(|\mathbf{E}|^2)}{2}}\|_{\frac{3}{3-q}} \le c\,J_p(\mathbf{v})^{\frac{q}{2}}\|(1+|\mathbf{D}(\mathbf{v})|^2)^{\frac{p(|\mathbf{E}|^2)}{2}}\|_1^{\frac{2-q}{2}} + c\,\|(1+|\mathbf{D}(\mathbf{v})|^2)^{\frac{p(|\mathbf{E}|^2)}{2}}\|_s^s\,. \tag{2.13}$$

Proof. One easily computes for $1\le q<2$, $p_\infty<2$,

$$\begin{aligned}\|\nabla\mathbf{D}(\mathbf{v})\|_q^q &\le c\int_\Omega\{(1+|\mathbf{D}|^2)^{\frac{p_\infty-2}{2}}\mathbf{D}(\nabla\mathbf{v})\cdot\mathbf{D}(\nabla\mathbf{v})\}^{\frac{q}{2}}(1+|\mathbf{D}|^2)^{\frac{2-p_\infty}{2}\frac{q}{2}}\,dx\\ &\le c\,J_p(\mathbf{v})^{\frac{q}{2}}\big(1+\|\nabla\mathbf{v}\|_{\frac{2-p_\infty}{2-q}q}^{\frac{2-p_\infty}{2}q}\big)\,.\end{aligned}$$

Choosing q such that $\frac{2-p_\infty}{2-q}q=\frac{3q}{3-q}$ and using the embedding $W^{1,q}(\Omega)\hookrightarrow L^{\frac{3q}{3-q}}(\Omega)$ we easily conclude that

$$1+\|\mathbf{D}(\nabla\mathbf{v})\|_q^q \le c\big(1+J_p(\mathbf{v})\big)^{\frac{q}{2}}\big(1+\|\mathbf{D}(\nabla\mathbf{v})\|_q^q\big)^{\frac{3(2-q)}{2(3-q)}}\,,$$

with $q=\frac{3p_\infty}{p_\infty+1}$, which yields (2.11). For $p_\infty\ge 2$ we use that $1\le(1+|\mathbf{D}|^2)^{\frac{p(|\mathbf{E}|^2)-2}{2}}$ and (2.12) follows. In both cases we have used Korn's inequality. Using the inequality

$$\ln(1+x^2)\le c(\alpha)(1+x^2)^{\alpha/2}\,, \tag{2.14}$$

which holds for all $\alpha > 0$ one deduces that, for $s > 1$,

$$\|\nabla(1+|\mathbf{D}(\mathbf{v})|^2)^{\frac{p(|\mathbf{E}|^2)}{2q}}\|_q^q \le c\int_\Omega (1+|\mathbf{D}(\mathbf{v})|^2)^{\frac{p(|\mathbf{E}|^2)}{2q}s}(1+|\mathbf{E}|^2)^{\frac{q}{2}}|\nabla\mathbf{E}|^q\,dx$$
$$+c\int_\Omega (1+|\mathbf{D}(\mathbf{v})|^2)^{\frac{p(|\mathbf{E}|^2)-q}{2}}|\mathbf{D}(\nabla\mathbf{v})|^q\,dx$$
$$\le c\|(1+|\mathbf{D}(\mathbf{v})|^2)^{\frac{p(|\mathbf{E}|^2)}{2}}\|_s^s + J_p(\mathbf{v})^{\frac{q}{2}}\|(1+|\mathbf{D}(\mathbf{v})|^2)^{\frac{p(|\mathbf{E}|^2)}{2}}\|_1^{\frac{2-q}{2}}$$

which together with the embedding $W^{1,q}(\Omega) \hookrightarrow L^{\frac{3q}{3-q}}(\Omega)$ and the equivalence of the $W^{1,q}$-norm with the $(\|\cdot\|_{sq} + \|\nabla\cdot\|_q)$-norm gives (2.13). ■

From (1.3) one easily computes that

$$\left|\frac{\partial\mathbf{S}(\mathbf{D},\mathbf{E})}{\partial E_k}\right| \le c_2|\mathbf{E}|(1+|\mathbf{E}|^2)(1+|\mathbf{D}|^2)^{\frac{p(|\mathbf{E}|^2)-1}{2}}(1+\ln(1+|\mathbf{D}|^2))\,.$$

Using (2.14) we can bound the first term on the right-hand side of (2.8) by

$$c(\mathbf{E})\int_\Omega \{(1+|\mathbf{D}(\mathbf{v}^N)|^2)^{\frac{p(|\mathbf{E}|^2)-2}{2}}|\mathbf{D}(\nabla\mathbf{v}^N)|^2\}^{\frac{1}{2}}(1+|\mathbf{D}(\mathbf{v}^N)|^2)^{\frac{p(|\mathbf{E}|^2)s}{2}\frac{1}{2}}\,dx$$
$$\le \frac{c_1}{4}J_p(\mathbf{v}^N(t)) + c(\mathbf{E})\|(1+|\mathbf{D}(\mathbf{v}^N)|^2)^{\frac{p(|\mathbf{E}|^2)}{2}}\|_s^s\,. \tag{2.15}$$

The right-hand side of (2.10) is bounded by

$$\frac{c_1}{4}J_p(\mathbf{v}^N(t)) + c\int_\Omega |\nabla\mathbf{E}|^2(1+|\mathbf{D}(\mathbf{v}^N)|^2)^{\frac{2-p(|\mathbf{E}|^2)}{2}}\,dx$$
$$\le \frac{c_1}{4}J_p(\mathbf{v}^N(t)) + c(\mathbf{E})\int_\Omega (1+|\mathbf{D}(\mathbf{v}^N)|^2)^{\frac{p(|\mathbf{E}|^2)}{2}}\,dx\,, \tag{2.16}$$

while the second term on the right-hand side of (2.7) can be estimated by (cf. the definition of $J_p(\mathbf{v}(t))$ and $\|\nabla^2\mathbf{v}\|_{r'} \le c\|\mathbf{D}(\nabla\mathbf{v})\|_{r'}$)

$$\|\mathbf{f}\|_r\|\nabla^2\mathbf{v}^N\|_{r'} \le \|\mathbf{f}\|_r\, J_p(\mathbf{v}^N(t))^{\frac{1}{2}}\|(1+|\mathbf{D}(\mathbf{v}^N)|^2)^{\frac{p(|\mathbf{E}|^2)}{2}}\|_1^{\frac{2-r'}{2r'}} \tag{2.17}$$
$$\le \frac{c_1}{4}J_p(\mathbf{v}^N(t)) + c\,\|\mathbf{f}\|_r^r + c\,\|(1+|\mathbf{D}(\mathbf{v}^N)|^2)^{\frac{p(|\mathbf{E}|^2)}{2}}\|_1$$

From (2.7)–(2.10), (2.15)–(2.17) we therefore deduce (put $q = 2$ in (2.13) and use Young's inequality for the last term in (2.17))

$$\frac{1}{2}\frac{d}{dt}(1+\|\nabla\mathbf{v}^N(t)\|_2^2) + \frac{c_1}{4}J_{p(|\mathbf{E}(t)|^2)}(\mathbf{v}^N(t)) \tag{2.18}$$
$$\le c(1+\|\nabla\mathbf{v}^N(t)\|_3^3) + c(\mathbf{E})\|(1+|\mathbf{D}(\mathbf{v}^N(t))|^2)^{\frac{p(|\mathbf{E}|^2)}{2}}\|_s^s + c\,\|\mathbf{f}(t)\|_r^r\,.$$

For brevity we denote

$$H_N(t) \equiv (1 + |\mathbf{D}(\mathbf{v}^N(t))|^2)^{\frac{p(|\mathbf{E}(t)|^2)}{2}}. \tag{2.19}$$

From (1.4) and Korn's inequality we obtain that for all $q \in [1, \infty)$

$$c \, \|\nabla \mathbf{v}^N(t)\|_{qp_\infty}^{p_\infty} \le \|H_N(t)\|_q \le c\big(1 + \|\nabla \mathbf{v}^N(t)\|_{qp_0}^{p_0}\big). \tag{2.20}$$

Moreover we can rewrite (2.13), for $q = 2$, as

$$\|H_N(t)\|_3 \le c \, J_{p(|\mathbf{E}(t)|^2)}(\mathbf{v}^N(t)) + c \, \|H_N(t)\|_s^s \tag{2.21}$$

and from (2.4) we see that

$$\int_0^T \|H_N(t)\|_1 \, dt \le c(\mathbf{f}, \mathbf{v}_0, \mathbf{E}). \tag{2.22}$$

Using (2.19) and (2.21) in (2.18) yields

$$\begin{aligned} &\frac{d}{dt}(1 + \|\nabla \mathbf{v}^N(t)\|_2^2) + J_{p(|\mathbf{E}(t)|^2)}(\mathbf{v}^N(t)) + \|H_N(t)\|_3 \\ &\le c\big(1 + \|\nabla \mathbf{v}^N(t)\|_3^3 + \|H_N(t)\|_s^s + \|\mathbf{f}(t)\|_r^r\big). \end{aligned} \tag{2.23}$$

Let us distinguish the cases when (a) $p_\infty \ge 3$ and (b) $p_\infty < 3$.

(a) $p_\infty \ge 3$

In this case we see, using (2.20), that the right-hand side of (2.23) is bounded from above by

$$c\big(1 + \|H_N(t)\|_s^s + \|\mathbf{f}(t)\|_r^r\big). \tag{2.24}$$

In order to bound the L^s-norm of $H_N(t)$ we use the following interpolation inequalities

$$\begin{aligned} \|g\|_s^s &\le \|g\|_1^{\frac{3-s}{2}} \, \|g\|_3^{\frac{3(s-1)}{2}}, \\ \|g\|_{sp_0}^{sp_0} &\le \|g\|_2^{2\frac{3p_\infty - sp_0}{3p_\infty - 2}} \, \|g\|_{3p_\infty}^{3p_\infty \frac{sp_0 - 2}{3p_\infty - 2}}, \end{aligned} \tag{2.25}$$

which hold for $2 \le p_0 < 3p_\infty$. Using the splitting $s = s\alpha + s(1-\alpha)$, $\alpha \in (0,1)$ and (2.20), (2.25) we obtain

$$\begin{aligned} \|H_N(t)\|_s^s &\le \|H_N(t)\|_1^{(1-\alpha)\frac{3-s}{2}} \, \|H_N(t)\|_3^{(1-\alpha)\frac{3(s-1)}{2}} \times \\ &\quad \times \|H_N(t)\|_3^{3\alpha\frac{sp_0-2}{3p_\infty-2}} (1 + \|\nabla \mathbf{v}^N(t)\|_2^2)^{\alpha\frac{3p_\infty - sp_0}{3p_\infty - 2}} \\ &\le \frac{1}{2}\|H_N(t)\|_3 + c\|H_N(t)\|_1^{(1-\alpha)\frac{3-s}{2}\delta'} (1 + \|\nabla \mathbf{v}^N(t)\|_2^2)^{\alpha\frac{3p_\infty - sp_0}{3p_\infty - 2}\delta'}, \end{aligned} \tag{2.26}$$

where we used Young's inequality with

$$\frac{1}{\delta}+\frac{1}{\delta'}=1\,,\qquad \Big((1-\alpha)\frac{3(s-1)}{2}+3\alpha\,\frac{sp_0-2}{3p_\infty-2}\Big)\delta=1\,.$$

Now we require that

$$(1-\alpha)\frac{3-s}{2}\delta'=1\,,$$

which enables us to compute α, δ and δ'. This yields

$$\alpha=\frac{(s-1)(3p_\infty-2)}{6-2s-3s(p_0-p_\infty)}\,,\qquad \delta'=\frac{2(6-2s-3s(p_0-p_\infty))}{(3-s)(4-3sp_0+3p_\infty)}\,. \tag{2.27}$$

Note, that $\alpha\in(0,1)$ and $\delta'\in(1,\infty)$ if

$$p_0<p_\infty+4/3 \tag{2.28}$$

for s chosen appropriately near 1. Putting the above calculation together we deduce from (2.23), (2.24) and (2.26)

$$\begin{aligned}&\frac{d}{dt}(1+\|\nabla\mathbf{v}^N(t)\|_2^2)+J_{p(|\mathbf{E}(t)|^2)}(\mathbf{v}^N(t))+\|H_N(t)\|_3\\&\le c\big(1+\|\mathbf{f}\|_r^r+\|H_N(t)\|_1(1+\|\nabla\mathbf{v}^N(t)\|_2^2)^{\lambda_1}\big)\,,\end{aligned} \tag{2.29}$$

where

$$\lambda_1=2\frac{(s-1)(3p_\infty-sp_0)}{(3-s)(4-3sp_0+3p_\infty)}\,. \tag{2.30}$$

Note, that we can chose λ_1 arbitrarily small for $s\to 1$. Dividing (2.29) by $(1+\|\nabla\mathbf{v}^N(t)\|_2^2)^{\lambda_1}$ and integrating over $(0,t)$ yields

$$\begin{aligned}&\frac{1}{1-\lambda_1}(1+\|\nabla\mathbf{v}^N(t)\|_2^2)^{1-\lambda_1}\\&+\int_0^t\big(J_{p(|\mathbf{E}(\tau)|^2)}(\mathbf{v}^N(\tau))+\|H_N(\tau)\|_3\big)(1+\|\nabla\mathbf{v}^N(\tau)\|_2^2)^{-\lambda_1}\,d\tau\le c,\end{aligned} \tag{2.31}$$

where $c=c(\mathbf{f},\mathbf{v}_0,\mathbf{E})$ and where we used the notation (2.22). Taking the supremum over $t\in(0,T)$ we obtain that

$$\mathbf{v}^N \text{ is bounded in } L^\infty(I,W^{1,2}(\Omega)) \tag{2.32}$$

and consequently also (cf. (2.12), (2.19), (2.20))

$$\mathbf{v}^N \text{ is bounded in } L^2(I,W^{2,2}(\Omega))\cap L^{p_\infty}(I,V_{3p_\infty})\,. \tag{2.33}$$

(b) $p_\infty \le 3$

In this case we also have to handle the first term on the right-hand side of (2.23). We have the interpolation inequalities

$$\begin{aligned} \|g\|_3 &\le \|g\|_{p_\infty}^{\frac{p_\infty-1}{2}} \|g\|_{3p_\infty}^{\frac{3-p_\infty}{2}}, \\ \|g\|_3 &\le \|g\|_2^{\frac{2(p_\infty-1)}{3p_\infty-2}} \|g\|_{3p_\infty}^{\frac{p_\infty}{3p_\infty-2}}, \end{aligned} \tag{2.34}$$

and thus we obtain using (2.20)

$$\begin{aligned} 1+\|\nabla \mathbf{v}^N(t)\|_3^3 &\le (1+\|\nabla \mathbf{v}^N(t)\|_2^2)^{(1-\beta)\frac{3(p_\infty-1)}{3p_\infty-2}} \|H_N(t)\|_1^{3\beta\frac{p_\infty-1}{2p_\infty}} \times \\ &\quad \times \|H_N(t)\|_3^{3\frac{1-\beta}{3p_\infty-2}+3\beta\frac{3-p_\infty}{2p_\infty}} \\ &\le \frac{1}{4}\|H_N(t)\|_3 + c\|H_N(t)\|_1^{3\beta\frac{p_\infty-1}{2p_\infty}\gamma'} (1+\|\nabla \mathbf{v}^N(t)\|_2^2)^{(1-\beta)\frac{3(p_\infty-1)}{3p_\infty-2}\gamma'}, \end{aligned} \tag{2.35}$$

where we used Young's inequality with

$$\frac{1}{\gamma}+\frac{1}{\gamma'}=1, \qquad 3\Big(\frac{1-\beta}{3p_\infty-2}+\beta\frac{3-p_\infty}{2p_\infty}\Big)\gamma = 1.$$

If we now require

$$3\beta\frac{p_\infty-1}{2p_\infty}\gamma' = 1,$$

we can compute β, γ and γ'. We obtain

$$\beta = \frac{p_\infty(3p_\infty-5)}{6(p_\infty-1)}, \qquad \gamma' = \frac{4}{3p_\infty-5} \tag{2.36}$$

and one sees that $\beta \in (0,1)$ and $\gamma' \in (1,\infty)$ if $p_\infty > 5/3$. From (2.23)–(2.30) and (2.34)–(2.36) we conclude

$$\begin{aligned} &\frac{d}{dt}(1+\|\nabla \mathbf{v}^N(t)\|_2^2) + J_{p(|\mathbf{E}(t)|^2)}(\mathbf{v}^N(t)) + \|H_N(t)\|_3 \\ &\le c\big(1+\|\mathbf{f}\|_r^r + \|H_N(t)\|_1(1+\|\nabla \mathbf{v}^N(t)\|_2^2)^{\lambda_1} + (1+\|\nabla \mathbf{v}^N(t)\|_2^2)^{\lambda_2}\big) \\ &\le c\big(1+\|\mathbf{f}\|_r^r + \|H_N(t)\|_1(1+\|\nabla \mathbf{v}^N(t)\|_2^2)^{\lambda_2}\big) \end{aligned} \tag{2.37}$$

with

$$\lambda_2 = \frac{3-p_\infty}{3p_\infty-5}. \tag{2.38}$$

We used that λ_1 can be chosen arbitrarily small. Dividing (2.37) by $(1+\|\nabla \mathbf{v}^N(t)\|_2^2)^{\lambda_2}$ and integrating over $(0,t)$ yields[5]

$$\frac{1}{1-\lambda_2}(1+\|\nabla \mathbf{v}^N(t)\|_2^2)^{1-\lambda_2} \tag{2.39}$$

$$+\int_0^t \big(J_{p(|\mathbf{E}(t)|^2)}(\mathbf{v}^N(t))+\|H_N(t)\|_3\big)(1+\|\nabla \mathbf{v}^N(t)\|_2^2)^{-\lambda_2}\,dt \le c\,,$$

where $c=c(\mathbf{f},\mathbf{v}_0,\mathbf{E})$. The first term in (2.39) gives only an information if $\lambda_2 \le 1$, which is equivalent to the requirement

$$p_\infty \ge 11/5\,, \tag{2.40}$$

which is the lower bound in the second part of Theorem 1.1. If (2.40) is satisfied we again obtain (2.32), (2.33) as in the case (a). In the other case, i.e. $p_\infty < 11/5$, the first term in (2.39) is negative, but it can be moved to the right-hand side and estimated there by $(\lambda-1)^{-1}$. In this case we obtain from (2.39) that

$$\int_0^T J_{p(|\mathbf{E}(t)|^2)}(\mathbf{v}^N(t))(1+\|\nabla \mathbf{v}^N(t)\|_2^2)^{-\lambda_2}\,dt \le c(\mathbf{v}_0,\mathbf{f},\mathbf{E})\,. \tag{2.41}$$

In order to derive the final estimate, which we need, we must distinguish the following two cases $(\alpha)\, 2 \le p_\infty < 11/5$ and $(\beta)\, p_\infty < 2$.

(α) $2 \le p_\infty < 11/5$

Using estimate (2.41) and inequality (2.12) we show that, for $\gamma = \frac{p_\infty(3p_\infty-5)}{3(p_\infty^2-3p_\infty+4)}$,

$$\int_0^T \|\nabla^2 \mathbf{v}^N(t)\|_2^{2\gamma}\,dt \le c(\mathbf{f},\mathbf{v}_0,\mathbf{E})\,. \tag{2.42}$$

Indeed, the quantity on the left-hand side of (2.42) is bounded by

$$\int_0^T \Big(\frac{J_p(|\mathbf{E}(t)|^2)^{\mathbf{v}^N(t)})}{(1+\|\nabla \mathbf{v}^N(t)\|_2^2)^{\lambda_2}}\Big)^\gamma (1+\|\nabla \mathbf{v}^N(t)\|_2^2)^{\lambda_2\gamma}\,dt$$

$$\le \Big(\int_0^T \frac{J_p(|\mathbf{E}(t)|^2)^{\mathbf{v}^N(t)})}{(1+\|\nabla \mathbf{v}^N(t)\|_2^2)^{\lambda_2}}\,dt\Big)^\gamma \Big(\int_0^T (1+\|\nabla \mathbf{v}^N(t)\|_2^2)^{\lambda_2\frac{\gamma}{1-\gamma}}\,dt\Big)^{1-\gamma}$$

$$\le c(\mathbf{f},\mathbf{v}_0,\mathbf{E})\,,$$

provided that $0<\gamma<1$ and

$$2\lambda_2\frac{\gamma}{1-\gamma} = p_\infty\,,$$

[5]Note, that the first term in (2.39) should be replaced by $\ln(1+\|\nabla \mathbf{v}^N(t)\|_2^2)$ if $\lambda_2 = 1$. Let us have this in mind in all following considerations.

which immediately gives the above formula for γ. To proceed further we use the interpolation inequality

$$\|\mathbf{v}\|_{1+\sigma,q} \le \|\mathbf{v}\|_{1,p_\infty}^{1-\sigma} \|\mathbf{v}\|_{2,2}^{\sigma},$$

which holds for $0 < \sigma < 1$ and $q = \frac{2p_\infty}{2+\sigma(p_\infty-2)}$. We chose $r \in (2\gamma, p_\infty)$ and compute that, using (2.4) and (2.42),

$$\begin{aligned}\int_0^T \|\mathbf{v}^N\|_{1+\sigma,q}^r \,dt &\le \int_0^T \|\mathbf{v}^N\|_{1,p_\infty}^{r(1-\sigma)} \|\mathbf{v}^N\|_{2,2}^{r\sigma}\,dt \qquad (2.43)\\ &\le c\Big(\int_0^T \|\nabla\mathbf{v}^N(t)\|_{p_\infty}^{p_\infty}\,dt\Big)^{1/\delta}\Big(\int_0^T \|\nabla^2\mathbf{v}^N(t)\|_2^{2\gamma}\,dt\Big)^{1/\delta'}\\ &\le c(\mathbf{f},\mathbf{v}_0,\mathbf{E}),\end{aligned}$$

provided that

$$1 = \frac{1}{\delta} + \frac{1}{\delta'} = \frac{(1-\sigma)r}{p_\infty} + \frac{\sigma r}{2\beta},$$

which implies that σ is given by $\sigma = \frac{2\gamma(p_\infty - r)}{r(p_\infty-2\gamma)}$. The estimate (2.43) implies that

$$\mathbf{v}^N \text{ is bounded in } L^r(I, W^{1+\sigma,q}(\Omega)), \qquad (2.44)$$

where

$$r \in (2\gamma, p_\infty), \qquad \sigma = \frac{2\gamma(p_\infty - r)}{r(p_\infty - 2\gamma)}, \qquad q = \frac{2p_\infty}{2+\sigma(p_\infty-2)}. \qquad (2.45)$$

(β) $p_\infty < 2$

We proceed similarly as in the case (α), and show that, for $\gamma = \frac{5p_\infty-9}{3(p_\infty-1)}$,

$$\int_0^T \|\nabla^2\mathbf{v}^N(t)\|_{\frac{3p_\infty}{p_\infty+1}}^{p_\infty\gamma}\,dt \le c(\mathbf{f},\mathbf{v}_0,\mathbf{E}). \qquad (2.46)$$

Since we want that $\gamma > 0$, the lower bound for p_∞ in case (i) of Theorem 1.1 appears. Using (2.11) we have

$$\begin{aligned}&\int_0^T \|\nabla^2\mathbf{v}^N(t)\|_{\frac{3p_\infty}{p_\infty+1}}^{p_\infty\gamma}\,dt \qquad (2.47)\\ &\le c + c\Big(\int_0^T \frac{J_{p(|\mathbf{E}(t)|^2)}(\mathbf{v}^N(t))}{(1+\|\nabla\mathbf{v}^N(t)\|_2^2)^{\lambda_2}}\,dt\Big)^{\gamma}\Big(\int_0^T (1+\|\nabla\mathbf{v}^N(t)\|_2^2)^{\lambda_2\frac{\gamma}{1-\gamma}}\,dt\Big)^{1-\gamma}.\end{aligned}$$

The interpolation inequality

$$\|\mathbf{v}\|_2^2 \le \|\mathbf{v}\|_{p_\infty}^{\frac{3p_\infty-2}{2}} \|\mathbf{v}\|_{3p_\infty}^{\frac{3(2-p_\infty)}{2}}$$

together with the embedding $W^{1,\frac{3p_\infty}{p_\infty+1}}(\Omega) \hookrightarrow L^{3p_\infty}(\Omega)$ implies that

$$\begin{aligned}
&\int_0^T (1+\|\nabla \mathbf{v}^N(t)\|_2^2)^{\lambda_2 \frac{\gamma}{1-\gamma}}\,dt \qquad (2.48)\\
&\le c + c\int_0^T \Big(\|\nabla \mathbf{v}^N(t)\|_{p_\infty}^{\frac{3p_\infty-2}{2}} \|\nabla^2 \mathbf{v}^N(t)\|_{\frac{3p_\infty}{p_\infty+1}}^{\frac{3(2-p_\infty)}{2}}\Big)^{\lambda_2\frac{\gamma}{1-\gamma}}\,dt\\
&\le c + c\Big(\int_0^T \|\nabla \mathbf{v}^N(t)\|_{p_\infty}^{p_\infty}\,dt\Big)^{1/\delta}\Big(\int_0^T \|\nabla^2 \mathbf{v}^N(t)\|_{\frac{3p_\infty}{p_\infty+1}}^{p_\infty\gamma}\,dt\Big)^{1/\delta'},
\end{aligned}$$

provided that

$$1 = \frac{1}{\delta} + \frac{1}{\delta'} = \lambda_2 \frac{\gamma}{1-\gamma}\frac{3p_\infty-2}{2p_\infty} + \lambda_2 \frac{3(2-p_\infty)}{2p_\infty(1-\gamma)},$$

which implies the above formula for γ. The estimate (2.46) now follows from (2.47), (2.48) and (2.41) since

$$\frac{1-\gamma}{\delta'} = \frac{3(2-p_\infty)(3-p_\infty)}{p_\infty(3p_\infty-5)} < 1$$

holds exactly if $p_\infty > 9/5$. From (2.46) and the interpolation inequality

$$\|\mathbf{v}\|_{1+\sigma,q} \le \|\mathbf{v}\|_{1,p_\infty}^{1-\sigma}\|\mathbf{v}\|_{2,\frac{3p_\infty}{p_\infty+1}}^{\sigma},$$

which holds for $0<\sigma<1$ and $q = \frac{3p_\infty}{3+\sigma(p_\infty-2)}$, we deduce as in the case (α) that

$$\int_0^T \|\mathbf{v}^N(t)\|_{1+\sigma,q}^r\,dt \le c(\mathbf{f},\mathbf{v}_0,\mathbf{E}),$$

provided that

$$r\in(1,p_\infty), \quad \sigma = \frac{\gamma(p_\infty-r)}{r(1-\gamma)}, \quad q = \frac{3p_\infty}{3+\sigma(p_\infty-2)}. \qquad (2.49)$$

Therefore we obtain also in the case $9/5 < p_\infty < 2$ the information (2.44), with r,σ and q now satisfying (2.49). Now, it remains to derive estimates for $\frac{\partial \mathbf{v}^N}{\partial t}$ in order to conclude our proof.

(iii) Estimate of the time derivative

Let us start with some observations for the Stokes operator and the projection P_N. We define the operator $A^\alpha \mathbf{u} = \sum_{j=1}^\infty \lambda_j^\alpha(\mathbf{u},\omega_j)\omega_j$, $\alpha\in\mathbb{R}$, with domain of definition $\mathcal{D}(A^\alpha) = \{\mathbf{u}\in H, \sum_{j=1}^\infty \lambda_j^{2\alpha}(\omega_j,\mathbf{u})^2 < \infty\}$.

Note, that for space periodic boundary conditions we can characterize $\mathcal{D}(A^\alpha)$ as $W^{2\alpha,2}(\Omega) \cap V_2$. In the following we shall need the inequality

$$\|P_N \mathbf{u}\|_{3,2} \le c\|\mathbf{u}\|_{3,2} \qquad \forall \mathbf{u} \in \mathcal{D}(A^{3/2}). \tag{2.50}$$

In order to show this we first realize that for $\mathbf{u} \in \mathcal{D}(A^\alpha)$

$$A^\alpha P_N \mathbf{u} = P_N A^\alpha \mathbf{u}, \tag{2.51}$$

which follows easily from the definition of A^α and P_N. Moreover, based on the regularity properties of the Stokes system one can show that for all $\mathbf{u} \in \mathcal{D}(A^{3/2})$

$$c\,\|\mathbf{u}\|_{3,2} \le \|A^{3/2}\mathbf{u}\|_2 \le \tilde{c}\,\|\mathbf{u}\|_{3,2}. \tag{2.52}$$

The crucial point in (2.52) is that for $\mathbf{u} \in \mathcal{D}(A^{3/2})$ holds

$$\|\nabla A\mathbf{u}\|_2 = -\int_\Omega A\mathbf{u} \cdot \Delta A\mathbf{u}\,dx = \int_\Omega A\mathbf{u} \cdot A^2\mathbf{u}\,dx = \|A^{3/2}\mathbf{u}\|_2,$$

where we used the definition of A^α, $P = P^2$ and that $P(-\Delta) = -\Delta P$ is valid for space periodic boundary conditions. The second inequality in (2.52) is trivial. Now, (2.50) follows, since

$$\|P_N\mathbf{u}\|_{3,2} \le c\,\|A^{3/2}P_N\mathbf{u}\|_2 = c\,\|P_N A^{3/2}\mathbf{u}\|_2 \le c\,\|A^{3/2}\mathbf{u}\|_2 \le c\,\|\mathbf{u}\|_{3,2},$$

where we used (2.51), the continuity of P_N in $L^2(\Omega)$ and (2.52).

Since we have different information in the cases $p_\infty \ge 11/5$ and $9/5 < p_\infty < 11/5$, we distinguish these two cases.

(α) $9/5 < p_\infty < 11/5$

From the Galerkin system (2.3) we obtain for all $\varphi \in W^{3,2}(\Omega) \cap V_2$

$$\begin{aligned} |\int_\Omega \frac{\partial \mathbf{v}^N}{\partial t} \cdot \varphi\,dx| &= \int_\Omega \frac{\partial \mathbf{v}^N}{\partial t} \cdot P_N\varphi\,dx| \\ &\le |\int_\Omega [\nabla \mathbf{v}^N]\mathbf{v}^N \cdot P_N\varphi\,dx| + \int_\Omega |\mathbf{S}(\mathbf{D}(\mathbf{v}^N),\mathbf{E}) \cdot \mathbf{D}(P_N\varphi)|\,dx \\ &\quad + \int_\Omega |\mathbf{f} \cdot P_N\varphi|\,dx + |\chi^E| \int_\Omega |\mathbf{E} \otimes \mathbf{E} \cdot \mathbf{D}(P_N\varphi)|\,dx = I_1 + \ldots + I_4. \end{aligned} \tag{2.53}$$

Using the embedding $W^{3,2}(\Omega) \hookrightarrow W^{1,\infty}(\Omega)$ we easily see that

$$\int_0^T I_3 + I_4\,dt \le c\big(\|\mathbf{f}\|_{L^2(Q_T)} + \|\mathbf{E}\|^2_{L^\infty(Q_T)}\big)\|\varphi\|_{L^2(I,W^{3,2}(\Omega))} \tag{2.54}$$

Moreover, we have

$$\begin{aligned} \int_0^T I_2\,dt &\le c(\mathbf{E}) \int_0^T \int_\Omega (1 + |\mathbf{D}(\mathbf{v}^N)|)^{p_0(2.52)-1} |\nabla P_N\varphi\,dx\,dt \\ &\le c\big(1 + \|\mathbf{D}(\mathbf{v}^N)\|^{p_0-1}_{L^{p_\infty}(Q_T)}\big)\|\varphi\|_{L^{\frac{p_\infty}{p_\infty+1-p_0}}(I,W^{3,2}(\Omega))} \end{aligned} \tag{2.55}$$

provided that

$$p_0 < p_\infty + 1\,, \tag{2.56}$$

which gives the upper bound in the first part of the theorem. For the remaining term we get

$$\begin{aligned}\int_0^T I_1\,dt &= \int_0^T \Big|\int_\Omega \mathbf{v}^N \otimes \mathbf{v}^N \cdot \nabla P_N\varphi\,dx\,\Big|\,dt \\ &\le c\|\mathbf{v}^N\|^2_{L^\infty(I,L^2(\Omega))}\|\varphi\|_{L^1(I,W^{3,2}(\Omega))}\,.\end{aligned} \tag{2.57}$$

From (2.53)–(2.57), (2.4) and the assumptions on the data we conclude that

$$\frac{\partial \mathbf{v}^N}{\partial t} \text{ is bounded in } L^s(I,(W^{3,2}(\Omega)\cap V_2)^*)\,, \tag{2.58}$$

provided

$$s = \frac{p_\infty}{p_0 - 1}\,, \qquad \text{and} \qquad p_0 < p_\infty + 1\,. \tag{2.59}$$

(β) $11/5 \le p_\infty$

In this case we can also use (2.32), (2.33), which implies via the interpolation of $L^{p_\infty+4/3}(\Omega)$ between $L^2(\Omega)$ and $L^{3p_\infty}(\Omega)$ that

$$\nabla\mathbf{v}^N \text{ is bounded in } L^{\frac{3p_\infty+4}{3}}(Q_T)\,. \tag{2.60}$$

The terms I_1, I_3 and I_4 in (2.53) are treated as in the case (α). For the remaining term we have

$$\begin{aligned}\int_0^T I_2\,dt &\le c(\mathbf{E})\int_0^T\int_\Omega (1+|\mathbf{D}(\mathbf{v}^N)|)^{p_0-1}|\nabla P_N\varphi|\,dx\,dt \\ &\le c\big(1+\|\mathbf{D}(\mathbf{v}^N)\|^{p_0-1}_{L^{\frac{3p_\infty+4}{3}}(Q_T)}\big)\|\varphi\|_{L^{\frac{3p_\infty+4}{3(p_\infty-p_0)+7}}(I,W^{3,2}(\Omega))}\,,\end{aligned} \tag{2.61}$$

where we used Hölder's inequality with $\delta = \frac{3p_\infty+4}{3(p_0-1)}$. Note, that $\delta > 1$ for $p_0 < p_\infty + 4/3$, which is the upper bound in the second part of the theorem. Therefore, we obtain from (2.7), (2.54), (2.61), (2.57), (2.60) and (2.4) that (2.58) holds with

$$s = \frac{3p_\infty + 4}{3(p_0 - 1)}\,, \qquad \text{and} \qquad p_0 < p_\infty + 4/3\,. \tag{2.62}$$

(iv) Limiting process $N \to \infty$

In the preceding paragraphs we have shown that our sequence of Galerkin solutions $\mathbf{v}^N$ fulfills (2.6), (2.58), (2.59) resp. (2.62). Moreover, we have that (2.44), (2.45) resp. (2.49) is satisfied for $p_\infty \in (9/5, 11/5)$ and that (2.32), (2.33) and (2.60) hold for $p_\infty \geq 11/5$. This information implies that we can chose a subsequence, still labeled $\mathbf{v}^N$, such that

$$\begin{aligned} \mathbf{v}^N &\rightharpoonup \mathbf{v} && \text{weakly in } L^{p_\infty}(I, V_{p_\infty}), \\ \mathbf{v}^N &\rightharpoonup^* \mathbf{v} && \text{weakly}^* \text{ in } L^\infty(I, L^2(\Omega)), \\ \frac{\partial \mathbf{v}^N}{\partial t} &\rightharpoonup \frac{\partial \mathbf{v}}{\partial t} && \text{weakly in } L^s(I, (W^{3,2}(\Omega) \cap V_2)^*), \\ \mathbf{v}^N &\to \mathbf{v} && \text{strongly in } L^{p_\infty}(I, L^2(\Omega)), \end{aligned} \tag{2.63}$$

where the last line is a consequence of (2.6), (2.58) and the Aubin-Lions lemma. In the case $p_\infty \geq 11/5$ we additionally can ensure that

$$\begin{aligned} \mathbf{v}^N &\rightharpoonup \mathbf{v} && \text{weakly in } L^2(I, W^{2,2}(\Omega)) \cap L^{p_\infty}(I, V_{3p_\infty}), \\ \mathbf{v}^N &\rightharpoonup^* \mathbf{v} && \text{weakly}^* \text{ in } L^\infty(I, W^{1,2}(\Omega)). \end{aligned} \tag{2.64}$$

The convergence indicated in (2.63) is sufficient for the limiting process $N \to \infty$ in all terms of the weak formulation of (1.2), which holds for all $\varphi \in \mathcal{D}(-\infty, T, \mathcal{V} \cap H_N)$

$$\begin{aligned} &\int_0^T \int_\Omega -\mathbf{v}^N \cdot \frac{\partial \varphi}{\partial t} + \mathbf{S}(\mathbf{D}(\mathbf{v}^N), \mathbf{E}) \cdot \mathbf{D}(\varphi) + [\nabla \mathbf{v}^N] \mathbf{v}^N \cdot \varphi \, dx \, dt \\ &= \int_0^T \int_\Omega \mathbf{f} \cdot \varphi - \chi^E \mathbf{E} \otimes \mathbf{E} \cdot \mathbf{D}(\varphi) \, dx \, dt + \int_\Omega \mathbf{v}_0 \cdot \varphi \, dx \end{aligned} \tag{2.65}$$

except the nonlinear extra stress tensor. The details are omitted here, since the treatment is exactly the same as in the case of generalized Newtonian fluids (cf. Málek, Nečas, Rokyta, Růžička [4] Sections 5.2, 5.3). Finally, from (2.44), (2.45) resp. (2.49) in the case $p_\infty \in (9/5, 11/5)$ and (2.32) in the case $p_\infty \geq 11/5$, the Aubin-Lions lemma and the compact embedding

$$W^{1+\sigma,q}(\Omega) \hookrightarrow\hookrightarrow W^{1,p_\infty}(\Omega)$$

resp.

$$W^{2,2}(\Omega) \hookrightarrow\hookrightarrow W^{1,2}(\Omega)$$

we deduce that

$$\nabla \mathbf{v}^N \to \nabla \mathbf{v} \qquad \text{strongly in } L^\alpha(Q_T),$$

where $\alpha = 2$ if $p_\infty \geq 11/5$ and $\alpha = r$ with r given by (2.45) resp. (2.49). This in turn implies that $\nabla \mathbf{v}^N \to \nabla \mathbf{v}$ almost everywhere in Q_T and thus

$$\mathbf{S}(\mathbf{D}(\mathbf{v}^N), \mathbf{E}) \to \mathbf{S}(\mathbf{D}(\mathbf{v}), \mathbf{E}) \qquad \text{a.e. in } Q_T. \tag{2.66}$$

Moreover, we have that for all measurable sets $M \subseteq Q_T$

$$\begin{aligned}\int_M |\mathbf{S}(\mathbf{D}(\mathbf{v}^N),\mathbf{E})|\,dx\,dt &\le c_1(\mathbf{E})\int_M (1+|\mathbf{D}(\mathbf{v}^N)|)^{p_0-1}\,dx\,dt \\ &\le c\Big(\int_0^T\int_\Omega (1+|\mathbf{D}(\mathbf{v}^N)|)^{p_\infty}\,dx\,dt\Big)^{\frac{p_0-1}{p_\infty}}|M|^{\frac{1}{s'}} \\ &\le c|M|^{1/s'},\end{aligned} \tag{2.67}$$

with s given in (2.59) resp. (2.62). Vitali's convergence theorem together with (2.66) and (2.67) yields that also the limiting element for the nonlinear extra stress tensor can be identified (cf. (1.13)). Finally, one deduces from (2.65) the weak formulation (1.10) in a standard way. The uniqueness in the second part of Theorem 1.1 follows exactly as in the case of generalized Newtonian fluids (cf. Málek, Nečas, Rokyta, Růžička [4]). The proof of Theorem 1.1 is finished.

(v) Steady case

The proof in this situation follows the lines indicated in the previous paragraphs with considerable simplifications due to the fact that we do not have to guard the time integrability. Let us outline the basic steps of the proof of Theorem 1.2. Again we consider a solution $\mathbf{v}^N$ of the Galerkin system for the steady system

$$\begin{aligned}&\int_\Omega \mathbf{S}(\mathbf{D}(\mathbf{v}^N),\mathbf{E})\cdot\mathbf{D}(\boldsymbol{\omega}^r)\,dx + \int_\Omega [\nabla\mathbf{v}^N]\mathbf{v}^N\cdot\boldsymbol{\omega}^r\,dx \\ &= \int_\Omega \mathbf{f}\cdot\boldsymbol{\omega}^r\,dx - \chi^{\mathbf{E}}\int_\Omega \mathbf{E}\otimes\mathbf{E}\cdot\mathbf{D}(\boldsymbol{\omega}^r)\,dx \qquad r=1,\cdots,N.\end{aligned} \tag{2.68}$$

The energy estimate now reads (cf. (2.4))

$$\int_\Omega |\mathbf{D}(\mathbf{v}^N)|^{p(|\mathbf{E}|^2)} + |\mathbf{D}(\mathbf{v})|^{p_\infty}\,dx \le c(\mathbf{f},\mathbf{E}), \tag{2.69}$$

which implies that

$$\begin{aligned}\mathbf{v}^N &\text{ is bounded in } V_{p_\infty}, \\ \mathbf{D}(\mathbf{v}^N) &\text{ is bounded in } L^{p(|\mathbf{E}|^2)}(\Omega).\end{aligned} \tag{2.70}$$

Proceeding exactly as in the derivation of (2.23) we obtain

$$J_{p(|\mathbf{E}|^2)}(\mathbf{v}^N) + \|H_N\|_3 \le c\big(1+\|\nabla\mathbf{v}^N\|_3^3 + \|H_N\|_s^s + \|\mathbf{f}\|_r^r\big), \tag{2.71}$$

where $H_N = (1+|\mathbf{D}(\mathbf{v}^N)|^2)^{\frac{p(|\mathbf{E}|^2)}{2}}$ and $r = \max(2, p'_\infty)$. To obtain the desired information from (2.71) is now easier, since we do not need to split the powers in the terms on the right-hand side into $\alpha + (1-\alpha)$. We distinguish again the cases $(a)\, p_\infty \geq 3$ and $(b)\, p_\infty < 3$.

(a) $p_\infty \geq 3$

We obtain using (2.24), $(2.25)_1$, (2.69) and (2.20) that the right-hand side of (2.71) is bounded by

$$c\big(1 + \|\mathbf{f}\|_r^r + \|H_N\|_1^{\frac{3-s}{2}} \|H_N\|_3^{\frac{3(s-1)}{2}}\big) \leq c(\mathbf{f},\mathbf{E})\big(1 + \|H_N\|_3^{\frac{3(s-1)}{2}}\big).$$

For $s \in (1, 5/3)$ the right-hand side in this inequality can be absorbed in the left-hand side of (2.71) and thus we obtain

$$J_{p(|\mathbf{E}|^2)}(\mathbf{v}^N) + \|H_N\|_3 \leq c(\mathbf{f},\mathbf{E}), \tag{2.72}$$

which implies that (cf. Lemma 2.1)

$$\mathbf{v}^N \text{ is bounded in } W^{2,2}(\Omega) \cap V_{3p_\infty}. \tag{2.73}$$

(b) $p_\infty < 3$

In this case we treat the second term on the right-hand side of (2.71) as in the case (a) and the first term we bound using $(2.34)_1$, (2.20). Thus we have

$$J_{p(|\mathbf{E}|^2)}(\mathbf{v}^N) + \|H_N\|_3 \leq c(\mathbf{f},\mathbf{E})\big(1 + \|H_N\|_3^{\frac{3(s-1)}{2}} + \|H_N\|_3^{3\frac{3-p_\infty}{2p_\infty}}\big).$$

The exponents on the right-hand side are smaller than 1 if

$$s \in (1, 5/3) \quad \text{and} \quad p_\infty > 9/5, \tag{2.74}$$

which gives the lower bound in Theorem 1.2. Thus we obtain also in this case that (2.72) and (2.73) hold. From (2.70) and (2.73) we deduce, that we can choose a subsequence such that

$$\begin{aligned} \mathbf{v}^N &\rightharpoonup \mathbf{v} && \text{weakly in } W^{2,2}(\Omega) \cap V_{3p_\infty}, \\ \nabla\mathbf{v}^N &\to \nabla\mathbf{v} && \text{strongly in } L^2(\Omega), \\ \mathbf{D}(\mathbf{v}^N) &\rightharpoonup \mathbf{D}(\mathbf{v}) && \text{weakly in } L^{p(|\mathbf{E}|^2)}(\Omega). \end{aligned} \tag{2.75}$$

This information is sufficient for the limiting process in all terms in the weak formulation of the steady problem (2.68) if we can show the analogue of (2.67). Indeed, for all measurable sets $M \subseteq \Omega$ we have

$$\begin{aligned} \int_M |\mathbf{S}(\mathbf{D}(\mathbf{v}^N),\mathbf{E})|\, dx &\leq c_1(\mathbf{E}) \int_M (1 + |\mathbf{D}(\mathbf{v}^N)|)^{p_0-1}\, dx \\ &\leq c(1 + \|\mathbf{D}(\mathbf{v}^N)\|_{3p_\infty}^{p_0-1})|M|^{\frac{3p_\infty+1-p_0}{3p_\infty}}, \end{aligned}$$

provided that

$$p_0 < 3p_\infty + 1\,, \tag{2.76}$$

which is the upper bound from the theorem. The proof of Theorem 1.2 is finished.

Acknowledgments

I would like to thank L. Diening for his careful reading of the manuscript.

References

[1] Bellout, H., Bloom, F. and Nečas, J. (1994). Young Measure-valued Solutions for Non-Newtonian Incompressible Fluids. *Comm. Partial Differential Equations*, 19:1763–1803.

[2] Constantin, P. and Foias, C. (1980). *Navier-Stokes Equations*. University of Chicago Press.

[3] Kováčik, O. and Rákosník, J. (1991). On Spaces $L^{p(x)}$ and $W^{k,p(x)}$. *Czechoslovak Math. J.*, 41:592–618.

[4] Málek, J., Nečas, J., Rokyta, M. and Růžička, M. (1996). *Weak and Measure-valued Solutions to Evolutionary Partial Differential Equations*. Applied Mathematics and Mathematical Computation, vol. 13 Chapman and Hall, London.

[5] Málek, J., Nečas, J. and Růžička, M. (1993). On the Non-Newtonian Incompressible Fluids. *Math. Models Methods Appl. Sci*, 3:35–63.

[6] Málek, J., Nečas, J. and Růžička, M. (1999). On Weak Solutions to a Class of Non-Newtonian Incompressible Fluids in Bounded Three-dimensional Domains. The Case $p \geq 2$, Preprint SFB 256 no. 481, Adv. Differential Equations. To appear.

[7] Málek, J., Rajagopal, K.R. and Růžička, M. (1995). Existence and Regularity of Solutions and the Stability of the Rest State forFluids with Shear Dependent Viscosity. *Math. Models Methods Appl. Sci*, 5:789–812.

[8] Nečas, J. (1991). Privat Communication.

[9] Rajagopal, K.R. and Růžička, M. (1996). On the Modeling of Electrorheological Materials. *Mech. Res. Comm.*, 23:401–407.

[10] Rajagopal, K.R. and Růžička, M. (1997). Mathematical Modeling of Electrorheological Materials. *Int. J. Engng. Sci.*, submitted.

[11] Růžička, M. (1998). *Electrorheological Fluids: Modeling and Mathematical Theory*, Habilitationsschrift University Bonn.

ON DECAY OF SOLUTIONS TO THE NAVIER-STOKES EQUATIONS

Maria Elena Schonbek

Abstract: We first recall results on space-time decay of solutions to the Navier-Stokes equation in the whole space $\mathbb{R}^n$ which were developed in [9] and [1]. Next we give an example of a solution with radial vorticity to the Navier-Stokes equations in 2D, where the space-time decay rate can be computed explicitly.

Keywords: Navier-Stokes equations, space-time decay in $\mathbb{R}^n$, solutions with radial vorticity in $\mathbb{R}^2$.

1. INTRODUCTION

In this note we discuss the pointwise space time decay of solutions to the Navier-Stokes equations in the whole space $\mathbb{R}^n$, with $2 \leq n \leq 5$. We present some results that show the interplay between the space and time decay of the solutions and give an example of an explicit solution were this relation is clear. This kind of interplay is already present at the level of the solutions to the Heat equation. In particular for the Heat kernel sharp rates can be established.

The space time decay for solutions to the Navier-Stokes equations is algebraic and seems not to be as fast as for the heat kernel itself. This raises questions of what causes the loss of decay. The proofs developed in [1] and [9] for solutions to the Navier-Stokes equations will naturally also work for solutions to the Heat equation. The question stands if for solutions to the Heat equations and solutions to the Navier-Stokes equations depending on the data one could refine our results to obtain faster decay. The example we give at the end seems to indicate that this rates could be optimal.

We present only the results. For details on the proofs we refer the reader to our joint papers with T. P. Schonbek [9] and with Amrouche, Girault and T.P. Schonbek[1]. Questions of time decay of solutions

Applied Nonlinear Analysis, edited by Sequeira *et al.*
Kluwer Academic / Plenum Publishers, New York, 1999.

to the Navier-Stokes equations in different norms have been studied, among many others, by R. Kajikiya and T. Miyakawa [2], T. Kato [3], H. Kozono [4], H. Kozono and T. Ogawa [5], M.E. Schonbek [7], [8], M. Wiegner [13], and Zhang-Linghai [14]. In the direction of space-time decay of particular interest in the are the results by Takahashi [12]. In this reference, Takahashi studies the pointwise decay in space and time of the solutions, and their first derivatives, to the Navier-Stokes equations with zero initial data and an external force which decays at an algebraic rate in both space and time. In our case the data is nonzero and the external force vanishes. Our results follow by moment estimates combined with a Gagliardo-Nirenberg estimate. Specifically in [1] we show that strong solutions to the Navier-Stokes equations with data in appropriate spaces for $0 \leq k \leq n/2$:

$$|D^\alpha u(x,t)| \leq C_{k,m} \frac{1}{(t+1)^{\rho_0}(1+|x|^2)^{k/2}}$$

where $\rho_O = (1-2k/n)(m/2+\mu+n/4)$, $|\alpha| = m$, $\mu > \frac{n}{4}$ and where μ is the L^2 time rate of decay of the solution. We recall that this decay depends only on norms of the data [6], [7], [13].

In this paper we first recall the results obtained in the papers we mention above, the we discuss questions of optimality related to the rates we obtained. Finally we analyze an explicit example. This example is a solution to the Navier-Stokes in 2 dimensions with radial vorticity, which turns out to be simultaneously a solution to the Heat equation, [10] with very special data which depends on the initial vorticity. Extensions of these types of solutions can be constructed in all even space dimensions [10].

We use the notation
Let $\alpha = (\alpha_1, \ldots, \alpha_n)$ be a multi-index with $\alpha_i \geq 0$.

$$D^\alpha = \frac{\partial^{|\alpha|}}{\partial x_1^{\alpha_1} \ldots \partial x_n^{\alpha_n}}, \tag{1.1}$$

where

$$|\alpha| = \alpha_1 + \ldots + \alpha_n, \tag{1.2}$$

and

$$D_i = \frac{\partial}{\partial x_i}. \tag{1.3}$$

For any integer $m \geq 0$, we set

$$D^m f(x) = \left(\sum_{|\alpha|=m} |D^\alpha f(x)|^2 \right)^{1/2},$$

where $x = (x_1, \ldots, x_n)$. The L^2 norm (or energy norm) will be denoted by

$$\|u\| = \|u(.,t)\|_2 = [\int_{I\!R^n} |u(x,t)|^2 dx]^{1/2}, \tag{1.4}$$

where $dx = dx_1 \ldots dx_n$. More generally we denote the L^p norm for $1 \leq p < \infty$ by

$$\|u(.,t)\|_p = [\int_{I\!R^n} |u(x,t)|^p dx]^{1/p}, \tag{1.5}$$

and the L^∞ norm by

$$\|u(.,t)\|_\infty = \text{ess sup}_x |u(x,t)|. \tag{1.6}$$

The H^m norm is defined by

$$\|u(.,t)\|_{H^m} = [\int_{I\!R^n} \sum_{|\alpha| \leq m} |D^\alpha u(x,t)|^2 dx]^{1/2}. \tag{1.7}$$

For $s = 0, 1, 2, \ldots$, we define the (s, α) moments

$$M_{s,\alpha}(t) = \int_{I\!R^n} |x|^s |D^\alpha u(x,t)|^2 \, dx,$$

and in particular for $s \geq 0$, $t \geq 0$, we define the moment of order s of u by

$$M_s((u)(t)) = M_{s,0}(t) = \int_{I\!R^n} |x|^s |u(x,t)|^2 \, dx = \left(\|u(t)\|_{L^2_{s/2}} \right)^2.$$

2. POINTWISE DECAY

The main difficulty in establishing spatial-time decay is to obtain a time independent estimate for the moments of the solution and their derivatives. In the presence of such a bound the time decay of the moments is straightforward. Once the estimates on the moments are established we use a Gagliardo-Nirenberg's estimate to obtain an L^∞ algebraic time decay for $v(x,t) = (1 + |x|^2)^{k/2} D^\alpha u(x,t)$. From where the space time decay will follow. Specifically we use Gagliardo-Nirenberg inequality to show

$$|(1 + |x|^2)^{k/2} D^\alpha u(x,t)| \leq \|v(\cdot,t)\|_\infty \leq \|v(\cdot,t)\|_2^{1-a} \|D^s v(\cdot,t)\|_2^a. \tag{2.1}$$

We note that the L^2 norms on the right are nothing else than energy norms of the moments of the solution and the moments of their derivatives. Thus the decay is a consequence of the following theorem.

For its statement, we need to introduce the real numbers ν, q, r and r_1 which satisfy the relations

$$0 \le s < n, \quad 2 \le r_1 \le r, \quad 1 \le q \le \infty, \quad r > n\,, \tag{2.2}$$

Theorem 2.1. *Let $u_0 \in W^{m,r} \cap W^{m,r_1} \cap H^1(\mathbb{R}^n)^n$ with r, s, r_1 as above. Let u be a strong solution of the Navier-Stokes equations with data u_0, satisfying*

$$\|u(t)\|_2 \le C(t+1)^{-\mu} \quad \textit{where } \mu > n/4\,. \tag{2.3}$$

Then

$$\tilde{M}_{s,m}(t) \le C(t+1)^{-(2\mu+m)(1-\frac{s}{n})}\,,$$

for $m = 0, 1, 2, \ldots,$ $s = 0, 1, \ldots, n$.

Proof. For a proof we refer the reader to [1]. As remarked above that the main step in this proof is a uniform bound on the moments of the solution and derivatives. The decay will follow by an appropriately chosen Hölder inequality.

This theorem combined with (2.1) yields

Theorem 2.2. *Let $2 \le n \le 5$. We retain the assumptions of the last theorem and we consider a strong solution u of the Navier-Stokes equations with data u_0. Let $k \le n/2$. Then*

$$|D^\alpha u(x,t)| \le C_{k,m} \frac{1}{(t+1)^{\rho_0}(1+|x|^2)^{k/2}}\,, \tag{2.4}$$

where $\rho_0 = (\mu + m/2 + n/4)(1 - 2k/n)$ and $|\alpha| = m$.

Proof. See [1] and [9]. The restriction to dimensions $n = 2, 3, 4, 5$ is due to the fact that we are using decay results for the derivatives of the solutions to Navier-Stokes which were established under those conditions, [11].

3. QUESTIONS ON OPTIMALITY OF THE DECAY

In order to understand the interplay between the time and space decay of our solutions we compare the situation with solutions to the Heat equation. In particular we first recall the behavior of the Heat Kernel. It is easy to show that the fundamental solution of the Heat equation,

$$E(x,t) = (4\pi t)^{-n/2} e^{-|x|^2/4t}$$

has the following asymptotic behavior:

$$|D^\alpha E(x,t)| \le c_0 |x|^{-a} t^{-b},$$

where $a + 2b = n + m$, with $m = |\alpha|$. The proof of the last fact follows by a simple induction argument on the order of derivation. On the other hand depending on the data, solutions to the Heat equation will decay at different algebraic order. If the data is compact it is easy to show that the solution decays at the same rates as the Heat Kernel. On the other extreme if the data is constant there is no decay. A simplified version of the proofs used for solutions to the Navier-Stokes equations can also be used for solutions to the Heat equation. The question that remains is if given appropriate data this decay rate for the solutions to the Navier-Stokes equations can be improved. Even if the data is compact our method will only show the decay obtained in Theorem 2.2.

For solutions to the Navier-Stokes equations, the interplay between the time and space decay can be described as follows. Let $\mu = n/4 + \gamma$ be such that

$$\|u(t)\|_{L^2} \le C(t+1)^{\mu}.$$

Note that such type of decay in the L^2 norm can be obtained easily when the appropriate data is given. See [6], [7], [13].

The relation that holds between the space and time decay which follows by Theorem 2.2 is

$$2\rho_0 + 2k = m + n + 2\gamma - \frac{2km}{n} - \frac{4\gamma m}{n} . \tag{3.1}$$

For $k = 0$, we recover the same decay of the Heat equation, but this only gives decay in time, [11]. If $m = 0$ we have the relation $2\rho_0 + 2k = n + 2\alpha(1 - 2k/n)$, For $m \ge 0$ the decay improves since $k \le n/2$. To have the same interplay between the space and time decay as for the Heat Kernel,

$$2\rho_0 + k = m + n$$

we would need $\gamma = k/2(\frac{n+2m}{n-2k})$ and this would imply that $\gamma \to \infty$ as $k \to n/2$. Since γ determines the order of the time decay of the L^2 norm of the solution, this would be equivalent to require that there is exponential L^2 time decay for the solutions. The above comments leave open the question of optimality of the decay rates.

4. A SPECIAL EXAMPLE

We will first give an example in two spatial dimensions and then mention how to extend it to all even spatial dimensions. The main purpose of this 2D example is to show explicitly the interplay between the time and space decay. In this case we compute directly the space-time decay of the solution and show that it agrees with the one obtained in the general theorem. We are going to require less conditions on the data and thus our resulting decay will only be for the solution and not for the derivatives.

Let u(x,t) be a solution to the 2-D Navier-Stokes equation with radial vorticity. Suppose that $u(x,0) = u_o \in L^2 \cap L^1$ and $\nabla u_o \in L^2$. Let $\omega_o = curl u_o$ and $\omega_o \in L^1$. It is well known that a solution can be expressed as

$$u(x,t) = 1/r^2 \int_o^r s\omega(s)ds Ax,$$

where $x = (x_1, x_2)$, $r = {x_1}^2 + {x_2}^2$, ω is the vorticity and

$$A = \begin{pmatrix} 0 & -1 \\ 1 & 0 \end{pmatrix}.$$

One can show that in this case the nonlinear term of the solution $u\nabla u$ is a gradient of some function p. (See [10]). Thus u is a solution to both the Navier-Stokes equation and the Heat equation with data

$$u(x,0) = 1/r^2 \int_o^r s\omega_o(s)ds Ax,$$

We can bound the solution pointwise in the following manner.
Let $u = (u_1, u_2)$, then

$$|u_1(x,t)|^2 + |u_2(x,t)|^2 \leq \frac{1}{r^2}\left(\int_o^r s\omega(s)ds\right)^2 = G(r) \tag{4.1}$$

Now let $\alpha + \beta = 2$, then the right hand side of equation (4.1) can be bounded as follows

$$G(r) \leq \tag{4.2}$$

$$\leq \frac{1}{r^2}\left[\frac{1}{2\pi}\left(\int_o^{2\pi}\int_o^r s|\omega(s)|ds\right)^{\alpha}\left(\int_o^{2\pi}\int_o^r s|\omega(s)|ds\right)^{\beta}\right] \leq$$

$$\leq \frac{1}{r^2}\frac{1}{2\pi}\left(\int_{I\!R^2}|w|dxdy\right)^{\alpha}\left(\int_o^{2\pi}\int_o^r sds\right)^{(\beta)/2}\left(\int_o^{2\pi}\int_o^r s|\omega(s)|^2 ds\right)^{(\beta)/2}$$

Combining equations (4.1) and (4.2) yields

$$|u_1(x,t)|^2 + |u_2(x,t)|^2 \leq$$

$$\frac{1}{r^2}\frac{1}{2\pi}\left(\int_{I\!R^2}|w|dxdy\right)^{\alpha} r^{\beta}\left(\int_{I\!R^2}|w|^2 dxdy\right)^{(\beta)/2} \tag{4.3}$$

Since ω is the radial vorticity of a solution to the 2D Navier-Stokes equation it satisfies the Heat equation, which in our case has data in L^1. Thus we know that the vorticity is bounded in L^1, since $\nabla u_o \in L^1$. Moreover since the data $u_o \in L^2 \cap L^1$ and $\nabla u_o \in L^2$ it follows that ω decays in L^2 at a rate of $(1+t)^{-n/4-1/2}$, with $n = 2$, [11]. Thus estimate (4.3) yields

$$|u(x,t)| \leq \left(|u_1(x,t)|^2 + |u_2(x,t)|^2\right)^{1/2} \leq$$
$$Cr^{-\alpha/2}(1+t)^{-\beta/2(n/4+1/2)}$$

From the last equation we have a clear interplay between the time and space decay. Moreover we can check the relation we had obtained in (3.1). In our case we have $m = 0, \gamma = 0, \mu = n/4$ and (3.1) thus reduces to

$$2\rho_0 + 2k = n$$

This relation holds in our example, since we have $\rho_0 = \beta/2(n/4+1/2)$ and $k = \alpha/2$ thus the last relation translates for $n = 2$ into

$$\alpha + \beta = 2$$

which follows by the definition of α and β.

To extend this example to all even dimensions we use the solution constructed in [10]. We quote the theorem that gives the extension.

Theorem 4.1. *Suppose n is even and let 1. $\omega : [0,\infty) \times \mathbb{R} \to \mathbb{R}$ be such that the function $v(x,t) = \omega(|x|,t)$ is a solution of the Heat equation $v_t = \Delta v$; so*

$$\omega_t = \omega_{rr} + \frac{n-1}{r}\omega_r;$$

$$2.g(r,t) = r^{-n}\int_o^r s^{n-1}w(s,t)ds;$$

$3.A = (a_{ij})$ *is an* $n \times n$ *matrix with real entries such that* $A^2 = \lambda I$ *for some* $\lambda \in \mathbb{R}$, $x^t Ax = O$ *for all* $x \in \mathbb{R}^n$

Then the function $u(x,t) = g(|x|,t)Ax$ satisfies

a) $u_t = \Delta u$.

b) There exists a function p such that (u,p) is a solution to the incompressible Navier-Stokes equations.

Proof. See [10].

We note that the matrix A will be an $n \times n$ and will have block in the diagonal of order 2×2 which will coincide with the block for the 2D case. The rest of the matrix will have zeroes.

Since the solutions constructed in Theorem 4.1 is of similar structure as the 2D solutions with radial vorticity one can compute the space-time decay as in the example above.

References

[1] Amrouche, C., Girault, V., Schonbek, M.E. and Schonbek, T.P. *Pointwise decay of solutions and higher derivatives to Navier-Stokes equations.* Preprint.

[2] Kajikiya, R. and Miyakawa, T. (1986). *On the L^2 decay of weak solutions of the Navier-Stokes equations in $\mathbb{R}^n$.* Math. Z., 192:135–148.

[3] Kato, T. (1982). *Strong L^p solutions of the Navier-Stokes equations with applications to weak solutions.* Math. Z., 187:471–480.

[4] Kozono, H. (1989). *Weak and classical solutions of the two-dimensional magneto-hydrodynamic equations.* Tohoku Math. J., 41:471–488.

[5] Kozono, H. and Ogawa, T. (1993). *Two dimensional Navier-Stokes equations in unbounded domains.* Math. Ann., 297:1–31.

[6] Schonbek, M.E. (1985). *L^2 decay of weak solutions of the Navier-Stokes equations.* Arch. Rational Mech. Anal., 88:209–222.

[7] Schonbek, M.E. (1986). *Large time behavior of solutions to the Navier-Stokes equations.* Comm. Partial Differential Equations, 11:733–763.

[8] Schonbek, M.E. (1991). *Lower bounds of rates of decay for solutions to the Navier-Stokes equations.* J. Amer. Math. Soc., 4:423–449.

[9] Schonbek, M.E. and Schonbek, T.P. *On the boundedness and decay of moments of solutions of the Navier-Stokes equations.* Preprint.

[10] Schonbek, M.E., Schonbek, T.P. and Süli, E. (1996). *Large-time behavior of solutions to the Magneto-Hydrodynamic equations.* Math. Ann., 304(4):717–756.

[11] Schonbek M.E. and Wiegner, M. (1996). *On the decay of higher order norms of the solutions of Navier-Stokes equations.* Proceedings of the Royal Society of Edinburgh, section A -Mathematics, 126:677–685.

[12] Takahashi, S. *A weighted equation approach to decay rate estimates for the Navier-Stokes equations.* Preprint.

[13] Wiegner, M. (1987). *Decay results for weak solutions to the Navier-Stokes equations in $\mathbb{R}^n$.* J. London Math. Soc. (2), 35:303–313.

[14] Zhang Linghai. *Sharp rates of decay of global solutions to 2-dimensional Navier-Stokes equations.* Preprint.

CONVEXITY CONDITIONS FOR ROTATIONALLY INVARIANT FUNCTIONS IN TWO DIMENSIONS

Miloslav Šilhavý

Abstract: Rotationally invariant functions can be represented as functions of the (signed) singular values of their tensor arguments. In two dimensions, the paper expresses the ordinary convexity, polyconvexity, and rank 1 convexity of the rotationally invariant function in terms of its representation, with the emphasis on the functions invariant only with respect to the proper orthogonal group.

Keywords: Rank one convexity, isotropy, stored energies.

1. INTRODUCTION

Let L denote the set of all second-order tensors on a two-dimensional space, which may be identified with the set of all 2 by 2 matrices. A function $f : U \to \mathsf{R}$, $U \subset \mathsf{L}$, is said to be rotationally invariant (briefly, invariant) if for every $\mathbf{A} \in U$ and every $\mathbf{Q}, \mathbf{R}$ proper orthogonal tensors we have $\mathbf{QAR} \in U$ and

$$f(\mathbf{QAR}) = f(\mathbf{A}). \tag{1.1}$$

The representation theorem for invariant functions (Proposition 2.1) says that

$$f(\mathbf{A}) = \tilde{f}(w_1, w_2),$$

where (w_1, w_2) are the signed singular values of $\mathbf{A}$, defined as the eigenvalues (v_1, v_2), $v_1 \geq v_2$, of $\sqrt{\mathbf{AA}^T}$ if $\det \mathbf{A} \geq 0$ and as $(v_1, -v_2)$ if $\det \mathbf{A} < 0$, and where $\tilde{f}$ is a function defined on a suitable domain $W \subset \mathsf{R}^2$, symmetric, and even if $U = \mathsf{L}$. The function f is fully invariant if (1.1) holds for every orthogonal $\mathbf{Q}, \mathbf{R}$, even improper. To distinguish an invariant function from a fully invariant one, the term proper invariant is used for the former occasionally.

Applied Nonlinear Analysis, edited by Sequeira *et al.*
Kluwer Academic / Plenum Publishers, New York, 1999.

The paper gives equivalent conditions for the convexity, polyconvexity, and rank 1 convexity of f in terms of $\tilde{f}$ when either $U = \mathrm{L}^+$, the set of all second-order tensors with positive determinant, or $U = \mathrm{L}$. The reader is referred to [4] or [11] for the motivations of these concepts in the calculus of variations and elasticity; the definitions are given below. The reader is also referred to [6] for treating the convexity conditions of proper invariant functions.

In Section 3, necessary and sufficient conditions are given in terms of $\tilde{f}$ for f to be convex. The case $U = \mathrm{L}$ and f fully invariant is well known (see [3; Theorem 5.1]), the case $U = \mathrm{L}^+$, is the content of [11; Proposition 18.3.5(2)], and thus only the case $U = \mathrm{L}$ and f proper invariant is proved here. In Section 4 a necessary and sufficient condition for f to be globally polyconvex is given, which is less restrictive than BALL'S [3] sufficient condition. The difference is connected with the distinction between invariant and fully invariant functions (see the remark following Proposition 4.2). For invariant functions f on $U = \mathrm{L}^+$ the condition equivalent to the rank 1 convexity in terms of $\tilde{f}$ has been given by AUBERT [1], and this condition is extended in Section 5 to include the cases $U = \mathrm{L}$ and f proper invariant or f fully invariant. The form is somewhat different from AUBERT'S; see (5.16).

One might be tempted to think that the rank 1 convexity translates into the ordinary convexity provided appropriate variables are used. The results of Section 6 show that this is (only) partly true. Namely, it turns out that (5.16) in the case $z_1 z_2 > 0$ simplifies into the ordinary convexity in terms of $x = (x_1, x_2)$, $x_1 := w_1 + w_2$, $x_2 := w_1 w_2$ (but only on a certain angle with vertex at x, see Figure 1). The case $z_1 z_2 < 0$ in (5.16) translates into the ordinary convexity in terms of $y = (y_1, y_2)$, $y_1 := w_1 - w_2$, $y_2 := w_1 w_2$ (again only on a certain angle with vertex at y, see Figure 2). One is thus led to consider simultaneously two more representations $\tilde{f}^{\pm}$, of f, $f(\mathbf{A}) = \tilde{f}^+(x) = \tilde{f}^-(y)$. For the global rank 1 convexity this gives a rather satisfactory result: the global rank 1 convexity translates into following three conditions, to be satisfied simultaneously: **(1)** the Baker-Ericksen inequality, **(2)** ordinary convexity of $\tilde{f}^+$ along specific lines in the domain of definition of $\tilde{f}^+$, and **(3)** ordinary convexity of $\tilde{f}^-$ along specific lines in the domain of definition of $\tilde{f}^-$.

To simplify the treatment, it is mostly assumed that the functions are continuously differentiable; only where the differentiability is not under control, less stringent assumptions are made. Most of the results can be generalized to functions with lower degree of smoothness.

2. SIGNED SINGULAR VALUES

Throughout, direct vector and tensor notation is used [11]. The symbol V denotes a 2-dimensional real vector space with scalar product $\mathbf{u}\cdot\mathbf{v}$. We write $|\mathbf{u}| := (\mathbf{u}\cdot\mathbf{u})^{1/2}$ and $\mathbf{u}^2 = |\mathbf{u}|^2$. A second-order tensor $\mathbf{A}$ is a linear transformation on V with the product $\mathbf{AB}$ of two second order tensors the composition of the transformations. The scalar product of tensors is defined by $\mathbf{A}\cdot\mathbf{B} = \mathrm{tr}(\mathbf{AB}^T)$ and the associated euclidean norm is denoted $|\cdot|$. In addition to the terminology and notation of Introduction, S and S^+ denote the sets of symmetric and positive definite symmetric tensors, respectively, C^1 is the unit sphere in V and $\mathsf{G}^2 = \{x \in \mathsf{R}^2 : x_1 \geq |x_2|\}$. We furthermore denote by $\mathsf{R}_+ (\mathsf{R}_{++})$ the nonnegative (positive) half-axis; R^2_+ and R^2_{++} have obvious meanings. The direct notation is used for the derivatives of real-valued functions f on an open subset U of L; for $\mathbf{A} \in U$, the derivative $\partial_{\mathbf{A}} f(\mathbf{A})$ is in L. We use the index notation for the partial derivatives of a real-valued function $\tilde{f}$ on on an open subset U of R^2.

We denote by $(\hat{v}_1(\mathbf{A}), \hat{v}_2(\mathbf{A}))$ the ordered pair of *singular values* and by $(\hat{w}_1(\mathbf{A}), \hat{w}_2(\mathbf{A})) \in \mathsf{G}^2$ the ordered pair of *signed singular values* of $\mathbf{A} \in \mathrm{L}$, respectively, as defined in Introduction. If we say that w are the signed singular values of $\mathbf{A}$ it is always understood that $w \in \mathsf{G}^2$. The same convention applies to the (ordinary) singular values.

Let $P : \mathsf{R}^2 \to \mathsf{R}^2$ denote the linear transformation $P(w_1, w_2) = (w_2, w_1)$. An $f : D \to \mathsf{R}$, $D \subset \mathsf{R}^2$, is said to be *symmetric* if $P(D) = D$ and $f(Pw) = f(w)$ for every $w \in D$. f is said to be *even* if for every $w \in D$ also $-w \in D$ and $f(-w) = f(w)$ for every $w \in D$. f is said to be *fully even* if it is even and for every $w \in D$ also $(-w_1, w_2) \in D$ and $f(w_1, w_2) = f(-w_1, w_2)$.

2.1 Proposition. *Let $f : U \to \mathsf{R}$, where $U = \mathrm{L}^+$ or $U = \mathrm{L}$, be an invariant function. Then there exists a unique $\tilde{f} : W \to \mathsf{R}$, where $W = \mathsf{R}^2_{++}$ if $U = \mathrm{L}^+$ and $W = \mathsf{R}^2$ if $U = \mathrm{L}$, such that the following two conditions hold:*

(1) *for every $\mathbf{A} \in U$ we have $f(\mathbf{A}) = \tilde{f}(w)$ where w are the signed singular values of $\mathbf{A}$;*

(2) *$\tilde{f}$ is symmetric if $U = \mathrm{L}^+$ and $\tilde{f}$ is symmetric and even if $U = \mathrm{L}$.*

The function $\tilde{f}$ satisfying **(1)**, **(2)** *has the following additional properties:*

(3) *If $U = \mathrm{L}$ then f is fully invariant if and only if $\tilde{f}$ is fully even;*

(4) *if f is continuously differentiable then $\tilde{f}$ is continuously differentiable; if $\mathbf{A} = \mathrm{diag}(w)$, where $w \in W$, then $f(\mathbf{A}) = \tilde{f}(w)$ and $\mathbf{S} := \partial_{\mathbf{A}} f(\mathbf{A}) = \mathrm{diag}(s)$, where $s_i = \tilde{f}_i(w)$, $i = 1, 2$.*

The function $\tilde{f}$ is called the *representation* of f.

Proof. In the case $U = \mathsf{L}^+$ this is the familiar representation theorem for objective-isotropic functions, see, e.g., [11; Chapter 8], and a generalization to $U = \mathsf{L}$ is immediate. □

3. INVARIANT CONVEX FUNCTIONS

Let $\tilde{f} : D \to \mathsf{R}$ a function on a possibly nonconvex subset D of a finite-dimensional real vector space X. The vector $s \in \mathrm{X}$ is said to be a *subgradient* of $\tilde{f}$ at $w \in D$ if

$$\tilde{f}(u) - \tilde{f}(w) \geq s \cdot (u - w) \tag{3.1}$$

for every $u \in D$ [8]. The function f is said to be convex on D if f has a convex extension $f^{co} : \mathrm{co}(D) \to \mathsf{R}$ to a convex hull $\mathrm{co}(D)$ of D. A necessary and sufficient condition that f is a convex function is that f has a subgradient at every point x of D.

3.1 Proposition. *Let $f : \mathsf{R}^2_{++} \to \mathsf{R}$ be a symmetric convex function. A necessary and sufficient condition that f have a convex, symmetric, and even extension $\tilde{f} : \mathsf{R}^2 \to \mathsf{R}$ is that for each $w \in \mathsf{R}^2_{++}$ there exists a subgradient $s \in \mathsf{R}^2$ of f at w such that*

$$s_1 + s_2 \geq 0. \tag{3.2}$$

Proof. The necessity: Let s be a subgradient of the extension $\tilde{f}$ of f at $w \in \mathsf{R}^2_{++}$. Setting $u = (-w_2, -w_1)$ in (3.1) we obtain (3.2). Sufficiency: Assume that (3.2) holds. Set $D = (-\infty, 0)^2 \cup (0, \infty)^2$, denote by $\bar{f} : D \to \mathsf{R}$ the even extension of f and prove that $\bar{f}$ is a convex function on a nonconvex domain. Let $w \in \mathsf{R}^2_+$. The convexity inequality

$$\bar{f}(u) - \bar{f}(w) \geq s_1(u_1 - w_1) + s_2(u_2 - w_2) \tag{3.3}$$

holds if $u \in \mathsf{R}^2_+$ as a consequence of the hypothesis on f. To prove (3.3) when $u \in (-\infty, 0]^2$, let $z = (-u_2, -u_1)$ and note

$$\begin{aligned}\bar{f}(u) - \bar{f}(w) &= f(z) - f(w)\\ &\geq s_1 \cdot (u_1 - w_1) + s_2 \cdot (u_2 - w_2) - (s_1 + s_2)(u_1 + u_2)\\ &\geq s_1 \cdot (u_1 - w_1) + s_2 \cdot (u_2 - w_2)\end{aligned}$$

using (3.2) and $u_1 + u_2 \leq 0$. The even character of $\bar{f}$ implies (3.3) also for any $w \in (-\infty, 0]^2$ and $u \in D$. Let $\tilde{f} : \mathsf{R}^2 \to \mathsf{R}$ be the lower convex hull of $\bar{f}$ which is an extension of $\bar{f}$ and hence also of f since $\bar{f}$ is convex on D. One easily finds that $\tilde{f}$ is symmetric and even as a consequence of the properties of $\bar{f}$. □

3.2 Proposition. *Let $f : \mathsf{L} \to \mathsf{R}$ be an invariant continuously differentiable function and $\tilde{f}$ its representation. Then f is convex if and only if $\tilde{f}$ is convex.*

By mollification, the result holds also without the continuous differentiability assumption. If f is convex, then as a consequence of the convexity, symmetry, and even nature of $\tilde{f}$, $\tilde{f}_i = \tilde{f}_i(u)$ satisfy

$$(\tilde{f}_1 - \tilde{f}_2)(u_1 - u_2) \ge 0 \quad \text{and} \quad (\tilde{f}_1 + \tilde{f}_2)(u_1 + u_2) \ge 0 \tag{3.4}$$

at every $u \in \mathsf{R}^2$; if additionally f is fully invariant then $\tilde{f}$ is fully even and

$$\tilde{f}_1 u_1 \ge 0, \quad \tilde{f}_2 u_2 \ge 0. \tag{3.5}$$

Proof. That the convexity of f implies the convexity of $\tilde{f}$ is immediate. Assume that $\tilde{f}$ and verify that

$$f(\mathbf{G}) - f(\mathbf{F}) \ge \mathbf{S} \cdot (\mathbf{G} - \mathbf{F}) \quad \text{where} \quad \mathbf{S} = \partial_{\mathbf{A}} f(\mathbf{F}) \tag{3.6}$$

for every $\mathbf{F}, \mathbf{G} \in \mathsf{L}$. By the invariance it suffices to consider the case when $\mathbf{F} = \mathbf{U}$ is symmetric with the eigenvalues $u \in \mathsf{G}^2$. Let w be the signed singular values of $\mathbf{G}$ and set $s_i = \tilde{f}_i(u)$, $i = 1, 2$; hence $\mathbf{S} \cdot \mathbf{F} = s_1 u_1 + s_2 u_2$. As $u \in \mathsf{G}^2$ we have $u_1 + u_2 \ge 0, u_1 - u_2 \ge 0$. If both these inequalities are strict, then (3.4) imply

$$s_1 + s_2 \ge 0, \quad s_1 \ge s_2 \tag{3.7}$$

if any of the inequalities $u_1 + u_2 \ge 0, u_1 - u_2 \ge 0$ is nonstrict, then (3.7) still holds by continuity. We have generally that if (3.7) holds then

$$s \cdot w \ge \mathbf{S} \cdot \mathbf{A}. \tag{3.8}$$

If $\det \mathbf{A} \ge 0, \det \mathbf{S} \ge 0$ or $\det \mathbf{A} \le 0, \det \mathbf{S} \le 0$ then w, s are the (ordinary) singular values of $\mathbf{A}, \mathbf{S}$, respectively and (3.8) follows from the trace inequality of von Neumann (see [3; Lemmas 5.1 & 5.2]). (ii) If $\det \mathbf{S} < 0, \det \mathbf{A} > 0$ then $\mathbf{A} = \mathbf{R}_0 \mathbf{V}$ where $\mathbf{R}_0$ is proper orthogonal, $\mathbf{V} \in \mathsf{S}^+$ and w are the ordered eigenvalues of $\mathbf{V}$. Hence $\mathbf{S} \cdot \mathbf{A} = \mathbf{S} \cdot (\mathbf{R}_0 \mathbf{V})$ and the inequality $\mathbf{S} \cdot (\mathbf{R}_0 \mathbf{V}) \le s \cdot w$ is the content of the direct implication of [11; Proposition 18.3.2(**2**)]. (iii) the case $\det \mathbf{S} > 0, \det \mathbf{A} < 0$ is treated similarly. By (3.8),

$$f(\mathbf{G}) - f(\mathbf{F}) - \mathbf{S} \cdot (\mathbf{G} - \mathbf{F}) \ge \tilde{f}(w) - \tilde{f}(u) - s_1(w_1 - u_1) - s_2(w_2 - u_2)$$

and the last term is nonnegative by the convexity of $\tilde{f}$. □

3.3 Proposition. *Let $f : \mathsf{L}^+ \to \mathsf{R}$ be an invariant function and $\tilde{f} : \mathsf{R}^2_{++} \to \mathsf{R}$ its representation. Then f is convex if and only if $\tilde{f}$ is convex and for every $w \in \mathsf{R}^2_{++} \cap \mathsf{G}^2$ there exists a subgradient s of $\tilde{f}$ at w such that* (3.7) *hold.*

The result is a specialization of [11; Proposition 18.3.5(**2**)] to the case $\dim \mathsf{V} = 2$.

4. INVARIANT POLYCONVEX FUNCTIONS

A function $f : U \to \mathsf{R}$, where $U \subset \mathsf{L}$, is said to be *polyconvex at* $\mathbf{F} \in U$ if there exists a $\mathbf{C} \in \mathsf{L}$ and a $c \in \mathsf{R}$ such that

$$f(\mathbf{G}) - f(\mathbf{F}) \geq \mathbf{C}\cdot(\mathbf{G}-\mathbf{F}) + c(\det\mathbf{G} - \det\mathbf{F}) \equiv \bar{\mathbf{C}}\cdot(\mathbf{G}-\mathbf{F}) + c\det(\mathbf{G}-\mathbf{F}) \tag{4.1}$$

for every $\mathbf{G} \in U$ where $\bar{\mathbf{C}} = \mathbf{C} + c\operatorname{cof}\mathbf{F}$. The function f is said to be *polyconvex* if it is polyconvex at every point of its domain. If f is differentiable at $\mathbf{F}$, one obtains

$$\mathbf{S} = \mathbf{C} + c\operatorname{cof}\mathbf{F} = \bar{\mathbf{C}}, \tag{4.2}$$

where $\mathbf{S} = \mathbf{S}(\mathbf{F}) = \partial_{\mathbf{F}} f(\mathbf{F})$. If additionally f is invariant with the representation $\tilde{f}$ and $\mathbf{F} = \mathbf{U} \in \mathsf{S}$ has the eigenvalues $w = (w_1, w_2) \in \mathsf{G}^2$, then $\mathbf{U}$ and $\mathbf{S}$ are diagonal in the basis of eigenvectors of $\mathbf{U}$ and (4.2) implies that also $\mathbf{C}$ is diagonal. Denoting $\mathbf{U} = \operatorname{diag}(w)$, $\mathbf{S} = \operatorname{diag}(s)$, $\mathbf{C} = \operatorname{diag}(c)$, $s_i = \tilde{f}_i(w)$, $i = 1, 2$, we have

$$s_1 = c_1 + cw_2, \quad s_2 = c_2 + cw_1. \tag{4.3}$$

4.1 Proposition. *Let* $f : U \to \mathsf{R}$, *where* $U = \mathsf{L}^+$ *or* $U = \mathsf{L}$, *be an invariant continuously differentiable function with the representation* $\tilde{f} : W \to \mathsf{R}$, *where* $W = \mathsf{R}^2_{++}$ *if* $U = \mathsf{L}^+$ *and* $W = \mathsf{R}^2$ *if* $U = \mathsf{L}$, *let* $\mathbf{F} \in U$ *have the signed singular values* w, *and let* $s_i = \tilde{f}_i(w)$, $i = 1, 2$. *Then* f *is polyconvex at* $\mathbf{F}$ *if and only if there exists a constant* $c \in \mathsf{R}$ *satisfying*

$$s_1 + s_2 \geq c(w_1 + w_2), \quad c(w_1 - w_2) \geq s_2 - s_1 \tag{4.4}$$

such that

$$\tilde{f}(\bar{w}) - \tilde{f}(w) \geq s_1 z_1 + s_2 z_2 + c z_1 z_2 \tag{4.5}$$

for every $\bar{w} \in W$, *where* $z = \bar{w} - w$. *If additionally* f *is fully invariant, then* (4.5) *implies that*

$$s_1 w_1 \geq 0, \quad s_2 w_2 \geq 0. \tag{4.6}$$

If $w_1 + w_2 \neq 0, w_1 - w_2 \neq 0$, (4.4) can be rewritten as

$$\frac{s_1 + s_2}{w_1 + w_2} \geq c \geq -\frac{s_1 - s_2}{w_1 - w_2}. \tag{4.7}$$

As a consequence we have the Baker-Ericksen inequality $s_1 w_1 - s_2 w_2 \geq 0$.

Proof. It suffices to consider the case $\mathbf{F} = \mathbf{U}$ symmetric with eigenvalues w. Let f be polyconvex at $\mathbf{U}$; the discussion preceding the proposition shows that $\mathbf{C}$ in (4.1) is diagonal with diagonal elements c_1, c_2; let finally c be as in (4.1). Testing (4.1) on $\mathbf{G} = \operatorname{diag}(-w_2, -w_1)$, $\mathbf{G} = \operatorname{diag}(w_2, w_1)$, and using that $f(\mathbf{G}) = f(\mathbf{U})$, one finds that $(c_1 + c_2)(w_1 + w_2) \geq 0$, $(c_1 - c_2)(w_1 - w_2) \geq 0$, respectively. We have $w_1 + w_2 \geq 0$, $w_1 - w_2 \geq 0$ as a consequence of $w \in \mathsf{G}^2$. If the first or the second of these inequalities is strict then

$$c_1 + c_2 \geq 0, \quad c_1 \geq c_2, \tag{4.8}$$

respectively and (4.3) gives (4.4). If if $w_1 = -w_2$ or $w_1 = w_2$ then (4.4) holds again by the symmetry of the first derivatives of symmetric functions. Let $\bar{w} \in W$, set $\mathbf{G} = \operatorname{diag}(\bar{w})$ and write (4.1) for this $\mathbf{G}$ to obtain

$$\tilde{f}(\bar{w}) - \tilde{f}(w) \geq c_1(\bar{w}_1 - w_1) + c_2(\bar{w}_2 - w_2) + c(\bar{w}_1\bar{w}_2 - w_1 w_2). \tag{4.9}$$

Using (4.3), this can be rewritten as (4.5), which completes the proof of the direct implication. Conversely, assume that there exists a c such that (4.4) and (4.5) hold. Defining c_1, c_2 by (4.3) we find that (4.4) imply (4.8); moreover (4.5) may be rewritten as (4.9). Define $\mathbf{C} = \operatorname{diag}(c)$ and prove (4.1). Given a $\mathbf{G} \in U$ with singular values $\bar{w}$, (3.8) and (4.8) say $\mathbf{C} \cdot \mathbf{G} \leq c_1 \bar{w}_1 + c_2 \bar{w}_2$ and

$$\begin{aligned} f(\mathbf{G}) - f(\mathbf{U}) &\geq c_1(\bar{w}_1 - w_1) + c_2(\bar{w}_2 - w_2) + c(\bar{w}_1\bar{w}_2 - w_1 w_2) \\ &\geq \mathbf{C} \cdot (\mathbf{G} - \mathbf{U}) + c(\det \mathbf{G} - \det \mathbf{U}). \end{aligned}$$

This completes the proof of the main statement. If additionally f is fully invariant, then the application of (4.5) to $\bar{w} = (-w_1, w_2)$ leads to $(4.6)_1$. Inequality $(4.6)_2$ is proved similarly. □

Let us now proceed to the global polyconvexity. Let $f : U \to \mathsf{R}$, $U \subset \mathsf{L}$ be polyconvex, define $V = \{(\mathbf{F}, \delta) \in \mathsf{L} \times \mathsf{R} : \mathbf{F} \in U, \delta = \det \mathbf{F}\}$ and $g : V \to \mathsf{R}$ by $g(\mathbf{F}, \det \mathbf{F}) = f(\mathbf{F})$, $\mathbf{F} \in U$. If $\mathbf{F} \in U$ and $\mathbf{C}, c$ are as in (4.1), then $(\mathbf{C}, c) \in \mathsf{L} \times \mathsf{R}$ is a subgradient of g at $(\mathbf{F}, \det \mathbf{F})$, and hence g is a convex function on a nonconvex domain. Thus g has a convex extension $h : Z \to \mathsf{R}$ where Z is the convex hull of V. Hence the function f is polyconvex if and only if there exists a convex function $h : Z \to \mathsf{R}$ defined on the convex hull of V such that $f(\mathbf{F}) = h(\mathbf{F}, \det \mathbf{F})$ for every $\mathbf{F} \in U$. If $U = \mathsf{L}^+$ then $Z = \mathsf{L} \times \mathsf{R}^2_{++}$ and if $U = \mathsf{L}$ then $Z = \mathsf{L} \times \mathsf{R}$ (see [3; Theorem 4.3]).

4.2 Proposition. *Let $f : U \to \mathsf{R}$, where $U = \mathsf{L}^+$ or $U = \mathsf{L}$, be an invariant continuously differentiable function. Then f is polyconvex if and only if there exists a convex function $\tilde{h} : T \to \mathsf{R}$, where $T = \mathsf{R}^3_{++}$ if $U = \mathsf{L}^+$ and $T = \mathsf{R}^3$ if $U = \mathsf{L}$, such that the following three conditions hold simultaneously:*

(1) *$\tilde{h}(w_1, w_2, \delta) = \tilde{h}(w_2, w_1, \delta)$ for every $(w_1, w_2, \delta) \in T$; furthermore, if $U = \mathsf{L}$ then $\tilde{h}(-w_1, -w_2, \delta) = \tilde{h}(w_1, w_2, \delta)$ for every $(w_1, w_2, \delta) \in T$;*

(2) *for every* $\mathbf{F} \in U$, $f(\mathbf{F}) = \tilde{h}(w_1, w_2, \det \mathbf{F})$ *where* w *are the signed singular values of* $\mathbf{F}$;

(3) *if* $U = \mathsf{L}^+$, *then at every* $(w_1, w_2, \delta) \in T$, $\tilde{h}$ *has a subgradient* $(s_1, s_2, \gamma) \in \mathsf{R}^3$ *satisfying* (3.2).

In the case $U = \mathsf{L}^+$, BALL [3; Theorem 5.2] gave a sufficient condition for polyconvexity which amounts to **(1)**–**(3)** above with (3.2) strengthened to assert that $s_1 \geq 0$, $s_2 \geq 0$.

Proof. Let f be polyconvex, let g, V be as in the discussion preceding the proposition, and let $h : Z \to \mathsf{R}$ be the lower convex hull of g, which by the polyconvexity hypothesis is an extension of g. Hence $f(\mathbf{F}) = h(\mathbf{F}, \det \mathbf{F})$, $\mathbf{F} \in U$. One easily finds that $h(\cdot, \delta)$ is an invariant function depending parametrically on δ. Moreover, h is convex. The invariant function $h : Z \to \mathsf{R}$ has by Proposition 2.1 a representation $\bar{h} : X \to \mathsf{R}$, where $X = \mathsf{R}^2 \times \mathsf{R}_{++}$ if $Z = \mathsf{L} \times \mathsf{R}_{++}$ and $X = \mathsf{R}^3$ if $Z = \mathsf{L} \times \mathsf{R}$, such that $h(\mathbf{F}, \delta) = \bar{h}(w_1, w_2, \delta)$ for each $(\mathbf{F}, \delta) \in Z$, where (w_1, w_2) are the signed singular values of $\mathbf{F}$, and $\bar{h}$ is symmetric and even with respect to the first two arguments. If $U = \mathsf{L}^+$ let $\tilde{h}$ be the restriction of $\bar{h}$ to R^3_{++} and if $U = \mathsf{L}$ let $\tilde{h} = \bar{h}$. Then $\tilde{h}$ is convex and satisfies **(1)**–**(3)**. Here the convexity of $\tilde{h}$ and **(3)** follow from the convexity of h and Proposition 3.3 if $U = \mathsf{L}^+$ and from Proposition 3.2 if $U = \mathsf{L}$, using that **(3)** is vacuous in this case. It is also noted that the validity of Propositions 3.3 and 3.2 extends to the present case of convex functions depending jointly on $\mathbf{F}$ and the scalar parameter δ. Furthermore, **(1)** is implied by the representation theorem and **(2)** follows from the construction. Assume conversely that there exists a convex function $\tilde{h}$ satisfying **(1)**–**(3)** and prove that f is polyconvex. If $U = \mathsf{L}^+$ one first extends $\tilde{h} : \mathsf{R}^3_{++}$ into a convex, symmetric and even function $\bar{h} : \mathsf{R}^2 \times \mathsf{R}_{++} \to \mathsf{R}$ using Proposition 3.1. Here the symmetry and the even nature concerns the dependence on the first two arguments when the third is held fixed (as in **(1)**) while the convexity is joint with respect to all arguments. It is again noted that Proposition 3.1 extends to the present case. If $U = \mathsf{L}$ set $\bar{h} = \tilde{h}$. The function $\bar{h}$ defines, via the representation theorem 2.1, a function $h : Z \to \mathsf{R}$ which is invariant (with respect to the first argument) and convex by Proposition 3.2. This function satisfies $f(\mathbf{F}) = h(\mathbf{F}, \det \mathbf{F})$ for every $\mathbf{F} \in U$ and thus f is polyconvex. □

5. INVARIANT RANK 1 CONVEX FUNCTIONS

Throughout the rest of the paper, we set $I_1 = w_1 + w_2$, $I_2 = w_1 w_2$, $\bar{I}_1 = \bar{w}_1 + \bar{w}_2$, $\bar{I}_2 = \bar{w}_1 \bar{w}_2$ where $w_1, w_2, \bar{w}_1, \bar{w}_2$ have the current local meaning specified by the surrounding text.

5.1 Remark. *Let* $\mathbf{U} \in \mathrm{S}$ *have the eigenvalues* $w \in \mathrm{G}^2$, *let* $\mathbf{n} \in \mathrm{C}^1, \mathbf{a} \in \mathrm{V}$, *set* $\mathbf{F} := \mathbf{U} + \mathbf{a} \otimes \mathbf{n}$ *and let* $\bar{w} \in \mathrm{G}^2$. *Then* $\bar{w} \in \mathrm{G}^2$ *are the signed singular values of* $\mathbf{F}$ *if and only if*

$$|I_1 \mathbf{n} + \mathbf{a}| = \bar{I}_1, \quad (\operatorname{cof} \mathbf{U})\mathbf{n} \cdot \mathbf{a} = \bar{I}_2 - I_2. \tag{5.1}$$

Proof. Equation $(5.1)_2$ is immediate. Evaluating $\mathbf{F}\mathbf{F}^T$ one finds $\bar{w}_1^2 + \bar{w}_2^2 = w_1^2 + w_2^2 + 2\mathbf{U}\mathbf{a} \cdot \mathbf{n} + \mathbf{a}^2$ and the Cayley-Hamilton theorem gives $\mathbf{U}\mathbf{a} \cdot \mathbf{n} - I_1 \mathbf{a} \cdot \mathbf{n} + (\operatorname{cof} \mathbf{U})\mathbf{a} \cdot \mathbf{n} = 0$. This gives $(5.1)_1$. The converse implication follows by reversing the steps in the above argument. □

5.2 Remark. *Let* $\mathbf{U} \in \mathrm{S}$ *have the eigenvalues* $w \in \mathrm{G}^2$. *Then* $\bar{w} \in \mathrm{G}^2$ *are the signed singular values of some rank 1 perturbation of* $\mathbf{U}$ *if and only if*

$$|w_2| \le \bar{w}_1 < \infty, \quad |\bar{w}_2| \le w_1. \tag{5.2}$$

This is well-known [2]. The set of all $\bar{w} \in \mathrm{G}^2$ satisfying (5.2) is the union of the two shaded regions in Figures 1(a) and 2(a).

We shall now exhibit special rank 1 perturbations.

5.3 Proposition. *Let* $\mathbf{U} \in \mathrm{S}$ *have the eigenvalues* $w \in \mathrm{G}^2$, *let* $\bar{w} \in \mathrm{G}^2$, $\bar{w} \ne w$, *satisfy inequalities* (5.2), *set*

$$z_1 = \bar{w}_1 - w_1, \quad z_2 = \bar{w}_2 - w_2, \tag{5.3}$$

and let $(\mathbf{a}, \mathbf{n}) \in \mathrm{V} \times \mathrm{C}^1$ *be any pair of vectors satisfying*

$$\left.\begin{aligned} n_1^2 &= \frac{z_1 z_2}{(z_1 + z_2)(w_2 - w_1)} + \frac{z_1}{z_1 + z_2}, \\ n_2^2 &= \frac{z_1 z_2}{(z_1 + z_2)(w_1 - w_2)} + \frac{z_2}{z_1 + z_2}, \\ \mathbf{a} &= (z_1 + z_2)\mathbf{n} \end{aligned}\right\} \text{ if } z_1 z_2 > 0, \tag{5.4}$$

$$n_1^2 = \frac{z_1}{z_1 + z_2}, \quad n_2^2 = \frac{z_2}{z_1 + z_2}, \quad \mathbf{a} = (z_1 + z_2)\mathbf{n} \quad \text{if} \quad z_1 z_2 = 0, \tag{5.5}$$

and

$$\left.\begin{aligned} n_1^2 &= \frac{z_1 z_2}{(z_1 - z_2)(w_1 + w_2)} + \frac{z_1}{z_1 - z_2}, \\ n_2^2 &= \frac{z_1 z_2}{(z_2 - z_1)(w_1 + w_2)} + \frac{z_2}{z_2 - z_1}, \\ \mathbf{a} &= (z_1 - z_2)(n_1, -n_2) \end{aligned}\right\} \text{ if } z_1 z_2 < 0, \tag{5.6}$$

where n_1, n_2 *denote the components of* $\mathbf{n}$ *in any basis of eigenvectors of* U. *Then* $\mathbf{U} + \mathbf{a} \otimes \mathbf{n}$ *has the signed singular values* $\bar{w}$.

It is a part of the assertion that the denominators in (5.4)–(5.6) are nonzero, that the expressions giving the squares of n_1, n_2 are nonnegative, and that these expressions give a unit vector. Note, e.g., that if $w_1 = w_2 > 0$, then (5.2) implies that $z_1 \geq 0, z_2 \leq 0$, hence $z_1 z_2 \leq 0$; moreover, combining with $\bar{w} \neq w$ we find that $z_1 - z_2 \neq 0$. Inequality $z_1 z_2 \leq 0$ implies that only (5.6) applies in this case, and thus the formulas (5.4) with vanishing denominators do not intervene.

Proof. The proof is left to the reader. See also [13; Section 4] for the analogue if $n = \dim \mathrm{V}$ is arbitrary. □

5.4 Lemma. *Let* $\mathbf{U} \in \mathrm{S}$ *have the eigenvalues* $w \in \mathrm{G}^2$, *let* $\bar{w} \in \mathrm{G}^2$ *satisfy inequalities* (5.2), *and set*

$$M := \max\{\mathbf{a} \cdot \mathbf{n} : \hat{w}_1(\mathbf{U} + \mathbf{a} \otimes \mathbf{n}) = \bar{w}_1, \hat{w}_2(\mathbf{U} + \mathbf{a} \otimes \mathbf{n}) = \bar{w}_2\}. \tag{5.7}$$

Then

$$M = \begin{cases} z_1 + z_2 & \text{if } z_1 z_2 \geq 0, \\ \dfrac{2 z_1 z_2}{w_1 + w_2} + z_1 + z_2 & \text{if } z_1 z_2 < 0, \end{cases} \tag{5.8}$$

where $z = \bar{w} - w$. *If* $\bar{w} \neq w$, *the maximum is realized on any pair* $(\mathbf{a}, \mathbf{n}) \in \mathrm{V} \times \mathrm{C}^1$ *described in Proposition 5.3.*

That the denominator in (5.8) is nonzero is part of the assertion: if $w_1 + w_2 = 0$ then the case $z_1 z_2 < 0$ cannot occur as a consequence of (5.2). We denote by $R(z_1, z_2)$ the right-hand side of (5.8) considered as function of z_1, z_2 arising from those $\bar{w}, w$ which satisfy (5.2).

Proof. By Remark 5.1, M is the maximum of $\mathbf{a} \cdot \mathbf{n}$ over all $(\mathbf{a}, \mathbf{n}) \in \mathrm{V} \times \mathrm{C}^1$ which satisfy (5.1). Let us first prove that

$$\mathbf{a} \cdot \mathbf{n} \leq z_1 + z_2 \tag{5.9}$$

and if additionally $w_1 + w_2 \neq 0$ then also

$$\mathbf{a} \cdot \mathbf{n} \leq \frac{2 z_1 z_2}{w_1 + w_2} + z_1 + z_2 \tag{5.10}$$

for every $(\mathbf{a}, \mathbf{n}) \in \mathsf{V} \times \mathsf{C}^1$ satisfying (5.1). Taking the square of $(5.1)_1$,

$$I_1^2 + 2I_1 \mathbf{a} \cdot \mathbf{n} + \mathbf{a}^2 = \bar{I}_1^2 \tag{5.11}$$

and combining with $(\mathbf{a} \cdot \mathbf{n})^2 \leq \mathbf{a}^2$ we obtain (5.9). To prove (5.10), we prove preliminarily

$$\hat{\mathbf{a}} \cdot \mathbf{n} \leq z_1 - z_2, \tag{5.12}$$

where in the basis of eigenvectors of $\mathbf{U}$, $\hat{\mathbf{a}} := (a_1, -a_2)$ provided $\mathbf{a} = (a_1, a_2)$. The constraint equation $(5.1)_2$ can be written in the following two equivalent forms:

$$w_2 \mathbf{a} \cdot \mathbf{n} + (w_1 - w_2) a_2 n_2 = \bar{I}_2 - I_2, \quad w_1 \mathbf{a} \cdot \mathbf{n} - (w_1 - w_2) a_1 n_1 = \bar{I}_2 - I_2,$$

from which

$$(w_1 + w_2) \mathbf{a} \cdot \mathbf{n} = 2\bar{I}_2 - 2I_2 + (w_1 - w_2) \hat{\mathbf{a}} \cdot \mathbf{n}. \tag{5.13}$$

Eliminating $\mathbf{a} \cdot \mathbf{n}$ via (5.11) we obtain

$$(w_1 - w_2)^2 + 2(w_1 - w_2) \hat{\mathbf{a}} \cdot \mathbf{n} + \mathbf{a}^2 = (\bar{w}_1 - \bar{w}_2)^2.$$

Combining with $(\hat{\mathbf{a}} \cdot \mathbf{n})^2 \leq \hat{\mathbf{a}}^2 = \mathbf{a}^2$ and taking the square root gives $w_1 - w_2 + \hat{\mathbf{a}} \cdot \mathbf{n} \leq \bar{w}_1 - \bar{w}_2$ which yields (5.12). An elimination of $\hat{\mathbf{a}} \cdot \mathbf{n}$ from (5.13) via (5.12) leads to (5.10). Combining inequalities (5.9) and (5.10) with the definition of R, we see that we have proved that $R(z_1, z_2)$ is an upper bound for M. A calculation that is left to the reader shows that this bound is attained on any pair $(\mathbf{a}, \mathbf{n})$ of vectors described in Proposition 5.3. □

A continuously differentiable function $f : U \to \mathsf{R}$ defined on an open subset U of L is said to be *rank 1 convex* at $\mathbf{F} \in U$ if

$$f(\mathbf{F} + \mathbf{a} \otimes \mathbf{n}) \geq f(\mathbf{F}) + \mathbf{S}\mathbf{n} \cdot \mathbf{a} \tag{5.14}$$

for every $\mathbf{a} \in \mathsf{V}, \mathbf{n} \in \mathsf{C}^1$ such that $\mathbf{F} + \mathbf{a} \otimes \mathbf{n} \in U$; here $\mathbf{S} = \partial_{\mathbf{F}} f(\mathbf{F})$. The function f is said to be *rank 1 convex* if it is rank 1 convex at every point of its domain.

5.5 Proposition. *Let $f : U \to \mathsf{R}$, where $U = \mathsf{L}^+$ or $U = \mathsf{L}$, be a continuously differentiable invariant function and $\tilde{f} : W \to \mathsf{R}$ its representation, where $W = \mathsf{R}^2_{++}$ if $U = \mathsf{L}^+$ and $W = \mathsf{R}^2$ if $U = \mathsf{L}$. Let $\mathbf{F} \in U$ have the signed singular values $w \in \mathsf{G}^2$ and set $s_i = \tilde{f}_i(w)$, $i = 1, 2$. Then f is rank 1 convex at $\mathbf{F}$ if and only if the following two conditions hold simultaneously:*

(1) $s_1 w_1 - s_2 w_2 \geq 0$;

(2) *for every $\bar{w} \in W \cap \mathsf{G}^2$ satisfying*

$$|w_2| \leq \bar{w}_1 < \infty, \quad |\bar{w}_2| \leq w_1 \tag{5.15}$$

we have the inequality

$$\tilde{f}(\bar{w}) - \tilde{f}(w) \geq \begin{cases} s_1 z_1 + s_2 z_2 - \dfrac{s_1 - s_2}{w_1 - w_2} z_1 z_2 & \text{if } z_1 z_2 > 0, \\ s_1 z_1 + s_2 z_2 & \text{if } z_1 z_2 = 0, \\ s_1 z_1 + s_2 z_2 + \dfrac{s_1 + s_2}{w_1 + w_2} z_1 z_2 & \text{if } z_1 z_2 < 0 \end{cases} \tag{5.16}$$

where $z = \bar{w} - w$.

5.6 Remarks.

(1) The case $U = \mathrm{L}^+$, $w_1 \neq w_2$, is due to AUBERT [1; Theorems 3.1 & 3.2].

(2) As in Proposition 5.3 and Lemma 5.4, the expressions with vanishing denominators in (5.16) never intervene because of the constraints imposed on z_1, z_2 by (5.15).

(3) If $w_1 \neq |w_2|$, (5.16) can be written equivalently

$$\tilde{f}(\bar{w}) - \tilde{f}(w) \geq s_1 z_1 + s_2 z_2 + z_1 z_2 \frac{s_2 w_1 - s_1 w_2}{w_1^2 - w_2^2} - |z_1 z_2| \frac{s_1 w_1 - s_2 w_2}{w_1^2 - w_2^2}. \tag{5.17}$$

If $w_1 = w_2 \equiv w$, we have from (5.15) $z_1 z_2 \leq 0$, and the second and the third cases of (5.16) give

$$f(\bar{w}) - f(w) \geq s(z_1 + z_2) + s z_1 z_2 / w \tag{5.18}$$

where $s_1 = s_2 \equiv s$. If $w_1 = -w_2 \equiv w > 0$, which can happen only if $U = \mathrm{L}$, then $z_1 z_2 \geq 0$, and (5.16) reduces to

$$\tilde{f}(\bar{w}) - \tilde{f}(w) \geq s(z_1 - z_2) - s z_1 z_2 / w$$

where $s_1 = -s_2 \equiv s$. If $w_1 = 0$ and hence $w_2 = 0$, inequalities (5.15) imply $z_2 = 0$ and (5.16) reduces to $\tilde{f}(\bar{w}) \geq \tilde{f}(w)$.

(4) Higher dimensional generalizations of Proposition 5.5 are treated in [13].

Proof. It suffices to consider the case when $\mathbf{F} = \mathbf{U}$ is symmetric with eigenvalues w. Let f be rank 1 convex at $\mathbf{U}$. **(1)**: The Baker-Ericksen inequality is well-known, see, e.g., [7]. **(2)**: Let $\bar{w} \in W \cap \mathrm{G}^2$ satisfy (5.15), and let us apply the rank 1 convexity inequality

$$f(\mathbf{U} + \mathbf{a} \otimes \mathbf{n}) \geq f(\mathbf{U}) + \mathbf{Sn} \cdot \mathbf{a} \tag{5.19}$$

to any $\mathbf{a}, \mathbf{n}$ such that $\mathbf{U} + \mathbf{a} \otimes \mathbf{n}$ has the signed singular values $\bar{w}$. The invariance of f ensures that the left-hand side of (5.19) is constant and equal to $\tilde{f}(\bar{w}) - \tilde{f}(w)$. Taking the maximum over all $\mathbf{a}, \mathbf{n}$ we are led to the maximum

$$N := \max\{\mathbf{Sn} \cdot \mathbf{a} : \hat{w}_1(\mathbf{U} + \mathbf{a} \otimes \mathbf{n}) = \bar{w}_1, \hat{w}_2(\mathbf{U} + \mathbf{a} \otimes \mathbf{n}) = \bar{w}_2\}.$$

Assume first that $w_1 > w_2$. In the basis of eigenvectors of $\mathbf{U}$ we have $\mathbf{S} = \operatorname{diag}(s_1, s_2)$, and one finds that

$$\mathbf{S} = c_1 \mathbf{1} + c_2 \operatorname{cof} \mathbf{U} \tag{5.20}$$

where

$$c_1 = \frac{s_1 w_1 - s_2 w_2}{w_1 - w_2}, \quad c_2 = -\frac{s_1 - s_2}{w_1 - w_2}.$$

Hence

$$\mathbf{Sn} \cdot \mathbf{a} = \frac{s_1 w_1 - s_2 w_2}{w_1 - w_2} \mathbf{a} \cdot \mathbf{n} - \frac{s_1 - s_2}{w_1 - w_2} (\bar{I}_2 - I_2), \tag{5.21}$$

where we have used $(5.1)_2$. Since the coefficient in front of $\mathbf{a} \cdot \mathbf{n}$ is nonnegative by **(1)**, we have

$$N = \frac{s_1 w_1 - s_2 w_2}{w_1 - w_2} M - \frac{s_1 - s_2}{w_1 - w_2} (\bar{I}_2 - I_2),$$

where M is given by (5.7), and Lemma 5.4 leads to the left-hand side of (5.16), which completes the proof of **(2)** in this case. If finally $w_1 = w_2 \equiv v$, one also has $s_1 = s_2 \equiv s$ and $\mathbf{Sn} \cdot \mathbf{a} = s \mathbf{a} \cdot \mathbf{n}$; invoking Lemma 5.4, we find that (5.16) again holds (cf. the text preceding the proof). This completes the proof of the direct implication. The converse implication is proved by reversing the direction of the arguments. □

For fully invariant functions on L the rank 1 convexity implies that $s_1 w_1 \geq 0$, $s_2 w_2 \geq 0$ and if this is known, it suffices to verify (5.16) only on the nonnegative part of the space of signed singular values:

5.7 Proposition. *Let $f : \mathsf{L} \to \mathsf{R}$ be a continuously differentiable fully invariant function and $\tilde{f} : \mathsf{R}^2 \to \mathsf{R}$ its representation. Let $\mathbf{F} \in \mathsf{L}$ have the (ordinary) singular values $w \in \mathsf{G}^2 \cap \mathsf{R}^2_+$ and let $s_i = \tilde{f}_i(w)$, $i = 1, 2$. Then f is rank 1 convex at $\mathbf{F}$ if and only if the following three conditions hold simultaneously:*

(1) $s_1 \geq 0$, $s_2 \geq 0$;

(2) $s_1 w_1 - s_2 w_2 \geq 0$;

(3) *for each $\bar{w} \in \mathsf{G}^2 \cap \mathsf{R}^2_+$ satisfying*

$$w_2 \leq \bar{w}_1 < \infty, \quad 0 \leq \bar{w}_2 \leq w_1$$

inequality (5.16) *holds.*

If $\det \mathbf{F} > 0$, *then equivalently f is rank 1 convex at $\mathbf{F}$ if and only if the restriction of f to L^+ is rank 1 convex and Item* **(1)** *above holds.*

Proof. Let first $\det \mathbf{F} \neq 0$ and prove the equivalent assertion at the end of the proposition. In view of the full invariance it suffices to treat the case $\mathbf{F} = \mathbf{U} \in \mathsf{S}^+$. Let f be rank 1 convex at $\mathbf{U}$ and prove **(1)**. In the basis of eigenvectors of $\mathbf{U}$ we have $\mathbf{U} = \operatorname{diag}(w_1, w_2)$, and let

$\mathbf{G} = \mathrm{diag}(-w_1, w_2)$. Note that $\mathbf{G}$ is a rank 1 perturbation of $\mathbf{U}$ (with negative determinant), the full invariance implies that $f(\mathbf{U}) = f(\mathbf{G})$ and the rank 1 convexity inequality (5.14) gives $s_1 \geq 0$. The inequality $s_2 \geq 0$ is proved similarly. Let now the restriction of f to L^+ be rank 1 convex at $\mathbf{U}$, let **(1)** hold and prove that f is rank 1 convex at $\mathbf{U}$. By hypothesis, the rank 1 convexity inequality (5.19) holds for all rank 1 perturbations of $\mathbf{U}+\mathbf{a}\otimes\mathbf{n}$ with positive determinant. Assume that $\mathbf{G} = \mathbf{U}+\mathbf{b}\otimes\mathbf{n}$ is a rank 1 perturbation of $\mathbf{U}$ with negative determinant and with the singular values $(\bar{w}_1, \bar{w}_2)$. Let $\mathbf{a}$ be given by $\mathbf{a} = \mathbf{b}+2\bar{w}_1\bar{w}_2\mathbf{U}^{-1}\mathbf{n}/(I_2|\mathbf{U}^{-1}\mathbf{n}|^2)$. Then $\mathbf{V} := \mathbf{U}+\mathbf{a}\otimes\mathbf{n}$ is a rank 1 perturbation of $\mathbf{U}$ with positive determinant and with the singular values $(\bar{w}_1, \bar{w}_2)$. The full invariance gives $f(\mathbf{G}) = f(\mathbf{V})$. We have $\mathbf{Sn}\cdot\mathbf{b} = \mathbf{Sn}\cdot\mathbf{a} - 2\bar{I}_2\mathbf{S}\mathbf{U}^{-1}\mathbf{n}\cdot\mathbf{n}/(I_2|\mathbf{U}^{-1}\mathbf{n}|^2)$ By Item **(1)**, $\mathbf{S}\mathbf{U}^{-1}\mathbf{n}\cdot\mathbf{n} \geq 0$ and hence $\mathbf{Sn}\cdot\mathbf{b} \leq \mathbf{Sn}\cdot\mathbf{a} \leq f(\mathbf{V}) - f(\mathbf{U}) = f(\mathbf{G}) - f(\mathbf{U})$ which proves the rank 1 convexity inequality for the rank 1 perturbations with negative determinant. Finally, the rank 1 perturbations with determinant 0 are obtained as limits. □

6. RANK 1 CONVEXITY IN TERMS OF INVARIANTS

Let $\mathsf{X}_\pm : \mathsf{G}^2 \to \mathsf{R}^2$ be defined by $\mathsf{X}_+(w) = x$, $\mathsf{X}_-(w) = y$ where

$$x_1 = w_1 + w_2, \quad y_1 = w_1 - w_2, \quad x_2 = y_2 = w_1 w_2. \tag{6.1}$$

The values x_1, x_2 are called the *invariants* of w and the values y_1, y_2 are called the *signed invariants* of w in this paper. The mappings $\mathsf{X}_\pm$ map G^2 bijectively onto $\mathsf{Q}_\pm$ where

$$\mathsf{Q}_+ := \{(x_1, x_2) \in \mathsf{R}^2 : x_1 \geq 0,\ x_2 \leq \tfrac{1}{4}x_1^2\},$$

$$\mathsf{Q}_- := \{(y_1, y_2) \in \mathsf{R}^2 : y_1 \geq 0, y_2 \geq -\tfrac{1}{4}y_1^2\},$$

with the inverses $\mathsf{X}_+^{-1}(x) = w$, $\mathsf{X}_-^{-1}(y) = w$ where

$$w_1 = \tfrac{1}{2}x_1 + \tfrac{1}{2}\sqrt{x_1^2 - 4x_2}, \quad w_2 = \tfrac{1}{2}x_1 - \tfrac{1}{2}\sqrt{x_1^2 - 4x_2},$$

$$w_1 = \tfrac{1}{2}y_1 + \tfrac{1}{2}\sqrt{y_1^2 + 4y_2}, \quad w_2 = -\tfrac{1}{2}y_1 + \tfrac{1}{2}\sqrt{y_1^2 + 4y_2}$$

on $\mathsf{Q}_\pm$. Furthermore, $\mathsf{X}_\pm$ maps the intersection of G^2 with the upper (lower) halfplane to the intersection of $\mathsf{Q}_\pm$ with the upper (lower) halfplane. The Jacobian of $\mathsf{X}_\pm$ at $w \in \mathsf{G}^2$ is $J = w_1 \mp w_2$, and thus $\mathsf{X}_\pm$ fails to be a local diffeomorphism on the lines $\mathsf{S}^\pm := \{(w_1, w_2) \in \mathsf{G}^2 : w_1 = \pm w_2\}$ on the boundary of G^2. Because of the specific form of (6.1), $\mathsf{X}_\pm$ map the horizontal and vertical line segments in G^2 to line segments in $\mathsf{Q}_\pm$.

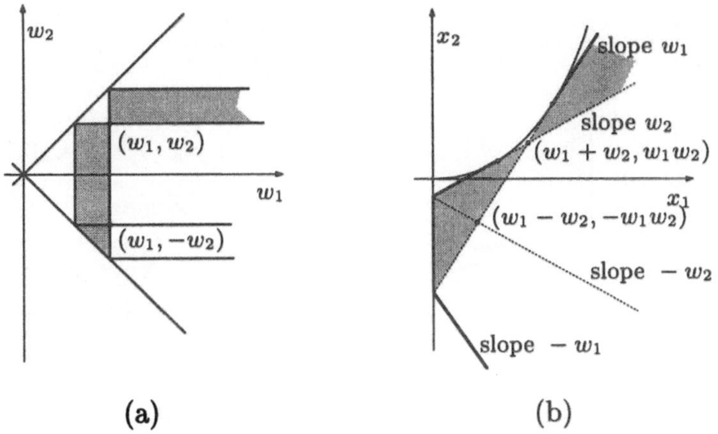

Figure 1 The image of A_+ under X_+

Let $x \in \mathsf{R}^2$ and $s_1, s_2 \in \mathsf{R}$ satisfy $s_1 \leq s_2$. We denote by $\mathsf{K}(x; s_1, s_2) \subset \mathsf{R}^2$ the union of all straight lines through x whose slope s satisfies $s_1 \leq s \leq s_2$. Recall from Proposition 5.5 the sets of signed singular values of the rank 1 perturbations of a given tensor and the subsets of this set satisfying $z_1 z_2 \geq 0$ and $z_1 z_2 \leq 0$, respectively. The following lemma, whose immediate proof is omitted, determines the images of these sets under $\mathsf{X}_\pm$.

6.1 Lemma. *Let* $w \in \mathsf{G}^2$ *and denote* $x = \mathsf{X}_+(w)$, $y = \mathsf{X}_-(w)$,

$$\mathsf{A}_\pm := \{\bar{w} \in \mathsf{G}^2 : |w_2| \leq \bar{w}_1 < \infty,\ |\bar{w}_2| \leq w_1,\ \pm z_1 z_2 \geq 0 \text{ where } z = \bar{w} - w\}.$$

Then

$$\mathsf{X}_+(\mathsf{A}_+) = \mathsf{K}(x; w_2, w_1) \cap \mathsf{Q}_+, \quad \mathsf{X}_-(\mathsf{A}_-) = \mathsf{K}(y; -w_1, w_2) \cap \mathsf{Q}_-.$$

See Figures 1 and 2. The boundary of $\mathsf{K}(x; w_2, w_1)$ consists of the two tangents to the parabola $\{x \in \mathsf{R}^2 : x_2 = \frac{1}{4} x_1^2\}$ which pass through x. The boundary of $\mathsf{K}(y; -w_1, w_2)$ consists of the two tangents to $\{y \in \mathsf{R}^2 : y_2 = -\frac{1}{4} y_1^2\}$ which pass through y.

Let $f : U \to \mathsf{R}$, where $U = \mathrm{L}^+$ or L, be a continuously differentiable invariant function with the representation $\tilde{f} : W \to \mathsf{R}$, where $W = \mathsf{R}^2_{++}$ or $W = \mathsf{R}^2$, respectively. We introduce two more representations $\tilde{f}^\pm : \mathsf{D}^\pm \to \mathsf{R}$ of f in terms of the invariants x_1, x_2 and the signed invariants y_1, y_2,

$$\tilde{f}(w) = \tilde{f}^+(x) = \tilde{f}^-(y) \tag{6.2}$$

for every $w \in W \cap \mathsf{G}^2$, where $x, y \in \mathsf{R}^2$ are given by (6.1) and

$$\mathsf{D}^\pm := \mathsf{Q}_\pm \cap (\mathsf{R} \times \mathsf{R}_{++}) \text{ if } U = \mathrm{L}^+ \text{ and } \mathsf{D}^\pm = \mathsf{Q}_\pm \text{ if } U = \mathrm{L}, \tag{6.3}$$

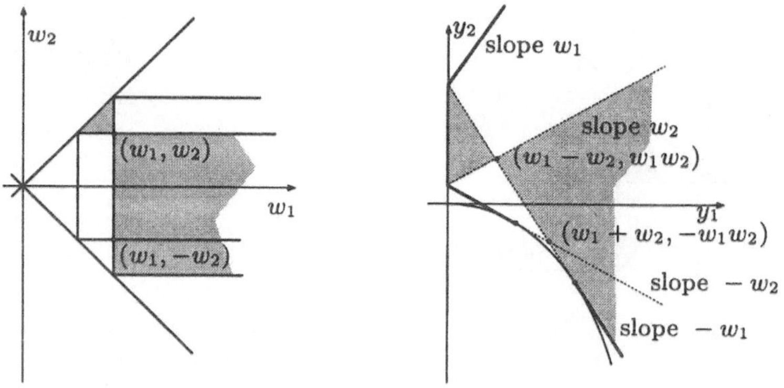

(a) (b)

Figure 2 The image of A_- under X_-

or explicitly,

$$\mathsf{D}^+ = \{x \in \mathsf{R}^2_{++} : x_2 \le \tfrac{1}{4}x_1^2\}, \quad \mathsf{D}^- = \mathsf{R}_+ \times \mathsf{R}_{++} \quad \text{if} \quad U = \mathsf{L}^+. \tag{6.4}$$

The mappings $\mathsf{X}_\pm$ have their Jacobians different from 0 everywhere in G^2 except on the lines $\mathsf{S}^\pm := \{(w_1, w_2) \in \mathsf{G}^2 : w_1 = \pm w_2\}$ on the boundary of G^2. Thus the continuous differentiability of $\tilde{f}$ implies the continuous differentiability of $\tilde{f}^\pm$ in the interior of $\mathsf{D}^\pm$, and a calculation gives

$$\tilde{f}_1^+ = \frac{s_1 w_1 - s_2 w_2}{w_1 - w_2}, \quad \tilde{f}_2^+ = -\frac{s_1 - s_2}{w_1 - w_2}, \quad \tilde{f}_1^- = \frac{s_1 w_1 - s_2 w_2}{w_1 + w_2}, \quad \tilde{f}_2^- = \frac{s_1 + s_2}{w_1 + w_2} \tag{6.5}$$

at the points related by $\mathsf{X}_\pm$, where $s_i = f_i$. The formulas are meaningful everywhere except the lines $\mathsf{S}^\pm$ where the Jacobian of $\mathsf{X}_\pm$ vanishes. We extend the meaning of the partial derivatives of $\tilde{f}^\pm$ to interpret them consistently as the usual partial derivatives in the interiors of the domains of definition of $\tilde{f}^\pm$ and as the expressions given by (6.5) on those parts of the boundary of the domains of definition of $\tilde{f}^\pm$ where the formulas are meaningful. Note also that if $U = \mathsf{L}$ then f is fully invariant if and only if $\tilde{f}^-(y_1, y_2) = \tilde{f}^+(y_1, -y_2)$ for every $(y_1, y_2) \in \mathsf{Q}_-$.

6.2 Proposition. *Let $f : U \to \mathsf{R}$, where $U = \mathsf{L}^+$ or $U = \mathsf{L}$, be a continuously differentiable invariant function with $\tilde{f}^\pm : \mathsf{D}^\pm \to \mathsf{R}$ introduced above (see (6.1)–(6.4)), let $\mathbf{F} \in U$ have the signed singular values w, and let x, y be given by (6.1). Then f is rank 1 convex at $\mathbf{F}$ if and only if the following three conditions hold simultaneously:*

(1) $s_1 w_1 - s_2 w_2 \ge 0$ *(if $w_1 \ne w_2$ then equivalently $\tilde{f}_1^+(x) \ge 0$; if $w_1 \ne -w_2$ then equivalently, $\tilde{f}_1^-(y) \ge 0$);*

(2) *if* $w_1 \neq w_2$, *then*

$$\tilde{f}^+(\bar{x}) - \tilde{f}^+(x) \geq \tilde{f}_1^+(x)(\bar{x}_1 - x_1) + \tilde{f}_2^+(x)(\bar{x}_2 - x_2) \tag{6.6}$$

for every $\bar{x} \in \mathsf{K}(x; w_2, w_1) \cap \mathsf{D}^+$;

(3) $w_1 \neq -w_2$, *then*

$$\tilde{f}^-(\bar{y}) - \tilde{f}^-(y) \geq \tilde{f}_1^-(y)(\bar{y}_1 - y_1) + \tilde{f}_2^-(y)(\bar{y}_2 - y_2) \tag{6.7}$$

for every $\bar{y} \in \mathsf{K}(y; -w_1, w_2) \cap \mathsf{D}^-$.

If $w_1 = w_2$, then **(2)** is not in force, while the derivatives of $\tilde{f}^-$ in **(3)** must be interpreted in the sense of (6.5) in this case; in $w_1 = -w_2$, then **(3)** is not in force while the derivatives of $\tilde{f}^-$ in **(2)** must again be interpreted in the sense of (6.5).

Proof. This is just a reformulation of Proposition 5.5 using (6.5) and Lemma 6.1. □

Let us now proceed to the conditions of global rank 1 convexity in terms of the representations $\tilde{f}^\pm$.

6.3 Lemma. *Consider the straight line* L *given parametrically by*

$$x_1 = \alpha t + \beta, \quad x_2 = t, \quad t \in \mathsf{R}, \tag{6.8}$$

where $\alpha, \beta \in \mathsf{R}$ *are constants. Then*

$$\mathsf{L} \cap \{x \in \mathsf{R} : x_2 > \tfrac{1}{4}x_1^2\} = \emptyset \;\Leftrightarrow\; \alpha\beta \geq 1,$$

$$\mathsf{L} \cap \{y \in \mathsf{R} : y_2 < -\tfrac{1}{4}y_1^2\} = \emptyset \;\Leftrightarrow\; \alpha\beta \leq -1.$$

6.4 Proposition. *Let* $f : U \to \mathsf{R}$, *where* $U = \mathrm{L}^+$ *or* $U = \mathrm{L}$, *be a continuously differentiable invariant function with* $\tilde{f}^\pm : \mathsf{D}^\pm \to \mathsf{R}$ *introduced above (see (6.1)–(6.4)). Then f is rank 1 convex if and only if the following three conditions hold simultaneously:*

(1) $\tilde{f}_1^+ \geq 0$ *on* D^+ *(equivalently,* $\tilde{f}_1^- \geq 0$ *on* D^-*)*;

(2) *if* L *is a line (6.8) such that* $\alpha\beta \geq 1$ *then the restriction of* $\tilde{f}^+$ *to* $\mathsf{L} \cap \mathsf{D}^+$ *is convex;*

(3) *if* L *is a line (6.8) such that* $\alpha\beta \leq -1$ *then the restriction of* $\tilde{f}^-$ *to* $\mathsf{L} \cap \mathsf{D}^-$ *is convex.*

If, moreover, $U = \mathrm{L}$ *and f is fully invariant, then conditions* **(2)** *and* **(3)** *are equivalent and thus f is rank 1 convex if and only if conditions* **(1)** *and* **(2)** *above hold.*

By Lemma 6.3, the lines L in **(2)**, **(3)** are exactly those which do not intersect the parabolas $\{x \in \mathsf{R} : x_2 = \frac{1}{4}x_1^2\}$ and $\{y \in \mathsf{R} : y_2 = -\frac{1}{4}y_1^2\}$, respectively; they can at most touch them tangentially.

Other characterizations of global rank 1 convexity as a convexity condition (in different representations) are given in [5; Proposition 1.1] and [10; Proposition 4.1].

Proof. This follows from Proposition 6.2 and Lemma 6.3. □

Remark. The preprint [12] was written without knowing the paper of AUBERT [1], to which I would have referred had I seen it in time. Since this work was completed, I became aware of a related work by ROSAKIS [9] which contains the special case of Proposition 4.2 with $U = \mathrm{L}^+$ without the continuous differentiability assumption and Proposition 3.2 for arbitrary dimension without the continuous differentiability assumption.

Acknowledgments

This work was supported by Grant No. A2019603 of the AV ČR.

References

[1] Aubert, G. (1995). Necessary and sufficient conditions for isotropic rank-one convex functions in dimension 2. *J. Elasticity*, 39:31–46.

[2] Aubert, G. and Tahraoui, R. (1980). Sur la faible fermeture de certains ensembles de contrainte en élasticité non-linéaire plane. *C. R. Acad. Sci. Paris*, 290:537–540.

[3] Ball, J.M. (1977). Convexity conditions and existence theorems in nonlinear elasticity. *Arch. Rational Mech. Anal.*, 63:337–403.

[4] Dacorogna, B. (1990). *Direct methods in the calculus of variations.* Berlin, Springer.

[5] Dacorogna, B., Douchet, J., Gangbo, W. and Rappaz, J. (1990). Some examples of rank one convex functions in dimension two. *Proc. Roy. Soc. Edinburgh*, 114A:135–150.

[6] Dacorogna, B. and Koshigoe, H. (1993). On the different notions of convexity for rotationally invariant functions. *Ann. Fac. Sci. Toulouse*, 2:163–184.

[7] Knowles, J.K. and Sternberg, E. (1977). On the failure of ellipticity of the equations for finite elastostatic plane strain. *Arch. Rational Mech. Anal.*, 63:321–326.

[8] Rockafellar, R.T. (1970). *Convex analysis.* Princeton, Princeton University Press.

[9] Rosakis, P. (1997). Characterization of convex isotropic functions. *Preprint,*

[10] Rosakis, P. and Simpson, H. (1995). On the relation between polyconvexity and rank-one convexity in nonlinear elasticity. *J. Elasticity*, 37:113–137.

[11] Šilhavý, M. (1997). *The Mechanics and Thermodynamics of Continuous Media.* Berlin, Springer.

[12] Šilhavý, M. (1997). Convexity conditions for rotationally invariant functions in two dimensions. Preprint.

[13] Šilhavý, M. (1998). On isotropic rank 1 convex functions. Preprint. Proc. Roy Soc. Edinburgh. To appear.

[14] Šilhavý, M. Rank 1 perturbations of deformation gradients. Int. J. Solids Structures. To appear.

HÖLDER CONTINUITY OF WEAK SOLUTIONS TO CERTAIN NONLINEAR PARABOLIC SYSTEMS IN TWO SPACE DIMENSIONS

Joerg Wolf

Abstract: In the present paper we prove the Hölder continuity of weak solutions to a nonlinear parabolic system in two space dimensions

$$\frac{\partial u^i}{\partial t} - D_\alpha a_i^\alpha(x,t,\nabla u) = B_i(x,t,u,\nabla u) \quad \text{in } Q \quad (i = 1,\dots,N)$$

($Q = \Omega \times (0,T), \Omega \subset I\!R^2$), where the coefficients $a_i^\alpha(x,t,\xi)(\alpha = 1,2;$ $i = 1,\dots,N)$ are measurable in x, continuous in t, and Lipschitz continuous in ξ, whereas the right hand side B_i satisfies the controlled growth condition.

Keywords: Nonlinear parabolic systems, controlled growth, Hölder continuity.

1. INTRODUCTION

Let $\Omega \subset I\!R^2$ be an open and bounded set. Given $0 < T < +\infty$ by Q we denote the cylinder $\Omega \times (0,T)$. Let $N \in I\!N \geq 1$. We consider the following parabolic system of PDE's:

$$\frac{\partial u^i}{\partial t} - D_\alpha a_i^\alpha(x,t,\nabla u) = B_i(x,t,u,\nabla u) \text{ [1]in} Q \quad (i = 1,\dots,N), \quad (1.1)$$

where

$$u = \{u^1,\dots,u^N\}.$$

We assume that $a_i^\alpha : Q \times I\!R^{2N} \to I\!R$ and $B_i : Q \times I\!R^{3N} \to I\!R\,(\alpha = 1,2; i = 1,\dots,N)$ are Carathéodory functions satisfying the following

[1] Here ∇u denotes the matrix of spatial derivatives $\{D_\alpha u^i\}$ ($D_\alpha u^i = \partial u^i/\partial x_\alpha$, $\alpha = 1,2$; $i = 1,\dots,N$). Throughout the paper repeated Greek indices (Latin indices resp.) imply summation over $1,2$ ($i = 1,\dots,N$ resp.).

conditions:

$$\left.\begin{aligned}&|a_i^\alpha(x,s,\eta) - a_i^\alpha(x,t,\xi)| \le \omega(|s-t|)(1+|\eta|+|\xi|) + c_0|\eta-\xi|\\&\forall\, x\in\Omega, \forall\,(s,\eta),(t,\xi)\in(0,T)\times I\!R^{2N} \quad (c_0=\text{const}),\\&\omega \text{ is positive and nondecreasing with } \lim_{h\to0}\omega(h)=0;\end{aligned}\right\}\tag{1.2}$$

$$\left.\begin{aligned}&(a_i^\alpha(x,t,\xi) - a_i^\alpha(x,t,\eta))(\xi_\alpha^i-\eta_\alpha^i) \ge \nu_0|\xi-\eta|^2\\&\forall\,(x,t)\in Q, \forall\,\xi,\eta\in I\!R^{2N} \quad (\nu_0=\text{const}>0);\end{aligned}\right\}\tag{1.3}$$

$$|a_i^\alpha(x,t,\xi)| \le c_1(1+|\xi|) \quad \forall\,(x,t,\xi)\in Q\times I\!R^{2N} \ (c_1=\text{const}),\tag{1.4}$$

and for the right hand side we assume controlled grows,

$$\left.\begin{aligned}&|B_i(x,t,u,\xi)| \le c_2(1+|u|^3+|\xi|^{3/2})\\&\forall\,(x,t,u,\xi)\in Q\times I\!R^N\times I\!R^{2N} \ (c_2=\text{const})\end{aligned}\right\}\tag{1.5}$$

$(\alpha=1,2; i=1,\ldots,N)$.

Next, define

$$\begin{aligned}\mathrm{W}_2^{1,0}(Q;I\!R^N) &= \left\{v\in L^2(Q;I\!R^N)\,\middle|\,\frac{\partial v}{\partial x_\alpha}\in L^2(Q;I\!R^N)\quad(\alpha=1,2)\right\},\\\mathrm{W}_2^{1,1}(Q;I\!R^N) &= \left\{v\in \mathrm{W}_2^{1,0}(Q;I\!R^N)\,\middle|\,\frac{\partial v}{\partial t}\in L^2(Q;I\!R^N)\right\}\cong\\&\cong \mathrm{W}_2^1(Q;I\!R^N),\\\mathrm{V}_2^{1,0}(Q;I\!R^N) &= \left\{v\in \mathrm{W}_2^{1,0}(Q;I\!R^N)\,\middle|\,\operatorname*{ess\,sup}_{(0,T)}\int_\Omega|v(x,t)|^2dx<+\infty\right\}.\end{aligned}$$

Let X be any Banach space whose norm is denoted by $\|\cdot\|_X$. Then by $L^p(0,T;X)$ $(1\le p\le+\infty)$ we denote the space of all (classes of equivalent) Bochner measurable functions $v:(0,T)\to X$ such that,

$$\|v\|_{L^p(0,T;X)} = \left(\int_0^T\|v(t)\|_X^p\,dt\right)^{1/p} < +\infty \qquad \text{if } 1\le p<+\infty$$

$$\|v\|_{L^\infty(0,T;X)} = \operatorname*{ess\,sup}_{(0,T)}\|v(t)\|_X < +\infty \qquad \text{if } p=+\infty.$$

By $\mathrm{W}_p^1(Q;I\!R^N)$ ($\overset{\circ}{\mathrm{W}}{}_p^1(Q;I\!R^N)$ resp.) $(1\le p\le\infty)$ we denote the usual Sobolev spaces.

Definition 1.1. Let (1.4) and (1.5) be fulfilled. A vector valued function $u \in \mathrm{V}_2^{1,0}(Q; \mathbb{R}^N)$ is said to be a *weak solution* to the system (1.1) if the following integral identity is valid for all $\varphi \in \mathrm{C}_c^\infty(Q; \mathbb{R}^N)$

$$- \int_Q u^i \frac{\partial \varphi^i}{\partial t} dxdt + \int_Q a_i^\alpha(x,t,\nabla u) D_\alpha \varphi^i dxdt = \int_Q B_i(x,t,u,\nabla u)\varphi^i dxdt. \tag{1.6}$$

REMARKS: 1.) In the particular case $N = 1$ it is well known that weak solutions to the system (1.1) are Hölder continuous. This result has been proved first by De Giorgi and Nash for weak solutions to a linear elliptic PDE with bounded coefficients (cf. [2]). The same result has been obtained by Moser using the so called *Harnack inequality* and later these result has been generalized by Ladyzenskaya and Ural'ceva both for the elliptic case (cf. in [8]) and for the parabolic case (cf. in [9]).

In contrary, it is well known that in case of systems (i.e. if $N > 1$) the results stated above doesn't continue to hold. Indeed, in recent time several counterexamples have been constructed for example by De Giorgi, Giusti, Giaquinta, Campanato, Nečas (cf. [4]). However, in the case of two dimensions no discontinuous counterexample are known. Therefore it is our conjecture that one shall obtain regularity of the weak solution in this particular case. For elliptic systems such a result has been proved by Giaquinta and Modica having employed the result of higher integrability via *Reverse Hölder Inequality* which is a generalization of an idea introduced by Gehring (cf.[3]).

2.) Studying the linear parabolic case, with measurable bounded coefficients being either continuous in t or in x the Hölder continuity of the weak solution has been proved in [6].

In Nečas and Šverák [14] the Hölder continuity of the weak solution of the system (1.1) in two, three and four space dimensions has been obtained for the case, that the coefficients a_i^α depend on the gradient matrix only. In recent time the Hölder continuity of the weak solution of the system (1.1) $(B_i \equiv 0)$ has been obtained in [12] under the additional assumption that the coefficients a_i^α are Hölder continuous with respect to t sufficiently near to 1.

The aim of the present work is to establish an analogous result as in [12] under weaker assumptions on the coefficients. Our main result is the following

Theorem 1.2. *Assume* (1.2), (1.3), (1.4) *and* (1.5). *Let* $\mathrm{V}_2^{1,0}(Q; \mathbb{R}^N)$ *be a weak solution to the system* (1.1). *Then there exists some* $0 < \gamma < 1$,

$$u \in \mathrm{C}^{\gamma,\gamma/2}(\overline{Q}'; \mathbb{R}^N) \qquad \forall Q' \subset\subset Q.^2 \tag{1.7}$$

The paper is organized as follows. In Section 2 we prove some lemmas which will be used in the sequence of the work.

Section 3 is dedicated to a fundamental estimate (cf. Theorem 3.1) for weak solutions to (1.1), where $B_i \equiv 0$ and where the coefficients a_i^α does not depend on t. The proof of this fundamental estimate are based on the so called Caccioppoli inequalities and the reverse inequality via Giaquinta, Modica (cf. [5]).

Finally, in Section 4 we complete the proof of Theorem 1.2 by freezing the coefficients a_i^α in t and by making use of the fundamental estimate provided in Section 3.

2. PRELIMINARIES

Let $X, Y (X \subset Y)$ denote some Banach spaces. Let $-\infty < a < b < +\infty$ be given. For $v \in L^2(a, b; Y)$ we define the mean value

$$v_{t^0,r} = \frac{1}{r} \int_{t^0-r}^{t^0} v(t)dt \qquad (t^0 \in (a,b),\ 0 < r < t^0 - a).$$

Lemma 2.1. *Let $v \in H^1(a, b; Y)$. Then for all $t^0 \in (a,b)$ and $0 < r < t^0 - a$ we have*

$$\int_{t^0-r}^{t^0} \|v(t) - v_{t^0,r}\|_Y^2 \, dt \le 4r^2 \int_{t^0-r}^{t^0} \left\| \frac{dv}{dt}(t) \right\|_Y^2 dt. \text{ [3]} \tag{2.1}$$

Proof. 1°) First, assume that $v : [a, b] \to Y$ is continuously differentiable. Clearly, by the mean value theorem we evaluate for all $a \le s < t \le b$,

$$v(s) - v(t) = \int_0^1 \frac{dv}{dt}(t + \tau(s-t))(s-t)d\tau.$$

Next, using Hölder's inequality and Fubini's theorem, we estimate,

$$\int_{t^0-r}^{t^0} \|v(t) - v_{t^0,r}\|_Y^2 \, dt \le \frac{1}{r} \int_{t^0-r}^{t^0} \int_{t^0-r}^{t^0} \|v(s) - v(t)\|_Y^2 \, dsdt \le$$

---

[2] A function v belongs to $C^{\gamma,\gamma/2}(\overline{Q})$ if there exists a constant c depending on v such that $|v(x,t) - v(y,s)| \le c(|x-y|^\gamma + |s-t|^{\gamma/2})\ \forall (x,t),(y,s) \in \overline{Q}$. - The expression $Q' \subset\subset Q$ says that $\overline{Q'} \subset Q$.

[3] Remark that dv/dt has to be understood in sense of distribution.

$$\leq r \int_{t^0-r}^{t^0} \int_{t^0-r}^{t^0} \int_0^1 \left\| \frac{dv}{dt}(t+\tau(s-t)) \right\|_Y^2 d\tau ds dt =$$

$$= r \int_{t^0-r}^{t^0} \int_{1/2}^{1} \int_{t^0-r}^{t^0} \left\| \frac{dv}{dt}(t+\tau(s-t)) \right\|_Y^2 ds d\tau dt +$$

$$+ r \int_{t^0-r}^{t^0} \int_{0}^{1/2} \int_{t^0-r}^{t^0} \left\| \frac{dv}{dt}(t+\tau(s-t)) \right\|_Y^2 dt d\tau ds.$$

Then the assertion (2.1) easily follows from the estimate above after having applied the transformation formula of the Lebesgue integral.

2°) Secondly, in the general case the inequality (2.1) is obtained by an approximation argument which is well known in the literature (for instance cf. [11]) . This concludes the proof of the lemma. ∎

The following result is due to Campanato (cf. [1]).

Lemma 2.2. *Let $v \in L^2(a,b;X)$. If there are some constants $0 < K < +\infty$ and $0 < \gamma < 1$ such that for each $t^0 \in (a,b]$:*

$$\int_{t^0-r}^{t^0} \|v(r) - v_{t^0,r}\|_X^2 \, dt \leq K r^{1+2\gamma} \quad \forall 0 < r < t^0 - a, \tag{2.2}$$

then v belongs to $C^\gamma([a,b];X)$. In addition, we have

$$[v]^2_{C^\gamma([a,b];X)} \leq cK \qquad (c = \text{const}). \tag{2.3}$$

Next we are going to introduce an interpolation result whose proof is based on the two lemmas stated above.

Lemma 2.3. *Let $v \in H^1(a,b;Y) \cap L^2(a,b;X)$. Assume that the interpolation space $X_\theta = [X,Y]_{1-\theta}$ $(0 < \theta < 1/2)$ exists* (cf. Triebel [15]).[4] *Then we have*

$$v \in C^{1/2-\theta}([a,b];X_\theta). \tag{2.4}$$

In particular, the following inequality holds true for all $t^0 \in (a,b]$ and $0 < r < \sqrt{t^0 - a}$,

$$\begin{aligned} \operatorname*{ess\,sup}_{(t^0-r^2,t^0)} \|v(t)\|^2_{X_\theta} \; &\leq \; c r^{2-4\theta} \left\| \frac{dv}{dt} \right\|^{2(1-\theta)}_{L^2(t^0-r^2,t^0;Y)} \|v\|^{2\theta}_{L^2(t^0-r^2,t^0;X)} + \\ &+ \; c r^{-2} \|v\|^{2(1-\theta)}_{L^2(t^0-r^2,t^0;Y)} \|v\|^{2\theta}_{L^2(t^0-r^2,t^0;X)} . \end{aligned} \tag{2.5}$$

[4] Notice, $\exists c = \text{const} : \|\varphi\|_{X_\theta} \leq c \|\varphi\|_Y^{1-\theta} \|\varphi\|_X^\theta \;\; \forall \varphi \in X$.

Proof. Let t^0 and $0 < r < t^0 - a$ be arbitrarily chosen. With help of Hölder's inequality and lemma 2.1 we estimate

$$\int\limits_{t^0-r}^{t^0} \|v(t) - v_{t^0,r}\|_{X_\theta}^2 \, dt \le$$

$$\le c\Bigg(\int\limits_{t^0-r}^{t^0} \|v(t) - v_{t^0,r}\|_Y^2 \, dt\Bigg)^{1-\theta} \Bigg(\int\limits_{t^0-r}^{t^0} \|v(t) - v_{t^0,r}\|_X^2 \, dt\Bigg)^{\theta} \quad (2.6)$$

$$\le cr^{2(1-\theta)} \left\|\frac{dv}{dt}\right\|_{L^2(a,b;Y)}^{2(1-\theta)} \|v\|_{L^2(a,b;X)}^{2\theta} \cdot$$

Now, making use of lemma 2.2 we easily verify that

$$v \in \mathrm{C}^{1/2-\theta}([a,b]; X_\theta).$$

Moreover, combining (2.3) and (2.6) we get

$$[v]^2_{\mathrm{C}^{1/2-\theta}([a,b];X_\theta)} \le c \left\|\frac{dv}{dt}\right\|_{L^2(a,b;Y)}^{2(1-\theta)} \|v\|_{L^2(a,b;X)}^{2\theta} \quad (2.7)$$

(c=const).

To prove the second part of the theorem we first consider the case $t^0 = 0$ and $r = 1$. By (2.7) we estimate for a.a. $s \in (-1, 0)$,

$$\|v(s)\|_{X_\theta}^2 \le 2\,\|v(s) - v_{0,1}\|_{X_\theta}^2 + 2\int\limits_{-1}^{0} \|v(t)\|_{X_\theta}^2 \, dt \le$$

$$\le 2\int\limits_{-1}^{0} \|v(s) - v(t)\|_{X_\theta}^2 \, dt + c\,\|v\|_{L^2(-1,0;Y)}^{2(1-\theta)} \|v\|_{L^2(-1,0;X)}^{2\theta} \le$$

$$\le c\left\|\frac{dv}{dt}\right\|_{L^2(-1,0;Y)}^{2(1-\theta)} \|v\|_{L^2(-1,0;X)}^{2\theta} + c\,\|v\|_{L^2(-1,0;Y)}^{2(1-\theta)} \|v\|_{L^2(-1,0;X)}^{2\theta} \cdot$$

Whence (2.5).

In order to verify assertion (2.5) for the general case we consider the change of variables $t' = (t - t^0)/r^2$ and apply the transformation formula of the Lebesgue integral. ■

Next, given $X^0 = (x^0, t^0) \in \mathbb{R}^3$, $0 < R < +\infty$ the parabolic cylinder is defined by

$$Q_R(X^0) = B_R(x^0) \times (t^0 - R^2, t^0)$$

$(B_R(x^0) = \{x \in \mathbb{R}^2 \,|\, |x - x^0| < R\})$.

For the sake of simplicity if no confusion is possible to arise we will use the shorter notation Q_R (B_R resp.) instead of $Q_R(X^0)$ ($B_R(x^0)$ resp.).

Now, we shall introduce a well known result of higher integrability which is based on a method developed by Gehring, Giuquinta, Modica (cf. [5],[13]).

Lemma 2.4. *Assume* (1.2), (1.4) *and* (1.5). *Let* $u \in V_2^{1,0}(Q, I\!R^N)$ *be a weak solution to the system* (1.1). *Then there exists some real number* $2 < q_0 < +\infty$ *such that* $\nabla u \in L^q_{\text{loc}}(Q; I\!R^{2N})$ $\forall q \in [2, q_0)$. *Moreover, for each pair of concentric subcylinder* $Q_{R/2} \subset Q_R \subset Q$ *we have,*

$$\left(\fint_{Q_{R/2}} (1 + |u|^4 + |\nabla u|^2)^{q/2} dxdt \right)^{2/q} \leq \tag{2.8}$$

$$\leq c \fint_{Q_R} (1 + |u|^4 + |\nabla u|^2) dxdt, [5]$$

where $c = \text{const} > 0$ *depending only on* ν_0, c_0, c_1, c_2 *and* N.

We close this section with a technical lemma which can be found in [4].

Lemma 2.5. *Let* f *be a nonnegative and bounded function defined on the interval* $[a, b]$ $(-\infty < a < b < +\infty)$. *Furthermore, let* A, B, θ *and* $0 < \varepsilon < 1$ *be positive constants such that for all* $a \leq r < \overline{r} \leq b$,

$$f(r) \leq A(\overline{r} - r)^{-\theta} + B + \varepsilon f(\overline{r}). \tag{2.9}$$

Then there exists a positive constant $c = c(\theta, \varepsilon)$ *such that for all* $a \leq r < \overline{r} \leq b$,

$$f(r) \leq c \left(A(\overline{r} - r)^{-\theta} + B \right). [6] \tag{2.10}$$

3. THE CASE $a_i^\alpha(x, \xi)$

In this section with aid of suitable Caccioppoli inequalities we shall derive a fundamental estimate which plays the main role in proving Theorem 1.2. An inspection of the proofs below shows that (with a slightly modification) all the results continue to hold even in the case of more than two space dimensions, whereas the Hölder continuity of

---

[5] For $v \in L^1(Q_r)$ $(0 < r < +\infty)$ we define the mean value $\fint_{Q_r} v dxdt = \frac{1}{\text{meas}(Q_r)} \int_{Q_r} v dxdt$.

[6] Notice that $c = c(\theta, \varepsilon) = (1 - \varepsilon^{1/(\theta+1)})^{-(\theta+1)}$ is the best constant in (2.10))

the weak solution to the system (1.1) is verified only in case of two dimensions.

We start with a differentiability result which one may easily obtain by using the method provided in [12]. We have

Lemma 3.1. *Let $v \in V_2^{1,0}(Q;\mathbb{R}^N)$ be a weak solution of the following parabolic system,*

$$\frac{\partial v^i}{\partial t} - D_\alpha a_i^\alpha(x,\nabla v) = 0 \quad \text{in} \quad Q \qquad (i=1,\dots,N), \tag{3.1}$$

where $a_i^\alpha(x,\xi)$ $(\alpha = 1,2; i = 1,\dots,N)$ are Carathéodory functions satisfying (1.2), (1.3) *and* (1.4). *Then we have*

$$\frac{\partial v}{\partial t},\ \frac{\partial}{\partial t}D_\alpha v \in L^2(Q';\mathbb{R}^N) \quad \forall Q' \subset\subset Q \qquad (\alpha = 1,2). \tag{3.2}$$

Next, we are going to prove the following Caccioppoli type inequalities.

Lemma 3.2. *Assume* (1.2), (1.3) *and* (1.4). *Let $v \in V_2^{1,0}(Q;\mathbb{R}^N)$ be a weak solution of the system* (3.1). *Then for each pair of concentric subcylinder $Q_{R/2} \subset Q_R \subset Q$ we have*

$$\int_{Q_{R/2}} \left|\frac{\partial v}{\partial t}\right|^2 dxdt \le \frac{c}{R^2}\int_{Q_R}(1+|\nabla v|^2)dxdt, \tag{3.3}$$

$$\int_{Q_{R/2}} \left|\frac{\partial}{\partial t}\nabla v\right|^2 dxdt \le \frac{c}{R^4}\int_{Q_R}(1+|\nabla v|^2)dxdt. \tag{3.4}$$

where $c = \text{const} > 0$ depending on ν_0, c_0, c_1 and N only.

Proof. First of all, we observe that the weak solution v to (3.1) obeys the identity

$$\left.\begin{array}{l} \displaystyle\int_Q v^i\frac{\partial \varphi^i}{\partial t}dxdt + \int_Q a_i^\alpha(x,\nabla v)D_\alpha\varphi^i dxdt = 0 \\ \forall \varphi \in W_2^{1,1}(Q;\mathbb{R}^N),\ \mathrm{supp}(\varphi) \subset Q. \end{array}\right\} \tag{3.5}$$

Let $\Omega' \subset\subset \Omega$ and $0 < t_0 < T$. Then, making use of the Steklov mean we may localize (3.5) with respect to t (cf. [13]),

$$\left.\begin{array}{l} \displaystyle -\int_\Omega \frac{\partial v_h^i}{\partial t}(x,t)\psi^i(x)dx + \int_\Omega (a_i^\alpha)_h(x,t)D_\alpha\psi^i(x)dx = 0 \\ \text{for a.a. } t \in (0,t_0),\ \forall 0 < h < T-t_0,\ \forall \psi \in \overset{\circ}{W}{}_2^1(\Omega';\mathbb{R}^N). \end{array}\right\} \tag{3.6}$$

[7]

Next, given $X^0 = (x^0, t^0) \in Q, 0 < R < \min\{\text{dist}(x^0, \partial\Omega), \sqrt{t^0}\}$, $0 < r < \bar{r} \le R$ and $0 < h < T - t^0$ we may insert $\psi(x) = (\Delta_h v)(x,t)\zeta^2$ ($(x,t) \in \Omega \times (t^0 - \bar{r}^2, t^0)$) into (3.6), where $\zeta \in C_c^\infty(B_{\bar{r}})$ denotes a cut-off function such that $0 \le \zeta \le 1$ in $B_{\bar{r}}$, $\zeta \equiv 1$ on B_r and $|\nabla\zeta| \le c(\bar{r} - r)^{-1}$ ($c =$ (const)). Integrating over the interval $(t^0 - \bar{r}^2, t^0)$ and employing Hölder's inequality gives

$$\frac{1}{h} \int_{t^0-\bar{r}^2}^{t^0} \int_{B_{\bar{r}}} |\Delta_h v|^2 \zeta^2 dx dt =$$

$$= \int_{t^0-\bar{r}^2}^{t^0} \int_{B_{\bar{r}}} (a_i^\alpha)_h ((\Delta_h D_\alpha v^i)\zeta^2 + 2(\Delta_h v^i)\zeta D_\alpha \zeta) dx dt \le$$

$$\le \left(\int_{t^0-\bar{r}^2}^{t^0+h} \int_{B_{\bar{r}}} (a_i^\alpha)_h^2 dx dt \right)^{1/2} \times$$

$$\times \left(\int_{t^0-\bar{r}^2}^{t^0} \int_{B_{\bar{r}}} \left[(\Delta_h D_\alpha v^i)\zeta^2 + 2(\Delta_h v^i)\zeta D_\alpha \zeta \right]^2 dx dt \right)^{1/2}.$$

Then, observing (1.4) applying Nirenberg's lemma and making use of Young's and Hölder's inequality gives,

$$\frac{1}{h^2} \int_{Q_r} |\Delta_h v|^2 dx dt \le \frac{c}{(\bar{r} - r)^2} \int_{t^0-\bar{r}^2}^{t^0+h} \int_{B_{\bar{r}}} (1 + |\nabla v|^2) dx dt + \tag{3.7}$$

$$+ c \left(\int_{t^0-\bar{r}^2}^{t^0+h} \int_{B_{\bar{r}}} (1 + |\nabla v|^2) dx dt \right)^{1/2} \left(\int_{t^0-\bar{r}^2}^{t^0+h} \int_{B_{\bar{r}}} \left| \frac{\partial}{\partial t} \nabla v \right|^2 dx dt \right)^{1/2}.$$

By virtue of reflexivity there exists an infinitesimal subsequence $\{h_j\}$ ($j = 1, 2, \ldots$),

$$(\Delta_{h_j} v)/h_j \to \partial v/\partial t \quad \text{weakly in} \quad L^2(Q_r; \mathbb{R}^N) \quad \text{as} \quad j \to +\infty.$$

[7] Here $v_h(t)$ denotes the Steklov mean $\frac{1}{h}\int_t^{t+h} v(\tau) d\tau$ and $(\Delta_h v)(x,t)$ denotes the difference $v(x, t+h) - v(x,t)$. For the properties of the Steklov mean $v_{(h)}$ and the differences $(\Delta_h v)(x,t)$ see in [13]. - Notice that the set of measure zero of those t for which (3.6) fails, does not depend on h.

Hence, in (3.7) passing to the limit $h_j \to 0$ implies

$$\int_{Q_r} \left|\frac{\partial v}{\partial t}\right|^2 dxdt \le \frac{c}{(\overline{r}-r)^2}\int_{Q_{\overline{r}}}(1+|\nabla v|^2)dxdt + \\ +c\left(\int_{Q_{\overline{r}}}(1+|\nabla v|^2)dxdt\right)^{1/2}\left(\int_{Q_{\overline{r}}}\left|\frac{\partial}{\partial t}\nabla v\right|^2 dxdt\right)^{1/2} \tag{3.8}$$

($c = \text{const} > 0$ doesn't depend on $r, \overline{r}$ and v).

Next, let $0 < h < T - t^0$ and form the difference Δ_h in (3.1). It follows that

$$\left.\begin{aligned} &\int_0^{T-h}\int_\Omega (\Delta_h v^i)\frac{\partial \varphi^i}{\partial t}dxdt = \\ &= \int_0^{T-h}\int_\Omega (a_i^\alpha(x,\nabla v(x,t+h)) - a_i^\alpha(x,\nabla v(x,t)))D_\alpha\varphi^i dxdt \\ &\forall \varphi \in \mathrm{W}_2^{1,1}(Q;\mathbb{R}^N),\ \operatorname{supp}(\varphi)\subset \Omega\times(0,T-h). \end{aligned}\right\} \tag{3.9}$$

Let $0 < r < \overline{r} \le R$ be arbitrarily chosen. Then we may localize (3.9) with respect to t arguing as in [13] we derive the following Caccioppoli inequality

$$\int_{Q_r}|\Delta_h \nabla v|^2 dxdt \le \frac{c}{(\overline{r}-r)^2}\int_{Q_{\overline{r}}}|\Delta_h v|^2 dxdt. \tag{3.10}$$

Then, from (3.10) letting $h \to 0$, we obtain,

$$\int_{Q_r}\left|\frac{\partial}{\partial t}\nabla v\right|^2 dxdt \le \frac{c}{(\overline{r}-r)^2}\int_{Q_{\overline{r}}}\left|\frac{\partial v}{\partial t}\right|^2 dxdt. \tag{3.11}$$

Inserting (3.11) into (3.8) and using Young's inequality gives,

$$\int_{Q_r}\left|\frac{\partial v}{\partial t}\right|^2 dxdt \le \frac{c}{(\overline{r}-r)^2}\int_{Q_{\overline{r}}}(1+|\nabla v|^2)dxdt + \frac{1}{2}\int_{Q_{\overline{r}}}\left|\frac{\partial v}{\partial t}\right|^2 dxdt$$

for all $0 < r < \overline{r} \le R$.

Now, after having applied the technical Lemma 2.5 we have for every $0 < r < \overline{r} \le R$

$$\int_{Q_r}\left|\frac{\partial v}{\partial t}\right|^2 dxdt \le \frac{c}{(\overline{r}-r)^2}\int_{Q_{\overline{r}}}(1+|\nabla v|^2)dxdt, \tag{3.12}$$

where c = const does not depend on $r, \bar{r}$ and v. Hence, from (3.12) putting $r = R/2$ and $\bar{r} = R$ therein we get (3.1). Finally, the second inequality (3.2) we receive by combining (3.12) and (3.11). ∎

The next result includes a fundamental estimate, which is crucial to prove the Hölder continuity of weak solutions of (1.1).

Theorem 3.1. *Assume* (1.2), (1.3) *and* (1.4). *Then there exists some real number* $0 < \gamma < 1$ *and a constant* $0 < A < +\infty$ *depending on* ν_0, c_0, c_1 *and* N *such that whenever* $v \in V_2^{1,0}(Q; \mathbb{R}^N)$ *is a weak solution to* (3.1) *the following fundamental estimate holds true for any subcylinder* $Q_R \subset Q$ *and for each* $0 < \tau < 1$,

$$\int_{Q_{\tau R}} (1 + |v|^4 + |\nabla v|^2) dxdt \le A\tau^{2+2\gamma} \int_{Q_R} (1 + |v|^4 + |\nabla v|^2) dxdt. \tag{3.13}$$

Proof. Let $X^0 = (x^0, t^0)$ and $0 < R < \min\{\text{dist}(x^0, \partial\Omega), \sqrt{t^0}\}$ be arbitrarily fixed. Since in the case $1/4 < \tau < 1$ the inequality (3.13) is trivially fulfilled without loss of generality we may assume that $0 < \tau \le 1/4$.

1° Firstly, having set $p = 8q/(3q+2)$, $\gamma = (q-2)/4q$ $(2 < q < q_0)$ [8] with help of Hölder's inequality and (2.5) (cf. Lemma 2.3) putting $X = L^q(B_{R/2}), Y = L^2(B_{R/2})$ and $\theta = \frac{1}{4}$ therein we estimate,

$$\begin{aligned}\int_{Q_{\tau R}} |\nabla v|^2 dxdt &\le c(\tau R)^{2+2\gamma} \operatorname*{ess\,sup}_{(t^0 - R/4, t^0)} \|\nabla v(\cdot, t)\|^2_{L^p(B_{R/2})} \le \\ &\le c(\tau R)^{2+2\gamma} R \left\| \frac{\partial}{\partial t} \nabla v \right\|^{3/2}_{L^2(Q_{R/2})} \|\nabla v\|^{1/2}_{L^2(t^0 - R^2/4, t^0; L^q(B_{R/2}))} + \\ &+ c(\tau R)^{2+2\gamma} R^{-2} \|\nabla v\|^{3/2}_{L^2(Q_{R/2})} \|\nabla v\|^{1/2}_{L^2(t^0 - R^2/4, t^0; L^q(B_{R/2}))} .^{9}\end{aligned} \tag{3.14}$$

Using Hölder's inequality from (2.8) we get

$$\|\nabla v\|^{1/2}_{L^2(t^0 - R^2/4, t^0; L^q(B_{R/2}))} \le cR^{-2\gamma} \left(\int_{Q_R} (1 + |\nabla v|^2) dxdt \right)^{1/4}. \tag{3.15}$$

[8] Here q refers to the exponent which occurs in the reverse inequality (2.8) (cf. Lemma 2.4).
[9] Here we have identified the spaces $L^2(t^0 - R^2, t^0; L^2(B_R; \mathbb{R}^N))$ and $L^2(Q_R; \mathbb{R}^N)$ by the linear isometry $L^2(t^0 - R^2, t^0; L^2(B_R; \mathbb{R}^N)) \cong L^2(Q_R; \mathbb{R}^N)$.

Now, the right hand side of (3.14) may be estimated by (3.4) and (3.15). Thus,

$$\int_{Q_{\tau R}} |\nabla v|^2 dxdt \le c\tau^{2+2\gamma} \int_{Q_R} (1+|\nabla v|^2) dxdt, \tag{3.16}$$

where $c = \text{const} > 0$ depending only on ν_0, c_0, c_1 and N.

2° Secondly, since $v, D_\alpha v \in V_2^{1,0}(Q_{R/2}; I\!R^N)$ $(\alpha = 1,2)$ by virtue of Sobolev's imbedding theorem we surely have $v \in L^{8/(1-\gamma)}(Q_{R/2}; I\!R^N)$. Then applying Hölder's inequality making use of (3.3) and (3.4) we get

$$\int_{Q_{\tau R}} |v|^4 dxdt \le \tag{3.17}$$

$$\le c\tau^{2+2\gamma}\Big\{ \int_{Q_R} (1+|v|^4) dxdt + \Big(\int_{Q_{R/2}} |\nabla v|^2 dxdt \Big)^2 \Big\}.$$

To estimate the second integral on the right hand side of (3.17) we argue as in [13] and apply Hölder's inequality. Then it follows,

$$\int_{Q_{\tau R}} |v|^4 dxdt \le c\tau^{2+2\gamma} \int_{Q_R} (1+|v|^4) dxdt. \tag{3.18}$$

Hence, the desired estimate (3.13) follows after having combined (3.16) and (3.18). ∎

4. PROOF OF THEOREM 1.2

Let $u \in V_2^{1,0}(Q; I\!R^N)$ be a weak solution of (4.1). Let $Q_R \subset Q$ be any fixed subcylinder. As it has been shown in [10] there exists an uniquely defined weak solution $v \in V_2^{1,0}(Q_R; I\!R^N)$ of the following Dirichlét problem:

$$\left.\begin{array}{l} \dfrac{\partial v^i}{\partial t} - D_\alpha a_i^\alpha(t^0, x, \nabla v) = 0 \quad \text{in} \quad Q_R \quad (i = 1, \dots, N) \\ v = u \quad \text{on } \Gamma = \partial B_R(x^0) \times (t^0 - R^2, t^0) \cup B_R(x^0) \times \{t^0 - R^2\}. \end{array}\right\} \tag{4.1}$$

Having defined

$$\Phi(R) = \int_{Q_R} (1 + |u|^4 + |\nabla u|^2) dxdt,$$

$$w = u - v$$

with help of (3.13) (cf. theorem 2) by means of the triangle inequality we easily deduce that

$$\Phi(\tau R) \le 16A\tau^{2+2\gamma}\Phi(R) + c(\tau)\int_{Q_R}(|w|^4 + |\nabla w|^2)dxdt. \qquad (4.2)$$

To estimate the integral of the right hand side in (4.2) we proceed as follows. Combining (4.1) and (1.1) we get the following integral identity for any $\varphi \in \mathrm{W}_2^{1,1}(Q_R;\mathbb{R}^N)$ with $\mathrm{supp}(\varphi) \subset Q_R$:

$$-\int_{Q_R} w^i\frac{\partial\varphi^i}{\partial t}dxdt + \int_{Q_R}(a_i^\alpha(x,t,\nabla u) - a_i^\alpha(x,t^0,\nabla v))D_\alpha\varphi^i dxdt \qquad (4.3)$$
$$= \int_{Q_R} B_i(x,t,u,\nabla u)\,\varphi^i dxdt.$$

Next, by $\overline{w}$ we denote the extension of w to $Q^* = B_R \times (-\infty, t^0)$ by zero. Furthermore, we set,

$$\overline{A}_i^\alpha(x,t) = \begin{cases} (a_i^\alpha(\cdot,\cdot,\nabla u) - a_i^\alpha(\cdot,t^0,\nabla v))(x,t) & \\ & \text{f.a.a. } (x,t) \in B_R \times (t^0 - R^2, t^0) \\ 0 & \text{f.a.a. } (x,t) \in B_R \times (-\infty, t^0 - R^2] \end{cases}$$

$$\overline{B}_i(x,t) = \begin{cases} B_i(\cdot,\cdot,u,\nabla u)(x,t) & \text{f.a.a. } (x,t) \in B_R \times (t^0 - R^2, t^0) \\ 0 & \text{f.a.a. } (x,t) \in B_R \times (-\infty, t^0 - R^2]. \end{cases}$$

$(\alpha = 1,2; i = 1,\dots,N)$.

Observing that $w \in \mathrm{H}^{1/2}_{(t^0-R^2)}(t^0 - R^2, t^0; L^2(B_R;\mathbb{R}^N))$ [10] we may extend (4.3) via Kaplan's method provided in [7]. Thus, with the notation introduced above we have

$$\left.\begin{array}{l} -\displaystyle\int_{Q^*}\overline{w}^i\frac{\partial\varphi^i}{\partial t}dxdt + \int_{Q^*}\overline{A}_i^\alpha D_\alpha\varphi^i dxdt = \int_{Q^*}\overline{B}_i\,\varphi^i dxdt \\ \forall\varphi \in \mathrm{W}_2^{1,1}(Q^*;\mathbb{R}^N),\ \mathrm{supp}(\varphi) \subset Q^*. \end{array}\right\} \qquad (4.4)$$

[10] By $\mathrm{H}_a^{1/2}(a,b;X)$ $(-\infty < a < b < +\infty)$ we denote the space of all $v \in L^2(a,b;X)$ such that

$$\int_a^b\int_a^b \frac{\|v(s)-v(t)\|_X^2}{|s-t|^2}dsdt + \int_a^b \frac{\|v(t)\|_X^2}{|t-a|}dt < +\infty.$$

Let $t' \in (t^0 - R^2, t^0)$ be any real number. Then, we may localize (4.4) with respect to t using the argument mentioned before. It follows,

$$\left.\begin{aligned}&\int_{B_R} \frac{\partial \overline{w}^i_\lambda}{\partial t}(x,s)\psi^i(x)dx + \int_{B_R} (\overline{A}^\alpha_i)_\lambda(x,s)D_\alpha\psi^i(x)dx = \\ &\quad = \int_{B_R} (\overline{B}_i)_\lambda(x,s)\psi^i(x)dx \\ &\text{for a.a. } s \in (-\infty, t'),\ \forall 0 < \lambda < t^0 - t',\ \forall \psi \in \overset{\circ}{W}{}^1_2(B_R; \mathbb{R}^N).\ ^{11}\end{aligned}\right\} \quad (4.5)$$

Now, for a.a. $(x,s) \in B_R \times (-\infty, t')$ we insert $\psi(x) = \overline{w}_\lambda(x,s)$ $(0 < \lambda < t^0 - t')$ into (4.5), integrate over the interval $(t^0 - R^2, t)$ $(t \in (t^0 - R^2, t'))$ and let tend $\lambda \to 0$. It follows that

$$\begin{aligned}&\frac{1}{2}\int_{B_R} |w(x,t)|^2 dx + \\ &\quad + \int_{t^0-R^2}^{t}\int_{B_R} (a^\alpha_i(x,t^0,\nabla u) - a^\alpha_i(x,t^0,\nabla v))D_\alpha w^i dxds = \\ &= \int_{t^0-R^2}^{t}\int_{B_R} (a^\alpha_i(x,t^0,\nabla u) - a^\alpha_i(x,s,\nabla u))D_\alpha w^i dxds + \\ &\quad + \int_{t^0-R^2}^{t}\int_{B_R} B_i w^i dxds.\end{aligned} \quad (4.6)$$

Observing (1.2), (1.3) and (1.5), letting tend $t' \to t^0$ using Hölder's and Young's inequality (with $0 < \varepsilon < 1$) gives,

$$\begin{aligned}&\frac{1}{2}\operatorname*{ess\,sup}_{(t^0-R^2,t^0)}\int_{B_R} |w(x,t)|^2 dx + \nu_0 \int_{Q_R} |\nabla w|^2 dxdt \le \\ &\le c\omega^2(R)\int_{Q_R} (1 + |\nabla u|^2)dxdt + \\ &\quad + \frac{c}{\varepsilon}\left(\int_{Q_R} (1 + |u|^4 + |\nabla u|^2)dxdt\right)^{3/2} + \varepsilon\left(\int_{Q_R} |w|^4 dxdt\right)^{1/2}.\end{aligned} \quad (4.7)$$

---

[11] Remind that the set of measure zero of those t for which (4.5) fails does not depend on λ.

On the other hand, by virtue of the imbedding $V_2^{1,0}(Q_R; \mathbb{R}^N) \subset L^4(Q_R; \mathbb{R}^N)$ which is continuous (cf. [9]) we have

$$\left(\int_{Q_R} |w|^4 dxdt \right)^{1/2} \leq \tag{4.8}$$

$$\leq c \operatorname*{ess\,sup}_{(t^0-R^2,t^0)} \int_{B_R} |w(x,t)|^2 dx + c \int_{Q_R} |\nabla w|^2 dxdt.$$

Combining (4.8) and (4.7) having chosen ε sufficiently small implies,

$$\int_{Q_R} (|w|^4 + |\nabla w|^2) dxdt \leq \mathcal{O}(R)\, \Phi(R), \tag{4.9}$$

where $\mathcal{O}(R)$ denotes some nondecreasing positive function which tends to zero as $R \to 0$. Thus, combining (4.9) and (4.2) implies

$$\Phi(\tau R) \leq \left\{ 16A\tau^{2+2\gamma} + c(\tau)\, \mathcal{O}(R) \right\} \Phi(R). \tag{4.10}$$

Now, choosing $0 < \tau < 1$ and $0 < R_1 \leq R$ such that

$$16A\tau^{2+2\gamma} + c(\tau)\, \mathcal{O}(R_1) \leq \tau^{2+\gamma}$$

from (4.10) we infer,

$$\Phi(\tau\sigma) \leq \tau^{2+\gamma}\Phi(\sigma) \qquad \forall\, 0 < \sigma \leq R_1. \tag{4.11}$$

After having applied (4.11) iteratively we get

$$\Phi(\tau^k R_1) \leq \tau^{k(2+\gamma)}\Phi(R_1) \qquad (k = 1, 2, \ldots).$$

Consequently,

$$\left.\begin{aligned} &\int_{Q_\sigma} (1 + |u|^4 + |\nabla u|^2) dxdt \leq \\ &\qquad \leq c \left(\frac{\sigma}{R_1} \right)^{2+\gamma} \int_{Q} (1 + |u|^4 + |\nabla u|^2) dxdt \\ &\forall\, 0 < \sigma \leq R_1 \quad (c = \text{const}). \end{aligned}\right\} \tag{4.12}$$

Finally, using the Poincaré inequality for weak solutions (cf. [13]) from (4.12) we deduce that for all $0 < \sigma \leq R_1$ we have

$$\int_{Q_\sigma} |u - u_{Q_\sigma}|^2 dxdt \leq C\, \sigma^{4+\gamma} \int_{Q} (1 + |u|^4 + |\nabla u|^2) dxdt,$$

where

$$u_{Q_\sigma} = \frac{1}{\text{meas}(Q_\sigma)} \int\limits_{Q_\sigma} u\, dxdt$$

(the constant C depends on ν_0, c_0, c_1, c_2, N and $1/R_1$ only). Hence, the Hölder continuity of u follows by Campanato's theory (cf. [1]). ∎

References

[1] Campanato, S. (1966). *Equazioni parabolice del secondo ordine e spazi* $\mathcal{L}^{2,\theta}(\Omega,\delta)$. Ann. Mat. Pura Appl., Vol. 73.

[2] De Giorgi, E. (1957). *Sulla differentiabilitá e l'analiticitá delle estremali deli integrali multipli.* Mem. Accad. Sci. Torino, Cl. Sci. Fis. Mat. Nat., 3(3):25–43.

[3] Gehring, F.W. (1973). *The* L^p*-integrability of partial derivatives of a quasiconformal mapping.* Acta. Math., 130:265–277.

[4] Giuaquinta, M. (1983). *Multiple integrals in calculus of variations and nonlinear elliptic systems.* Princeton Univ. Press, Princeton, New Jersey.

[5] Giuaquinta, M. and Modica, G. (1979). *Almost- everywhere regularity results for solutions of nonlinear elliptic systems.* Manuscripta Math., 28:109–158.

[6] John, O. and Stará, J. *On the regularity of weak solutions to parabolic systems in two dimensions.* To appear.

[7] Kaplan, S. (1966). *Abstract boundary value problems for linear parabolic equations.* Ann. Scuola Norm. Sup. Pisa, 20(3):395–419.

[8] Ladyzenskaya, O.A. and Ural'ceva, N.N. (1968). *Linear and quasilinear elliptic equations.* Academic Press.

[9] Ladyzenskaya, O.A., Solonnikov, V.A. and Ural'ceva, N.N. *Linear and quasilinear equations of parabolic type.* Trans. Math. Monographs 28, Amer. Math. Soc., Providence, R.I..

[10] Lions, J.L. (1969) *Quelques méthodes de résolutions des problèmes aux limites non linéaires.* Paris.

[11] Lions, J.L. and Magenes, E. (1968) *Problemes aux limites non homogenes et applications*, Dunot.

[12] Naumann, J., Wolff, M. and Wolf, J. *On the Hölder continuity of weak solutions to nonlinear parabolic systems in two space dimensions.* To appear.

[13] Naumann, J. and Wolff, M. *Interior integral estimates on weak solutions of nonlinear parabolic systems.* Humboldt Univ. Berlin, FB. Math. Preprint 94–12.

[14] Nečas, J. and Šverák, V. (1991). *On regularity of nonlinear parabolic systems.* Annali Scuola Normale Superiore Pisa, 18(4):1–11.

[15] Triebel, H. (1995). *Interpolation theory, function spaces, differential operators.* J.A. Barth, Heidelberg, Leipzig.

INDEX